T0188365

Nonlinear Photonics

Suitable for both graduate and senior undergraduate students, this textbook offers a logical progression through the underlying principles and practical applications of nonlinear photonics. Building up from essential physics, general concepts, and fundamental mathematical formulations, it provides a robust introduction to nonlinear optical processes and phenomena, and their practical applications in real-world devices and systems. Over 45 worked problems illustrate key concepts and provide hands-on models for students, and over 160 end-of-chapter exercises supply students with plenty of scope to master the material. Accompanied by a complete solutions manual for instructors, including detailed explanations of each result, and drawing on the author's 35 years of teaching experience, this is the ideal introduction to nonlinear photonics for students in electrical engineering.

Jia-Ming Liu is Distinguished Professor of Electrical and Computer Engineering, Northrop Grumman Optoelectronics Chair in Electrical Engineering, and Associate Dean of the Henry Samueli School of Engineering at the University of California, Los Angeles. He is the author of *Photonic Devices* (2005) and *Principles of Photonics* (2016), a co-author of *Graphene Photonics* (2019), and a Fellow of Optica, the American Physical Society, the IEEE, and the Guggenheim Foundation.

Nonlinear Photonics

JIA-MING LIU

University of California, Los Angeles

CAMBRIDGE
UNIVERSITY PRESS

CAMBRIDGE
UNIVERSITY PRESS

University Printing House, Cambridge CB2 8BS, United Kingdom

One Liberty Plaza, 20th Floor, New York, NY 10006, USA

477 Williamstown Road, Port Melbourne, VIC 3207, Australia

314–321, 3rd Floor, Plot 3, Splendor Forum, Jasola District Centre, New Delhi – 110025, India

103 Penang Road, #05–06/07, Visioncrest Commercial, Singapore 238467

Cambridge University Press is part of the University of Cambridge.

It furthers the University's mission by disseminating knowledge in the pursuit of
education, learning, and research at the highest international levels of excellence.

www.cambridge.org
Information on this title: www.cambridge.org/highereducation/isbn/9781316512524
DOI: 10.1017/9781009067652

© Jia-Ming Liu 2022

First published 2022

Printed in the United Kingdom by TJ Books Limited, Padstow Cornwall

A catalogue record for this publication is available from the British Library.

Library of Congress Cataloging-in-Publication Data
Names: Liu, Jia-Ming, 1953– author.
Title: Nonlinear photonics / Jia Ming Liu, University of California, Los Angeles.
Description: Cambridge, UK ; New York, NY : Cambridge University Press, 2022.
Identifiers: LCCN 2021025537 (print) | LCCN 2021025538 (ebook) | ISBN 9781316512524 (hardback)
| ISBN 9781009067652 (ebook)
Subjects: LCSH: Nonlinear optics.
Classification: LCC QC446.2 .L58 2022 (print) | LCC QC446.2 (ebook) | DDC 535/.2–dc23
LC record available at https://lccn.loc.gov/2021025537
LC ebook record available at https://lccn.loc.gov/2021025538

ISBN 978-1-316-51252-4 Hardback

To Vida and Janelle

CONTENTS

19   Supercontinuum Generation   545

PREFACE

The field of nonlinear optics was born in 1961, when Franken and co-workers experimentally observed second-harmonic generation, followed by the groundbreaking theoretical work in 1962 by Bloembergen and co-workers. In the first two decades, this field was mainly developed as a sub-discipline of optical physics, with its primary interest focused on exploring the physical mechanisms of various nonlinear optical phenomena but with only limited technological applications. As nonlinear optics became a mature field of research in physics, it began to find increasing applications in a wide range of scientific and engineering areas, ranging from chemistry and biology to electrical engineering and materials. Over the past two decades, it has developed into an important sub-discipline of photonics as the photonics field matured into an important discipline of modern engineering and technology. This new sub-discipline is nonlinear photonics, which covers both the underlining physics of nonlinear optical processes and the technological applications of various nonlinear optical devices and systems.

Today, nonlinear photonics is an essential course for all graduate students in any program that is related to optics, photonics, optoelectronics, or optical engineering. Such a program can be found in the physics department, the electrical engineering department, the mechanical engineering department, the materials science department, or the photonics department. A graduate student in this program is required to take the nonlinear photonics course. Aside from graduate students in such a program, many graduate students from other programs of science or engineering often take this course as an elective. Many researchers and technologists in a diverse range of fields regularly use the knowledge and the techniques of nonlinear photonics in their work. This book is written primarily for a course on nonlinear photonics. It can also be used as a reference or self-study book for researchers and technologists who work in various related fields.

At the electrical and computer engineering department of UCLA, I began teaching a graduate course on nonlinear optics in 1986 by using various books and my own notes. This course gradually evolved into nonlinear photonics about 15 years ago. The key challenge to the students and instructors in this course has been the lack of a book that is self-contained and sufficiently complete for them to simply follow through in the course. Most of the existing books are nonlinear optics books targeting physics students; as such, they fall short of the needs of a course on nonlinear photonics. This book solves the problem.

Although this book is different from existing books on nonlinear optics, in that it emphasizes photonics applications, it actually goes into the basic physics in depth, sometimes more so than existing nonlinear optics books. The book is logically structured in this manner: (1) It first introduces the essential physics and general concepts; (2) the mathematical formulations are then presented through concise, but rigorous, derivations; (3) various nonlinear optical processes and phenomena are described; and (4) devices and systems resulting from the applications of the nonlinear optical processes are covered. This structural approach is followed across the entire book, as well as within each chapter where applicable.

This book is self-contained. It covers in depth both the physics, which is generally covered by a book on nonlinear optics, and the practical photonics applications, which are generally neglected or only casually mentioned in a book on nonlinear optics. As such, this book is ideal for a course on nonlinear photonics in any engineering program. It is also an excellent choice for a modern course on nonlinear optics in a science program, because the learning of science can be enhanced by the illustration of its applications.

There are two prerequisites for a course based on this book: (1) electromagnetics and (2) quantum mechanics. These two requirements are generally fulfilled by a student who is enrolled in a graduate program that offers a course on nonlinear photonics. In addition, many graduate students also have a background in photonics at the level of *Principles of Photonics*, such as one taught through my book of that title published in 2016 by Cambridge University Press.

One concern an instructor usually has is the use of quantum mechanics and density matrix, because many engineering students are weak on this subject. For this reason, quantum mechanics is to be avoided if possible. However, quantum-mechanical formulations using density matrix are necessary for a rigorous treatment of some fundamental concepts. It is thus necessary to find a proper trade-off between these two demands. My approach is to lay out a rigorous treatment of the fundamental concepts by using quantum mechanics, but then to use a macroscopic, nonquantum-mechanical approach for the treatment of nonlinear optical processes. In this book, quantum-mechanical formulations involving density matrix appear only in Sections 1.2–1.8 and Section 3.1. It is possible for a reader who finds the quantum-mechanical treatment challenging to completely avoid quantum mechanics when reading this book by simply taking the results obtained in these sections for granted. By contrast, a reader who has sufficient background in quantum mechanics will find the quantum-mechanical formulations in these sections interesting and useful.

The book starts in Chapter 1 with the fundamental physical processes of light–matter interaction and their formulations. This chapter is unique and deeper than that covered in existing nonlinear optics books. The general concepts and formulations of optical nonlinearity and optical susceptibilities are then introduced in Chapters 2 and 3. Chapter 4 covers the essential background on the propagation of optical waves in various optical media. In Chapter 5, a general discussion is given on all second- and third-order nonlinear processes to give the reader an overview understanding and a complete picture of these processes. The essential mathematical formulations that are necessary for the analysis of nonlinear optical interactions in various situations are then derived and presented in Chapters 6–8. The concepts and techniques of phase matching required for efficient nonlinear optical interactions are discussed in Chapter 9. These nine chapters complete the foundation of the subjects regarding the general concepts, the basic physics, and the mathematical formulations. Chapters 10–19 then address various phenomena and applications. The choice of these topics is based on both their importance and general interest in the photonics field.

This book is written with a pedagogical purpose in mind. The structured approach as described above is one important pedagogical feature. The book has 150 carefully prepared figures to illustrate relevant concepts. Important data or concepts are summarized in 11 tables.

The book contains 46 examples that are properly placed in the text. The examples are treated as illustrative teaching materials that utilize realistic numerical data and have detailed explanations. A reader is thus strongly advised to study these examples as an essential step to gain a clear understanding of the text. The book has a total of 161 problems. A solution manual that contains detailed solutions and explanations for these problems is available to instructors.

The materials in this book were developed through my teaching of the graduate course on nonlinear optics, and then on nonlinear photonics, at UCLA over the past 35 years. I would first like to express my gratitude to my colleagues and my students, who have given me valuable feedback during these years. The book was written from February to October 2020 during the lockdown due to the COVID-19 pandemic. I would like to thank my editors, Julie Lancashire and Lisa Pinto, for getting this book published. I am most grateful to the content manager, Charles Howell, and the copy-editor, Gary Smith, for their help during the publication process. Thanks are also due to Kuan-Yuan Chang, who helped me complete the figures for soliton propagation. This book is dedicated to my wife, Vida, who gave me constant support and created an original oil painting for the cover art, and to my daughter, Janelle, who took a daily interest in my progress during the writing of this book.

1 Light–Matter Interaction

1.1 OPTICAL FIELDS

Optical nonlinearity emerges from nonlinear interaction of light with matter. In this chapter, the basic concept and formulation of light–matter interaction are discussed through a semiclassical approach. In this semiclassical treatment of the interaction between an optical field and a material, the behavior of the optical field is classically described by *Maxwell's equations*, but the state of the material is quantum mechanically described by a wave function that is governed by the *Hamiltonian* of the material.

The behavior of a spatially and temporally varying electromagnetic field is governed by Maxwell's equations:

$$\nabla \times \boldsymbol{E} = -\frac{\partial \boldsymbol{B}}{\partial t}, \qquad \text{Faraday's law;} \tag{1.1}$$

$$\nabla \times \boldsymbol{H} = \frac{\partial \boldsymbol{D}}{\partial t} + \boldsymbol{J}, \qquad \text{Ampere's law;} \tag{1.2}$$

$$\nabla \cdot \boldsymbol{D} = \rho, \qquad \text{Gauss' law, Coulomb's law;} \tag{1.3}$$

$$\nabla \cdot \boldsymbol{B} = 0, \qquad \text{absence of magnetic monopoles.} \tag{1.4}$$

The *electric field* $\boldsymbol{E}(\mathbf{r},t)$ and the *magnetic induction* $\boldsymbol{B}(\mathbf{r},t)$ are the macroscopic forms of the microscopic fields seen by the charge and current densities in a medium. The *polarization* $\boldsymbol{P}(\mathbf{r},t)$ and the *magnetization* $\boldsymbol{M}(\mathbf{r},t)$ are, respectively, the macroscopically averaged densities of the microscopic electric dipole moments and magnetic dipole moments that are induced by the presence of the electromagnetic field in the medium. These macroscopic forms are obtained by averaging over a volume that is small compared to the dimension of the optical wavelength but is large compared to the atomic dimension. The *electric displacement* $\boldsymbol{D}(\mathbf{r},t)$ and the *magnetic field* $\boldsymbol{H}(\mathbf{r},t)$ are macroscopic fields that are defined as

$$\boldsymbol{D}(\mathbf{r},t) = \epsilon_0 \boldsymbol{E}(\mathbf{r},t) + \boldsymbol{P}(\mathbf{r},t), \tag{1.5}$$

and

$$\boldsymbol{H}(\mathbf{r},t) = \frac{1}{\mu_0} \boldsymbol{B}(\mathbf{r},t) - \boldsymbol{M}(\mathbf{r},t), \tag{1.6}$$

where $\epsilon_0 \approx \frac{1}{36\pi} \times 10^{-9} \, \mathrm{F\,m^{-1}} = 8.854 \times 10^{-12} \, \mathrm{F\,m^{-1}}$ is the *electric permittivity* of free space and $\mu_0 = 4\pi \times 10^{-7} \, \mathrm{H\,m^{-1}}$ is the *magnetic permeability* of free space. The current density $\boldsymbol{J}$ and the charge density ρ are constrained by the *continuity equation*:

$$\mathbf{V} \cdot \mathbf{J} + \frac{\partial \rho}{\partial t} = 0, \qquad \text{conservation of charge.} \qquad (1.7)$$

1.1.1 Coulomb Gauge

According to the two Maxwell's equations in (1.1) and (1.4), there exist a *vector potential* $A(\mathbf{r}, t)$ and a *scalar potential* $\varphi(\mathbf{r}, t)$ such that the electric field and the magnetic induction can be expressed as

$$\boldsymbol{E} = -\mathbf{V}\varphi - \frac{\partial A}{\partial t}, \qquad (1.8)$$

$$\boldsymbol{B} = \mathbf{V} \times \boldsymbol{A}. \qquad (1.9)$$

By explicitly accounting for all current contributions to the optical field with $\boldsymbol{J}$ and all charge contributions with ρ, we effectively use the microscopic form of Maxwell's equations by setting $\boldsymbol{P} = 0$ and $\boldsymbol{M} = 0$ so that $\boldsymbol{D} = \epsilon_0 \boldsymbol{E}$ and $\boldsymbol{B} = \mu_0 \boldsymbol{H}$ for the other two Maxwell's equations that are given in (1.2) and (1.3). Then, by using (1.8) and (1.9), the two Maxwell's equations in (1.2) and (1.3) can be expressed in terms of $A(\mathbf{r}, t)$ and $\varphi(\mathbf{r}, t)$ as

$$\nabla^2 A - \frac{1}{c^2} \frac{\partial^2 A}{\partial t^2} - \mathbf{V}\left(\mathbf{V} \cdot \boldsymbol{A} + \frac{1}{c^2} \frac{\partial \varphi}{\partial t}\right) = -\mu_0 \boldsymbol{J}, \qquad (1.10)$$

$$\nabla^2 \varphi + \frac{\partial}{\partial t}(\mathbf{V} \cdot \boldsymbol{A}) = -\frac{\rho}{\epsilon_0}. \qquad (1.11)$$

An optical field is completely defined when both $\boldsymbol{E}$ and $\boldsymbol{B}$ are fully defined, which have six field components because each of the vector fields $\boldsymbol{E}$ and $\boldsymbol{B}$ has three components. By contrast, from (1.8) and (1.9), $\boldsymbol{E}$ and $\boldsymbol{B}$ are fully defined by the vector potential A and the scalar potential φ, which together have only four components. For this reason, there are multiple combinations of A and φ for a given combination of $\boldsymbol{E}$ and $\boldsymbol{B}$ that are determined by (1.8) and (1.9). In other words, A and φ of a given optical field need not be unique, though $\boldsymbol{E}$ and $\boldsymbol{B}$ for the given optical field have to be unique. It can be easily shown that for a given combination of A and φ that defines the unique set of $\boldsymbol{E}$ and $\boldsymbol{B}$ through (1.8) and (1.9), a different set A' and φ' of the forms

$$A' = A + \mathbf{V}\Lambda \quad \text{and} \quad \varphi' = \varphi - \frac{\partial \Lambda}{\partial t}, \qquad (1.12)$$

with $\Lambda(\mathbf{r}, t)$ being an arbitrary smooth scalar function of space and time, also defines the same set of $\boldsymbol{E}$ and $\boldsymbol{B}$:

$$\boldsymbol{E} = -\mathbf{V}\varphi - \frac{\partial A}{\partial t} = -\mathbf{V}\varphi' - \frac{\partial A'}{\partial t}, \qquad (1.13)$$

$$\boldsymbol{B} = \mathbf{V} \times \boldsymbol{A} = \mathbf{V} \times A'. \qquad (1.14)$$

The transformation from A and φ to A' and φ' through (1.12) without changing $\boldsymbol{E}$ and $\boldsymbol{B}$ is known as *gauge transformation*. Maxwell's equations are invariant under a gauge transformation as defined in (1.12). The choice of $\Lambda(\mathbf{r}, t)$ for a particular gauge depends on the convenience of formulating a problem.

For the semiclassical treatment of light–matter interaction, it is convenient to use the *Coulomb gauge*, for which we choose a proper $\Lambda(\mathbf{r}, t)$ such that

$$\mathbf{\nabla} \cdot \mathbf{A} = 0. \tag{1.15}$$

The current density $\mathbf{J} = \mathbf{J}_\mathrm{L} + \mathbf{J}_\mathrm{T}$ can be generally decomposed, as any vectorial function of space can be decomposed, into an irrotational longitudinal component $\mathbf{J}_\mathrm{L}$ with $\mathbf{\nabla} \times \mathbf{J}_\mathrm{L} = 0$ and a solenoidal transverse component $\mathbf{J}_\mathrm{T}$ with $\mathbf{\nabla} \cdot \mathbf{J}_\mathrm{T} = 0$. It can be easily shown by using the continuity equation given in (1.7) and the condition given in (1.15) for the Coulomb gauge that $\frac{1}{c^2} \mathbf{\nabla} \frac{\partial \varphi}{\partial t} = \mu_0 \mathbf{J}_\mathrm{L}$. Then, in the Coulomb gauge, (1.10) and (1.11) reduce to the forms:

$$\nabla^2 \mathbf{A} - \frac{1}{c^2} \frac{\partial^2 \mathbf{A}}{\partial t^2} = -\mu_0 \mathbf{J}_\mathrm{T}, \tag{1.16}$$

$$\nabla^2 \varphi = -\frac{\rho}{\epsilon_0}. \tag{1.17}$$

1.1.2 Source-Free Maxwell's Equations

The total current density $\mathbf{J}$ consists of both the *induced current*, which is induced by the optical field, and the *external current*, which is contributed by an external current source; thus, $\mathbf{J} = \mathbf{J}_\mathrm{ind} + \mathbf{J}_\mathrm{ext}$. The total charge density ρ also consists of both the *induced charges* and the *external charges*; thus, $\rho = \rho_\mathrm{ind} + \rho_\mathrm{ext}$. For an optical field that is away from its source, the external current and external charges are both absent; thus, $\mathbf{J}_\mathrm{ext} = 0$ and $\rho_\mathrm{ext} = 0$. For optical interaction, charge conservation requires that there is no net macroscopic induced charge density; thus, $\rho_\mathrm{ind} = 0$ and $\rho = 0$. By contrast, an optical field can cause a net induced current density $\mathbf{J}_\mathrm{ind}$ either in the form of a *displacement current* from the field-induced movement of bound electrons or in the form of a *conduction current* from the field-induced flow of free electrons, or both. In any event, an induced current can be either explicitly expressed as a separate term, as seen in (1.2), or implicitly included in the $\partial \mathbf{D}/\partial t$ term so that $\mathbf{J} = 0$. If the latter choice is taken and the fact that $\rho = 0$ is considered for an optical field, then Maxwell's equations for an optical field can be expressed as

$$\mathbf{\nabla} \times \mathbf{E} = -\frac{\partial \mathbf{B}}{\partial t}, \tag{1.18}$$

$$\mathbf{\nabla} \times \mathbf{H} = \frac{\partial \mathbf{D}}{\partial t}, \tag{1.19}$$

$$\mathbf{\nabla} \cdot \mathbf{D} = 0, \tag{1.20}$$

$$\mathbf{\nabla} \cdot \mathbf{B} = 0. \tag{1.21}$$

1.1.3 Symmetry Properties of the Fields

There are three vector fields, E, D, and P, of the electric type, and three vector fields, B, H, and M, of the magnetic type. The electric-type fields and the magnetic-type fields are fundamentally different types of vectors, and they have different spatial and temporal symmetry properties. Besides the above six fields, there are also the current density vector J, the charge density ρ, the vector potential A, and the scalar potential φ. For these four fields, J and A have the same symmetry properties, and ρ and φ have the same symmetry properties, but they are all different from those of the electric-type fields and those of the magnetic-type fields.

The three fields of the electric type are *polar vectors*. They change sign under *space inversion* through the $\mathbf{r} \to -\mathbf{r}$ space transformation:

$$E(\mathbf{r},t) \xrightarrow{\mathbf{r}\to-\mathbf{r}} -E(-\mathbf{r},t),\ D(\mathbf{r},t) \xrightarrow{\mathbf{r}\to-\mathbf{r}} -D(-\mathbf{r},t),\ P(\mathbf{r},t) \xrightarrow{\mathbf{r}\to-\mathbf{r}} -P(-\mathbf{r},t); \quad (1.22)$$

but they do not change sign under *time reversal* through the $t \to -t$ time transformation:

$$E(\mathbf{r},t) \xrightarrow{t\to-t} E(\mathbf{r},-t),\ D(\mathbf{r},t) \xrightarrow{t\to-t} D(\mathbf{r},-t),\ P(\mathbf{r},t) \xrightarrow{t\to-t} P(\mathbf{r},-t). \quad (1.23)$$

The three fields of the magnetic type are *axial vectors*. They do not change sign under space inversion through the $\mathbf{r} \to -\mathbf{r}$ space transformation:

$$B(\mathbf{r},t) \xrightarrow{\mathbf{r}\to-\mathbf{r}} B(-\mathbf{r},t),\ H(\mathbf{r},t) \xrightarrow{\mathbf{r}\to-\mathbf{r}} H(-\mathbf{r},t),\ M(\mathbf{r},t) \xrightarrow{\mathbf{r}\to-\mathbf{r}} M(-\mathbf{r},t); \quad (1.24)$$

but they change sign under time reversal through the $t \to -t$ time transformation:

$$B(\mathbf{r},t) \xrightarrow{t\to-t} -B(\mathbf{r},-t),\ H(\mathbf{r},t) \xrightarrow{t\to-t} -H(\mathbf{r},-t),\ M(\mathbf{r},t) \xrightarrow{t\to-t} -M(\mathbf{r},-t). \quad (1.25)$$

The current density vector and the vector potential have the same symmetry properties. Both of them change sign under either space inversion or time reversal:

$$J(\mathbf{r},t) \xrightarrow{\mathbf{r}\to-\mathbf{r}} -J(-\mathbf{r},t),\ A(\mathbf{r},t) \xrightarrow{\mathbf{r}\to-\mathbf{r}} -A(-\mathbf{r},t); \quad (1.26)$$

$$J(\mathbf{r},t) \xrightarrow{t\to-t} -J(\mathbf{r},-t),\ A(\mathbf{r},t) \xrightarrow{t\to-t} -A(\mathbf{r},-t). \quad (1.27)$$

The charge density and the scalar potential have the same symmetry properties. Both of them do not change sign under either space inversion or time reversal:

$$\rho(\mathbf{r},t) \xrightarrow{\mathbf{r}\to-\mathbf{r}} \rho(-\mathbf{r},t),\ \varphi(\mathbf{r},t) \xrightarrow{\mathbf{r}\to-\mathbf{r}} \varphi(-\mathbf{r},t); \quad (1.28)$$

$$\rho(\mathbf{r},t) \xrightarrow{t\to-t} \rho(\mathbf{r},-t),\ \varphi(\mathbf{r},t) \xrightarrow{t\to-t} \varphi(\mathbf{r},-t). \quad (1.29)$$

Though the electric-type fields and the magnetic-type fields have different transformation properties, Maxwell's equations are invariant under space inversion or time reversal, or both together. We also find that (1.10) and (1.11), and their forms given in (1.16) and (1.17) under the Coulomb gauge, are also invariant under space inversion or time reversal, or both together.

1.1.4 Polarization and Magnetization

Polarization and magnetization are generated in a medium by the response of the medium to the electric field and the magnetic induction, respectively: $P(\mathbf{r}, t)$ depends on $E(\mathbf{r}, t)$, and $M(\mathbf{r}, t)$ depends on $B(\mathbf{r}, t)$. For a natural material, *the magnetization vanishes*, $M \approx 0$, at an *optical frequency* [1–3]. The only exceptions of an enhanced magnetic response that leads to a perceptible magnetization M at an optical frequency appear in a specifically designed artificial *metamaterial* [4–6] or *photonic nanostructure* [3, 7], or for a *magnetic dipole transition* [8, 9], which is associated with a magnetic dipole interaction as discussed in Subsection 1.4.5. Even in such exceptions, at an optical frequency the nonvanishing magnetization M is very small compared to the polarization P. Therefore, except when the effect of a magnetic dipole transition or the property of a specific metamaterial is the focus of interest, it is generally true for an optical field that $M \approx 0$ so that

$$B(\mathbf{r}, t) = \mu_0 H(\mathbf{r}, t). \tag{1.30}$$

Because μ_0 is a constant that is independent of the medium, the magnetic induction $B(\mathbf{r}, t)$ can be replaced by $\mu_0 H(\mathbf{r}, t)$ for any equations that describe optical fields, including Maxwell's equations, thus effectively eliminating one field variable. Note that this is not true at DC or low frequencies, however, because a nonzero DC or low-frequency magnetization, $M \neq 0$, can exist in any material.

In the common situation, with the exception of dealing with a magnetic dipole transition or a specifically designed metamaterial of a significant magnetization, the optical properties of a material are completely determined by the relation between $P(\mathbf{r}, t)$ and $E(\mathbf{r}, t)$. This relation is generally characterized by an *electric susceptibility tensor*, χ, through the definition for the *electric polarization* in the real space and time domain:

$$P(\mathbf{r}, t) = \epsilon_0 \int_{-\infty}^{t} \iiint_{\text{all } \mathbf{r}'} \chi(\mathbf{r} - \mathbf{r}', t - t') \cdot E(\mathbf{r}', t') d\mathbf{r}' dt'. \tag{1.31}$$

The relation between $D(\mathbf{r}, t)$ and $E(\mathbf{r}, t)$ is characterized by the *electric permittivity tensor*, ϵ, also known as the *dielectric permittivity tensor*, of the medium:

$$D(\mathbf{r}, t) = \epsilon_0 E(\mathbf{r}, t) + P(\mathbf{r}, t) = \int_{-\infty}^{t} \iiint_{\text{all } \mathbf{r}'} \epsilon(\mathbf{r} - \mathbf{r}', t - t') \cdot E(\mathbf{r}', t') d\mathbf{r}' dt'. \tag{1.32}$$

From (1.31) and (1.32), the relation between χ and ϵ in the real space and time domain is

$$\epsilon(\mathbf{r}, t) = \epsilon_0 [\mathbf{I}\delta(\mathbf{r})\delta(t) + \chi(\mathbf{r}, t)], \tag{1.33}$$

where $\mathbf{I}$ is the identity tensor that has the form of a 3×3 unit matrix and the delta functions are Dirac delta functions: $\iiint_{\text{all } \mathbf{r}} \delta(\mathbf{r}) d\mathbf{r} = 1$ and $\int_{-\infty}^{\infty} \delta(t) dt = 1$. The relation in (1.33) indicates that χ and ϵ contain exactly the same information about the medium: one is known when the other is known. Because χ and, equivalently, ϵ represent the response of a medium to an optical field and thus completely characterize the macroscopic electromagnetic properties of the medium, (1.31) and (1.32) can be regarded as the definitions of $P(\mathbf{r}, t)$ and $D(\mathbf{r}, t)$, respectively.

1.1.5 Wave Equation

By taking $\nabla \times$ on (1.18), followed by using (1.19) and (1.30), we can combine Maxwell's equations to get the *wave equation*:

$$\nabla \times \nabla \times E + \mu_0 \frac{\partial^2 D}{\partial t^2} = 0, \tag{1.34}$$

which can also be expressed as

$$\nabla \times \nabla \times E + \frac{1}{c^2} \frac{\partial^2 E}{\partial t^2} = -\mu_0 \frac{\partial^2 P}{\partial t^2}, \tag{1.35}$$

where $c = 1/\sqrt{\mu_0 \epsilon_0} \approx 3 \times 10^8 \text{ m s}^{-1}$ is the *speed of light* in free space. We see from (1.35) that we can view the electric polarization P as a driving force for the propagation of the optical wave by expressing it on the right-hand side of the wave equation. By contrast, in (1.34) P is included in the electric displacement D as the optical property of the medium. The separation of P from D is flexible and somewhat arbitrary because it does not change the behavior of the optical wave. Therefore, it is only a matter of convenience in treating the wave behavior.

In practice, a useful alternative is to include only a background polarization in D by dividing the total electric polarization P into two parts as $P = P_b + \Delta P$, where P_b is the background polarization of the medium and ΔP consists of all of the rest. We can then define $D(\mathbf{r}, t) = \epsilon_0 E(\mathbf{r}, t) + P_b(\mathbf{r}, t)$ and write the wave equation in the form:

$$\nabla \times \nabla \times E + \mu_0 \frac{\partial^2 D}{\partial t^2} = -\mu_0 \frac{\partial^2 \Delta P}{\partial t^2}. \tag{1.36}$$

In this form, we can view ΔP as a force that causes changes in the behavior of the optical wave as it propagates in the medium. It is a polarization on top of the background polarization of the medium. Its choice simply depends on our convenience and purpose. For example, it can be the resonant polarization between two energy levels, as described in Subsection 1.7.1, or a nonlinear polarization, as described in Section 6.1.

1.1.6 Complex Harmonic Fields

Optical fields are harmonic fields that vary sinusoidally with time. The fields defined in the preceding subsections are all *real* harmonic fields. They include the following 10 vector or scalar field quantities: $E(\mathbf{r}, t)$, $B(\mathbf{r}, t)$, $D(\mathbf{r}, t)$, $H(\mathbf{r}, t)$, $P(\mathbf{r}, t)$, $M(\mathbf{r}, t)$, $\rho(\mathbf{r}, t)$, $J(\mathbf{r}, t)$, $\varphi(\mathbf{r}, t)$, and $A(\mathbf{r}, t)$. For harmonic fields, it is always convenient to use *complex fields*. We define the spatially and temporally varying complex electric field, $\mathbf{E}(\mathbf{r}, t)$, through its relation to the real electric field, $\boldsymbol{E}(\mathbf{r}, t)$, as[1]

$$\boldsymbol{E}(\mathbf{r}, t) = \mathbf{E}(\mathbf{r}, t) + \mathbf{E}^*(\mathbf{r}, t) = \mathbf{E}(\mathbf{r}, t) + \text{c.c.}, \tag{1.37}$$

where c.c. means the complex conjugate. In our convention, $\mathbf{E}(\mathbf{r}, t)$ contains the complex field components that vary with time as $\exp(-i\omega t)$ with ω having a positive value, while $\mathbf{E}^*(\mathbf{r}, t)$ contains those varying with time as $\exp(i\omega t)$ with positive ω. Each of the 10 real harmonic fields listed above has a complex field defined in the same manner: $\mathbf{E}(\mathbf{r}, t)$, $\mathbf{B}(\mathbf{r}, t)$, $\mathbf{D}(\mathbf{r}, t)$, $\mathbf{H}(\mathbf{r}, t)$, $\mathbf{P}(\mathbf{r}, t)$, $\mathbf{M}(\mathbf{r}, t)$, $\rho(\mathbf{r}, t)$, $\mathbf{J}(\mathbf{r}, t)$, $\varphi(\mathbf{r}, t)$, and $\mathbf{A}(\mathbf{r}, t)$.

With this definition for the complex fields, all of the linear field equations retain their forms, but this is not true for nonlinear field equations, as we shall see in later chapters. Because Maxwell's equations are linear equations, they retain their forms whether they are expressed in real fields or in complex fields. For example, in terms of complex optical fields, the source-free Maxwell's equations that are given in (1.18)–(1.21) take the forms:

$$\mathbf{\nabla} \times \mathbf{E} = -\frac{\partial \mathbf{B}}{\partial t}, \tag{1.38}$$

$$\mathbf{\nabla} \times \mathbf{H} = \frac{\partial \mathbf{D}}{\partial t}, \tag{1.39}$$

$$\mathbf{\nabla} \cdot \mathbf{D} = 0, \tag{1.40}$$

$$\mathbf{\nabla} \cdot \mathbf{B} = 0. \tag{1.41}$$

The electric polarization defined in (1.31) and the relation for the electric displacement given in (1.32) are both linear field equations. They also retain their forms when expressed in terms of complex fields:

$$\mathbf{P}(\mathbf{r}, t) = \epsilon_0 \int_{-\infty}^{t} \iiint_{\text{all } \mathbf{r}'} \boldsymbol{\chi}(\mathbf{r} - \mathbf{r}', t - t') \cdot \mathbf{E}(\mathbf{r}', t') d\mathbf{r}' dt', \tag{1.42}$$

$$\mathbf{D}(\mathbf{r}, t) = \epsilon_0 \mathbf{E}(\mathbf{r}, t) + \mathbf{P}(\mathbf{r}, t) = \int_{-\infty}^{t} \iiint_{\text{all } \mathbf{r}'} \boldsymbol{\epsilon}(\mathbf{r} - \mathbf{r}', t - t') \cdot \mathbf{E}(\mathbf{r}', t') d\mathbf{r}' dt'. \tag{1.43}$$

Note that the electric susceptibility tensor $\boldsymbol{\chi}(\mathbf{r} - \mathbf{r}', t - t')$ in (1.42) and the electric permittivity tensor $\boldsymbol{\epsilon}(\mathbf{r} - \mathbf{r}', t - t')$ in (1.43) remain the same real functions of space and time as those in (1.31) and (1.32). They are not harmonic fields and therefore cannot be decomposed as (1.37) for a harmonic field.

The three different forms of wave equations expressed in (1.34)–(1.36) can each be expressed in a corresponding form in terms of complex fields. For example, (1.36) can be expressed as

---

[1] In some literature, the complex field is defined through a relation with the real field as $\boldsymbol{E}(\mathbf{r}, t) = \left[\mathbf{E}(\mathbf{r}, t) + \mathbf{E}^*(\mathbf{r}, t) \right]/2$, which differs from our definition expressed in (1.37) by the factor $1/2$. The magnitude of the complex field defined through this alternative relation is twice that of the complex field that is defined through (1.37). As a result, expressions for many quantities may be different under the two different definitions. Our definition is particularly convenient for the expressions of nonlinear optical polarizations, as we shall see in later chapters.

$$\mathbf{\nabla} \times \mathbf{\nabla} \times \mathbf{E} + \mu_0 \frac{\partial^2 \mathbf{D}}{\partial t^2} = -\mu_0 \frac{\partial^2 \Delta \mathbf{P}}{\partial t^2}. \tag{1.44}$$

The complex electric field of a harmonic optical field that has a carrier wavevector of $\mathbf{k}$ and a carrier angular frequency of ω can be further expressed as

$$\mathbf{E}(\mathbf{r}, t) = \boldsymbol{\mathcal{E}}(\mathbf{r}, t) \exp(\mathrm{i} \mathbf{k} \cdot \mathbf{r} - \mathrm{i} \omega t) = \hat{e} \mathcal{E}(\mathbf{r}, t) \exp(\mathrm{i} \mathbf{k} \cdot \mathbf{r} - \mathrm{i} \omega t), \tag{1.45}$$

where $\boldsymbol{\mathcal{E}}(\mathbf{r}, t)$ is the spatially and temporally varying amplitude of the field, and $\hat{e}$ is the unit polarization vector of the field. The *vectorial field amplitude* $\boldsymbol{\mathcal{E}}(\mathbf{r}, t)$ is generally a complex vectorial quantity that has a magnitude, a phase, and a polarization. Other complex field quantities, such as $\mathbf{D}(\mathbf{r}, t)$, $\mathbf{P}(\mathbf{r}, t)$, $\mathbf{B}(\mathbf{r}, t)$, and $\mathbf{H}(\mathbf{r}, t)$, can be similarly expressed. The space- and time-dependent phase factor in (1.45) indicates the direction of wave propagation:

$\mathrm{i} \mathbf{k} \cdot \mathbf{r} - \mathrm{i} \omega t$: propagating in the $\mathbf{k}$ direction;
$-\mathrm{i} \mathbf{k} \cdot \mathbf{r} - \mathrm{i} \omega t$: propagating in the $-\mathbf{k}$ direction.

For harmonic optical fields, it is often useful to consider the complex fields in the momentum space and frequency domain defined by the Fourier-transform relations:

$$\mathbf{E}(\mathbf{k}, \omega) = \int_{-\infty}^{\infty} \iiint_{\text{all } \mathbf{r}} \mathbf{E}(\mathbf{r}, t) \exp(-\mathrm{i} \mathbf{k} \cdot \mathbf{r} + \mathrm{i} \omega t) \mathrm{d} \mathbf{r} \mathrm{d} t, \quad \text{for } \omega > 0, \tag{1.46}$$

$$\mathbf{E}(\mathbf{r}, t) = \frac{1}{(2\pi)^4} \int_{0}^{\infty} \iiint_{\text{all } \mathbf{k}} \mathbf{E}(\mathbf{k}, \omega) \exp(\mathrm{i} \mathbf{k} \cdot \mathbf{r} - \mathrm{i} \omega t) \mathrm{d} \mathbf{k} \mathrm{d} \omega. \tag{1.47}$$

Note that $\mathbf{E}(\mathbf{k}, \omega)$ in (1.46) is only defined for $\omega > 0$; therefore, the integral in (1.47) over the frequency for obtaining $\mathbf{E}(\mathbf{r}, t)$ through the inverse Fourier transform only extends over positive values of ω. This is in accordance with the convention we use in (1.37) to define the complex field $\mathbf{E}(\mathbf{r}, t)$. All other space- and time-dependent quantities, including other field vectors and the permittivity and susceptibility tensors, are transformed in a similar manner.

Through the Fourier transform, the convolution integrals in the real space and time domain become simple products in the momentum space and frequency domain. Consequently, we have

$$\mathbf{P}(\mathbf{k}, \omega) = \epsilon_0 \boldsymbol{\chi}(\mathbf{k}, \omega) \cdot \mathbf{E}(\mathbf{k}, \omega), \tag{1.48}$$

and

$$\mathbf{D}(\mathbf{k}, \omega) = \epsilon_0 [\mathbf{I} + \boldsymbol{\chi}(\mathbf{k}, \omega)] \cdot \mathbf{E}(\mathbf{k}, \omega) = \boldsymbol{\epsilon}(\mathbf{k}, \omega) \cdot \mathbf{E}(\mathbf{k}, \omega). \tag{1.49}$$

Note that in the real space and time domain $\mathbf{P}(\mathbf{r}, t)$ and $\mathbf{D}(\mathbf{r}, t)$ are connected to $\mathbf{E}(\mathbf{r}, t)$ through convolution integrals in space and time, whereas in the momentum space and frequency domain $\mathbf{P}(\mathbf{k}, \omega)$ and $\mathbf{D}(\mathbf{k}, \omega)$ are connected to $\mathbf{E}(\mathbf{k}, \omega)$ through direct products.

1.2 INTERACTION HAMILTONIANS

Light–matter interaction is characterized by an *interaction Hamiltonian*. An optical field interacts with a material through its interaction with the charged particles in the material, which include the negatively charged electrons and the positively charged atomic nuclei. Because the mass of an atomic nucleus is three to four orders of magnitude larger than that of an electron, the response of an atomic nucleus to an optical field is negligible compared to the response of an electron unless specific nuclear resonances are the focus of attention. For this reason, we only consider the interaction of an optical field with an electron, which has a negative charge of $q = -e$ with $e = 1.602 \times 10^{-19}$ C being the elementary charge. In the following discussion, we take the Coulomb gauge for the optical field, as discussed in Subsection 1.1.1.

An electron interacts with an optical field through its interaction with both the electric and the magnetic fields. The interaction of an electron with the electric and magnetic fields leads to an addition of a term of $q\varphi = -e\varphi$ to the Hamiltonian of the electron and the replacement of the momentum operator $\hat{\mathbf{p}}$ of the electron with $\hat{\mathbf{p}} - qA = \hat{\mathbf{p}} + eA$, where φ is the scalar potential and A is the vector potential, both defined in Subsection 1.1.1. The interaction of the electron with the magnetic field also depends on the band structure of the system [10, 11] and further on the spin of the electron [12]. Therefore, it cannot be generalized for all material systems. In the following, we consider two common systems:

1. *Schrödinger electron.* A Schrödinger electron is nonrelativistic with a nonzero mass. It can be an electron in an atom, wherein the electron mass is its rest mass m_0, or an electron in a parabolic band structure, such as that of a semiconductor, wherein the electron has an effective mass of m^*.
2. *Dirac electron.* A Dirac electron is relativistic with a zero mass. It is an electron in a linear band structure, such as that of monolayer graphene, wherein the electron behaves as a massless electron.

1.2.1 Schrödinger Electron

The quantum-mechanical behavior of a nonrelativistic electron that has a nonzero mass can be described by the *Schrödinger equation*. This applies to an electron in a parabolic energy band and an electron in an atom. For an electron of an effective mass of m^* in a parabolic band structure and a rest mass of m_0, the total Hamiltonian in the presence of an optical field is [13]

$$\hat{H} = \frac{1}{2m^*} (\hat{\mathbf{p}} + eA)^2 + \frac{g_{\mathrm{s}} e}{2m_0 \hbar} i\hat{\mathbf{S}} \cdot (\hat{\mathbf{p}} \times A + A \times \hat{\mathbf{p}}) - e\varphi + V, \tag{1.50}$$

where A and φ are the vector and scalar potentials, respectively, of the light field, $\hat{\mathbf{p}}$ is the *linear momentum operator* of the electron, $\hat{\mathbf{S}}$ is the *spin angular momentum operator* of the electron, $g_{\mathrm{s}} = 2.0023$ is the *electronic spin gyromagnetic factor*, commonly known as the *electronic spin g-factor*, and V is some additional potential in the system that is independent of the optical field. Note that because the Hamiltonian has to be a Hermitian operator, it has to be expressed in terms of the real fields A, φ, and V, but not in terms of complex fields. By contrast, the quantum-mechanical operators $\hat{\mathbf{p}}$ and $\hat{\mathbf{S}}$ need not be real but have to be Hermitian operators. Note that in (1.50) the effective mass m^* is used in the first term for the motion of the electron in a parabolic band [10, 11], but the electron rest mass is used in the second term for the interaction of the electron spin with the optical field [12, 13]. For an electron in an atom, we can take $m^* = m_0$.

The electron spin angular momentum operator $\hat{\mathbf{S}}$ cannot be expressed as a function of space and time because spin is an intrinsic property of an electron that is independent of the spatial variable. It can be expressed in the matrix form:

$$\hat{\mathbf{S}} = \frac{\hbar}{2}\boldsymbol{\sigma}, \tag{1.51}$$

where $\boldsymbol{\sigma}$ is the *Pauli vector*:

$$\boldsymbol{\sigma} = \begin{bmatrix} 0 & 1 \\ 1 & 0 \end{bmatrix}\hat{x} + \begin{bmatrix} 0 & -i \\ i & 0 \end{bmatrix}\hat{y} + \begin{bmatrix} 1 & 0 \\ 0 & -1 \end{bmatrix}\hat{z}. \tag{1.52}$$

The total Hamiltonian in the presence of the optical field, given in (1.50), can be expressed as the combination of two parts:

$$\hat{H} = \hat{H}_0 + \hat{H}_{\text{int}}, \tag{1.53}$$

where

$$\hat{H}_0 = \frac{\hat{p}^2}{2m^*} + V \tag{1.54}$$

is the Hamiltonian of the electron in the absence of the optical field, and

$$\hat{H}_{\text{int}} = \frac{1}{2m^*}\left(e\hat{\mathbf{p}}\cdot\mathbf{A} + e\mathbf{A}\cdot\hat{\mathbf{p}} + e^2A^2\right) + \frac{g_s e}{2m_0\hbar}i\hat{\mathbf{S}}\cdot(\hat{\mathbf{p}}\times\mathbf{A} + \mathbf{A}\times\hat{\mathbf{p}}) - e\varphi \tag{1.55}$$

is the *interaction Hamiltonian* that describes the interaction between the optical field and the electron. When the interaction takes place in a source-free region, the charge density is absent, such that $\rho = 0$ and the scalar potential can be taken as $\varphi = 0$. Then, [14]

$$\hat{H}_{\text{int}} = \frac{1}{2m^*}\left(e\hat{\mathbf{p}}\cdot\mathbf{A} + e\mathbf{A}\cdot\hat{\mathbf{p}} + e^2A^2\right) + \frac{g_s e}{2m_0\hbar}i\hat{\mathbf{S}}\cdot(\hat{\mathbf{p}}\times\mathbf{A} + \mathbf{A}\times\hat{\mathbf{p}}). \tag{1.56}$$

In (1.56), the first compound term, which consists of three terms of dot products, describes the interaction of the optical field with the electron in real space. This interaction causes the motion and displacement of the electron in real space; thus, it depends on the effective mass m^* that is determined by the band structure of the material in which the electron resides [10, 11]. The second compound term, which consists of two terms of cross products, describes the interaction of the optical field with the spin of the electron [12, 13]. This interaction might cause the electron to make a transition between different spin states. This interaction does not depend on the effective mass, but only on the electron rest mass m_0 because it does not depend on the band structure of the material.

In the Coulomb gauge taken here, $\mathbf{V} \cdot A = 0$. Then, we find that $\hat{\mathbf{p}}$ and A commute:

$$\hat{\mathbf{p}} \cdot A = \frac{\hbar}{i} \mathbf{V} \cdot A + A \cdot \hat{\mathbf{p}} = A \cdot \hat{\mathbf{p}}. \tag{1.57}$$

We also find that

$$\hat{\mathbf{p}} \times A = \frac{\hbar}{i} (\mathbf{V} \times A) - A \times \hat{\mathbf{p}}, \tag{1.58}$$

so that the interaction of the spin of the electron with the optical field can be expressed as

$$\frac{g_s e}{2m_0 \hbar} i \hat{\mathbf{S}} \cdot (\hat{\mathbf{p}} \times A + A \times \hat{\mathbf{p}}) = \frac{g_s e}{2m_0} \hat{\mathbf{S}} \cdot (\mathbf{V} \times A) = \frac{g_s e}{2m_0} \hat{\mathbf{S}} \cdot B = \frac{g_s \mu_B}{\hbar} \hat{\mathbf{S}} \cdot B, \tag{1.59}$$

where

$$\mu_B = \frac{e\hbar}{2m_0} \tag{1.60}$$

is the *Bohr magneton*.

By using (1.57) and (1.59), the interaction Hamiltonian given in (1.56) can be expressed as

$$\hat{H}_{int} = \frac{e}{m^*} A \cdot \hat{\mathbf{p}} + \frac{e^2}{2m^*} A^2 + \frac{g_s \mu_B}{\hbar} \hat{\mathbf{S}} \cdot B. \tag{1.61}$$

In (1.61), the second term, $e^2 A^2 / 2m^*$, is known as the *diamagnetic interaction* [15], which is generally much smaller than the first term, $eA \cdot \hat{\mathbf{p}} / m^*$, except for an interaction in a very high field; thus, it is often neglected to write the interaction Hamiltonian approximately as

$$\hat{H}_{int} \approx \frac{e}{m^*} A \cdot \hat{\mathbf{p}} + \frac{g_s \mu_B}{\hbar} \hat{\mathbf{S}} \cdot B. \tag{1.62}$$

For an electron in a parabolic energy band, we take m^* to be its effective mass in this band. For an electron in an atom, we take m^* to be the electron rest mass, $m^* = m_0$.

1.2.2 Dirac Electron

The quantum-mechanical behavior of a relativistic electron is described by the *Dirac equation*. An electron in a linear band structure has a zero effective mass; it behaves like a *massless Dirac electron*, though it might not be moving at a relativistic speed [16]. Here we specifically consider the case of an electron in intrinsic monolayer graphene, which is a two-dimensional material that has linear conduction and valence bands [16]. The total Hamiltonian of an electron in intrinsic monolayer graphene in the presence of an optical field is [16]

$$\hat{H} = v_{\mathrm{F}}\boldsymbol{\sigma} \cdot (\hat{\mathbf{p}} + e\boldsymbol{A}) - e\varphi, \tag{1.63}$$

where $\boldsymbol{A}$ and φ are the vector and scalar potentials, respectively, of the light field, $v_{\mathrm{F}} \approx 1.0 \times 10^6 \, \mathrm{m\,s}^{-1}$ is the *Fermi velocity* in graphene, and $\boldsymbol{\sigma}$ is the Pauli vector given in (1.52).

The total Hamiltonian of an electron in graphene in the presence of the optical field, given in (1.63), can be expressed as the combination of two parts:

$$\hat{H} = \hat{H}_0 + \hat{H}_{\mathrm{int}}, \tag{1.64}$$

where

$$\hat{H}_0 = v_{\mathrm{F}}\boldsymbol{\sigma} \cdot \hat{\mathbf{p}} \tag{1.65}$$

is the Hamiltonian of the electron in intrinsic monolayer graphene in the absence of the optical field [16], and

$$\hat{H}_{\mathrm{int}} = v_{\mathrm{F}}e\boldsymbol{\sigma} \cdot \boldsymbol{A} - e\varphi \tag{1.66}$$

is the *interaction Hamiltonian* that describes the interaction between the optical field and the electron in intrinsic monolayer graphene [16]. When the interaction takes place in a source-free region, the charge density is absent such that $\rho = 0$ and $\varphi = 0$. Then,

$$\hat{H}_{\mathrm{int}} = v_{\mathrm{F}}e\boldsymbol{\sigma} \cdot \boldsymbol{A}. \tag{1.67}$$

1.3 TRANSFORMATION OF HAMILTONIANS

The total Hamiltonian and the interaction Hamiltonian of an electron in the presence of an optical field can be transformed into different forms for the convenience of expression and analysis without changing the physics by maintaining the invariance of physical laws in the process of the transformation. For a quantum-mechanical system, such a transformation has to be a *unitary transformation* that maintains the invariance of the time-dependent Schrödinger equation of the system. For the total Hamiltonian and the interaction Hamiltonian that involve light–matter interaction, such a transformation can be performed simply through the transformation of $\boldsymbol{A}$ and φ as given in (1.12) by choosing any smooth scalar function $\Lambda(\mathbf{r}, t)$. The transformation given in (1.12) for both the total Hamiltonian and the interaction Hamiltonian can be formally shown by performing a proper unitary transformation on the total Hamiltonian, followed by subtracting $\hat{H}_0$ from the transformed total Hamiltonian to find the transformed interaction Hamiltonian.

Consider the time-dependent Schrödinger equation:

$$\hat{H}\psi = i\hbar \frac{\partial \psi}{\partial t}, \tag{1.68}$$

where $\hat{H}$ is the total Hamiltonian as discussed in the preceding section and ψ is the space- and time-dependent wave function $\psi(\mathbf{r}, t)$ of an electron in a system under consideration. Our task is to find a proper unitary transformation that simultaneously transforms $\hat{H}$ and ψ by using a unitary matrix U without changing the form of (1.68) to maintain the invariance of the Schrödinger equation, which is a physical law. Because U is unitary, $UU^\dagger = U^\dagger U = I$, where $U^\dagger$ is the Hermitian conjugate of U and I is the identity matrix. By multiplying U on both sides of (1.68), we find

$$
\begin{aligned}
&U\hat{H}\psi = i\hbar U \frac{\partial}{\partial t}\psi \\
\Rightarrow \quad &U\hat{H}U^\dagger U\psi = i\hbar U \frac{\partial}{\partial t}(U^\dagger U\psi) = i\hbar U\left(\frac{\partial U^\dagger}{\partial t} + U^\dagger \frac{\partial}{\partial t}\right)(U\psi) \\
\Rightarrow \quad &\left(U\hat{H}U^\dagger - i\hbar U \frac{\partial U^\dagger}{\partial t}\right)(U\psi) = i\hbar \frac{\partial}{\partial t}(U\psi) \\
\Rightarrow \quad &\hat{H}'\psi' = i\hbar \frac{\partial}{\partial t}\psi'.
\end{aligned}
\tag{1.69}
$$

Therefore, the unitary transformation of $\hat{H}$ and ψ that keeps the time-dependent Schrödinger equation invariant is

$$
\hat{H} \;\rightarrow\; \hat{H}' = U\hat{H}U^\dagger - i\hbar U\frac{\partial U^\dagger}{\partial t} \quad \text{and} \quad \psi \;\rightarrow\; \psi' = U\psi.
\tag{1.70}
$$

To make certain that U is a unitary matrix, we choose it in the form:

$$
U = \exp\left(-i\frac{e\Lambda}{\hbar}\right) \quad \text{and} \quad U^\dagger = \exp\left(i\frac{e\Lambda}{\hbar}\right),
\tag{1.71}
$$

where $\Lambda(\mathbf{r}, t)$ is any smooth scalar function of space and time.

The specific form of the total Hamiltonian $\hat{H}$ and that of the interaction Hamiltonian $\hat{H}_{\text{int}}$ both depend on the band structure. Therefore, their specific forms $\hat{H}'$ and $\hat{H}'_{\text{int}}$ after the unitary transformation of (1.70) also depend on the band structure. Nonetheless, as shown below, one conclusion is true for any system irrespective of its band structure: The forms of $\hat{H}'$ and $\hat{H}'_{\text{int}}$ can be written out by simply carrying out the transformation of A and φ as given in (1.12). In other words, the transformation of A and φ given in (1.12) gives the correct forms of $\hat{H}'$ and $\hat{H}'_{\text{int}}$ that are obtained through a unitary transformation of the system, without the need to go through the process of a formal unitary transformation as described above.

1.3.1 Schrödinger Electron

For the interaction of an optical field with an electron in a parabolic band, we consider the total Hamiltonian given in (1.50), but use (1.59) for the interaction of the electron spin with the optical field:

$$\hat{H} = \frac{1}{2m^*} (\hat{\mathbf{p}} + e\mathbf{A})^2 + \frac{g_s \mu_B}{\hbar} \hat{\mathbf{S}} \cdot \mathbf{B} - e\varphi + V. \tag{1.72}$$

Then, the interaction Hamiltonian given in (1.55) can be expressed as

$$\hat{H}_{int} = \frac{e}{m^*} \mathbf{A} \cdot \hat{\mathbf{p}} + \frac{e^2}{2m^*} A^2 + \frac{g_s \mu_B}{\hbar} \hat{\mathbf{S}} \cdot \mathbf{B} - e\varphi, \tag{1.73}$$

where $\hat{\mathbf{p}} \cdot \mathbf{A} = \mathbf{A} \cdot \hat{\mathbf{p}}$, as given in (1.57), is used.

Because $\mathbf{A}$, $\mathbf{B}$, φ, and V are functions of only space and time, and $\hat{\mathbf{S}}$ is independent of either space or time, they all commute with the scalar function $\Lambda(\mathbf{r}, t)$ of only space and time. In (1.72), only the momentum operator $\hat{\mathbf{p}}$ does not commute with $\Lambda(\mathbf{r}, t)$ and, therefore, does not commute with U, nor with $U^\dagger$, which are defined in (1.71). Thus,

$$U\hat{\mathbf{p}}U^\dagger = e^{-ie\Lambda/\hbar} \left(\frac{\hbar}{i} \mathbf{\nabla} \right) e^{ie\Lambda/\hbar} = \hat{\mathbf{p}} + e\mathbf{\nabla}\Lambda. \tag{1.74}$$

Then, from (1.70), we find that

$$
\begin{aligned}
\hat{H}' &= U\hat{H}U^\dagger - i\hbar U \frac{\partial U^\dagger}{\partial t} \\
&= \frac{1}{2m^*} (\hat{\mathbf{p}} + e\mathbf{\nabla}\Lambda + e\mathbf{A})^2 + \frac{g_s \mu_B}{\hbar} \hat{\mathbf{S}} \cdot \mathbf{B} - e\varphi + V + e\frac{\partial \Lambda}{\partial t} \\
&= \frac{1}{2m^*} [\hat{\mathbf{p}} + e(\mathbf{A} + \mathbf{\nabla}\Lambda)]^2 + \frac{g_s \mu_B}{\hbar} \hat{\mathbf{S}} \cdot \mathbf{B} - e\left(\varphi - \frac{\partial \Lambda}{\partial t} \right) + V \\
&= \frac{1}{2m^*} (\hat{\mathbf{p}} + e\mathbf{A}')^2 + \frac{g_s \mu_B}{\hbar} \hat{\mathbf{S}} \cdot \mathbf{B} - e\varphi' + V.
\end{aligned}
\tag{1.75}
$$

By taking $\hat{H}_0$ from (1.54), the transformed interaction Hamiltonian has the form:

$$
\begin{aligned}
\hat{H}'_{int} &= \hat{H}' - \hat{H}_0 \\
&= \frac{e}{m^*} (\mathbf{A} + \mathbf{\nabla}\Lambda) \cdot \hat{\mathbf{p}} + \frac{e^2}{2m^*} (\mathbf{A} + \mathbf{\nabla}\Lambda)^2 + \frac{g_s \mu_B}{\hbar} \hat{\mathbf{S}} \cdot \mathbf{B} - e\left(\varphi - \frac{\partial \Lambda}{\partial t} \right) \\
&= \frac{e}{m^*} \mathbf{A}' \cdot \hat{\mathbf{p}} + \frac{e^2}{2m^*} A'^2 + \frac{g_s \mu_B}{\hbar} \hat{\mathbf{S}} \cdot \mathbf{B} - e\varphi'.
\end{aligned}
\tag{1.76}
$$

We see from (1.75) and (1.76) that both $\hat{H}'$ and $\hat{H}'_{int}$ can be obtained by directly taking the transformation that is given in (1.12) for $\mathbf{A}$ and φ.

1.3.2 Dirac Electron

For the interaction of an optical field with an electron in a linear band, we consider the total Hamiltonian given in (1.63) for an electron in intrinsic monolayer graphene in the presence of an optical field:

$$\hat{H} = v_F \boldsymbol{\sigma} \cdot (\hat{\mathbf{p}} + e\mathbf{A}) - e\varphi. \tag{1.77}$$

Then, the interaction Hamiltonian is that given in (1.66):

$$\hat{H}_{\text{int}} = v_{\text{F}} e \boldsymbol{\sigma} \cdot \boldsymbol{A} - e\varphi. \tag{1.78}$$

Because $\boldsymbol{A}$ and φ are functions of only space and time, and $\boldsymbol{\sigma}$ is independent of either space or time, they all commute with the scalar function $\Lambda(\mathbf{r}, t)$ of only space and time. In (1.77), only the momentum operator $\hat{\mathbf{p}}$ does not commute with $\Lambda(\mathbf{r}, t)$ and, therefore, does not commute with U, nor with $U^\dagger$. Then, by using (1.74), we find from (1.70) that

$$\begin{aligned}
\hat{H}' &= U\hat{H}U^\dagger - i\hbar U \frac{\partial U^\dagger}{\partial t} \\
&= v_{\text{F}} \boldsymbol{\sigma} \cdot (\hat{\mathbf{p}} + e\boldsymbol{\nabla}\Lambda + e\boldsymbol{A}) - e\varphi + e\frac{\partial \Lambda}{\partial t} \\
&= v_{\text{F}} \boldsymbol{\sigma} \cdot [\hat{\mathbf{p}} + e(\boldsymbol{A} + \boldsymbol{\nabla}\Lambda)] - e\left(\varphi - \frac{\partial \Lambda}{\partial t}\right) \\
&= v_{\text{F}} \boldsymbol{\sigma} \cdot (\hat{\mathbf{p}} + e\boldsymbol{A}') - e\varphi'.
\end{aligned} \tag{1.79}$$

By taking $\hat{H}_0$ from (1.65), the interaction Hamiltonian is transformed as

$$\begin{aligned}
\hat{H}'_{\text{int}} &= \hat{H}' - \hat{H}_0 \\
&= v_{\text{F}} e \boldsymbol{\sigma} \cdot (\boldsymbol{A} + \boldsymbol{\nabla}\Lambda) - e\left(\varphi - \frac{\partial \Lambda}{\partial t}\right) \\
&= v_{\text{F}} e \boldsymbol{\sigma} \cdot \boldsymbol{A}' - e\varphi'.
\end{aligned} \tag{1.80}$$

We see again that both $\hat{H}'$ and $\hat{H}'_{\text{int}}$ can be obtained by directly taking the transformation that is given in (1.12) for $\boldsymbol{A}$ and φ.

1.4 MULTIPOLE EXPANSION

The interaction Hamiltonian obtained in (1.61), or its approximate form given in (1.62), for an electron in an atom or in a parabolic band, and that obtained in (1.67) for an electron in the linear band of monolayer graphene, can be directly used to solve a problem of light–matter interaction in a source-free region where the scalar potential can be taken as $\varphi = 0$. However, because they are expressed in terms of the vector potential $\boldsymbol{A}$ but $\boldsymbol{A}$ is not uniquely defined for a given optical field, the usefulness of these expressions depends on the proper choice of the vector potential. By contrast, the $\boldsymbol{E}$ and $\boldsymbol{B}$ fields of an optical field are uniquely defined and are most commonly used to describe the optical field. Therefore, it is usually more convenient to express an interaction Hamiltonian in terms of $\boldsymbol{E}$ and $\boldsymbol{B}$ instead of $\boldsymbol{A}$. For this purpose, an interaction Hamiltonian is expanded as a series of electric and magnetic *multipole interactions*.

As shown in the preceding section, the form of an interaction Hamiltonian that is found through a proper unitary transformation of a system can be obtained through the transformation of $\boldsymbol{A}$ and φ, as given in (1.12), by properly choosing a smooth scalar function $\Lambda(\mathbf{r}, t)$ of space and time. Because the interaction Hamiltonians for parabolic and linear bands are different, the multipole expansions of them are different, though the transformation of $\boldsymbol{A}$ and φ through the same $\Lambda(\mathbf{r}, t)$ applies to both cases, as seen below.

From (1.12) and the results found in the preceding section, $\boldsymbol{A}$ and φ are transformed as

$$A' = A + \nabla\Lambda \quad \text{and} \quad \varphi' = \varphi - \frac{\partial\Lambda}{\partial t}. \tag{1.81}$$

For light–matter interaction in a source-free region to be considered in the following, $\varphi = 0$ but $\varphi' = -\partial\Lambda/\partial t \neq 0$. Therefore, for multipole expansion, we have to find A' and φ' by properly choosing a smooth scalar function $\Lambda(\mathbf{r}, t)$ for the expansion. Because $A(\mathbf{r}, t)$ and $\Lambda(\mathbf{r}, t)$ are both functions of the space variable $\mathbf{r}$, we shall expand them with respect to a center of motion for the electron, which is commonly the center of charge. For example, in an atom, the center of motion is commonly taken to be the location of the nucleus. For a center of motion that is located at $\mathbf{R}$, the location $\mathbf{r}$ of the electron can be expressed as

$$\mathbf{r} = \mathbf{R} + \xi, \tag{1.82}$$

where ξ is the relative spatial displacement of the electron with respect to the center of motion that is located at $\mathbf{R}$, as shown in Fig. 1.1(a).

With $\xi = \mathbf{r} - \mathbf{R}$, as given in (1.82), $A(\mathbf{r}, t)$ can be expanded with respect to $\mathbf{R}$ as

$$A(\mathbf{r}, t) = \sum_{k=0}^{\infty} \frac{1}{k!} (\xi \cdot \nabla_{\mathbf{R}})^k A(\mathbf{R}, t), \tag{1.83}$$

where $\nabla_{\mathbf{R}}$ takes the vectorial spatial derivative with respect to the variable $\mathbf{R}$. For multipole expansion based on the series expansion of $A(\mathbf{r}, t)$ given above, the proper choice of $\Lambda(\mathbf{r}, t)$ is

$$\Lambda(\mathbf{r}, t) = -\int_0^1 \xi \cdot A(\mathbf{R} + u\xi, t)du = -\sum_{k=0}^{\infty} \frac{1}{(k+1)!} \xi^{k+1} \vdots \nabla_{\mathbf{R}}^k A(\mathbf{R}, t). \tag{1.84}$$

With $\varphi = 0$ for the interaction in a source-free region, we have $E = -\partial A/\partial t$ and $B = \nabla \times A$ from (1.8) and (1.9), respectively. Then, by using (1.12), we find after some tedious vector algebra that

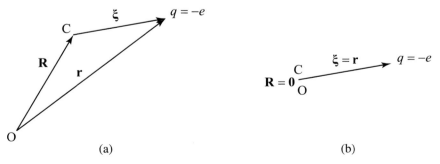

(a)   (b)

Figure 1.1 An electron at the location $\mathbf{r}$ moving around its center of motion located at $\mathbf{R}$ for a relative spatial displacement of ξ. The point O is the origin of the spatial coordinate, and the point C is the center of motion for the electron. (a) $\mathbf{R} \neq \mathbf{0}$ so that $\xi \neq \mathbf{r}$. In this case, C is located away from the origin O. (b) $\mathbf{R} = \mathbf{0}$ so that $\xi = \mathbf{r}$. In this case, the origin O is chosen to be at the center of motion C.

$$A'(\mathbf{r}, t) = A(\mathbf{r}, t) + \nabla\Lambda(\mathbf{r}, t)$$

$$= \sum_{k=0}^{\infty} \frac{1}{k!} (\boldsymbol{\xi} \cdot \nabla_{\mathbf{R}})^k A(\mathbf{R}, t) - \nabla \sum_{k=0}^{\infty} \frac{1}{(k+1)!} \boldsymbol{\xi}^{k+1} \vdots \nabla_{\mathbf{R}}^k A(\mathbf{R}, t)$$

$$= \sum_{k=0}^{\infty} \frac{1}{k!} (\boldsymbol{\xi} \cdot \nabla_{\mathbf{R}})^k A - \sum_{k=0}^{\infty} \frac{1}{(k+1)!} (\boldsymbol{\xi} \cdot \nabla_{\mathbf{R}})^k A - \sum_{k=0}^{\infty} \frac{k}{(k+1)!} (\boldsymbol{\xi} \cdot \nabla_{\mathbf{R}})^{k-1} \nabla_{\mathbf{R}} (\boldsymbol{\xi} \cdot A)$$

$$= \sum_{k=0}^{\infty} \frac{k}{(k+1)!} (\boldsymbol{\xi} \cdot \nabla_{\mathbf{R}})^k A - \sum_{k=0}^{\infty} \frac{k}{(k+1)!} (\boldsymbol{\xi} \cdot \nabla_{\mathbf{R}})^{k-1} \nabla_{\mathbf{R}} (\boldsymbol{\xi} \cdot A) \qquad (1.85)$$

$$= \sum_{k=0}^{\infty} \frac{k}{(k+1)!} (\boldsymbol{\xi} \cdot \nabla_{\mathbf{R}})^{k-1} [\boldsymbol{\xi} \cdot \nabla_{\mathbf{R}} A - \nabla_{\mathbf{R}} (\boldsymbol{\xi} \cdot A)]$$

$$= -\sum_{k=0}^{\infty} \frac{k+1}{(k+2)!} (\boldsymbol{\xi} \cdot \nabla_{\mathbf{R}})^k \boldsymbol{\xi} \times [\nabla_{\mathbf{R}} \times A(\mathbf{R}, t)]$$

$$= -\sum_{k=0}^{\infty} \frac{k+1}{(k+2)!} (\boldsymbol{\xi} \cdot \nabla_{\mathbf{R}})^k \boldsymbol{\xi} \times B(\mathbf{R}, t),$$

and

$$\varphi'(\mathbf{r}, t) = -\frac{\partial \Lambda(\mathbf{r}, t)}{\partial t}$$

$$= \sum_{k=0}^{\infty} \frac{1}{(k+1)!} \boldsymbol{\xi}^{k+1} \vdots \nabla_{\mathbf{R}}^k \frac{\partial A(\mathbf{R}, t)}{\partial t} \qquad (1.86)$$

$$= -\sum_{k=0}^{\infty} \frac{1}{(k+1)!} \boldsymbol{\xi}^{k+1} \vdots \nabla_{\mathbf{R}}^k E(\mathbf{R}, t).$$

1.4.1 Schrödinger Electron

For an electron in a parabolic band or in an atom, we find by using (1.85) and (1.86) for (1.76) that the interaction Hamiltonian can be expressed as [17]

$$\hat{H}'_{\text{int}} = \frac{e}{m^*} A' \cdot \hat{\mathbf{p}} + \frac{e^2}{2m^*} A'^2 + \frac{g_s \mu_B}{\hbar} \hat{S} \cdot B - e\varphi'$$

$$= -\frac{e}{m^*} \sum_{k=0}^{\infty} \frac{k+1}{(k+2)!} (\boldsymbol{\xi} \cdot \nabla_{\mathbf{R}})^k \boldsymbol{\xi} \times B \cdot \hat{\mathbf{p}} + \frac{e^2}{2m^*} A'^2 + \frac{g_s \mu_B}{\hbar} \hat{S} \cdot B + e\sum_{k=0}^{\infty} \frac{1}{(k+1)!} \boldsymbol{\xi}^{k+1} \vdots \nabla_{\mathbf{R}}^k E$$

$$= e\boldsymbol{\xi} \cdot E + \frac{e}{2} \boldsymbol{\xi}\boldsymbol{\xi} : \nabla_{\mathbf{R}} E + \cdots + \frac{e}{2m^*} \boldsymbol{\xi} \times \hat{\mathbf{p}} \cdot B + \frac{g_s \mu_B}{\hbar} \hat{S} \cdot B + \cdots. \qquad (1.87)$$

By resetting the origin of the spatial coordinate at the center of motion so that $\mathbf{R} = 0$ and $\boldsymbol{\xi} = \mathbf{r}$, as seen in Fig. 1.1(b), the interaction Hamiltonian can be expressed as

$$\hat{H}_{\text{int}} = e\mathbf{r} \cdot E(\mathbf{0}, t) + \frac{e}{2} \mathbf{rr} : \nabla E(\mathbf{0}, t) + \cdots + \frac{e}{2m^*} \mathbf{r} \times \hat{\mathbf{p}} \cdot B(\mathbf{0}, t) + \frac{g_s \mu_B}{\hbar} \hat{S} \cdot B(\mathbf{0}, t) + \cdots$$

$$= -\hat{\mathbf{p}} \cdot E(\mathbf{0}, t) - \hat{Q} : \nabla E(\mathbf{0}, t) + \cdots - \hat{\mathbf{m}}_1 \cdot B(\mathbf{0}, t) - \hat{\mathbf{m}}_s \cdot B(\mathbf{0}, t) + \cdots \qquad (1.88)$$

$$= -\hat{\mathbf{p}} \cdot E(\mathbf{0}, t) - \hat{Q} : \nabla E(\mathbf{0}, t) + \cdots - \hat{\mathbf{m}} \cdot B(\mathbf{0}, t) + \cdots,$$

where the strengths and the gradients of the electric field and the magnetic induction are all evaluated at $\mathbf{R} = \mathbf{0}$,

$$\hat{p} = q\mathbf{r} = -e\mathbf{r} \quad \text{and} \quad \hat{Q} = \frac{q}{2}\mathbf{rr} = -\frac{e}{2}\mathbf{rr} \tag{1.89}$$

are the *electric dipole moment* and the *electric quadrupole moment* of an electron of charge $q = -e$, respectively, and

$$\hat{m}_1 = \frac{q}{2m^*}\mathbf{r} \times \hat{p} = -\frac{e}{2m^*}\mathbf{r} \times \hat{p} = -\frac{g_1\mu_B}{\hbar}\hat{\mathbf{L}} \quad \text{and} \quad \hat{m}_s = -\frac{g_s\mu_B}{\hbar}\hat{\mathbf{S}} \tag{1.90}$$

are the *orbital magnetic dipole moment* and the *spin magnetic dipole moment* of an electron of charge $q = -e$, respectively, and

$$\hat{m} = \hat{m}_1 + \hat{m}_s = -\frac{\mu_B}{\hbar}(g_1\hat{\mathbf{L}} + g_s\hat{\mathbf{S}}) \tag{1.91}$$

is the total *magnetic dipole moment* of the electron. In (1.91),

$$\hat{\mathbf{L}} = \mathbf{r} \times \hat{p} \tag{1.92}$$

is the *orbital angular momentum operator* of the electron, and $g_1 = m_0/m^*$ is the *effective orbital g-factor* of the electron [10]. In an atom, $g_1 = 1$. In a semiconductor, the value of g_1 depends on the band structure in which the electron resides.

1.4.2 Dirac Electron

For an electron in a linear band such as that of monolayer graphene, we find by using (1.85) and (1.86) for (1.80) that the interaction Hamiltonian can be expressed as [17]

$$\begin{aligned}\hat{H}'_{\text{int}} &= v_F e\boldsymbol{\sigma} \cdot \mathbf{A}' - e\varphi' \\ &= -v_F e\boldsymbol{\sigma} \cdot \sum_{k=0}^{\infty}\frac{k+1}{(k+2)!}(\boldsymbol{\xi} \cdot \nabla_{\mathbf{R}})^k\boldsymbol{\xi} \times \mathbf{B} + e\sum_{k=0}^{\infty}\frac{1}{(k+1)!}\boldsymbol{\xi}^{k+1}\vdots\nabla_{\mathbf{R}}^k\mathbf{E} \\ &= e\boldsymbol{\xi} \cdot \mathbf{E} + \frac{e}{2}\boldsymbol{\xi}\boldsymbol{\xi} : \nabla_{\mathbf{R}}\mathbf{E} + \cdots + \frac{v_F e}{2}\boldsymbol{\xi} \times \boldsymbol{\sigma} \cdot \mathbf{B} + \cdots.\end{aligned} \tag{1.93}$$

By resetting the origin of the spatial coordinate at the center of motion, as shown in Fig. 1.1(b), so that $\mathbf{R} = \mathbf{0}$ and $\boldsymbol{\xi} = \mathbf{r}$, the interaction Hamiltonian can be expressed as

$$\begin{aligned}\hat{H}_{\text{int}} &= e\mathbf{r} \cdot \mathbf{E}(\mathbf{0}, t) + \frac{e}{2}\mathbf{rr} : \nabla\mathbf{E}(\mathbf{0}, t) + \cdots + \frac{v_F e}{2}\mathbf{r} \times \boldsymbol{\sigma} \cdot \mathbf{B}(\mathbf{0}, t) + \cdots \\ &= -\hat{p} \cdot \mathbf{E}(\mathbf{0}, t) - \hat{Q} : \nabla\mathbf{E}(\mathbf{0}, t) + \cdots - \hat{m}_\sigma \cdot \mathbf{B}(\mathbf{0}, t) + \cdots,\end{aligned} \tag{1.94}$$

where the strengths and the gradients of the electric field and the magnetic induction are all evaluated at $\mathbf{R} = \mathbf{0}$; the electric dipole moment, $\hat{p} = q\mathbf{r} = -e\mathbf{r}$, and the electric quadrupole moment, $\hat{Q} = q\mathbf{rr}/2 = -e\mathbf{rr}/2$, are those of an electron of charge $q = -e$ as given in (1.89); and

$$\hat{\boldsymbol{m}}_\sigma = \frac{v_F q}{2}\mathbf{r} \times \boldsymbol{\sigma} = -\frac{v_F e}{2}\mathbf{r} \times \boldsymbol{\sigma} \qquad (1.95)$$

is the *relativistic magnetic dipole moment* of a massless Dirac electron of charge $q = -e$ in the linear band of monolayer graphene.

1.4.3 Electric Dipole Interaction

From the preceding two subsections, we find that, for both cases of nonrelativistic and relativistic electrons, the first term in the multipole expansion of the interaction Hamiltonian is the *electric dipole interaction*:

$$\hat{H}_{\mathrm{ED}} = -\hat{\boldsymbol{p}} \cdot \boldsymbol{E}(\mathbf{0}, t) = e\mathbf{r} \cdot \boldsymbol{E}(\mathbf{0}, t). \qquad (1.96)$$

As we shall see later, this is the most important term in light–matter interaction. It dominates most optical processes, including most nonlinear optical processes. We can estimate its magnitude by taking the displacement of an electron to be the order of an atom by using the *Bohr radius*, a_0, for the radius of an atom:

$$|\mathbf{r}| = r \approx a_0 = \frac{\hbar}{m_0 c \alpha}, \qquad (1.97)$$

where α is the *fine structure constant*:

$$\alpha = \frac{1}{4\pi\epsilon_0}\frac{e^2}{\hbar c} \approx \frac{1}{137}. \qquad (1.98)$$

The electric dipole has a magnitude of $p \approx er \approx ea_0$. Thus, we find that the electric dipole interaction has a magnitude of the order:

$$H_{\mathrm{ED}} \approx ea_0 E. \qquad (1.99)$$

The strength of the electric dipole interaction is linearly proportional to the strength of the electric field.

1.4.4 Electric Quadrupole Interaction

For both cases of nonrelativistic and relativistic electrons, the second term in the multipole expansion of the interaction Hamiltonian is the *electric quadrupole interaction*:

$$\hat{H}_{\mathrm{EQ}} = -\hat{\boldsymbol{Q}} : \boldsymbol{\nabla}\boldsymbol{E}(\mathbf{0}, t) = \frac{e}{2}\mathbf{r}\mathbf{r} : \boldsymbol{\nabla}\boldsymbol{E}(\mathbf{0}, t). \qquad (1.100)$$

The electric quadrupole has a magnitude of $Q \approx er^2/2 \approx ea_0^2/2$. The gradient of the electric field depends on the focus of the optical field and on the structure of the material; it is largest at a junction or an interface of a material. It has the order of magnitude of $|\boldsymbol{\nabla}\boldsymbol{E}| \approx E/\lambda$, where λ is the wavelength of the optical field. Thus, in the ordinary situation,

$$H_{EQ} \approx \frac{ea_0^2}{2} \frac{E}{\lambda}. \tag{1.101}$$

By comparing the electric quadrupole interaction to the electric dipole interaction, we find that in the ordinary situation, the electric quadrupole interaction is four to five orders of magnitude weaker than the electric dipole interaction:

$$\frac{H_{EQ}}{H_{ED}} \approx \frac{a_0}{2\lambda} \approx \frac{5.29 \times 10^{-11} \text{ m}}{2 \times 500 \times 10^{-9} \text{ m}} \approx 5 \times 10^{-5}, \tag{1.102}$$

where the Bohr radius is $a_0 = 5.29 \times 10^{-11}$ m and the optical wavelength is taken to be $\lambda = 500$ nm in the middle of the visible spectral range for this estimate. The electric quadrupole interaction can be observed in certain special circumstances, such as in a tightly confined interface [18] or nanostructure where the gradient of the electric field is strongly enhanced or in a material where the electric dipole interaction for a certain nonlinear process vanishes due to the symmetry of the material.

1.4.5 Magnetic Dipole Interaction

There is a *magnetic dipole interaction* in the multipole expansion of the interaction Hamiltonian, but it takes different forms for nonrelativistic and relativistic electrons. For a nonrelativistic Schrödinger electron, the magnetic dipole interaction has the form:

$$\hat{H}_{MD}^{S} = -\hat{\boldsymbol{m}} \cdot \boldsymbol{B}(\boldsymbol{0}, t) = -(\hat{\boldsymbol{m}}_l + \hat{\boldsymbol{m}}_s) \cdot \boldsymbol{B}(\boldsymbol{0}, t) = -\frac{\mu_B}{\hbar} (g_l \hat{\boldsymbol{L}} + g_s \hat{\boldsymbol{S}}) \cdot \boldsymbol{B}(\boldsymbol{0}, t). \tag{1.103}$$

For a massless Dirac electron in monolayer graphene, the magnetic dipole interaction has the form:

$$\hat{H}_{MD}^{D} = -\hat{\boldsymbol{m}}_{\sigma} \cdot \boldsymbol{B}(\boldsymbol{0}, t) = \frac{v_F e}{2} \boldsymbol{r} \times \boldsymbol{\sigma} \cdot \boldsymbol{B}(\boldsymbol{0}, t). \tag{1.104}$$

By taking $|\boldsymbol{r}| = r \approx a_0$, we can estimate the magnitude of the magnetic dipole interaction for both cases:

$$H_{MD}^{S} \approx \mu_B B \quad \text{and} \quad H_{MD}^{D} \approx \frac{v_F e}{2} a_0 B. \tag{1.105}$$

For an electromagnetic field, the magnitudes of the electric field and the magnetic induction have the relationship $E/B = c$ from Faraday's law given in (1.1). By using the relations in (1.60) for the Bohr magneton μ_B, (1.97) for the Bohr radius a_0, (1.98) for the fine structure constant α, and $v_F \approx 1 \times 10^6$ m s^{-1} for the Fermi velocity of an electron in monolayer graphene, we compare the magnetic dipole interaction to the electric dipole interaction:

$$\frac{H_{MD}^{S}}{H_{ED}} \approx \frac{\mu_B}{ea_0} \frac{B}{E} = \frac{\alpha}{2} \approx \frac{1}{274} \quad \text{and} \quad \frac{H_{MD}^{D}}{H_{ED}} \approx \frac{v_F e a_0 B/2}{ea_0 E} = \frac{v_F}{2c} \approx \frac{1}{600}. \tag{1.106}$$

We see for both Schrödinger and Dirac electrons that the magnetic dipole interaction is two to three orders of magnitude weaker than the electric dipole interaction. Therefore, its effect is not significant in the ordinary situation. The magnetic dipole interaction can be enhanced in the case of resonant *magnetic dipole transition* [8] when the optical frequency is in resonance with the transition.

1.5 DENSITY MATRIX FORMULATION

From the above discussions, the total Hamiltonian of an atomic, molecular, or electronic system that interacts with an optical field includes the *unperturbed Hamiltonian* $\hat{H}_0$ and the interaction Hamiltonian $\hat{H}_{int}$. In addition, the system also stochastically interacts with its surroundings through various random processes, such as the interactions with incoherent radiation, spontaneous emission, phonon absorption and emission, and elastic and inelastic collisions [19]. Stochastic processes always happen in a realistic system because they are fundamentally required by thermodynamic laws. Therefore, a *random-interaction Hamiltonian* $\hat{H}_{rand}$ that characterizes the stochastic interactions has to be included in the total Hamiltonian:

$$\hat{H} = \hat{H}_0 + \hat{H}_{int} + \hat{H}_{rand}. \tag{1.107}$$

The unperturbed Hamiltonian $\hat{H}_0$ of the system is time-independent; it defines the time-independent *eigenstates*, $|a\rangle$, and the corresponding *eigenenergies*, E_a, of the system through the solutions of the *time-independent Schrödinger equation*:

$$\hat{H}_0|a\rangle = E_a|a\rangle. \tag{1.108}$$

The interaction Hamiltonian $\hat{H}_{int}$ acts as a coherent perturbation to the system. As discussed in the preceding sections, it can be expressed in terms of the vector potential as given in (1.62) or (1.67). For direct coupling to the electric and magnetic fields that are used in expressing Maxwell's equations, it is most convenient to express the interaction Hamiltonian in the form of multipole expansion:

$$\hat{H}_{int} = -\hat{\boldsymbol{p}} \cdot \boldsymbol{E}(\boldsymbol{0}, t) - \hat{\boldsymbol{Q}} : \boldsymbol{\nabla}\boldsymbol{E}(\boldsymbol{0}, t) - \hat{\boldsymbol{m}} \cdot \boldsymbol{B}(\boldsymbol{0}, t) + \cdots, \tag{1.109}$$

or

$$\hat{H}_{int} = -\hat{\boldsymbol{p}} \cdot \boldsymbol{E}(\boldsymbol{0}, t) - \hat{\boldsymbol{Q}} : \boldsymbol{\nabla}\boldsymbol{E}(\boldsymbol{0}, t) - \hat{\boldsymbol{m}}_\sigma \cdot \boldsymbol{B}(\boldsymbol{0}, t) + \cdots. \tag{1.110}$$

By contrast, the random-interaction Hamiltonian $\hat{H}_{rand}$ cannot be generally expressed as an operator in an analytical manner because of its stochastic nature. Nevertheless, its effect can be accounted for through an ensemble average. For this reason, it is not convenient to solve for the wave function of the system through the Schrödinger equation. Instead, it is convenient to use the density matrix formulation.

1.5.1 Density Matrix Operator

The total Hamiltonian $\hat{H}$ is a function of time because both $\hat{H}_{int}$ and $\hat{H}_{rand}$ vary with time. The temporal evolution of the system is described by a temporally varying *state vector* $|\psi(t)\rangle$ that is governed by the *time-dependent Schrödinger equation* in the general form:

$$\hat{H}|\psi(t)\rangle = i\hbar \frac{\partial}{\partial t}|\psi(t)\rangle. \tag{1.111}$$

The time-dependent state vector $|\psi(t)\rangle$ can be expressed as a linear superposition of the eigenstates of the system with time-dependent coefficients:

$$|\psi(t)\rangle = \sum_a C_a(t)|a\rangle. \tag{1.112}$$

The problem of dealing with an ensemble of identical systems that all have the same total Hamiltonian $\hat{H}$ can be formulated with the *density matrix operator* that is defined as

$$\rho = \overline{|\psi\rangle\langle\psi|}, \tag{1.113}$$

which is an ensemble average of the product of the ket and the bra state vectors, as indicated by the overhead bar. The diagonal element $\rho_{aa} = \overline{C_a C_a^*}$ of the density matrix is the probability for a system in the ensemble to be in the level $|a\rangle$, thus describing the population distribution of the ensemble. The off-diagonal element $\rho_{ab} = \overline{C_a C_b^*}$ describes the relative coherence of the phase between levels $|a\rangle$ and $|b\rangle$ in the ensemble. A density matrix is a Hermitian matrix with a unity trace; its elements satisfy the relations:

$$\sum_a \rho_{aa} = 1 \quad \text{and} \quad \rho_{ab} = \rho_{ba}^*. \tag{1.114}$$

The density matrix can be used to calculate the ensemble average of the expectation value of a physical observable A that is represented by an operator $\hat{A}$ as

$$\overline{\langle A \rangle} = \text{Tr}\left(\rho\hat{A}\right) = \sum_a \left(\rho\hat{A}\right)_{aa} = \sum_{a,b} \rho_{ab}A_{ba}, \tag{1.115}$$

where Tr is the trace of the matrix – that is, the sum of the diagonal elements of the matrix. For example, the electric polarization of an optical medium can be found by using the density matrix ρ that describes the state of the medium and the operator $\hat{P}$ of the electric polarization, as

$$\overline{\langle P \rangle} = \text{Tr}\left(\rho\hat{P}\right). \tag{1.116}$$

1.5.2 Equation of Motion

To find the temporal evolution of the density matrix elements for the purpose of using (1.115) to find the temporal evolution of a physical observable, it is necessary to solve the *equation of motion for the density matrix*:

$$\frac{\partial \rho}{\partial t} = \frac{1}{i\hbar}[\hat{H}, \rho] = \frac{1}{i\hbar}[\hat{H}_0, \rho] + \frac{1}{i\hbar}[\hat{H}_{\text{int}}, \rho] + \frac{1}{i\hbar}[\hat{H}_{\text{rand}}, \rho], \tag{1.117}$$

which describes the temporal evolution of the ensemble through its interactions with the coherent radiation and with the surrounding random thermal reservoir. The equation of motion for the density matrix given above can be obtained from the time-dependent Schrödinger equation given in (1.111) by using the definition of the density matrix given in (1.113). Therefore, it is equivalent to the time-dependent Schrödinger equation.

There are three commutator terms on the right-hand side of (1.117). The first term contains the unperturbed Hamiltonian $\hat{H}_0$ of the system; the density matrix elements of this term can be easily calculated by using (1.108):

$$\frac{1}{i\hbar}[\hat{H}_0, \rho]_{aa} = \frac{1}{i\hbar}(\hat{H}_0\rho - \rho\hat{H}_0)_{aa} = \frac{1}{i\hbar}(E_a\rho_{aa} - \rho_{aa}E_a) = 0, \tag{1.118}$$

$$\frac{1}{i\hbar}[\hat{H}_0, \rho]_{ab} = \frac{1}{i\hbar}(\hat{H}_0\rho - \rho\hat{H}_0)_{ab} = \frac{1}{i\hbar}(E_a\rho_{ab} - \rho_{ab}E_b) = -i\omega_{ab}\rho_{ab}, \tag{1.119}$$

where

$$\omega_{ab} = \frac{E_a - E_b}{\hbar} = -\frac{E_b - E_a}{\hbar} = -\omega_{ba}. \tag{1.120}$$

It is clear from (1.120) that $\omega_{ab} > 0$ and $\omega_{ba} < 0$ if $E_a > E_b$, $\omega_{ab} < 0$ and $\omega_{ba} > 0$ if $E_a < E_b$, and $\omega_{aa} = 0$. The positive value $|\omega_{ab}| = |\omega_{ba}|$ is the *resonance frequency* between energy levels $|a\rangle$ and $|b\rangle$. The second term on the right-hand side of (1.117) contains the interaction Hamiltonian; it cannot be readily computed without solving the equation of motion after specifying its form of interaction. The third term is the stochastic term that cannot be computed by using an analytical operator but can be treated through a phenomenological approach discussed below.

The effect of the random interaction of the system with the thermal reservoir of its surroundings is basically to bring the system into thermal equilibrium with its surroundings. In thermal equilibrium, the population distribution among the energy levels is characterized by the thermodynamic equilibrium distribution, while the phase relationships among them are randomized. The density matrix operator, $\rho^{(0)}$, in thermal equilibrium is a diagonal matrix that has diagonal elements $\rho_{aa}^{(0)}$ given by the equilibrium population distribution function, such as the *Boltzmann distribution* for atomic and molecular systems and the *Fermi distribution* for electrons in a semiconductor or a metal, and vanishing off-diagonal elements $\rho_{ab}^{(0)} = 0$. The effect of $\hat{H}_{\text{rand}}$ is to thermalize the ensemble and to bring ρ back to $\rho^{(0)}$ from any deviation from thermal equilibrium. We can thus write

$$\frac{1}{i\hbar}[\hat{H}_{\text{rand}}, \rho] = \left(\frac{\partial \rho}{\partial t}\right)_{\text{rand}} \tag{1.121}$$

and

$$\frac{\partial \rho^{(0)}}{\partial t} = \left(\frac{\partial \rho^{(0)}}{\partial t}\right)_{\text{rand}} = 0. \tag{1.122}$$

The population relaxation caused by the interaction of the system with the surrounding thermal reservoir can be described as

$$\left(\frac{\partial \rho_{aa}}{\partial t}\right)_{\text{rand}} = \sum_{\substack{b \\ b \neq a}} W_{ba}^{\text{th}} \rho_{bb} - \sum_{\substack{b \\ b \neq a}} W_{ab}^{\text{th}} \rho_{aa}, \tag{1.123}$$

where $W_{ab}^{\text{th}} = W_{a \to b}^{\text{th}}$ is the total thermal transition rate from level $|a\rangle$ to level $|b\rangle$ and $W_{ba}^{\text{th}} = W_{b \to a}^{\text{th}}$ is the transition rate from $|b\rangle$ to $|a\rangle$.[2] These rates include all thermally induced transitions, including thermal excitation and thermal relaxation and including radiative and nonradiative processes [19]. They are determined by the temperature of the thermal reservoir through the temperature-dependent population distribution in thermal equilibrium:

$$W_{ba}^{\text{th}} \rho_{bb}^{(0)} = W_{ab}^{\text{th}} \rho_{aa}^{(0)} \tag{1.124}$$

from (1.122) and (1.123). For example, in an atomic or molecular system,

$$\rho_{aa}^{(0)} = \rho_{bb}^{(0)} \exp\left(-\frac{\hbar \omega_{ab}}{k_{\text{B}} T}\right) \quad \text{and} \quad W_{ba}^{\text{th}} = W_{ab}^{\text{th}} \exp\left(-\frac{\hbar \omega_{ab}}{k_{\text{B}} T}\right). \tag{1.125}$$

Thermal relaxation randomizes the phase between two energy levels among identical systems in the ensemble. Because the coherence of this phase is characterized by an off-diagonal element of the density matrix, we can write, for $b \neq a$,

$$\left(\frac{\partial \rho_{ab}}{\partial t}\right)_{\text{rand}} = -\gamma_{ab} \rho_{ab}, \tag{1.126}$$

where $\gamma_{ab} = \gamma_{ba}$ is the *phase relaxation rate* for the coherence between energy levels $|a\rangle$ and $|b\rangle$. The phase relaxation rate can have contributions from many physical mechanisms [19]; here we treat it as a phenomenological parameter without further discussion of such mechanisms.

By treating the terms containing $\hat{H}_0$ and $\hat{H}_{\text{rand}}$ in the manner discussed above, the equation of motion for the density matrix given in (1.117) can be expressed in terms of its diagonal and off-diagonal elements as

$$\frac{\partial \rho_{aa}}{\partial t} = \sum_{\substack{b \\ b \neq a}} W_{ba}^{\text{th}} \rho_{bb} - \sum_{\substack{b \\ b \neq a}} W_{ab}^{\text{th}} \rho_{aa} + \frac{1}{i\hbar} [\hat{H}_{\text{int}}, \rho]_{aa}, \tag{1.127}$$

$$\frac{\partial \rho_{ab}}{\partial t} = -i\omega_{ab} \rho_{ab} - \gamma_{ab} \rho_{ab} + \frac{1}{i\hbar} [\hat{H}_{\text{int}}, \rho]_{ab}, \tag{1.128}$$

---

[2] The subscript notation for the transition rate is not consistent in the literature. Some use W_{ab} to denote the transition from $|a\rangle$ to $|b\rangle$. Others use W_{ba}. Here we use the former notation.

where

$$\omega_{ab} = -\omega_{ba} \quad \text{and} \quad \gamma_{ab} = \gamma_{ba}, \tag{1.129}$$

but no simple relation exists for W_{ab}^{th} and W_{ba}^{th} except that both are real parameters.

1.5.3 Perturbation Expansion

Only in special cases can the equation of motion for the density matrix as described in (1.127) and (1.128) be solved in a closed form. In most cases, it is generally not possible. Then, there are two useful approaches:

1. In the case that the interaction Hamiltonian H_{int} is small compared to the Hamiltonian of the system such that $H_{\text{int}} \ll H_0 + H_{\text{rand}}$, it can be treated as a perturbation to the system by expanding the density matrix in a perturbation series.
2. In the case that the interaction Hamiltonian H_{int} is not small compared to the Hamiltonian of the system, it cannot be treated as a perturbation. Then the equation of motion has to be numerically solved if it does not have an analytical solution.

In this subsection, we consider the case that $H_{\text{int}} \ll H_0 + H_{\text{rand}}$ such that the interaction Hamiltonian can be considered as a perturbation and the density matrix can be expanded as a perturbation series:

$$\hat{H} = \hat{H}_0 + \hat{H}_{\text{rand}} + \hat{H}_{\text{int}}, \tag{1.130}$$

$$\rho = \rho^{(0)} + \rho^{(1)} + \rho^{(2)} + \cdots. \tag{1.131}$$

By writing $\hat{H} = \hat{H}_0 + \hat{H}_{\text{rand}} + \lambda \hat{H}_{\text{int}}$ and $\rho = \rho^{(0)} + \lambda \rho^{(1)} + \lambda^2 \rho^{(2)} + \cdots$ with a parameter λ to carry out the perturbation expansion, we obtain a series of *perturbation equations of motion*:

$$
\begin{aligned}
\frac{\partial \rho^{(0)}}{\partial t} &= \left(\frac{\partial \rho^{(0)}}{\partial t} \right)_{\text{rand}} + \frac{1}{i\hbar} \left[\hat{H}_0, \rho^{(0)} \right], \\
\frac{\partial \rho^{(1)}}{\partial t} &= \left(\frac{\partial \rho^{(1)}}{\partial t} \right)_{\text{rand}} + \frac{1}{i\hbar} \left[\hat{H}_0, \rho^{(1)} \right] + \frac{1}{i\hbar} \left[\hat{H}_{\text{int}}, \rho^{(0)} \right], \\
&\vdots \\
\frac{\partial \rho^{(n)}}{\partial t} &= \left(\frac{\partial \rho^{(n)}}{\partial t} \right)_{\text{rand}} + \frac{1}{i\hbar} \left[\hat{H}_0, \rho^{(n)} \right] + \frac{1}{i\hbar} \left[\hat{H}_{\text{int}}, \rho^{(n-1)} \right],
\end{aligned}
\tag{1.132}
$$

where we have, from (1.118), (1.119), (1.123), and (1.126),

$$\frac{1}{i\hbar} [\hat{H}_0, \rho^{(n)}]_{aa} = 0, \tag{1.133}$$

$$\frac{1}{i\hbar} [\hat{H}_0, \rho^{(n)}]_{ab} = -i\omega_{ab} \rho_{ab}^{(n)}, \tag{1.134}$$

$$\left(\frac{\partial \rho_{aa}^{(n)}}{\partial t}\right)_{rand} = \sum_b W_{ba}^{th}\, \rho_{bb}^{(n)} - \sum_b W_{ab}^{th}\, \rho_{aa}^{(n)}, \tag{1.135}$$

$$\left(\frac{\partial \rho_{ab}^{(n)}}{\partial t}\right)_{rand} = -\gamma_{ab}\, \rho_{ab}^{(n)}. \tag{1.136}$$

By using (1.133)–(1.136), we find from (1.132) that [14]

$$\frac{\partial \rho_{aa}^{(n)}}{\partial t} = \sum_b W_{ba}^{th}\, \rho_{bb}^{(n)} - \sum_b W_{ab}^{th}\, \rho_{aa}^{(n)} + \frac{1}{i\hbar}\left[\hat{H}_{int}, \rho^{(n-1)}\right]_{aa}, \tag{1.137}$$

$$\frac{\partial \rho_{ab}^{(n)}}{\partial t} = -(i\omega_{ab} + \gamma_{ab})\rho_{ab}^{(n)} + \frac{1}{i\hbar}\left[\hat{H}_{int}, \rho^{(n-1)}\right]_{ab}. \tag{1.138}$$

In most cases, we are interested in the off-diagonal elements $\rho_{ab}^{(n)}$, with $a \neq b$, of a density matrix $\rho^{(n)}$ because they couple different states through an interaction. Equation (1.138) can be integrated to explicitly express $\rho_{ab}^{(n)}$ in the form:

$$\rho_{ab}^{(n)}(t) = e^{-(i\omega_{ab}+\gamma_{ab})t} \int_{-\infty}^{t} \frac{1}{i\hbar}\left[\hat{H}_{int}(t'), \rho^{(n-1)}(t')\right]_{ab} e^{(i\omega_{ab}+\gamma_{ab})t'}\, dt'. \tag{1.139}$$

Therefore, the off-diagonal elements of a high-order density matrix can be found from the elements of a lower-order density matrix through this relation.

The expectation value of a physical observable can then be written in a perturbation series. For example, the electric polarization can be written as

$$\overline{\langle \boldsymbol{P} \rangle} = \mathrm{Tr}\,(\rho \hat{\boldsymbol{P}}) = \overline{\langle \boldsymbol{P}^{(0)} \rangle} + \overline{\langle \boldsymbol{P}^{(1)} \rangle} + \overline{\langle \boldsymbol{P}^{(2)} \rangle} + \overline{\langle \boldsymbol{P}^{(3)} \rangle} + \cdots, \tag{1.140}$$

where

$$\overline{\langle \boldsymbol{P}^{(n)} \rangle} = \mathrm{Tr}\,\left(\rho^{(n)} \hat{\boldsymbol{P}}\right). \tag{1.141}$$

1.6 ELECTRIC POLARIZATION

The electric polarization of a material is the *polarization density* in the material, which is the total electric dipole moments per unit volume of the material. The operator of the electric dipole moment for an electron, which has a charge of $q = -e$, is given in (1.89):

$$\hat{\boldsymbol{p}} = q\mathbf{r} = -e\mathbf{r}. \tag{1.142}$$

The matrix of the electric dipole operator is a symmetric matrix that has vanishing diagonal elements and real off-diagonal elements:

$$p_{aa} = \langle a | \hat{p} | a \rangle = 0 \quad \text{and} \quad p_{ab} = \langle a | \hat{p} | b \rangle = \langle b | \hat{p} | a \rangle = p_{ba} = -e\mathbf{r}_{ba} = -e\mathbf{r}_{ab}. \quad (1.143)$$

The operator $\hat{\mathbf{P}}$ of the electric polarization that appears in (1.116) and (1.140) is obtained by summing all of the electric dipole moments in a unit volume. Then, the electric polarization $\mathbf{P}$ is found by using (1.116), or (1.140), for $\mathbf{P} = \overline{\langle \mathbf{P} \rangle}$. In a material where the energy levels involved in the optical interaction form continuous energy bands, such as those in a semiconductor, (1.116) has to be evaluated by integrating over the band structure in the k space:

$$\mathbf{P} = \frac{1}{(2\pi)^3} \iiint \mathrm{Tr}\,(\rho\hat{p})\mathrm{dk} = \frac{1}{(2\pi)^3} \iiint \sum_{a,b} \rho_{ab}\mathbf{p}_{ba}\mathrm{dk} = \frac{1}{(2\pi)^3} \iiint \sum_{a,b} \rho_{ab}\mathbf{p}_{ab}\mathrm{dk}$$

$$(1.144)$$

$$= \frac{1}{(2\pi)^3} \sum_{\substack{a,b \\ a > b}} \mathbf{p}_{ab} \iiint (\rho_{ab} + \rho_{ab}^*)\mathrm{dk} = \frac{2}{(2\pi)^3} \sum_{\substack{a,b \\ a > b}} \mathbf{p}_{ab} \iiint \mathrm{Re}\,(\rho_{ab})\mathrm{dk},$$

where the relations as stated in (1.143) are used and (1.114) is used for $\rho_{ab} + \rho_{ba} = \rho_{ab} + \rho_{ab}^* = 2\mathrm{Re}\,(\rho_{ab})$. For an ensemble of separate atoms or molecules that have a total density of N_{total} atoms or molecules per unit volume, we simply have $\hat{\mathbf{P}} = N_{\mathrm{total}}\,\hat{p}$; thus

$$\mathbf{P} = N_{\mathrm{total}}\,\mathrm{Tr}\,(\rho\hat{p}) = N_{\mathrm{total}}\sum_{a,b}\rho_{ab}\mathbf{p}_{ba} = N_{\mathrm{total}}\sum_{\substack{a,b \\ a > b}}\mathbf{p}_{ab}(\rho_{ab} + \rho_{ab}^*) = 2N_{\mathrm{total}}\sum_{\substack{a,b \\ a > b}}\mathbf{p}_{ab}\mathrm{Re}\,(\rho_{ab}).$$

$$(1.145)$$

We shall use this simple form in this book. Similar results can be easily obtained with (1.144) in the case of a material in which the energy levels form a band structure.

As stated in the preceding section, in most cases we are interested in the off-diagonal elements of a density matrix. As seen in (1.144) and (1.145), only the off-diagonal elements ρ_{ab}, with $a \neq b$, appear in the calculation of the electric polarization. The reason is that the diagonal elements of the dipole moment operator are zero, as expressed in (1.143).

To find the electric polarization $\mathbf{P}$ from (1.144) or (1.145), we need the following information:

1. The off-diagonal elements of the electric dipole moment: $\mathbf{p}_{ab} = -e\mathbf{r}_{ab}$. For this purpose, we have to find the eigenstates of the material that are involved in an optical interaction so that $\mathbf{r}_{ab} = \langle a | \mathbf{r} | b \rangle$ can be calculated.
2. The interaction Hamiltonian: $\hat{H}_{\mathrm{int}}$. We have to identify the proper form of the interaction Hamiltonian $\hat{H}_{\mathrm{int}}$ according to the discussions in the preceding sections, and those in the following sections.
3. The off-diagonal elements of the density matrix: ρ_{ab}, with $a \neq b$. Once the form of $\hat{H}_{\mathrm{int}}$ is determined, ρ_{ab} can be found by solving (1.128) together with (1.127).
4. Numerical solution. Depending on the complexity of the problem, the numerical solution for ρ_{ab} might be needed in solving (1.128).
5. Perturbation solution. In most of the cases of interest in nonlinear optics, the perturbation expansion discussed in the preceding section is carried out to find $\rho_{ab}^{(n)}$ from (1.138), or from

its integral form given in (1.139). Then the nth-order polarization, $\boldsymbol{P}^{(n)}$, is found by applying (1.141) to (1.145):

$$\boldsymbol{P}^{(n)} = N_{\text{total}} \text{Tr}\left(\rho^{(n)}\hat{\boldsymbol{p}}\right) = N_{\text{total}}\sum_{a,b}\rho^{(n)}_{ab}\boldsymbol{P}_{ba} = N_{\text{total}}\sum_{\substack{a,b \\ a>b}}\boldsymbol{P}_{ab}\left(\rho^{(n)}_{ab} + \rho^{(n)*}_{ab}\right)$$

$$= 2N_{\text{total}}\sum_{\substack{a,b \\ a>b}}\boldsymbol{P}_{ab}\,\text{Re}\left(\rho^{(n)}_{ab}\right). \tag{1.146}$$

1.7 ELECTRIC DIPOLE APPROXIMATION

In Section 1.4, we found that the interaction Hamiltonian can be expressed as multipole expansion. The first three terms are electric dipole, electric quadrupole, and magnetic dipole interactions; all other high-order terms are much smaller than these terms and thus are negligible. Among the three leading terms, we also found in Section 1.4 that the electric dipole interaction dominates each of the other two terms by a few orders of magnitude. The electric quadrupole interaction is normally about four to five orders of magnitude smaller than the electric dipole interaction, except in certain structures where a large gradient of the electric field is found or where the effect of the electric dipole interaction vanishes. The magnetic dipole interaction is two to three orders of magnitude smaller than the electric dipole interaction except when magnetic resonance occurs.

In most applications, it is generally justified to keep only the electric dipole term for the interaction Hamiltonian by taking the *electric dipole approximation*:

$$\hat{H}_{\text{int}} \approx \hat{H}_{\text{ED}} = -\hat{\boldsymbol{p}}\cdot\boldsymbol{E}(\boldsymbol{0},t) = e\mathbf{r}\cdot\boldsymbol{E}(\boldsymbol{0},t). \tag{1.147}$$

The electric dipole approximation expressed above is a widely used approximation. It is a very good approximation because all other terms in the multipole expansion of the interaction Hamiltonian are normally orders of magnitude smaller and because the electric dipole interaction is expressed in terms of the electric field $\boldsymbol{E}(\boldsymbol{0},t)$ at the center of motion, which is uniquely defined for an optical field.

For the rest of this book, we shall take the electric dipole approximation as expressed in (1.147) for the Hamiltonian of light–matter interaction. Nevertheless, it is always important to bear in mind that it is still an *approximation*, no matter how good and how widely applicable it is. An approximation fails when it is used beyond the limit of the conditions under which the approximation is made. The conditions for using the electric dipole approximation are:

1. The electric dipole interaction has a nonvanishing contribution to the effect of interest.
2. All contributions to the effect of interest from high-order terms, including the electric quadrupole interaction and the magnetic dipole interaction, are negligibly small compared to that from the electric dipole interaction.

When one of these conditions fails, as in one of the possible situations mentioned in the beginning of this section, the electric quadrupole interaction or the magnetic dipole interaction has to be considered for the interaction Hamiltonian.

1.7.1　Electric Polarization of a Two-Level System

We consider the interaction between an optical field of a frequency ω with a two-level system that has a lower energy level $|1\rangle$ of an eigenenergy E_1 and an upper energy level $|2\rangle$ of an eigenenergy E_2, as shown in Fig. 1.2. The *resonance frequency* between these energy levels is

$$\omega_{21} = \frac{E_2 - E_1}{\hbar} > 0, \tag{1.148}$$

as is generally defined in (1.120) for that between two energy levels. From (1.143), the matrix of the electric dipole operator for the two-level system is a 2×2 matrix that has zero diagonal elements and real off-diagonal elements:

$$\boldsymbol{p}_{11} = \boldsymbol{p}_{22} = \boldsymbol{0} \quad \text{and} \quad \boldsymbol{p}_{12} = -e\mathbf{r}_{12} = -e\mathbf{r}_{21} = \boldsymbol{p}_{21}. \tag{1.149}$$

From (1.145), we find that the electric polarization for the two-level system is

$$\boldsymbol{P} = N_{\text{total}}\text{Tr}(\rho\hat{\boldsymbol{p}}) = N_{\text{total}}\boldsymbol{p}_{12}(\rho_{12} + \rho_{12}^*) = 2N_{\text{total}}\boldsymbol{p}_{12}\text{Re}(\rho_{12}) = 2N_{\text{total}}\boldsymbol{p}_{12}\text{Re}(\rho_{21}). \tag{1.150}$$

To find the electric polarization $\boldsymbol{P}$ from (1.150), it is only necessary to find the off-diagonal density matrix element ρ_{21}. The diagonal elements ρ_{11} and ρ_{22} are the population probabilities of energy levels $|1\rangle$ and $|2\rangle$ with $\rho_{11} + \rho_{22} = 1$. We take the electric dipole approximation as given in (1.147) for the interaction Hamiltonian:

$$\hat{H}_{\text{int}} \approx -\hat{\boldsymbol{p}} \cdot \boldsymbol{E}(\boldsymbol{0}, t) = e\mathbf{r} \cdot \boldsymbol{E}(\boldsymbol{0}, t). \tag{1.151}$$

From (1.149), we find that the matrix elements of this electric dipole interaction Hamiltonian has the properties:

$$(\hat{H}_{\text{int}})_{11} = (\hat{H}_{\text{int}})_{22} = 0 \quad \text{and} \quad (\hat{H}_{\text{int}})_{12} = -\boldsymbol{p}_{12} \cdot \boldsymbol{E}(\boldsymbol{0}, t) = -\boldsymbol{p}_{21} \cdot \boldsymbol{E}(\boldsymbol{0}, t) = (\hat{H}_{\text{int}})_{21}. \tag{1.152}$$

We then find from (1.127) and (1.128) the coupled equations for the density matrix elements of the two-level system under electric dipole interaction:

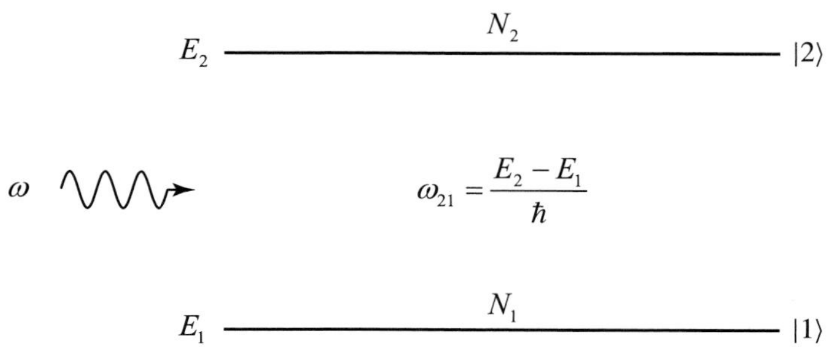

Figure 1.2 Interaction between an optical field of a frequency ω with a two-level system that has a lower level $|1\rangle$ of an eigenenergy E_1 and an upper level $|2\rangle$ of an eigenenergy E_2. The resonance frequency of the two-level system is ω_{21}.

$$\frac{\partial \rho_{11}}{\partial t} = \gamma_2 \rho_{22} + \frac{i}{\hbar} \boldsymbol{p}_{21} \cdot \boldsymbol{E}(0, t)(\rho_{21} - \rho_{12}), \tag{1.153}$$

$$\frac{\partial \rho_{22}}{\partial t} = -\gamma_2 \rho_{22} - \frac{i}{\hbar} \boldsymbol{p}_{21} \cdot \boldsymbol{E}(0, t)(\rho_{21} - \rho_{12}), \tag{1.154}$$

$$\frac{\partial \rho_{21}}{\partial t} = -i\omega_{21}\rho_{21} - \gamma_{21}\rho_{21} - \frac{i}{\hbar} \boldsymbol{p}_{21} \cdot \boldsymbol{E}(0, t)(\rho_{22} - \rho_{11}), \tag{1.155}$$

$$\rho_{12} = \rho_{21}^*, \tag{1.156}$$

where $\gamma_2 = W_{21}^{\text{th}}$ is the *population relaxation rate* from level $|2\rangle$ to level $|1\rangle$, γ_{21} is the *phase relaxation rate* for the coherence between the two levels, and we take $W_{12}^{\text{th}} \approx 0$ by assuming that the thermal excitation rate from level $|1\rangle$ to level $|2\rangle$ is negligibly small when the resonance frequency ω_{21} corresponding to the energy separation of these two levels is in the optical region. We see from (1.153) and (1.154) that $\partial(\rho_{11} + \rho_{22})/\partial t = 0$ as expected because $\rho_{11} + \rho_{22} = 1$ for a two-level system.

The solution of the coupled equations (1.153)–(1.156) depends on the strength of the electric field $\boldsymbol{E}$. The solution is a nonlinear function of the electric field. It cannot be expressed in a simple analytic form. To gain some insight, we can integrate (1.155) to obtain the relation:

$$\rho_{21}(t) = \rho_{12}^*(t) = \frac{i}{\hbar} \int_{-\infty}^{t} \boldsymbol{p}_{21} \cdot \boldsymbol{E}(t') \left[\rho_{11}(t') - \rho_{22}(t')\right] e^{-(i\omega_{21}+\gamma_{21})(t-t')} dt'. \tag{1.157}$$

By using (1.150), we can express the electric polarization as

$$\boldsymbol{P}(t) = \frac{2}{\hbar} \int_{-\infty}^{t} \boldsymbol{p}_{12}\boldsymbol{p}_{21} \cdot \boldsymbol{E}(t') \left[N_1(t') - N_2(t')\right] \sin\omega_{21}(t - t') e^{-\gamma_{21}(t-t')} dt'$$

$$= \frac{2}{\hbar} \int_{0}^{\infty} \boldsymbol{p}_{12}\boldsymbol{p}_{21} \cdot \boldsymbol{E}(t - t') \left[N_1(t - t') - N_2(t - t')\right] \sin\omega_{21}t' e^{-\gamma_{21}t'} dt', \tag{1.158}$$

where $N_1 = \rho_{11}N_{\text{total}}$ and $N_2 = \rho_{22}N_{\text{total}}$ are the *population densities* in levels $|1\rangle$ and $|2\rangle$, respectively, and N_{total} is the total density of active atoms in the system. It can be seen from (1.153) and (1.154) that in general the solutions of ρ_{11} and ρ_{22} depend on the electric field $\boldsymbol{E}$; therefore, the population densities N_1 and N_2 in (1.158) are generally functions of $\boldsymbol{E}$. Because of this field dependence of the population densities of the energy levels that are involved in the optical interaction, it can be seen from (1.158) that the polarization $\boldsymbol{P}$ is generally a nonlinear function of the electric field $\boldsymbol{E}$.

From (1.31), the electric polarization is defined as the response of a system to the electric field through an electric susceptibility tensor as

$$\boldsymbol{P}(t) = \epsilon_0 \int_{-\infty}^{t} \boldsymbol{\chi}(t - t') \cdot \boldsymbol{E}(t') dt' = \epsilon_0 \int_{0}^{\infty} \boldsymbol{\chi}(t; t') \cdot \boldsymbol{E}(t - t') dt'. \tag{1.159}$$

By comparing (1.159) with (1.158), we can express the electric susceptibility tensor in the time domain for the two-level system as

$$\chi(t; t') = \frac{2}{\epsilon_0 \hbar} \, \boldsymbol{p}_{12} \, \boldsymbol{p}_{21} \left[N_1(t - t') - N_2(t - t') \right] \sin \omega_{21} t' e^{-\gamma_{21} t'}. \tag{1.160}$$

Though we can formally express the electric susceptibility tensor χ as (1.160), it cannot be evaluated without the knowledge of the dependence of the population densities on the electric field because the population densities N_1 and N_2 in (1.160) are generally functions of $\boldsymbol{E}$. In other words, the electric susceptibility tensor χ is generally a function of the electric field $\boldsymbol{E}$ – that is, $\chi(t; t'; \boldsymbol{E})$. *This is the origin of optical nonlinearity.*

1.7.2 Linear Susceptibility

We consider the situation that $H_{\text{int}} \approx |\hat{\boldsymbol{p}} \cdot \boldsymbol{E}(\boldsymbol{0}, t)| \ll H_0 + H_{\text{rand}}$ so that the perturbation expansion of the density matrix discussed in Subsection 1.5.3 is valid. The population probabilities of the energy levels $|1\rangle$ and $|2\rangle$ of the two-level system that is in thermal equilibrium with its surroundings are $\rho_{11}^{(0)}$ and $\rho_{22}^{(0)}$, with $\rho_{11}^{(0)} + \rho_{22}^{(0)} = 1$. Then, the population densities in thermal equilibrium are $N_1^{(0)} = \rho_{11}^{(0)} N_{\text{total}}$ and $N_2^{(0)} = \rho_{22}^{(0)} N_{\text{total}}$, with $N_1^{(0)} + N_2^{(0)} = N_{\text{total}}$. By solving (1.138) or by directly applying (1.139), we find the first-order solution for the off-diagonal elements of the density matrix:

$$\rho_{21}^{(1)}(t) = \rho_{12}^{(1)*}(t) = \frac{\mathrm{i}}{\hbar} \left(\rho_{11}^{(0)} - \rho_{22}^{(0)} \right) \int_{-\infty}^{t} \boldsymbol{p}_{21} \cdot \boldsymbol{E}(t') e^{-(\mathrm{i}\omega_{21} + \gamma_{21})(t - t')} \mathrm{d}t'. \tag{1.161}$$

Note that the difference between (1.161) and (1.157) is that $\rho_{11}^{(0)} - \rho_{22}^{(0)}$ in (1.161) is independent of the electric field $\boldsymbol{E}$ and thus independent of time, whereas $\rho_{11}(t') - \rho_{22}(t')$ in (1.157) is a function of $\boldsymbol{E}$ and thus a function of time.

By following the procedure from (1.158) to (1.160), we find the *linear polarization*:

$$\boldsymbol{P}^{(1)}(t) = \frac{2}{\hbar} \left(N_1^{(0)} - N_2^{(0)} \right) \int_{-\infty}^{t} \boldsymbol{p}_{12} \, \boldsymbol{p}_{21} \cdot \boldsymbol{E}(t') \sin \omega_{21}(t - t') e^{-\gamma_{21}(t - t')} \mathrm{d}t'$$

$$= \frac{2}{\hbar} \left(N_1^{(0)} - N_2^{(0)} \right) \int_{0}^{\infty} \boldsymbol{p}_{12} \, \boldsymbol{p}_{21} \cdot \boldsymbol{E}(t - t') \sin \omega_{21} t' e^{-\gamma_{21} t'} \mathrm{d}t', \tag{1.162}$$

which is linearly related to the electric field $\boldsymbol{E}$ through a *linear susceptibility tensor* $\chi^{(1)}(t)$ as

$$\boldsymbol{P}^{(1)}(t) = \epsilon_0 \int_{-\infty}^{t} \chi^{(1)}(t - t') \cdot \boldsymbol{E}(t') \mathrm{d}t' = \epsilon_0 \int_{0}^{\infty} \chi^{(1)}(t') \cdot \boldsymbol{E}(t - t') \mathrm{d}t'. \tag{1.163}$$

Thus, the linear susceptibility tensor can be found as

$$\chi^{(1)}(t) = \frac{2}{\epsilon_0 \hbar} \boldsymbol{p}_{12}\boldsymbol{p}_{21} \left(N_1^{(0)} - N_2^{(0)}\right) \sin \omega_{21} t \, e^{-\gamma_{21} t}. \tag{1.164}$$

Because $N_1^{(0)}$ and $N_2^{(0)}$ are equilibrium population densities that are independent of the electric field $\boldsymbol{E}$, the linear susceptibility tensor $\chi^{(1)}$ as given in (1.164) is independent of the electric field. *This is the susceptibility that is generally considered in linear optics*, where the total polarization $\boldsymbol{P}$ is assumed to be linearly proportional to the electric field $\boldsymbol{E}$ and the total susceptibility χ is assumed to be independent of $\boldsymbol{E}$. *This assumption is invalid in nonlinear optics*, as we shall discuss throughout this book.

1.8 ROTATING-WAVE APPROXIMATION

We can find the characteristics of the linear susceptibility in the frequency domain by taking the Fourier transform of (1.164):

$$\begin{aligned}
\chi^{(1)}(\omega) &= \int_{-\infty}^{\infty} \chi^{(1)}(t) e^{i\omega t} dt \\
&= \frac{2}{\epsilon_0 \hbar} \boldsymbol{p}_{12}\boldsymbol{p}_{21} \left(N_1^{(0)} - N_2^{(0)}\right) \int_{-\infty}^{\infty} \sin \omega_{21} t \, e^{-\gamma_{21} t} e^{i\omega t} dt \\
&= \frac{1}{\epsilon_0 \hbar} \boldsymbol{p}_{12}\boldsymbol{p}_{21} \left(N_1^{(0)} - N_2^{(0)}\right) \left(\frac{1}{\omega + \omega_{21} + i\gamma_{21}} - \frac{1}{\omega - \omega_{21} + i\gamma_{21}}\right).
\end{aligned} \tag{1.165}$$

In an optical system, ω_{21} of the system is of the order of the optical frequency ω, and the phase relaxation rate γ_{21} is at least a few orders of magnitude smaller than the optical frequency such that $\omega_{21} \gg \gamma_{21}$. The optical frequency can be tuned such that it is in resonance with the optical transition with $\omega = \omega_{12}$, or near resonance such that $|\omega - \omega_{12}| \approx \gamma_{21}$. In these situations, it is clear that $|\omega + \omega_{21} + i\gamma_{21}|$ is orders of magnitude larger than $|\omega - \omega_{21} + i\gamma_{21}|$ so that the resonant term in (1.165), which has the frequency dependence of $(\omega - \omega_{21} + i\gamma_{21})^{-1}$, is orders of magnitude larger than the nonresonant term, which has the frequency dependence of $(\omega + \omega_{21} + i\gamma_{21})^{-1}$. Therefore, for resonant interactions, the nonresonant term can usually be neglected by approximating the susceptibility in the frequency domain as

$$\begin{aligned}
\chi^{(1)}(\omega) &\approx \frac{1}{\epsilon_0 \hbar} \boldsymbol{p}_{12}\boldsymbol{p}_{21} \left(N_1^{(0)} - N_2^{(0)}\right) \left(-\frac{1}{\omega - \omega_{21} + i\gamma_{21}}\right) \\
&= \frac{1}{\epsilon_0 \hbar} \boldsymbol{p}_{12}\boldsymbol{p}_{21} \left(N_1^{(0)} - N_2^{(0)}\right) \mathcal{L}(\omega; \omega_{21}, \gamma_{21}),
\end{aligned} \tag{1.166}$$

where

$$\mathcal{L}(\omega; \omega_{21}, \gamma_{21}) = -\frac{1}{\omega - \omega_{21} + i\gamma_{21}} = -\frac{\omega - \omega_{21}}{(\omega - \omega_{21})^2 + \gamma_{21}^2} + i\frac{\gamma_{21}}{(\omega - \omega_{21})^2 + \gamma_{21}^2} \tag{1.167}$$

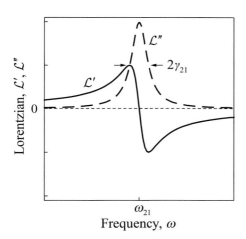

Figure 1.3 Real part, $\mathcal{L}'$, and imaginary part, $\mathcal{L}''$, of the complex Lorentzian lineshape function near a resonance frequency of ω_{21} with a relaxation rate of γ_{21}.

is the *complex Lorentzian lineshape function*, which is plotted in Fig. 1.3. The approximation of neglecting the nonresonant term as described above is known as the *rotating-wave approximation*. It is valid only for a nearly resonant interaction but is invalid at frequencies far away from the resonance frequency.

The rotating-wave approximation as discussed above is performed in the frequency domain. It can equally be carried out in the time domain. From the definition of the complex field given in (1.37), the real electric polarization P can be expressed in terms of the sum of the complex electric polarization and its complex conjugate as

$$\boldsymbol{P} = \mathbf{P} + \mathbf{P}^*. \tag{1.168}$$

By using the above relation with (1.150), we have

$$\boldsymbol{P} = N_{\text{total}} \text{Tr} \left(\rho \hat{\boldsymbol{p}} \right) = N_{\text{total}} \, \boldsymbol{p}_{12} (\rho_{21} + \rho_{21}^*) = \mathbf{P} + \mathbf{P}^*. \tag{1.169}$$

Therefore, the complex electric polarization is

$$\mathbf{P} = N_{\text{total}} \, \boldsymbol{p}_{12} \rho_{21}. \tag{1.170}$$

From (1.155) and (1.157), we know that ρ_{21} and ρ_{12} respectively vary with time as

$$\rho_{21} \propto \mathrm{i}\mathrm{e}^{-\mathrm{i}\omega_{21}t - \gamma_{21}t} \quad \text{and} \quad \rho_{12} = \rho_{21}^* \propto -\mathrm{i}\mathrm{e}^{\mathrm{i}\omega_{21}t - \gamma_{21}t} \tag{1.171}$$

so that

$$\rho_{21} + \rho_{21}^* = 2 \, \text{Re}(\rho_{21}) \propto 2 \sin \omega_{21}t \, \mathrm{e}^{-\gamma_{21}t}. \tag{1.172}$$

From (1.169) and (1.170), we then find that

$$\mathbf{P} \propto \mathrm{i}\mathrm{e}^{-\mathrm{i}\omega_{21}t - \gamma_{21}t}, \quad \mathbf{P}^* \propto -\mathrm{i}\mathrm{e}^{\mathrm{i}\omega_{21}t - \gamma_{21}t}, \quad \text{and} \quad \boldsymbol{P} \propto 2 \sin \omega_{21}t \, \mathrm{e}^{-\gamma_{21}t}. \tag{1.173}$$

This is the reason why $2 \sin \omega_{21} t \, e^{-\gamma_{21} t}$ appears in the expressions of $\boldsymbol{P}$ in (1.158) and (1.162), which leads to its appearance in $\chi^{(1)}(t)$ in (1.164) and the two frequency-dependent terms of $\chi^{(1)}(\omega)$ in (1.165). By neglecting the nonresonant term to take the rotating-wave approximation in the frequency domain, we are dropping the term that depends on frequency as $\omega + \omega_{21}$ while keeping the term that depends on $\omega - \omega_{21}$. In the time domain, this process is equivalent to dropping $\boldsymbol{P}^*$ while keeping $\boldsymbol{P}$ for $\boldsymbol{P}$, or dropping $\rho_{12} = \rho_{21}^*$ while keeping ρ_{21} for $\rho_{21} + \rho_{12}$. Consequently, the rotating-wave approximation is taken in the time domain by carrying out the approximation:

$$\rho_{21} + \rho_{12} \approx \rho_{21} \propto \mathrm{i} e^{-\mathrm{i}\omega_{21} t - \gamma_{21} t} \quad \text{and} \quad \boldsymbol{P} = \mathbf{P} + \mathbf{P}^* \approx \mathbf{P} \propto \mathrm{i} e^{-\mathrm{i}\omega_{21} t - \gamma_{21} t}. \qquad (1.174)$$

The rotating-wave approximation is often taken in dealing with laser transitions because such transitions are resonant transitions. However, it has to be avoided in calculating nonlinear optical susceptibilities because the energy levels that are involved in a nonlinear optical interaction are generally not in resonance with the frequency of the optical field. Even in the special case that some energy levels are in resonance with the optical frequency in a certain nonlinear process, not all energy levels involved are in resonance. Consequently, *the rotating-wave approximation is generally not taken in treating a nonlinear optical interaction.* By comparison, the electric dipole approximation is generally a good approximation for nonlinear optical interactions except for certain special situations, as discussed in the preceding section.

Problem Set

1.1.1 Show that for a given combination of vector and scalar potentials A and φ that defines a unique set of E and B fields through (1.8) and (1.9), a different set A' and φ' of the form given in (1.12) with $\Lambda(\mathbf{r}, t)$ being an arbitrary smooth scalar function of space and time also defines the same set of E and B fields.

1.1.2 Show that Maxwell's equations are invariant under a gauge transformation of A and φ as defined in (1.12).

1.1.3 Show that in the Coulomb gauge defined in (1.15), (1.10) and (1.11) reduce to (1.16) and (1.17), respectively.

1.1.4 Show that Maxwell's equations are invariant under space inversion or time reversal, or both together.

1.5.1 Show that the equation of motion for the density matrix given in (1.117) can be obtained from the time-dependent Schrödinger equation given in (1.111).

1.5.2 Show that the off-diagonal elements $\rho_{ab}^{(n)}$, with $a \neq b$, of the nth-order perturbation term of a density matrix is governed by (1.138). Then show that it can be integrated to the form shown in (1.139).

1.7.1 Show that the coupled equations for the density matrix elements of a two-level system under the electric dipole interaction are those given in (1.153)–(1.156).

1.7.2 Show that the electric polarization of a two-level system can be expressed in the form of (1.158) and the electric susceptibility tensor in the form of (1.160).

1.7.3 Show that the linear electric polarization and the linear electric susceptibility of a two-level system are those given in (1.162) and (1.164), respectively.

References

[1] L. D. Landau and E. M. Lifshitz, *Electrodynamics of Continuous Media*. New York: Pergamon Press, 1960.

[2] H. Giessen and R. Vogelgesang, "Glimpsing the weak magnetic field of light," *Science*, vol. 326, pp. 529–530, 2009.

[3] M. Burresi, D. van Oosten, T. Kampfrath, H. Schoenmaker, R. Heideman, A. Leinse, *et al.*, "Probing the magnetic field of light at optical frequencies," *Science*, vol. 326, pp. 550–553, 2009.

[4] S. Linden, C. Enkrich, G. Dolling, M. W. Klein, J. Zhou, T. Koschny, *et al.*, "Photonic metamaterials: magnetism at optical frequencies," *IEEE Journal of Selected Topics in Quantum Electronics*, vol. 12, pp. 1097–1105, 2006.

[5] R. Merlin, "Metamaterials and the Landau–Lifshitz permeability argument: large permittivity begets high-frequency magnetism," *Proceedings of the National Academy of Sciences*, vol. 106, pp. 1693–1698, 2009.

[6] V. M. Shalaev, "Optical negative-index metamaterials," *Nature Photonics*, vol. 1, pp. 41–48, 2007.

[7] M. Burresi, T. Kampfrath, D. van Oosten, J. C. Prangsma, B. S. Song, S. Noda, *et al.*, "Magnetic light–matter interactions in a photonic crystal nanocavity," *Physical Review Letters*, vol. 105, p. 123901, 2010.

[8] M. Kasperczyk, S. Person, D. Ananias, L. D. Carlos, and L. Novotny, "Excitation of magnetic dipole transitions at optical frequencies," *Physical Review Letters*, vol. 114, p. 163903, 2015.

[9] N. R. Brewer, Z. N. Buckholtz, Z. J. Simmons, E. A. Mueller, and D. D. Yavuz, "Coherent magnetic response at optical frequencies using atomic transitions," *Physical Review X*, vol. 7, p. 011005, 2017.

[10] G. W. Winkler, D. Varjas, R. Skolasinski, A. A. Soluyanov, M. Troyer, and M. Wimmer, "Orbital contributions to the electron g factor in semiconductor nanowires," *Physical Review Letters*, vol. 119, p. 037701, 2017.

[11] A. Mielnik-Pyszczorski, K. Gawarecki, and P. Machnikowski, "Limited accuracy of conduction band effective mass equations for semiconductor quantum dots," *Scientific Reports*, vol. 8, p. 2873, 2018.

[12] C. Rudowicz and S. K. Misra, "Spin-Hamiltonian formalisms in electron magnetic resonance (EMR) and related spectroscopies," *Applied Spectroscopy Reviews*, vol. 36, pp. 11–63, 2001.

[13] R. E. Raab, "Magnetic multipole moments," *Molecular Physics*, vol. 29, pp. 1323–1331, 1975.

[14] N. Bloembergen, *Nonlinear Optics*. Reading, MA: W. A. Benjamin, Inc., 1965.

[15] G. A. Aucar, T. Saue, L. Visscher, and H. J. A. Jensen, "On the origin and contribution of the diamagnetic term in four-component relativistic calculations of magnetic properties," *The Journal of Chemical Physics*, vol. 110, pp. 6208–6218, 1999.

[16] J. M. Liu and I. T. Lin, *Graphene Photonics*. Cambridge: Cambridge University Press, 2019.

[17] T. A. Marian, "Higher-order multipole expansion in the Dirac equation," *Physical Review A*, vol. 53, pp. 1992–1999, 1996.

[18] N. Bloembergen and P. S. Pershan, "Light waves at the boundary of nonlinear media," *Physical Review*, vol. 128, pp. 606–622, 1962.

[19] J. M. Liu, *Photonic Devices*. Cambridge: Cambridge University Press, 2005.

2 Optical Nonlinearity

2.1 NONLINEAR OPTICAL POLARIZATION

Optical nonlinearity manifests nonlinear interaction of an optical field with a material. The origin of optical nonlinearity is the nonlinear response of electrons in a material to an optical field as the strength of the field increases. Microscopically, optical nonlinearity is caused by the field dependence of the population distribution of the energy levels that are involved in the interaction, as discussed in Subsection 1.7.1, as well as the field dependence of the coupling among these energy levels, as seen below. Macroscopically, the nonlinear optical response of a material is described by a polarization that is a nonlinear function of the optical field, as also discussed in Subsection 1.7.1. In general, such nonlinear dependence on the optical field can take a variety of forms. In particular, it can be very complicated when the optical field becomes extremely strong.

As discussed in Section 1.7, the electric dipole approximation is generally a good approximation for nonlinear optical interactions except for certain special situations. For this reason, in the discussions throughout this book, we shall take the electric dipole approximation [1–3]:

$$\hat{H}_{\text{int}} \approx \hat{H}_{\text{ED}} = -\hat{\boldsymbol{p}} \cdot \boldsymbol{E} = e\mathbf{r} \cdot \boldsymbol{E}. \tag{2.1}$$

By contrast, the rotating-wave approximation is generally not taken for nonlinear optical interactions because it is good only for resonant transitions. Therefore, we shall not take the rotating-wave approximation except when it is justified and specifically stated.

Under the electric dipole approximation, the diagonal elements, ρ_{aa}, and the off-diagonal elements, ρ_{ab} with $a \neq b$, of the density matrix of a system that is involved in an optical interaction are determined by (1.127) and (1.128) with the interaction Hamiltonian given in (2.1) for electric dipole interaction [1, 3]:

$$\frac{\partial \rho_{aa}}{\partial t} = \sum_{\substack{b \\ b \neq a}} W_{ba}^{\text{th}} \rho_{bb} - \sum_{\substack{b \\ b \neq a}} W_{ab}^{\text{th}} \rho_{aa} + \frac{\mathrm{i}}{\hbar} \left[\hat{\boldsymbol{p}} \cdot \boldsymbol{E}, \rho \right]_{aa}, \tag{2.2}$$

$$\frac{\partial \rho_{ab}}{\partial t} = -\mathrm{i}\omega_{ab}\rho_{ab} - \gamma_{ab}\rho_{ab} + \frac{\mathrm{i}}{\hbar} \left[\hat{\boldsymbol{p}} \cdot \boldsymbol{E}, \rho \right]_{ab}, \tag{2.3}$$

where

$$\omega_{ab} = -\omega_{ba} \quad \text{and} \quad \gamma_{ab} = \gamma_{ba} > 0. \tag{2.4}$$

Once the off-diagonal elements of the density matrix are found, the electric polarization can be found by using (1.145) as

$$P = N_{\text{total}} \text{Tr}(\rho \hat{p}) = N_{\text{total}} \sum_{a,b} \rho_{ab} \, p_{ba} = N_{\text{total}} \sum_{\substack{a,b \\ a>b}} p_{ab}(\rho_{ab} + \rho_{ab}^*) = 2N_{\text{total}} \sum_{\substack{a,b \\ a>b}} p_{ab} \, \text{Re}(\rho_{ab}).$$

(2.5)

The matrix elements of the electric dipole moment $\hat{p}$ that appears in (2.2), (2.3), and (2.5) have the properties stated in (1.143):

$$p_{aa} = 0 \quad \text{and} \quad p_{ab} = -e\mathbf{r}_{ab} = -e\mathbf{r}_{ba} = p_{ba}.$$

(2.6)

We can integrate (2.3) to express the off-diagonal elements of the density matrix in the form:

$$\rho_{ab}(t) = \rho_{ba}^*(t) = \frac{i}{\hbar} \int_{-\infty}^{t} \sum_{c} \left[p_{ac}\rho_{cb}(t') - \rho_{ac}(t')p_{cb} \right] \cdot E(t') e^{-(i\omega_{ab}+\gamma_{ab})(t-t')} dt',$$

(2.7)

which gives (1.157) for a two-level system that was considered in Subsection 1.7.1. We find from (2.7) that the expression given in (1.157) for the off-diagonal elements, $\rho_{21}(t) = \rho_{12}^*(t)$, in an integral form in terms of only the diagonal elements, ρ_{11} and ρ_{22}, is exceptional only for a two-level system under the electric dipole approximation. Such expression of an off-diagonal element in terms of only the diagonal elements, but not other off-diagonal elements, cannot be generalized to a multilevel system. As (2.7) indicates, for a multilevel system, each off-diagonal element is related to all other off-diagonal elements besides all diagonal elements. Similarly, from (2.2), each diagonal element is also related to all off-diagonal elements and all other diagonal elements. Therefore, for a multilevel system, all diagonal and off-diagonal elements of the density matrix have to be solved together by simultaneously solving the coupled equations of (2.2) and (2.3) with the relations given in (2.4) and (2.6). Although a general analytic solution for these coupled equations is not possible, it can be easily seen from the forms of these coupled equations that the solutions for the elements of the density matrix are all functions of the electric field E; consequently, from (2.7), it follows that they are all nonlinear functions of E. Hence, it is clear that the electric polarization obtained from (2.5) is generally a nonlinear function of E.

From the above discussions, nonlinear response of a material in its interaction with an optical field is the norm rather than the exception. In other words, all materials respond to optical fields nonlinearly. This is the basic reason why a general analytic solution for the coupled equations of (2.2) and (2.3) is not possible. The question is therefore not whether an optical interaction is nonlinear or not, but how nonlinear such an interaction is. Though an exact solution is generally not possible, good and useful solutions can be found for most of the realistic situations.

2.2   PERTURBATION EXPANSION

As discussed in the preceding section, the optical polarization P is generally a nonlinear function of the electric field E because the elements of the density matrix ρ are nonlinear functions of the electric field. Except for some special cases, an exact solution of P as a nonlinear function of E is generally not possible. However, as discussed in Subsection 1.5.3, in the case that the interaction Hamiltonian H_{int} is small compared to the Hamiltonian of the system such that $H_{\text{int}} \approx |\mathbf{p} \cdot \mathbf{E}| \ll H_0 + H_{\text{rand}}$ under the electric dipole approximation, as considered here and below, it can be treated as a perturbation to the system by expanding the density matrix in a perturbation series as given in (1.131):

$$\rho = \rho^{(0)} + \rho^{(1)} + \rho^{(2)} + \cdots. \tag{2.8}$$

When this perturbation expansion of the density matrix is valid, the total optical polarization can be expanded as (1.140) in terms of a series of polarizations:

$$\mathbf{P}(\mathbf{r}, t) = \text{Tr}(\rho\hat{\mathbf{P}}) = \mathbf{P}^{(0)}(\mathbf{r}, t) + \mathbf{P}^{(1)}(\mathbf{r}, t) + \mathbf{P}^{(2)}(\mathbf{r}, t) + \mathbf{P}^{(3)}(\mathbf{r}, t) + \cdots, \tag{2.9}$$

with

$$\mathbf{P}^{(n)}(\mathbf{r}, t) = \text{Tr}\left(\rho^{(n)}\hat{\mathbf{P}}\right), \tag{2.10}$$

where $\mathbf{P}^{(0)}$ is a *permanent electric polarization*, also known as *spontaneous electric polarization*, that spontaneously appears in the absence of an optical field, thus independent of $\mathbf{E}$, $\mathbf{P}^{(1)}$ is the *linear polarization* that is linearly proportional to the optical field $\mathbf{E}$, $\mathbf{P}^{(2)}$ is the *second-order nonlinear polarization* that is a quadratic function of the optical field $\mathbf{E}$, and $\mathbf{P}^{(3)}$ is the *third-order nonlinear polarization* that is a cubic function of the optical field $\mathbf{E}$. The spontaneous electric polarization appears only in ferroelectric materials; we take $\mathbf{P}^{(0)} = 0$ in this book because it does not exist in most materials. Except in some special cases, nonlinear polarizations of the fourth and higher orders are usually not important and thus can be ignored.

In most nonlinear optical processes of interest, with the exceptions mentioned in Section 2.3, the perturbation expansion of the polarization is valid, and only the three terms of linear, second-order, and third-order polarizations are significant. In the real space and time domain, these polarizations can be macroscopically defined as convolution integrals of space and time [4, 5]:

$$\mathbf{P}^{(1)}(\mathbf{r}, t) = \epsilon_0 \int\limits_{-\infty}^{t} \int\limits_{-\infty}^{\infty} \chi^{(1)}(\mathbf{r} - \mathbf{r}', t - t') \cdot \mathbf{E}(\mathbf{r}', t') d\mathbf{r}' dt', \tag{2.11}$$

$$\mathbf{P}^{(2)}(\mathbf{r}, t) = \epsilon_0 \int\limits_{-\infty}^{t} \int\limits_{-\infty}^{t} \int\limits_{-\infty}^{\infty} \int\limits_{-\infty}^{\infty} \chi^{(2)}(\mathbf{r} - \mathbf{r}_1, t - t_1; \mathbf{r} - \mathbf{r}_2, t - t_2) : \mathbf{E}(\mathbf{r}_1, t_1) \mathbf{E}(\mathbf{r}_2, t_2) d\mathbf{r}_1 d\mathbf{r}_2 dt_1 dt_2,$$

$$\tag{2.12}$$

$$\mathbf{P}^{(3)}(\mathbf{r}, t) = \epsilon_0 \int\limits_{-\infty}^{t} \int\limits_{-\infty}^{t} \int\limits_{-\infty}^{t} \int\limits_{-\infty}^{\infty} \int\limits_{-\infty}^{\infty} \int\limits_{-\infty}^{\infty} \chi^{(3)}(\mathbf{r} - \mathbf{r}_1, t - t_1; \mathbf{r} - \mathbf{r}_2, t - t_2; \mathbf{r} - \mathbf{r}_3, t - t_3) \vdots$$

$$\tag{2.13}$$

$$\mathbf{E}(\mathbf{r}_1, t_1) \mathbf{E}(\mathbf{r}_2, t_2) \mathbf{E}(\mathbf{r}_3, t_3) d\mathbf{r}_1 d\mathbf{r}_2 d\mathbf{r}_3 dt_1 dt_2 dt_3,$$

where $\chi^{(1)}$ is the *linear susceptibility*, and $\chi^{(2)}$ and $\chi^{(3)}$ are the second- and third-order *nonlinear susceptibilities*, respectively. The linear susceptibility $\chi^{(1)}$ is that of linear optics; it is a second-order tensor:

$$\chi^{(1)} = \left[\chi^{(1)}_{ij} \right]. \tag{2.14}$$

The second-order nonlinear susceptibility $\chi^{(2)}$ and the third-order nonlinear susceptibility $\chi^{(3)}$ are, respectively, third-order and fourth-order tensors:

$$\chi^{(2)} = \left[\chi^{(2)}_{ijk} \right], \tag{2.15}$$

and

$$\chi^{(3)} = \left[\chi^{(3)}_{ijkl} \right]. \tag{2.16}$$

The linear susceptibility, $\chi^{(1)}$, characterizes the linear optical property, whereas the nonlinear susceptibilities, $\chi^{(2)}$ and $\chi^{(3)}$, characterize the second-order and third-order nonlinear optical properties of a material. Therefore, the linear relation given in (2.11) defines the linear response of a material to an optical field, whereas the relations in (2.12) and (2.13), respectively, define the second-order and third-order nonlinear polarizations that describe the nonlinear responses of the material to an optical field. Because of the generally anisotropic nature of nonlinear susceptibility tensors, the nonlinear polarizations $P^{(2)}$ and $P^{(3)}$ that are defined in (2.12) and (2.13) are expressed in the form of high-order tensor products of the nonlinear susceptibilities and the optical field. The response of a material to an optical field can be nonlocal in space and noninstantaneous in time. This statement is true for both linear and nonlinear responses. For this reason, the linear polarization, $P^{(1)}$ defined in (2.11), and the nonlinear polarizations, $P^{(2)}$ and $P^{(3)}$ defined in (2.12) and (2.13), are generally expressed as convolution integrals of both space and time.

For material responses that are local in space but not instantaneous in time, the susceptibilities can be expressed as

$$\chi^{(n)}(\mathbf{r} - \mathbf{r}_1, t - t_1; \mathbf{r} - \mathbf{r}_2, t - t_2; \cdots) = \chi^{(n)}(t - t_1, t - t_2, \cdots)\delta(\mathbf{r} - \mathbf{r}_1)\delta(\mathbf{r} - \mathbf{r}_2) \cdots . \tag{2.17}$$

Then, the linear and nonlinear polarizations can be expressed as

$$P^{(1)}(\mathbf{r}, t) = \epsilon_0 \int_{-\infty}^{t} \chi^{(1)}(t - t') \cdot E(\mathbf{r}, t')dt', \tag{2.18}$$

$$P^{(2)}(\mathbf{r}, t) = \epsilon_0 \int_{-\infty}^{t} \int_{-\infty}^{t} \chi^{(2)}(t - t_1, t - t_2) : E(\mathbf{r}, t_1)E(\mathbf{r}, t_2)dt_1 dt_2, \tag{2.19}$$

$$P^{(3)}(\mathbf{r}, t) = \epsilon_0 \int_{-\infty}^{t} \int_{-\infty}^{t} \int_{-\infty}^{t} \chi^{(3)}(t - t_1, t - t_2, t - t_3) \vdots E(\mathbf{r}, t_1)E(\mathbf{r}, t_2)E(\mathbf{r}, t_3)dt_1 dt_2 dt_3. \tag{2.20}$$

In the momentum space and frequency domain, spatially local but temporally noninstantaneous responses are characterized by linear and nonlinear susceptibilities that are functions of optical frequencies but are independent of optical wavevectors:

$$\chi^{(n)}(\omega_1, \cdots, \omega_n) = \int_{-\infty}^{t} \cdots \int_{-\infty}^{t} \chi^{(n)}(t_1, \cdots, t_n) e^{i\omega_1 t_1 + \cdots + i\omega_n t_n} dt_1 \cdots dt_n. \tag{2.21}$$

This situation applies to the interactions discussed in this chapter and the following chapter. Therefore, the discussions in these two chapters are restricted to spatially local interactions where the linear and nonlinear polarizations can be expressed in the form of (2.18)–(2.20), and the linear and nonlinear susceptibilities in the momentum space and frequency domain are functions of only optical frequencies.

The polarizations defined above in (2.18)–(2.20) are expressed in terms of real fields, just as all basic definitions of electromagnetic field quantities are. However, it is generally convenient to deal with optical fields in terms of complex fields because optical fields are harmonic fields. As seen in Subsection 1.1.6, the conversion to expressions in terms of complex fields is quite straightforward. All linear equations, such as Maxwell's equations as expressed in (1.38)–(1.41), retain their general forms after the conversion. The complex field, $\mathbf{E}(\mathbf{r}, t)$, is defined in (1.37) through its relation to the real field, $\boldsymbol{E}(\mathbf{r}, t)$, as

$$\boldsymbol{E}(\mathbf{r}, t) = \mathbf{E}(\mathbf{r}, t) + \mathbf{E}^*(\mathbf{r}, t) = \mathbf{E}(\mathbf{r}, t) + \text{c.c.} \tag{2.22}$$

By following this definition, a complex nonlinear polarization, $\mathbf{P}^{(n)}(\mathbf{r}, t)$, in the real space and time domain can be defined through its relation to the real nonlinear polarization, $\boldsymbol{P}^{(n)}(\mathbf{r}, t)$, as

$$\boldsymbol{P}^{(n)}(\mathbf{r}, t) = \mathbf{P}^{(n)}(\mathbf{r}, t) + \mathbf{P}^{(n)*}(\mathbf{r}, t) = \mathbf{P}^{(n)}(\mathbf{r}, t) + \text{c.c.} \tag{2.23}$$

Note that in our convention, $\mathbf{E}(\mathbf{r}, t)$ and $\mathbf{P}^{(n)}(\mathbf{r}, t)$ contain components that vary with time as $\exp(-i\omega t)$ with positive values of ω, whereas $\mathbf{E}^*(\mathbf{r}, t)$ and $\mathbf{P}^{(n)*}(\mathbf{r}, t)$ contain those varying with time as $\exp(i\omega t)$ with positive values of ω or, equivalently, $\exp(-i\omega t)$ with ω assuming negative values. By substituting (2.22) and (2.23) in (2.18)–(2.20), the expressions of complex polarizations in terms of complex fields can be obtained. The relation between the complex linear polarization $\mathbf{P}^{(1)}(\mathbf{r}, t)$ and the complex field $\mathbf{E}(\mathbf{r}, t)$ has the same form as (2.18), as shown in (1.42). However, the complex nonlinear polarizations $\mathbf{P}^{(2)}(\mathbf{r}, t)$ and $\mathbf{P}^{(3)}(\mathbf{r}, t)$ contain products of $\mathbf{E}(\mathbf{r}, t)$ and $\mathbf{E}^*(\mathbf{r}, t)$ in addition to those of $\mathbf{E}(\mathbf{r}, t)$ alone. Consequently, they have more complicated expressions than those of the real polarizations seen in (2.19) and (2.20).

An optical field that is involved in a nonlinear interaction usually contains multiple, distinct frequency components. Such a field can be expanded in terms of its frequency components as

$$\mathbf{E}(\mathbf{r}, t) = \sum_q \mathbf{E}_q(\mathbf{r}) \exp(-i\omega_q t) = \sum_q \boldsymbol{\mathcal{E}}_q(\mathbf{r}) \exp(i\mathbf{k}_q \cdot \mathbf{r} - i\omega_q t), \tag{2.24}$$

where $\boldsymbol{\mathcal{E}}_q(\mathbf{r})$ is the slowly varying amplitude and $\mathbf{k}_q$ is the wavevector of the frequency component ω_q, and

$$\mathbf{E}_q(\mathbf{r}) = \boldsymbol{\mathcal{E}}_q(\mathbf{r}) \exp(i\mathbf{k}_q \cdot \mathbf{r}). \tag{2.25}$$

The nonlinear polarizations also contain multiple frequency components and can be expanded as

$$\mathbf{P}^{(n)}(\mathbf{r}, t) = \sum_q \mathbf{P}_q^{(n)}(\mathbf{r}) \exp(-i\omega_q t). \tag{2.26}$$

Note that here we do not attempt to further express $\mathbf{P}_q^{(n)}(\mathbf{r})$ in terms of a slowly varying polarization amplitude $\mathcal{P}_q^{(n)}(\mathbf{r})$ that is defined through the relation $\mathbf{P}_q(\mathbf{r}) = \mathcal{P}_q(\mathbf{r}) \exp(i\mathbf{k}_q \cdot \mathbf{r})$, as is done for $\mathbf{E}_q(\mathbf{r})$ in (2.25). The reason is that, as we shall see later, *the wavevector that characterizes the fast-varying spatial phase of a nonlinear polarization $\mathbf{P}_q^{(n)}(\mathbf{r})$ is not simply determined by the frequency ω_q but is determined by the fields involved in a particular nonlinear interaction of interest.* Therefore, $\mathcal{P}_q^{(n)}(\mathbf{r})$ is not necessarily slowly varying in space because the wavevector of $\mathbf{P}_q^{(n)}(\mathbf{r})$ is not necessarily $\mathbf{k}_q$. Nonetheless, it is still convenient to use $\mathcal{P}_q^{(n)}(\mathbf{r})$ in some cases, as will be seen in Chapter 8.

In the discussions of nonlinear polarizations, we also use the notations $\mathbf{E}(\omega_q)$ and $\mathbf{P}^{(n)}(\omega_q)$, which are respectively defined as

$$\mathbf{E}(\omega_q) = \mathbf{E}_q(\mathbf{r}) \quad \text{and} \quad \mathbf{P}^{(n)}(\omega_q) = \mathbf{P}_q^{(n)}(\mathbf{r}). \tag{2.27}$$

Field and polarization components with negative frequencies are interpreted as

$$\mathbf{E}(-\omega_q) = \mathbf{E}^*(\omega_q) \quad \text{and} \quad \mathbf{P}^{(n)}(-\omega_q) = \mathbf{P}^{(n)*}(\omega_q). \tag{2.28}$$

The following notation for nonlinear susceptibilities is also used:

$$\chi^{(n)}(\omega_q = \omega_1 + \cdots + \omega_n) = \chi^{(n)}(\omega_q; \omega_1, \cdots, \omega_n) = \chi^{(n)}(\omega_1, \cdots, \omega_n), \tag{2.29}$$

for $\omega_1 + \cdots + \omega_n = \omega_q$, where $\chi^{(n)}(\omega_1, \cdots, \omega_n)$ is the frequency-domain susceptibility that is defined in (2.21).

By using the definitions of the complex fields and the complex polarizations given in (2.22) and (2.23), as well as their expansions given in (2.24) and (2.26), we obtain, by taking the Fourier transform of (2.18)–(2.20), the relations [2, 4, 5]:

$$\mathbf{P}^{(1)}(\omega_q) = \epsilon_0 \chi^{(1)}(\omega_q = \omega_q) \cdot \mathbf{E}(\omega_q), \tag{2.30}$$

$$\mathbf{P}^{(2)}(\omega_q) = \epsilon_0 \sum_{m,n} \chi^{(2)}(\omega_q = \omega_m + \omega_n) : \mathbf{E}(\omega_m)\mathbf{E}(\omega_n), \tag{2.31}$$

$$\mathbf{P}^{(3)}(\omega_q) = \epsilon_0 \sum_{m,n,p} \chi^{(3)}(\omega_q = \omega_m + \omega_n + \omega_p) \vdots \mathbf{E}(\omega_m)\mathbf{E}(\omega_n)\mathbf{E}(\omega_p). \tag{2.32}$$

The summation is performed over all *positive and negative* values of frequencies that, for a given ω_q, satisfy the constraint of $\omega_m + \omega_n = \omega_q$ in the case of (2.31) and the constraint of $\omega_m + \omega_n + \omega_p = \omega_q$ in the case of (2.32). More explicitly, by performing the summation over *only positive frequencies* and by expanding each tensor product, we have [4, 6]

$$P_i^{(1)}(\omega_q) = \epsilon_0 \sum_j \chi_{ij}^{(1)}(\omega_q = \omega_q)E_j(\omega_q), \tag{2.33}$$

$$\begin{aligned} P_i^{(2)}(\omega_q) = \epsilon_0 \sum_{j,k} \sum_{\omega_m,\omega_n>0} \Big[&\chi_{ijk}^{(2)}(\omega_q = \omega_m + \omega_n)E_j(\omega_m)E_k(\omega_n) \\ &+ \chi_{ijk}^{(2)}(\omega_q = \omega_m - \omega_n)E_j(\omega_m)E_k^*(\omega_n) \\ &+ \chi_{ijk}^{(2)}(\omega_q = -\omega_m + \omega_n)E_j^*(\omega_m)E_k(\omega_n) \Big], \end{aligned} \tag{2.34}$$

$$\begin{aligned} P_i^{(3)}(\omega_q) = \epsilon_0 \sum_{j,k,l} \sum_{\omega_m,\omega_n,\omega_p>0} \Big[&\chi_{ijkl}^{(3)}(\omega_q = \omega_m + \omega_n + \omega_p)E_j(\omega_m)E_k(\omega_n)E_l(\omega_p) \\ &+ \chi_{ijkl}^{(3)}(\omega_q = \omega_m + \omega_n - \omega_p)E_j(\omega_m)E_k(\omega_n)E_l^*(\omega_p) \\ &+ \chi_{ijkl}^{(3)}(\omega_q = \omega_m - \omega_n + \omega_p)E_j(\omega_m)E_k^*(\omega_n)E_l(\omega_p) \\ &+ \chi_{ijkl}^{(3)}(\omega_q = -\omega_m + \omega_n + \omega_p)E_j^*(\omega_m)E_k(\omega_n)E_l(\omega_p) \\ &+ \chi_{ijkl}^{(3)}(\omega_q = \omega_m - \omega_n - \omega_p)E_j(\omega_m)E_k^*(\omega_n)E_l^*(\omega_p) \\ &+ \chi_{ijkl}^{(3)}(\omega_q = -\omega_m + \omega_n - \omega_p)E_j^*(\omega_m)E_k(\omega_n)E_l^*(\omega_p) \\ &+ \chi_{ijkl}^{(3)}(\omega_q = -\omega_m - \omega_n + \omega_p)E_j^*(\omega_m)E_k^*(\omega_n)E_l(\omega_p) \Big]. \end{aligned} \tag{2.35}$$

Usually only a limited number of frequencies participate in a given nonlinear optical interaction. Consequently, only one or a few terms among those listed in (2.34) or (2.35) contribute to a specific nonlinear polarization of interest.

EXAMPLE 2.1

Three optical fields at the wavelengths of $\lambda_1 = 300$ nm, $\lambda_2 = 750$ nm, and $\lambda_3 = 1500$ nm, corresponding to the frequencies of $\omega_1 = 2\pi c/\lambda_1$, $\omega_2 = 2\pi c/\lambda_2$, and $\omega_3 = 2\pi c/\lambda_3$, respectively, are involved in second-order nonlinear optical interactions. The optical fields at the three frequencies are $\mathbf{E}(\omega_1) = E_1(\hat{x} + \hat{y})/\sqrt{2}$, $\mathbf{E}(\omega_2) = E_2(\hat{y} + \hat{z})/\sqrt{2}$, and $\mathbf{E}(\omega_3) = E_3\hat{z}$, where $\hat{x}$, $\hat{y}$, and $\hat{z}$ are the principal x, y, and z axes of the nonlinear crystal. Find the nonlinear polarization $\mathbf{P}^{(2)}(\omega_4)$ at the frequency of $\omega_4 = 2\pi c/\lambda_4$ where $\lambda_4 = 375$ nm. Express the components of $\mathbf{P}^{(2)}(\omega_4)$ explicitly in terms of the elements of $\chi^{(2)}$ and the given magnitudes, E_1, E_2, and E_3, of the three optical fields.

Solution

Because $\lambda_1^{-1} - \lambda_3^{-1} = \lambda_2^{-1} + \lambda_2^{-1} = \lambda_4^{-1}$, we find that $\omega_4 = \omega_1 - \omega_3 = \omega_2 + \omega_2$. The second-order nonlinear polarization at the frequency ω_4 is

$$\begin{aligned} \mathbf{P}^{(2)}(\omega_4) = \epsilon_0 \Big[&\chi^{(2)}(\omega_4 = \omega_1 - \omega_3) : \mathbf{E}(\omega_1)\mathbf{E}^*(\omega_3) + \chi^{(2)}(\omega_4 = -\omega_3 + \omega_1) : \mathbf{E}^*(\omega_3)\mathbf{E}(\omega_1) \\ &+ \chi^{(2)}(\omega_4 = \omega_2 + \omega_2) : \mathbf{E}(\omega_2)\mathbf{E}(\omega_2) \Big]. \end{aligned}$$

Note that there are two terms from the mixing of ω_1 and ω_3 because of permutation, but there is only one term from ω_2 mixing with itself. By using the given fields at the three frequencies of ω_1, ω_2, and ω_3, we can express the components of $\mathbf{P}^{(2)}(\omega_4)$ as

$$P_x^{(2)}(\omega_4) = \epsilon_0 \left[\chi_{xxz}^{(2)}(\omega_4 = \omega_1 - \omega_3) \frac{E_1 E_3^*}{\sqrt{2}} + \chi_{xyz}^{(2)}(\omega_4 = \omega_1 - \omega_3) \frac{E_1 E_3^*}{\sqrt{2}} \right.$$

$$+ \chi_{xzx}^{(2)}(\omega_4 = -\omega_3 + \omega_1) \frac{E_3^* E_1}{\sqrt{2}} + \chi_{xzy}^{(2)}(\omega_4 = -\omega_3 + \omega_1) \frac{E_3^* E_1}{\sqrt{2}}$$

$$+ \chi_{xyy}^{(2)}(\omega_4 = \omega_2 + \omega_2) \frac{E_2^2}{2} + \chi_{xyz}^{(2)}(\omega_4 = \omega_2 + \omega_2) \frac{E_2^2}{2}$$

$$\left. + \chi_{xzy}^{(2)}(\omega_4 = \omega_2 + \omega_2) \frac{E_2^2}{2} + \chi_{xzz}^{(2)}(\omega_4 = \omega_2 + \omega_2) \frac{E_2^2}{2} \right],$$

$$P_y^{(2)}(\omega_4) = \epsilon_0 \left[\chi_{yxz}^{(2)}(\omega_4 = \omega_1 - \omega_3) \frac{E_1 E_3^*}{\sqrt{2}} + \chi_{yyz}^{(2)}(\omega_4 = \omega_1 - \omega_3) \frac{E_1 E_3^*}{\sqrt{2}} \right.$$

$$+ \chi_{yzx}^{(2)}(\omega_4 = -\omega_3 + \omega_1) \frac{E_3^* E_1}{\sqrt{2}} + \chi_{yzy}^{(2)}(\omega_4 = -\omega_3 + \omega_1) \frac{E_3^* E_1}{\sqrt{2}}$$

$$+ \chi_{yyy}^{(2)}(\omega_4 = \omega_2 + \omega_2) \frac{E_2^2}{2} + \chi_{yyz}^{(2)}(\omega_4 = \omega_2 + \omega_2) \frac{E_2^2}{2}$$

$$\left. + \chi_{yzy}^{(2)}(\omega_4 = \omega_2 + \omega_2) \frac{E_2^2}{2} + \chi_{yzz}^{(2)}(\omega_4 = \omega_2 + \omega_2) \frac{E_2^2}{2} \right],$$

$$P_z^{(2)}(\omega_4) = \epsilon_0 \left[\chi_{zxz}^{(2)}(\omega_4 = \omega_1 - \omega_3) \frac{E_1 E_3^*}{\sqrt{2}} + \chi_{zyz}^{(2)}(\omega_4 = \omega_1 - \omega_3) \frac{E_1 E_3^*}{\sqrt{2}} \right.$$

$$+ \chi_{zzx}^{(2)}(\omega_4 = -\omega_3 + \omega_1) \frac{E_3^* E_1}{\sqrt{2}} + \chi_{zzy}^{(2)}(\omega_4 = -\omega_3 + \omega_1) \frac{E_3^* E_1}{\sqrt{2}}$$

$$+ \chi_{zyy}^{(2)}(\omega_4 = \omega_2 + \omega_2) \frac{E_2^2}{2} + \chi_{zyz}^{(2)}(\omega_4 = \omega_2 + \omega_2) \frac{E_2^2}{2}$$

$$\left. + \chi_{zzy}^{(2)}(\omega_4 = \omega_2 + \omega_2) \frac{E_2^2}{2} + \chi_{zzz}^{(2)}(\omega_4 = \omega_2 + \omega_2) \frac{E_2^2}{2} \right].$$

2.3 HIGHLY NONLINEAR OPTICAL PROCESSES

For most cases of interest in nonlinear optics, the perturbation expansion described in the preceding section is applicable. There are two significant exceptions that are discussed in this book: (1) *high-order harmonic generation* and (2) *optical saturation*. These processes are highly nonlinear so that the perturbation expansion completely fails. Consequently, a full analysis that is specific for a particular nonlinear process is required for each case. High-order harmonic generation will be discussed in Section 10.8. Optical saturation will be discussed in Chapter 15.

When the perturbation expansion is valid for a nonlinear process, it is usually sufficient to consider only the second-order and third-order nonlinear susceptibilities, as described in the preceding section. Furthermore, usually only the second-order susceptibility or the third-order susceptibility is involved in a specific process because the second-order and the third-order susceptibilities have very different symmetry properties, as discussed in the following chapter.

In some cases, a high-order process that can be described by the perturbation approach is of interest. One example is the process of *three-photon absorption*, which is a fifth-order nonlinear optical process. Three-photon absorption will be discussed in Section 14.3.

Problem Set

2.1.1 Show that the off-diagonal elements of the density matrix of a multilevel system can be expressed in the form of (2.7).

2.2.1 Three optical fields at the wavelengths of $\lambda_1 = 750$ nm, $\lambda_2 = 600$ nm, and $\lambda_3 = 500$ nm, corresponding to the frequencies of $\omega_1 = 2\pi c/\lambda_1$, $\omega_2 = 2\pi c/\lambda_2$, and $\omega_3 = 2\pi c/\lambda_3$, respectively, are involved in second-order nonlinear optical interactions. The optical fields at the three frequencies are $\mathbf{E}(\omega_1) = E_1 \hat{x}$, $\mathbf{E}(\omega_2) = E_2(\hat{x} + \hat{y})/\sqrt{2}$, and $\mathbf{E}(\omega_3) = E_3(\hat{y} + \hat{z})/\sqrt{2}$, where $\hat{x}$, $\hat{y}$, and $\hat{z}$ are aligned with the principal x, y, and z axes of the nonlinear crystal. Find the nonlinear polarization $\mathbf{P}^{(2)}$ at the frequency of $\omega_4 = 2\pi c/\lambda_4$, where $\lambda_4 = 300$ nm. Express each of the components of $\mathbf{P}^{(2)}(\omega_4)$ explicitly in terms of the elements of $\chi^{(2)}$ and the given magnitudes, E_1, E_2, and E_3, of the three optical fields.

2.2.2 Answer the questions in Problem 2.2.1 for the nonlinear polarization $\mathbf{P}^{(2)}$ at the frequency of $\omega_5 = 2\pi c/\lambda_5$, where $\lambda_5 = 3$ μm.

2.2.3 Three optical fields at the wavelengths of $\lambda_1 = 750$ nm, $\lambda_2 = 500$ nm, and $\lambda_3 = 375$ nm, corresponding to the frequencies of $\omega_1 = 2\pi c/\lambda_1$, $\omega_2 = 2\pi c/\lambda_2$, and $\omega_3 = 2\pi c/\lambda_3$, respectively, are involved in second-order nonlinear optical interactions. The optical fields at the three frequencies are $\mathbf{E}(\omega_1) = E_1 \hat{y}$, $\mathbf{E}(\omega_2) = E_2(\hat{y} + \hat{z})/\sqrt{2}$, and $\mathbf{E}(\omega_3) = E_3 \hat{z}$, where $\hat{x}$, $\hat{y}$, and $\hat{z}$ are the principal x, y, and z axes of the nonlinear crystal. Find the nonlinear polarization $\mathbf{P}^{(2)}(\omega_4)$ at the frequency of $\omega_4 = 2\pi c/\lambda_4$, where $\lambda_4 = 1500$ nm. Express the components of $\mathbf{P}^{(2)}(\omega_4)$ explicitly in terms of the elements of $\chi^{(2)}$ and the given magnitudes, E_1, E_2, and E_3, of the three optical fields.

2.2.4 Answer the questions in Problem 2.2.1 for the nonlinear polarization $\mathbf{P}^{(2)}$ at the frequency of $\omega_5 = 2\pi c/\lambda_5$, where $\lambda_5 = 250$ nm.

References

[1] N. Bloembergen, *Nonlinear Optics*. Reading, MA: W. A. Benjamin, Inc., 1965.

[2] J. A. Armstrong, N. Bloembergen, J. Ducuing, and P. S. Pershan, "Interactions between light waves in a nonlinear dielectric," *Physical Review*, vol. 127, pp. 1918–1939, 1962.

[3] N. Bloembergen and Y. R. Shen, "Quantum-theoretical comparison of nonlinear susceptibilities in parametric media, lasers, and Raman lasers," *Physical Review*, vol. 133, pp. A37–A49, 1964.

[4] J. M. Liu, *Photonic Devices*. Cambridge: Cambridge University Press, 2005.

[5] Y. R. Shen, *The Principles of Nonlinear Optics*. New York: Wiley-Interscience, 1984.

[6] R. W. Boyd, *Nonlinear Optics*, 4th ed. New York: Academic Press, 2020.

3 Optical Susceptibilities

3.1 PERTURBATION FORMS OF OPTICAL SUSCEPTIBILITIES

The response of a material to an optical field is generally described by an electric polarization through an optical susceptibility. In general, the total optical susceptibility is a function of the optical field because the optical response is generally nonlinear, as discussed in the preceding chapter. When the electric polarization can be expressed as a perturbation series of linear and nonlinear polarizations, field-independent linear and nonlinear susceptibilities can be defined, such as done in (2.18)–(2.20) in the real space and time domain, and in (2.30)–(2.32) in the frequency domain. Thus, the nonlinear optical response of a certain order is described by a nonlinear polarization of the order through a field-independent nonlinear optical susceptibility of the order. The overall nonlinear optical properties of a material are then characterized by its nonlinear optical susceptibilities. In this chapter, the general properties of linear and nonlinear optical susceptibilities are discussed.

To find a linear or nonlinear optical susceptibility, it is necessary to first find the corresponding linear or nonlinear polarization, usually through the density matrix of the system. The electric dipole approximation is valid for most optical interactions, as discussed in Section 2.1. Except for high-order harmonic generation and optical saturation, which were mentioned in Section 2.3, perturbation expansion of the polarization can be done. In this chapter, we take the electric dipole approximation for the interaction Hamiltonian, which is given in (2.1), and follow the perturbation approach discussed in Section 2.2. For generality, we consider a multilevel system. For simplicity, we assume that the energy levels are discrete and the total density of atoms is N_{total}. In case the energy levels form energy bands, integration of the states over the energy bands is required.

The linear and nonlinear susceptibilities are defined in the real space and time domain through convolution integrals of space and time, as seen in (2.18)–(2.20). These susceptibilities $\chi^{(n)}(\mathbf{r} - \mathbf{r}_1, t - t_1; \cdots; \mathbf{r} - \mathbf{r}_n, t - t_n)$ in the real space and time domain are real tensors because $E(\mathbf{r}_1, t_1), \cdots, E(\mathbf{r}_n, t_n)$, and $P^{(n)}(\mathbf{r}, t)$ are real field vectors. Through the Fourier transform, the convolution integrals in the real space and time domain are transformed into direct products in the momentum space and frequency domain, as seen in (2.30)–(2.32), which are represented in the frequency domain. Because the fields $\mathbf{E}(\omega_m)$, $\mathbf{E}(\omega_n)$, $\mathbf{E}(\omega_p)$, $\mathbf{E}(\omega_q)$, and the polarization $\mathbf{P}^{(n)}(\omega_q)$ in (2.30)–(2.32) are all complex field vectors, the susceptibilities $\chi^{(n)}(\omega_q = \omega_1 + \omega_2 + \cdots + \omega_n)$ in the frequency domain are generally complex – that is, tensors with complex elements. This characteristic is common to linear and nonlinear susceptibilities. Also common to linear and nonlinear susceptibilities is the fact that *the imaginary part of an element of a frequency-domain susceptibility signifies the presence of loss or gain in an optical medium, meaning that there is a net exchange of energy between the medium and one or multiple optical fields that are involved in the interaction that is described by this susceptibility element. Moreover, the sign of this imaginary part tells whether it is a loss or a gain for an optical field.* These facts are seen in optical absorption, in which energy flows from the optical field to the medium, and optical amplification, in which energy flows from the medium to the optical field. *The real part of a frequency-dependent susceptibility, irrespective of whether it is linear or nonlinear, does not cause a net energy exchange between the medium and the optical field.*

The linear and nonlinear optical properties of a given material are connected to each other. Indeed, there are close relations, at both microscopic and macroscopic levels, between the linear and the nonlinear optical susceptibilities of the same material. The reason for such connections is simply that both linear and nonlinear optical properties of a material have their roots in the same microscopic material properties, including the atomic compositions, the energy levels, the resonance frequencies, and the relaxation rates, that determine the optical responses of the material.

For the rest of this chapter and throughout the rest of this book, we generally consider the susceptibilities in the frequency domain by representing them as functions of the frequencies of the interacting optical fields. The elements of these susceptibilities are generally complex, as discussed above. An interaction of one or multiple optical fields with an optical medium through the real part of a susceptibility tensor element does not cause energy exchange between the optical fields and the medium, though there can be energy exchange among the optical fields; such an optical process with no net energy exchange with the optical medium is known as a *parametric optical process*. The existence of the imaginary part of a susceptibility tensor element indicates energy exchange between the optical medium and one or multiple optical fields for an optical process that is coupled through this imaginary part; such an optical process with a net energy exchange with the optical medium is known as a *nonparametric optical process*.

To find the nth-order susceptibility, $\chi^{(n)}$, it is necessary to find the nth-order density matrix, $\rho^{(n)}$, and then, through (2.10), the nth-order polarization, $\boldsymbol{P}^{(n)}$ [1]. Because $\rho^{(n)}$ depends on $\rho^{(n-1)}$, in order to find $\rho^{(n)}$ it is necessary to find all orders of the density matrix below the nth order in the perturbation expansion, starting with the zeroth order, $\rho^{(0)}$. We take the zeroth order to be the state of thermal equilibrium for the system in the absence of optical interaction, as described by the equation of motion for $\rho^{(0)}$ in (1.122). Then the population distribution among the energy levels is determined by the thermal statistics, such as the Boltzmann distribution or the Fermi distribution, of the system; meanwhile, the phases among different energy levels are completely thermalized – that is, completely randomized without coherence. Therefore,

$$\sum_a \rho_{aa}^{(0)} = 1, \quad \text{and} \quad \rho_{ab}^{(0)} = 0 \quad \text{for} \quad a \neq b. \tag{3.1}$$

The nth-order density matrix can be found from the $(n-1)$th order by using (1.139):

$$\rho_{ab}^{(n)}(t) = e^{-(i\omega_{ab}+\gamma_{ab})t} \int_{-\infty}^{t} \frac{1}{i\hbar} \left[\hat{H}_{int}(t'), \rho^{(n-1)}(t') \right]_{ab} e^{(i\omega_{ab}+\gamma_{ab})t'} dt'. \tag{3.2}$$

Under the electric dipole approximation that is taken here, $\hat{H}_{int} \approx -\hat{\boldsymbol{p}} \cdot \boldsymbol{E}$, and

$$\frac{1}{i\hbar} \left[\hat{H}_{int}(t'), \rho^{(n-1)}(t') \right]_{ab} \approx \frac{i}{\hbar} \sum_c \left[\boldsymbol{p}_{ac} \rho_{cb}^{(n-1)}(t') - \rho_{ac}^{(n-1)}(t') \boldsymbol{p}_{cb} \right] \cdot \boldsymbol{E}(t'). \tag{3.3}$$

Note that $\boldsymbol{E}(t')$ in (3.3) is the real electric field. By using (2.22), (2.24), (2.27), and (2.28), it can be expressed as

$$\boldsymbol{E}(t') = \boldsymbol{E}(t') + \boldsymbol{E}^*(t') = \sum_{\substack{q \\ \omega_q > 0}} \left[\boldsymbol{E}(\omega_q) e^{-i\omega_q t'} + \boldsymbol{E}^*(\omega_q) e^{i\omega_q t'} \right] = \sum_q \boldsymbol{E}(\omega_q) e^{-i\omega_q t'}, \tag{3.4}$$

where the last summation covers both positive and negative frequencies with $\boldsymbol{E}(-\omega_q) = \boldsymbol{E}^*(\omega_q)$ according to (2.28). Then, (3.2) takes the form:

$$\rho_{ab}^{(n)}(t) = e^{-(i\omega_{ab}+\gamma_{ab})t} \frac{i}{\hbar} \sum_c \int_{-\infty}^{t} \left[\boldsymbol{p}_{ac} \rho_{cb}^{(n-1)}(t') - \rho_{ac}^{(n-1)}(t') \boldsymbol{p}_{cb} \right] \cdot \boldsymbol{E}(t') e^{(i\omega_{ab}+\gamma_{ab})t'} dt'$$

$$= e^{-(i\omega_{ab}+\gamma_{ab})t} \frac{i}{\hbar} \sum_q \sum_c \int_{-\infty}^{t} \left[\boldsymbol{p}_{ac} \rho_{cb}^{(n-1)}(t') - \rho_{ac}^{(n-1)}(t') \boldsymbol{p}_{cb} \right] \cdot \boldsymbol{E}(\omega_q) e^{-i\omega_q t'} e^{(i\omega_{ab}+\gamma_{ab})t'} dt'. \tag{3.5}$$

3.1.1 Linear Susceptibility

By using (3.1) for the zeroth-order density matrix, such that $\rho_{ab}^{(0)} = 0$ for $a \neq b$, and the fact that the electric dipole operator has vanishing diagonal elements, $\boldsymbol{p}_{aa} = 0$, the off-diagonal elements of the first-order density matrix can be readily found by applying (3.5) to the first order:

$$\rho_{ab}^{(1)}(t) = e^{-(i\omega_{ab}+\gamma_{ab})t} \frac{i}{\hbar} \left(\rho_{bb}^{(0)} - \rho_{aa}^{(0)} \right) \sum_q \int_{-\infty}^{t} \boldsymbol{p}_{ab} \cdot \boldsymbol{E}(\omega_q) e^{-i\omega_q t'} e^{(i\omega_{ab}+\gamma_{ab})t'} dt'$$

$$= -\frac{1}{\hbar} \left(\rho_{bb}^{(0)} - \rho_{aa}^{(0)} \right) \sum_q \frac{\boldsymbol{p}_{ab} \cdot \boldsymbol{E}(\omega_q) e^{-i\omega_q t}}{\omega_q - \omega_{ab} + i\gamma_{ab}} \tag{3.6}$$

$$= \frac{1}{\hbar} \left(\rho_{aa}^{(0)} - \rho_{bb}^{(0)} \right) \sum_q \frac{\boldsymbol{p}_{ab} \cdot \boldsymbol{E}(\omega_q) e^{-i\omega_q t}}{\omega_q - \omega_{ab} + i\gamma_{ab}}.$$

Then, by using (2.10) for $n = 1$, with $\hat{\boldsymbol{P}} = N_{\text{total}}\,\hat{\boldsymbol{p}}$ for the relation between the polarization operator and the electric dipole operator, we find the first-order electric polarization:

$$\boldsymbol{P}^{(1)}(t) = \text{Tr}\left(\rho^{(1)}\hat{\boldsymbol{P}}\right) = N_{\text{total}}\,\text{Tr}\left(\rho^{(1)}\hat{\boldsymbol{p}}\right) = \frac{N_{\text{total}}}{\hbar}\sum_{a,b}\left(\rho_{aa}^{(0)} - \rho_{bb}^{(0)}\right)\sum_{\substack{q \\ \pm\omega_q}}\frac{\boldsymbol{p}_{ab}\cdot\mathbf{E}(\omega_q)e^{-i\omega_q t}}{\omega_q - \omega_{ab} + i\gamma_{ab}}\boldsymbol{p}_{ba}.$$

$$(3.7)$$

By comparing (3.7) with the relation that

$$\boldsymbol{P}^{(1)}(t) = \mathbf{P}^{(1)}(t) + \mathbf{P}^{(1)*}(t) = \sum_q \mathbf{P}^{(1)}(\omega_q)e^{-i\omega_q t} = \epsilon_0\sum_q \chi^{(1)}(\omega_q)\cdot\mathbf{E}(\omega_q)e^{-i\omega_q t}, \quad (3.8)$$

we find the linear susceptibility in the frequency domain:

$$\chi^{(1)}(\omega_q) = \frac{N_{\text{total}}}{\epsilon_0\hbar}\sum_{a,b}\left(\rho_{aa}^{(0)} - \rho_{bb}^{(0)}\right)\frac{\boldsymbol{p}_{ba}\,\boldsymbol{p}_{ab}}{\omega_q - \omega_{ab} + i\gamma_{ab}}. \quad (3.9)$$

This second-order tensor can be expressed in terms of its tensor elements as

$$\chi_{ij}^{(1)}(\omega_q) = \frac{N_{\text{total}}}{\epsilon_0\hbar}\sum_{a,b}\left(\rho_{aa}^{(0)} - \rho_{bb}^{(0)}\right)\frac{p_{ba}^i\,p_{ab}^j}{\omega_q - \omega_{ab} + i\gamma_{ab}}. \quad (3.10)$$

By rearranging the dummy indices a and b, the linear susceptibility in (3.10) can be expressed as

$$\chi_{ij}^{(1)}(\omega_q) = \frac{N_{\text{total}}}{\epsilon_0\hbar}\sum_{a,b}\rho_{aa}^{(0)}\left[\frac{p_{ba}^i\,p_{ab}^j}{\omega_q - \omega_{ab} + i\gamma_{ab}} - \frac{p_{ab}^i\,p_{ba}^j}{\omega_q + \omega_{ab} + i\gamma_{ab}}\right]$$

$$= \frac{N_{\text{total}}}{\epsilon_0\hbar}\sum_{a,b}\rho_{aa}^{(0)}\left[-\frac{p_{ab}^i\,p_{ba}^j}{\omega_q + \omega_{ab} + i\gamma_{ab}} + \frac{p_{ab}^j\,p_{ba}^i}{\omega_q - \omega_{ab} + i\gamma_{ab}}\right],$$

$$(3.11)$$

where the optical frequency ω_q can take either positive or negative values; the transition frequencies ω_{ab} can also be positive or negative, with $\omega_{ab} > 0$ for $E_a > E_b$ and $\omega_{ab} < 0$ for $E_a < E_b$; but the phase relaxation rates γ_{ab} are always positive, with $\gamma_{ab} = \gamma_{ba} > 0$.

3.1.2 Second-Order Nonlinear Susceptibility

The second-order nonlinear susceptibility is found by applying (3.5) to the second order to find the off-diagonal elements $\rho_{ab}^{(2)}$ of the second-order density matrix, followed by using (2.10) for $n = 2$ to find $\boldsymbol{P}^{(2)}$. The procedure is similar to that outlined above for finding the linear susceptibility. Also similar to the linear susceptibility, the second-order susceptibility $\chi^{(2)}(\omega_q = \omega_m + \omega_n)$ and its tensor elements $\chi_{ijk}^{(2)}(\omega_q = \omega_m + \omega_n)$ can be expressed in a few different, but equivalent, forms. Without going through the detailed mathematics, we express it in a symmetric form with eight terms as [1–4]

$$\chi_{ijk}^{(2)}(\omega_q = \omega_m + \omega_n) = \frac{N_{\text{total}}}{2\epsilon_0\hbar^2}\sum_{a,b,c}\rho_{aa}^{(0)}\times$$

$$\left[\frac{p_{ab}^i p_{bc}^j p_{ca}^k}{(\omega_m + \omega_n + \omega_{ab} + i\gamma_{ab})(\omega_n - \omega_{ca} + i\gamma_{ca})} + \frac{p_{ab}^i p_{bc}^k p_{ca}^j}{(\omega_m + \omega_n + \omega_{ab} + i\gamma_{ab})(\omega_m - \omega_{ca} + i\gamma_{ca})}\right.$$

$$-\frac{p_{ab}^k p_{bc}^i p_{ca}^j}{(\omega_m + \omega_n + \omega_{bc} + i\gamma_{bc})(\omega_n - \omega_{ab} + i\gamma_{ab})} - \frac{p_{ab}^j p_{bc}^i p_{ca}^k}{(\omega_m + \omega_n + \omega_{bc} + i\gamma_{bc})(\omega_m - \omega_{ab} + i\gamma_{ab})}$$

$$-\frac{p_{ab}^j p_{bc}^i p_{ca}^k}{(\omega_m + \omega_n + \omega_{bc} + i\gamma_{bc})(\omega_n - \omega_{ca} + i\gamma_{ca})} - \frac{p_{ab}^k p_{bc}^i p_{ca}^j}{(\omega_m + \omega_n + \omega_{bc} + i\gamma_{bc})(\omega_m - \omega_{ca} + i\gamma_{ca})}$$

$$+\frac{p_{ab}^k p_{bc}^j p_{ca}^i}{(\omega_m + \omega_n + \omega_{ca} + i\gamma_{ca})(\omega_n - \omega_{ab} + i\gamma_{ab})} + \left.\frac{p_{ab}^j p_{bc}^k p_{ca}^i}{(\omega_m + \omega_n + \omega_{ca} + i\gamma_{ca})(\omega_m - \omega_{ab} + i\gamma_{ab})}\right],$$

$$(3.12)$$

where the optical frequencies ω_m, ω_n, and $\omega_q = \omega_m + \omega_n$ can take either positive or negative values; the transition frequencies ω_{ab} can also be positive or negative, with $\omega_{ab} > 0$ for $E_a > E_b$ and $\omega_{ab} < 0$ for $E_a < E_b$; but the phase relaxation rates γ_{ab} are always positive, with $\gamma_{ab} = \gamma_{ba} > 0$.

As discussed in Subsection 3.4.1, the frequency dependence of a nonlinear susceptibility has the *intrinsic permutation symmetry*, which is purely a result of the convention that is used for the representation of frequency-dependent nonlinear susceptibilities. This symmetry allows the optical frequencies ω_m and ω_n to be freely permutated in the expression for $\chi_{ijk}^{(2)}(\omega_q = \omega_m + \omega_n)$ without changing its value if the corresponding Cartesian coordinate indices j and k, which respectively represent the polarization directions of the excitation fields at the frequencies ω_m and ω_n, are simultaneously permutated. By performing this intrinsic permutation on (3.12), it can be seen that this eight-term form of $\chi_{ijk}^{(2)}(\omega_q = \omega_m + \omega_n)$ indeed possesses the intrinsic symmetry.

3.1.3 Third-Order Nonlinear Susceptibility

The third-order nonlinear susceptibility is found by applying (3.5) to the third order to find the off-diagonal elements $\rho_{ab}^{(3)}$ of the third-order density matrix, followed by using (2.10) for $n = 3$ to find $\boldsymbol{P}^{(3)}$. The procedure is similar to that outlined above for finding the linear susceptibility and the second-order susceptibility. Also similar to the linear susceptibility and the second-order susceptibility, the third-order susceptibility $\chi^{(3)}(\omega_q = \omega_m + \omega_n + \omega_p)$ and its tensor elements $\chi_{ijkl}^{(3)}(\omega_q = \omega_m + \omega_n + \omega_p)$ can be expressed in a few different, but equivalent, forms. Without going through the detailed mathematics, we express it in a form that has eight terms with an intrinsic permutation operator $\hat{P}_i$ as [4]

$$\chi_{ijkl}^{(3)}\left(\omega_q = \omega_m + \omega_n + \omega_p\right) = \frac{N_{\text{total}}}{\epsilon_0 \hbar^3} \hat{P}_{\text{i}} \sum_{a,b,c,d} \rho_{aa}^{(0)} \times$$

$$\left[-\frac{p_{ab}^i p_{bc}^j p_{cd}^k p_{da}^l}{(\omega_m + \omega_n + \omega_p + \omega_{ab} + i\gamma_{ab})(\omega_m + \omega_n + \omega_{ac} + i\gamma_{ac})(\omega_m + \omega_{ad} + i\gamma_{ad})} \right.$$

$$+\frac{p_{ab}^l p_{bc}^i p_{cd}^j p_{da}^k}{(\omega_m + \omega_n + \omega_p + \omega_{bc} + i\gamma_{bc})(\omega_m + \omega_n + \omega_{bd} + i\gamma_{bd})(\omega_m - \omega_{ab} + i\gamma_{ab})}$$

$$+\frac{p_{ab}^k p_{bc}^i p_{cd}^j p_{da}^l}{(\omega_m + \omega_n + \omega_p + \omega_{bc} + i\gamma_{bc})(\omega_m + \omega_n + \omega_{bd} + i\gamma_{bd})(\omega_m + \omega_{ad} + i\gamma_{ad})}$$

$$-\frac{p_{ab}^l p_{bc}^k p_{cd}^i p_{da}^j}{(\omega_m + \omega_n + \omega_p + \omega_{cd} + i\gamma_{cd})(\omega_m + \omega_n - \omega_{ac} + i\gamma_{ac})(\omega_m - \omega_{ab} + i\gamma_{ab})} \qquad (3.13)$$

$$+\frac{p_{ab}^j p_{bc}^i p_{cd}^k p_{da}^l}{(\omega_m + \omega_n + \omega_p + \omega_{bc} + i\gamma_{bc})(\omega_m + \omega_n + \omega_{ac} + i\gamma_{ac})(\omega_m + \omega_{ad} + i\gamma_{ad})}$$

$$-\frac{p_{ab}^l p_{bc}^j p_{cd}^i p_{da}^k}{(\omega_m + \omega_n + \omega_p + \omega_{cd} + i\gamma_{cd})(\omega_m + \omega_n + \omega_{bd} + i\gamma_{bd})(\omega_m - \omega_{ab} + i\gamma_{ab})}$$

$$-\frac{p_{ab}^k p_{bc}^j p_{cd}^i p_{da}^l}{(\omega_m + \omega_n + \omega_p + \omega_{cd} + i\gamma_{cd})(\omega_m + \omega_n + \omega_{bd} + i\gamma_{bd})(\omega_m + \omega_{ad} + i\gamma_{ad})}$$

$$\left. +\frac{p_{ab}^l p_{bc}^k p_{cd}^j p_{da}^i}{(\omega_m + \omega_n + \omega_p - \omega_{ad} + i\gamma_{ad})(\omega_m + \omega_n - \omega_{ac} + i\gamma_{ac})(\omega_m - \omega_{ab} + i\gamma_{ab})} \right].$$

In (3.13), the optical frequencies ω_m, ω_n, ω_p, and $\omega_q = \omega_m + \omega_n + \omega_p$ can take either positive or negative values; the transition frequencies ω_{ab} can also be positive or negative, with $\omega_{ab} > 0$ for $E_a > E_b$ and $\omega_{ab} < 0$ for $E_a < E_b$; but the phase relaxation rates γ_{ab} are always positive, with $\gamma_{ab} = \gamma_{ba} > 0$.

The intrinsic permutation symmetry for $\chi_{ijkl}^{(3)}(\omega_q = \omega_m + \omega_n + \omega_p)$ allows the optical frequencies ω_m, ω_n, and ω_p to be freely permutated on the right-hand side of the equals sign without changing the value of the nonlinear susceptibility if the corresponding Cartesian coordinate indices j, k, and l that respectively represent the polarizations of the fields at the three frequencies are simultaneously permuted. Though this is true for $\chi_{ijkl}^{(3)}(\omega_q = \omega_m + \omega_n + \omega_p)$, this symmetry is not formally seen in the eight-term form that is expressed in (3.13). The expression of $\chi_{ijkl}^{(3)}(\omega_q = \omega_m + \omega_n + \omega_p)$ can be made formally symmetric for the three frequencies ω_m, ω_n, and ω_p without changing the value of $\chi_{ijkl}^{(3)}(\omega_q = \omega_m + \omega_n + \omega_p)$ by applying an intrinsic permutation operator $\hat{P}_{\text{i}}$ to it, as shown in (3.13). The operator $\hat{P}_{\text{i}}$ averages everything to its right-hand side by summing over all possible permutations of the input frequencies ω_m, ω_n, and ω_p, which are permuted together with the corresponding coordinate indices j, k, and l, and then by dividing the sum by the number of possible intrinsic permutations among these frequencies [4]. In other words, $\hat{P}_{\text{i}}$ performs the operation:

$$\hat{P}_{\mathrm{i}} f(\omega_m, \omega_n, \omega_p) = \frac{1}{\text{Number of permutations}} \sum_{\substack{\omega_m, \omega_n, \omega_p \\ \omega_m + \omega_n + \omega_p = \omega_q}} f(\omega_m, \omega_n, \omega_p). \qquad (3.14)$$

Because there are six permutations for three different frequencies, the formally symmetric form of $\chi_{ijkl}^{(3)}(\omega_q = \omega_m + \omega_n + \omega_p)$ has 48 terms [2] instead of the eight terms listed in (3.13).

3.2 REALITY CONDITION

The susceptibility tensors in the real space and time domain, including the total susceptibility, $\chi(\mathbf{r}, t)$, the linear susceptibility, $\chi^{(1)}(\mathbf{r}, t)$, and the nonlinear susceptibilities, $\chi^{(2)}(\mathbf{r}_1, t_1; \mathbf{r}_2, t_2)$ and $\chi^{(3)}(\mathbf{r}_1, t_1; \mathbf{r}_2, t_2; \mathbf{r}_3, t_3)$, are all real tensors with real elements. The optical field vectors, including the electric field $\mathbf{E}(\mathbf{r}, t)$, the electric polarization $\mathbf{P}(\mathbf{r}, t)$, and the nonlinear polarizations $\mathbf{P}^{(n)}(\mathbf{r}, t)$, in the real space and time domain are real vectors. Because these fields are all harmonic fields, complex fields $\mathbf{E}(\mathbf{r}, t)$, $\mathbf{P}(\mathbf{r}, t)$, and $\mathbf{P}^{(n)}(\mathbf{r}, t)$ in the real space and time domain can be defined. By contrast, no complex susceptibilities are defined in the real space and time domain; each element of a susceptibility in the real space and time domain is always a real quantity. Even in a relation that is expressed in terms of complex fields, such as (1.42), the susceptibility tensor $\chi(\mathbf{r} - \mathbf{r}', t - t')$ is still a real tensor. This is also true for a permittivity tensor $\epsilon(\mathbf{r}, t)$ in the real space and time domain. *This reality property applies to each susceptibility tensor in the real space and time domain and every permittivity tensor in the real space and time domain* because the susceptibility and permittivity in the real space and time domain describe the response of a medium to an optical field in reality. It is true even in the presence of an optical loss or optical gain in the medium when a susceptibility in the *frequency domain* is complex and the sign of its imaginary part indicates whether there is a loss or a gain for the optical field, as discussed in Section 3.1.

Both the fields and the susceptibilities, as well as the permittivity, can be transformed to the momentum space and frequency domain through Fourier transform. In the real space and time domain, the relation between a polarization and an optical field is a convolution integral, as seen in (2.18)–(2.20), whereas in the momentum space and frequency domain, it is a direct tensor product, as seen in (2.30)–(2.32). In the real space and time domain, the susceptibilities $\chi^{(n)}(\mathbf{r}_1, t_1; \cdots; \mathbf{r}_n, t_n)$ are all real tensors though the polarizations and the fields can be defined in the complex form as $\mathbf{P}^{(n)}(\mathbf{r}, t)$ and $\mathbf{E}(\mathbf{r}, t)$. In the momentum space and frequency domain, however, the susceptibilities $\chi^{(n)}(\mathbf{k}_1, \omega_1; \cdots; \mathbf{k}_n, \omega_n)$ are generally complex, and the polarizations $\mathbf{P}^{(n)}(\mathbf{k}, \omega)$ and the electric fields $\mathbf{E}(\mathbf{k}, \omega)$ are also generally complex. The difference is that $\chi^{(n)}(\mathbf{k}_1, \omega_1; \cdots; \mathbf{k}_n, \omega_n)$ is the Fourier transform of a real susceptibility $\chi^{(n)}(\mathbf{r}_1, t_1; \cdots; \mathbf{r}_n, t_n)$ of space and time, whereas in our convention $\mathbf{P}^{(n)}(\mathbf{k}, \omega)$ and $\mathbf{E}(\mathbf{k}, \omega)$ are the Fourier transforms of the complex fields $\mathbf{P}^{(n)}(\mathbf{r}, t)$ and $\mathbf{E}(\mathbf{r}, t)$ of space and time, as defined in (1.46). Thus, the reality property of the susceptibility tensor in the real space and time domain, as discussed above, has to be reflected in the momentum space and time domain, but no such reality property applies to the complex polarizations and fields.

The mathematical fact is that the real part of the Fourier transform of a real function in real space and time is an even function when $\mathbf{k}$ and ω both change sign together, whereas the imaginary part is an odd function. In other words, because $\chi_{ij}(\mathbf{r}, t)$ is a real function of space and time, $\chi'_{ij}(\mathbf{k}, \omega) = \chi'_{ij}(-\mathbf{k}, -\omega)$ and $\chi''_{ij}(\mathbf{k}, \omega) = -\chi''_{ij}(-\mathbf{k}, -\omega)$ so that [5]

$$\chi^*(\mathbf{k}, \omega) = \chi(-\mathbf{k}, -\omega). \qquad (3.15)$$

This *reality condition* in the momentum space and frequency domain also applies to the permittivity:

$$\epsilon^*(\mathbf{k}, \omega) = \epsilon(-\mathbf{k}, -\omega). \qquad (3.16)$$

The reality condition for a nonlinear susceptibility in the momentum space and frequency domain has the form:

$$\chi^{(n)*}(\mathbf{k}_1, \omega_1; \cdots; \mathbf{k}_n, \omega_n) = \chi^{(n)}(-\mathbf{k}_1, -\omega_1; \cdots; -\mathbf{k}_n, -\omega_n). \qquad (3.17)$$

In the case of spatially local interaction when the relation in (2.17) is valid, we can use (2.29) to write the reality condition for nonlinear susceptibilities in the form:

$$\chi^{(n)*}(\omega_q = \omega_1 + \cdots + \omega_n) = \chi^{(n)}(-\omega_q = -\omega_1 - \cdots - \omega_n). \qquad (3.18)$$

3.3 ELEMENTS OF SUSCEPTIBILITY TENSORS

To gain a general perspective of the susceptibility tensor elements, we first review the properties of the linear susceptibility tensor $\chi^{(1)}$. Because $\chi^{(1)} = \left[\chi_{ij}^{(1)}\right]$ is a second-order tensor, it consists of nine tensor elements. Because the linear susceptibility is a function of a single frequency, only one frequency, ω, needs to be specified. When both $\chi^{(1)}(\omega)$ and $\chi^{(1)}(-\omega)$ are considered, the number of elements doubles. The reality condition, by stating that the elements of $\chi^{(1)}(-\omega)$ are completely determined by those of $\chi^{(1)}(\omega)$, reduces the maximum number of independent elements back to nine. The linear susceptibility tensor $\chi^{(1)}(\omega)$ of a material can always be diagonalized, thus further reducing the nine tensor elements to only three diagonal elements. Therefore, the maximum number of independent $\chi^{(1)}$ tensor elements is three.

Similar concepts apply in the consideration of the properties of nonlinear susceptibilities. However, the complexity dramatically increases due to the fact that the nonlinear susceptibilities are high-order tensors and are functions of multiple frequencies. Being a third-order tensor, the second-order nonlinear susceptibility, $\chi^{(2)} = \left[\chi_{ijk}^{(2)}\right]$, has 27 tensor elements. In the most general situation, three different frequencies are involved in a second-order nonlinear process that is characterized by $\chi^{(2)}$. The three frequencies, say ω_1, ω_2, and ω_3, are not independent of one another, but are subject to the condition $\omega_3 = \omega_1 + \omega_2$, assuming that $\omega_3 > \omega_1, \omega_2$. For each tensor element $\chi_{ijk}^{(2)}$, there are 3! different permutations of the three frequencies, resulting in the six different frequency dependences:

$$\chi_{ijk}^{(2)}(\omega_3 = \omega_1 + \omega_2), \quad \chi_{ijk}^{(2)}(\omega_2 = \omega_3 - \omega_1), \quad \chi_{ijk}^{(2)}(\omega_1 = -\omega_2 + \omega_3),$$
$$\chi_{ijk}^{(2)}(\omega_3 = \omega_2 + \omega_1), \quad \chi_{ijk}^{(2)}(\omega_2 = -\omega_1 + \omega_3), \quad \chi_{ijk}^{(2)}(\omega_1 = \omega_3 - \omega_2). \tag{3.19}$$

The signs of the frequencies in each element in (3.19) can be simultaneously changed to have elements such as $\chi_{ijk}^{(2)}(-\omega_3 = -\omega_1 - \omega_2)$, and so on. Fortunately, because of the reality condition, as expressed in (3.18), this sign change does not result in additional susceptibility elements needed for describing a nonlinear process. Therefore, the maximum number of frequency-dependent $\chi^{(2)}$ tensor elements that are needed to completely describe a second-order nonlinear interaction among three different optical frequencies is $27 \times 3! = 162$.

The fourth-order tensor of the third-order nonlinear susceptibility tensor, $\chi^{(3)} = \left[\chi_{ijkl}^{(3)}\right]$, has 81 tensor elements. For a third-order nonlinear process that is characterized by $\chi^{(3)}$, there can in general be four different frequencies involved in the interaction. Similar to what is discussed above for $\chi^{(1)}$ and $\chi^{(2)}$, simultaneous sign change of all frequencies does not increase the number of independent $\chi^{(3)}$ elements because of the reality condition. Therefore, the maximum number of frequency-dependent $\chi^{(3)}$ tensor elements that are possibly independent of one another is $81 \times 4! = 1,944$.

In most situations of practical interest, the number of independent elements of a nonlinear susceptibility tensor that has to be considered in a particular nonlinear interaction can be greatly reduced by applying the symmetry considerations discussed in the following.

3.4 PERMUTATION SYMMETRY

As we saw from the preceding section, permutations of the participating optical frequencies in a nonlinear optical interaction can significantly increase the number of the susceptibility elements that have to be considered for the interaction. This is not an issue of concern for the linear susceptibility because there is only one frequency variable for a linear susceptibility, thus no permutations of frequencies to be considered. A nonlinear susceptibility is generally a function of multiple frequencies. The number of possible permutations of the participating optical frequencies increases as the number of possible frequencies increases with the order of the nonlinear susceptibility, thus increasing the number of frequency-dependent elements of the nonlinear susceptibility. The reality condition discussed in Section 3.2 already connects each pair of susceptibility elements for which each corresponding frequency variable has opposite signs, thus cutting the number of possibly independent susceptibility elements to half. Furthermore, many of the remaining elements are not independent of one another because of the permutation symmetries discussed in this section.

3.4.1 Intrinsic Permutation Symmetry

Intrinsic permutation symmetry is purely a matter of the convention of the notation used for frequency-dependent nonlinear susceptibilities. It does not imply any physical change in a nonlinear interaction. For this reason, it is always valid for a nonlinear susceptibility.

As an example, we consider, in the case of $\omega_3 = \omega_1 + \omega_2$, a nonlinear polarization $P_x^{(2)}(\omega_3)$ that is generated by two orthogonally polarized optical field components $E_y(\omega_1)$ and $E_z(\omega_2)$ through a second-order nonlinear process. According to (2.34), we have

$$P_x^{(2)}(\omega_3) = \epsilon_0 \left[\chi_{xyz}^{(2)}(\omega_3 = \omega_1 + \omega_2)E_y(\omega_1)E_z(\omega_2) + \chi_{xzy}^{(2)}(\omega_3 = \omega_2 + \omega_1)E_z(\omega_2)E_y(\omega_1) \right].$$
(3.20)

Both terms on the right-hand side of (3.20) are needed because of the convention used in (2.34) for expanding the product in (2.31). However, they are equal in magnitude because they represent the same physical process of the nonlinear mixing of $E_y(\omega_1)$ and $E_z(\omega_2)$ to generate $P_x^{(2)}(\omega_3)$. Therefore, $\chi_{xyz}^{(2)}(\omega_3 = \omega_1 + \omega_2) = \chi_{xzy}^{(2)}(\omega_3 = \omega_2 + \omega_1)$. Generalization of this result leads to the intrinsic permutation symmetry [3–5]:

$$\chi_{ijk}^{(2)}(\omega_3 = \omega_1 + \omega_2) = \chi_{ikj}^{(2)}(\omega_3 = \omega_2 + \omega_1).$$
(3.21)

This intrinsic permutation symmetry permits free permutation of *only the frequencies on the right-hand side* of the equals sign in the argument of a nonlinear susceptibility *if the corresponding Cartesian coordinate indices are simultaneously permuted*. It applies to the elements of $\chi^{(3)}$ as well.

It can be seen that that the symmetric form of $\chi^{(2)}$ that is expressed in (3.12) possesses the intrinsic permutation symmetry as expected. It can also be seen that the form of $\chi^{(3)}$ that is expressed in (3.13) does not formally show the intrinsic permutation symmetry before applying the intrinsic permutation operator $\hat{P}_i$; the operator $\hat{P}_i$ makes it symmetric for intrinsic permutation. The intrinsic permutation symmetry reduces the number of independent $\chi^{(2)}$ elements from 162 to 81 and that of independent $\chi^{(3)}$ elements from 1,944 to 324 without imposing any qualifying physical conditions.

EXAMPLE 3.1
Simplify the expressions for $\mathbf{P}^{(2)}(\omega_4)$ and $P_x^{(2)}(\omega_4)$ in Example 2.1 by using the intrinsic permutation symmetry of $\chi^{(2)}$. Write out the simplified expressions for $P_y^{(2)}(\omega_4)$ and $P_z^{(2)}(\omega_4)$.

Solution
The intrinsic permutation symmetry requires that

$$\chi^{(2)}(\omega_4 = \omega_1 - \omega_3) : \mathbf{E}(\omega_1)\mathbf{E}^*(\omega_3) = \chi^{(2)}(\omega_4 = -\omega_3 + \omega_1) : \mathbf{E}^*(\omega_3)\mathbf{E}(\omega_1).$$

Therefore, the first two terms in $\mathbf{P}^{(2)}(\omega_4)$ can be combined to have the expression:

$$\mathbf{P}^{(2)}(\omega_4) = \epsilon_0 \left[2\boldsymbol{\chi}^{(2)}(\omega_4 = \omega_1 - \omega_3) : \mathbf{E}(\omega_1)\mathbf{E}^*(\omega_3) + \boldsymbol{\chi}^{(2)}(\omega_4 = \omega_2 + \omega_2) : \mathbf{E}(\omega_2)\mathbf{E}(\omega_2) \right].$$

By explicitly applying the intrinsic permutation symmetry to the elements of $\boldsymbol{\chi}^{(2)}$, we can use the relations $\chi^{(2)}_{xxz}(\omega_4 = \omega_1 - \omega_3) = \chi^{(2)}_{xzx}(\omega_4 = -\omega_3 + \omega_1)$, $\chi^{(2)}_{xyz}(\omega_4 = \omega_1 - \omega_3) = \chi^{(2)}_{xzy}(\omega_4 = -\omega_3 + \omega_1)$, and $\chi^{(2)}_{xyz}(\omega_4 = \omega_2 + \omega_2) = \chi^{(2)}_{xzy}(\omega_4 = \omega_2 + \omega_2)$ to express the x component of $\mathbf{P}^{(2)}(\omega_4)$ as

$$P^{(2)}_x(\omega_4) = \epsilon_0 \left[\sqrt{2}\chi^{(2)}_{xxz}(\omega_4 = \omega_1 - \omega_3)E_1 E_3^* + \sqrt{2}\chi^{(2)}_{xyz}(\omega_4 = \omega_1 - \omega_3)E_1 E_3^* \right.$$
$$\left. + \chi^{(2)}_{xyy}(\omega_4 = \omega_2 + \omega_2)\frac{E_2^2}{2} + \chi^{(2)}_{xyz}(\omega_4 = \omega_2 + \omega_2)E_2^2 + \chi^{(2)}_{xzz}(\omega_4 = \omega_2 + \omega_2)\frac{E_2^2}{2} \right].$$

The y and z components of $\mathbf{P}^{(2)}(\omega_4)$ are, respectively,

$$P^{(2)}_y(\omega_4) = \epsilon_0 \left[\sqrt{2}\chi^{(2)}_{yxz}(\omega_4 = \omega_1 - \omega_3)E_1 E_3^* + \sqrt{2}\chi^{(2)}_{yyz}(\omega_4 = \omega_1 - \omega_3)E_1 E_3^* \right.$$
$$\left. + \chi^{(2)}_{yyy}(\omega_4 = \omega_2 + \omega_2)\frac{E_2^2}{2} + \chi^{(2)}_{yyz}(\omega_4 = \omega_2 + \omega_2)E_2^2 + \chi^{(2)}_{yzz}(\omega_4 = \omega_2 + \omega_2)\frac{E_2^2}{2} \right]$$

and

$$P^{(2)}_z(\omega_4) = \epsilon_0 \left[\sqrt{2}\chi^{(2)}_{zxz}(\omega_4 = \omega_1 - \omega_3)E_1 E_3^* + \sqrt{2}\chi^{(2)}_{zyz}(\omega_4 = \omega_1 - \omega_3)E_1 E_3^* \right.$$
$$\left. + \chi^{(2)}_{zyy}(\omega_4 = \omega_2 + \omega_2)\frac{E_2^2}{2} + \chi^{(2)}_{zyz}(\omega_4 = \omega_2 + \omega_2)E_2^2 + \chi^{(2)}_{zzz}(\omega_4 = \omega_2 + \omega_2)\frac{E_2^2}{2} \right].$$

3.4.2 Full Permutation Symmetry

A full permutation symmetry exists when all of the optical frequencies that are contained in a susceptibility, as well as all combinations of these optical frequencies, are far away from any resonance frequencies of a material so that the material causes no loss or gain to the optical field at those frequencies. Therefore, the full permutation symmetry is valid when the imaginary part of a susceptibility is negligibly small. It breaks down in a nonparametric process, where the imaginary part of the susceptibility is significant. *The full permutation symmetry allows all of the frequencies in a nonlinear susceptibility to be freely permuted if the Cartesian coordinate indices are also permuted accordingly.* It permits the interchange of the frequency on the left-hand side of the equals sign in the argument of a nonlinear susceptibility with any one on the right-hand side, which is not permitted by the intrinsic permutation symmetry. However, the sign of a frequency has to be changed at the time when it is moved across the equals sign in a permutation. For example, $\chi^{(2)}_{ijk}(\omega_3 = \omega_1 + \omega_2) = \chi^{(2)}_{jik}(-\omega_1 = -\omega_3 + \omega_2)$, and so on. By applying the reality condition given in (3.18) and the fact that the susceptibility is necessarily real when the full permutation symmetry is valid, we then have [3–5]

$$\chi^{(2)}_{ijk}(\omega_3 = \omega_1 + \omega_2) = \chi^{(2)}_{jik}(\omega_1 = \omega_3 - \omega_2) = \chi^{(2)}_{kij}(\omega_2 = \omega_3 - \omega_1). \qquad (3.22)$$

Similar relations can be written for the $\chi^{(3)}$ elements. This full permutation symmetry further reduces the maximum number of independent $\chi^{(2)}$ elements from 81 to 27 and that of independent $\chi^{(3)}$ elements from 324 to 81.

3.4.3 Kleinman's Symmetry Condition

If, in addition to being lossless so that the full permutation symmetry is valid, a medium is also nondispersive in the entire spectral range that covers all of the frequencies that are contained in a nonlinear susceptibility, *the frequencies in the susceptibility can be freely permuted independently of the Cartesian coordinate indices*. Similarly, *the Cartesian coordinate indices can also be permuted independently of the frequencies*. This permutation symmetry is known as *Kleinman's symmetry condition*. Under this condition, we have [4, 5]

$$
\begin{aligned}
\chi_{ijk}^{(2)}(\omega_3 = \omega_1 + \omega_2) &= \chi_{ijk}^{(2)}(\omega_1 = \omega_3 - \omega_2) = \chi_{ijk}^{(2)}(\omega_2 = \omega_3 - \omega_1) \\
&= \chi_{jik}^{(2)}(\omega_3 = \omega_1 + \omega_2) = \chi_{kij}^{(2)}(\omega_3 = \omega_1 + \omega_2) \\
&= \cdots
\end{aligned}
\tag{3.23}
$$

and so on. Kleinman's symmetry condition, when applicable, further reduces the number of independent $\chi^{(2)}$ elements from 27 to a maximum of 10 and that of independent $\chi^{(3)}$ elements from 81 to a maximum of 15.

3.5 SPATIAL SYMMETRY

The forms of the linear and nonlinear susceptibility tensors of a material reflect the spatial symmetry property of the material structure. As a result, some elements in a nonlinear susceptibility tensor may be zero and others may be related in one way or another, greatly reducing the total number of independent tensor elements. However, as discussed below, *the linear susceptibility tensor has its form determined only by the crystal system of a material*, whereas the *form of a nonlinear susceptibility tensor further depends on the point group of the material*. There are seven crystal systems for dielectric crystals, which are listed in Table 3.1 [5].

Within the seven crystal systems, there are 32 point groups. Among the 32 point groups, 21 are noncentrosymmetric and 11 are centrosymmetric. The 21 noncentrosymmetric point groups

Table 3.1 **Linear optical properties of crystals**

Crystal system	Linear optical property
Cubic	Isotropic: $\chi_x^{(1)} = \chi_y^{(1)} = \chi_z^{(1)}$, $n_x = n_y = n_z$
Hexagonal, trigonal, tetragonal	Uniaxial: $\chi_x^{(1)} = \chi_y^{(1)} \neq \chi_z^{(1)}$, $n_x = n_y \neq n_z$
Orthorhombic, monoclinic, triclinic	Biaxial: $\chi_x^{(1)} \neq \chi_y^{(1)} \neq \chi_z^{(1)}$, $n_x \neq n_y \neq n_z$

are triclinic 1, monoclinic 2 and m, orthorhombic 222 and $mm2$, tetragonal 4, $\bar{4}$, 422, $4mm$, and $\bar{4}2m$, trigonal 3, 32, and $3m$, hexagonal 6, $\bar{6}$, 622, $6mm$, and $\bar{6}m2$, and cubic 432, 23, and $\bar{4}3m$. The 11 centrosymmetric point groups are triclinic $\bar{1}$, monoclinic $2/m$, orthorhombic mmm, tetragonal $4/m$ and $4/mmm$, trigonal $\bar{3}$ and $\bar{3}m$, hexagonal $6/m$ and $6/mmm$, and cubic $m3$ and $m3m$.

Many materials, including gases, liquids, amorphous solids, and all crystals that belong to the 11 centrosymmetric point groups, are centrosymmetric because they possess space-inversion symmetry. *In the electric dipole approximation, optical effects of all even orders, but not those of the odd orders, vanish identically in a centrosymmetric material. Therefore, $\chi^{(2)}$ contributed by electric dipole interaction is identically zero in a centrosymmetric material, whereas a nonzero $\chi^{(1)}$ and a nonzero $\chi^{(3)}$ exists in any material.* This fact can be easily verified by considering the effect of space inversion on the linear and nonlinear polarizations $\boldsymbol{P}^{(1)}$, $\boldsymbol{P}^{(2)}$, and $\boldsymbol{P}^{(3)}$, which are given in (2.11), (2.12), and (2.13), respectively. The space-inversion transformation can be performed on a centrosymmetric material without changing the properties of the material. Being polar vectors, $\boldsymbol{P}^{(1)}$, $\boldsymbol{P}^{(2)}$, $\boldsymbol{P}^{(3)}$, and $\boldsymbol{E}$ all change sign under such a transformation. From (2.12), we then find that $\boldsymbol{P}^{(2)} = -\boldsymbol{P}^{(2)}$. Therefore, $\boldsymbol{P}^{(2)}$ cannot exist, and $\chi^{(2)}$ has to vanish identically in a centrosymmetric material. No such conclusion is drawn for $\boldsymbol{P}^{(1)}$ and $\boldsymbol{P}^{(3)}$ as we examine (2.11) and (2.13) by following the same procedure. Therefore, nonvanishing $\chi^{(1)}$ and $\chi^{(3)}$ generally exist in any material. Note that the above only applies to electric dipole interaction but not to electric quadrupole interaction or magnetic dipole interaction. Therefore, even-order optical susceptibilities that are contributed by electric quadrupole interaction or magnetic dipole interaction do not vanish in a centrosymmetric material.

3.5.1 Linear Susceptibility

Because $\chi^{(1)} = [\chi_{ij}^{(1)}]$ is a second-order tensor, it consists of nine tensor elements:

$$\chi^{(1)} = \begin{bmatrix} \chi_{11}^{(1)} & \chi_{12}^{(1)} & \chi_{13}^{(1)} \\ \chi_{21}^{(1)} & \chi_{22}^{(1)} & \chi_{23}^{(1)} \\ \chi_{31}^{(1)} & \chi_{32}^{(1)} & \chi_{33}^{(1)} \end{bmatrix}, \tag{3.24}$$

where 1, 2, and 3 are arbitrarily chosen rectilinear coordinate directions. Because the linear susceptibility is a function of a single frequency, only one frequency, ω, needs to be specified. The linear susceptibility tensor $\chi^{(1)}(\omega)$ of a material can always be diagonalized. Furthermore, for a nonmagnetic dielectric material, the susceptibility tensor is a symmetric tensor – that is, $\chi_{ij}^{(1)} = \chi_{ji}^{(1)}$. The eigenvectors of a symmetric matrix are real vectors. This means that the linear susceptibility tensor $\chi^{(1)}(\omega)$ can be diagonalized in a rectilinear coordinate system, thus further reducing the nine tensor elements to only three diagonal elements that represent the eigenvalues of the tensor:

$$\chi^{(1)} = \begin{bmatrix} \chi_x^{(1)} & 0 & 0 \\ 0 & \chi_y^{(1)} & 0 \\ 0 & 0 & \chi_z^{(1)} \end{bmatrix}, \tag{3.25}$$

where $\hat{x}$, $\hat{y}$, and $\hat{z}$ are the unique set of rectilinear coordinate axes, known as the *principal dielectric axes*, or simply the *principal axes*, that diagonalize $\chi^{(1)}$, and the corresponding eigenvalues $\chi_x^{(1)}$, $\chi_y^{(1)}$, and $\chi_z^{(1)}$ are known as the *principal dielectric susceptibilities*.

The principal axes also diagonalize the permittivity tensor $\boldsymbol{\epsilon}$ with the eigenvalues ϵ_x, ϵ_y, and ϵ_z, known as the *principal dielectric permittivities*. The values ϵ_x/ϵ_0, ϵ_y/ϵ_0, and ϵ_z/ϵ_0 are the eigenvalues of the *dielectric constant tensor*, $\boldsymbol{\epsilon}/\epsilon_0$, and are known as the *principal dielectric constants*. They define three *principal indices of refraction*:

$$n_x = \sqrt{\frac{\epsilon_x}{\epsilon_0}}, \quad n_y = \sqrt{\frac{\epsilon_y}{\epsilon_0}}, \quad n_z = \sqrt{\frac{\epsilon_z}{\epsilon_0}}. \tag{3.26}$$

The principal axes $\hat{x}$, $\hat{y}$, and $\hat{z}$ are unique and specific for a given material because they are determined by the structure of the material, and so are the corresponding principal susceptibilities, the principal dielectric permittivities, the principal dielectric constants, and the principal indices of refraction.

Depending on the spatial symmetry of a material, the number of independent linear susceptibility elements that are needed for characterizing the linear optical properties of the material can be further reduced from three to two or one, as summarized in Table 3.1 in terms of the relations among the three principal susceptibilities and among the three principal indices of refraction. As seen in Table 3.1, the linear optical property of a crystal only depends on its crystal system.

Noncrystalline materials, including gases, liquids, and amorphous solids, generally have isotropic linear optical properties. A cubic crystal also has an isotropic linear optical property. A material that has an isotropic linear optical property has only one principal susceptibility, $\chi_x^{(1)} = \chi_y^{(1)} = \chi_z^{(1)}$, thus one index of refraction, $n_x = n_y = n_z$. For a noncrystalline isotropic material, the three principal axes can be any three mutually orthogonal axes that are arbitrarily chosen for convenience. For a cubic crystal, however, the three principal axes cannot be arbitrarily chosen because there is an underlining crystal structure. Though its linear optical property does not depend on the principal axes, its nonlinear optical properties do depend on the principal axes, as we shall see in the following two subsections.

A crystal of the hexagonal, trigonal, or tetragonal crystal system is a *uniaxial* crystal, for which one unique axis, which is designated as the z axis by convention, has a principal susceptibility that is different from the other two, $\chi_x^{(1)} = \chi_y^{(1)} \neq \chi_z^{(1)}$, thus a principal index of refraction that is different from the other two, $n_x = n_y \neq n_z$. Therefore, a uniaxial crystal has only two distinct values of indices of refraction, which are known as the *ordinary index of refraction*, $n_o = n_x = n_y$, and the *extraordinary index of refraction*, $n_e = n_z$. A crystal is *positive uniaxial* if $n_e > n_o$ and is *negative uniaxial* if $n_e < n_o$. An optical wave that propagates along the z axis has to be polarized in the xy plane; it sees an index of refraction that is independent of its polarization because $n_x = n_y$. Such an axis along which an optical wave propagates with an index of refraction that is independent of its polarization is known as an *optical axis* of the crystal. In a uniaxial crystal, there is only one optical axis, which coincides with the z principal axis of the crystal.

A crystal of the orthorhombic, monoclinic, or triclinic crystal system is a *biaxial crystal* because it has two optical axes. A biaxial crystal has three distinct values of principal suscepti- bilities, $\chi_x^{(1)} \neq \chi_y^{(1)} \neq \chi_z^{(1)}$, thus three different values of principal indices of refraction, $n_x \neq n_y \neq n_z$. By convention, *the principal axis of the largest index of refraction is designated as the z axis, and that of the smallest index of refraction is designated as the x axis.* Thus, $n_z > n_y > n_x$. If $n_z - n_y > n_y - n_x$, the crystal is called *positive biaxial*; if $n_z - n_y < n_y - n_x$, the crystal is called *negative biaxial.* None of the three principal axes of a biaxial crystal is an optical axis. The two optical axes lie on the *zx principal plane* on either side of the z axis at an angle of θ_{OA} with respect to the z axis:

$$\theta_{OA} = \sin^{-1} \left(\frac{n_x^{-2} - n_y^{-2}}{n_x^{-2} - n_z^{-2}} \right)^{1/2}, \tag{3.27}$$

which will be further discussed in Subsection 4.3.3 and shown in Fig. 4.7(c). Equivalently, they are also on either side of the x axis at an angle of $\pi/2 - \theta_{OA}$ with respect to the x axis. The directions of the two optical axes are determined by the relations among the three indices of refraction.

The existence of two or three different values among the three principal indices cause light of different polarizations to propagate with different propagation constants when it is not propagating along an optical axis. This phenomenon is known as *birefringence.* Optically isotropic materials, including cubic crystals, are *nonbirefringent.* Uniaxial and biaxial crystals are *birefringent crystals.* The birefringence of a crystal is quantified as the largest difference between two different indices of refraction. For a uniaxial crystal, the birefringence is

$$B = \Delta n = n_e - n_o, \tag{3.28}$$

which has a positive value for a positive uniaxial crystal, and a negative value for a negative uniaxial crystal. For a biaxial crystal, the birefringence is

$$B = \Delta n = n_z - n_x, \tag{3.29}$$

which is defined to have a positive value by the convention that $n_z > n_y > n_x$ for both positive and negative biaxial crystals.

3.5.2 Second-Order Nonlinear Susceptibility

Though the linear optical property of a material is determined only by the crystal system of the lattice structure, the nonlinear optical properties are further determined by the point group of the crystal structure. The elements of the second-order susceptibility tensor due to electric dipole interaction are listed in Table 3.2 for the 32 point groups. As discussed above, $\chi^{(2)}$ contributed by electric dipole interaction vanishes for centrosymmetric materials, including

Table 3.2 **Elements of the second-order nonlinear susceptibility tensor from electric dipole interaction**

Crystal system	Point group	Nonvanishing tensor elements
Triclinic	1	All 27 elements are independent and nonvanishing
	$\bar{1}$	Centrosymmetric; all elements vanish
Monoclinic	2	$xyz, xzy, xxy, xyx, yxx, yyy, yzz, yzx, yxz, zyz, zzy, zxy, zyx$ (two-fold axis parallel to $\hat{y}$)
	m	$xxx, xyy, xzz, xzx, xxz, yyz, yzy, yxy, yyx, zxx, zyy, zzz, zzx, zxz$ (mirror plane perpendicular to $\hat{y}$)
	$2/m$	Centrosymmetric; all elements vanish
Orthorhombic	222	$xyz, xzy, yzx, yxz, zxy, zyx$
	$mm2$	$xzx, xxz, yyz, yzy, zxx, zyy, zzz$
	mmm	Centrosymmetric; all elements vanish
Tetragonal	4	$xyz = -yxz, \; xzy = -yzx, xzx = yzy, xxz = yyz,$ $zxx = zyy, zzz, zxy = -zyx$
	$\bar{4}$	$xyz = yxz, xzy = yzx, xzx = -yzy, xxz = -yyz, zxx = -zyy, zxy = zyx$
	422	$xyz = -yxz, xzy = -yzx, zxy = -zyx$
	$4mm$	$xzx = yzy, xxz = yyz, zxx = zyy, zzz$
	$\bar{4}2m$	$xyz = yxz, xzy = yzx, zxy = zyx$
	$4/m$	Centrosymmetric; all elements vanish
	$4/mmm$	Centrosymmetric; all elements vanish
Trigonal	3	$xxx = -xyy = -yxy = -yyx, xyz = -yxz, xzy = -yzx,$ $xzx = yzy, xxz = yyz, yyy = -yxx = -xxy = -xyx,$ $zxx = zyy, zzz, zxy = -zyx$
	32	$xxx = -xyy = -yxy = -yyx, xyz = -yxz, xzy = -yzx, zxy = -zyx$
	$3m$	$xzx = yzy, xxz = yyz, yyy = -yxx = -xxy = -xyx, zxx = zyy, zzz$ (mirror plane perpendicular to $\hat{x}$)
	$\bar{3}$	Centrosymmetric; all elements vanish
	$\bar{3}m$	Centrosymmetric; all elements vanish
Hexagonal	6	$xyz = -yxz, xzy = -yzx, xzx = yzy, xxz = yyz,$ $zxx = zyy, zzz, zxy = -zyx$

Table 3.2 (*cont.*)

Crystal system	Point group	Nonvanishing tensor elements
	$\bar{6}$	$xxx = -xyy = -yxy = -yyx, yyy = -yxx = -xxy = -xyx$
	622	$xyz = -yxz, xzy = -yzx, zxy = -zyx$
	$6mm$	$xzx = yzy, xxz = yyz, zxx = zyy, zzz$
	$\bar{6}m2$	$yyy = -yxx = -xxy = -xyx$
	$6/m$	Centrosymmetric; all elements vanish
	$6/mmm$	Centrosymmetric; all elements vanish
Cubic	432	$xyz = yzx = zxy = -xzy = -yxz = -zyx$
	23	$xyz = yzx = zxy, xzy = yxz = zyx$
	$\bar{4}3m$	$xyz = yzx = zxy = xzy = yxz = zyx$
	$m3$	Centrosymmetric; all elements vanish
	$m3m$	Centrosymmetric; all elements vanish

crystals that belong to the 11 centrosymmetric point groups and noncrystalline material such as amorphous solids, liquids, and gases. Therefore, only crystals that belong to the 21 noncentrosymmetric point groups have nonvanishing $\chi^{(2)}$ elements. It is clear that though not all noncentrosymmetric crystals are useful, any material that can support a second-order nonlinear process through electric dipole interaction is necessarily a noncentrosymmetric crystal. The nonvanishing $\chi^{(2)}$ tensor elements and the relations among them for each of the 21 noncentrosymmetric point groups are listed in Table 3.2.

The discussion above about the vanishing $\chi^{(2)}$ for a centrosymmetric material is valid only for the bulk nonlinear optical property of the material due to electric dipole interaction. Nonlinear optical effects of even orders that are contributed by magnetic dipole interaction or electric quadrupole interaction can still exist in a centrosymmetric material, resulting in a nonvanishing $\chi^{(2)}$ from magnetic dipole interaction or electric quadrupole interaction. A nonvanishing $\chi^{(2)}$ from electric dipole interaction can also be found on the surface or at an interface of the material. Centrosymmetry is broken on the surface of any material or at an interface between two different materials even when the materials themselves are centrosymmetric. Therefore, $\chi^{(2)}$ contributed by electric dipole interaction exists at any material surface or interface. As a result, second-order nonlinear processes that normally do not occur in the bulk of a certain material, such as silicon, which is centrosymmetric, can take place on its surface or interface. The surface $\chi^{(2)}$ also depends on the structure of the material surface.

EXAMPLE 3.2

The $\hat{x}$, $\hat{y}$, and $\hat{z}$ directional unit vectors used to define the electric field polarizations in Examples 2.1 and 3.1 are aligned with the principal x, y, and z axes of a crystal. (a) Use the result obtained in Example 3.1 to find the nonvanishing terms in the three components of $\mathbf{P}^{(2)}(\omega_4)$ if the nonlinear interaction takes place in a $\overline{4}3m$ crystal, such as GaAs. (b) Find the nonvanishing terms in the case of a crystal of the $3m$ point group, such as LiNbO$_3$.

Solution

(a) From Table 3.2, the only nonvanishing $\chi^{(2)}$ elements for the $\overline{4}3m$ point group are $\chi_{xyz}^{(2)} = \chi_{yzx}^{(2)} = \chi_{zxy}^{(2)} = \chi_{xzy}^{(2)} = \chi_{yxz}^{(2)} = \chi_{zyx}^{(2)}$. From the expressions for the components of $\mathbf{P}^{(2)}(\omega_4)$ obtained in Example 3.1, we have, for a $\overline{4}3m$ crystal,

$$P_x^{(2)}(\omega_4) = \epsilon_0 \left[\sqrt{2} \chi_{xyz}^{(2)}(\omega_4 = \omega_1 - \omega_3) E_1 E_3^* + \chi_{xyz}^{(2)}(\omega_4 = \omega_2 + \omega_2) E_2^2 \right],$$

$$P_y^{(2)}(\omega_4) = \sqrt{2} \epsilon_0 \chi_{yxz}^{(2)}(\omega_4 = \omega_1 - \omega_3) E_1 E_3^*,$$

$$P_z^{(2)}(\omega_4) = 0.$$

(b) For the $3m$ point group, the only nonvanishing $\chi^{(2)}$ elements are $\chi_{xzx}^{(2)} = \chi_{yzy}^{(2)}$, $\chi_{xxz}^{(2)} = \chi_{yyz}^{(2)}$, $\chi_{yyy}^{(2)} = -\chi_{yxx}^{(2)} = -\chi_{xxy}^{(2)} = -\chi_{xyx}^{(2)}$, $\chi_{zxx}^{(2)} = \chi_{zyy}^{(2)}$, and $\chi_{zzz}^{(2)}$, according to Table 3.2. Then, the expressions for the components of $\mathbf{P}^{(2)}(\omega_4)$ obtained in Example 3.1 reduce to the forms:

$$P_x^{(2)}(\omega_4) = \sqrt{2} \epsilon_0 \chi_{xxz}^{(2)}(\omega_4 = \omega_1 - \omega_3) E_1 E_3^*,$$

$$P_y^{(2)}(\omega_4) = \epsilon_0 \left[\sqrt{2} \chi_{yyz}^{(2)}(\omega_4 = \omega_1 - \omega_3) E_1 E_3^* + \chi_{yyy}^{(2)}(\omega_4 = \omega_2 + \omega_2) \frac{E_2^2}{2} \right.$$
$$\left. + \chi_{yyz}^{(2)}(\omega_4 = \omega_2 + \omega_2) E_2^2 \right],$$

$$P_z^{(2)}(\omega_4) = \epsilon_0 \left[\chi_{zyy}^{(2)}(\omega_4 = \omega_2 + \omega_2) \frac{E_2^2}{2} + \chi_{zzz}^{(2)}(\omega_4 = \omega_2 + \omega_2) \frac{E_2^2}{2} \right].$$

3.5.3 Third-Order Nonlinear Susceptibility

All materials, including centrosymmetric crystals and isotropic materials, have nonvanishing $\chi^{(3)}$. Nonetheless, the nonvanishing tensor elements and their relations are determined by the symmetry property of a material. The nonvanishing tensor elements and the relations among them for crystals of all point groups and for isotropic materials are listed in Table 3.3.

Because $\chi^{(3)}$ exists in all materials, the materials used for the devices that are based on third-order nonlinear optical processes are usually isotropic noncrystalline materials, such as glasses, or cubic crystals, such as the III–V semiconductors. Only occasionally are noncubic crystals used for such devices. Third-order nonlinear processes are particularly important for isotropic materials because $\chi^{(2)}$ vanishes identically so that $\chi^{(3)}$ becomes the leading nonlinear

Table 3.3 **Elements of the third-order nonlinear susceptibility tensor from electric dipole interaction**

Crystal system	Point group	Nonvanishing tensor elements
Triclinic	$1, \overline{1}$	All 81 elements are independent and nonvanishing
Monoclinic	$2, m, 2/m$	$xxxx, yyyy, zzzz,$ $xxyy, xyxy, xyyx, yyzz, yzyz, yzzy, zzxx, zxzx, zxxz,$ $yyxx, yxyx, yxxy, zzyy, zyzy, zyyz, xxzz, xzxz, xzzx,$ $yyxz, yyzx, xyyz, zyyx, xzyy, zxyy, xyzy, zyxy, yxyz, yzyx, yzxy, yxzy,$ $xxxz, xxzx, xzxx, zxxx, zzzx, zzxz, zxzz, xzzz$
Orthorhombic	$222, mm2, mmm$	$xxxx, yyyy, zzzz,$ $xxyy, xyxy, xyyx, yyzz, yzyz, yzzy, zzxx, zxzx, zxxz,$ $yyxx, yxyx, yxxy, zzyy, zyzy, zyyz, xxzz, xzxz, xzzx$
Tetragonal	$4, \overline{4}, 4/m$	$xxxx = yyyy, \ zzzz,$ $xxyy = yyxx, \ xyxy = yxyx, \ xyyx = yxxy,$ $yyzz = xxzz, \ yzyz = xzxz, \ yzzy = xzzx,$ $zzyy = zzxx, \ zyzy = zxzx, \ zyyz = zxxz,$ $xxxy = -yyyx, \ xxyx = -yyxy, \ xyxx = -yxyy, \ yxxx = -xyyy,$ $xyzz = -yxzz, \ xzyz = -yzxz, \ xzzy = -yzzx,$ $zzxy = -zzyx, \ zxzy = -zyzx, \ zxyz = -zyxz$
	$422, 4mm, \overline{4}2m, 4/mmm$	$xxxx = yyyy, \ zzzz,$ $xxyy = yyxx, \ xyxy = yxyx, \ xyyx = yxxy,$ $yyzz = xxzz, \ yzyz = xzxz, \ yzzy = xzzx,$ $zzyy = zzxx, \ zyzy = zxzx, \ zyyz = zxxz$
Trigonal	$3, \overline{3}$	$xxxx = yyyy = xxyy + xyxy + xyyx, \ zzzz,$ $xxyy = yyxx, \ xyxy = yxyx, \ xyyx = yxxy,$ $yyzz = xxzz, \ yzyz = xzxz, \ yzzy = xzzx,$ $zzyy = zzxx, \ zyzy = zxzx, \ zyyz = zxxz,$ $xyzz = -yxzz, \ xzyz = -yzxz, \ xzzy = -yzzx,$ $zzxy = -zzyx, \ zxzy = -zyzx, \ zxyz = -zyxz$ $xxxy = -yyyx = yyxy + yxyy + xyyy,$ $yyxy = -xxyx, \ yxyy = -xyxx, \ xyyy = -yxxx,$ $yyyz = -yxxz = -xyxz = -xxyz, \ yyzy = -yxzx = -xyzx = -xxzy,$ $yzyy = -yzxx = -xzyx = -xzxy, \ zyyy = -zyxx = -zxyx = -zxxy,$ $xxxz = -xyyz = -yxyz = -yyxz, \ xxzx = -xyzy = -yxzy = -yyzx,$ $xzxx = -yzxy = -yzyx = -xzyy, \ zxxx = -zxyy = -zyxy = -zyyx$
	$32, 3m, \overline{3}m$	$xxxx = yyyy = xxyy + xyxy + xyyx, \ zzzz,$ $xxyy = yyxx, \ xyxy = yxyx, \ xyyx = yxxy,$ $yyzz = xxzz, \ yzyz = xzxz, \ yzzy = xzzx,$

Table 3.3 (*cont.*)

Crystal system	Point group	Nonvanishing tensor elements
		$zzyy = zzxx,\ zyzy = zxzx,\ zyyz = zxxz,$
		$xxxz = -xyyz = -yxyz = -yyxz,\ xxzx = -xyzy = -yxzy = -yyzx,$
		$xzxx = -xzyy = -yzxy = -yzyx,\ zxxx = -zxyy = -zyxy = -zyyx$
Hexagonal	$6, \overline{6}, 6/m$	$xxxx = yyyy = xxyy + xyxy + xyyx,\ zzzz,$
		$xxyy = yyxx,\ xyxy = yxyx,\ xyyx = yxxy,$
		$yyzz = xxzz,\ yzyz = xzxz,\ yzzy = xzzx,$
		$zzyy = zzxx,\ zyzy = zxzx,\ zyyz = zxxz,$
		$xyzz = -yxzz,\ xzyz = -yzxz,\ xzzy = -yzzx,$
		$zzxy = -zzyx,\ zxzy = -zyzx,\ zxyz = -zyxz$
		$xxxy = -yyyx = yyxy + yxyy + xyyy,$
		$yyxy = -xxyx,\ yxyy = -xyxx,\ xyyy = -yxxx$
	$622, 6mm, \overline{6}m2, 6/mmm$	
		$xxxx = yyyy = xxyy + xyxy + xyyx,\ zzzz,$
		$xxyy = yyxx,\ xyxy = yxyx,\ xyyx = yxxy,$
		$yyzz = xxzz,\ yzyz = xzxz,\ yzzy = xzzx,$
		$zzyy = zzxx,\ zyzy = zxzx,\ zyyz = zxxz$
Cubic	$23, m3$	$xxxx = yyyy = zzzz,$
		$xxyy = yyzz = zzxx,\ yyxx = zzyy = xxzz,$
		$xyxy = yzyz = zxzx,\ yxyx = zyzy = xzxz,$
		$xyyx = yzzy = zxxz,\ yxxy = zyyz = xzzx$
	$432, \overline{4}3m, m3m$	$xxxx = yyyy = zzzz,$
		$xxyy = yyxx = yyzz = zzyy = zzxx = xxzz,$
		$xyxy = yxyx = yzyz = zyzy = zxzx = xzxz,$
		$xyyx = yxxy = yzzy = zyyz = zxxz = xzzx$
Isotropic		$xxxx = yyyy = zzzz = xxyy + xyxy + xyyx,$
		$xxyy = yyxx = yyzz = zzyy = zzxx = xxzz,$
		$xyxy = yxyx = yzyz = zyzy = zxzx = xzxz,$
		$xyyx = yxxy = yzzy = zyyz = zxxz = xzzx$

susceptibility of such materials. It can be seen that for all of the point groups in the cubic system and for isotropic materials, there are only 21 nonvanishing $\chi^{(3)}$ tensor elements. For the 23 and $m3$ point groups, there are seven independent $\chi^{(3)}$ elements. For the 432, $\overline{4}3m$, and $m3m$ point groups, there are only four independent elements of the types $\chi^{(3)}_{1111}$, $\chi^{(3)}_{1122}$, $\chi^{(3)}_{1212}$, $\chi^{(3)}_{1221}$. If Kleinman's symmetry condition is valid, the number of independent elements reduces to two of the types $\chi^{(3)}_{1111}$ and $\chi^{(3)}_{1122} = \chi^{(3)}_{1212} = \chi^{(3)}_{1221}$ for all point groups in the cubic system. For an isotropic material, there are only three independent $\chi^{(3)}$ elements among the four types of nonvanishing elements because $\chi_{1111}^{(3)} = \chi_{1122}^{(3)} + \chi_{1212}^{(3)} + \chi_{1221}^{(3)}$. If Kleinman's symmetry condition is valid in an isotropic material, we have

$$\chi_{1122}^{(3)} = \chi_{1212}^{(3)} = \chi_{1221}^{(3)} = \frac{\chi_{1111}^{(3)}}{3},$$ (3.30)

thus reducing the number of independent $\chi^{(3)}$ elements to only one.

3.5.4 Nonlinear Optical *d* Coefficients

In the literature, experimentally measured values of the second-order nonlinear susceptibilities of a material are commonly quoted in terms of nonlinear coefficients d_{ijk}, or $d_{i\alpha}$ under *index contraction*, defined below in (3.32). The relation between the *d* coefficients and the $\chi^{(2)}$ elements is simply

$$d_{ijk} = \frac{1}{2}\chi_{ijk}^{(2)}$$ (3.31)

if neither index *j* nor index *k* is associated with a DC field.

3.5.5 Index Contraction

In certain situations, *the rule of index contraction* given below can be applied to $\chi_{ijk}^{(2)}$ and d_{ijk} on the last two indices *j* and *k* by replacing *jk* with α:

$$\begin{array}{ccccccc} jk: & 11 & 22 & 33 & 23, 32 & 31, 13 & 12, 21 \\ \text{or } jk: & xx & yy & zz & yz, zy & zx, xz & xy, yx. \\ \alpha: & 1 & 2 & 3 & 4 & 5 & 6 \end{array}$$ (3.32)

Then the 27 elements of $\chi_{ijk}^{(2)}$, or d_{ijk}, are reduced to the 18 elements of $\chi_{i\alpha}^{(2)}$, or $d_{i\alpha}$, for $i = 1, 2, 3$ and $\alpha = 1, 2, \cdots, 6$. Clearly, the condition for index contraction to be applicable is that there is no physical significance in interchanging the last two indices *j* and *k* independently of the frequencies in $\chi_{i\alpha}^{(2)}$.

For $\chi^{(2)}(\omega_3 = \omega_1 + \omega_2)$ in general, index contraction applies only when Kleinman's symmetry condition is valid so that $\chi_{ijk}^{(2)}(\omega_3 = \omega_1 + \omega_2) = \chi_{ikj}^{(2)}(\omega_3 = \omega_1 + \omega_2)$. However, index contraction applies without the requirement of Kleinman's symmetry condition in the special cases of $\chi^{(2)}(2\omega = \omega + \omega)$ and $\chi^{(2)}(0 = \omega - \omega)$.

For $\chi^{(2)}(2\omega = \omega + \omega)$, which characterizes the process of *second-harmonic generation*, index contraction always applies because $\chi_{ijk}^{(2)}(2\omega = \omega + \omega) = \chi_{ikj}^{(2)}(2\omega = \omega + \omega)$ by the definition of the intrinsic permutation symmetry.

For $\chi^{(2)}(0 = \omega - \omega)$, which characterizes the process of *optical rectification* for the generation of a DC electric field by an optical field, index contraction applies only when the medium is lossless at the frequency ω so that $\chi_{ijk}^{(2)}(0 = \omega - \omega) = \chi_{ikj}^{(2)}(0 = -\omega + \omega) =$

$\chi_{ikj}^{(2)}(0 = \omega - \omega) = \chi_{ikj}^{(2)}(0 = -\omega + \omega)$ due to the reality condition and the intrinsic permutation symmetry. Note that Kleinman's symmetry condition is never valid for $\chi^{(2)}(0 = \omega - \omega)$ because no material can be completely nondispersive in the entire spectral range from DC to the optical frequencies.

With index contraction, the second-order nonlinear susceptibilities $\chi_{ia}^{(2)}$ and, correspondingly, the nonlinear coefficients d_{ia} can be expressed in the form of a 3×6 matrix. If Kleinman's symmetry condition is valid, the matrix form of $\chi_{ia}^{(2)}$ and d_{ia} is further simplified to result in a maximum of only 10 independent parameters. For example, $d_{14} = d_{25} = d_{36}$ under Kleinman's symmetry condition. We also find that under Kleinman's symmetry condition the three independent nonlinear coefficients d_{14}, d_{25}, and d_{36} for the 222 point group reduce to one identical parameter, and the two independent parameters $d_{14} = d_{25}$ and d_{36} for the $\overline{4}2m$ point group also reduce to a single parameter. The properties of some important nonlinear crystals are listed in Table 3.4.

By using (2.31) and (3.31), the second-order nonlinear polarization can be expressed in terms of the d_{ia} matrix. In the general case of $\omega_1 + \omega_2 = \omega_3$ with $\omega_1 \neq \omega_2$, we have

$$\begin{bmatrix} P_x^{(2)}(\omega_3) \\ P_y^{(2)}(\omega_3) \\ P_z^{(2)}(\omega_3) \end{bmatrix} = 4\epsilon_0 \begin{bmatrix} d_{11} & d_{12} & d_{13} & d_{14} & d_{15} & d_{16} \\ d_{21} & d_{22} & d_{23} & d_{24} & d_{25} & d_{26} \\ d_{31} & d_{32} & d_{33} & d_{34} & d_{35} & d_{36} \end{bmatrix} \begin{bmatrix} E_x(\omega_1)E_x(\omega_2) \\ E_y(\omega_1)E_y(\omega_2) \\ E_z(\omega_1)E_z(\omega_2) \\ E_y(\omega_1)E_z(\omega_2) + E_z(\omega_1)E_y(\omega_2) \\ E_z(\omega_1)E_x(\omega_2) + E_x(\omega_1)E_z(\omega_2) \\ E_x(\omega_1)E_y(\omega_2) + E_y(\omega_1)E_x(\omega_2) \end{bmatrix}.$$

$$(3.33)$$

In the case of second-harmonic generation with $\omega_1 = \omega_2 = \omega$ and $\omega_3 = 2\omega$, we have

$$\begin{bmatrix} P_x^{(2)}(2\omega) \\ P_y^{(2)}(2\omega) \\ P_z^{(2)}(2\omega) \end{bmatrix} = 2\epsilon_0 \begin{bmatrix} d_{11} & d_{12} & d_{13} & d_{14} & d_{15} & d_{16} \\ d_{21} & d_{22} & d_{23} & d_{24} & d_{25} & d_{26} \\ d_{31} & d_{32} & d_{33} & d_{34} & d_{35} & d_{36} \end{bmatrix} \begin{bmatrix} E_x^2(\omega) \\ E_y^2(\omega) \\ E_z^2(\omega) \\ 2E_y(\omega)E_z(\omega) \\ 2E_z(\omega)E_x(\omega) \\ 2E_x(\omega)E_y(\omega) \end{bmatrix}.$$

$$(3.34)$$

Note that each d coefficient in (3.33) has the same value as the corresponding one in (3.34) if we ignore dispersion due to the frequency differences between (3.33) and (3.34). It is true that in the case that $\omega_1 \neq \omega_2$, $\mathbf{P}^{(2)}(\omega_3) = 2\mathbf{P}^{(2)}(2\omega)$ if $\mathbf{E}(\omega_1) = \mathbf{E}(\omega_2) = \mathbf{E}(\omega)$, as is seen by comparing (3.33) and (3.34).

3.6 UNIT CONVERSION

The SI system, which is essentially the MKSA system, is used consistently in this book. Nevertheless, the Gaussian system is also used quite often in the literature. In the Gaussian

Table 3.4 Properties of representative nonlinear crystals[a]

Crystal	LiNbO$_3$ (LN)	β-BaB$_2$O$_4$ (BBO)	LiB$_3$O$_5$ (LBO)	KTiOPO$_4$ (KTP)	KTiOAsO$_4$ (KTA)	AgGaS$_2$	AgGaSe$_2$	ZnGeP$_2$	LiIO$_3$
Point group	3m	3m	mm2	mm2	mm2	$\bar{4}$2m	$\bar{4}$2m	$\bar{4}$2m	6
n_x (n_o): see Note b									
A	4.913	2.735 9	2.454 3	3.006 48	3.142 40	5.728	6.854 40	4.473 30	3.414
B	0.118 8	0.018 78	0.011 413	0.038 81	0.045 89	0.241 07	0.408 15	5.265 76	0.046 29
C	0.045 97	0.018 22	0.009 498 1	0.043 52	0.044 48	0.087 03	0.192 64	0.133 81	0.037 11
D	0.027 78	0.013 54	0.013 900	0.013 20	0.010 60	0.002 10	0.001 283 8	1.490 85	0.007 603 2
E								662.55	
n_y: see Note b									
A	Same as	Same as	2.538 2	3.030 42	3.167 90	Same as	Same as	Same as	Same as
B	n_x	n_x	0.012 83	0.041 76	0.044 05	n_x	n_x	n_x	n_x
C			0.011 387	0.047 53	0.056 55				
D			0.017 034	0.013 27	0.015 00				
E									
n_z (n_e): see Note b									
A	4.579 8	2.375 3	2.585 4	3.313 72	3.448 30	5.497	6.691 30	4.633 18	2.922 82
B	0.099 4	0.012 24	0.013 065	0.056 80	0.063 53	0.202 59	0.395 83	5.342 15	0.032 87
C	0.042 35	0.016 67	0.011 617	0.056 79	0.057 70	0.130 70	0.283 65	0.142 55	0.033 335
D	0.022 4	0.015 16	0.018 146	0.016 79	0.017 40	0.002 33	0.001 343 0	0.145 795	0.004 262 3
E								662.55	
d_{31} (pm V^{-1})	−4.4	0.04	−1.09	3.7	2.8	0	0	0	−4.64
d_{32}	$= d_{31}$	$= d_{31}$	1.17	2.2	4.2	0	0	0	$= d_{31}$
d_{33}	−25.2	—	0.065	14.6	16.2	0	0	0	−4.84
d_{24}	$= d_{31}$	$= d_{31}$	−1.1	1.9	4.24	0	0	0	—
d_{15}	$= d_{24}$	$= d_{24}$	1.00	3.7	2.24	0	0	0	$= d_{24}$
d_{36}	0	0	0	0	0	28.7	49.3	75	0

Table 3.4 (cont.)

Crystal	LiNbO$_3$ (LN)	β-BaB$_2$O$_4$ (BBO)	LiB$_3$O$_5$ (LBO)	KTiOPO$_4$ (KTP)	KTiOAsO$_4$ (KTA)	AgGaS$_2$	AgGaSe$_2$	ZnGeP$_2$	LiIO$_3$
d_{14}	0	0	0	0	0	$= d_{36}$	$= d_{36}$	—	0.31
d_{25}	0	0	0	0	0	$= d_{14}$	$= d_{14}$	—	$= -d_{14}$
$d_{22} = -d_{21} = -d_{16}$	2.4	2.22	0	0	0	0	0	0	0
Transparency (nm)	400–5,000	190–3,000	160–2,600	350–4,500	400–5,000	500–13,000	780–18,000	700–12,000	300–5,500
Damage threshold (GW cm^{-2})	1	13	25	1	1.2	0.02	0.02	>1	0.5
dn_x/dT ($\times 10^{-5}$ K^{-1})	2.0	−1.7	−0.19	1.1		5	4	14.3	−8.9
dn_y/dT ($\times 10^{-5}$ K^{-1})	2.0	−1.7	−1.3	1.3		5	4	14.3	−8.9
dn_z/dT ($\times 10^{-5}$ K^{-1})	7.6	−0.93	−0.83	1.6		5	4	15.0	−7.8

[a] The data are collected from various sources in the literature. Many of the crystal properties are constantly being revised. The refractive indices of some crystals vary with crystal growth and preparation procedures. The nonlinear susceptibilities can vary with optical wavelength and temperature, too.

[b] The parameters listed here for the refractive indices are those at 300 K. For all crystals except ZnGeP$_2$, the refractive indices are calculated using the following Sellmeier equation:

$$n^2 = A + \frac{B}{\lambda^2 - C} - D\lambda^2.$$

For ZnGeP$_2$, the following Sellmeier equation has to be used:

$$n^2 = A + \frac{B}{1 - C/\lambda^2} + \frac{D}{1 - E/\lambda^2}.$$

In both formulas, λ is in micrometers.

system, cgs units are used, but the electric field quantities and the susceptibilities are normally given in the units of esu, meaning electrostatic units, without explicitly spelling out their true dimensions. In the SI system, the units are explicit. Unit conversion for susceptibilities between the SI and the Gaussian systems follows the relations:

$$\text{SI} \qquad\qquad\qquad \text{Gaussian}$$

$$\chi^{(1)}(\text{dimensionless}) = 4\pi\chi^{(1)}(\text{dimensionless}), \tag{3.35}$$

$$\chi^{(2)}(\text{m V}^{-1}) = \frac{4\pi}{3 \times 10^4}\chi^{(2)}(\text{esu}), \tag{3.36}$$

$$\chi^{(3)}(\text{m}^2\, \text{V}^{-2}) = \frac{4\pi}{9 \times 10^8}\chi^{(3)}(\text{esu}). \tag{3.37}$$

Unit conversion for the d coefficient is the same as that for $\chi^{(2)}$ given in (3.36) because the relation in (3.31) is independent of the unit system used.

Problem Set

3.1.1 Show that the frequency dependence of the second-order nonlinear susceptibility element $\chi^{(2)}_{ijk}(\omega_q = \omega_m + \omega_n)$ is that given in (3.12) by finding the second-order polarization $\mathbf{P}^{(2)}(t)$ through the second-order density matrix $\rho^{(2)}(t)$.

3.1.2 Show that the frequency dependence of the third-order nonlinear susceptibility element $\chi^{(3)}_{ijkl}(\omega_q = \omega_m + \omega_n + \omega_p)$ is that given in (3.13) by finding the third-order polarization $\mathbf{P}^{(3)}(t)$ through the third-order density matrix $\rho^{(3)}(t)$.

3.1.3 In this problem, we calculate the linear and nonlinear susceptibilities of a material that contains N_{total} valence electrons per unit volume by using a classical one-dimensional anharmonic oscillator model. Each of these electrons, with a charge of $q = -e$ and a mass of m_0, oscillates in an anharmonic potential of the form:

$$V(x) = \frac{1}{2}m_0\omega_0^2 x^2 + \frac{1}{3}m_0 ax^3 \tag{3.38}$$

with a damping constant of γ so that its motion in response to externally applied optical fields can be described by an equation of motion of the form:

$$\frac{d^2 x}{dt^2} + 2\gamma\frac{dx}{dt} + \omega_0^2 x + ax^2 = \frac{F}{m_0}, \tag{3.39}$$

where $F = -e(E_m\, e^{-i\omega_m t} + E_n\, e^{-i\omega_n t} + E_p\, e^{-i\omega_p t} + \cdots) + \text{c.c.}$ We are interested in the linear and nonlinear susceptibilities that are contributed by electric dipole interactions. The electric dipole polarization is defined as

$$P(t) = -N_{\text{total}}ex(t). \tag{3.40}$$

For the material response at a particular frequency ω_q, $P(t) = P(\omega_q)e^{-i\omega_q t} + \text{c.c.}$ and $x(t) = x(\omega_q)e^{-i\omega_q t} + \text{c.c.}$ so that $P(\omega_q) = -N_{\text{total}}ex(\omega_q)$. The linear and nonlinear susceptibilities can be found by solving (3.39) by using the classical perturbation method. In this approach, $x(t)$ is expanded in a perturbation series as $x = x^{(1)} + x^{(2)} + x^{(3)} + \cdots$. Each order of $x^{(n)}$ is solved successively from (3.39). The polarizations of different orders are then defined as $P^{(n)}(\omega_q) = -N_{\text{total}}ex^{(n)}(\omega_q)$ for $n = 1, 2, 3, \ldots$.

(a) Is this material centrosymmetric or noncentrosymmetric?

(b) Find the linear susceptibility $\chi^{(1)}(\omega)$.

(c) Find the second-order susceptibilities: $\chi^{(2)}(\omega = \omega_1 + \omega_2)$, $\chi^{(2)}(2\omega_1 = \omega_1 + \omega_1)$, and $\chi^{(2)}(0 = \omega_1 - \omega_1)$, where $\omega_1 \neq \omega_2$.

(d) Find the third-order susceptibilities: $\chi^{(3)}(\omega = \omega_1 + \omega_2 + \omega_3)$, $\chi^{(3)}(3\omega_1 = \omega_1 + \omega_1 + \omega_1)$, and $\chi^{(3)}(\omega_1 = \omega_1 + \omega_1 - \omega_1)$, where $\omega_1 \neq \omega_2 \neq \omega_3$.

3.1.4 There is a relationship between the second-order susceptibility $\chi_{ijk}^{(2)}(\omega_3 = \omega_1 + \omega_2)$ and the linear susceptibilities $\chi_{ii}^{(1)}(\omega_3)$, $\chi_{jj}^{(1)}(\omega_1)$, and $\chi_{kk}^{(1)}(\omega_2)$. It is known as *Miller's rule* and states that the ratio

$$\Delta_{ijk}^{(2)} = \frac{\chi_{ijk}^{(2)}(\omega_3 = \omega_1 + \omega_2)}{\chi_{ii}^{(1)}(\omega_3)\chi_{jj}^{(1)}(\omega_1)\chi_{kk}^{(1)}(\omega_2)} \tag{3.41}$$

is nearly a constant for all noncentrosymmetric crystals.

(a) Use the results in Problem 3.1.3 to find the constant $\Delta^{(2)}$ for a one-dimensional case.

(b) Estimate the value of the constant $\Delta^{(2)}$ for the typical noncentrosymmetric solid crystal by taking these typical values: $N_{\text{total}} \approx 10^{29}$ m^{-3} and $\omega_0 \approx 10^{16}$ rad s^{-1}. We also take $|a|x^2 \approx \omega_0^2 x$ and $x \approx N_{\text{total}}^{-1/3}$ so that $|a| \approx \omega_0^2 N_{\text{total}}^{1/3}$. Find the typical range of $\chi^{(2)}$ by considering the fact that $\chi^{(1)}$ falls in the range of 1–10 for most crystals.

(c) Find a similar rule for the third-order susceptibility $\chi_{ijkl}^{(3)}(\omega_4 = \omega_1 + \omega_2 + \omega_3)$. What is the constant $\Delta^{(3)}$ for a one-dimensional case in this situation?

(d) Estimate the value of the constant $\Delta^{(3)}$ for the typical noncentrosymmetric crystal by taking the parameters used in (b). Find the typical range of $\chi^{(3)}$ by considering the fact that $\chi^{(1)}$ falls in the range of 1–10 for most crystals.

(e) What are the physical implications of Miller's rule and the similar rule for $\chi^{(3)}$?

3.1.5 By using the quantum-mechanical forms of $\chi_{ijk}^{(2)}(\omega_q = \omega_m + \omega_n)$ as given in (3.12) and $\chi_{ij}^{(1)}(\omega_q)$ as given in (3.11), find the form of the ratio $\Delta_{ijk}^{(2)}$ that is defined in (3.41) for Miller's rule. Estimate the magnitude of $\Delta^{(2)}$ by taking these typical values: $N_{\text{total}} \approx 10^{29}$ m^{-3}, $\hbar\omega = 1$ eV for the photon energy, and $p \approx ea_0$ for the magnitude of the electric dipole moment, where $a_0 = 5.29 \times 10^{-11}$ m is the Bohr radius. Compare this magnitude with that found in Problem 3.1.4(b) from the classical forms of the linear and nonlinear susceptibilities.

3.1.6 In this problem, we calculate the linear and nonlinear susceptibilities of a material that contains N_{total} valence electrons per unit volume by using a classical one-dimensional anharmonic oscillator model that is different from the one considered in Problem 3.1.3. Each of these electrons, with a charge of $q = -e$ and a mass of m_0, oscillates in an anharmonic potential of the form:

$$V(x) = \frac{1}{2}m_0\omega_0^2 x^2 + \frac{1}{4}m_0 b x^4 \tag{3.42}$$

with a damping constant of γ so that its motion in response to externally applied optical fields can be described by an equation of motion of the form:

$$\frac{d^2 x}{dt^2} + 2\gamma \frac{dx}{dt} + \omega_0^2 x + bx^3 = \frac{F}{m_0}, \tag{3.43}$$

where $F = -e(E_m e^{-i\omega_m t} + E_n e^{-i\omega_n t} + E_p e^{-i\omega_p t} + \cdots) + \text{c.c.}$

(a) Is this material centrosymmetric or noncentrosymmetric?

(b) Use the perturbation method and the definition for the electric dipole polarization given in Problem 3.1.3 to find $\chi^{(1)}(\omega)$, $\chi^{(2)}(2\omega = \omega + \omega)$, and $\chi^{(3)}(3\omega = \omega + \omega + \omega)$ that are contributed by electric dipole interactions.

(c) Explain why $\chi^{(2)} = 0$ in this problem.

3.1.7 For centrosymmetric materials, there is a relation between $\chi^{(3)}$ and $\chi^{(1)}$ similar to Miller's rule, discussed in Problem 3.1.4.

(a) Use the results obtained in Problem 3.1.6(b) to show that there is a constant $\Delta^{(3)}$ that relates $\chi^{(3)}(3\omega = \omega + \omega + \omega)$ and the linear susceptibilities $\chi^{(1)}(\omega)$ and $\chi^{(1)}(3\omega)$ for centrosymmetric materials. Find this constant. Compare it with that found in Problem 3.1.4(c) for noncentrosymmetric crystals.

(b) Estimate the value of the constant $\Delta^{(3)}$ for the typical centrosymmetric solid by taking the following typical values: $N_{total} \approx 10^{29}$ m^{-3} and $\omega_0 \approx 10^{16}$ rad s^{-1}. We also take $|b|x^3 \approx \omega_0^2 x$ and $x \approx N_{total}^{-1/3}$ so that $|b| \approx \omega_0^2 N_{total}^{2/3}$. Find the typical range of $\chi^{(3)}$ by considering the fact that $\chi^{(1)}$ falls in the range of 1–10 for most solids. Compare the numerical results obtained here with those found in Problem 3.1.4(d).

3.1.8 Consider an isotropic two-level absorbing medium with a single resonance frequency at ω_0. Follow the procedure used in Problem 3.1.6 to find $\chi^{(2)}(0 = \omega - \omega)$ and $\chi^{(3)}(\omega = \omega + \omega - \omega)$.

3.2.1 Verify the reality condition given in (3.17) for nonlinear susceptibilities.

3.4.1 Simplify the expressions for $\mathbf{P}^{(2)}(\omega_4)$ and $P_x^{(2)}(\omega_4)$ in Problem 2.2.1 by using the intrinsic permutation symmetry of $\chi^{(2)}$. Write out the simplified expressions for $P_y^{(2)}(\omega_4)$ and $P_z^{(2)}(\omega_4)$.

3.4.2 Simplify the expressions for $\mathbf{P}^{(2)}(\omega_5)$ and $P_x^{(2)}(\omega_5)$ in Problem 2.2.2 by using the intrinsic permutation symmetry of $\chi^{(2)}$. Write out the simplified expressions for $P_y^{(2)}(\omega_5)$ and $P_z^{(2)}(\omega_5)$.

3.4.3 Simplify the expressions for $\mathbf{P}^{(2)}(\omega_4)$ and $P_x^{(2)}(\omega_4)$ in Problem 2.2.3 by using the intrinsic permutation symmetry of $\chi^{(2)}$. Write out the simplified expressions for $P_y^{(2)}(\omega_4)$ and $P_z^{(2)}(\omega_4)$.

3.4.4 Simplify the expressions for $\mathbf{P}^{(2)}(\omega_5)$ and $P_x^{(2)}(\omega_5)$ in Problem 2.2.4 by using the intrinsic permutation symmetry of $\chi^{(2)}$. Write out the simplified expressions for $P_y^{(2)}(\omega_5)$ and $P_z^{(2)}(\omega_5)$.

3.4.5 Use the results from Problem 3.1.3 to answer the following questions.

(a) Show the permutation symmetry of $\chi^{(2)}(\omega_3 = \omega_1 + \omega_2)$ for a lossless medium when $\omega_1 \neq \omega_2$.

(b) What is the permutation symmetry relation of $\chi^{(2)}$ for a lossless medium in the case of frequency degeneracy, $\omega_1 = \omega_2$?

(c) Without calculation, write down similar permutation symmetry relations of $\chi^{(3)}(\omega_4 = \omega_1 + \omega_2 + \omega_3)$ for a lossless medium in the cases of no frequency degeneracy, two-frequency degeneracy, and three-frequency degeneracy, respectively.

3.4.6 Spell out explicitly the relations among the frequency-dependent elements of the $\chi^{(3)}$ tensor that characterize the interaction of four frequencies ω_1, ω_2, ω_3, and ω_4 for $\omega_4 = \omega_1 + \omega_2 + \omega_3$ under (a) intrinsic permutation symmetry, (b) full permutation symmetry, and (c) Kleinman's symmetry condition, respectively.

3.5.1 The $\hat{x}$, $\hat{y}$, and $\hat{z}$ directional unit vectors used to define the electric field polarizations in Problems 2.2.1 and 3.4.1 are aligned with the principal x, y, and z axes of a crystal. Use the result obtained in Problem 3.4.1 to find the nonvanishing terms in the three components of $\mathbf{P}^{(2)}(\omega_4)$ if the nonlinear interaction takes place in a crystal of the $3m$ point group, such as LiNbO$_3$, for which the only nonvanishing $\chi^{(2)}$ elements are $\chi_{xzx}^{(2)} = \chi_{yzy}^{(2)}$, $\chi_{xxz}^{(2)} = \chi_{yyz}^{(2)}$, $\chi_{yyy}^{(2)} = -\chi_{yxx}^{(2)} = -\chi_{xxy}^{(2)} = -\chi_{xyx}^{(2)}$, $\chi_{zxx}^{(2)} = \chi_{zyy}^{(2)}$, and $\chi_{zzz}^{(2)}$.

3.5.2 The $\hat{x}$, $\hat{y}$, and $\hat{z}$ directional unit vectors used to define the electric field polarizations in Problems 2.2.2 and 3.4.2 are aligned with the principal x, y, and z axes of a crystal. Use the result obtained in Problem 3.4.2 to find the nonvanishing terms in the three components of $\mathbf{P}^{(2)}(\omega_5)$ if the nonlinear interaction takes place in a crystal of the $mm2$ point group, such as KTP, for which the only nonvanishing $\chi^{(2)}$ elements are $\chi_{xzx}^{(2)}, \chi_{xxz}^{(2)}, \chi_{yyz}^{(2)}, \chi_{yzy}^{(2)}, \chi_{zxx}^{(2)}, \chi_{zyy}^{(2)}$, and $\chi_{zzz}^{(2)}$.

3.5.3 The $\hat{x}$, $\hat{y}$, and $\hat{z}$ directional unit vectors used to define the electric field polarizations in Problems 2.2.3 and 3.4.3 are aligned with the principal x, y, and z axes of a crystal. Use the result obtained in Problem 3.4.3 to find the nonvanishing terms in the three components of $\mathbf{P}^{(2)}(\omega_4)$ if the nonlinear interaction takes place in a crystal of the $3m$ point group, such as LiNbO$_3$, for which the only nonvanishing $\chi^{(2)}$ elements are $\chi_{xzx}^{(2)} = \chi_{yzy}^{(2)}$, $\chi_{xxz}^{(2)} = \chi_{yyz}^{(2)}$, $\chi_{yyy}^{(2)} = -\chi_{yxx}^{(2)} = -\chi_{xxy}^{(2)} = -\chi_{xyx}^{(2)}$, $\chi_{zxx}^{(2)} = \chi_{zyy}^{(2)}$, and $\chi_{zzz}^{(2)}$.

3.5.4 The $\hat{x}$, $\hat{y}$, and $\hat{z}$ directional unit vectors used to define the electric field polarizations in Problems 2.2.4 and 3.4.4 are aligned with the principal x, y, and z axes of a crystal. Use the result obtained in Problem 3.4.4 to find the nonvanishing terms in the three components of $\mathbf{P}^{(2)}(\omega_5)$ if the nonlinear interaction takes place in a crystal of the *mm*2 point group, such as KTP, for which the only nonvanishing $\chi^{(2)}$ elements are $\chi^{(2)}_{xzx}$, $\chi^{(2)}_{xxz}$, $\chi^{(2)}_{yyz}$, $\chi^{(2)}_{yzy}$, $\chi^{(2)}_{zxx}$, $\chi^{(2)}_{zyy}$, and $\chi^{(2)}_{zzz}$.

3.5.5 Show that $\chi^{(2)}$ contributed by electric dipole interaction is identically zero in a centrosymmetric material, whereas a nonzero $\chi^{(3)}$ exists in any material.

3.5.6 Answer the following questions regarding nonlinear optical susceptibilities.

(a) Does a second-order nonlinear optical effect exist on the surface of a centrosymmetric material? Why?

(b) In general, would you expect a highly refractory material to have larger or smaller nonlinear optical susceptibilities than a less refractory material? Explain.

(c) Given 10 nonlinear crystals without any knowledge of their $\chi^{(2)}$, but with the refractive indices checked out from a handbook, how do you make an intelligent guess at which ones are likely to have a large $\chi^{(2)}$ before taking any measurements? What do you base your guess on?

(d) How does the effect of quantum confinement, such as that in the quantum-well structures of a semiconductor, enhance nonlinear susceptibilities?

References

[1] N. Bloembergen and Y. R. Shen, "Quantum-theoretical comparison of nonlinear susceptibilities in parametric media, lasers, and Raman lasers," *Physical Review*, vol. 133, pp. A37–A49, 1964.
[2] N. Bloembergen, H. Lotem, and R. T. Lynch, "Lineshapes in coherent resonant Raman scattering," *Indian Journal of Pure and Applied Physics*, vol. 16, pp. 151–158, 1978.
[3] Y. R. Shen, *The Principles of Nonlinear Optics*. New York: Wiley-Interscience, 1984.
[4] R. W. Boyd, *Nonlinear Optics*, 4th ed. New York: Academic Press, 2020.
[5] J. M. Liu, *Photonic Devices*. Cambridge: Cambridge University Press, 2005.

4 Propagation of Optical Waves

4.1 OPTICAL ENERGY FLOW

The propagation of an optical wave is governed by Maxwell's equations. The flow of optical power and energy is a consequence of wave propagation and is thus also governed by Maxwell's equations. For generality, we begin our discussion with the general form of Maxwell's equations given in (1.1)–(1.4).

By using (1.1) and (1.2) with the vector identity $\mathbf{V} \cdot (\mathbf{E} \times \mathbf{H}) = \mathbf{H} \cdot (\mathbf{V} \times \mathbf{E}) - \mathbf{E} \cdot (\mathbf{V} \times \mathbf{H})$, we obtain the relation:

$$\mathbf{V} \cdot (\mathbf{E} \times \mathbf{H}) = -\mathbf{H} \cdot \frac{\partial \mathbf{B}}{\partial t} - \mathbf{E} \cdot \frac{\partial \mathbf{D}}{\partial t} - \mathbf{E} \cdot \mathbf{J}. \tag{4.1}$$

By using the relations $\mathbf{D} = \epsilon_0 \mathbf{E} + \mathbf{P}$ from (1.5) and $\mathbf{B} = \mu_0 \mathbf{H} + \mu_0 \mathbf{M}$ from (1.6), we can rearrange (4.1) as

$$\mathbf{E} \cdot \mathbf{J} = -\mathbf{V} \cdot (\mathbf{E} \times \mathbf{H}) - \frac{\partial}{\partial t}\left(\frac{\epsilon_0}{2}\left|\mathbf{E}\right|^2 + \frac{\mu_0}{2}\left|\mathbf{H}\right|^2 \right) - \left(\mathbf{E} \cdot \frac{\partial \mathbf{P}}{\partial t} + \mu_0 \mathbf{H} \cdot \frac{\partial \mathbf{M}}{\partial t} \right). \tag{4.2}$$

Each term in (4.2) has the unit of power density (i.e., power per unit volume). The term on the left-hand side, $\mathbf{E} \cdot \mathbf{J}$, is the power expended by the field $\mathbf{E}$ to drive the current $\mathbf{J}$. It is similar to the resistive power loss in an electric circuit. This power is lost as heat to the surroundings. The vector quantity

$$\mathbf{S} = \mathbf{E} \times \mathbf{H} \tag{4.3}$$

is called the *Poynting vector* of the optical field. It represents the *instantaneous magnitude and direction of the power flow* of the field. The scalar quantity

$$u_0 = \frac{\epsilon_0}{2}\left|\mathbf{E}\right|^2 + \frac{\mu_0}{2}\left|\mathbf{H}\right|^2 \tag{4.4}$$

has the unit of energy per unit volume and is the *energy density that is stored in the propagating field*. It consists of two components, thus accounting for the energies stored in both the electric field and the magnetic field at any instant of time. The last term in (4.2) also has two components that are associated with the electric field and the magnetic field, respectively. The quantity

$$W_{\mathrm{p}} = \boldsymbol{E} \cdot \frac{\partial \boldsymbol{P}}{\partial t} \tag{4.5}$$

is the *power density that is expended by the electromagnetic field on the polarization.* It is the rate of energy transfer from the optical field to the medium by inducing an electric polarization in the medium. Similarly, the quantity

$$W_{\mathrm{m}} = \mu_0 \boldsymbol{H} \cdot \frac{\partial \boldsymbol{M}}{\partial t} \tag{4.6}$$

is the *power density that is expended by the electromagnetic field on the magnetization.* For an optical wave, $W_{\mathrm{m}} \approx 0$ because $\boldsymbol{M} \approx 0$, as discussed in Subsection 1.1.4. With these physical meanings attached to these terms, (4.2) simply states the law of conservation of energy in the propagation of an electromagnetic wave in a medium.

4.1.1 Light Intensity

The *light intensity*, also known as the *irradiance*, is the power density of an optical field. It can be calculated by time averaging the Poynting vector over one wave cycle of a time period T:

$$\overline{\boldsymbol{S}} = \frac{1}{T} \int_0^T \boldsymbol{S} \mathrm{d}t = \frac{1}{T} \int_0^T \boldsymbol{E} \times \boldsymbol{H} \mathrm{d}t = 2 \operatorname{Re} \left(\mathbf{E} \times \mathbf{H}^* \right), \tag{4.7}$$

where $\operatorname{Re}(\cdot)$ means taking the real part. Note that the Poynting vector $\boldsymbol{S}$ is a real vector that is defined in terms of the real fields $\boldsymbol{E}$ and $\boldsymbol{H}$. We can define a *complex Poynting vector* in terms of the complex fields $\mathbf{E}$ and $\mathbf{H}$ as

$$\mathbf{S} = \mathbf{E} \times \mathbf{H}^*, \tag{4.8}$$

so that

$$\overline{\boldsymbol{S}} = \mathbf{S} + \mathbf{S}^*, \tag{4.9}$$

which has the same form as the relation between the real field and the complex field defined in (1.37), except that the real Poynting vector $\overline{\boldsymbol{S}}$ that appears in this relation is time averaged. *The direction of the Poynting vector is the direction of the power flow.* The *light intensity, I,* that is projected on a surface depends on the angle between the Poynting vector and the normal direction, $\hat{n}$, of the surface:

$$I = \left| \overline{\boldsymbol{S}} \cdot \hat{n} \right| = \left| \left(\mathbf{S} + \mathbf{S}^* \right) \cdot \hat{n} \right|, \tag{4.10}$$

where I has the unit of watt per square meter. The maximum intensity is the intensity projected on a surface that is normal to the Poynting vector such that $\overline{\boldsymbol{S}} \| \hat{n}$. Thus, the maximum intensity is simply the magnitude of the real time-averaged Poynting vector:

$$I_{\max} = \left| \overline{\mathbf{S}} \right| = \left| \mathbf{S} + \mathbf{S}^* \right|. \tag{4.11}$$

4.1.2 Power Exchange

The power density W_p that is expended by an optical field E on a polarization P quantifies the power transfer between an optical field and an optical medium. If $W_p > 0$, the optical field transfers energy to the optical medium to drive the polarization P while the optical field diminishes. If $W_p < 0$, the optical medium transfers energy to the optical field through the polarization P by either amplifying the optical field or radiating the optical field. By using the definition of the complex field for $E(\mathbf{r}, t) = \mathbf{E}(\mathbf{r})e^{-i\omega t} + \mathbf{E}^*(\mathbf{r})e^{i\omega t}$ and $P(\mathbf{r}, t) = \mathbf{P}(\mathbf{r})e^{-i\omega t} + \mathbf{P}^*(\mathbf{r})e^{i\omega t}$, we find from (4.5) the *time-averaged power density that is expended by the optical field on driving the polarization* to be

$$\overline{W}_p(\mathbf{r}) = 2\omega \left| \mathbf{E}^*(\mathbf{r}) \cdot \mathbf{P}(\mathbf{r}) \right| \sin \Delta\varphi(\mathbf{r}), \tag{4.12}$$

where

$$\Delta\varphi(\mathbf{r}) = \varphi_{\mathbf{P}(\mathbf{r})} - \varphi_{\mathbf{E}(\mathbf{r})} \tag{4.13}$$

is the phase of the polarization $\mathbf{P}(\mathbf{r})$ with respect to that of the electric field $\mathbf{E}(\mathbf{r})$ – that is, it is the phase of $\mathbf{E}^*(\mathbf{r}) \cdot \mathbf{P}(\mathbf{r})$. This phase can be a function of space because the polarization and the optical field are not necessarily phase synchronized in space – that is, they are not necessarily *phase matched*, as we shall see in later chapters.

The relation given in (4.12) is not limited to a linear polarization. It can in general be applied to any polarizations, including nonlinear polarizations. For power exchange between an optical field at a frequency of ω_q and an optical medium through a nonlinear polarization $\mathbf{P}_q^{(n)}$, the power density that is expended by the optical field $\mathbf{E}_q$ is

$$\overline{W}_q^{(n)}(\mathbf{r}) = 2\omega_q \left| \mathbf{E}_q^*(\mathbf{r}) \cdot \mathbf{P}_q^{(n)}(\mathbf{r}) \right| \sin \Delta\varphi_q(\mathbf{r}), \tag{4.14}$$

where

$$\Delta\varphi_q(\mathbf{r}) = \varphi_{\mathbf{P}_q^{(n)}(\mathbf{r})} - \varphi_{\mathbf{E}_q(\mathbf{r})} \tag{4.15}$$

is the phase of the polarization $\mathbf{P}_q^{(n)}(\mathbf{r})$ with respect to that of the electric field $\mathbf{E}_q(\mathbf{r})$ – that is, it is the phase of $\mathbf{E}_q^*(\mathbf{r}) \cdot \mathbf{P}_q^{(n)}(\mathbf{r})$. When $\pi > \Delta\varphi_q(\mathbf{r}) > 0$, power is transferred from the optical field to the nonlinear polarization, and the maximum transfer in this direction occurs when $\Delta\varphi_q(\mathbf{r}) = \pi/2$. When $-\pi < \Delta\varphi_q(\mathbf{r}) < 0$, power is transferred from the nonlinear polarization to the optical field, and the maximum transfer in this direction occurs when $\Delta\varphi_q(\mathbf{r}) = -\pi/2$.

4.2 PROPAGATION IN AN ISOTROPIC MEDIUM

The propagation of an optical wave depends on the optical property and physical structure of the medium. It also depends on the makeup of the optical wave, such as its frequency contents and its temporal characteristics. In this and the following sections, we consider the basic characteristics of the propagation of a monochromatic plane optical wave in an infinite isotropic and structurally homogeneous medium. We also consider only the linear optical property of the medium so that ϵ includes only the field-independent $\chi^{(1)}$.

For a monochromatic plane wave, there is only one value of $\mathbf{k}$ and one value of ω. Its complex electric field is that given by (1.45), but with the field amplitude $\mathcal{E}$ being a constant of space and time – that is, it is independent of $\mathbf{r}$ and t:

$$\mathbf{E}(\mathbf{r}, t) = \mathcal{E} \exp(i\mathbf{k} \cdot \mathbf{r} - i\omega t) = \hat{e}\mathcal{E} \exp(i\mathbf{k} \cdot \mathbf{r} - i\omega t). \tag{4.16}$$

Therefore, we have the relations:

$$\mathbf{P}^{(1)}(\mathbf{r}, t) = \epsilon_0 \boldsymbol{\chi}^{(1)}(\mathbf{k}, \omega) \cdot \mathbf{E}(\mathbf{r}, t) \tag{4.17}$$

and

$$\mathbf{D}(\mathbf{r}, t) = \boldsymbol{\epsilon}(\mathbf{k}, \omega) \cdot \mathbf{E}(\mathbf{r}, t). \tag{4.18}$$

We also assume no spatial nonlocality in the medium, thus neglecting the $\mathbf{k}$ dependence of $\boldsymbol{\chi}^{(1)}$ and $\boldsymbol{\epsilon}$. Then,

$$\mathbf{P}^{(1)}(\mathbf{r}, t) = \epsilon_0 \boldsymbol{\chi}^{(1)}(\omega) \cdot \mathbf{E}(\mathbf{r}, t) \tag{4.19}$$

and

$$\mathbf{D}(\mathbf{r}, t) = \boldsymbol{\epsilon}(\omega) \cdot \mathbf{E}(\mathbf{r}, t). \tag{4.20}$$

In this section, we consider the propagation of a monochromatic optical wave in an isotropic and structurally homogeneous medium. For a monochromatic wave of a frequency ω, the wave equation is simply

$$\boldsymbol{\nabla} \times \boldsymbol{\nabla} \times \mathbf{E} + \mu_0 \boldsymbol{\epsilon}(\omega) \cdot \frac{\partial^2 \mathbf{E}}{\partial t^2} = 0. \tag{4.21}$$

In an isotropic and structurally homogeneous medium, the permittivity tensor $\boldsymbol{\epsilon}(\omega)$ reduces to a scalar permittivity $\epsilon(\omega)$, and

$$\boldsymbol{\nabla} \cdot \mathbf{E} = \frac{1}{\epsilon(\omega)} \boldsymbol{\nabla} \cdot \mathbf{D}(\mathbf{r}, t) = 0 \tag{4.22}$$

from (1.40). Then, by using the vector identity $\boldsymbol{\nabla} \times \boldsymbol{\nabla} \times = \boldsymbol{\nabla}\boldsymbol{\nabla} \cdot - \nabla^2$, the wave equation in (4.21) reduces to the form:

$$\nabla^2 \mathbf{E} - \mu_0 \epsilon(\omega) \cdot \frac{\partial^2 \mathbf{E}}{\partial t^2} = 0. \tag{4.23}$$

For a medium that is anisotropic, such as a birefringent crystal, or structurally inhomogeneous, such as a waveguide, (4.23) is generally not valid because (4.22) does not hold in such a medium.

With the field amplitude $\mathcal{E}$ of the complex field $\mathbf{E}(\mathbf{r}, t)$ being a constant independent of $\mathbf{r}$ and t for a monochromatic plane wave, as expressed in (4.16) for $\mathbf{E}(\mathbf{r}, t)$ and similarly for the magnetic field $\mathbf{H}(\mathbf{r}, t)$, we can make the replacement for the operators when operating on $\mathbf{E}$ of the form in (4.16) or on $\mathbf{H}$ of a similar form:

$$\mathbf{\nabla} \rightarrow i\mathbf{k}, \quad \frac{\partial}{\partial t} \rightarrow -i\omega. \tag{4.24}$$

4.2.1 Free Space

In free space, $\mathbf{P} = 0$ and $\epsilon(\omega)$ reduces to ϵ_0. Substitution of (4.24) in (4.23) then yields

$$k^2 = \omega^2 \mu_0 \epsilon_0. \tag{4.25}$$

The propagation constant in free space is

$$k = \omega\sqrt{\mu_0\epsilon_0} = \frac{\omega}{c} = \frac{2\pi v}{c} = \frac{2\pi}{\lambda}, \tag{4.26}$$

where $c = 1/\sqrt{\mu_0\epsilon_0} \approx 3 \times 10^8 \text{ m s}^{-1}$ is the speed of light in free space, v is the frequency of the optical wave, and λ is its wavelength. Because k is proportional to $1/\lambda$, it is also known as the *wavenumber*.

By using (4.24) and noting that $\mathbf{D} = \epsilon_0 \mathbf{E}$ and $\mathbf{B} = \mu_0 \mathbf{H}$ in free space, the Maxwell's equations of the forms given in (1.38)–(1.41) reduce to the forms:

$$\mathbf{k} \times \mathbf{E} = \omega\mu_0\mathbf{H}, \tag{4.27}$$

$$\mathbf{k} \times \mathbf{H} = -\omega\epsilon_0\mathbf{E}, \tag{4.28}$$

$$\mathbf{k} \cdot \mathbf{E} = 0, \tag{4.29}$$

$$\mathbf{k} \cdot \mathbf{H} = 0. \tag{4.30}$$

From (4.27) and (4.28), we find that

$$\mathbf{E} \cdot \mathbf{H} = 0. \tag{4.31}$$

The relationships in (4.27)–(4.30) indicate that the three vectors $\mathbf{E}$, $\mathbf{H}$, and $\mathbf{k}$ are mutually orthogonal. These relationships also imply that

$$\mathbf{S} \parallel \mathbf{k}. \tag{4.32}$$

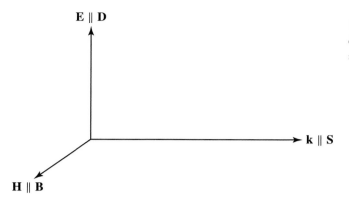

E ∥ D

k ∥ S

H ∥ B

Figure 4.1 Relationships among the directions of **E**, **D**, **B**, **H**, **k**, and **S** in free space or in an isotropic medium.

In free space, the light intensity that is projected on a surface normal to the propagation direction can be expressed as

$$I = \overline{\mathbf{S}} \cdot \hat{k} = \frac{2k}{\omega\mu_0}|\mathbf{E}|^2 = 2c\epsilon_0|\mathbf{E}|^2 = \frac{2k}{\omega\epsilon_0}|\mathbf{H}|^2 = 2c\mu_0|\mathbf{H}|^2. \tag{4.33}$$

The relationships among the directions of these vectors are shown in Fig. 4.1.

4.2.2 Medium without Optical Loss or Gain

For an isotropic and structurally homogeneous medium that does not exhibit an optical loss or gain, the permittivity is a positive real scalar $\epsilon(\omega)$ that is different from ϵ_0. All of the results obtained for free space remain valid, except that ϵ_0 is replaced by $\epsilon(\omega)$. This change of the electric permittivity from that of a vacuum to that of a material is measured by the relative electric permittivity, ϵ/ϵ_0, which is a dimensionless quantity also known as the *dielectric constant* of the material. Therefore, the propagation constant in the medium is

$$k = \omega\sqrt{\mu_0\epsilon} = \frac{n\omega}{c} = \frac{2\pi n v}{c} = \frac{2\pi n}{\lambda}, \tag{4.34}$$

where

$$n = \sqrt{\frac{\epsilon}{\epsilon_0}} = (\text{dielectric constant})^{1/2} \tag{4.35}$$

is the *index of refraction*, or the *refractive index*, of the medium. In a medium that has a refractive index of n, the optical frequency is still v, but the optical wavelength is λ/n and the speed of light is $v = c/n$. Because $n(\omega)$ in a medium is generally frequency dependent, the speed of light in a medium is also frequency dependent. This frequency dependence results in various dispersive phenomena, as discussed in Section 4.4. The light intensity is then

$$I = \overline{\mathbf{S}} \cdot \hat{k} = \frac{2k}{\omega\mu_0}\left|\mathbf{E}\right|^2 = 2c\epsilon_0 n\left|\mathbf{E}\right|^2 = \frac{2k}{\omega\epsilon}\left|\mathbf{H}\right|^2 = \frac{2c\mu_0}{n}\left|\mathbf{H}\right|^2. \tag{4.36}$$

The three Maxwell's equations that are given in (4.27), (4.29), and (4.30) remain the same, but the one given in (4.28) has to be modified by replacing ϵ_0 with ϵ. The relationships among **E**, **D**, **B**, **H**, **k**, and **S** as shown in Fig. 4.1 are still valid.

4.2.3 Medium with Optical Loss or Gain

As mentioned in Section 3.1, $\chi^{(1)}(\omega)$ is complex for a medium that has an optical loss, when $\chi^{(1)''} > 0$, or an optical gain, when $\chi^{(1)''} < 0$. This fact can be seen by examining the direction of power transfer between the optical field and the medium. By using the relations $E(\mathbf{r}, t) = \mathbf{E}(\mathbf{r})e^{-i\omega t} + \mathbf{E}^*(\mathbf{r})e^{i\omega t}$ and $P^{(1)}(\mathbf{r}, t) = \mathbf{P}^{(1)}(\mathbf{r})e^{-i\omega t} + \mathbf{P}^{(1)*}(\mathbf{r})e^{i\omega t}$, with $\mathbf{P}^{(1)}(\mathbf{r}) = \epsilon_0 \chi^{(1)}(\omega)\mathbf{E}(\mathbf{r})$, we find from (4.5), or equivalently from (4.14) for $n = 1$, that the power that is expended by the optical field and absorbed by the medium is

$$\overline{W}_{\mathrm{p}}^{(1)}(\mathbf{r}) = 2\omega\epsilon_0\chi^{(1)''}\left|\mathbf{E}(\mathbf{r})\right|^2. \tag{4.37}$$

Therefore, when $\chi^{(1)''} > 0$, power is transferred from the optical field to the medium, resulting in an optical loss. When $\chi^{(1)''} < 0$, power is transferred from the medium to the optical field, resulting in an optical gain.

Because $\epsilon(\omega) = \epsilon_0\left[1 + \chi^{(1)}(\omega)\right]$ for the linear optical permittivity, $\epsilon(\omega)$ is also complex when $\chi^{(1)}(\omega)$ is complex, with $\epsilon'' > 0$ for an optical loss and $\epsilon'' < 0$ for an optical gain. Therefore,

$$k^2 = \omega^2\mu_0\epsilon = \omega^2\mu_0(\epsilon' + i\epsilon''), \tag{4.38}$$

and the propagation constant k becomes complex:

$$k = k' + ik'' = \beta + i\frac{\alpha}{2} = \beta - i\frac{g}{2}, \tag{4.39}$$

where α is the *absorption coefficient*, also known as the *attenuation coefficient*, and g is the *gain coefficient*, also known as the *amplification coefficient*. The index of refraction also becomes complex:

$$n = \sqrt{\frac{\epsilon' + i\epsilon''}{\epsilon_0}} = n' + in''. \tag{4.40}$$

The relation between k and n as expressed in (4.34) is still valid. However, the light intensity I is not simply given by (4.36) but is given by the real part of it:

$$I = \overline{\mathbf{S}} \cdot \hat{k} = \frac{2k'}{\omega\mu_0}\left|\mathbf{E}\right|^2 = 2c\epsilon_0 n'\left|\mathbf{E}\right|^2 \approx \frac{2k'}{\omega\epsilon'}\left|\mathbf{H}\right|^2 \approx \frac{2c\mu_0}{n'}\left|\mathbf{H}\right|^2. \tag{4.41}$$

The three Maxwell's equations that are given in (4.27), (4.29), and (4.30) remain the same, but the one given in (4.28) has to be modified by replacing ϵ_0 with ϵ. The relations among **E**, **D**, **B**, **H**, **k**, and **S** as shown in Fig. 4.1 are still valid. However, with k and ϵ being complex, it can be seen that **E** and **H** are no longer in phase.

It can be shown that if we choose β to be positive, as is generally the case, the sign of α is the same as that of ϵ'', which has the same sign as that of $\chi^{(1)''}$. In this case, k' and n' are also positive, and k'' and n'' also have the same sign as ϵ'' and $\chi^{(1)''}$. If we consider as an example an optical wave that propagates in the z direction, then $\hat{k} = \hat{z}$ and, from (4.16) and (4.39), the complex electric field is

$$\mathbf{E}(\mathbf{r}, t) = \boldsymbol{\mathcal{E}} e^{-\alpha z/2} \exp(i\beta z - i\omega t) = \hat{e}\mathcal{E} e^{-\alpha z/2} \exp(i\beta z - i\omega t). \tag{4.42}$$

It can be seen that the wave has a phase that varies sinusoidally with a period of $1/\beta$ in the z direction; however, its amplitude is not constant but varies exponentially with z. Thus, the light intensity also varies exponentially with z:

$$I = I_0 e^{-\alpha z} \quad \text{or} \quad I = I_0 e^{gz}. \tag{4.43}$$

Clearly, β is the wavenumber in this case, and the sign of α determines the attenuation or amplification of the optical wave. If $\chi^{(1)''} > 0$, then $\epsilon'' > 0$; the medium absorbs light with $\alpha > 0$ and $g < 0$. If $\chi^{(1)''} < 0$, then $\epsilon'' < 0$; the medium amplifies light with $\alpha < 0$ and $g > 0$.

EXAMPLE 4.1

At the optical wavelength of $\lambda = 532$ nm, the refractive index of silicon has a real part of $n' = 4.142$ and an imaginary part of $n'' = 0.0324$. (a) Find the linear susceptibility $\chi^{(1)}$ and the permittivity ϵ of silicon at this wavelength. (b) Find the propagation constant β and the absorption coefficient α for an optical wave at this wavelength in silicon. (c) What is the distance that an optical wave at this wavelength can propagate in silicon before 99% of its energy is absorbed?

Solution

(a) By using the relation that $n = (\epsilon/\epsilon_0)^{1/2} = \left(1 + \chi^{(1)}\right)^{1/2}$, we find that

$$\chi^{(1)} = n^2 - 1 = (4.142 + i0.0324)^2 - 1 = 16.155 + i0.268$$

and

$$\epsilon = n^2\epsilon_0 = (4.142 + i0.0324)^2\epsilon_0 = (17.155 + i0.268)\epsilon_0.$$

(b) By using the relation that

$$k = \frac{n\omega}{c} = \frac{2\pi}{\lambda}(n' + in'') = \beta + i\frac{\alpha}{2},$$

we find that

$$\beta = \frac{2\pi n'}{\lambda} = \frac{2\pi \times 4.142}{532 \times 10^{-9}} \text{ m}^{-1} = 4.892 \times 10^7 \text{ m}^{-1}$$

and

$$\alpha = \frac{4\pi n''}{\lambda} = \frac{4\pi \times 0.0324}{532 \times 10^{-9}} \; m^{-1} = 7.653 \times 10^5 \; m^{-1}.$$

(c) The distance l for 99% of the optical energy to be absorbed is found by solving the relation that $1 - e^{-\alpha l} = 0.99$. Therefore,

$$l = -\frac{\ln(1 - 0.99)}{\alpha} = -\frac{\ln 0.01}{7.653 \times 10^5} \; m = 6.0 \; \mu m.$$

4.3 PROPAGATION IN A BIREFRINGENT CRYSTAL

As shown in Table 3.1 and discussed in Subsection 3.5.1, the linear optical property of a crystal is only determined by its crystal system. Among the seven crystal systems, only the cubic system has an isotropic linear optical property with $n_x = n_y = n_z$. Note that though $n_x = n_y = n_z$ for a cubic crystal, $\hat{x}$, $\hat{y}$, and $\hat{z}$ are not arbitrary axes but are the principal axes that are determined by the crystal structure. Three crystal systems, hexagonal, trigonal, and tetragonal, are uniaxial with $n_x = n_y \neq n_z$, which can be either $n_z > n_y = n_x$ for a positive uniaxial crystal or $n_z < n_y = n_x$ for a negative uniaxial crystal. The other three crystal systems, orthorhombic, monoclinic, and triclinic, are biaxial with $n_x \neq n_y \neq n_z$, which are designated as $n_z > n_y > n_x$ by convention.

The principal axes can be found by diagonalizing the tensor $\chi^{(1)}$, as illustrated in Section 3.5.1, or equivalently by diagonalizing the tensor $\boldsymbol{\epsilon}$:

$$\boldsymbol{\epsilon} = \begin{bmatrix} \epsilon_{11} & \epsilon_{12} & \epsilon_{13} \\ \epsilon_{21} & \epsilon_{22} & \epsilon_{23} \\ \epsilon_{31} & \epsilon_{32} & \epsilon_{33} \end{bmatrix} \xrightarrow{\text{diagonalize}} \boldsymbol{\epsilon} = \begin{bmatrix} \epsilon_x & 0 & 0 \\ 0 & \epsilon_y & 0 \\ 0 & 0 & \epsilon_z \end{bmatrix}. \tag{4.44}$$

The components of **D** and **E** along the principal axes have the simple relations:

$$D_x = \epsilon_x E_x, \quad D_y = \epsilon_y E_y, \quad D_z = \epsilon_z E_z. \tag{4.45}$$

In a birefringent crystal, either $\epsilon_x = \epsilon_y \neq \epsilon_z$, as in the case of a uniaxial crystal, or $\epsilon_x \neq \epsilon_y \neq \epsilon_z$, as in the case of a biaxial crystal. Therefore, the electric displacement and the electric field are not generally parallel to each other in a birefringent crystal because $\boldsymbol{\nabla} \cdot \mathbf{D} = 0$ but $\boldsymbol{\nabla} \cdot \mathbf{E} \neq 0$.

EXAMPLE 4.2
At the optical wavelength of 1.064 μm, the permittivity tensor of the LiNbO$_3$ crystal represented in a rectilinear coordinate system defined by $\hat{x}_1$, $\hat{x}_2$, and $\hat{x}_3$ is found to be

$$\boldsymbol{\epsilon} = \epsilon_0 \begin{bmatrix} 4.9909 & 0 & 0 \\ 0 & 4.9046 & 0.1495 \\ 0 & 0.1495 & 4.7320 \end{bmatrix}.$$

Find the principal axes and the corresponding principal indices for this crystal. What is the birefringence of the crystal?

Solution

The permittivity tensor $\boldsymbol{\epsilon}$ is represented by a symmetric matrix because LiNbO$_3$ is a dielectric crystal. Diagonalization of this matrix yields the eigenvalues and the corresponding eigenvectors:

$$\epsilon_x = 4.9909\epsilon_0, \ \hat{x} = \hat{x}_1;$$

$$\epsilon_y = 4.9909\epsilon_0, \ \hat{y} = \frac{\sqrt{3}}{2}\hat{x}_2 + \frac{1}{2}\hat{x}_3;$$

$$\epsilon_z = 4.6457\epsilon_0, \ \hat{z} = -\frac{1}{2}\hat{x}_2 + \frac{\sqrt{3}}{2}\hat{x}_3.$$

Therefore, the principal axes of the crystal are $\hat{x}$, $\hat{y}$, and $\hat{z}$, given above, and the principal indices of refraction are

$$n_x = \sqrt{4.9907} = 2.2340, \quad n_y = \sqrt{4.9907} = 2.2340, \quad n_z = \sqrt{4.6457} = 2.1554.$$

Because $n_x = n_y > n_z$, the LiNbO$_3$ crystal is negative uniaxial with $n_o = 2.2340 > n_e = 2.1554$. The birefringence is $B = n_e - n_o = -0.0786$.

The inverse of the dielectric constant tensor is the *relative impermeability tensor*:

$$\boldsymbol{\eta} = [\eta_{ij}] = \left(\frac{\boldsymbol{\epsilon}}{\epsilon_0}\right)^{-1}, \tag{4.46}$$

where i and j are spatial coordinate indices. In a general rectilinear coordinate system (x_1, x_2, x_3), the ellipsoid that is defined by the equation

$$\sum_{i,j} x_i \eta_{ij} x_j = 1 \tag{4.47}$$

is called the *index ellipsoid* or the *optical indicatrix*. For a nonmagnetic dielectric material, $\boldsymbol{\eta}$ is a symmetric tensor – that is, $\eta_{ij} = \eta_{ji}$ – because $\boldsymbol{\epsilon}$ is symmetric. Therefore, (4.47) can be expressed as

$$\eta_{11}x_1^2 + \eta_{22}x_2^2 + \eta_{33}x_3^2 + 2\eta_{23}x_2x_3 + 2\eta_{31}x_3x_1 + 2\eta_{12}x_1x_2 = 1. \tag{4.48}$$

This equation is usually written as

$$\eta_1 x_1^2 + \eta_2 x_2^2 + \eta_3 x_3^2 + 2\eta_4 x_2 x_3 + 2\eta_5 x_3 x_1 + 2\eta_6 x_1 x_2 = 1 \tag{4.49}$$

to reduce the double index ij of η_{ij} to the single index α of η_α by using the index contraction rule defined in (3.32).

The index ellipsoid equation is invariant with respect to coordinate rotation. When a coordinate system with its axes aligned with the principal dielectric axes of the crystal is chosen, $\boldsymbol{\epsilon}$ is diagonalized. Thus, the tensor $\boldsymbol{\eta}$ is also diagonalized with the eigenvalues:

$$\eta_x = \frac{\epsilon_0}{\epsilon_x} = \frac{1}{n_x^2}, \quad \eta_y = \frac{\epsilon_0}{\epsilon_y} = \frac{1}{n_y^2}, \quad \eta_z = \frac{\epsilon_0}{\epsilon_z} = \frac{1}{n_z^2}. \tag{4.50}$$

In this coordinate system, the index ellipsoid takes the simple form:

$$\frac{x^2}{n_x^2} + \frac{y^2}{n_y^2} + \frac{z^2}{n_z^2} = 1. \tag{4.51}$$

By comparing (4.51) with (4.49), we find that the terms that contain cross products of different coordinates are eliminated when the coordinate system of the principal dielectric axes is used. In this coordinate system, the principal axes of the index ellipsoid coincide with the principal dielectric axes of the crystal, and the principal indices of refraction of the crystal are given by the semiaxes of the index ellipsoid, as illustrated in Fig. 4.2. Therefore, a coordinate transformation by rotation to eliminate cross-product terms in the index ellipsoid equation is equivalent to diagonalization of the $\boldsymbol{\epsilon}$ tensor. The principal dielectric axes and their corresponding principal indices of refraction can be found through either approach. Between the two approaches, however, diagonalization of the $\boldsymbol{\epsilon}$ tensor is better because it is more systematic and is easier to carry out.

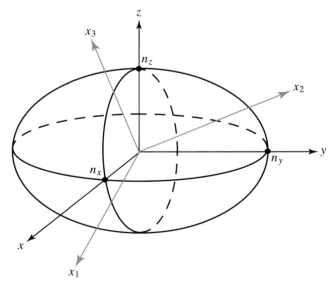

Figure 4.2 Index ellipsoid and its relationship with the coordinate system. Here (x, y, z) is the coordinate system aligned with the principal axes of the crystal, whereas (x_1, x_2, x_3) is an arbitrary coordinate system.

EXAMPLE 4.3

By using the index ellipsoid instead of diagonalizing the ϵ tensor as done in Example 4.2, find the principal axes and their corresponding principal indices for the LiNbO$_3$ crystal given in Example 4.2 at the optical wavelength of 1.064 µm. Compare the two approaches.

Solution

By inverting the ϵ tensor given in Example 4.2, the relative impermeability tensor in the (x_1, x_2, x_3) coordinate system can be found:

$$\boldsymbol{\eta} = \left(\frac{\epsilon}{\epsilon_0}\right)^{-1} = \begin{bmatrix} 4.9909 & 0 & 0 \\ 0 & 4.9046 & 0.1495 \\ 0 & 0.1495 & 4.7320 \end{bmatrix}^{-1} \approx \begin{bmatrix} \dfrac{1}{4.9909} & 0 & 0 \\ 0 & \dfrac{1}{4.8999} & -0.00645 \\ 0 & -0.00645 & \dfrac{1}{4.7274} \end{bmatrix}.$$

In the (x_1, x_2, x_3) coordinate system, the index ellipsoid is thus described by the equation:

$$\frac{x_1^2}{4.9909} + \frac{x_2^2}{4.8999} + \frac{x_3^2}{4.7274} - 0.0129 x_2 x_3 = 1.$$

To find the principal axes and their principal indices of refraction, the cross-product term has to be eliminated by rotating the coordinates. From Example 4.2, we know that this can be done by taking the coordinate transformation:

$$x_1 = x, \quad x_2 = \frac{\sqrt{3}}{2}y - \frac{1}{2}z, \quad x_3 = \frac{1}{2}y + \frac{\sqrt{3}}{2}z.$$

Substitution of these relations into the above index ellipsoid equation transforms it into the equation for the index ellipsoid in the (x, y, z) coordinate system:

$$\frac{x^2}{4.9909} + \frac{y^2}{4.9909} + \frac{z^2}{4.6457} = 1.$$

Thus, the principal indices are $n_x = \sqrt{4.9907} = 2.2340$, $n_y = \sqrt{4.9907} = 2.2340$, and $n_z = \sqrt{4.6457} = 2.1554$.

By comparing the two approaches illustrated in this example and in Example 4.2, it is clear that they are equivalent to each other. It is also clear that the method of diagonalizing ϵ described in Example 4.2 is more systematic and straightforward than that of eliminating the cross-product terms in the equation for the index ellipsoid, particularly when there is more than one cross-product term.

4.3.1 Polarization Normal Modes

An optical wave that propagates in a homogeneous medium with a wavevector **k** has two *polarization normal modes* that are orthogonally polarized, with their polarizations characterized by two unit eigenvectors $\hat{e}_1$ and $\hat{e}_2$ of the orthonormal relations:

$$\hat{e}_1 \cdot \hat{e}_1^* = \hat{e}_2 \cdot \hat{e}_2^* = 1 \quad \text{and} \quad \hat{e}_1 \cdot \hat{e}_2^* = \hat{e}_2 \cdot \hat{e}_1^* = 0. \tag{4.52}$$

The polarization eigenvectors $\hat{e}_1$ and $\hat{e}_2$ can be both real, representing two linearly polarized normal modes, or both complex, representing two circularly polarized normal modes or two elliptically polarized normal modes. For the propagation of an optical wave in a dielectric medium that is not optically active, it is generally convenient to choose linearly polarized normal modes because the principal axes of a dielectric medium are linear axes. However, for an arbitrary propagation direction $\hat{k}$, the two polarization directions $\hat{e}_1$ and $\hat{e}_2$ do not generally line up with any of the principal axes, but they are always orthogonal with the propagation direction:

$$\hat{e}_1 \cdot \hat{k} = \hat{e}_2 \cdot \hat{k} = 0. \tag{4.53}$$

Note that the polarization eigenvectors $\hat{e}_1$ and $\hat{e}_2$ represent the directions of the electric displacements $\mathbf{D}_1$ and $\mathbf{D}_2$ of the two polarization normal modes but not necessarily those of the electric fields $\mathbf{E}_1$ and $\mathbf{E}_2$. The reason is that the relation $\mathbf{D} \cdot \mathbf{k} = 0$ is always true in any source-free medium but the relation $\mathbf{E} \cdot \mathbf{k} = 0$ is not necessarily true in an anisotropic medium, as we shall see below. Therefore, we can write

$$\mathbf{D}_1 = D_1\hat{e}_1 \quad \text{and} \quad \mathbf{D}_2 = D_2\hat{e}_2, \tag{4.54}$$

but *not necessarily* $\mathbf{E}_1 = E_1\hat{e}_1$ or $\mathbf{E}_2 = E_2\hat{e}_2$.

For an optical wave that propagates in a nonbirefringent medium, including a cubic crystal, which has an isotropic linear optical property, any set of two unit polarization vectors $\hat{e}_1$ and $\hat{e}_2$ that satisfy the conditions given in (4.52) and (4.53) can be chosen as polarization normal modes. Because the relations $\mathbf{D} \cdot \mathbf{k} = 0$ and $\mathbf{E} \cdot \mathbf{k} = 0$ are both valid in an isotropic and structurally homogeneous medium, we also have $\mathbf{E}_1 = E_1\hat{e}_1$ and $\mathbf{E}_2 = E_2\hat{e}_2$ besides $\mathbf{D}_1 = D_1\hat{e}_1$ and $\mathbf{D}_2 = D_2\hat{e}_2$. The polarization of an optical wave remains unchanged as the wave propagates through an isotropic and structurally homogeneous medium no matter what its initial state of polarization is.

By contrast, the state of polarization of an optical wave generally varies along its path of propagation through a birefringent crystal unless it is linearly polarized in the direction of one of the two polarization normal modes. From the above discussions, we also know that the two polarization normal modes cannot be arbitrarily chosen for the propagation in a birefringent crystal because they are determined by both the direction of propagation, $\hat{k}$, and the properties of the birefringent crystal. Furthermore, the eigenvector of a polarization normal mode represents the direction of the electric displacement $\mathbf{D}$, but not necessarily that of the electric field $\mathbf{E}$, of the normal mode.

4.3.2 Propagation along a Principal Axis

We first consider the simple case that an optical wave propagates along one of the principal axes, say $\hat{z}$. Then the field can be decomposed into two normal modes, each of which is polarized along one of the other two principal axes, $\hat{x}$ or $\hat{y}$. We see from (4.45) and (3.26) that each field component along a principal axis has a characteristic index of refraction of $n_i = \sqrt{\epsilon_i/\epsilon_0}$, meaning that it has a characteristic propagation constant of $k^i = n_i\omega/c$, which is determined by the polarization of the field but not by the direction of wave propagation. For a wave that propagates along $\hat{z}$, the electric field can be expressed as

$$\mathbf{E} = \hat{x}\mathcal{E}_x \, e^{ik^x z - i\omega t} + \hat{y}\mathcal{E}_y \, e^{ik^y z - i\omega t} = \left[\hat{x}\mathcal{E}_x + \hat{y}\mathcal{E}_y \, e^{i(k^y - k^x)z}\right] e^{ik^x z - i\omega t}. \tag{4.55}$$

Because the wave propagates in the z direction, the wavevectors are $\mathbf{k}^x = k^x \hat{z}$ for the x-polarized field and $\mathbf{k}^y = k^y \hat{z}$ for the y-polarized field. Note that $k^x = n_x \omega / c$ and $k^y = n_y \omega / c$ are the propagation constants of the x- and y-polarized fields, respectively, not to be confused with the x and y components of a wavevector $\mathbf{k}$, which are normally expressed as k_x and k_y. The field that is expressed in (4.55) has these propagation characteristics:

1. If it is originally linearly polarized along one of the principal axes, it remains linearly polarized in the same direction.
2. If it is originally linearly polarized at an angle of $\theta = \tan^{-1}(\mathcal{E}_y / \mathcal{E}_x)$ with respect to the x axis, its polarization state varies periodically along z with a period of $2\pi / |k^y - k^x|$. In general, its polarization follows a sequence of variations from linear polarization to elliptical polarization to linear polarization in the first half-period and then reverses the sequence back to linear polarization in the second half-period. At the half-period position, it is linearly polarized at an angle of θ on the other side of the x axis, as shown in Fig. 4.3(a). Thus, the polarization is rotated by an angle of 2θ from the original direction. In the special case that $\theta = 45°$, the wave is circularly polarized at the quarter-period point and is linearly polarized at the half-period point with its polarization rotated by $90°$ from the original direction, as shown in Fig. 4.3(b).

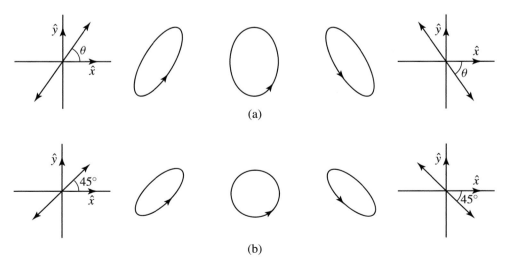

(a)

(b)

Figure 4.3 Evolution of the polarization state of an optical wave propagating along the principal axis $\hat{z}$ of an anisotropic crystal that has $n_x \neq n_y$. Only the evolution over one half-period is shown here. (a) The optical wave is initially linearly polarized at an arbitrary angle of θ with respect to the principal axis $\hat{x}$. (b) The optical wave is initially polarized at $45°$ with respect to $\hat{x}$.

These characteristics have useful applications. A plate of a birefringent crystal that has a quarter-period thickness of

$$l_{\lambda/4} = \frac{1}{4} \cdot \frac{2\pi}{|k^y - k^x|} = \frac{\lambda}{4|n_y - n_x|} \qquad (4.56)$$

is called a *quarter-wave plate*. It can be used to convert a linearly polarized wave to circular or elliptic polarization, and vice versa. A plate of a thickness of any odd integral multiple of $l_{\lambda/4}$, such as $3l_{\lambda/4}$ or $5l_{\lambda/4}$, also has the same function. By contrast, a plate of a half-period thickness of

$$l_{\lambda/2} = \frac{1}{2} \cdot \frac{2\pi}{|k^y - k^x|} = \frac{\lambda}{2|n_y - n_x|} \qquad (4.57)$$

is called a *half-wave plate*. It can be used to rotate the polarization direction of a linearly polarized wave by any angular amount by properly choosing the angle θ of the incident polarization with respect to the principal axis $\hat{x}$, or $\hat{y}$, of the crystal. A plate of a thickness that is any odd integral multiple of $l_{\lambda/2}$ has the same function.

Note that though the output from a quarter-wave or half-wave plate can be linearly polarized, the wave plates are not polarizers. They are based on different principles and have completely different functions. For the quarter-wave and half-wave plates discussed here, $n_x \neq n_y$. Between the two principal axes $\hat{x}$ and $\hat{y}$, the one with the smaller index is called the *fast axis*, whereas the other, with the larger index, is the *slow axis*.

4.3.3 Propagation in a General Direction

When an optical wave propagates along an optical axis of a birefringent crystal, both of its polarization normal modes see the same index of refraction. In this case, the normal modes of polarization $\hat{e}_1$ and $\hat{e}_2$ can be arbitrarily chosen as long as they are orthogonal to each other and both are orthogonal to the propagation direction $\hat{k}$. The propagation characteristics of a wave that propagates along an optical axis is similar to those of a wave that propagates in a nonbirefringent medium, as discussed in the preceding section. When a wave propagates in a general direction that is neither along a principal axis nor along an optical axis, it still has two normal modes of polarization $\hat{e}_1$ and $\hat{e}_2$ that are mutually orthogonal and are both orthogonal to $\hat{k}$. In this situation, the directions of $\hat{e}_1$ and $\hat{e}_2$ are determined by whether the crystal is uniaxial or biaxial and by the birefringence of the crystal. In the following, the simpler case of general propagation in a uniaxial crystal is discussed in detail. General propagation in a biaxial crystal is more complicated and is only briefly discussed.

When an optical wave propagates in a uniaxial crystal in a direction that is not along the optical axis, the index of refraction depends on the direction of its polarization. In this situation, there exist two normal modes of linearly polarized waves, each of which sees a unique index of refraction. One of them is the polarization that is perpendicular to the optical axis. This normal mode is called the *ordinary wave*. We use $\hat{e}_o$ to represent the direction of polarization of the ordinary wave. The other normal mode is clearly one that is perpendicular to $\hat{e}_o$ because the two normal-mode polarizations are orthogonal to each other. This normal mode is called the *extraordinary wave*, and its direction of polarization is represented by $\hat{e}_e$.

Note that $\hat{e}_o$ and $\hat{e}_e$ are the directions of **D** rather than those of **E**, as discussed in the preceding subsection. For the ordinary wave, $\hat{e}_o \| \mathbf{D}_o \| \mathbf{E}_o$. For the extraordinary wave, however, $\hat{e}_e \| \mathbf{D}_e \nparallel \mathbf{E}_e$ except when $\hat{e}_e$ is parallel to a principal axis. Both $\hat{e}_o$ and $\hat{e}_e$, being the unit vectors of $\mathbf{D}_o$ and $\mathbf{D}_e$, are perpendicular to the direction of wave propagation, $\hat{k}$, because **D** is always perpendicular to $\hat{k}$. From this understanding, both $\hat{e}_o$ and $\hat{e}_e$ can be found if both $\hat{k}$ and the optical axis are known.

For a uniaxial crystal with its optical axis designated as $\hat{z}$ by convention, we have

$$\hat{e}_o = \frac{1}{\sin\theta}\hat{k}\times\hat{z}, \qquad \hat{e}_e = \hat{e}_o \times \hat{k}, \tag{4.58}$$

if the vector $\hat{k}$ is in a direction that is at an angle of θ with respect to the optical axis $\hat{z}$ and an angle of ϕ on the xy plane with respect to the principal axis $\hat{x}$. Therefore, we have

$$\hat{k} = \hat{x}\sin\theta\cos\phi + \hat{y}\sin\theta\sin\phi + \hat{z}\cos\theta, \tag{4.59}$$

$$\hat{e}_o = \hat{x}\sin\phi - \hat{y}\cos\phi, \tag{4.60}$$

$$\hat{e}_e = -\hat{x}\cos\theta\cos\phi - \hat{y}\cos\theta\sin\phi + \hat{z}\sin\theta. \tag{4.61}$$

The relationships among these vectors are illustrated in Fig. 4.4.

The indices of refraction of the ordinary and extraordinary waves can be found by using the index ellipsoid given in (4.51), with $n_x = n_y = n_o$ and $n_z = n_e$ for a uniaxial crystal, as is shown in Fig. 4.5. The intersection of the index ellipsoid and the plane that is normal to $\hat{k}$ at the origin of the ellipsoid defines an *index ellipse*. The principal axes of this index ellipse are in the directions of $\hat{e}_o$ and $\hat{e}_e$, and their half-lengths are the corresponding indices of refraction. For

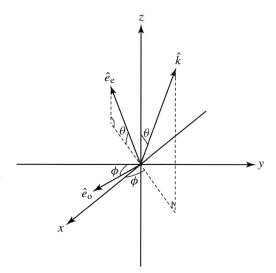

Figure 4.4 Relationships among the direction of wave propagation and the polarization directions of the ordinary and extraordinary waves in a uniaxial crystal, for which $\hat{x}, \hat{y}, \hat{z}$ are the principal axes, and $\hat{z}$ is the optical axis.

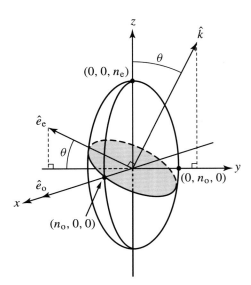

Figure 4.5 Determination of the refractive indices of the ordinary and extraordinary waves in a positive uniaxial crystal using index ellipsoid, for which $n_e > n_o$. If the crystal is negative uniaxial, for which $n_e < n_o$, then the ellipsoid has a shorter semiaxis on the z axis than those on the x and y axes.

a uniaxial crystal, the index of refraction of the ordinary wave is simply n_o. The index of refraction of the extraordinary wave depends on the angle θ, but not on the angle ϕ; it is given by the relation:

$$\frac{1}{n_e^2(\theta)} = \frac{\cos^2\theta}{n_o^2} + \frac{\sin^2\theta}{n_e^2},\qquad(4.62)$$

which can be seen from Fig. 4.5.

The refractive indices of the two normal modes as functions of the propagation direction $\hat{k}$ describe a two-sheet *refractive index surface*. This refractive index surface is also known as the *normal surface* or the *k-vector surface*. The refractive index surface of a birefringent crystal has two sheets for the two normal modes. The indices of the two normal modes for a given $\hat{k}$ vector are found at the two intersections of the $\hat{k}$ vector with the two-sheet refractive index surface.

For a uniaxial crystal, one sheet is a sphere that has a radius of n_o described by the relation $n(\theta, \phi) = n_o$ for the ordinary wave. The other sheet is an ellipsoid that has two semiaxes of n_o and one semiaxis of n_e defined by the relation $n(\theta, \phi) = n_e(\theta)$, which is given in (4.62), for the extraordinary wave. These two sheets intersect on the z axis, where the optical axis is found. Figure 4.6 shows the intersections of this two-sheet refractive index surface with the three principal planes for a positive uniaxial crystal. Though the sheet of refractive index surface for the extraordinary wave of a uniaxial crystal is also an ellipsoid, it is different from the index ellipsoid. The *ellipsoid sheet of the refractive index surface* seen in Fig. 4.6 is described by (4.62), whereas *the index ellipsoid* plotted in Fig. 4.5 is defined by (4.51). For example, the intersection of the index ellipsoid with the z axis is $(0, 0, n_e)$ (i.e., it has a value of n_e on the z axis), as seen in Fig. 4.5, but the intersection of the refractive index surface that is described by (4.62) has a value of $n_e(0°) = n_o$ on the z axis, as seen in Fig. 4.6.

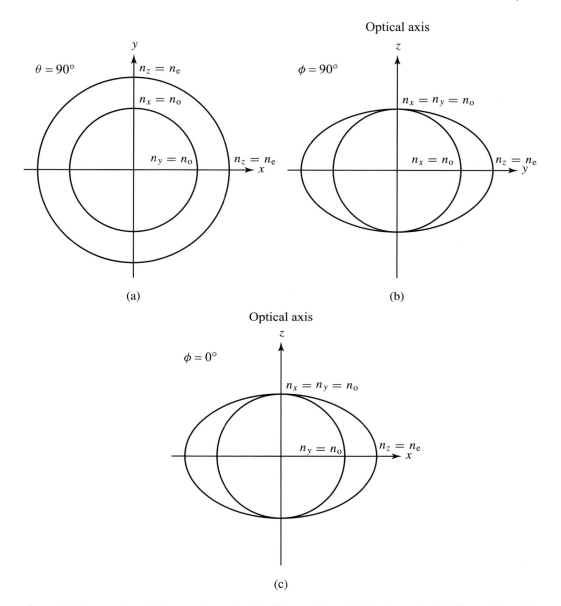

Figure 4.6 Intersections of the two-sheet refractive index surface with the three principal planes of a positive uniaxial crystal, for which $n_e > n_o$. The optical axis is found on the z axis, where the two sheets of the refractive index intersect. If the crystal is negative uniaxial, for which $n_e < n_o$, then in (a) the n_e circle is inside the n_o circle, and in (b) and (c) the ellipse is encircled by the n_o circle.

Because **D** is orthogonal to $\hat{k}$ and can be decomposed into $\mathbf{D}_o$ and $\mathbf{D}_e$ components, we have

$$\mathbf{D} = \hat{e}_o \mathcal{D}_o e^{ik_o\hat{k}\cdot\mathbf{r}-i\omega t} + \hat{e}_e \mathcal{D}_e e^{ik_e\hat{k}\cdot\mathbf{r}-i\omega t}, \tag{4.63}$$

where $k_o = n_o\omega/c$ and $k_e = n_e\omega/c$. In general, **E** cannot be written in the form of (4.63) because its longitudinal component along the wave propagation direction $\hat{k}$ does not vanish except when

$\theta = 0°$ or $90°$. We see that $n_e(0°) = n_o$ and $n_e(90°) = n_e$. The special case that the wave propagates along one of the principal axes discussed earlier belongs to one of these situations.

EXAMPLE 4.4

From the preceding two examples, we find that LiNbO$_3$ is a uniaxial crystal with $\hat{z}$ being its optical axis because $n_x = n_y \neq n_z$. (a) Is it positive or negative uniaxial? (b) For an optical wave at the wavelength of 1.064 μm, which propagates in LiNbO$_3$ along a direction $\hat{k}$ that makes an angle θ with respect to the optical axis $\hat{z}$, find the values of the refractive index of the extraordinary wave for $\theta = 0°$, $30°$, $45°$, $60°$, and $90°$. What are the values of the refractive index of the ordinary wave for these angles?

Solution

(a) From Examples 4.2 and 4.3, we find that $n_x = n_y = 2.2340$ and $n_z = 2.1554$ for LiNbO$_3$ at the wavelength of 1.064 μm. Therefore, LiNbO$_3$ is a negative uniaxial crystal because $n_e = 2.1554 < n_o = 2.2340$.

(b) The refractive index of the extraordinary wave varies with the angle θ according to (4.62) such that $n_o = 2.2340 > n_e(\theta) > n_e = 2.1554$ for $0° < \theta < 90°$. Therefore,

$$n_e(0°) = \left(\frac{\cos^2 0°}{n_o^2} + \frac{\sin^2 0°}{n_e^2} \right)^{-1/2} = n_o = 2.2340,$$

$$n_e(30°) = \left(\frac{\cos^2 30°}{n_o^2} + \frac{\sin^2 30°}{n_e^2} \right)^{-1/2} = 2.2136,$$

$$n_e(45°) = \left(\frac{\cos^2 45°}{n_o^2} + \frac{\sin^2 45°}{n_e^2} \right)^{-1/2} = 2.1936,$$

$$n_e(60°) = \left(\frac{\cos^2 60°}{n_o^2} + \frac{\sin^2 60°}{n_e^2} \right)^{-1/2} = 2.1743,$$

$$n_e(90°) = \left(\frac{\cos^2 90°}{n_o^2} + \frac{\sin^2 90°}{n_e^2} \right)^{-1/2} = n_e = 2.1554.$$

The refractive index of the ordinary wave is independent of the angle θ such that $n_o = 2.2340$ for any value of θ.

When an optical wave propagates in a biaxial crystal in a direction that is not along one of the two optic axes, it has two normal modes of polarization, a *fast wave* $\hat{e}_f$ of an index of refraction n_f and a *slow wave* $\hat{e}_s$ of an index of refraction n_s, which have the relation that $n_f < n_s$. The directions of the two polarizations $\hat{e}_f$ and $\hat{e}_s$ are determined by the propagation direction $\hat{k}$ with respect to the two optical axes through the index ellipsoid of the biaxial crystal by following a procedure similar to that for a uniaxial crystal shown in Fig. 4.5. The index ellipsoid defined by

(4.51) for a biaxial crystal has three different semiaxes because $n_x < n_y < n_z$. The intersection of the index ellipsoid and the plane normal to $\hat{k}$ at the origin of the ellipsoid still defines an index ellipse. The principal axes of this index ellipse give the directions of the two normal modes of polarization, $\hat{e}_f$ and $\hat{e}_s$, and their half-lengths are the corresponding indices of refraction, with $n_f < n_s$. Experimentally, the polarization directions of these two normal modes can be found for a biaxial crystal by using the *Biot–Fresnel law* [1, 2].

The indices of refraction n_f and n_s of the two modes are functions of the angles θ and ϕ for the general propagation direction $\hat{k} = \hat{x} \sin\theta \cos\phi + \hat{y} \sin\theta \sin\phi + \hat{z} \cos\theta$, as given in (4.59) and shown in Fig. 4.4, where θ is the angle between $\hat{k}$ and $\hat{z}$, and ϕ is the angle between the projection of $\hat{k}$ on the xy plane and $\hat{x}$. They are the two positive solutions of the equation known as *Fresnel's equation* [1, 2]:

$$\frac{k_x^2}{n^{-2} - n_x^{-2}} + \frac{k_y^2}{n^{-2} - n_y^{-2}} + \frac{k_z^2}{n^{-2} - n_z^{-2}} = 0, \tag{4.64}$$

where $k_x = \sin\theta \cos\phi$, $k_y = \sin\theta \sin\phi$, and $k_z = \cos\theta$ are the x, y, and z components of the unit vector $\hat{k}$. In terms of the angles θ and ϕ, Fresnel's equation can be expressed as

$$\frac{\sin^2\theta \cos^2\phi}{n^{-2} - n_x^{-2}} + \frac{\sin^2\theta \sin^2\phi}{n^{-2} - n_y^{-2}} + \frac{\cos^2\theta}{n^{-2} - n_z^{-2}} = 0. \tag{4.65}$$

Of the two solutions of Fresnel's equation for n, the smaller one is the fast index n_f, and the larger one is the slow index n_s.

It can be easily verified that (4.65) reduces to (4.62) when $n_x = n_y = n_o$ and $n_z = n_e$ for a uniaxial crystal. For a biaxial crystal with $n_z > n_y > n_x$ by convention, however, (4.65) does not describe an ellipsoid for one sheet of the refractive index surface as (4.62) does for a uniaxial crystal. Instead, it describes a two-sheet refractive index surface that has intersections with the three principal planes as shown in Fig. 4.7. As stated above, this refractive index surface is not the same as the index ellipsoid. This difference is clear for a biaxial crystal because the refractive index surface of a biaxial crystal is not an ellipsoid, whereas the relation given in (4.51) still defines an index ellipsoid for the biaxial crystal. As also stated above, the two intersections of a given $\hat{k}$ vector with the two-sheet refractive index surface give the indices of the two normal modes for the $\hat{k}$ vector. The two optical axes of a biaxial crystal are found in the two directions where the two sheets intersect, as shown in Fig. 4.7(c). The two sheets of the refractive index surface do not intersect on the z axis, nor on the x or y axis, because none of the principal axes is an optical axis of a biaxial crystal. It can be shown by using (4.65) that the two optical axes lie on the zx principal plane, which are at an angle of θ_{OA}, given by (3.27), with respect to the z axis on either side of the z axis:

$$\theta_{OA} = \sin^{-1} \left(\frac{n_x^{-2} - n_y^{-2}}{n_x^{-2} - n_z^{-2}} \right)^{1/2}. \tag{4.66}$$

Thus, the two optical axes are in these directions:

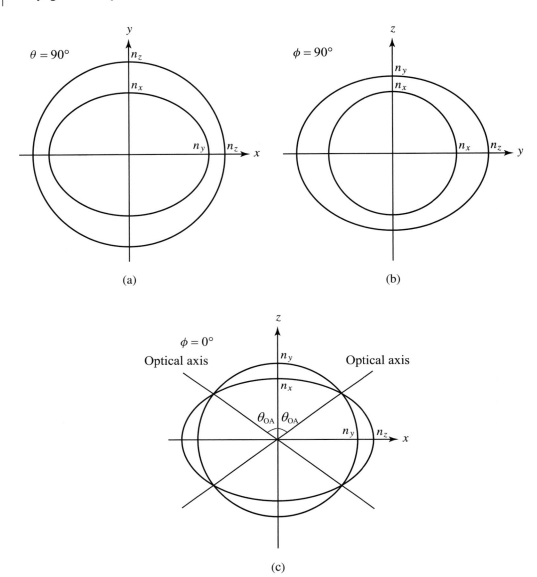

Figure 4.7 Intersections of the two-sheet refractive index surface with the three principal planes of a biaxial crystal, for which $n_z > n_y > n_x$ by convention. The two sheets of the refractive index surface intersect at four points on the zx principal plane, through which the two optical axes are found. Each of the two optical axes is at an angle of θ_{OA} with respect to the z axis on either side of the z axis.

$$\hat{k}_{OA1} = \hat{x}\sin\theta_{OA} + \hat{z}\cos\theta_{OA} = \left(\frac{n_x^{-2} - n_y^{-2}}{n_x^{-2} - n_z^{-2}}\right)^{1/2}\hat{x} + \left(\frac{n_y^{-2} - n_z^{-2}}{n_x^{-2} - n_z^{-2}}\right)^{1/2}\hat{z} \qquad (4.67)$$

and

$$\hat{k}_{OA2} = -\hat{x}\sin\theta_{OA} + \hat{z}\cos\theta_{OA} = -\left(\frac{n_x^{-2} - n_y^{-2}}{n_x^{-2} - n_z^{-2}}\right)^{1/2}\hat{x} + \left(\frac{n_y^{-2} - n_z^{-2}}{n_x^{-2} - n_z^{-2}}\right)^{1/2}\hat{z}. \qquad (4.68)$$

EXAMPLE 4.5

KTP is a positive biaxial crystal. Its principal indices of refraction at the 1.064 μm wavelength are $n_x = 1.7399$, $n_y = 1.7475$, and $n_z = 1.8296$. Find the two optical axes of KTP at 1.064 μm.

Solution

With the given principal indices of refraction, we find by using (4.66) that

$$\theta_{OA} = \sin^{-1}\left(\frac{n_x^{-2} - n_y^{-2}}{n_x^{-2} - n_z^{-2}}\right)^{1/2} = \sin^{-1}\left(\frac{1.7399^{-2} - 1.7475^{-2}}{1.7399^{-2} - 1.8296^{-2}}\right)^{1/2} = 17.5°.$$

Thus, the two optical axes are

$$\hat{k}_{OA1} = \hat{x}\sin\theta_{OA} + \hat{z}\cos\theta_{OA} = 0.301\hat{x} + 0.954\hat{z}$$

and

$$\hat{k}_{OA2} = -\hat{x}\sin\theta_{OA} + \hat{z}\cos\theta_{OA} = -0.301\hat{x} + 0.954\hat{z}.$$

4.3.4 Spatial Beam Walk-Off

Each of the normal modes of polarization has a well-defined propagation constant. For an optical wave that propagates in a birefringent crystal, the Poynting vectors of the two normal modes might point in different directions even though the two modes have the same propagation direction that is defined by the $\hat{k}$ vector. Thus, the power flows of the two modes might be separated though their wavefronts travel in the same $\hat{k}$ direction. This results in *spatial walk-off* of the two modes. In the following, the simpler case of spatial beam walk-off in a uniaxial crystal is discussed in detail. Spatial beam walk-off in a biaxial crystal is similar but more complicated, and is only briefly discussed at the end of this subsection.

The fields of monochromatic ordinary and extraordinary waves in a uniaxial crystal can be separately written in the form of (4.16), with $\mathbf{k} = \mathbf{k}_o$ for the ordinary way and $\mathbf{k} = \mathbf{k}_e$ for the extraordinary way. By using (4.24), we reduce Maxwell's equations for a normal mode, either ordinary or extraordinary, to the forms:

$$\mathbf{k} \times \mathbf{E} = \omega\mu_0\mathbf{H}, \tag{4.69}$$

$$\mathbf{k} \times \mathbf{H} = -\omega\mathbf{D}, \tag{4.70}$$

$$\mathbf{k} \cdot \mathbf{D} = 0, \tag{4.71}$$

$$\mathbf{k} \cdot \mathbf{H} = 0. \tag{4.72}$$

Note that because $n_o \neq n_e$, these relations apply to the ordinary and the extraordinary normal modes separately with different values of $\mathbf{k}$, but they do not apply to a wave that mixes the two modes. At optical frequencies, $\mathbf{B} = \mu_0\mathbf{H}$ is true in a birefringent crystal. Therefore, (4.69) and (4.72) have the same forms as (4.27) and (4.30), respectively. However, (4.29) for a wave in an isotropic medium is now replaced by (4.71) for a normal mode in a birefringent crystal. Therefore, we have $\mathbf{D}\perp\mathbf{k}$ for both ordinary and extraordinary waves. For an ordinary wave, $\mathbf{E}_o\perp\mathbf{k}_o$ because $\mathbf{D}_o\|\mathbf{E}_o$. The relationships shown in Fig. 4.8(a) among the field vectors for an ordinary wave in birefringent crystal are the same as those shown in Fig. 4.1 for a wave in an isotropic medium. However, in general $\mathbf{E}_e\not\perp\mathbf{k}_e$ for an extraordinary wave, and the Poynting vector $\mathbf{S}_e$, which is perpendicular to $\mathbf{E}_e$, is not necessarily parallel to $\mathbf{k}_e$ because $\mathbf{D}_e\not\|\mathbf{E}_e$. The only exception is when $\hat{e}_e$ is parallel to a principal axis. As a result, the direction of power flow, which is that of $\mathbf{S}_e$, is not the same as the direction of wave propagation, which is normal to the planes of constant phase and is that of $\mathbf{k}_e$. This is shown in Fig. 4.8(b) together with the relationships among the directions of the field vectors. Note that $\mathbf{E}_e$, $\mathbf{D}_e$, $\mathbf{k}_e$, and $\mathbf{S}_e$ all lie in the plane that is normal to $\mathbf{H}_e$ because $\mathbf{B}_e\|\mathbf{H}_e$.

If the electric field of an extraordinary wave is not parallel to a principal axis, its Poynting vector is not parallel to its propagation direction because $\mathbf{E}_e$ is not parallel to $\mathbf{D}_e$ in this situation. As a result, its energy flows away from the direction of its wavefront propagation. This phenomenon is known as *spatial beam walk-off*. When this characteristic appears in one of the two normal modes of an optical wave that propagates in a birefringent crystal, the optical wave splits into two beams of parallel wavevectors but separate, nonparallel traces of energy flow.

For simplicity, let us consider the propagation of an optical wave in a uniaxial crystal with $\hat{k}$, for both ordinary and extraordinary waves, at an angle of θ with respect to the optical axis $\hat{z}$. Clearly, there is no walk-off for the ordinary wave because $\mathbf{E}_o\|\mathbf{D}_o$ and therefore $\mathbf{S}_o\|\hat{k}$. For the extraordinary wave, $\mathbf{S}_e$ is not parallel to $\hat{k}$ but points in a direction at an angle of ψ_e with respect to the optical axis. Figure 4.9(a) shows the relationships among these vectors. The angle α between $\mathbf{S}_e$ and $\hat{k}$, which is defined as $\alpha = \psi_e - \theta$, is called the *walk-off angle* of the extraordinary wave. Note that α is also the angle between $\mathbf{E}_e$ and $\mathbf{D}_e$, as is shown in Fig. 4.9(a). Because neither $\mathbf{E}_e$ nor $\mathbf{D}_e$ is parallel to any principal axis, their relationship is found through their projections on the principal axes: $D_z^e = n_e^2\epsilon_0 E_z^e$ and $D_{xy}^e = n_o^2\epsilon_0 E_{xy}^e$. By using these two relations

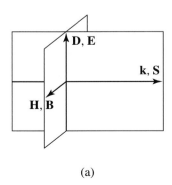

(a)

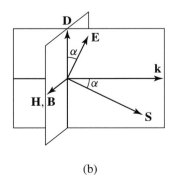

(b)

Figure 4.8 Relationships among the directions of E, D, B, H, k, and S in a birefringent crystal for (a) an ordinary wave and (b) an extraordinary wave. In both cases, the vectors E, D, k, and S all lie in the plane that is normal to B and H.

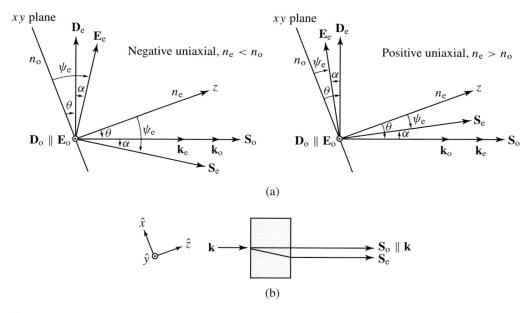

Figure 4.9 (a) Wave propagation and walk-off in a uniaxial crystal. (b) Birefringent plate acting as a polarizing beam splitter for a normally incident wave. The $\hat{x}$, $\hat{y}$, and $\hat{z}$ unit vectors indicate the principal axes of the birefringent plate.

and the definition of α in Figs. 4.8(b) and 4.9(a), it can be shown that the walk-off angle is given by the relation:

$$\alpha = \psi_e - \theta = \tan^{-1}\left(\frac{n_o^2}{n_e^2}\tan\theta\right) - \theta. \tag{4.73}$$

If the crystal is positive uniaxial, α as defined in Fig. 4.9(a) is negative. This means that $\mathbf{S}_e$ is between $\hat{k}$ and $\hat{z}$ for a positive uniaxial crystal. If the crystal is negative uniaxial, α is positive, and $\hat{k}$ is between $\mathbf{S}_e$ and $\hat{z}$. *No walk-off appears if an optical wave propagates along any of the principal axes of a crystal.*

A birefringent crystal can be used to construct a simple *polarizing beam splitter* by taking advantage of the walk-off phenomenon. For such a purpose, a uniaxial crystal can be cut into a plate whose surfaces are at an oblique angle with respect to the optical axis, as is shown in Fig. 4.9(b). When an optical wave is normally incident upon the plate, it splits into ordinary and extraordinary waves in the crystal if its original polarization contains components of both polarization modes. The extraordinary wave is then separated from the ordinary wave because of spatial walk-off, creating two orthogonally polarized beams. However, because of normal incidence, both $\mathbf{k}_e$ and $\mathbf{k}_o$ are parallel to the direction of $\hat{k}$ although they have different magnitudes. When both beams reach the other side of the plate, they are separated by a distance of $d = l \tan\alpha$, where l is the thickness of the plate. After leaving the plate, the two spatially separated beams propagate parallel to each other along the same direction $\hat{k}$ because the directions of their wavevectors have not changed, as is also shown in Fig. 4.9(b).

EXAMPLE 4.6

Find the spatial walk-off angle at the 1.064 μm wavelength for a few representative propagation directions in LiNbO$_3$. Design a polarizing beam splitter at this wavelength by using a LiNbO$_3$ crystal.

Solution

LiNbO$_3$ is a negative uniaxial crystal that has $n_o = 2.2340$ and $n_e = 2.1554$ at the wavelength of 1.064 μm, which are found in Examples 4.2 and 4.3. The spatial walk-off angle α of an extraordinary wave is a function of the angle θ between the wave propagation direction $\hat{k}$ and the optical axis $\hat{z}$ of the uniaxial crystal. Because LiNbO$_3$ is negative uniaxial, the walk-off angle α has a positive value according to (4.73) and Fig. 4.9(a). For example,

$$\alpha = \tan^{-1}\left(\frac{2.2340^2}{2.1554^2}\tan 0°\right) - 0° = 0°, \quad \text{for} \quad \theta = 0°;$$

$$\alpha = \tan^{-1}\left(\frac{2.2340^2}{2.1554^2}\tan 30°\right) - 30° = 1.81°, \quad \text{for} \quad \theta = 30°;$$

$$\alpha = \tan^{-1}\left(\frac{2.2340^2}{2.1554^2}\tan 45°\right) - 45° = 2.05°, \quad \text{for} \quad \theta = 45°;$$

$$\alpha = \tan^{-1}\left(\frac{2.2340^2}{2.1554^2}\tan 60°\right) - 60° = 1.74°, \quad \text{for} \quad \theta = 60°;$$

$$\alpha = \tan^{-1}\left(\frac{2.2340^2}{2.1554^2}\tan 90°\right) - 90° = 0°, \quad \text{for} \quad \theta = 90°.$$

From these numerical examples, we find that the walk-off angle does not vary monotonically with θ. There is no beam walk-off at either $\theta = 0°$ or $\theta = 90°$, and the largest walk-off angle is found near $\theta = 45°$ at $\theta = 43.97°$.

A polarizing beam splitter can be made by cutting a LiNbO$_3$ crystal at an angle, such as 45°, with respect to its optical axis for a parallel plate that has a thickness of l. A beam at the 1.064 μm wavelength that consists of a mixture of extraordinary and ordinary polarizations is normally incident on the plate for $\theta = 45°$ and $\alpha = 2.05°$. Because the ordinary wave does not have walk-off, the Poynting vectors of the extraordinary and ordinary components of the beam separate at an angle of $\alpha = 2.05°$. If a minimum spatial separation of $d = 100$ μm between the extraordinary and ordinary components is desired on the exit surface of the LiNbO$_3$ plate, the minimum thickness of the plate has to be $l > d/\tan \alpha = 2.8$ mm.

Spatial beam walk-off in a biaxial crystal is similar to that in a uniaxial crystal, except that the Poynting vectors $\mathbf{S}_f$ and $\mathbf{S}_s$ for both fast and slow waves are not parallel to the propagation direction $\hat{k}$

when the wave propagates in a general direction that is neither along an optical axis nor along a principal axis. Furthermore, $\mathbf{S}_f$ and $\mathbf{S}_s$ are not parallel to each other because $n_f \neq n_s$. Consequently, all of the three vectors $\mathbf{S}_f$, $\mathbf{S}_s$, and $\hat{k}$ point in different directions, creating a complicated walk-off situation that depends on the birefringence of the crystal and the specific direction of propagation.

4.4 DISPERSION

For a monochromatic plane optical wave that propagates in the z direction, $\mathbf{k} = k\hat{z}$ so that the complex electric field given in (4.16) can be written as

$$\mathbf{E}(z, t) = \boldsymbol{\mathcal{E}} \exp(ikz - i\omega t), \tag{4.74}$$

where $\boldsymbol{\mathcal{E}}$ is a constant vector that is independent of space and time. This represents a sinusoidal wave whose phase varies with z and t as

$$\varphi = kz - \omega t. \tag{4.75}$$

For a point of constant phase on the spatially and temporally varying field, $\varphi = $ constant and thus $k\,dz - \omega\,dt = 0$. If we track this point of constant phase, we find that it travels with a velocity of

$$v_p = \frac{dz}{dt} = \frac{\omega}{k}. \tag{4.76}$$

This is called the *phase velocity* of the wave.

Note that the phase velocity is a function of the optical frequency because the refractive index of a medium is a function of frequency. There is *phase-velocity dispersion* due to the fact that $dn/d\omega \neq 0$. In the case of *normal dispersion*,

$$\frac{dn}{d\omega} > 0 \quad \text{and} \quad \frac{dn}{d\lambda} < 0. \tag{4.77}$$

In the case of *anomalous dispersion*,

$$\frac{dn}{d\omega} < 0 \quad \text{and} \quad \frac{dn}{d\lambda} > 0. \tag{4.78}$$

In real circumstances, a propagating optical wave rarely contains only one frequency. It usually consists of many frequency components that are grouped around some center frequency of ω_0 with a corresponding center propagation constant of k_0. For the simplicity of illustration, we consider a wave packet traveling in the z direction that is composed of two plane waves of equal real amplitude $\mathcal{E}$. The frequencies and propagation constants of the two component plane waves are

$$\begin{aligned} \omega_1 = \omega_0 + d\omega, \quad k_1 = k_0 + dk; \\ \omega_2 = \omega_0 - d\omega, \quad k_2 = k_0 - dk. \end{aligned} \tag{4.79}$$

The spatially and temporally varying total real field of the wave packet is then given by the expression:

$$E(z,t) = \mathcal{E}\exp\left(ik_1 z - i\omega_1 t\right) + \text{c.c.} + \mathcal{E}\exp\left(ik_2 z - i\omega_2 t\right) + \text{c.c.}$$
$$= 4\mathcal{E}\cos\left(\mathrm{d}kz - \mathrm{d}\omega t\right)\cos\left(k_0 z - \omega_0 t\right). \tag{4.80}$$

We find that the resultant wave packet has a *carrier*, which has a frequency of ω_0 and a propagation constant of k_0, and an *envelope*, which varies with space and time as $\cos\left(\mathrm{d}kz - \mathrm{d}\omega t\right)$. This is illustrated in Fig. 4.10. Therefore, a fixed point on the envelope is defined by the condition that $\mathrm{d}kz - \mathrm{d}\omega t = \text{constant}$, and it travels with a velocity of

$$v_g = \frac{\mathrm{d}z}{\mathrm{d}t} = \frac{\mathrm{d}\omega}{\mathrm{d}k}. \tag{4.81}$$

This is called the *group velocity* because it is the velocity of the wave packet. Because the energy of a harmonic wave is proportional to the square of its field amplitude, the energy carried by a wave packet that is composed of many frequency components is concentrated in regions where the amplitude of the envelope is large. Therefore, the energy in a wave packet is transported at the group velocity v_g. *The constant-phase wavefront propagates at the phase velocity, but the group velocity is the velocity at which energy and information propagate.*

In reality, the group velocity is usually a function of optical frequency. Then,

$$\frac{\mathrm{d}^2 k}{\mathrm{d}\omega^2} = \frac{\mathrm{d}}{\mathrm{d}\omega}v_g^{-1} \neq 0. \tag{4.82}$$

Therefore, $\mathrm{d}^2 k/\mathrm{d}\omega^2$ represents the *group-velocity dispersion*. A *dimensionless* coefficient for the group-velocity dispersion can be defined as

$$D = c\omega\frac{\mathrm{d}^2 k}{\mathrm{d}\omega^2} = \frac{2\pi c^2}{\lambda}\frac{\mathrm{d}^2 k}{\mathrm{d}\omega^2}. \tag{4.83}$$

When measuring the transmission delay or the broadening of optical pulses due to group-velocity dispersion in an optical fiber, another group-velocity dispersion coefficient defined as

ω_1 component

ω_2 component

Total

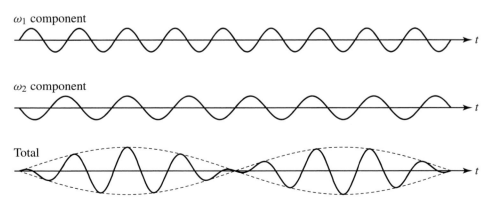

Figure 4.10 Wave packet composed of two frequency components showing the carrier and the envelope. The carrier travels at the phase velocity, whereas the envelope travels at the group velocity.

$$D_\lambda = \frac{2\pi c}{\lambda^2} \frac{\mathrm{d}^2 k}{\mathrm{d}\omega^2} = -\frac{D}{c\lambda} \tag{4.84}$$

is usually used. This coefficient is generally expressed as a function of wavelength in the unit of *picosecond per kilometer per nanometer*. It is a direct measure of the chromatic pulse transmission delay over a unit transmission length.

Group-velocity dispersion is an important consideration in the propagation of optical pulses. It can cause the broadening of an individual pulse, as well as changes in the time delay between pulses of different frequencies. The sign of the group-velocity dispersion can be either positive or negative. A long-wavelength, or low-frequency, pulse travels faster than a short-wavelength, or high-frequency, pulse in the case of *positive group-velocity dispersion*:

$$\frac{\mathrm{d}^2 k}{\mathrm{d}\omega^2} > 0, \quad D > 0, \quad \text{and} \quad D_\lambda < 0. \tag{4.85}$$

By contrast, a short-wavelength pulse travels faster than a long-wavelength pulse in the case of *negative group-velocity dispersion*:

$$\frac{\mathrm{d}^2 k}{\mathrm{d}\omega^2} < 0, \quad D < 0, \quad \text{and} \quad D_\lambda > 0. \tag{4.86}$$

In a given material, the sign of D generally depends on the spectral region of concern. Group-velocity dispersion and phase-velocity dispersion discussed earlier have different meanings. They should not be confused with each other.

In general, both $\epsilon(\omega)$ and $n(\omega)$ of an optical medium are frequency dependent, and the propagation constant is

$$k = \frac{\omega}{c} n(\omega). \tag{4.87}$$

Therefore, we have

$$v_\mathrm{p} = \frac{c}{n} \tag{4.88}$$

and

$$v_\mathrm{g} = \frac{c}{N}, \tag{4.89}$$

where

$$N = n + \omega \frac{\mathrm{d}n}{\mathrm{d}\omega} = n - \lambda \frac{\mathrm{d}n}{\mathrm{d}\lambda} \tag{4.90}$$

is called the *group index*. By using (4.83) and (4.84), we also have

$$D(\lambda) = \lambda^2 \frac{\mathrm{d}^2 n}{\mathrm{d}\lambda^2} \quad \text{and} \quad D_\lambda(\lambda) = -\frac{\lambda}{c} \frac{\mathrm{d}^2 n}{\mathrm{d}\lambda^2}. \tag{4.91}$$

4.4.1 Material Dispersion

The root cause of optical dispersion is the fact that the response of a material to an optical excitation does not decay instantaneously. This delayed response in the time domain leads to a frequency dependence of the susceptibility, thus a frequency dependence of the permittivity and that of the index of refraction, in the frequency domain. This effect is particularly significant around a transition resonance frequency, as can be seen in (1.166) and Fig. 1.3.

As discussed in Subsection 4.2.3, $\chi^{(1)''}(\omega) > 0$ for an optical material so that it absorbs light at the frequency ω when the material is in its normal state in thermal equilibrium with its surroundings. In this state, the lower energy level is more populated than the upper energy level. This characteristic can be seen from (1.166) for $N_1^{(0)} > N_2^{(0)}$. In this situation, we find from (1.166) that $d\chi^{(1)'}/d\omega < 0$ so that $dn/d\omega < 0$ for $|\omega - \omega_{21}| < \gamma_{21}$, but $d\chi^{(1)'}/d\omega > 0$ so that $dn/d\omega > 0$ for $|\omega - \omega_{21}| > \gamma_{21}$. This feature can be seen in Fig. 1.3. As is indicated in Fig. 1.3, the full width at half-maximum of the absorption peak that is associated with the resonance frequency ω_{21} is $2\gamma_{21}$. Therefore, *in the normal state in thermal equilibrium, an optical material has anomalous dispersion at frequencies around a resonance frequency within the full width at half-maximum of its absorption peak, but it has normal dispersion at frequencies that are away from any transition resonance frequency.*

Also discussed in Subsection 4.2.3, an optical material can have an optical gain at the frequency ω when it is pumped to reach population inversion so that $\chi^{(1)''}(\omega) < 0$. This aspect can be seen from (1.166) for $N_1^{(0)} < N_2^{(0)}$. Also from (1.166), we see that $\chi^{(1)'}(\omega)$ changes sign together with $\chi^{(1)''}(\omega)$. Therefore, we find that when an optical material has population inversion for the resonant transition at ω_{21} so that $\chi^{(1)''}(\omega_{21}) < 0$, we have $d\chi^{(1)'}/d\omega > 0$ so that $dn/d\omega > 0$ for $|\omega - \omega_{21}| < \gamma_{21}$, but $d\chi^{(1)'}/d\omega < 0$ so that $dn/d\omega < 0$ for $|\omega - \omega_{21}| > \gamma_{21}$. Therefore, *in the state of population inversion for an optical gain at a resonance frequency, an optical material has normal dispersion at frequencies around that resonance frequency within the full width at half-maximum of its gain peak, but it has anomalous dispersion at frequencies that are immediately outside of that full width at half-maximum.*

An optical material generally has many transition resonance frequencies. Each of these resonances contributes an absorption spectrum; the entire absorption spectrum of the material is the superposition of all of these individual spectra, which can either be separated from one another or overlap on one another. Figure 4.11 shows, as an example, such a spectrum that covers three resonance frequencies; for simplicity we take them to be well separated so that their spectra do not overlap. If one of the resonances is pumped to reach population inversion, the spectrum immediately around that resonance frequency is inverted. Note that it is not possible to pump a material for all of its resonances to be population inverted. Therefore, an inverted spectrum is generally localized around one or at most a few resonance frequencies. In most transparent materials, such as water and glass, light is not absorbed in the visible spectral region; thus, there are no transition resonances in this region. For this reason, they have normal dispersion in the visible spectral region, which often extends to the near-infrared and near-ultraviolet regions.

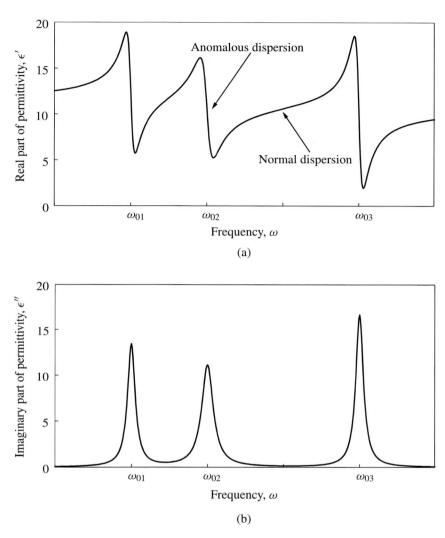

Figure 4.11 Real and imaginary parts of the permittivity as a function of frequency for a material in its normal state over a spectral region covering three separated resonance frequencies.

4.4.2 Modal Dispersion

As we have seen in Section 4.2, the propagation of a monochromatic plane wave in an isotropic and structurally homogeneous medium is characterized by a single propagation constant, $k = n\omega/c$. Except when the optical medium is isotropic and structurally homogeneous, however, an optical wave generally does not propagate as a plane wave with a single propagation constant. Instead, it propagates as a superposition of normal modes that are defined by the properties and structures of the medium. We have already seen in Section 4.3 that the birefringence of a crystal leads to ordinary and extraordinary waves of different propagation constants. This is a clear example of *polarization mode dispersion*, or simply called *polarization dispersion*, that different polarizations of the same frequency and the same propagation direction have different propagation constants.

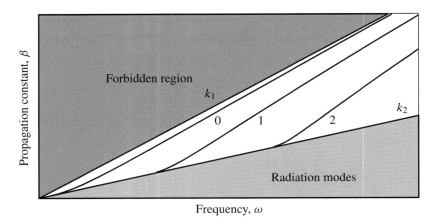

Figure 4.12 Propagation constants as a function of mode number and frequency for a step-index planar dielectric waveguide that has a core of the refractive index n_1, for $k_1 = n_1 \omega / c$, and a substrate of the refractive index n_2, for $k_2 = n_2 \omega / c$.

Besides birefringence, an optical structure, such as a waveguide, also creates normal modes of propagation that have different propagation constants, which have different dispersion properties. Figure 4.12 shows as an example three guided normal modes, with the mode indices of $m = 0$, 1, and 2, of a step-index planar dielectric waveguide. It can be seen that these three modes have different propagation constants, and different frequency dependences for their propagation constants. This phenomenon is known as *modal dispersion*.

For waveguide modes, the fundamental mode, which has an index of $m = 0$ for a planar waveguide, appears first and has the largest propagation constant when other modes also exist. A low-order mode always appears before a high-order mode and has a larger propagation constant than a high-order mode at the same frequency:

$$\beta_m > \beta_n \quad \text{for} \quad m < n. \tag{4.92}$$

These characteristics are clearly seen in Fig. 4.12. Waveguide modes also have different polarizations. The polarization mode dispersion exists between two waveguide modes of different polarizations even if the waveguide material is nonbirefringent and the two modes are of the same order. A planar dielectric waveguide only supports TE and TM modes. Between a TE mode and a TM mode of the same order at the same frequency, the TE mode has a larger propagation constant:

$$\beta_m^{\text{TE}} > \beta_m^{\text{TM}}. \tag{4.93}$$

A nonplanar dielectric waveguide, such as an optical fiber, supports modes of four different polarization characteristics. The polarization mode dispersion among such modes is more complicated.

Problem Set

4.1.1 Show, by using (4.5), that the time-averaged power density that is expended by an optical field on driving a polarization is that given in (4.12) in terms of the complex field, the complex polarization, and the phase of the polarization with respect to the optical field given in (4.13).

4.2.1 (a) Show, by using (4.5), that the time-averaged power density that is expended by an optical field and absorbed by the medium is that given in (4.37).

(b) Show also that the same result is obtained from (4.14).

4.2.2 Show that the wavenumber β and the attenuation coefficient α defined in (4.39) for the propagation of an optical wave in an absorptive medium can be expressed in terms of the real part, χ', and the imaginary part, χ'', of the electric susceptibility χ of the medium as

$$\beta = \frac{\omega}{c} \frac{\left[(1+\chi') + |1+\chi| \right]^{1/2}}{2^{1/2}} \quad \text{and} \quad \alpha = -\frac{\omega}{c} \frac{2^{1/2}\chi''}{\left[(1+\chi') + |1+\chi| \right]^{1/2}}. \quad (4.94)$$

Show that when $|\chi''| \ll \chi'$,

$$\beta \approx \frac{\omega}{c}(1+\chi')^{1/2} \approx \frac{\omega}{c}n' \quad \text{and} \quad \alpha \approx \beta\frac{\chi''}{n'^2} \approx \frac{\omega \chi''}{c\,n'}. \quad (4.95)$$

4.2.3 In an intrinsic silicon crystal, an optical wave at the wavelength of $\lambda = 266$ nm has a propagation constant of $\beta = 4.718 \times 10^7$ m^{-1} and an absorption coefficient of $\alpha = 2.099 \times 10^8$ m^{-1}, and an optical wave at the wavelength of $\lambda = 1.064$ µm has a propagation constant of $\beta = 2.097 \times 10^7$ m^{-1} and an absorption coefficient of $\alpha = 986$ m^{-1}.

(a) Find the refractive index n, the linear susceptibility $\chi^{(1)}$, and the permittivity ϵ of silicon at each wavelength.

(b) What is the distance that an optical wave at each wavelength can propagate in silicon before 99% of its energy is absorbed?

4.3.1 In the (x, y, z) coordinate system, the electric permittivity tensor of a particular crystal has the form:

$$\epsilon = \epsilon_0 \begin{bmatrix} 2.25 & 0 & 0 \\ 0 & 2.13 & 0 \\ 0 & 0 & 2.02 \end{bmatrix}.$$

A linearly polarized optical wave that has a free-space wavelength of $\lambda = 600$ nm is sent through the crystal. Find the propagation constant and the wavelength of the wave in the crystal for each of these arrangements:

(a) The wave is polarized along $\hat{x}$ and propagates along $\hat{z}$.

(b) The wave is polarized along $\hat{y}$ and propagates along $\hat{z}$.

(c) The wave is polarized along $\hat{x}$ and propagates along $\hat{y}$.

(d) The wave is polarized along $\hat{z}$ and propagates along $\hat{y}$.

4.3.2 The electric permittivity of a crystal measured at $\lambda = 1$ μm with respect to an arbitrary rectilinear coordinate system defined by $\hat{x}_1$, $\hat{x}_2$, and $\hat{x}_3$ is found to be given by the tensor:

$$\epsilon = \epsilon_0 \begin{bmatrix} 4.786 & 0 & 0.168 \\ 0 & 5.010 & 0 \\ 0.168 & 0 & 4.884 \end{bmatrix}.$$

(a) Find the principal dielectric axes $\hat{x}$, $\hat{y}$, and $\hat{z}$ of the crystal and their corresponding principal indices of refraction. Identify the crystal from the values of the principal indices of refraction found at $\lambda = 1$ μm.

(b) Write down the equation that describes the index ellipsoid of the crystal in the original coordinate system. What is the equation for the index ellipsoid in the coordinate system defined by the principal axes?

(c) Is the crystal uniaxial or biaxial? Find its optical axis if it is uniaxial or its optical axes if it is biaxial.

(d) Is the crystal positive or negative uniaxial, or positive or negative biaxial? What is the birefringence of the crystal at $\lambda = 1$ μm?

(e) How do you arrange an optical wave to propagate in such a crystal so that the polarization of the wave remains unchanged throughout the entire path if the wave is linearly polarized? How about if the wave is circularly polarized?

(f) Make a quarter-wave plate by using this crystal for the optical wave at $\lambda = 1$ μm. What is the thickness of the plate?

4.3.3 The electric permittivity of a crystal measured at $\lambda = 850$ nm with respect to an arbitrary rectilinear coordinate system defined by $\hat{x}_1$, $\hat{x}_2$, and $\hat{x}_3$ is found to be given by the tensor:

$$\epsilon = \epsilon_0 \begin{bmatrix} 3.0684 & 0.0143 & 0 \\ 0.0143 & 3.0684 & 0 \\ 0 & 0 & 3.3869 \end{bmatrix}.$$

(a) Find the principal dielectric axes $\hat{x}$, $\hat{y}$, and $\hat{z}$ of the crystal and their corresponding principal indices of refraction. Identify the crystal from the values of the principal indices of refraction found at $\lambda = 850$ nm.

(b) Write down the equation that describes the index ellipsoid of the crystal in the original coordinate system. What is the equation for the index ellipsoid in the coordinate system defined by the principal axes?

(c) Is the crystal uniaxial or biaxial? Find its optical axis if it is uniaxial or its optical axes if it is biaxial.

(d) Is the crystal positive or negative uniaxial, or positive or negative biaxial? What is the birefringence of the crystal at $\lambda = 850$ nm?

4.3.4 (a) For a linearly polarized optical wave, under what condition can the polarization remain unchanged as the wave propagates through a birefringent crystal for any distance of propagation?

(b) Answer the question for an optical wave that is not linearly polarized.

4.3.5 A linearly polarized wave at a wavelength of $\lambda = 800$ nm propagates in the z direction through a crystal that has $n_x = 1.65$ and $n_y = 1.63$. How far must it propagate before its polarization is changed into each of the following states? In answering these questions, explain by showing the arrangements with sketches.

(a) It is made circularly polarized.

(b) It remains linearly polarized but with its polarization rotated by 90°.

(c) It remains linearly polarized but with its polarization rotated by 60°.

4.3.6 Rutile (TiO$_2$) is a uniaxial crystal. Its ordinary and extraordinary indices of refraction as functions of wavelength are given by the equations:

$$n_o^2 = 5.913 + \frac{0.2441}{\lambda^2 - 0.083}, \tag{4.96}$$

$$n_e^2 = 7.197 + \frac{0.3322}{\lambda^2 - 0.0843}, \tag{4.97}$$

where λ is in micrometers. A rutile plate of a thickness l is cut in such a way that its surface normal is perpendicular to its optical axis.

(a) If the plate is to be used as a first-order half-wave plate at an optical wavelength of $\lambda = 1$ μm, what should its thickness l be? How do you arrange the plate with respect to the polarization of the incident beam if the polarization of a linearly polarized input beam is to be rotated 60° after passing through the plate?

(b) With the thickness of the plate obtained in (a), find another wavelength at which the plate also functions as a half-wave plate. Find a wavelength at which it functions as a quarter-wave plate.

4.3.7 Consider wave propagation in a uniaxial crystal whose optical axis is $\hat{z}$.

(a) By using the relationships among $\hat{k}$, $\hat{e}_o$, and $\hat{e}_e$ given in (4.58), verify that the unit vectors $\hat{e}_o$ and $\hat{e}_e$ are given by the expressions in (4.60) and (4.61), respectively.

(b) By examining the index ellipsoid, show that $n_e(\theta)$ for the extraordinary wave is given by (4.62).

4.3.8 Show that the walk-off angle as defined in Fig. 4.9(a) is given by (4.73). Given n_e and n_o for a uniaxial crystal, find the angle θ for the propagation direction $\hat{k}$ that results in the largest walk-off for an extraordinary wave.

4.3.9 An extraordinary optical wave propagates in a uniaxial crystal with its wavevector **k** making an angle θ with respect to the optical axis, $\hat{z}$, of the crystal. In the case that $0° < \theta < 90°$ such that $\theta \neq 0°$ and $\theta \neq 90°$, the Poynting vector, **S**, of the wave is not parallel to **k**. The angle α between **S** and **k** is the same as that between **E** and **D**.

(a) Show that the vector **S** lies between **k** and the optical axis if the crystal is positive uniaxial, whereas **k** lies between **S** and $\hat{z}$ if it is negative uniaxial. What is the relationship among **E**, **D**, and $\hat{z}$ in either case?

(b) Show that the walk-off angle given by (4.73) can also be expressed as

$$\alpha = \tan^{-1}\left[\frac{n_e^2(\theta)}{2}\left(\frac{1}{n_e^2} - \frac{1}{n_o^2}\right)\sin 2\theta\right], \tag{4.98}$$

where $n_e(\theta)$ is that given by (4.62).

(c) Show that the maximum walk-off between **S** and **k** occurs at

$$\theta = \tan^{-1}\frac{n_e}{n_o} \tag{4.99}$$

for

$$\alpha = \tan^{-1}\frac{n_o}{n_e} - \tan^{-1}\frac{n_e}{n_o}. \tag{4.100}$$

4.3.10 Rutile (TiO_2) is a uniaxial crystal. Its ordinary and extraordinary indices of refraction as functions of wavelength are given by (4.96) and (4.97), respectively. A rutile plate of a thickness l is cut in such a way that its surface normal is at an angle of $\theta = 30°$ with respect to its optical axis. If this plate is used as a polarizing beam splitter for normally incident light at $\lambda = 800$ nm, what is the separation between the two orthogonally polarized beams leaving the plate? If the spot size of a collimated incident beam is 100 μm in diameter, what is the minimum value of l for the two orthogonally polarized beams at the exit to be completely separated without spatial overlap?

4.3.11 Show, by using the wave equation given in (4.21) for a monochromatic plane wave and Gauss's law $\mathbf{V} \cdot \mathbf{D} = 0$ in the charge-free region, that the refractive index of a crystal is described by Fresnel's equation given in (4.64), or (4.65).

4.3.12 Show by using (4.65) that the two optical axes of a biaxial crystal lie on the zx principal plane, each at an angle θ_{OA}, given by (4.66), with respect to the z axis on either side of the z axis. Then show that these two optical axes are in the directions given in (4.67) and (4.68).

References

[1] M. Born and E. Wolf, *Principles of Optics*, 6th ed. Oxford: Pergamon Press, 1980.

[2] H. Ito, H. Naito, and H. Inaba, "Generalized study on angular dependence of induced second-order nonlinear optical polarizations and phase matching in biaxial crystals," *Journal of Applied Physics*, vol. 46, pp. 3992–3998, 1975.

5 Nonlinear Optical Interactions

5.1 PARAMETRIC AND NONPARAMETRIC PROCESSES

An optical interaction always takes place in a material or through a material because an optical field interacts with electrons, and optical fields interact with one another through their interactions with electrons. Optical interactions can generally be categorized into two types: *parametric processes* and *nonparametric processes*. *A parametric process does not cause any change in the quantum-mechanical state of the material.* Therefore, the process does not cause any net exchange of energy, nor exchange of momentum, between an optical field and the optical medium. Both the total energy of the optical fields and the state of the material in the medium remain unchanged in a purely parametric process. By contrast, *a nonparametric process causes some changes in the quantum-mechanical state of the material*, resulting in an exchange of energy, accompanied by an exchange of momentum, between an optical field and the medium. As a consequence, the total optical energy changes in a nonparametric process. Because a change of the quantum-mechanical state of a material is associated with a transition between two different states, a parametric process does not involve any optical transition in the material, whereas a nonparametric process is connected to one or multiple optical transitions in the material. This feature was already discussed in Subsection 4.2.3 for optical absorption and optical amplification, which clearly are nonparametric processes.

As we have seen in the preceding chapter, linear and nonlinear optical susceptibilities in the frequency domain are generally complex quantities. It is noted that *to interpret the meaning of the real or imaginary part of a susceptibility, it is necessary to first represent the susceptibility in the principal coordinates of the material.* For the linear susceptibility, this means that we consider only the principal susceptibilities: $\chi_i^{(1)} = \chi_i^{(1)\prime} + i\chi_i^{(1)\prime\prime}$, where $i = x$, y, or z, with the real part being $\chi_i^{(1)\prime}$ and the imaginary part being $\chi_i^{(1)\prime\prime}$. To clearly define their real and imaginary parts, complex nonlinear susceptibilities in the frequency domain are expressed as $\chi_{ijk}^{(2)} = \chi_{ijk}^{(2)\prime} + i\chi_{ijk}^{(2)\prime\prime}$ and $\chi_{ijkl}^{(3)} = \chi_{ijkl}^{(3)\prime} + i\chi_{ijkl}^{(3)\prime\prime}$, with i, j, k, and l each representing one of the three principal coordinates, x, y, and z.

By examining $\chi_{ij}^{(1)}(\omega_q)$ given in (3.11), we can see that the linear susceptibility tensor is almost purely real if the optical frequency ω_q is very far away from any transition frequency ω_{ab} such that $|\omega_q - \omega_{ab}| \gg \gamma_{ab}$ for all of the relaxation rates to be negligible in (3.11). By contrast, the linear susceptibility becomes complex with a significant imaginary part near one or multiple resonances such that $|\omega_q - \omega_{ab}| < \gamma_{ab}$ or $|\omega_q - \omega_{ab}| \approx \gamma_{ab}$. Therefore, the imaginary part $\chi_i^{(1)\prime\prime}$ of a principal susceptibility is directly connected with one or multiple resonant transitions in the material. Consequently, it signifies an energy exchange between the optical field and the material, and its sign indicates the direction of energy flow, with $\chi_i^{(1)''} > 0$ for a loss of the optical energy and $\chi_i^{(1)''} < 0$ for an optical gain, as already discussed in Subsection 4.2.3.

Similar features can be seen, and similar conclusions can be drawn, by examining the second-order nonlinear susceptibility $\chi_{ijk}^{(2)}(\omega_q = \omega_m + \omega_n)$ that is given in (3.12) and the third-order nonlinear susceptibility $\chi_{ijkl}^{(3)}(\omega_q = \omega_m + \omega_n + \omega_p)$ that is given in (3.13). The imaginary part of a nonlinear susceptibility is strongly coupled to the transition resonances of a material in a way similar to the dependence of the imaginary part of the linear susceptibility on the material resonances. The sign of the imaginary part of a nonlinear susceptibility also indicates the direction of energy flow in the nonlinear interaction. The real and imaginary parts of a linear or nonlinear susceptibility are not independent of each other, but are fundamentally related to each other through the Kramers–Kronig relation, because physically they together represent one optical process, and mathematically they are the two parts of one complex quantity that represent a physical property of a material.

From the above discussion, it is clear that the real part of a linear or nonlinear susceptibility is not involved in an exchange of energy between an optical field and a material; thus, it represents a parametric process. By contrast, the imaginary part of a linear or nonlinear susceptibility is connected with one or multiple resonant transitions in the material, with energy exchange between an optical field and the material as a consequence; therefore, it represents a nonparametric process. For this reason, an optical interaction that is characterized by a real susceptibility in the frequency domain, or by the real part of a complex susceptibility, is generally classified as parametric, whereas one that is associated with the imaginary part of a complex frequency-dependent susceptibility is nonparametric.

In a *parametric linear process*, the energy of any given optical frequency component is conserved because a linear process does not couple optical fields of different frequencies, as can be seen from the fact that the linear susceptibility $\chi_i^{(1)}(\omega)$ is a function of a single frequency. In a *parametric nonlinear process*, energy exchange among different optical frequency components caused by nonlinear coupling among them can occur, though the sum of the energies from all of the interacting frequency components is conserved. In such a process, the material only facilitates the nonlinear optical interaction without being involved through any optical transition. For example, in a parametric process that is characterized by a real $\chi_{ijk}^{(2)}(\omega_3 = \omega_1 + \omega_2)$, optical energy can be transferred from the frequency components at ω_1 and ω_2 to the component at ω_3, or vice versa, without any energy exchange with the material. Therefore, the energy in each individual frequency component may change, but the total optical energy contained in all three frequency components is conserved because there is no net exchange of energy between the optical fields and the optical medium in a parametric process.

There are two features that are unique to nonlinear optical processes: one is *optical frequency conversion*, and the other is *field-dependent modification of a certain material property*. All nonlinear optical processes exhibit at least one, though not necessarily both, of these two features. All functional nonlinear optical devices take advantage of one or both of these two unique features. Neither feature is uniquely parametric or nonparametric; in other words, either feature can be a consequence of a parametric or nonparametric process. For example, optical frequency conversion through harmonic generation or parametric frequency conversion is a parametric process, but that through stimulated Raman or Brillouin scattering is a nonparametric process. Self-phase modulation and the optical Kerr effect are parametric processes that cause field-induced modification of a material property, whereas absorption saturation and gain saturation are nonparametric processes that lead to field-induced changes of a material property.

5.1.1 Phase-Matching Condition

A very important condition for efficient coupling among optical waves or among optical modes is *phase matching*, regardless of which physical mechanism is responsible for the coupling or whether it is linear or nonlinear. As we shall see in later chapters, phase matching is also most important for efficient nonlinear optical interactions. From the wave-optics point of view, the condition $\omega_q = \omega_1 + \cdots + \omega_n$ for the frequencies of the interacting optical fields is the synchronization of the temporal phases of these fields throughout the interaction process. The accompanying condition is

$$\mathbf{k}_q = \mathbf{k}_1 + \cdots + \mathbf{k}_n \tag{5.1}$$

for the match of the spatial phases of the interacting fields; thus, it is the condition for phase matching. From the photon point of view, the condition $\omega_q = \omega_1 + \cdots + \omega_n$ implies that $\hbar\omega_q = \hbar\omega_1 + \cdots + \hbar\omega_n$, which is conservation of total energy, including that in the photons and that in the material, as required by any process. The phase-matching condition $\mathbf{k}_q = \mathbf{k}_1 + \cdots + \mathbf{k}_n$ then implies the conservation of the total momentum, $\hbar\mathbf{k}_q = \hbar\mathbf{k}_1 + \cdots + \hbar\mathbf{k}_n$, among the interacting photons and the material. Specifically, the phase-matching condition for a second-order nonlinear interaction that is characterized by the relation $\omega_3 = \omega_1 + \omega_2$ among the interacting frequencies is

$$\mathbf{k}_3 = \mathbf{k}_1 + \mathbf{k}_2. \tag{5.2}$$

The phase-matching condition for a third-order interaction that is characterized by the relation $\omega_4 = \omega_1 + \omega_2 + \omega_3$ among the interacting frequencies is

$$\mathbf{k}_4 = \mathbf{k}_1 + \mathbf{k}_2 + \mathbf{k}_3. \tag{5.3}$$

When a particular frequency changes sign, the corresponding wavevector in the phase-matching condition also changes sign.

Note that the phase-matching conditions that are expressed in (5.1)–(5.3) are vector equations with vector sums on the right-hand side. For a second-order process, the condition given in (5.2) can be arranged on a plane, not necessarily in a line. For a third-order or high-order process, the conditions in (5.3) or (5.1) can be arranged in three dimensions, not necessarily on a plane or in a line. When all wavevectors fall in a line, the phase-matching process is *collinear*; otherwise, it is *noncollinear*. When they fall on a plane, the phase-matching process is *coplanar*; otherwise, it is *noncoplanar*.

The phase-matching condition can be automatically satisfied without the need for us to make an effort to satisfy it. One common situation for this automatic phase matching to happen is when the optical process is nonparametric. In a nonparametric process, any mismatched momentum among the interacting photons can be absorbed by the material because there is exergy exchange, and accompanying momentum exchange, between the photons and the material, and because the atoms in the materials are massive compared to the massless photons. Therefore, *phase matching among interacting optical fields is always automatically satisfied in a nonparametric process.* Another situation for automatic phase matching takes place in certain parametric processes. In general, *phase matching among interacting optical fields is not automatically satisfied in a parametric process; the exception is when the phase-matching condition is formally true irrespective of the values and directions of the wavevectors in the condition.* For example, the Pockels effect, which is characterized by $\chi^{(2)\prime}(\omega = \omega + 0)$, and cross-phase modulation, which is characterized by $\chi^{(3)\prime}(\omega = \omega + \omega' - \omega')$, are both parametric processes. Nonetheless, their phase-matching conditions, $\mathbf{k} = \mathbf{k} + \mathbf{0}$ for the Pockels effect and $\mathbf{k} = \mathbf{k} + \mathbf{k}' - \mathbf{k}'$ for cross-phase modulation, are automatically satisfied because both are formally true irrespective of the values and directions of the wavevectors in these relations.

5.2 SECOND-ORDER NONLINEAR OPTICAL PROCESSES

The second-order nonlinear optical processes are listed in Table 5.1. As indicated in this table, the contributing susceptibility for each of these processes is the real part, $\chi^{(2)\prime}$, of a frequency-dependent second-order susceptibility. Therefore, *all of these second-order processes are parametric in nature.* The optical frequencies that are involved in a second-order process of interest are generally far away from any resonance frequency of the nonlinear medium.

As discussed in Section 3.5, $\chi^{(2)}$ from electric dipole interaction vanishes for all centrosymmetric materials. Therefore, none of the processes listed in Table 5.1 are expected to occur in the bulk of a centrosymmetric material. They can only take place in a noncentrosymmetric crystal,

Table 5.1 **Second-order nonlinear optical processes**

Process	Susceptibility	Phase matching
Second-harmonic generation	$\chi^{(2)\prime}(2\omega = \omega + \omega)$	Required
Sum-frequency generation	$\chi^{(2)\prime}(\omega_3 = \omega_1 + \omega_2)$	Required
Difference-frequency generation	$\chi^{(2)\prime}(\omega_2 = \omega_3 - \omega_1)$	Required
Optical parametric generation	$\chi^{(2)\prime}(\omega_3 = \omega_1 + \omega_2)$	Required
Optical parametric amplification	$\chi^{(2)\prime}(\omega_1 = \omega_3 - \omega_2)$	Required
Optical rectification	$\chi^{(2)\prime}(0 = \omega - \omega)$	Automatic
Pockels effect	$\chi^{(2)\prime}(\omega = \omega + 0)$	Automatic

on the surface of a material, or at an interface between two different materials, where centro-symmetry is broken. For device applications that require a high efficiency, noncentrosymmetric crystals of large $\chi^{(2)}$ values are used.

Although each process listed in Table 5.1 represents a unique phenomenon and has its own specific applications, every one of them is basically a *parametric frequency-conversion process* that is generally characterized by a real susceptibility of the form $\chi^{(2)}(\omega_3 = \omega_1 + \omega_2)$. The differences among different processes come either from different experimental conditions or from different subjective purposes of application. The full permutation symmetry expressed in (3.22) applies to the real susceptibilities that characterize parametric second-order processes. Therefore, the processes of sum-frequency generation, difference-frequency generation, and optical parametric generation listed in Table 5.1 have the same nonlinear susceptibility. Second-harmonic generation is the degenerate case of sum-frequency generation with $\omega_1 = \omega_2 = \omega$ and $\omega_3 = 2\omega$. Optical rectification is a special case of difference-frequency generation with $\omega_3 = \omega_1 = \omega$ and $\omega_2 = 0$. The Pockels effect is a special case of sum-frequency generation with $\omega_3 = \omega_1 = \omega$ and $\omega_2 = 0$. In nondegenerate sum-frequency generation and difference-frequency generation, two optical waves at different frequencies have to be supplied at the input. In second-harmonic generation and optical rectification, only one optical wave is needed at the input. The Pockels effect also requires only one input optical wave, but it simultaneously requires a DC field.

5.2.1 Second-Harmonic Generation

Second-harmonic generation is the first experimentally observed nonlinear optical process [1]. Its observation in 1961 kicked off the field of nonlinear optics [2]. Second-harmonic generation performs the function of *frequency doubling* for an input optical wave at a *fundamental frequency* of ω to generate a harmonic wave at the *second-harmonic frequency* of 2ω, as shown in Fig. 5.1. Second-harmonic generation is the degenerate case of sum-frequency generation for the generation of $\omega_3 = \omega_1 + \omega_2$ with $\omega_1 = \omega_2 = \omega$ and $\omega_3 = 2\omega$. In the process, energy conservation requires that two photons at the fundamental frequency ω be simultan-eously annihilated to generate one photon at the second-harmonic frequency 2ω. The intensity at the fundamental frequency ω diminishes while that at the second-harmonic frequency 2ω grows, but the total intensity of the two frequencies remains constant because there is no energy exchange with the optical medium.

Second-harmonic generation is a parametric frequency-conversion process that uses the second-order susceptibility $\chi^{(2)}(2\omega = \omega + \omega)$, which has real elements. With only one

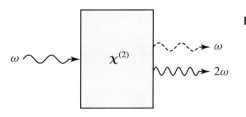

Figure 5.1 Second-harmonic generation.

input *fundamental wave* at the frequency of ω, the nonlinear polarization of the second-harmonic frequency at 2ω is

$$\mathbf{P}^{(2)}(2\omega) = \epsilon_0\chi^{(2)}(2\omega = \omega + \omega) : \mathbf{E}(\omega)\mathbf{E}(\omega). \tag{5.4}$$

Note that there is no permutation terms on the right-hand side of (5.4) because there is only one frequency ω for the input electric field. Meanwhile, the fundamental wave at the fundamental frequency ω also participates in the interaction. Its nonlinear polarization in this process is

$$\mathbf{P}^{(2)}(\omega) = \epsilon_0\left[\chi^{(2)}(\omega = 2\omega - \omega) : \mathbf{E}(2\omega)\mathbf{E}^*(\omega) + \chi^{(2)}(\omega = -\omega + 2\omega) : \mathbf{E}^*(\omega)\mathbf{E}(2\omega)\right]$$
$$= 2\epsilon_0\chi^{(2)}(\omega = 2\omega - \omega) : \mathbf{E}(2\omega)\mathbf{E}^*(\omega).$$
$$\tag{5.5}$$

There are two permutation terms on the right-hand side of (5.5) because there are two frequencies, 2ω and $-\omega$, for the participating electric fields. Note that ω and $-\omega$ are treated as two different frequencies for the permutation of frequencies in the nonlinear susceptibility. Note also that the intrinsic permutation symmetry is used to combine the two terms.

The phase-matching condition is

$$\mathbf{k}_{2\omega} = \mathbf{k}_\omega + \mathbf{k}_\omega = 2\mathbf{k}_\omega, \tag{5.6}$$

which is not automatically satisfied. Therefore, second-harmonic generation is not automatically phase matched but requires an effort on the phase-matching arrangement for efficiency. For device applications, nonlinear crystals that do not have centrosymmetry are commonly used. Second-harmonic generation is also observed on the surface of a material or at the interface between two different materials.

Second-harmonic generation is the most widely used process for the conversion of optical frequencies. The features and practical considerations of second-harmonic generation will be further discussed in Section 10.1.

5.2.2 Sum-Frequency Generation

Sum-frequency generation is a parametric frequency-conversion process that transfers optical energy among different optical frequencies but not between any optical wave and the optical medium. In sum-frequency generation, shown in Fig. 5.2, an optical wave at a high frequency of ω_3 is generated through the nonlinear interaction of two optical waves at the lower frequencies of ω_1 and ω_2 with the nonlinear medium. In sum-frequency generation, energy conservation, which is imposed by the relation $\omega_3 = \omega_1 + \omega_2$, requires that one photon at ω_1 and another at ω_2 be simultaneously annihilated when one photon at ω_3 is generated. The intensities of both input waves at ω_1 and ω_2 decrease as the intensity of the output wave at ω_3 grows. However, the total intensity of the three frequencies remains constant because of the parametric nature of this process.

Sum-frequency generation is a parametric process that uses the second-order susceptibility $\chi^{(2)}(\omega_3 = \omega_1 + \omega_2)$, which has real elements. In the general case that $\omega_1 \neq \omega_2$, the nonlinear polarization of the sum frequency at ω_3 is

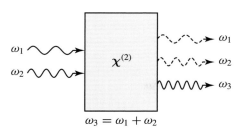

Figure 5.2 Sum-frequency generation.

$$\mathbf{P}^{(2)}(\omega_3) = \epsilon_0 \left[\chi^{(2)}(\omega_3 = \omega_1 + \omega_2) : \mathbf{E}(\omega_1)\mathbf{E}(\omega_2) + \chi^{(2)}(\omega_3 = \omega_2 + \omega_1) : \mathbf{E}(\omega_2)\mathbf{E}(\omega_1) \right]$$
$$= 2\epsilon_0 \chi^{(2)}(\omega_3 = \omega_1 + \omega_2) : \mathbf{E}(\omega_1)\mathbf{E}(\omega_2).$$

(5.7)

Note that there are two terms on the right-hand side for $\omega_1 \neq \omega_2$ because of frequency permutation. These two terms can be combined because of intrinsic permutation symmetry. The phase-matching condition is

$$\mathbf{k}_3 = \mathbf{k}_1 + \mathbf{k}_2,$$

(5.8)

which is not automatically satisfied. Therefore, sum-frequency generation is not automatically phase matched but requires an effort on the phase-matching arrangement for efficiency. For device applications, nonlinear crystals that do not have centrosymmetry are commonly used. Sum-frequency generation is also observed on the surface of a material or at the interface between two different materials.

Because second-harmonic generation is a degenerate case of sum-frequency generation, the two processes share many common features. For example, both are parametric frequency-conversion processes that require efforts on phase matching. Both transfer energy among different optical frequencies but not between any optical field and the optical medium. Nonetheless, there are technical and practical differences between the two processes. The most easily seen is that second-harmonic generation requires only one input wave, whereas sum-frequency generation requires two input waves. This difference is not superficial and has practical implications, which will be discussed in Section 10.2.

5.2.3 Difference-Frequency Generation

Difference-frequency generation is also a parametric frequency-conversion process that transfers optical energy among different optical frequencies but not between any optical wave and the optical medium. In difference-frequency generation, shown in Fig. 5.3, two optical waves at the frequencies of ω_3 and ω_1 interact with the nonlinear optical medium to generate an optical wave at the difference frequency of $\omega_2 = \omega_3 - \omega_1$. In the difference-frequency generation that is characterized by the relation $\omega_2 = \omega_3 - \omega_1$, however, one photon at the input frequency ω_1 is generated simultaneously with the generation of one photon at the difference frequency ω_2 when one photon at ω_3 is annihilated. Therefore, as the intensity of the output wave at ω_2 grows, the intensity of the low-frequency input wave at ω_1 also increases, though that of the high-

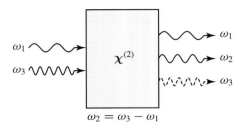

Figure 5.3 Difference-frequency generation and optical parametric amplification. For difference-frequency generation, the purpose is to generate the frequency at ω_2, which is absent at the input. The same arrangement can be used for optical parametric amplification if the purpose is to amplify the intensity of the input wave at ω_1.

frequency input wave at ω_3 diminishes. Nonetheless, because of the parametric nature of this process, the total intensity of the three frequencies remains constant, similar to sum-frequency generation.

Difference-frequency generation is a parametric process that uses the real second-order susceptibility $\chi^{(2)}(\omega_2 = \omega_3 - \omega_1)$. The nonlinear polarization of the difference frequency at ω_2 is

$$
\begin{aligned}
\mathbf{P}^{(2)}(\omega_2) &= \epsilon_0 \left[\chi^{(2)}(\omega_2 = \omega_3 - \omega_1) : \mathbf{E}(\omega_3)\mathbf{E}^*(\omega_1) + \chi^{(2)}(\omega_2 = -\omega_1 + \omega_3) : \mathbf{E}^*(\omega_1)\mathbf{E}(\omega_3) \right] \\
&= 2\epsilon_0 \chi^{(2)}(\omega_2 = \omega_3 - \omega_1) : \mathbf{E}(\omega_3)\mathbf{E}^*(\omega_1).
\end{aligned}
$$

(5.9)

The phase-matching condition is

$$
\mathbf{k}_2 = \mathbf{k}_3 - \mathbf{k}_1,
$$

(5.10)

which is not automatically satisfied. The phase-matching condition of (5.10) for difference-frequency generation is mathematically identical to that of (5.8) for sum-frequency generation. Therefore, difference-frequency generation is not automatically phase matched but requires an effort on the phase-matching arrangement for efficiency. For device applications, nonlinear crystals that do not have centrosymmetry are commonly used. Difference-frequency generation is also observed on the surface of a material or at the interface between two different materials.

Difference-frequency generation shares many common features with sum-frequency generation. Both are parametric frequency-conversion processes that require efforts on phase matching, and both transfer energy among different optical frequencies but not between any optical field and the optical medium. Nonetheless, there are technical and practical differences between the two processes. A clear difference is that in difference-frequency generation one of the two input waves grows in intensity, whereas in sum-frequency generation the intensities of both input waves diminish. Other practical differences will be discussed in Section 10.3.

5.2.4 Optical Parametric Generation

Optical parametric generation is the reverse process of sum-frequency generation. Therefore, just as sum-frequency generation, it is a parametric frequency-conversion process that transfers optical energy among different optical frequencies but not between any optical wave and the optical medium. In optical parametric generation, an optical wave at a high frequency of ω_3 interacts with the nonlinear optical medium to generate two optical waves at the lower frequencies of ω_1 and ω_2, as shown in Fig. 5.4. In this process, one photon at ω_3 is annihilated when two photons, one at ω_1 and another at ω_2, are simultaneously generated. This process can spontaneously occur in the form of *parametric fluorescence* without an input at ω_1 or ω_2; however, spontaneous parametric fluorescence is a linear optical process but not a nonlinear optical process [3]. In practical applications, optical parametric generation is usually carried out with a feedback or with another input wave at either ω_1 or ω_2. When a feedback is provided, usually by placing the nonlinear crystal in a resonant optical cavity, *optical parametric oscillation* is possible with only one input wave at ω_3. Nonetheless, optical parametric generation can occur with a good efficiency without feedback or another input wave at ω_1 or ω_2 when the input intensity at ω_3 is sufficiently high for the parametric fluorescence photons at ω_1 and ω_2 to be coherently amplified through the nonlinear process that is characterized by the real part of $\chi^{(2)}(\omega_3 = \omega_1 + \omega_2)$. Note that though parametric fluorescence is a linear process, coherent parametric amplification is a nonlinear optical process. As expected for any parametric process, when the intensity at ω_3 diminishes to generate and enhance those at ω_1 and ω_2, the total intensity of the three optical frequencies remain constant without energy exchange with the optical medium.

Being the reverse process of sum-frequency generation, optical parametric generation uses the same second-order real susceptibility $\chi^{(2)}(\omega_3 = \omega_1 + \omega_2)$ as that of sum-frequency generation. The nonlinear polarizations of the two parametrically generated frequencies at ω_1 and ω_2 are, respectively,

$$\mathbf{P}^{(2)}(\omega_1) = \epsilon_0 \left[\chi^{(2)}(\omega_1 = \omega_3 - \omega_2) : \mathbf{E}(\omega_3)\mathbf{E}^*(\omega_2) + \chi^{(2)}(\omega_1 = -\omega_2 + \omega_3) : \mathbf{E}^*(\omega_2)\mathbf{E}(\omega_3) \right]$$
$$= 2\epsilon_0 \chi^{(2)}(\omega_1 = \omega_3 - \omega_2) : \mathbf{E}(\omega_3)\mathbf{E}^*(\omega_2),$$

$$(5.11)$$

$$\mathbf{P}^{(2)}(\omega_2) = \epsilon_0 \left[\chi^{(2)}(\omega_2 = \omega_3 - \omega_1) : \mathbf{E}(\omega_3)\mathbf{E}^*(\omega_1) + \chi^{(2)}(\omega_2 = -\omega_1 + \omega_3) : \mathbf{E}^*(\omega_1)\mathbf{E}(\omega_3) \right]$$
$$= 2\epsilon_0 \chi^{(2)}(\omega_2 = \omega_3 - \omega_1) : \mathbf{E}(\omega_3)\mathbf{E}^*(\omega_1).$$

$$(5.12)$$

The phase-matching condition is

$$\mathbf{k}_1 = \mathbf{k}_3 - \mathbf{k}_2 , \text{ i.e., } \mathbf{k}_2 = \mathbf{k}_3 - \mathbf{k}_1 \text{ and } \mathbf{k}_3 = \mathbf{k}_1 + \mathbf{k}_2, \qquad (5.13)$$

which is not automatically satisfied. For device applications, it also uses the same noncentro-symmetric nonlinear crystals that are used for its corresponding reverse process of sum-

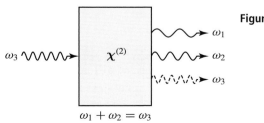

Figure 5.4 Optical parametric generation.

frequency generation. Basically, all technical requirements and arrangements are the same as the corresponding sum-frequency generation process. However, because optical parametric generation generally requires a sufficiently high efficiency for it to be useful, it is normally not performed on the surface of a material or at the interface between two different materials, though it is possible. Various forms of optical parametric conversion, generation, and oscillation will be discussed in Sections 10.4 and 10.6.

5.2.5 Optical Parametric Amplification

Optical parametric amplification, as shown in Fig. 5.3, is fundamentally the same as difference-frequency generation, except for its purpose. It is technically the same as optical parametric generation except for the possible arrangements on the input waves. By comparing Figs. 5.3 and 5.4, it can be seen that the physical difference between difference-frequency generation and optical parametric generation is that only one input wave at ω_3 is required for optical parametric generation, but two input waves, at ω_3 and ω_1, are required for difference-frequency generation. The difference in their purposes is that difference-frequency generation aims at generating the frequency at ω_2, which is absent at the input, whereas optical parametric generation aims at generating ω_1 and ω_2, which are both absent at the input. If an input wave at one of the parametric frequencies, say ω_1, is also provided for optical parametric generation, the process is basically the same as that of difference-frequency generation except that the purpose might not be the same. This process is called optical parametric amplification if the purpose is the amplification of the signal at the input frequency ω_1, which is present at the input, rather than the generation of the wave at the difference frequency ω_2, which is absent at the input.

From the above discussion, it is clear that optical parametric amplification uses the second-order real susceptibility $\chi^{(2)}(\omega_1 = \omega_3 - \omega_2)$ for ω_1 and the same second-order real susceptibility $\chi^{(2)}(\omega_2 = \omega_3 - \omega_1)$ as that of difference-frequency generation for ω_2. The nonlinear polarizations of these two parametric frequencies are those given in (5.11) and (5.12) for optical parametric generation. It has the same phase-matching condition as that for optical parametric generation:

$$\mathbf{k}_1 = \mathbf{k}_3 - \mathbf{k}_2, \text{ i.e., } \mathbf{k}_2 = \mathbf{k}_3 - \mathbf{k}_1 \text{ and } \mathbf{k}_3 = \mathbf{k}_1 + \mathbf{k}_2, \tag{5.14}$$

which is not automatically satisfied. The two processes use the same noncentrosymmetric crystal and the same phase-matching arrangement. Basically, all technical requirements and arrangements are the same for the two processes. However, like optical parametric generation, optical parametric amplification is normally not performed on the surface of a material or at the interface between two different materials, though it is possible. Optical parametric amplification will be further discussed in Section 10.5.

5.2.6 Optical Rectification

Optical rectification, shown in Fig. 5.5, is the degenerate case of difference-frequency generation for $\omega_3 = \omega_1 = \omega$ and $\omega_2 = 0$. Optical rectification generates a DC electric field in a nonlinear medium through the second-order interaction of an optical wave with the medium, which is represented by $\chi^{(2)}(0 = \omega - \omega)$. This is a parametric process that does not involve energy exchange between the optical field and the optical medium. A voltage is created across the medium by the DC field that is generated through this process, as shown in Fig. 5.5, but no current is generated. Therefore, there is no exchange of energy between the optical field and the medium. The DC field might cause a modification on the optical property of the medium, for example through the Pockels effect discussed in the following subsection, but that is caused by a secondary effect rather than directly by the optical field.

The nonlinear polarization of optical rectification is

$$\mathbf{P}^{(2)}(0) = \epsilon_0 \left[\chi^{(2)}(0 = \omega - \omega) : \mathbf{E}(\omega)\mathbf{E}^*(\omega) + \chi^{(2)}(0 = -\omega + \omega) : \mathbf{E}^*(\omega)\mathbf{E}(\omega) \right]$$
$$= 2\epsilon_0 \chi^{(2)}(0 = \omega - \omega) : \mathbf{E}(\omega)\mathbf{E}^*(\omega).$$
(5.15)

Note that ω and $-\omega$ are treated as two different frequencies in the nonlinear susceptibility. Therefore, the permutations of them have to be counted. The phase-matching condition for optical rectification is

$$0 = \mathbf{k} - \mathbf{k},$$
(5.16)

which is always mathematically true irrespective of the value or direction of $\mathbf{k}$. Therefore, optical rectification is automatically phase matched though it is a parametric process. Nonetheless, being a second-order process, it does not occur in a centrosymmetric material under the electric dipole interaction but only takes place in a noncentrosymmetric crystal.

5.2.7 Pockels Effect

In the *Pockels effect*, as shown in Fig. 5.5, a DC field at the zero frequency interacts with an optical field at a frequency of ω through the second-order real susceptibility $\chi^{(2)}(\omega = \omega + 0)$. It is a parametric process because it involves only the real part of a nonlinear susceptibility. The process can be considered to be a special case of sum-frequency generation with $\omega_1 = \omega$, $\omega_2 = 0$, and $\omega_3 = \omega$. The Pockels effect involves modification of the material property, but such a modification is caused by a DC, or low frequency, electric field, not by an optical field. The optical field that participates in the Pockels effect senses the change of the material property caused by the DC field. The special case of the Pockels effect can also be considered as parametric mixing of an optical field at ω with a DC or low-frequency electric field, though it is generally described in terms of a modification on the permittivity tensor of a crystal by the DC

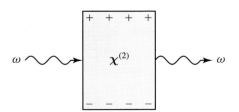

Figure 5.5 Optical rectification and Pockels effect. For optical rectification, the DC field across the medium is generated by the optical field. For the Pockels effect, the DC field is an externally applied field.

or low-frequency electric field. Being a parametric process, the Pockels effect does not cause exchange of energy between the optical field and the optical medium.

The nonlinear polarization of the Pockels effect is

$$\mathbf{P}^{(2)}(\omega) = \epsilon_0 \left[\chi^{(2)}(\omega = \omega + 0) : \mathbf{E}(\omega)\mathbf{E}(0) + \chi^{(2)}(\omega = 0 + \omega) : \mathbf{E}(0)\mathbf{E}(\omega) \right]$$
$$= 2\epsilon_0 \chi^{(2)}(\omega = \omega + 0) : \mathbf{E}(\omega)\mathbf{E}(0). \tag{5.17}$$

The phase-matching condition for the Pockels effect is

$$\mathbf{k} = \mathbf{k} + \mathbf{0}, \tag{5.18}$$

which is always mathematically true irrespective of the value or the direction of $\mathbf{k}$. Therefore, the Pockels effect is automatically phase matched though it is a parametric process. Being a second-order process, it does not occur in a centrosymmetric material under the electric dipole interaction but only takes place in a noncentrosymmetric crystal.

Similar to optical rectification, the Pockels effect involves an optical field at the frequency ω and a DC field at the zero frequency, as both illustrated in Fig. 5.5. Therefore, they share some common features. For example, both processes are automatically phase matched though both are parametric processes. However, the Pockels effect is not the reverse process of optical rectification, in the sense that sum-frequency generation is not the reverse process of difference-frequency generation. Electro-optic modulation based on the Pockels effect will be thoroughly discussed in Chapter 11.

EXAMPLE 5.1

By starting with a Nd:YAG laser that emits at a single wavelength of 1.064 μm, what different optical wavelengths can possibly be generated through various second-order nonlinear optical processes in a single step? What can be generated in two cascaded steps of second-order processes?

Solution

Starting from $\lambda_\omega = 1.064$ μm, we can generate its second harmonic at $\lambda_{2\omega} = \lambda_\omega/2 = 532$ nm by frequency doubling in one step through $\chi^{(2)}(2\omega = \omega + \omega)$ in a nonlinear crystal. We can also use optical parametric generation to generate a pair of tunable wavelengths at $\lambda_1, \lambda_2 > \lambda_\omega = 1.064$ μm that satisfies the condition $\lambda_1^{-1} + \lambda_2^{-1} = \lambda_\omega^{-1}$ in one step by using $\chi^{(2)}(\omega = \omega_1 + \omega_2)$ of a nonlinear crystal. Therefore, it is possible to obtain a visible wavelength at 532 nm and a range of tunable infrared wavelengths longer than 1.064 μm.

In two cascaded steps, many more wavelengths can be generated. The second harmonic can be further frequency doubled to the fourth harmonic at $\lambda_{4\omega} = \lambda_\omega/4 = 266$ nm. We can also mix the fundamental and the second harmonic in a nonlinear crystal to generate the third harmonic at $\lambda_{3\omega} = \lambda_\omega/3 = 354.7$ nm through sum-frequency generation by using $\chi^{(2)}(3\omega = \omega + 2\omega)$ of a nonlinear crystal. The second harmonic at 532 nm that is generated in the first step can be used to generate a range of tunable wavelengths longer than 532 nm through optical parametric generation. This range of tunable wavelengths longer than 532 nm can also be covered by frequency doubling the tunable infrared wavelengths that are generated through optical parametric generation in the first step, as well as by mixing the tunable infrared wavelengths with the fundamental at 1.064 μm through sum-frequency generation.

We can see further along this line that if we take merely three steps of second-order nonlinear optical mixing processes, either all in cascade or mixed in parallel/cascade, it is possible to cover a wide range of frequencies from the deep-ultraviolet to the far-infrared by starting with a single laser wavelength. The requirements for these processes to efficiently take place, as well as their limitations, will be discussed in Chapter 10.

5.3 THIRD-ORDER NONLINEAR OPTICAL PROCESSES

The third-order nonlinear optical processes of common interest are listed in Table 5.2. Different from the second-order processes considered in the preceding section, which are all parametric, the third-order processes considered in this section can be either parametric or nonparametric. Among these third-order processes, some, such as third-harmonic generation, parametric frequency conversion, and the optical Kerr effect, are parametric, whereas others, such as

Table 5.2 **Third-order nonlinear optical processes**

Process	Susceptibility	Phase matching
Third-harmonic generation	$\chi^{(3)\prime}(3\omega = \omega + \omega + \omega)$	Required
Parametric frequency conversion	$\begin{cases} \chi^{(3)\prime}(\omega_4 = \omega_1 + \omega_2 + \omega_3) \\ \chi^{(3)\prime}(\omega_4 = \omega_1 + \omega_2 - \omega_3) \\ \chi^{(3)\prime}(\omega_4 = \omega_1 - \omega_2 - \omega_3) \end{cases}$	Required
Optical Kerr effect	$\begin{cases} \chi^{(3)\prime}(\omega = \omega + \omega - \omega) \\ \chi^{(3)\prime}(\omega = \omega + \omega' - \omega') \end{cases}$	Automatic
Self-phase modulation	$\chi^{(3)\prime}(\omega = \omega + \omega - \omega)$	Automatic
Cross-phase modulation	$\chi^{(3)\prime}(\omega = \omega + \omega' - \omega')$	Automatic
Electro-optic Kerr effect	$\chi^{(3)\prime}(\omega = \omega + 0 - 0)$	Automatic
Absorption saturation	$\chi^{(3)\prime\prime}(\omega = \omega + \omega - \omega)$	Automatic
Gain saturation	$\chi^{(3)\prime\prime}(\omega = \omega + \omega - \omega)$	Automatic
Two-photon absorption	$\chi^{(3)\prime\prime}(\omega_1 = \omega_1 + \omega_2 - \omega_2)$	Automatic
Stimulated Raman scattering	$\chi^{(3)\prime\prime}(\omega_S = \omega_S + \omega_p - \omega_p)$	Automatic
Stimulated Brillouin scattering	$\chi^{(3)\prime\prime}(\omega_S = \omega_S + \omega_p - \omega_p)$	Automatic

absorption saturation, two-photon absorption, and stimulated Raman scattering, are nonparametric. As can be seen in Table 5.2, the contributing susceptibility of a parametric third-order process is the real part, $\chi^{(3)\prime}$, and that of a nonparametric process is the imaginary part, $\chi^{(3)\prime\prime}$, of a frequency-dependent third-order susceptibility.

We already know that if an optical frequency that is involved in an optical process is in resonance with a transition frequency of the optical medium, then the process is nonparametric. Therefore, all optical frequencies involved in a parametric process are far away from any resonance frequency of the medium. This is true for linear and second-order processes, but it is only necessary but not sufficient for a third-order process. For a linear process, there is only one optical frequency; therefore, the above condition for a parametric process is clearly necessary and sufficient. For a second-order process, there are a maximum of three participating frequencies satisfying the relation that $\omega_1 + \omega_2 = \omega_3$, where for generality each of the three frequencies can be positive or negative. It is clear from the relation that if a transition frequency is not in resonance with any of the three frequencies, then it is not in resonance with the combination of any two frequencies among ω_1, ω_2, and ω_3. This fact can be easily verified by examining the frequency dependence of $\chi^{(2)}_{ijk}$ given in (3.12). Therefore, the above condition for a parametric process is also necessary and sufficient for a second-order process.

However, the situation is different for a third-order process, which can have a maximum of four participating frequencies satisfying the relation that $\omega_1 + \omega_2 + \omega_3 = \omega_4$. It is true that if a transition frequency is not in resonance with any of the four frequencies, then it is not in resonance with the combination of any three frequencies among ω_1, ω_2, ω_3, and ω_4. Nonetheless, it is still possible for the transition frequency to be in resonance with a combination of two frequencies among the four frequencies; such resonance is sufficient to make the process nonparametric even when no transition is in resonance with any of the four frequencies ω_1, ω_2, ω_3, and ω_4. This fact can be verified by examining the frequency dependence of $\chi^{(3)}_{ijkl}$ given in (3.13). Examples for such resonances are seen in stimulated Raman scattering and two-photon absorption, both of which are nonparametric. Therefore, *the condition that all optical frequencies involved in a process are far away from any resonance frequency of the medium is only necessary but not sufficient for a parametric third-order process*. Besides the above necessary condition, *the sufficient condition for a third-order process also requires that the combination of any two frequencies among the four participating frequencies in the process be not in resonance with any transition frequency.*

Another fundamental difference between the second-order and the third-order processes is that $\chi^{(2)}$ from electric dipole interaction vanishes for all centrosymmetric materials, whereas $\chi^{(3)}$ from electric dipole interaction exists in any material. Therefore, it is not necessary to restrict a third-order process to a noncentrosymmetric crystal, though such a crystal can still be used for a third-order process. This is a great advantage for the third-order processes over the second-order processes. However, the disadvantage is that $\chi^{(3)}$ is generally orders of magnitude smaller than $\chi^{(2)}$ when $\chi^{(2)}$ does not vanish. Therefore, for a given purpose, a second-order process is generally chosen when a noncentrosymmetric crystal to facilitate it is available; otherwise, a third-order process takes over when a second-order process is not possible.

All of the processes listed in Table 5.2 are generally characterized by a third-order susceptibility of the form $\chi^{(3)}(\omega_4 = \omega_1 + \omega_2 + \omega_3)$, with each parametric process represented by the real part $\chi^{(3)\prime}$ of a $\chi^{(3)}$ and each nonparametric process represented by the imaginary part $\chi^{(3)\prime\prime}$ of a $\chi^{(3)}$. A nonlinear process that is characterized by the interaction of four optical waves is generally referred to as *four-wave mixing*. In *nondegenerate four-wave mixing*, all four optical waves have different frequencies. The process becomes partially degenerate if there are only two or three distinct frequencies. In fully *degenerate four-wave mixing*, all of the participating waves have the same frequency. With the exception of the *electro-optic Kerr effect*, all of the third-order nonlinear processes can be described as four-wave mixing processes. A nondegenerate or degenerate parametric four-wave mixing process is also often called a *four-photon mixing process* because four photons are involved in the third-order process regardless of whether the frequencies are degenerate or not.

5.3.1 Third-Harmonic Generation

Third-harmonic generation is the degenerate case of sum-frequency generation for the generation of $\omega_4 = \omega_1 + \omega_2 + \omega_3$ with $\omega_1 = \omega_2 = \omega_3 = \omega$ and $\omega_4 = 3\omega$. It performs the function of *frequency tripling* for an input optical wave at the *fundamental frequency* of ω to generate a harmonic wave at the *third-harmonic frequency* of 3ω, as shown in Fig. 5.6. Third-harmonic generation is a parametric frequency-conversion process represented by the real part of the third-order susceptibility $\chi^{(3)}(3\omega = \omega + \omega + \omega)$. In the process, three photons at the fundamental frequency ω are simultaneously annihilated to generate one photon at the third-harmonic frequency 3ω. Energy is converted from the three low-frequency photons to the high-frequency photon, but there is no energy exchange with the optical medium. The total intensity of the two frequencies remains constant while the intensity at the third-harmonic frequency grows at the expense of the intensity at the fundamental frequency.

The nonlinear polarization of third-harmonic generation with only one input fundamental wave at ω is

$$\mathbf{P}^{(3)}(3\omega) = \epsilon_0 \chi^{(3)}(3\omega = \omega + \omega + \omega) \vdots \mathbf{E}(\omega)\mathbf{E}(\omega)\mathbf{E}(\omega). \tag{5.19}$$

The phase-matching condition for third-harmonic generation is

$$\mathbf{k}_{3\omega} = \mathbf{k}_\omega + \mathbf{k}_\omega + \mathbf{k}_\omega = 3\mathbf{k}_\omega, \tag{5.20}$$

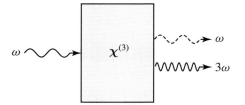

Figure 5.6 Third-harmonic generation.

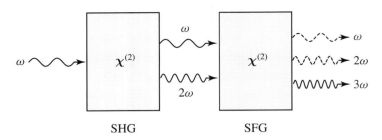

Figure 5.7 Third-harmonic generation through cascaded second-harmonic generation and second-order sum-frequency generation.

which is not automatically satisfied. Therefore, third-harmonic generation is not automatically phase matched but requires an effort on the phase-matching arrangement for efficiency. In principle, third-harmonic generation can be performed in any material because $\chi^{(3)}$ does not vanish in any material. In practice, however, efficient third-harmonic generation by using $\chi^{(3)}$ is difficult to accomplish. One reason is that $\chi^{(3)}$ is generally small in most materials. The other reason is that it is difficult to satisfy the phase-matching condition given in (5.20) for third-harmonic generation. For example, $\chi^{(3)}$ exists in any isotropic medium, but (5.20) is generally not satisfied in an isotropic medium due to material dispersion. Therefore, efficient third-harmonic generation for practical applications is usually carried out by using cascaded two-step second-order processes of second-harmonic generation for $\omega + \omega \rightarrow 2\omega$ and sum-frequency generation for $\omega + 2\omega \rightarrow 3\omega$, as shown in Fig. 5.7.

5.3.2 Parametric Frequency Conversion

As many as four different optical frequencies can participate in a third-order parametric frequency-conversion process. As shown in Fig. 5.8, there are a few different variations of this process. In a scenario of third-order sum-frequency generation, which is similar to second-order sum-frequency generation but with three input frequencies instead of two, three photons at the frequencies of ω_1, ω_2, and ω_3 combine to generate a photon at a higher frequency of $\omega_4 = \omega_1 + \omega_2 + \omega_3$ for $\omega_1 + \omega_2 + \omega_3 \rightarrow \omega_4$, as shown in Fig. 5.8(a). In another scenario, shown in Fig. 5.8(b), two photons at ω_1 and ω_2 combine to generate two other photons at ω_3 and ω_4 for $\omega_1 + \omega_2 \rightarrow \omega_3 + \omega_4$. This process has no analogous second-order process. In yet another scenario, which is similar to second-order difference-frequency generation but somewhat different, shown in Fig. 5.8(c), one photon at ω_4 breaks into three photons at ω_1, ω_2, and ω_3 through the interaction of the wave at ω_4 with another wave at a lower frequency, say ω_1, for $\omega_4 - \omega_1 \rightarrow \omega_2 + \omega_3$. In a process similar to parametric generation, it is also possible for a photon at ω_4 to break into three photons at ω_1, ω_2, and ω_3 without an input at any of the lower frequencies, for $\omega_4 \rightarrow \omega_1 + \omega_2 + \omega_3$, as shown in Fig. 5.8(d). Third-harmonic generation is simply the degenerate case of the first scenario for $\omega_1 = \omega_2 = \omega_3 = \omega$ and $\omega_4 = 3\omega$. Other partially degenerate cases also exist. For example, for any of the scenarios shown in Fig. 5.8, it is possible that $\omega_1 = \omega_2 = \omega$ but $\omega_3 \neq \omega$ and $\omega_3 \neq \omega_4$.

As a parametric process that is represented by a real third-order susceptibility, the process of parametric frequency conversion does not cause energy exchange between any optical field and the optical medium while it converts photon energies among different optical frequencies. In the

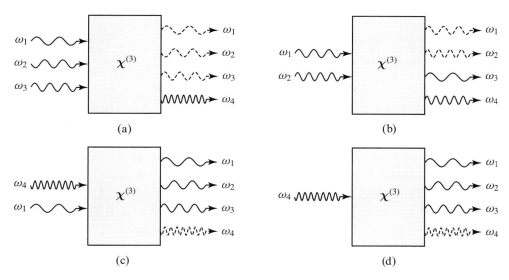

Figure 5.8 Third-order parametric frequency-conversion processes. (a) $\omega_1 + \omega_2 + \omega_3 \rightarrow \omega_4$. (b) $\omega_1 + \omega_2 \rightarrow \omega_3 + \omega_4$. (c) $\omega_4 - \omega_1 \rightarrow \omega_2 + \omega_3$. (d) $\omega_4 \rightarrow \omega_1 + \omega_2 + \omega_3$.

case that all four frequencies are different, $\omega_1 \neq \omega_2 \neq \omega_3 \neq \omega_4$, there is a six-fold permutation for each of the nonlinear polarizations of the four frequencies:

$$\mathbf{P}^{(3)}(\omega_1) = 6\epsilon_0 \chi^{(3)}(\omega_1 = \omega_4 - \omega_2 - \omega_3) \vdots \mathbf{E}(\omega_4)\mathbf{E}^*(\omega_2)\mathbf{E}^*(\omega_3), \tag{5.21}$$

$$\mathbf{P}^{(3)}(\omega_2) = 6\epsilon_0 \chi^{(3)}(\omega_2 = \omega_4 - \omega_3 - \omega_1) \vdots \mathbf{E}(\omega_4)\mathbf{E}^*(\omega_3)\mathbf{E}^*(\omega_1), \tag{5.22}$$

$$\mathbf{P}^{(3)}(\omega_3) = 6\epsilon_0 \chi^{(3)}(\omega_3 = \omega_4 - \omega_1 - \omega_2) \vdots \mathbf{E}(\omega_4)\mathbf{E}^*(\omega_1)\mathbf{E}^*(\omega_2), \tag{5.23}$$

$$\mathbf{P}^{(3)}(\omega_4) = 6\epsilon_0 \chi^{(3)}(\omega_4 = \omega_1 + \omega_2 + \omega_3) \vdots \mathbf{E}(\omega_1)\mathbf{E}(\omega_2)\mathbf{E}(\omega_3). \tag{5.24}$$

In the case that there is frequency degeneracy or degeneracies, the number of permutations is reduced. For example, in the case that $\omega_1 = \omega_2 \neq \omega_3 \neq \omega_4$, the nonlinear polarizations of the three different frequencies of $\omega_1 = \omega_2 = \omega$, ω_3, and ω_4 are

$$\mathbf{P}^{(3)}(\omega) = 6\epsilon_0 \chi^{(3)}(\omega = \omega_4 - \omega_3 - \omega) \vdots \mathbf{E}(\omega_4)\mathbf{E}^*(\omega_3)\mathbf{E}^*(\omega), \tag{5.25}$$

$$\mathbf{P}^{(3)}(\omega_3) = 3\epsilon_0 \chi^{(3)}(\omega_3 = \omega_4 - \omega - \omega) \vdots \mathbf{E}(\omega_4)\mathbf{E}^*(\omega)\mathbf{E}^*(\omega), \tag{5.26}$$

$$\mathbf{P}^{(3)}(\omega_4) = 3\epsilon_0 \chi^{(3)}(\omega_4 = \omega + \omega + \omega_3) \vdots \mathbf{E}(\omega)\mathbf{E}(\omega)\mathbf{E}(\omega_3). \tag{5.27}$$

In the case that $\omega_1 = \omega_2 = \omega_3 \neq \omega_4$, parametric frequency conversion is the same as third-harmonic generation or is the reverse of third-harmonic generation. The nonlinear polarizations of the two different frequencies of $\omega_1 = \omega_2 = \omega_3 = \omega$ and $\omega_4 = 3\omega$ are

$$\mathbf{P}^{(3)}(\omega) = 3\epsilon_0 \chi^{(3)}(\omega = 3\omega - \omega - \omega) \vdots \mathbf{E}(3\omega)\mathbf{E}^*(\omega)\mathbf{E}^*(\omega), \tag{5.28}$$

$$\mathbf{P}^{(3)}(3\omega) = \epsilon_0 \chi^{(3)}(3\omega = \omega + \omega + \omega) \vdots \mathbf{E}(\omega)\mathbf{E}(\omega)\mathbf{E}(\omega). \tag{5.29}$$

The phase-matching condition is

$$\mathbf{k}_4 = \mathbf{k}_1 + \mathbf{k}_2 + \mathbf{k}_3, \tag{5.30}$$

or its equivalent, such as $\mathbf{k}_1 = \mathbf{k}_4 - \mathbf{k}_2 - \mathbf{k}_3$. This phase-matching condition is not automatically satisfied. Just as third-harmonic generation through $\chi^{(3)}$ discussed in the preceding subsection, phase matching for third-order parametric frequency conversion is usually difficult to accomplish in common materials, particularly when at least one of the four frequencies is very different from the others so that the material dispersion is large among them. This is the scenario for the degenerate parametric frequency conversion of third-harmonic generation, in which the third-harmonic frequency is three times the fundamental frequency so that the material dispersion makes it difficult to satisfy the phase-matching condition of $\mathbf{k}_{3\omega} = 3\mathbf{k}_\omega$ that is required by (5.20).

In some other scenarios, the four participating frequencies in a parametric frequency-conversion process can be very close to one another, for example, $\omega_1 = \omega - \delta\omega$, $\omega_2 = \omega + \delta\omega$, $\omega_3 = -\omega$, and $\omega_4 = \omega$, with $\delta\omega \ll \omega$; or $\omega_1 = \omega_4 = \omega$, $\omega_2 = -\omega - \delta\omega$, and $\omega_3 = \omega + \delta\omega$, with $\delta\omega \ll \omega$. Then dispersion among the participating frequencies is small so that the phase mismatch is small or easily compensated. In this case, the efficiency of the third-order parametric frequency conversion can be significant, particularly when the interaction length is very large, such as that in an optical fiber. The analogous scenario for the second-order parametric process does not take place in an optical fiber because an optical fiber is made of a glass material, which is centrosymmetric and thus has a vanishing $\chi^{(2)}$. This is an example for a third-order parametric process not being dominated by a second-order process.

The possibility of phase matching for four-photon mixing in an optical fiber will be discussed in Section 9.4.1. Partially degenerate four-wave mixing in an optical fiber will be further discussed in Section 10.7.

5.3.3 Optical-Field-Induced Changes in Optical Properties

Except for third-harmonic generation and parametric frequency conversion, which were discussed in the preceding two subsections, and for stimulated Raman scattering and stimulated Brillouin scattering, which will be discussed below in Subsections 5.3.10 and 5.3.11, other third-order processes listed in Table 5.2 do not cause optical frequency conversion but have the characteristic of inducing *field-dependent changes* in the optical properties of a material. These processes are characterized by either the real or the imaginary part of a susceptibility of the form $\chi^{(3)}(\omega = \omega + \omega' - \omega')$. When $\omega' = \omega$, there can be only one beam in the interaction, as illustrated in Fig. 5.9(a), but there can still be two physically distinguishable beams of the same frequency, as illustrated in Fig. 5.9(b). When $\omega' \neq \omega$, there are always two optical beams in the interaction, as illustrated in Fig. 5.9(c).

In the case of one-beam interaction, we find that

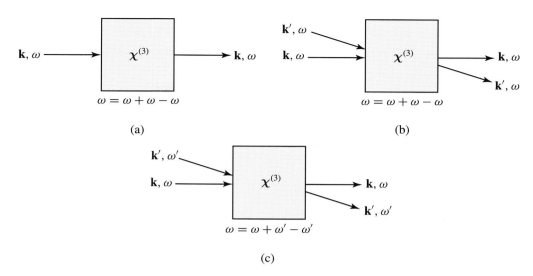

Figure 5.9 Third-order processes for field-induced susceptibility changes: (a) one-beam interaction, (b) interaction of two beams of the same frequency, and (c) interaction of two beams of different frequencies.

$$\mathbf{P}^{(3)}(\omega) = 3\epsilon_0\chi^{(3)}(\omega = \omega + \omega - \omega) \vdots \mathbf{E}(\omega)\mathbf{E}(\omega)\mathbf{E}^*(\omega) \tag{5.31}$$

by using (2.32) while taking into account frequency permutations and the intrinsic permutation symmetry. In the case of two-beam interaction, for either $\omega' = \omega$ or $\omega' \neq \omega$, we have

$$\begin{aligned}\mathbf{P}^{(3)}(\omega) = {} & 3\epsilon_0\chi^{(3)}(\omega = \omega + \omega - \omega) \vdots \mathbf{E}(\omega)\mathbf{E}(\omega)\mathbf{E}^*(\omega) \\ & + 6\epsilon_0\chi^{(3)}(\omega = \omega + \omega' - \omega') \vdots \mathbf{E}(\omega)\mathbf{E}(\omega')\mathbf{E}^*(\omega'),\end{aligned} \tag{5.32}$$

and a similar expression for $\mathbf{P}^{(3)}(\omega')$ in terms of $\chi^{(3)}(\omega' = \omega' + \omega' - \omega')$ and $\chi^{(3)}(\omega' = \omega' + \omega - \omega)$.

Each of the third-order processes discussed here and in the following uses a nonlinear susceptibility of the form of $\chi^{(3)}(\omega = \omega + \omega - \omega)$ or $\chi^{(3)}(\omega = \omega + \omega' - \omega')$. Its phase-matching condition is

$$\mathbf{k} = \mathbf{k} + \mathbf{k} - \mathbf{k} \quad \text{or} \quad \mathbf{k} = \mathbf{k} + \mathbf{k}' - \mathbf{k}', \tag{5.33}$$

which is always mathematically true irrespective of the values or directions of $\mathbf{k}$ and $\mathbf{k}'$. Therefore, such a process is automatically phase matched no matter whether it is parametric or nonparametric.

By identifying the total polarization at the frequency ω as $\mathbf{P}(\omega) = \mathbf{P}^{(1)}(\omega) + \mathbf{P}^{(3)}(\omega)$, we find that the total field-dependent permittivity tensor can be expressed as

$$\boldsymbol{\epsilon}(\omega, \mathbf{E}) = \boldsymbol{\epsilon}(\omega) + \Delta\boldsymbol{\epsilon}(\omega, \mathbf{E}), \tag{5.34}$$

where $\epsilon(\omega) = \epsilon_0 \left[\mathbf{I} + \chi^{(1)}(\omega) \right]$ represents the field-independent linear permittivity tensor of the medium and $\Delta\epsilon(\omega, \mathbf{E})$ accounts for the field-dependent changes that are induced by nonlinear optical interactions. For one-beam interaction,

$$\Delta\epsilon(\omega, \mathbf{E}) = 3\epsilon_0 \chi^{(3)}(\omega = \omega + \omega - \omega) : \mathbf{E}(\omega)\mathbf{E}^*(\omega), \text{ i.e.,}$$

$$\Delta\epsilon_{ij}(\omega, \mathbf{E}) = 3\epsilon_0 \sum_{k,l} \chi^{(3)}_{ijkl}(\omega = \omega + \omega - \omega) E_k(\omega) E_l^*(\omega). \tag{5.35}$$

For two-beam interaction,

$$\Delta\epsilon(\omega, \mathbf{E}) = 3\epsilon_0 \chi^{(3)}(\omega = \omega + \omega - \omega) : \mathbf{E}(\omega)\mathbf{E}^*(\omega)$$
$$+ 6\epsilon_0 \chi^{(3)}(\omega = \omega + \omega' - \omega') : \mathbf{E}(\omega')\mathbf{E}^*(\omega'), \text{ i.e.,}$$

$$\Delta\epsilon_{ij}(\omega, \mathbf{E}) = 3\epsilon_0 \sum_{k,l} \chi^{(3)}_{ijkl}(\omega = \omega + \omega - \omega) E_k(\omega) E_l^*(\omega)$$
$$+ 6\epsilon_0 \sum_{k,l} \chi^{(3)}_{ijkl}(\omega = \omega + \omega' - \omega') E_k(\omega') E_l^*(\omega'). \tag{5.36}$$

Note that $\Delta\epsilon(\omega, \mathbf{E})$ for the frequency at ω consists of two terms, one for the self-interaction of the field at ω, which is characterized by $\chi^{(3)}(\omega = \omega + \omega - \omega)$, and the other for the cross-interaction between the fields at ω and ω', which is characterized by $\chi^{(3)}(\omega = \omega + \omega' - \omega')$. The field-dependent permittivity of the form described here is the basis of many nonlinear optical phenomena that have important practical applications.

The nonlinear process discussed in this subsection generally leads to an *optical-field-induced birefringence* because $\Delta\epsilon$ is a tensor. The simplest case involves a single linearly polarized optical wave in an isotropic medium with the optical field polarized in any fixed direction, or in a cubic crystal with the optical field polarized along one of the principal axes. Then $\mathbf{P}^{(3)}$ is parallel to $\mathbf{E}$ of the optical field, and the only susceptibility element that contributes to this type of interaction is $\chi^{(3)}_{1111}(\omega = \omega + \omega - \omega)$. In this case, the permittivity seen by the optical field is

$$\epsilon(\omega, \mathbf{E}) = \epsilon(\omega) + 3\epsilon_0 \chi^{(3)}_{1111} \left| E(\omega) \right|^2 = \epsilon(\omega) + \frac{3\chi^{(3)}_{1111}}{2cn_0} I(\omega), \tag{5.37}$$

where n_0 is the field-independent linear refractive index of the medium and $I(\omega)$ is the intensity of the optical beam.

5.3.4 Optical Kerr Effect

We see from (5.37) that the real part of $\chi^{(3)}_{1111}(\omega = \omega + \omega - \omega)$ leads to the *intensity-dependent index of refraction*:[1]

$$n = n_0 + n_2 I(\omega), \tag{5.38}$$

---

[1] In the literature, a different expression of $n = n_0 + 2n_2|E|^2$ is also used. The value of n_2 defined by this expression is accordingly different from that defined in (5.38).

where

$$n_2 = \frac{3\chi_{1111}^{(3)\prime}}{4c\epsilon_0 n_0^2} \tag{5.39}$$

is the *nonlinear refractive index*. The value of n_2 for a given material varies with the optical wavelength, the impurities in the optical material, and temperature. In particular, it can be significantly enhanced by transition resonances in a manner like the resonant enhancement of the linear refractive index. Its value also depends on its response speed because n_2 of a given material can be contributed by many different physical mechanisms that have different relaxation times. For example, at room temperature the value of n_2 of a semiconductor, such as GaAs or AlGaAs, can range from the order of 1×10^{-17} m^2 W^{-1} for a wavelength that is far away from the bandgap to the order of 2×10^{-10} m^2 W^{-1} for a wavelength near the bandgap with exciton enhancement, and then further to the order of 1×10^{-8} m^2 W^{-1} with exciton enhancement near the band edge in GaAs/AlGaAs quantum wells.

This intensity-dependent index of refraction represents the simplest case of the *optical Kerr effect*. Various features of the optical Kerr effect and optical modulation based on this effect will be discussed in Chapter 12. Depending on the material properties and the experimental conditions, the optical Kerr effect leads to the phenomena of self-phase modulation, self-focusing, and self-defocusing, which will be discussed in Section 12.2.

EXAMPLE 5.2

Silica glass has an electronically contributed nonlinear susceptibility of $\chi_{1111}^{(3)\prime}$ $(\omega = \omega + \omega - \omega) = 1.8 \times 10^{-22}$ m^2 V^{-2} that causes an intensity-dependent index change in an optical fiber. Its linear refractive index is $n_0 \approx 1.45$ in the visible and near-infrared spectral regions for most applications of optical fibers. Find the value of n_2 for a silica fiber. For an ultrashort optical pulse that has a pulsewidth of the order of picoseconds or femtoseconds, the peak power can easily be a few kilowatts. Take a femtosecond pulse of a 10 kW peak power that propagates in a fiber of a 10 μm core diameter. What is the peak optical-field-induced index change seen by this pulse?

Solution

By using (5.39), we find that

$$n_2 = \frac{3 \times 1.8 \times 10^{-22}}{4 \times 3 \times 10^8 \times 8.85 \times 10^{-12} \times 1.45^2} \text{ m}^2 \text{ W}^{-1} = 2.4 \times 10^{-20} \text{ m}^2 \text{ W}^{-1}.$$

Therefore, $n_2 = 2.4 \times 10^{-20}$ m^2 W^{-1} for a silica fiber. The peak intensity of the pulse is

$$I_{pk} = \frac{10 \times 10^3}{\pi \times (10 \times 10^{-6}/2)^2} \text{ W m}^{-2} = 1.27 \times 10^{14} \text{ W m}^{-2}.$$

The peak optical-field-induced index change seen by the pulse is

$$\Delta n_{pk} = n_2 I_{pk} = 2.4 \times 10^{-20} \times 1.27 \times 10^{14} = 3.05 \times 10^{-6}.$$

We see that the optical-field-induced index change is very small even for a pulse of 10 kW peak power that is confined in a very small fiber core to reach a very high intensity. Nevertheless, this small index change can have very significant effects on the characteristics of the optical pulse through the process of self-phase modulation, and on other pulses through cross-phase modulation. It is the root cause of such fascinating phenomena as *pulse spectral broadening* and *optical solitons*, which will be discussed in Chapter 18. The value of n_2 for an optical fiber varies with the dopants in the fiber and with the optical wavelength. In the literature, the value of $n_2 = 3.2 \times 10^{-20} \text{ m}^2 \text{ W}^{-1}$ is often quoted for silica optical fibers. In Er-doped fibers, the value of n_2 at a resonance wavelength of 980 nm can be increased to the order of $1\text{–}3 \times 10^{-15} \text{ m}^2 \text{ W}^{-1}$, depending on the doping concentration, due to resonance enhancement.

5.3.5 Self-Phase Modulation

The spatially and temporally varying phase of an optical field that has a wavevector of **k** and a frequency of ω is

$$\varphi(\mathbf{r}, t) = \mathbf{k} \cdot \mathbf{r} - \omega t. \tag{5.40}$$

Consider the situation discussed in the preceding subsection for the optical Kerr effect, in which the optical field causes the index of refraction to change with its intensity as given in (5.38). Then, the phase of the optical field also depends on the intensity of the field:

$$
\begin{aligned}
\varphi(\mathbf{r}, t) &= \mathbf{k} \cdot \mathbf{r} - \omega t \\
&= \frac{\omega}{c} n r \hat{k} \cdot \hat{r} - \omega t \\
&= \frac{\omega}{c} \left[n_0 + n_2 I(\mathbf{r}, t) \right] r \hat{k} \cdot \hat{r} - \omega t.
\end{aligned}
\tag{5.41}
$$

In (5.41), there is a term that represents the intensity-dependent *Kerr phase change*:

$$\varphi_K(\mathbf{r}, t) = \frac{\omega}{c} n_2 r I(\mathbf{r}, t) \hat{k} \cdot \hat{r}. \tag{5.42}$$

This phase change is caused by the intensity of the optical field itself. It is known as *self-phase modulation* because it is self-induced. In (5.41) and (5.42), the intensity of the optical field is explicitly expressed as a function of space and time as it generally varies with space or time, or both, in the practical applications of self-phase modulation.

As discussed in Subsection 5.3.3, self-phase modulation is automatically phase matched. It originates from the optical Kerr effect; therefore, it is a parametric process associated with the real part of a third-order nonlinear susceptibility. The detailed features of self-phase modulation will be discussed in Subsection 12.1.1.

5.3.6 Cross-Phase Modulation

The optical Kerr effect discussed in Subsection 5.3.4 and the Kerr phase change obtained in Subsection 5.3.5 are induced by an optical field itself at a frequency of ω. As discussed in Subsection 5.3.3, a different field, which is at the same frequency ω but is spatially separate and distinguishable, or is at a different frequency ω', can induce similar changes on the phase of the optical field at ω. The resulting phase modulation due to this cross-interaction is known as *cross-phase modulation*.

Consider, for simplicity, the case that involves two linearly polarized optical waves in an isotropic medium with the optical field that is at a frequency of ω being polarized in any fixed direction, or in a cubic crystal with the optical field being polarized along one of the principal axes. The other field at a frequency of ω' is polarized in a direction that can be the same or different from that of the field at ω. From (5.35) and (5.36), we find that the effect of self-modulation is quantified by $\Delta\epsilon(\omega, \mathbf{E}) = 3\epsilon_0 \boldsymbol{\chi}^{(3)}(\omega = \omega + \omega - \omega) : \mathbf{E}(\omega)\mathbf{E}^*(\omega)$, which results in $\Delta\epsilon(\omega, \mathbf{E}) = 3\epsilon_0\chi_{1111}^{(3)\prime}(\omega = \omega + \omega - \omega)\left|E(\omega)\right|^2$ for an isotropic medium. By comparison, the effect of cross-modulation is quantified by $\Delta\epsilon(\omega, \mathbf{E}) = 6\epsilon_0\boldsymbol{\chi}^{(3)}(\omega = \omega + \omega' - \omega') : \mathbf{E}(\omega')\mathbf{E}^*(\omega')$, which for an isotropic medium results in $\Delta\epsilon(\omega, \mathbf{E}) = 6\epsilon_0\chi_{1111}^{(3)\prime}(\omega = \omega + \omega' - \omega')\left|E(\omega')\right|^2$ if $\mathbf{E}(\omega')\| \mathbf{E}(\omega)$ but $\Delta\epsilon(\omega, \mathbf{E}) = 6\epsilon_0\chi_{1122}^{(3)\prime}(\omega = \omega + \omega' - \omega')\left|E(\omega')\right|^2$ if $\mathbf{E}(\omega')\perp\mathbf{E}(\omega)$. Because $\chi_{1122}^{(3)\prime} = \chi_{1111}^{(3)\prime}\big/3$ for an isotropic material if Kleinman's symmetry condition is valid, we find that the effect of cross-phase modulation is 2/3 to 2 times that of self-phase modulation, depending on the relation between the polarization directions of the two optical waves. The details of this dependence of cross-phase modulation on the polarization directions of the interacting waves will be further discussed in Subsection 12.1.2.

Therefore, the cross-modulation permittivity that is seen by the optical field at the frequency ω due to the presence of the field at the frequency ω' is

$$\epsilon(\omega, \mathbf{E}) = \epsilon(\omega) + 3\epsilon_0\xi\chi_{1111}^{(3)\prime}\left|E(\omega')\right|^2 = \epsilon(\omega) + \frac{3\xi\chi_{1111}^{(3)\prime}}{2cn_0}I(\omega'), \tag{5.43}$$

where $I(\omega')$ is the intensity of the optical beam at the frequency ω'. In (5.43), we focus the attention on the cross-modulation by ignoring the self-induced change as expressed in (5.37). The factor ξ has a value that depends on the polarization direction of the field at the frequency ω' with respect to that at the frequency ω; it ranges from a value of 2/3 when the two are orthogonally polarized to a value of 2 when they are polarized in the same direction. Therefore, the *cross-intensity-dependent index of refraction* is

$$n(\omega) = n_0 + \xi n_2 I(\omega'), \tag{5.44}$$

which leads to a Kerr phase change for the field at the frequency ω due to cross-phase modulation from the intensity of the field at the frequency ω':

$$\varphi_K(\omega, \mathbf{r}, t) = \frac{\omega}{c}\xi n_2 r I(\omega', \mathbf{r}, t)\hat{k}\cdot\hat{r}. \tag{5.45}$$

Just as self-phase modulation and other effects that originate from the optical Kerr effect, cross-phase modulation is a third-order parametric process but is automatically phase matched. The detailed features of cross-phase modulation will be discussed in Subsection 12.1.2.

5.3.7 Electro-optic Kerr Effect

The *electro-optic Kerr effect* is similar to the Pockels effect in that it is an electro-optic effect in which a DC electric field induces a change of the property of a material at an optical frequency ω. Different from the Pockels effect, it is a third-order process that is associated with the third-order real susceptibility $\chi^{(3)}(\omega = \omega + 0 + 0)$. It can be viewed as a special case of cross-phase modulation of the optical Kerr effect for $\omega' = 0$. Indeed, the electro-optic Kerr effect was the effect that was discovered by Kerr in 1875 [4, 5], whereas the optical Kerr effect is its generalization after nonlinear optics was developed much later. Similar to the optical Kerr effect, the electro-optic Kerr effect is a third-order parametric process that does not cause energy exchange between the optical field and the optical medium.

The nonlinear polarization of the electro-optic Kerr effect is

$$\mathbf{P}^{(3)}(\omega) = 3\epsilon_0 \chi^{(3)}(\omega = \omega + 0 + 0) \vdots \mathbf{E}(\omega)\mathbf{E}(0)\mathbf{E}(0), \tag{5.46}$$

where the factor 3 accounts for frequency permutations with intrinsic permutation symmetry. The phase-matching condition is

$$\mathbf{k} = \mathbf{k} + \mathbf{0} + \mathbf{0}, \tag{5.47}$$

which is always mathematically true irrespective of the value or the direction of $\mathbf{k}$. Therefore, just as the optical Kerr effect discussed in Subsection 5.3.4, the electro-optic Kerr effect is automatically phase matched, though it is a parametric process. Being a third-order process, it exists in any materials, including centrosymmetric and noncentrosymmetric ones.

Because the Pockels effect vanishes in a centrosymmetric material, the electro-optic Kerr effect is the leading electro-optic effect in a centrosymmetric material. However, being a third-order process, the effect is normally too weak to be practically useful. In the case that the Pockels effect does not vanish, it completely overshadows the electro-optic Kerr effect.

5.3.8 Absorption Saturation and Gain Saturation

As seen in Section 1.8, the optical frequency ω for single-photon optical absorption and optical amplification is in resonance with an optical transition frequency between two energy levels, as shown in Fig. 5.10. Whether the medium absorbs or amplifies an incoming optical wave depends on the population difference between the two energy levels. It absorbs light when the lower level is more populated, and it amplifies light when the upper level is more populated. The unsaturated absorption coefficient, α_0, and the unsaturated gain coefficient, g_0, are proportional to the imaginary parts of the linear susceptibility that is associated with the resonant transition: $\alpha_0 \propto \chi_{\text{res}}^{(1)\prime\prime}$ and $g_0 \propto -\chi_{\text{res}}^{(1)\prime\prime}$.

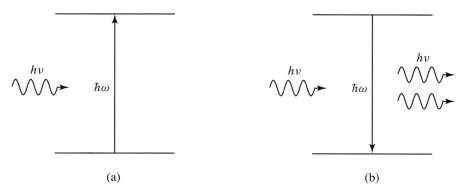

Figure 5.10 Resonant transitions for (a) single-photon optical absorption and (b) single-photon optical amplification.

As discussed in Section 15.1, in general, absorption saturation and gain saturation cannot be treated as merely a third-order effect through perturbation expansion because the light intensity can be higher than the saturation intensity of the system. The saturated absorption coefficient, α, and the saturated gain coefficient, g, have to be generally expressed in the forms:

$$\alpha = \frac{\alpha_0}{1 + I/I_{\text{sat}}} \quad \text{and} \quad g = \frac{g_0}{1 + I/I_{\text{sat}}}, \tag{5.48}$$

where I is the intensity of the optical field and I_{sat} is known as the *saturation intensity*. Optical absorption or amplification is a nonparametric process as it clearly involves energy exchange between the optical field and the optical medium.

It can be clearly seen that absorption saturation or gain saturation that is generally expressed in the forms given in (5.48) is not simply a third-order process. In the low-intensity limit such that the light intensity is much lower than the saturation intensity, $I \ll I_{\text{sat}}$, however, (5.48) can be expressed in the form of series expansion:

$$\alpha = \alpha_0 \left[1 - \frac{I}{I_{\text{sat}}} + \left(\frac{I}{I_{\text{sat}}}\right)^2 - \left(\frac{I}{I_{\text{sat}}}\right)^3 + \cdots \right] \text{ and } g = g_0 \left[1 - \frac{I}{I_{\text{sat}}} + \left(\frac{I}{I_{\text{sat}}}\right)^2 - \left(\frac{I}{I_{\text{sat}}}\right)^3 + \cdots \right]. \tag{5.49}$$

Then, the leading term is the unsaturated absorption coefficient or unsaturated gain coefficient that is independent of the optical intensity but is proportional to $\chi_{\text{res}}^{(1)''}(\omega)$. The second term of the series expansion is linearly proportional to the optical intensity and is also proportional to the imaginary part of the third-order resonant susceptibility $\chi_{\text{res}}^{(3)''}(\omega = \omega + \omega - \omega)$. Only in this low-intensity limit, can we consider absorption saturation or gain saturation as a third-order process. In this situation, the nonlinear polarization for low-intensity absorption saturation or gain saturation can be expressed in terms of the third-order susceptibility as

$$\mathbf{P}^{(3)}(\omega) = 3\epsilon_0 \chi^{(3)}(\omega = \omega + \omega - \omega) \vdots \mathbf{E}(\omega)\mathbf{E}(\omega)\mathbf{E}^*(\omega). \tag{5.50}$$

This process is automatically phase matched because it is nonparametric. Note that the nonlinear polarization given in (5.50) is identical in form to that given in (5.31) for any third-order one-beam interaction.

As an example, we consider the simplest general case discussed at the end of Subsection 5.3.3, which involves a single linearly polarized optical wave in an isotropic medium with the optical field polarized in any fixed direction, or in a cubic crystal with the optical field polarized along one of the principal axes. Without considering the details of the exact process, the field-induced change by such one-beam interaction leads to an intensity-dependent permittivity of the form given in (5.37). From (5.37), we see that the imaginary part of the total susceptibility is

$$\chi'' = \chi^{(1)''} + \frac{3\chi_{1111}^{(3)''}}{2c\epsilon_0 n_0} I(\omega). \tag{5.51}$$

Therefore, the imaginary part of $\chi_{1111}^{(3)}(\omega = \omega + \omega - \omega)$ leads to an intensity-dependent change in the loss or gain of a medium. In general, the sign of $\chi_{1111}^{(3)''}(\omega = \omega + \omega - \omega)$ that is contributed by a single-photon transition of a resonance frequency at or near ω is always the opposite of that of $\chi^{(1)''}(\omega)$. When $\chi^{(1)''}(\omega) > 0$, the medium has a linear loss. In this case, $\chi_{1111}^{(3)''}(\omega = \omega + \omega - \omega)$ is negative, and it contributes to an intensity-dependent reduction of the loss, resulting in absorption saturation. When $\chi^{(1)''}(\omega) < 0$, the medium has a gain. Then, $\chi_{1111}^{(3)''}(\omega = \omega + \omega - \omega)$ is positive, and it causes intensity-dependent gain saturation.

The differences among various one-beam interactions are not in their formal expressions of $\mathbf{P}^{(3)}$ but in the details of whether the real part or the imaginary part of $\chi^{(3)}$ contributes to a process and what kind of resonance, such as one-photon resonance or two-photon resonance, is involved in the process. Such differences are important for the two-photon absorption process and for stimulated Raman scattering and stimulated Brillouin scattering, as we shall see below. For absorption saturation and gain saturation, the resonance is with one-photon transitions. Various phenomena and applications of absorption saturation and gain saturation will be discussed in Chapter 15.

5.3.9 Two-Photon Absorption

As discussed above, the characteristics of a nonparametric third-order process are determined by the resonant transition that is responsible for the process. In Table 5.2, we see that the susceptibility for two-photon absorption has the same form as that for stimulated Raman scattering. However, the resonance frequency of the transition for stimulated Raman scattering is the difference, $\omega_p - \omega_S$, of two participating frequencies, whereas that for two-photon absorption is the sum, $\omega_1 + \omega_2$, of two participating frequencies, as shown in Fig. 5.11(a). In the special case that $\omega_1 = \omega_2 = \omega$, both the susceptibility for two-photon absorption and that

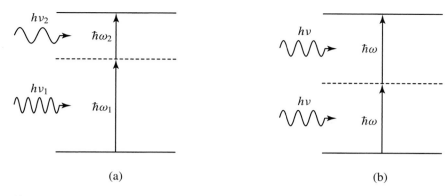

Figure 5.11 Resonant transitions for two-photon absorption at (a) $\omega_1 + \omega_2$ and (b) 2ω.

for absorption saturation or gain saturation have the form $\chi^{(3)''}(\omega = \omega + \omega - \omega)$. The difference between them is that the nonlinear susceptibility responsible for absorption saturation or gain saturation is resonant at ω but that for two-photon absorption is resonant at 2ω, as shown in Fig. 5.11(b).

The nonlinear polarizations for ω_1 and ω_2 when $\omega_1 \neq \omega_2$ are formally the same as those of the general cross-modulation term in the nonlinear polarization of two-beam interaction that is given in (5.32). In the case of two-photon absorption shown in Fig. 5.11(a), the resonance occurs at $\omega_1 + \omega_2$ but not at $2\omega_1$ or $2\omega_2$. The self-modulation term of the polarization for two-beam interaction is not in resonance and thus can be ignored. Then, we have, for two-photon absorption with $\omega_1 \neq \omega_2$, the nonlinear polarizations:

$$\mathbf{P}^{(3)}(\omega_1) = 6\epsilon_0\chi^{(3)}(\omega_1 = \omega_1 + \omega_2 - \omega_2) \vdots \mathbf{E}(\omega_1)\mathbf{E}(\omega_2)\mathbf{E}^*(\omega_2), \tag{5.52}$$

$$\mathbf{P}^{(3)}(\omega_2) = 6\epsilon_0\chi^{(3)}(\omega_2 = \omega_2 + \omega_1 - \omega_1) \vdots \mathbf{E}(\omega_2)\mathbf{E}(\omega_1)\mathbf{E}^*(\omega_1). \tag{5.53}$$

In the case that $\omega_1 = \omega_2 = \omega$, as shown in Fig. 5.12(b), the nonlinear polarization for two-photon absorption is formally the same as that for single-photon absorption saturation or gain saturation:

$$\mathbf{P}^{(3)}(\omega) = 3\epsilon_0\chi^{(3)}(\omega = \omega + \omega - \omega) \vdots \mathbf{E}(\omega)\mathbf{E}(\omega)\mathbf{E}^*(\omega). \tag{5.54}$$

As a nonparametric process, two-photon absorption is automatically phase matched.

The details of multiphoton absorption processes, including two-photon absorption and three-photon absorption, will be further discussed in Chapter 14.

5.3.10 Stimulated Raman Scattering

The parametric frequency-conversion processes are not the only third-order nonlinear processes that result in optical frequency conversion. Among the nonparametric processes, stimulated Raman scattering and stimulated Brillouin scattering also lead to optical frequency conversion. However, being nonparametric, these processes are connected to the intrinsic resonances of the optical medium and are dependent on the initial state of the material. This characteristic is shared with the processes of one-photon absorption or gain saturation and with two-photon absorption. The difference among them is the energy levels that are involved in the resonance and how the resonance takes place.

If the material is originally in the ground state of the relevant transition, annihilation of a photon at the pump frequency, ω_p, of the incident optical wave creates an excitation in the material and a photon at a down-shifted *Stokes frequency* of $\omega_S = \omega_p - \Omega$, as illustrated in Fig. 5.12(a). If the material is originally in an excited state, it is possible to create a photon at an up-shifted *anti-Stokes frequency* of $\omega_{AS} = \omega_p + \Omega$ while the material simultaneously makes a transition from the excited state to the ground state, as illustrated in Fig. 5.12(b). The *Stokes susceptibility* is $\chi^{(3)''}(\omega_S = \omega_S + \omega_p - \omega_p)$, whereas the *anti-Stokes susceptibility* is $\chi^{(3)''}(\omega_{AS} = \omega_{AS} + \omega_p - \omega_p)$. The amount of frequency shift, $\Omega = \omega_p - \omega_S = \omega_{AS} - \omega_p$, is determined by the excitation that is responsible for a given process and is a characteristic of the material. For stimulated Raman scattering, the excitation is at the molecular or atomic level, such as the optical phonons of a medium or the vibrational modes of molecules. The *Raman frequency shift* is typically in the range of 10–100 THz. The material is generally more populated in the ground state so that stimulated Raman scattering generates a Stokes signal at a down-shifted frequency, as shown in Fig. 5.13, not an anti-Stokes signal.

Because ω_p is always different from ω_S, the nonlinear polarizations of stimulated Raman scattering are formally

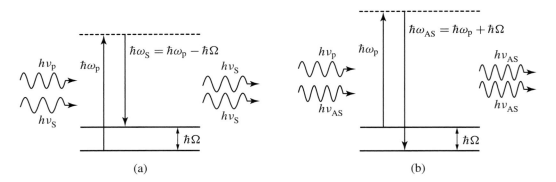

Figure 5.12 (a) Stokes and (b) anti-Stokes transitions for stimulated Raman scattering.

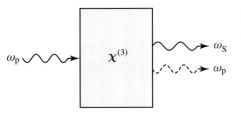

Figure 5.13 Stimulated Raman scattering. Most of the Stokes field propagates in the direction of the pump field.

$$\mathbf{P}^{(3)}(\omega_S) = 6\epsilon_0\chi^{(3)}(\omega_S = \omega_S + \omega_p - \omega_p) \vdots \mathbf{E}(\omega_S)\mathbf{E}(\omega_p)\mathbf{E}^*(\omega_p), \tag{5.55}$$

$$\mathbf{P}^{(3)}(\omega_p) = 6\epsilon_0\chi^{(3)}(\omega_p = \omega_p + \omega_S - \omega_S) \vdots \mathbf{E}(\omega_p)\mathbf{E}(\omega_S)\mathbf{E}^*(\omega_S). \tag{5.56}$$

As a nonparametric process, stimulated Raman scattering is automatically phase matched. Most of the Stokes optical field propagates in the same direction as the pump field, as shown in Fig. 5.13, though in principle it can propagate in any direction.

The process of stimulated Raman scattering can be utilized to construct optical amplifiers and generators based on Raman amplification and generation of the Stokes frequency. The features of the Raman processes and the details of the Raman amplifiers and generators will be discussed in Chapter 13.

5.3.11 Stimulated Brillouin Scattering

Stimulated Brillouin scattering is similar to stimulated Raman scattering in that it also involves material excitation, and it generates a Stokes signal at a down-shifted frequency of $\omega_S = \omega_p - \Omega$. The fundamental difference between a Raman process and a Brillouin process is the mode of excitation in the material that participates in the interaction. In stimulated Raman scattering, the interaction is associated with the excitation at the molecular or atomic level. In stimulated Brillouin scattering, by contrast, the interaction is associated with the long-range excitation that is characterized by acoustic phonons, or an acoustic wave, in a medium. The *Brillouin frequency shift* is typically in the range of 1–50 GHz in the hypersonic frequency region. Another difference is that the Stokes field that is generated by stimulated Brillouin scattering generally propagates in the backward direction opposite to that of the pump field.

The nonlinear polarizations of stimulated Brillouin scattering are formally the same as those given in (5.55) and (5.56) for stimulated Raman scattering. Again, the difference is in the detailed resonances of the nonlinear susceptibilities, not in the formal expressions of the nonlinear polarizations. As a nonparametric process, phase matching between the optical fields involved in the interaction is automatic, but phase matching among the optical waves and the acoustic wave determines the direction of the Stokes signal. In an isotropic medium, the only possible direction for this phase matching is for the Brillouin Stokes wave to propagate in the backward direction opposite to that of the pump wave, as shown in Fig. 5.14. In a birefringent crystal, other directions are possible, including the forward direction.

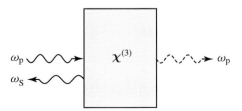

Figure 5.14 Stimulated Brillouin scattering. The Stokes field generally propagates in the backward direction of the pump field.

The process of stimulated Brillouin scattering can be utilized to construct optical amplifiers and generators based on Brillouin amplification and generation of the Stokes frequency. The features of the Brillouin processes and the details of the Brillouin amplifiers and generators will be discussed in Chapter 13.

Problem Set

5.1.1 Name the two unique features of nonlinear optical processes that set them apart from linear optical processes. For each of the second-order processes that are listed in Table 5.1, identify the unique feature or features it has. For each of the third-order processes that are listed in Table 5.2, identify the unique feature or features it has.

5.2.1 Most group IV crystals, including carbon in the diamond form, germanium, and silicon, and most III–V compound crystals, such as GaAs, GaAl, and InP, are cubic crystals. However, they belong to different point groups. Diamond, germanium, and silicon have the diamond structure of the $m3m$ point group, whereas GaAs, GaAl, and InP have the zinc blende structure of the $\overline{4}3m$ point group. They are all technologically important semiconductors for electronics applications, but they have very different nonlinear optical properties, and therefore limited photonics applications. Consider the applications of these crystals for the second-order nonlinear optical processes that are listed in Table 5.1 and discussed in Section 5.2.

(a) Name the processes for which the group IV cubic crystals in the bulk form are useful. Explain.

(b) Name the processes for which the III–V compound cubic crystals in the bulk form are useful. Explain.

5.2.2 The basic structure of a GaAs semiconductor laser is shown in Fig. 5.15(a), where the active waveguide core layer has junction planes perpendicular to the [001] crystal axis, which is taken to be the z axis. The laser light propagates in the [110] direction with $\hat{k} = (\hat{x} + \hat{y})/\sqrt{2}$. The field directions of the TE and TM modes of the waveguide are shown in Fig. 5.15(b). It is found that second-harmonic emission at a frequency that is twice the laser frequency can be observed at the laser facets when the laser field is a TE mode of the waveguide, but it disappears when the laser field is a TM mode of the waveguide.

(a) Explain why a TE laser mode can generate second-harmonic emission but a TM laser mode cannot.

(b) What is the polarization direction of the second-harmonic field that is generated by the TE-polarized laser mode?

(c) Explain why second-harmonic emission can be generated at the laser facets. Is there any contradiction to the discussions in Problem 5.2.1(b)?

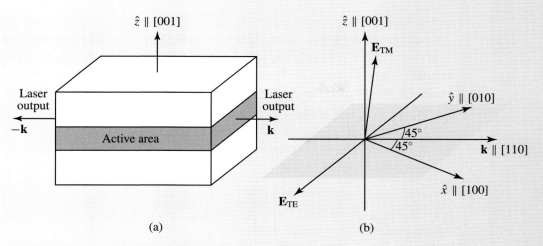

Figure 5.15 Crystal axes and field directions in a GaAs laser structure.

5.2.3 The only nonvanishing elements of the $\chi^{(2)}$ tensor for GaAs are $\chi^{(2)}_{14} = \chi^{(2)}_{25} = \chi^{(2)}_{36}$. A GaAs crystal is cleaved at the (011) surface so that the cleaved surface is normal to the [011] direction with the surface normal being $\hat{n} = (\hat{y} + \hat{z})/\sqrt{2}$. The bandgap energy of intrinsic GaAs is 1.424 eV, which gives an absorption edge at the wavelength of 870 nm. The crystal has a thickness of a few millimeters. It is placed in a nondispersive medium, such as a vacuum.

(a) A linearly polarized optical wave at $\lambda = 1$ μm is normally incident on the crystal surface with its polarization direction, $\hat{e}$, making an angle of ϕ with the [100] crystal axis, $\hat{x}$. Is it possible to generate substantial second-harmonic radiation in transmission? Why?

(b) It is possible to generate second-harmonic radiation in reflection. With the arrangement in (a), what is the dependence of the second-harmonic intensity in reflection on the polarization direction $\hat{e}$ as a function of angle ϕ? Discuss the relation between the polarization directions of the fundamental and second-harmonic fields.

(c) Another GaAs crystal is instead cleaved at the (001) surface with $\hat{n} = \hat{z}$, and the fundamental wave is incident at an angle of θ with respect to the surface normal and is TE polarized. If we rotate the crystal with respect to its normal axis $\hat{n}$ but with the incident fundamental beam fixed in space, how does the second-harmonic intensity in reflection vary with the rotation angle?

(d) In (c), what is the relation between the polarization directions of the fundamental field and the second-harmonic field in reflection? Does it vary with the rotation angle of the crystal?

(e) At a fixed rotational position of the crystal, what is the dependence of the second-harmonic intensity in reflection on the incident angle θ?

5.3.1 Most group IV crystals, including carbon in the diamond form, germanium, and silicon, and most III–V compound crystals, such as GaAs, GaAl, and InP, are cubic crystals. However, they belong to different point groups. Diamond, germanium, and silicon have the diamond structure of the $m3m$ point group, whereas GaAs, GaAl, and InP have the zinc blende structure of the $\bar{4}3m$ point group. The nonvanishing $\chi^{(3)}$ elements of these crystals have relatively large values. Consider the applications of these crystals for the third-order nonlinear optical processes that are listed in Table 5.2 and discussed in Section 5.3.

(a) Which processes can efficiently take place in a group IV cubic crystal in the bulk form? Explain.

(b) Which processes can efficiently take place in a III–V compound cubic crystal in the bulk form? Explain.

5.3.2 Show that irrespective of the difficulty for phase matching, a circularly polarized optical wave cannot generate third-harmonic radiation directly through a third-order nonlinear optical process in an isotropic medium.

5.3.3 The optical-field-induced birefringence with linearly polarized optical waves in an isotropic medium is considered in this problem.

(a) What is the third-order nonlinear polarization generated by a field that is linearly polarized along $\hat{x}$? What if the field is linearly polarized at an angle of $45°$ with respect to the x and y coordinate axes?

(b) Use the results obtained in (a) to show that $\chi^{(3)}_{1111} = \chi^{(3)}_{1122} + \chi^{(3)}_{1212} + \chi^{(3)}_{1221}$ for an isotropic medium. Show also that $\chi^{(3)}_{1111} = 3\chi^{(3)}_{1122}$ if Kleinman's symmetry condition is valid.

(c) What is the field-induced nonlinear index of refraction for a linearly polarized wave in an isotropic medium?

5.3.4 GaAs is a cubic crystal of the $\bar{4}3m$ point group, which has the isotropic linear optical property that $n_x = n_y = n_z = n$. However, it is not centrosymmetric. Therefore, its nonlinear optical properties are different from those of an isotropic material. Consider a linearly polarized optical wave at a frequency of ω that propagates along the z axis of the crystal. The optical field is polarized in a direction on the xy plane that makes an angle ϕ with respect to the x axis.

(a) The optical wave may change the index of refraction of the medium through optical-field-induced birefringence. Show that the direction of the optical field polarization is not changed by this effect only when the incident optical wave is polarized at $\phi = m\pi/4$, where $m = 0, \pm 1, \pm 2, \ldots$.

(b) What is the third-harmonic nonlinear polarization? For efficient harmonic generation, the optical frequencies involved have to be below the GaAs bandgap to avoid absorption. Is phase matching possible under this condition? Explain.

5.3.5 The third-order nonlinear susceptibility tensor $\chi^{(3)}$ of an isotropic material has 21 nonvanishing elements of the forms: $\chi^{(3)}_{1111}$, $\chi^{(3)}_{1122}$, $\chi^{(3)}_{1212}$, and $\chi^{(3)}_{1221}$, with $\chi^{(3)}_{1111} = \chi^{(3)}_{1122} + \chi^{(3)}_{1212} + \chi^{(3)}_{1221}$.

(a) Show that by applying a DC electric field it is possible to generate second-harmonic radiation of a linearly polarized fundamental optical wave in an isotropic medium.

(b) How do you apply this DC field and choose the polarization direction for the fundamental wave so that the second-harmonic field is polarized in a direction that is parallel to the fundamental field polarization?

(c) With an applied DC field, is it possible to generate the second harmonic of a circularly polarized fundamental wave? What is the polarization of this second-harmonic field if it is possible?

(d) Without the applied DC field, is it possible to generate second-harmonic emission in this medium? Explain.

5.3.6 In this problem, we consider phase-matched degenerate four-wave mixing of optical waves at an optical frequency of ω and a corresponding wavelength of λ in an isotropic medium. The arrangement involves two contra-propagating pump waves that have wavevectors of $\mathbf{k}_1$ and $\mathbf{k}_1' = -\mathbf{k}_1$, respectively, and a probe wave that has a wavevector of $\mathbf{k}_i$ for the generation of an output wave that has a wavevector of $\mathbf{k}_s = -\mathbf{k}_i$.

(a) Show that the third-order nonlinear polarization at the frequency ω is

$$\mathbf{P}_s^{(3)}(\mathbf{k}_s = -\mathbf{k}_i, \omega) = A(\mathbf{E}_1 \cdot \mathbf{E}_i^*)\mathbf{E}_1' + B(\mathbf{E}_1' \cdot \mathbf{E}_i^*)\mathbf{E}_1 + C(\mathbf{E}_1 \cdot \mathbf{E}_1')\mathbf{E}_i^*. \quad (5.57)$$

(b) If the angle between $\mathbf{k}_1$ and $\mathbf{k}_i$ is θ, what are the periods of the static gratings that are generated by the A and B terms, respectively? Express each in terms of the optical wavelength λ.

(c) If both pump waves are s polarized (normal to the plane formed by the wavevectors), what is the polarization of the output wave that is generated by an s-polarized probe wave? What is the polarization if the probe wave is p polarized (parallel to the plane formed by the wavevectors)? Indicate the contribution from each term in (5.57).

(d) Answer the questions in (c) for the situation that one pump wave is s polarized but the other is p polarized.

(e) In a gaseous medium, the static gratings may degrade with time because of atomic thermal motion and time-dependent interactions, if such gratings are created by short optical pulses. In such a situation, the degenerate four-wave mixing signal then depends on a parameter of $a = \tau/\Delta t_{ps}$, where τ is the atomic relaxation time constant and Δt_{ps} is the pulse duration. Based on this fact and the results obtained in (b)–(d), discuss an experimental approach that allows the deduction of information on the time constant τ by using variable pulse durations.

References

[1] P. A. Franken, A. E. Hill, C. W. Peters, and G. Weinreich, "Generation of optical harmonics," *Physical Review Letters*, vol. 7, pp. 118–119, 1961.

[2] J. A. Armstrong, N. Bloembergen, J. Ducuing, and P. S. Pershan, "Interactions between light waves in a nonlinear dielectric," *Physical Review*, vol. 127, pp. 1918–1939, 1962.

[3] K. Koch, E. C. Cheung, G. T. Moore, S. H. Chakmakjian, and J. M. Liu, "Hot spots in parametric fluorescence with a pump beam of finite cross section," *IEEE Journal of Quantum Electronics*, vol. 31, pp. 769–781, 1995.

[4] J. Kerr, "XL. A new relation between electricity and light: dielectrified media birefringent," *The London, Edinburgh, and Dublin Philosophical Magazine and Journal of Science*, vol. 50, pp. 337–348, 1875.

[5] J. Kerr, "LIV. A new relation between electricity and light: dielectrified media birefringent (second paper)," *The London, Edinburgh, and Dublin Philosophical Magazine and Journal of Science*, vol. 50, pp. 446–458, 1875.

6 Coupled-Wave Analysis

6.1 COUPLED-WAVE THEORY

The coupled-wave formalism [1, 2] deals with the coupling of waves of different frequencies, which takes place in nonlinear optical interactions. The coupled-mode formalism [3, 4] has to be used for the coupling between different spatial modes, such as different waveguide modes. For the coupling of waves of different frequencies in different waveguide modes, a combination of the coupled-wave and coupled-mode formalisms has to be used, which will be described in the next chapter. For simplicity, we consider in this chapter only the coupling among plane optical waves, but the formulation can be extended to nonplane waves, such as optical waves of Gaussian beam profiles.

The coupling mechanism can be generally described with a polarization $\Delta\mathbf{P}$, which causes different waves or different modes to couple, on top of the background polarization that represents the property of the optical medium in the absence of the coupling mechanism. Coupling among optical waves of different frequencies is possible only if the optical property of the medium in which the optical waves propagate is either temporally varying or optically nonlinear. Therefore, in the coupled-wave formalism, $\Delta\mathbf{P}$ can generally represent either a temporally varying modulation on the optical medium or a nonlinear interaction that takes place in the medium. By contrast, coupling between different spatial modes can be caused by a structural perturbation; in this case, $\Delta\mathbf{P}$ represents a spatial modulation that describes the structural perturbation. For the coupling of waves through a nonlinear optical interaction as discussed in this chapter, the coupling polarization is $\Delta\mathbf{P} = \mathbf{P}^{(n)}$.

In the absence of a coupling mechanism, the propagation of an optical wave in a medium is described by the linear wave equation that is given in (1.34),

$$\nabla \times \nabla \times \mathbf{E} + \mu_0 \frac{\partial^2 \mathbf{D}}{\partial t^2} = 0, \tag{6.1}$$

where $\mathbf{D}$ only accounts for the linear, static property of the medium. For a monochromatic optical wave of a constant amplitude at a frequency of ω, this equation reduces to the form:

$$\nabla \times \nabla \times \mathbf{E} - \omega^2 \mu_0 \epsilon(\mathbf{k}, \omega) \cdot \mathbf{E} = 0, \tag{6.2}$$

where $\epsilon(\mathbf{k}, \omega)$ describes the linear, time-independent optical property of the medium. Among the solutions of (6.2) are monochromatic plane waves and Gaussian waves. Here we consider only the plane waves, but the same concept discussed below applies to Gaussian waves as well.

An optical wave of a frequency ω that is governed by (6.2) propagates independently of waves of other frequencies. Therefore, optical waves of different frequencies do not couple if each of them is governed by (6.1), with **D** characterizing only the linear, static property of the medium. To describe the coupling through an nth-order nonlinear optical process, the nonlinear polarization $\mathbf{P}^{(n)}$ that characterizes the coupling mechanism is included in the wave equation as

$$\nabla \times \nabla \times \mathbf{E} + \mu_0 \frac{\partial^2 \mathbf{D}}{\partial t^2} = -\mu_0 \frac{\partial^2 \mathbf{P}^{(n)}}{\partial t^2}, \tag{6.3}$$

which is obtained from (1.44) for $\Delta \mathbf{P} = \mathbf{P}^{(n)}$. Because $\mathbf{P}^{(n)}$ couples waves of different frequencies, an optical wave of a given frequency ω that is governed by (6.3) does not propagate independently of waves of other frequencies any more. A monochromatic wave that is coupled to other frequencies cannot propagate without changing its amplitude, which includes magnitude, phase, and polarization. As the wave propagates, the nonlinear coupling to other waves results in changes in at least one of these parameters of its amplitude. Consequently, a monochromatic plane wave of a constant amplitude is not a solution of (6.3).

In most cases of interest, the condition

$$|\mathbf{P}^{(n)}| \ll |\mathbf{D}| \tag{6.4}$$

is valid; hence the wave-coupling mechanism can be considered as a perturbation on the linear, static property of the medium. Then, the total field of the waves being coupled can be expressed as a linear combination of plane waves of different frequencies, each of which has a spatially varying amplitude, as expressed in (2.24):

$$\mathbf{E}(\mathbf{r}, t) = \sum_q \mathbf{E}_q(\mathbf{r}) \exp(-\mathrm{i}\omega_q t) = \sum_q \boldsymbol{\mathcal{E}}_q(\mathbf{r}) \exp(\mathrm{i}\mathbf{k}_q \cdot \mathbf{r} - \mathrm{i}\omega_q t). \tag{6.5}$$

The field amplitude $\boldsymbol{\mathcal{E}}_q(\mathbf{r})$ is defined in (2.25) as

$$\mathbf{E}_q(\mathbf{r}) = \boldsymbol{\mathcal{E}}_q(\mathbf{r}) \exp(\mathrm{i}\mathbf{k}_q \cdot \mathbf{r}) = \hat{e}_q \mathcal{E}_q(\mathbf{r}) \exp(\mathrm{i}\mathbf{k}_q \cdot \mathbf{r}), \tag{6.6}$$

where $\hat{e}_q$ is the unit vector of the polarization of the field $\mathbf{E}_q$ with an orthonormalization relation of

$$\hat{e}_q \cdot \hat{e}_{q'}^* = \delta_{qq'}, \tag{6.7}$$

as given in (4.52). The nonlinear polarizations can be expanded in terms of a linear combination of multiple frequency components, as given in (2.26):

$$\mathbf{P}^{(n)}(\mathbf{r}, t) = \sum_q \mathbf{P}_q^{(n)}(\mathbf{r}) \exp(-\mathrm{i}\omega_q t). \tag{6.8}$$

Substitution of (6.5) and (6.8) in (6.3) yields the *coupled-wave equation*:

$$\nabla \times \nabla \times \mathbf{E}_q - \omega_q^2 \mu_0 \boldsymbol{\epsilon}(\mathbf{k}_q, \omega_q) \cdot \mathbf{E}_q = \omega_q^2 \mu_0 \mathbf{P}_q^{(n)}. \tag{6.9}$$

Note that $\mathbf{P}_q^{(n)}(\mathbf{r})$ is not generally proportional to $\mathbf{E}_q(\mathbf{r})$. It contains the fields of other frequencies to facilitate the coupling. Moreover, it does not necessarily contain a spatial phase factor of $\exp(\mathrm{i}\mathbf{k}_q \cdot \mathbf{r})$. In the special case that the spatial phase factor of $\mathbf{P}_q^{(n)}(\mathbf{r})$ is $\exp(\mathrm{i}\mathbf{k}_q \cdot \mathbf{r})$, the nonlinear interaction is most efficient and is *phase matched*.

6.1.1 Slowly Varying Amplitude Approximation

The coupled-wave equation expressed in (6.9) is a second-order differential equation. It can be reduced to a first-order differential equation by applying the *slowly varying amplitude approximation*, which assumes that the variation of the wave amplitude $\mathcal{E}_q(\mathbf{r})$ caused by the coupling to other frequencies is negligibly small over the distance of an optical wavelength. This approximation is valid in almost all situations of practical interest.

We first consider the situation in an isotropic medium for which the permittivity tensor $\boldsymbol{\epsilon}$ reduces to a scalar permittivity ϵ. In this case, $\nabla \cdot \mathbf{E}_q = 0$ so that $\nabla \times \nabla \times \mathbf{E}_q = -\nabla^2 \mathbf{E}_q$. Then, (6.9) becomes

$$\nabla^2 \mathbf{E}_q + \omega_q^2 \mu_0 \epsilon(\mathbf{k}_q, \omega_q) \mathbf{E}_q = -\omega_q^2 \mu_0 \mathbf{P}_q^{(n)}. \tag{6.10}$$

Substitution of the relation $\mathbf{E}_q = \mathcal{E}_q \exp(\mathrm{i}\mathbf{k}_q \cdot \mathbf{r})$ in (6.10), followed by applying the condition that $k_q^2 = \omega_q^2 \mu_0 \epsilon(\mathbf{k}_q, \omega_q)$, yields

$$\nabla^2 \mathcal{E}_q + 2\mathrm{i}(\mathbf{k}_q \cdot \nabla)\mathcal{E}_q = -\omega_q^2 \mu_0 \mathbf{P}_q^{(n)} \mathrm{e}^{-\mathrm{i}\mathbf{k}_q \cdot \mathbf{r}}. \tag{6.11}$$

Under the slowly varying amplitude approximation, we have

$$|\nabla^2 \mathcal{E}_q| \ll |(\mathbf{k}_q \cdot \nabla)\mathcal{E}_q|. \tag{6.12}$$

Consequently, the coupled-wave equation for wave propagation in an isotropic medium can be written as

$$(\mathbf{k}_q \cdot \nabla)\mathcal{E}_q \approx \frac{\mathrm{i}\omega_q^2 \mu_0}{2} \mathbf{P}_q^{(n)} \mathrm{e}^{-\mathrm{i}\mathbf{k}_q \cdot \mathbf{r}}. \tag{6.13}$$

In the special situation that the amplitudes of all waves being coupled vary only in a particular direction, say the z direction, we can write $\mathcal{E}_q(\mathbf{r}) = \mathcal{E}_q(z)$ even though $\mathbf{P}_q^{(n)}(\mathbf{r})$ might have variations in other directions so that it cannot necessarily be expressed as $\mathbf{P}_q^{(n)}(z)$. Then, the coupled-wave equation can be written as

$$\frac{\mathrm{d}\mathcal{E}_q(z)}{\mathrm{d}z} \approx \frac{\mathrm{i}\omega_q^2 \mu_0}{2k_{q,z}} \mathbf{P}_q^{(n)}(\mathbf{r}) \mathrm{e}^{-\mathrm{i}\mathbf{k}_q \cdot \mathbf{r}}. \tag{6.14}$$

If, furthermore, the interaction is collinear along the z direction, all participating waves have parallel or antiparallel wavevectors such that $\mathbf{k}_q = k_q \hat{z}$ for all q. In this situation, $\mathbf{P}_q^{(n)}$ can only have spatial variations along the z direction such that $\mathbf{P}_q^{(n)}(\mathbf{r}) = \mathbf{P}_q^{(n)}(z)$. Then (6.14) can be further simplified to the form:

$$\frac{d\mathcal{E}_q(z)}{dz} \approx \frac{i\omega_q^2 \mu_0}{2k_q} \mathbf{P}_q^{(n)}(z) \, e^{-ik_q z}.$$ (6.15)

For an optical wave that propagates in an anisotropic medium, $\mathbf{E}_q$ is not necessarily perpendicular to $\mathbf{k}_q$ so that, in general, $\nabla \cdot \mathbf{E}_q \neq 0$. Consequently, (6.10) and (6.13)–(6.15) are not generally valid in an anisotropic medium. In this situation, the field $\mathbf{E}_q$ that propagates in the $\mathbf{k}_q = k_q \hat{k}_q$ direction can be divided into a transverse component and a longitudinal component:

$$\mathbf{E}_q = \mathbf{E}_{q,\mathrm{T}} + \mathbf{E}_{q,\mathrm{L}},$$ (6.16)

where the transverse component is given by the relation:

$$\mathbf{E}_{q,\mathrm{T}} = \left(\hat{k}_q \times \mathbf{E}_q \right) \times \hat{k}_q,$$ (6.17)

and the longitudinal component is given by the relation:

$$\mathbf{E}_{q,\mathrm{L}} = \left(\hat{k}_q \cdot \mathbf{E}_q \right) \hat{k}_q.$$ (6.18)

Clearly, $\nabla \cdot \mathbf{E}_{q,\mathrm{T}} = 0$ but $\nabla \cdot \mathbf{E}_{q,\mathrm{L}} \neq 0$. Therefore, an equation similar to (6.10) can be written for the transverse component:

$$\nabla^2 \mathbf{E}_{q,\mathrm{T}} + \omega_q^2 \mu_0 \left[\boldsymbol{\epsilon}(\mathbf{k}_q, \omega_q) \cdot \mathbf{E}_q \right]_{\mathrm{T}} = -\omega_q^2 \mu_0 \mathbf{P}_{q,\mathrm{T}}^{(n)},$$ (6.19)

where $\mathbf{P}_{q,\mathrm{T}}^{(n)} = \left(\hat{k}_q \times \mathbf{P}_q^{(n)} \right) \times \hat{k}_q$. Note that $\mathbf{P}_{q,\mathrm{T}}^{(n)}$ can have contributions from the longitudinal field components of the interacting waves. By following the same procedure as that leading to (6.13), we find the coupled-wave equation for wave propagation in an anisotropic medium under the slowing varying amplitude approximation:

$$(\mathbf{k}_q \cdot \nabla)\mathcal{E}_{q,\mathrm{T}} \approx \frac{i\omega_q^2 \mu_0}{2} \mathbf{P}_{q,\mathrm{T}}^{(n)} \, e^{-i\mathbf{k}_q \cdot \mathbf{r}}.$$ (6.20)

6.1.2 Transverse Approximation

A nonlinear optical interaction often takes place in an anisotropic crystal, as can be expected from the fact that $\chi^{(2)}$ vanishes identically in the bulk of an isotropic, centrosymmetric medium under the electric dipole approximation. Even when a nonlinear interaction takes place in an isotropic medium, a longitudinal field component can sometimes be generated because of a field-dependent birefringence induced by a third-order nonlinear process such as the optical Kerr effect. For these reasons, under the slowly varying amplitude approximation, the correct coupled-wave equation to be used for the analysis of nonlinear optical interactions is the one given in (6.20). In the special situation that (6.13) can be reduced to (6.14) or (6.15), an equation similar to (6.14) or (6.15), but expressed in terms of the transverse field components, can be obtained from (6.20) for wave coupling in an anisotropic medium. Further simplification is possible, as discussed below.

In most cases of interest, the amplitudes of all of the interacting waves vary along the same direction, which is designated to be the z direction. Then, the coupled-wave equation can be written as

$$\frac{\mathrm{d}\mathcal{E}_{q,\mathrm{T}}(z)}{\mathrm{d}z} \approx \frac{\mathrm{i}\omega_q^2\mu_0}{2k_{q,z}}\mathbf{P}_{q,\mathrm{T}}^{(n)}(\mathbf{r})\,\mathrm{e}^{-\mathrm{i}\mathbf{k}_q\cdot\mathbf{r}}. \tag{6.21}$$

Note that the propagation direction of each wave, which is the direction normal to the wavefront and is defined by the wavevector $\mathbf{k}_q$, is not necessarily the same as the direction along which the field amplitude varies. In general, the transverse nonlinear polarization $\mathbf{P}_{q,\mathrm{T}}^{(n)}$ may also have variations along other directions.

Except in some unusual cases, the longitudinal field components of the interacting optical waves are small and unimportant, though they might exist. Then, we can make a *transverse approximation* by replacing $\mathcal{E}_{q,\mathrm{T}}$ and $\mathbf{P}_{q,\mathrm{T}}^{(n)}$ in (6.20) and (6.21) with $\mathcal{E}_q$ and $\mathbf{P}_q^{(n)}$, respectively, to further simplify the coupled-wave equation. When this simplification is done, we can multiply both sides of (6.21) by the unit vector $\hat{e}_q^*$ to write the coupled-wave equation as

$$\frac{\mathrm{d}\mathcal{E}_q}{\mathrm{d}z} \approx \frac{\mathrm{i}\omega_q^2\mu_0}{2k_{q,z}}\hat{e}_q^*\cdot\mathbf{P}_q^{(n)}\,\mathrm{e}^{-\mathrm{i}\mathbf{k}_q\cdot\mathbf{r}} = \frac{\mathrm{i}\omega_q}{2c\epsilon_0 n_{q,z}}\hat{e}_q^*\cdot\mathbf{P}_q^{(n)}\,\mathrm{e}^{-\mathrm{i}\mathbf{k}_q\cdot\mathbf{r}}, \tag{6.22}$$

where $n_{q,z} = ck_{q,z}/\omega_q$. This is the equation that is most commonly used in the coupled-wave analysis of nonlinear optical interactions. Note that it is obtained under the slowly varying amplitude approximation and the transverse approximation, which are usually very good approximations but still are not exact.

In the analysis of a nonlinear optical interaction, a coupled-wave equation is written for each of the interacting waves. The nonlinear polarization on the right-hand side of each equation couples the equations of different waves, resulting in an array of coupled nonlinear equations. It is clear from (6.22) that the nonlinear coupling of $\mathcal{E}_q$ to other waves is determined by the value and the characteristics of the factor $\hat{e}_q^*\cdot\mathbf{P}_q^{(n)}$, as we shall see below.

6.2 PARAMETRIC INTERACTIONS

To write the coupled-wave equations for a nonlinear interaction, we first identify the optical frequencies that are involved in the interaction. The next step is to write the nonlinear polarization $\mathbf{P}_q^{(n)}$ for each frequency ω_q. Then the equation for each frequency can be written according to (6.22). The equations for all participating frequencies form the complete set of coupled-wave equations. This general procedure applies to both parametric and nonparametric processes.

As we have seen in the preceding chapters, the permutations of the optical frequencies have to be accounted for in writing a nonlinear polarization. For all nonlinear interactions, including parametric and nonparametric processes, the intrinsic permutation symmetry is valid. Therefore, the terms from frequency permutations of the electric fields can be combined. For a parametric interaction, the nonlinear susceptibility involved is real so that the full permutation symmetry is also valid. Therefore, the frequency of the nonlinear polarization can be permutated with that of any participating electric field without changing the value of the nonlinear susceptibility, as we shall see below in this section. This is true only for a parametric interaction and is not true for a nonparametric interaction, as we shall see in Section 6.3.

In this section, we consider two second-order parametric interactions as examples. The general concepts and procedures apply to third-order parametric processes as well, which can have up to four frequencies. In any event, the number of coupled-wave equations that are required for a process does not depend on the order of the interaction, but on the number of distinct waves that are involved in the process. For a third-order process that involves four different frequencies, four coupled-wave equations are needed.

6.2.1 Three-Frequency Parametric Interaction

To illustrate several important concepts by using a concrete example, we first consider the coupled equations that describe a parametric second-order interaction of three different frequencies ω_1, ω_2, and ω_3 with the relation that $\omega_3 = \omega_1 + \omega_2$. We also take the approximations that allow us to use (6.22). By using (2.31) and the intrinsic permutation symmetry, or by directly using (5.7), (5.11), and (5.12), we find that

$$\hat{e}_3^* \cdot \mathbf{P}_3^{(2)} = 2\epsilon_0 \hat{e}_3^* \cdot \boldsymbol{\chi}^{(2)}(\omega_3 = \omega_1 + \omega_2) : \hat{e}_1 \hat{e}_2 \mathcal{E}_1 \mathcal{E}_2 \, \mathrm{e}^{\mathrm{i}(\mathbf{k}_1 + \mathbf{k}_2)\cdot\mathbf{r}}, \tag{6.23}$$

$$\hat{e}_1^* \cdot \mathbf{P}_1^{(2)} = 2\epsilon_0 \hat{e}_1^* \cdot \boldsymbol{\chi}^{(2)}(\omega_1 = \omega_3 - \omega_2) : \hat{e}_3 \hat{e}_2^* \mathcal{E}_3 \mathcal{E}_2^* \, \mathrm{e}^{\mathrm{i}(\mathbf{k}_3 - \mathbf{k}_2)\cdot\mathbf{r}}, \tag{6.24}$$

$$\hat{e}_2^* \cdot \mathbf{P}_2^{(2)} = 2\epsilon_0 \hat{e}_2^* \cdot \boldsymbol{\chi}^{(2)}(\omega_2 = \omega_3 - \omega_1) : \hat{e}_3 \hat{e}_1^* \mathcal{E}_3 \mathcal{E}_1^* \, \mathrm{e}^{\mathrm{i}(\mathbf{k}_3 - \mathbf{k}_1)\cdot\mathbf{r}}. \tag{6.25}$$

The full permutation symmetry is valid for the real $\boldsymbol{\chi}^{(2)}$ that characterizes the parametric process. Therefore, we can define an *effective nonlinear susceptibility* as

$$\begin{aligned}
\chi_{\mathrm{eff}} &= \hat{e}_3^* \cdot \boldsymbol{\chi}^{(2)}(\omega_3 = \omega_1 + \omega_2) : \hat{e}_1 \hat{e}_2 \\
&= \hat{e}_1 \cdot \boldsymbol{\chi}^{(2)*}(\omega_1 = \omega_3 - \omega_2) : \hat{e}_3^* \hat{e}_2 = \hat{e}_1 \cdot \boldsymbol{\chi}^{(2)}(\omega_1 = \omega_3 - \omega_2) : \hat{e}_3^* \hat{e}_2 \\
&= \hat{e}_2 \cdot \boldsymbol{\chi}^{(2)*}(\omega_2 = \omega_3 - \omega_1) : \hat{e}_3^* \hat{e}_1 = \hat{e}_2 \cdot \boldsymbol{\chi}^{(2)}(\omega_2 = \omega_3 - \omega_1) : \hat{e}_3^* \hat{e}_1.
\end{aligned} \tag{6.26}$$

Note that $\boldsymbol{\chi}^{(2)*} = \boldsymbol{\chi}^{(2)}$ in (6.26) because $\boldsymbol{\chi}^{(2)}$ is real. Note also that though $\boldsymbol{\chi}^{(2)}$ is real, χ_{eff} might be complex because the unit polarization vectors $\hat{e}_q$ of the optical fields can be complex, such as that for a circularly or elliptically polarized field. In terms of the d coefficient, the effective d coefficient for this interaction is simply $d_{\mathrm{eff}} = \chi_{\mathrm{eff}}/2$ by following the relation given in (3.31).

We then have the coupled equations for a parametric second-order interaction [1]:

$$\frac{\mathrm{d}\mathcal{E}_3}{\mathrm{d}z} = \frac{\mathrm{i}\omega_3^2}{c^2 k_{3,z}} \chi_{\mathrm{eff}} \mathcal{E}_1 \mathcal{E}_2 \, \mathrm{e}^{\mathrm{i}\Delta k z}, \tag{6.27}$$

$$\frac{\mathrm{d}\mathcal{E}_1}{\mathrm{d}z} = \frac{\mathrm{i}\omega_1^2}{c^2 k_{1,z}} \chi_{\mathrm{eff}}^* \mathcal{E}_3 \mathcal{E}_2^* \, \mathrm{e}^{-\mathrm{i}\Delta k z}, \tag{6.28}$$

$$\frac{d\mathcal{E}_2}{dz} = \frac{i\omega_2^2}{c^2 k_{2,z}} \chi_{\text{eff}}^* \mathcal{E}_3 \mathcal{E}_1^* e^{-i\Delta kz}, \tag{6.29}$$

where

$$\Delta \mathbf{k} = \mathbf{k}_1 + \mathbf{k}_2 - \mathbf{k}_3 = \Delta k \hat{z} \tag{6.30}$$

is the *phase mismatch*. If all three waves are linearly polarized, we have $\hat{e}_q^* = \hat{e}_q$, and thus $\chi_{\text{eff}}^* = \chi_{\text{eff}}$ is a real quantity. Otherwise, χ_{eff} can be complex, as stated above.

EXAMPLE 6.1

Take $\hat{x}$, $\hat{y}$, and $\hat{z}$ to be the principal axes of LiNbO$_3$. Find the effective nonlinear susceptibilities for second-harmonic generation in LiNbO$_3$ with (a) a linearly x-polarized fundamental wave that propagates in the z direction and (b) a linearly z-polarized fundamental wave that propagates in the x direction. The propagation direction of the second-harmonic wave is in practice determined by many factors. Here we make it the same as that of the fundamental wave.

Solution

For second-harmonic generation, we have the degenerate case of $\omega_1 = \omega_2 = \omega$ and $\omega_3 = 2\omega$. From Table 3.4, we find that the only nonvanishing second-order nonlinear susceptibility elements of LiNbO$_3$ are $d_{31} = d_{32} = d_{24} = d_{15} = -4.4$ pm V^{-1}, $d_{22} = -d_{21} = -d_{16} = 2.4$ pm V^{-1}, and $d_{33} = -25.2$ pm V^{-1}. We also know that $\chi_{ia}^{(2)} = 2d_{ia}$, according to (3.31).

In case (a), we have $\mathbf{k}_{2\omega} \| \mathbf{k}_\omega \| \hat{z}$ and $\hat{e}_\omega = \hat{x}$. We then find from the nonvanishing elements of $\chi^{(2)}$ for LiNbO$_3$ that $\mathbf{P}^{(2)}(2\omega)$ has only two components, $P_y^{(2)}$ and $P_z^{(2)}$, in the y and z directions, which are contributed by $\chi_{21}^{(2)}$ and $\chi_{31}^{(2)}$, respectively. However, because $\mathbf{k}_{2\omega}$ is forced to be in the z direction, $P_z^{(2)}$ is a longitudinal component that does not contribute to the propagation of the second-harmonic wave. In this situation, $\hat{e}_{2\omega} = \hat{y}$ because only the transverse component $P_y^{(2)}$ is useful for generating the second-harmonic wave. Therefore,

$$\chi_{\text{eff}} = \hat{e}_{2\omega}^* \cdot \chi^{(2)} : \hat{e}_\omega \hat{e}_\omega = \hat{y} \cdot \chi^{(2)} : \hat{x}\hat{x} = \chi_{21}^{(2)} = 2d_{21} = -4.8 \text{ pm V}^{-1}.$$

In case (b), we have $\mathbf{k}_{2\omega} \| \mathbf{k}_\omega \| \hat{x}$ and $\hat{e}_\omega = \hat{z}$. We find from the nonvanishing elements of $\chi^{(2)}$ for LiNbO$_3$ that $\mathbf{P}^{(2)}(2\omega)$ has only one component, $P_z^{(2)}$, in the z direction, which is contributed by $\chi_{33}^{(2)}$. Therefore, $\hat{e}_{2\omega} = \hat{z}$ and

$$\chi_{\text{eff}} = \hat{e}_{2\omega}^* \cdot \chi^{(2)} : \hat{e}_\omega \hat{e}_\omega = \hat{z} \cdot \chi^{(2)} : \hat{z}\hat{z} = \chi_{33}^{(2)} = 2d_{33} = -50.4 \text{ pm V}^{-1}.$$

We see from this example that the value of χ_{eff} can vary significantly depending on the polarization directions of the interacting waves, which in turn are constrained by the propagation directions of the interacting waves.

Note that though $\mathbf{k}_1$, $\mathbf{k}_2$, and $\mathbf{k}_3$ individually might not be parallel to $\hat{z}$, the phase mismatch $\Delta \mathbf{k}$ has to be parallel to $\hat{z}$ if the field amplitudes are to vary only along the z direction. This fact is mathematically required in (6.27)–(6.29) because $\mathcal{E}_1$, $\mathcal{E}_2$, and $\mathcal{E}_3$ are all functions of z only. Physically, the boundary conditions, which are dictated by Maxwell's equations, at the surface of a nonlinear crystal where the input waves enter the crystal require that the tangential component, but not the normal component, of $\mathbf{k}_1 + \mathbf{k}_2$ be equal to that of $\mathbf{k}_3$ for an interaction that is defined by the relation $\omega_3 = \omega_1 + \omega_2$. Therefore, any phase mismatch Δk occurs only in the direction that is normal to the input surface of the nonlinear crystal, as illustrated in Fig. 6.1(a). This condition can always be satisfied because only one or two of the interacting waves are provided at the input and only their wavevectors are initially given. For example, in sum-frequency generation, $\mathbf{k}_1$ and $\mathbf{k}_2$ are determined by the propagation directions of the input waves at the frequencies of ω_1 and ω_2, respectively. The propagation direction, $\mathbf{k}_3$, of the generated sum-frequency wave is then determined by two conditions:

1. Its magnitude, $k_3 = n_3\omega_3/c$, is determined by the dispersion and the birefringence of the nonlinear crystal.
2. Its projection on the crystal surface has to be equal to the projection of $\mathbf{k}_1 + \mathbf{k}_2$ on the crystal surface, as Fig. 6.1(a) shows.

Because $\Delta\mathbf{k} = \Delta k\hat{z}$, the z direction, along which the field amplitudes vary, is normal to the input surface of the nonlinear crystal. Figure 6.1(b) shows the fact that though the wavevector of each wave may not line up with $\hat{z}$, the magnitude of each wave varies only along the z direction.

According to (4.10), the intensity of a wave at a frequency of ω_q, projected on the plane of constant amplitude that is normal to the z direction, is given by the relation:

$$I_q = \left|\overline{\mathbf{S}}_q \cdot \hat{z}\right| \approx \frac{2k_{q,z}}{\omega_q \mu_0}|\mathcal{E}_q|^2 = 2c\epsilon_0 n_{q,z}|\mathcal{E}_q|^2, \tag{6.31}$$

where $\mathbf{S}_q$ is the Poynting vector defined in (4.3), $n_{q,z} = ck_{q,z}/\omega_q = n_q\cos\theta_q$, and θ_q is the angle between $\mathbf{k}_q$ and $\hat{z}$. In a birefringent crystal, a possible spatial walk-off between the vectors $\mathbf{S}_q$ and $\mathbf{k}_q$ is neglected by taking the approximation in (6.31). By using (6.31), we find that (6.27)–(6.29) lead to

$$\frac{dI_3}{dz} = -\frac{2\omega_3|\chi_{\text{eff}}|}{(2c^3\epsilon_0 n_{3,z}n_{1,z}n_{2,z})^{1/2}}I_3^{1/2}I_1^{1/2}I_2^{1/2}\sin\varphi, \tag{6.32}$$

$$\frac{dI_1}{dz} = \frac{2\omega_1|\chi_{\text{eff}}|}{(2c^3\epsilon_0 n_{3,z}n_{1,z}n_{2,z})^{1/2}}I_3^{1/2}I_1^{1/2}I_2^{1/2}\sin\varphi, \tag{6.33}$$

$$\frac{dI_2}{dz} = \frac{2\omega_2|\chi_{\text{eff}}|}{(2c^3\epsilon_0 n_{3,z}n_{1,z}n_{2,z})^{1/2}}I_3^{1/2}I_1^{1/2}I_2^{1/2}\sin\varphi, \tag{6.34}$$

where

$$\varphi = \varphi_\chi + \varphi_1 + \varphi_2 - \varphi_3 + \Delta kz; \tag{6.35}$$

φ_χ is the phase of χ_{eff}, defined as $\chi_{\text{eff}} = |\chi_{\text{eff}}|e^{i\varphi_\chi}$; and φ_1, φ_2, and φ_3 are the phases of $\mathcal{E}_1$, $\mathcal{E}_2$, and $\mathcal{E}_3$, respectively, defined as $\mathcal{E}_q = |\mathcal{E}_q|e^{i\varphi_q}$.

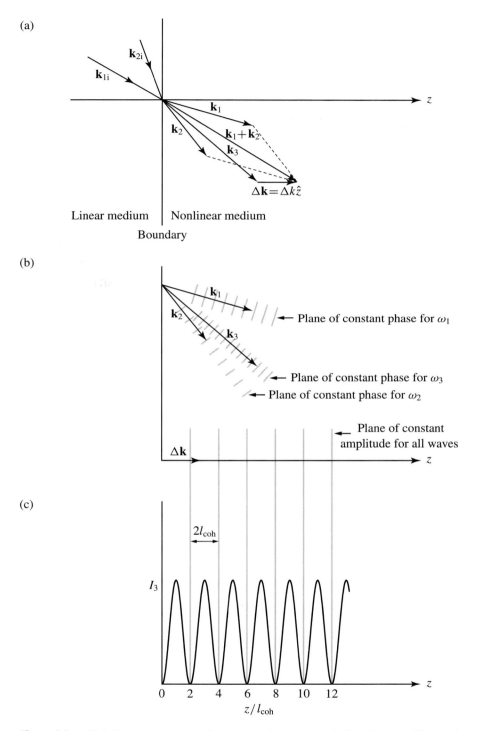

Figure 6.1 (a) Relations of wavevectors. In a parametric interaction, the boundary conditions at the input surface of a nonlinear crystal require that the phase mismatch $\Delta\mathbf{k}$, as well as the z direction, along which the field amplitudes vary, be normal to the input surface. (b) Wavefronts and the plane of constant field amplitude. The wavefront, defined by the plane of constant phase, of each wave is normal to its wavevector, but the plane of constant field amplitude is parallel to the input surface and is normal to $\hat{z}$. (c) Periodic intensity variation of a nonlinearly generated wave in the presence of a phase mismatch. The intensity varies periodically in the z direction.

By taking $n = 2$ for the second-order nonlinear interaction, we find from (4.14) that the time-averaged power that is expended by $\mathbf{E}(\omega_3)$ on $\mathbf{P}^{(2)}(\omega_3)$ is $\overline{W}_3^{(2)}(\mathbf{r}) = 2\omega_3|\mathbf{E}_3^*(\mathbf{r}) \cdot \mathbf{P}_3^{(3)}(\mathbf{r})|$ $\sin\Delta\varphi_3(\mathbf{r})$, where $\Delta\varphi_3(\mathbf{r}) = \varphi_{\mathbf{P}_3^{(3)}(\mathbf{r})} - \varphi_{\mathbf{E}_3(\mathbf{r})}$. Similarly, the power expended by $\mathbf{E}(\omega_1)$ on $\mathbf{P}^{(2)}(\omega_1)$ is $\overline{W}_1^{(2)}(\mathbf{r}) = 2\omega_1\left|\mathbf{E}_1^*(\mathbf{r}) \cdot \mathbf{P}_1^{(3)}(\mathbf{r})\right|\sin\Delta\varphi_1(\mathbf{r})$ with $\Delta\varphi_1(\mathbf{r}) = \varphi_{\mathbf{P}_1^{(3)}(\mathbf{r})} - \varphi_{\mathbf{E}_1(\mathbf{r})}$, and that by $\mathbf{E}(\omega_2)$ on $\mathbf{P}^{(2)}(\omega_2)$ is $\overline{W}_2^{(2)}(\mathbf{r}) = 2\omega_2\left|\mathbf{E}_2^*(\mathbf{r}) \cdot \mathbf{P}_2^{(3)}(\mathbf{r})\right|\sin\Delta\varphi_2(\mathbf{r})$ with $\Delta\varphi_2(\mathbf{r}) = \varphi_{\mathbf{P}_2^{(3)}(\mathbf{r})} - \varphi_{\mathbf{E}_2(\mathbf{r})}$. By using (5.7) and (6.23), it can be shown that $\varphi = \Delta\varphi_3(\mathbf{r}) = -\Delta\varphi_1(\mathbf{r}) = -\Delta\varphi_2(\mathbf{r})$, where φ is that defined in (6.35). The direction of energy flow in the interaction depends on the sign of $\sin\varphi$:

1. In the case that $\sin\varphi > 0$, we find that $\overline{W}_3^{(2)}(\mathbf{r}) > 0$, $\overline{W}_1^{(2)}(\mathbf{r}) < 0$, and $\overline{W}_2^{(2)}(\mathbf{r}) < 0$. Then, the optical energy in the field $\mathbf{E}(\omega_3)$ of the frequency ω_3 is expended to generate a nonlinear polarization $\mathbf{P}^{(2)}(\omega_3)$ because $\overline{W}_3^{(2)}(\mathbf{r}) > 0$. Because there is no energy exchange with the medium in this parametric interaction, the energy of $\mathbf{P}^{(2)}(\omega_3)$ is coupled to $\mathbf{P}^{(2)}(\omega_1)$ and $\mathbf{P}^{(2)}(\omega_2)$. The energies of $\mathbf{P}^{(2)}(\omega_1)$ and $\mathbf{P}^{(2)}(\omega_2)$ are subsequently transferred to $\mathbf{E}(\omega_1)$ and $\mathbf{E}(\omega_2)$, respectively, because $\overline{W}_1^{(2)}(\mathbf{r}) < 0$, and $\overline{W}_2^{(2)}(\mathbf{r}) < 0$.
2. In the case that $\sin\varphi < 0$, we find that $\overline{W}_3^{(2)}(\mathbf{r}) < 0$, $\overline{W}_1^{(2)}(\mathbf{r}) > 0$, and $\overline{W}_2^{(2)}(\mathbf{r}) > 0$. Then, the optical energies in $\mathbf{E}(\omega_1)$ and $\mathbf{E}(\omega_2)$ are expended to generate the nonlinear polarizations $\mathbf{P}^{(2)}(\omega_1)$ and $\mathbf{P}^{(2)}(\omega_2)$, respectively, because $\overline{W}_1^{(2)}(\mathbf{r}) > 0$ and $\overline{W}_2^{(2)}(\mathbf{r}) > 0$. The energies of $\mathbf{P}^{(2)}(\omega_1)$ and $\mathbf{P}^{(2)}(\omega_2)$ are coupled to $\mathbf{P}^{(2)}(\omega_3)$, which subsequently transfers its energy to $\mathbf{E}(\omega_3)$ because $\overline{W}_3^{(2)}(\mathbf{r}) < 0$.

The total intensity of the three interacting waves is $I = I_1 + I_2 + I_3$. By using the relation $\omega_3 = \omega_1 + \omega_2$, we find from (6.32)–(6.34) that

$$\frac{\mathrm{d}I}{\mathrm{d}z} = \frac{\mathrm{d}(I_1 + I_2 + I_3)}{\mathrm{d}z} = 0. \tag{6.36}$$

Consequently, the total optical energy is conserved in a parametric process, as is expected. In addition, we also find that

$$\frac{\mathrm{d}}{\mathrm{d}z}\left(\frac{I_1}{\omega_1}\right) = \frac{\mathrm{d}}{\mathrm{d}z}\left(\frac{I_2}{\omega_2}\right) = -\frac{\mathrm{d}}{\mathrm{d}z}\left(\frac{I_3}{\omega_3}\right). \tag{6.37}$$

Therefore, every time a photon at the frequency of ω_3 is annihilated, one photon at the frequency of ω_1 and another at the frequency of ω_2 are simultaneously generated, and vice versa. The relations in (6.36) and (6.37) are known as the *Manley–Rowe relations*.

The coupled equations and the Manley–Rowe relations formulated above apply to all parametric second-order interactions that involve three different optical frequencies. We see from (6.32)–(6.34) that optical energy is converted from ω_3 to ω_1 and ω_2 if $\sin\varphi > 0$, whereas it is converted from ω_1 and ω_2 to ω_3 if $\sin\varphi < 0$. If φ_χ, φ_1, φ_2, and φ_3 are fixed or vary slowly with z, as is normally the case, then $\sin\varphi$ changes sign periodically with z because of the phase mismatch Δk. This periodic change of sign in $\sin\varphi$ results in periodic reversal of a parametric process.

Therefore, in the presence of a phase mismatch, the maximum interaction length that a given frequency-conversion process can take without a reversal of the process is the *coherence length*:

$$l_{\text{coh}} = \frac{\pi}{|\Delta k|}. \tag{6.38}$$

From the above discussions and an examination of (6.32)–(6.34), we can see that the intensity of a wave that is generated by a parametric nonlinear process in the presence of a phase mismatch varies periodically *along the direction normal to the input surface of the nonlinear crystal* with a half period of l_{coh}, as illustrated in Fig. 6.1(c). The intensities of the other waves in the interaction also vary with the same period along this direction. Therefore, an interaction length larger than l_{coh} is not useful and can even be detrimental. Clearly, phase matching is very important for an efficient parametric interaction.

EXAMPLE 6.2

With a fundamental wave at $\lambda = 1.064$ μm, find the coherence length for each of the two cases of second-harmonic generation in LiNbO$_3$ discussed in Example 6.1. At room temperature, LiNbO$_3$ has $n_o = 2.2340$ and $n_e = 2.1554$ at the 1.064 μm wavelength and $n_o = 2.3251$ and $n_e = 2.2330$ at the 532 nm wavelength.

Solution

Because $\mathbf{k}_{2\omega} \| \mathbf{k}_\omega$, we have $\Delta k = 2k_\omega - k_{2\omega} = 4\pi (n_\omega - n_{2\omega})/\lambda$. In case (a), we have $n_\omega = n_o(\omega) = 2.2340$ and $n_{2\omega} = n_o(2\omega) = 2.3251$ because $\hat{e}_\omega = \hat{x}$ and $\hat{e}_{2\omega} = \hat{y}$. Then,

$$l_{\text{coh}} = \frac{\pi}{|\Delta k|} = \frac{\lambda}{4 \times \left| n_\omega^o - n_{2\omega}^o \right|} = \frac{1.064}{4 \times 0.0911} \text{ μm} = 2.92 \text{ μm}.$$

In case (b), we have $n_\omega = n_e(\omega) = 2.1554$ and $n_{2\omega} = n_e(2\omega) = 2.2330$ because $\hat{e}_\omega = \hat{z}$ and $\hat{e}_{2\omega} = \hat{z}$. Then,

$$l_{\text{coh}} = \frac{\pi}{|\Delta k|} = \frac{\lambda}{4 \times \left| n_\omega^e - n_{2\omega}^e \right|} = \frac{1.064}{4 \times 0.0776} \text{ μm} = 3.43 \text{ μm}.$$

We see from this example that the coherence length is very small for both cases. Clearly, the interaction would not be efficient. The reason for this undesirable situation is that we have arbitrarily chosen in Example 6.1 some convenient directions of propagation and polarization for the optical waves that are involved in the second-harmonic generation process without any consideration of the requirement for phase matching. These two examples together illustrate that it is possible to obtain a decent value of $|\chi_{\text{eff}}|$ while the interaction is still very inefficient because of phase mismatch. The methods for phase matching will be discussed in Chapter 9. Phase-matched second-harmonic generation processes in LiNbO$_3$ with properly chosen propagation and polarization directions of the optical waves will be demonstrated in Examples 9.1 and 9.3.

For a parametric interaction among linearly polarized waves in a homogeneous bulk crystal, $\varphi_\chi = 0$ or π, depending on the sign of χ_{eff}. Then, $\varphi = \varphi_1 + \varphi_2 - \varphi_3$ or $\varphi = \pi + \varphi_1 + \varphi_2 - \varphi_3$ in the case of *perfect phase matching* for $\Delta k = 0$. In this situation, it is possible for a frequency-conversion process to continue over the entire length of a crystal. Which parametric process occurs is determined completely by the value of φ. For sum-frequency generation, we need $\varphi = -\pi/2$ so that optical energy is converted most efficiently in the direction $\omega_1 + \omega_2 \to \omega_3$. For difference-frequency generation, optical parametric amplification, or optical parametric generation, $\varphi = \pi/2$ is needed to have the highest efficiency for the conversion of optical energy in the direction $\omega_3 \to \omega_1 + \omega_2$.

In real experimental settings, a desired process is controlled by the input conditions. Normally only one or two waves in a parametric three-wave interaction are supplied at the input; therefore, only one or two phases are set, and at least one phase is arbitrary. Consider the situation that χ_{eff} is real and positive so that $\varphi_\chi = 0$. If the input waves are at the frequencies of ω_1 and ω_2 and the phase-matching condition $\mathbf{k}_3 = \mathbf{k}_1 + \mathbf{k}_2$ is satisfied, sum-frequency generation occurs with the generation of a wave at the frequency of ω_3 that automatically picks a phase of $\varphi_3 = \varphi_1 + \varphi_2 + \pi/2$. If the same phase-matching condition is satisfied but the input waves are at the frequencies of ω_3 and ω_2, a wave at the frequency of ω_1 is generated with a phase of $\varphi_1 = \varphi_3 - \varphi_2 + \pi/2$, resulting in difference-frequency generation, or optical parametric amplification in the case that the amplification of the signal at the frequency of ω_2 is the objective. If only a wave at the frequency of ω_3 is supplied at the input, optical parametric generation occurs with $\varphi_1 + \varphi_2 = \varphi_3 + \pi/2$. In this situation, the values of ω_1 and ω_2 are determined by the phase-matching condition subject to the condition that $\omega_1 + \omega_2 = \omega_3$.

An interesting question is whether it is possible for other parametric processes, such as sum-frequency generation for the frequency of $\omega_1 + \omega_3$ and difference-frequency generation for the frequency of $\omega_1 - \omega_2$, and so on, to occur once all three waves at ω_1, ω_2, and ω_3 exist in a crystal, say, through a sum-frequency generation process of $\omega_1 + \omega_2 \to \omega_3$. From the above discussions, it is clear that any parametric process can occur if all three conditions are satisfied:

1. The nonlinear interaction has a nonvanishing χ_{eff}.
2. The process is phase matched.
3. The interaction has the correct initial value of the phase φ.

It is thus possible to have simultaneous multiple parametric processes if all of them satisfy the required conditions. In normal situations, however, it is highly unusual for two or more different processes to occur in a single experimental arrangement because of the difficulty of satisfying their respective phase-matching conditions all at once.

6.2.2   Second-Harmonic Generation

For second-harmonic generation, there are only two distinct frequencies, the fundamental frequency of ω and the second-harmonic frequency of 2ω, in the interaction. The nonlinear polarizations for the second-harmonic frequency and the fundamental frequency are given in (5.4) and (5.5), respectively. By using (5.4) and (5.5), we have

$$\hat{e}_{2\omega}^* \cdot \mathbf{P}_{2\omega}^{(2)} = \epsilon_0 \hat{e}_{2\omega}^* \cdot \boldsymbol{\chi}^{(2)} (2\omega = \omega + \omega) : \hat{e}_\omega \hat{e}_\omega \mathcal{E}_\omega \mathcal{E}_\omega \, e^{i2\mathbf{k}_\omega \cdot \mathbf{r}}, \tag{6.39}$$

$$\hat{e}_\omega^* \cdot \mathbf{P}_\omega^{(2)} = 2\epsilon_0 \hat{e}_\omega^* \cdot \boldsymbol{\chi}^{(2)} (\omega = 2\omega - \omega) : \hat{e}_{2\omega} \hat{e}_\omega^* \mathcal{E}_{2\omega} \mathcal{E}_\omega^* \, e^{i(\mathbf{k}_{2\omega} - \mathbf{k}_\omega) \cdot \mathbf{r}}. \tag{6.40}$$

The full permutation symmetry is valid for the real $\chi^{(2)}$ that characterizes the parametric process of second-harmonic generation so that we can define an effective nonlinear susceptibility as

$$\chi_{\text{eff}} = \hat{e}_{2\omega}^* \cdot \boldsymbol{\chi}^{(2)} (2\omega = \omega + \omega) : \hat{e}_\omega \hat{e}_\omega = \hat{e}_\omega \cdot \boldsymbol{\chi}^{(2)} (\omega = 2\omega - \omega) : \hat{e}_{2\omega}^* \hat{e}_\omega, \tag{6.41}$$

where the relation that $\chi^{(2)*} = \chi^{(2)}$ is used because $\chi^{(2)}$ is real. Again, as mentioned above for (6.26), though $\chi^{(2)}$ is real, χ_{eff} might be complex because the unit vectors $\hat{e}_q$ of the polarizations of the interacting fields can be complex, such as that for a circularly or elliptically polarized field.

The two coupled equations for second-harmonic generation are

$$\frac{d\mathcal{E}_{2\omega}}{dz} = \frac{i(2\omega)^2}{2c^2 k_{2\omega,z}} \chi_{\text{eff}} \mathcal{E}_\omega^2 \, e^{i\Delta kz} = \frac{i\omega}{cn_{2\omega,z}} \chi_{\text{eff}} \mathcal{E}_\omega^2 \, e^{i\Delta kz}, \tag{6.42}$$

$$\frac{d\mathcal{E}_\omega}{dz} = \frac{i\omega^2}{c^2 k_{\omega,z}} \chi_{\text{eff}}^* \mathcal{E}_{2\omega} \mathcal{E}_\omega^* \, e^{-i\Delta kz} = \frac{i\omega}{cn_{\omega,z}} \chi_{\text{eff}}^* \mathcal{E}_{2\omega} \mathcal{E}_\omega^* \, e^{-i\Delta kz}, \tag{6.43}$$

where

$$\Delta\mathbf{k} = 2\mathbf{k}_\omega - \mathbf{k}_{2\omega} = \Delta k \hat{z} \tag{6.44}$$

is the phase mismatch for second-harmonic generation. If both the fundamental and the second-harmonic waves are linearly polarized, we have $\hat{e}_\omega^* = \hat{e}_\omega$ and $\hat{e}_{2\omega}^* = \hat{e}_{2\omega}$, and thus $\chi_{\text{eff}}^* = \chi_{\text{eff}}$ is a real quantity. Otherwise, χ_{eff} can be complex, as stated above.

By using (6.31), the coupled-wave equations given in (6.42) and (6.43) lead to the equations in terms of the intensities of the fundamental and second-harmonic waves:

$$\frac{dI_{2\omega}}{dz} = -\frac{dI_\omega}{dz} = -\frac{2\omega |\chi_{\text{eff}}|}{(2c^3 \epsilon_0 n_{\omega,z}^2 n_{2\omega,z})^{1/2}} I_\omega I_{2\omega}^{1/2} \sin\varphi, \tag{6.45}$$

where

$$\varphi = \varphi_\chi + 2\varphi_\omega - \varphi_{2\omega} + \Delta kz. \tag{6.46}$$

From (6.45), we find the Manley–Rowe relations for second-harmonic generation:

$$\frac{dI}{dz} = \frac{d(I_\omega + I_{2\omega})}{dz} = 0 \tag{6.47}$$

and

$$\frac{d}{dz}\left(\frac{I_\omega}{\omega}\right) = -2\frac{d}{dz}\left(\frac{I_{2\omega}}{2\omega}\right). \tag{6.48}$$

Two photons at the fundamental frequency are annihilated to create one photon at the second-harmonic frequency.

The effects of the phase φ and the phase mismatch Δk on second-harmonic generation are similar to those discussed in the preceding subsection for a three-frequency parametric process. When $\varphi = -\pi/2$, optical energy is most efficiently converted from the fundamental frequency to the second-harmonic frequency in the direction $\omega + \omega \to 2\omega$. When $\varphi = \pi/2$, optical energy is most efficiently converted from the second-harmonic frequency to the fundamental frequency in the direction $2\omega \to \omega + \omega$, which is basically a degenerate parametric generation process.

For second-harmonic generation using linearly polarized waves in a homogeneous bulk crystal, $\varphi_\chi = 0$ or π, depending on the sign of χ_{eff}. Then, in the case of *perfect phase matching* for $\Delta k = 0$, $\varphi = \varphi_1 + \varphi_2 - \varphi_3$ when $\varphi_\chi = 0$, or $\varphi = \pi + \varphi_1 + \varphi_2 - \varphi_3$ when $\varphi_\chi = \pi$. In this situation, it is possible for the second-harmonic generation process to continue over the entire length of a crystal if the phase is fixed at $\varphi = -\pi/2$ so that optical energy is converted most efficiently throughout the crystal in the direction $\omega + \omega \to 2\omega$. When perfect phase matching is not accomplished such that $\Delta k \neq 0$, the phase φ changes sign periodically so that the second-harmonic intensity varies periodically with a half period of the coherence length as given in (6.38) and shown in Fig. 6.1(c).

6.3 NONPARAMETRIC INTERACTIONS

When writing the coupled-wave equations for any nonlinear process, it is important to first clearly understand the properties of the nonlinear susceptibility that characterizes the process under consideration. For a parametric process, the full permutation symmetry is valid for the susceptibility; this fact is used in defining the effective susceptibility that is given in (6.26) for a second-order parametric process. In general, the susceptibility for a nonparametric process does not have the full permutation symmetry because its imaginary part is significant for the process.

The susceptibilities for different nonparametric processes generally have different permutation properties because they are related to different resonant transitions in the material. For example, the susceptibility, $\chi^{(3)}(\omega_S = \omega_S + \omega_p - \omega_p)$, for the Stokes process of stimulated Raman scattering and the susceptibility, $\chi^{(3)}(\omega_1 = \omega_1 + \omega_2 - \omega_2)$, for two-photon absorption look the same, but they have different microscopic forms and thus very different properties because the former is resonant at $\omega_p - \omega_S$, whereas the latter is resonant at $\omega_1 + \omega_2$, as discussed in Subsections 5.3.10 and 5.3.9, respectively. If ω_p, ω_S, and $\omega_p + \omega_S$ are all far from any resonant transition frequencies, while $\omega_p - \omega_S$ is in resonance with a transition in the material, the Raman susceptibility has the permutation property:

$$
\begin{aligned}
\chi_{ijkl}^{(3)}(\omega_S = \omega_S + \omega_p - \omega_p) &= \chi_{klij}^{(3)*}(\omega_p = \omega_p + \omega_S - \omega_S) \\
&= \chi_{jilk}^{(3)}(\omega_S = \omega_S + \omega_p - \omega_p) \\
&= \chi_{lkji}^{(3)*}(\omega_p = \omega_p + \omega_S - \omega_S).
\end{aligned}
\tag{6.49}
$$

By contrast, the susceptibility for two-photon absorption has the permutation property:

$$\chi_{ijkl}^{(3)}(\omega_1 = \omega_1 + \omega_2 - \omega_2) = \chi_{klij}^{(3)}(\omega_2 = \omega_2 + \omega_1 - \omega_1)$$
$$= \chi_{jilk}^{(3)}(\omega_1 = \omega_1 + \omega_2 - \omega_2) \tag{6.50}$$
$$= \chi_{lkji}^{(3)}(\omega_2 = \omega_2 + \omega_1 - \omega_1)$$

if ω_1, ω_2, and $|\omega_1 - \omega_2|$ are all far from any resonant transition frequencies while $\omega_1 + \omega_2$ is in resonance with a transition. The difference between the relations in (6.49) and (6.50) is significant because the imaginary parts of these susceptibilities are responsible for the nonlinear processes under consideration.

As we shall see from the following discussions, the characteristics of a nonparametric process are completely determined by the state of the material. Phase matching for the interacting optical waves is automatically satisfied in a nonparametric process because the material can absorb any difference in the momenta of the interacting photons if there is energy exchange between the optical fields and the medium. For the same reason, the phase relationship among the interacting waves, which determines the direction of frequency conversion in a parametric process, plays no role in a nonparametric process.

6.3.1 Stimulated Raman Scattering

We first consider stimulated Raman scattering as an example of a nonparametric process. The coupled-wave equations for the process of stimulated Raman scattering are considered. By using (5.55) and (5.56), we can write

$$\hat{e}_S^* \cdot \mathbf{P}_S^{(3)} = 6\epsilon_0 \hat{e}_S^* \cdot \boldsymbol{\chi}^{(3)}(\omega_S = \omega_S + \omega_p - \omega_p) \vdots \hat{e}_S \hat{e}_p \hat{e}_p^* \mathcal{E}_S |\mathcal{E}_p|^2 e^{i\mathbf{k}_S \cdot \mathbf{r}}, \tag{6.51}$$

$$\hat{e}_p^* \cdot \mathbf{P}_p^{(3)} = 6\epsilon_0 \hat{e}_p^* \cdot \boldsymbol{\chi}^{(3)}(\omega_p = \omega_p + \omega_S - \omega_S) \vdots \hat{e}_p \hat{e}_S \hat{e}_S^* \mathcal{E}_p |\mathcal{E}_S|^2 e^{i\mathbf{k}_p \cdot \mathbf{r}}. \tag{6.52}$$

By applying the relation given in (6.49), we can define an *effective Raman susceptibility* as

$$\chi_R = \hat{e}_S^* \cdot \boldsymbol{\chi}^{(3)}(\omega_S = \omega_S + \omega_p - \omega_p) \vdots \hat{e}_S \hat{e}_p \hat{e}_p^* = \hat{e}_p \cdot \boldsymbol{\chi}^{(3)*}(\omega_p = \omega_p + \omega_S - \omega_S) \vdots \hat{e}_p^* \hat{e}_S^* \hat{e}_S. \tag{6.53}$$

The coupled-wave equations for stimulated Raman scattering are

$$\frac{d\mathcal{E}_S}{dz} = \frac{i3\omega_S^2}{c^2 k_{S,z}} \chi_R \mathcal{E}_S |\mathcal{E}_p|^2, \tag{6.54}$$

$$\frac{d\mathcal{E}_p}{dz} = \frac{i3\omega_p^2}{c^2 k_{p,z}} \chi_R^* \mathcal{E}_p |\mathcal{E}_S|^2. \tag{6.55}$$

By comparing these two equations to the three equations given in (6.27)–(6.29) for the parametric second-order interaction, we see clearly that the nonparametric process of stimulated Raman scattering is automatically phase matched, as discussed in Subsection 5.3.10.

The relation in (6.31) can be used to transform (6.54) and (6.55) into the forms:

$$\frac{dI_S}{dz} = -\frac{3\omega_S}{c^2\epsilon_0 n_{S,z} n_{p,z}} \chi_R'' I_S I_p,$$ (6.56)

$$\frac{dI_p}{dz} = \frac{3\omega_p}{c^2\epsilon_0 n_{S,z} n_{p,z}} \chi_R'' I_S I_p.$$ (6.57)

We find that the total light intensity, $I = I_S + I_p$, varies as

$$\frac{dI}{dz} = \frac{d(I_S + I_p)}{dz} = \frac{3(\omega_p - \omega_S)}{c^2\epsilon_0 n_{S,z} n_{p,z}} \chi_R'' I_S I_p.$$ (6.58)

Therefore, optical energy is not conserved in the nonparametric Raman process because there is energy exchange with the material due to the resonant transition at the frequency of $\Omega = \omega_p - \omega_S$. Nevertheless, one Stokes photon is created for every pump photon that is annihilated. Therefore, in the absence of other loss mechanisms, we still have the Manley–Rowe relation:

$$\frac{d}{dz}\left(\frac{I_S}{\omega_S}\right) = -\frac{d}{dz}\left(\frac{I_p}{\omega_p}\right).$$ (6.59)

We see from (6.56) and (6.57) that the direction of energy flow in the process of stimulated Raman scattering is determined by the sign of χ_R'', which depends on the state of the material. If the material is in the ground state of the Raman transition, the imaginary part of $\chi^{(3)}(\omega_S = \omega_S + \omega_p - \omega_p)$ is negative, resulting in $\chi_R'' < 0$ according to (6.53). In this situation, energy is converted from the pump wave to the Stokes wave. We also see from (6.58) that in a Stokes process, there is a net loss in the total optical intensity. The energy corresponding to this loss is absorbed by the material in making the Stokes Raman transition from the ground state to the excited state. In case the excited state of the Raman transition is more populated than the ground state, the imaginary part of $\chi^{(3)}(\omega_S = \omega_S + \omega_p - \omega_p)$ for $\Omega = \omega_p - \omega_S$ becomes positive. Then, $\chi_R'' > 0$. In this situation, the anti-Stokes process occurs with the wave at the frequency of ω_S acting as the pump wave and the wave at the frequency of ω_p acting as the anti-Stokes wave. In the anti-Stokes process, the total optical intensity has a net gain corresponding to the energy that is released by the material in making the anti-Stokes Raman transition from the excited state to the ground state.

6.3.2 Two-Photon Absorption

In this subsection, we consider the process of two-photon absorption as another example of a nonparametric process. The coupled-wave equations for this process are different from those of the process of stimulated Raman scattering because of the difference in the symmetry properties of their nonlinear susceptibilities. Consequently, the Manley–Rowe relations are also different for the two processes.

By using (5.52) and (5.53), we have,

$$\hat{e}_1^* \cdot \mathbf{P}_1^{(3)} = 6\epsilon_0 \hat{e}_1^* \cdot \boldsymbol{\chi}^{(3)}(\omega_1 = \omega_1 + \omega_2 - \omega_2) \vdots \hat{e}_1 \hat{e}_2 \hat{e}_2^* \mathcal{E}_1 |\mathcal{E}_2|^2 \, e^{i\mathbf{k}_1 \cdot \mathbf{r}}, \tag{6.60}$$

$$\hat{e}_2^* \cdot \mathbf{P}_2^{(3)} = 6\epsilon_0 \hat{e}_2^* \cdot \boldsymbol{\chi}^{(3)}(\omega_2 = \omega_2 + \omega_1 - \omega_1) \vdots \hat{e}_2 \hat{e}_1 \hat{e}_1^* \mathcal{E}_2 |\mathcal{E}_1|^2 \, e^{i\mathbf{k}_2 \cdot \mathbf{r}}, \tag{6.61}$$

which have the same forms as (6.51) and (6.52) for stimulated Raman scattering if we replace ω_S and ω_p in (6.51) and (6.52) with ω_1 and ω_2, respectively. However, by applying the relation given in (6.50) for the two-photon absorption process, we find an *effective two-photon absorption susceptibility* as

$$\chi_{\text{TPA}} = \hat{e}_1^* \cdot \boldsymbol{\chi}^{(3)}(\omega_1 = \omega_1 + \omega_2 - \omega_2) \vdots \hat{e}_1 \hat{e}_2 \hat{e}_2^* = \hat{e}_2^* \cdot \boldsymbol{\chi}^{(3)}(\omega_2 = \omega_2 + \omega_1 - \omega_1) \vdots \hat{e}_2 \hat{e}_1 \hat{e}_1^*. \tag{6.62}$$

By comparing (6.62) for two-photon absorption with (6.53) for stimulated Raman scattering, we find that χ_{TPA} given in (6.62) and χ_R given in (6.53) have different forms, though (6.60) and (6.61) have the same forms as those of (6.51) and (6.52), respectively. The difference is caused by the different symmetry properties of the susceptibilities for the two processes, as shown in (6.49) and (6.50). The different symmetry properties for the two susceptibilities originate from the difference in the resonances of the two processes, as discussed above.

With the effective susceptibility given in (6.62), the coupled-wave equations for two-photon absorption are

$$\frac{d\mathcal{E}_1}{dz} = \frac{i3\omega_1^2}{c^2 k_{1,z}} \chi_{\text{TPA}} \mathcal{E}_1 |\mathcal{E}_2|^2, \tag{6.63}$$

$$\frac{d\mathcal{E}_2}{dz} = \frac{i3\omega_2^2}{c^2 k_{2,z}} \chi_{\text{TPA}} \mathcal{E}_2 |\mathcal{E}_1|^2. \tag{6.64}$$

Similar to the process of stimulated Raman scattering, the nonparametric process of two-photon absorption is also automatically phase matched, as can be clearly seen in (6.63) and (6.64). Different from the stimulated Raman scattering, the two equations in (6.63) and (6.64) both use χ_{TPA}, but (6.54) and (6.55) for stimulated Raman scattering use χ_R and χ_R^*, respectively.

The relation in (6.31) can be used to transform (6.63) and (6.64) into the forms:

$$\frac{dI_1}{dz} = -\frac{3\omega_1}{c^2 \epsilon_0 n_{1,z} n_{2,z}} \chi_{\text{TPA}}'' I_1 I_2, \tag{6.65}$$

$$\frac{dI_2}{dz} = -\frac{3\omega_2}{c^2 \epsilon_0 n_{1,z} n_{2,z}} \chi_{\text{TPA}}'' I_1 I_2. \tag{6.66}$$

We see that (6.66) is different from (6.57) by a minus sign because of the difference between (6.64) and (6.55), as discussed above. We find that the total light intensity, $I = I_1 + I_2$, varies as

$$\frac{dI}{dz} = \frac{d(I_1 + I_2)}{dz} = -\frac{3(\omega_1 + \omega_2)}{c^2 \epsilon_0 n_{1,z} n_{2,z}} \chi_{\text{TPA}}'' I_1 I_2. \tag{6.67}$$

Optical energy is not conserved in the nonparametric two-photon absorption process but is absorbed by the medium through the resonant two-photon transition at the frequency of $\omega_1 + \omega_2$. In the absence of other loss mechanisms, we have the Manley–Rowe relation:

$$\frac{d}{dz}\left(\frac{I_1}{\omega_1}\right) = \frac{d}{dz}\left(\frac{I_2}{\omega_2}\right) = -\frac{3}{c^2\epsilon_0 n_{1,z} n_{2,z}}\chi''_{\text{TPA}} I_1 I_2. \tag{6.68}$$

This relation indicates that the two photons at the two frequencies of ω_1 and ω_2 are simultaneously absorbed in the two-photon absorption process. The absorption of one photon at one frequency is always accompanied by the simultaneous absorption of another photon at the other frequency.

Similar to the stimulated Raman scattering process discussed in the preceding subsection, we also see from (6.67) and (6.68) that the direction of energy flow in the two-photon transition process is determined by the sign of χ''_{TPA}, which depends on the state of the material. If the material is in the normal state in thermal equilibrium, $\chi''_{\text{TPA}} > 0$ so that two-photon absorption takes place. This is usually the case. In the unusual situation that the medium is pumped to population inversion between the two energy levels that are in resonance with the two-photon transition, then $\chi''_{\text{TPA}} < 0$ and stimulated two-photon emission becomes possible. In any event, a nonparametric process depends strongly on the quantum state of the optical medium.

Problem Set

6.1.1 Show, by expanding the field $\mathbf{E}(\mathbf{r}, t)$ and the nonlinear polarization $\mathbf{P}^{(n)}(\mathbf{r}, t)$ into the linear combinations of their frequency components as expressed in (6.5) and (6.8), respectively, that the general time-independent wave equation given in (6.3) reduces to the coupled-wave equation given in (6.9).

6.1.2 The coupled-wave equation given in (6.9) is valid for both isotropic and anisotropic media.

(a) Show that (6.9) reduces to the form given in (6.10) for wave propagation in an isotropic medium.

(b) Show that in an anisotropic medium, (6.9) leads to two equations, one for the propagating transverse component in the form that is given in (6.19) and another for the nonpropagating longitudinal component in the form:

$$\left[\epsilon(\mathbf{k}_q, \omega_q) \cdot \mathbf{E}_q(\mathbf{r})\right]_{\text{L}} = -\mathbf{P}^{(n)}_{q,\text{L}}(\mathbf{r}). \tag{6.69}$$

6.1.3 Show that the coupled-wave equation given in (6.19) for wave propagation in an anisotropic medium reduces to that given in (6.20) under the slowly varying amplitude approximation. Show also that it further reduces to (6.21) under proper conditions. What are the conditions that allow such reduction?

6.2.1 A BaTiO$_3$ crystal is a uniaxial crystal of the 4mm point group. The nonvanishing elements of the $\chi^{(2)}$ tensor are $\chi_{15}^{(2)} = \chi_{24}^{(2)}$, $\chi_{31}^{(2)} = \chi_{32}^{(2)}$, and $\chi_{33}^{(2)}$. The optical axis is the z principal axis.

   (a) An optical wave at a frequency of ω propagates through the crystal along the x principal axis with its electric field $\mathbf{E}(\omega)$ polarized in the yz plane making an angle of ϕ with respect to the y principal axis. Write down the second-harmonic nonlinear polarization $\mathbf{P}^{(2)}(2\omega)$ as a function of the angle ϕ and the nonvanishing elements of $\chi^{(2)}$. What is χ_{eff} if $\mathbf{E}(2\omega)$ is polarized at an angle of θ with respect to the y axis?

   (b) What should the direction of $\mathbf{E}(\omega)$ be so that $\mathbf{P}^{(2)}(2\omega) \| \mathbf{E}(\omega)$? What should it be so that $\mathbf{P}^{(2)}(2\omega) \perp \mathbf{E}(\omega)$?

   (c) How should the direction of beam propagation $\mathbf{k}$ and that of field polarization $\mathbf{E}(\omega)$ be arranged so that an optical rectification field can be generated along the direction of beam propagation? How should they be arranged so that an optical rectification field can be generated in a direction that is perpendicular to the direction of beam propagation?

6.2.2 A plane wave at a frequency of ω traverses an isotropic medium. It produces a nonlinear polarization at the third-harmonic frequency 3ω.

   (a) What is the direction of this third-harmonic polarization at 3ω if the fundamental wave at ω is linearly polarized? Find the effective nonlinear susceptibility, χ_{eff}, for this process in terms of an element of the nonlinear susceptibility tensor $\chi^{(3)}$. Write down the coupled-wave equations for this interaction.

   (b) Answer the questions in (a) for a circularly polarized fundamental wave.

6.3.1 Show that the coupled-wave equations for stimulated Raman scattering are those given in (6.54) and (6.55). Show also that they lead to the coupled equations in terms of light intensity as given in (6.56) and (6.57).

6.3.2 Show that the coupled-wave equations for two-photon absorption are those given in (6.63) and (6.64). Show also that they lead to the coupled equations in terms of light intensity as given in (6.65) and (6.66).

References

[1] J. A. Armstrong, N. Bloembergen, J. Ducuing, and P. S. Pershan, "Interactions between light waves in a nonlinear dielectric," *Physical Review*, vol. 127, pp. 1918–1939, 1962.

[2] N. Bloembergen, *Nonlinear Optics*. Reading, MA: W. A. Benjamin, Inc., 1965.

[3] A. Yariv, "Coupled-mode theory for guided-wave optics," *IEEE Journal of Quantum Electronics*, vol. 9, pp. 919–933, 1973.

[4] H. A. Haus and W. Huang, "Coupled-mode theory," *Proceedings of the IEEE*, vol. 79, pp. 1505–1518, 1991.

7 Nonlinearly Coupled Waveguide Modes

7.1 COUPLED-WAVE AND COUPLED-MODE FORMULATION

The efficiency of a nonlinear optical interaction generally increases with the intensities of the interacting optical waves and the interaction length. In a spatially homogeneous medium, the intensity of an optical wave can be increased by tightening the focus of the beam to reduce its cross-sectional spot size, but often at the expense of reducing the effective interaction length due to an increase in the beam divergence because of the decrease in the beam spot size. In an optical waveguide, however, an optical wave is guided and confined to a small cross-sectional area for the entire length of the waveguide. Because of optical confinement, a guided optical wave can maintain a high intensity over a long distance that is practically limited only by the length and the attenuation coefficient of the waveguide. Therefore, both high intensity and long interaction length that are desired for efficient nonlinear optical interactions can be simultaneously fulfilled in an optical waveguide. For example, in a low-loss optical fiber the effective interaction length can be tens of kilometers, and the optical intensity can be quite high at a modest power level because of the small core diameter of an optical fiber, particularly a single-mode optical fiber. This unique feature makes optical waveguides ideal media for efficient nonlinear optical devices.

The coupled-wave theory discussed in the preceding chapter is used in the analysis of the interactions among optical waves of different frequencies, including all of the nonlinear optical interactions discussed in Chapter 5. In the analysis of the coupling of waveguide modes, however, coupled-mode theory has to be used. In general, both the interaction among different optical frequencies and the characteristics of the waveguide modes have to be considered for a nonlinear optical interaction in an optical waveguide. Therefore, a *combination of coupled-wave and coupled-mode theories* has to be employed in the analysis of such an interaction.

7.1.1 Coupled-Mode Theory

A waveguide mode is a transverse field pattern that remains constant as the mode propagates through the waveguide along its longitudinal direction. We take the longitudinal direction of the waveguide to be the z direction; thus, the transverse directions are x and y. Then, the electric and magnetic field patterns, $\mathcal{E}_v$ and $\mathcal{H}_v$, respectively, are functions of only x and y, but not functions of z, nor functions of time – that is, $\mathcal{E}_v(x, y)$ and $\mathcal{H}_v(x, y)$, where v is the mode index. For a planar waveguide, the mode index v represents a single index, $v = m$, for only one transverse spatial variable. For a nonplanar waveguide, the mode index v represents two simple indices, $v = mn$, for the two transverse spatial variables.

Each waveguide mode has a characteristic propagation constant, β_ν. Mathematically, the field patterns $\mathcal{E}_\nu(x, y)$ and $\mathcal{H}_\nu(x, y)$ constitute the eigenfunctions, and the propagation constant β_ν is the eigenvalue, of the normal mode solution of the wave equation for wave propagation in the waveguide. Therefore, the complete spatial dependence of a mode field can be expressed as

$$\mathbf{E}_\nu(\mathbf{r}) = \mathcal{E}_\nu(x, y) \exp(i\beta_\nu z), \tag{7.1}$$

$$\mathbf{H}_\nu(\mathbf{r}) = \mathcal{H}_\nu(x, y) \exp(i\beta_\nu z). \tag{7.2}$$

The electric and magnetic fields of a mode at a frequency of ω can be expressed in terms of the mode field patterns as

$$\mathbf{E}_\nu(\mathbf{r}, t) = \mathbf{E}_\nu(\mathbf{r}) \exp(-i\omega t) = \mathcal{E}_\nu(x, y) \exp(i\beta_\nu z - i\omega t), \tag{7.3}$$

$$\mathbf{H}_\nu(\mathbf{r}, t) = \mathbf{H}_\nu(\mathbf{r}) \exp(-i\omega t) = \mathcal{H}_\nu(x, y) \exp(i\beta_\nu z - i\omega t). \tag{7.4}$$

These normal modes are characteristic solutions of the two propagation equations of Maxwell's equations:

$$\nabla \times \mathbf{E} = -\mu_0 \frac{\partial \mathbf{H}}{\partial t} = i\omega\mu_0 \mathbf{H}, \tag{7.5}$$

$$\nabla \times \mathbf{H} = \epsilon \cdot \frac{\partial \mathbf{E}}{\partial t} = -i\omega\epsilon \cdot \mathbf{E}, \tag{7.6}$$

where the spatial dependence of the permittivity $\epsilon(x, y)$ defines the waveguide structure. Note that $\epsilon(x, y)$ of an ideal waveguide does not vary with the longitudinal spatial variable z.

The normal modes of a waveguide are mutually orthogonal. They do not couple to one another unless the waveguide is subject to some perturbation, which can be a spatial modulation on the waveguide structure, a temporal modulation, or an optical nonlinearity. By using (4.10) for the light intensity, the intensity of a mode field can be expressed as

$$I_\nu = \bar{\mathbf{S}}_\nu \cdot \hat{z} = (\mathbf{S}_\nu + \mathbf{S}_\nu^*) \cdot \hat{z} = (\mathcal{E}_\nu \times \mathcal{H}_\nu^* + \mathcal{E}_\nu^* \times \mathcal{H}_\nu) \cdot \hat{z}, \tag{7.7}$$

where $\bar{\mathbf{S}}$ is the time average of the real Poynting vector, defined in (4.7), and $\mathbf{S}$ is the complex Poynting vector, defined in (4.8). Therefore, the mode fields satisfy the *orthogonality condition*:

$$\int\limits_{-\infty}^{\infty} \int\limits_{-\infty}^{\infty} (\mathcal{E}_\nu \times \mathcal{H}_\mu^* + \mathcal{E}_\mu^* \times \mathcal{H}_\nu) \cdot \hat{z}\mathrm{d}x\mathrm{d}y = \pm P_\nu \, \delta_{\nu\mu}, \tag{7.8}$$

where P_ν is the power of the waveguide mode ν. In (7.8), the plus sign is used for a forward-propagating mode that propagates in the z direction, whereas the minus sign is used for a backward-propagating mode that propagates in the $-z$ direction.

The *normalized mode field distributions* for the electric field $\hat{\mathcal{E}}_\nu$ and the corresponding magnetic field $\hat{\mathcal{H}}_\nu$ of the waveguide mode ν are defined through the *orthonormality relation*:

$$\int\limits_{-\infty}^{\infty} \int\limits_{-\infty}^{\infty} (\hat{\mathcal{E}}_\nu \times \hat{\mathcal{H}}_\mu^* + \hat{\mathcal{E}}_\mu^* \times \hat{\mathcal{H}}_\nu) \cdot \hat{z}\mathrm{d}x\mathrm{d}y = \pm \delta_{\nu\mu}, \tag{7.9}$$

which integrates over the transverse cross-section of the waveguide in the xy plane. The delta function $\delta_{\nu\mu}$ is the Kronecker delta function for discrete modes, which are guided modes. For a planar waveguide, $\nu = m$, $\mu = m'$, and $\delta_{\nu\mu} = \delta_{mm'}$. For a nonplanar waveguide, $\nu = mn$, $\mu = m'n'$, and $\delta_{\nu\mu} = \delta_{mm'}\delta_{nn'}$. For continuous modes, which are unguided, $\delta_{\nu\mu}$ has to be replaced by the Dirac delta function $\delta(\nu - \mu)$. For a planar waveguide, $\nu = a$, $\mu = a'$, and $\delta(\nu - \mu) = \delta(a - a')$. For a nonplanar waveguide, $\nu = ab$, $\mu = a'b'$, and $\delta(\nu - \mu) = \delta(a - a')\delta(b - b')$.

For TE and TM modes, the general orthonormality relation given in (7.9) in the cross-product form can be transformed to dot-product forms:

$$\frac{2\beta_\nu}{\omega\mu_0} \int_{-\infty}^{\infty}\int_{-\infty}^{\infty} \hat{\mathcal{E}}_\nu \cdot \hat{\mathcal{E}}_\mu^* dxdy = \delta_{\nu\mu} \tag{7.10}$$

among TE modes, and

$$\frac{2\beta_\nu}{\omega} \int_{-\infty}^{\infty}\int_{-\infty}^{\infty} \frac{1}{\epsilon(x,y)} \hat{\mathcal{H}}_\nu \cdot \hat{\mathcal{H}}_\mu^* dxdy = \delta_{\nu\mu} \tag{7.11}$$

among TM modes. The two relations in (7.10) and (7.11) become the same for TEM modes; thus, both apply to TEM modes, which exist in a metallic waveguide but not in a dielectric waveguide.

The normal modes of a waveguide form a basis for linear expansion of an optical field at a given frequency ω in the waveguide:

$$\mathbf{E}(\mathbf{r}) = \sum_\nu A_\nu \hat{\mathcal{E}}_\nu(x,y) \exp(\mathrm{i}\beta_\nu z), \tag{7.12}$$

$$\mathbf{H}(\mathbf{r}) = \sum_\nu A_\nu \hat{\mathcal{H}}_\nu(x,y) \exp(\mathrm{i}\beta_\nu z), \tag{7.13}$$

where A_ν is the expansion coefficient for mode ν, and $\hat{\mathcal{E}}_\nu$ and $\hat{\mathcal{H}}_\nu$ are the normalized mode fields defined in (7.9). Note that in the expansion, the electric and magnetic fields of mode ν have the same expansion coefficient A_ν because $\hat{\mathcal{E}}_\nu$ and $\hat{\mathcal{H}}_\nu$ together define mode ν. All expansion coefficients A_ν are constants that do not vary with space or time when the waveguide is not subject to any perturbation so that the normal modes are not coupled to one another.

When the waveguide is subject to a perturbation, represented by $\Delta\mathbf{P}$, the two Maxwell's equations given in (7.5) and (7.6) are changed as

$$\nabla \times \mathbf{E} = -\mu_0 \frac{\partial \mathbf{H}}{\partial t} = \mathrm{i}\,\omega\mu_0 \mathbf{H}, \tag{7.14}$$

$$\nabla \times \mathbf{H} = \boldsymbol{\epsilon} \cdot \frac{\partial \mathbf{E}}{\partial t} + \frac{\partial \Delta\mathbf{P}}{\partial t} = -\mathrm{i}\,\omega\boldsymbol{\epsilon} \cdot \mathbf{E} - \mathrm{i}\,\omega\Delta\mathbf{P}. \tag{7.15}$$

The normal modes of the unperturbed waveguide, defined with $\Delta\mathbf{P} = 0$, are not exact normal modes of the perturbed waveguide with $\Delta\mathbf{P} \neq 0$. Consequently, these modes can be coupled to one another to exchange power. If the perturbation is sufficiently weak so that the normal modes of the unperturbed waveguide remain valid as a basis for expanding the total field in the waveguide, then the total field can be expanded as

$$\mathbf{E}(\mathbf{r}) = \sum_v A_v(z)\hat{\boldsymbol{\mathcal{E}}}_v(x,y) \exp(\mathrm{i}\beta_v z), \tag{7.16}$$

$$\mathbf{H}(\mathbf{r}) = \sum_v A_v(z)\hat{\boldsymbol{\mathcal{H}}}_v(x,y) \exp(\mathrm{i}\beta_v z), \tag{7.17}$$

where the expansion coefficients $A_v(z)$ are now functions of the space variable z.

By using (7.8) and (7.9) for (7.12)–(7.17), we find that the power of mode v is

$$P_v = |A_v|^2 = A_v A_v^*. \tag{7.18}$$

When a waveguide is not perturbed, the mode power P_v is a constant that is independent of space and time. Thus, each mode propagates independently without coupling. When the waveguide is perturbed so that $A_v(z)$ varies with z, the mode power $P_v(z)$ also varies with z. Then the modes exchange power through mode coupling as they propagate through the waveguide.

The *coupled-mode equations* can be derived by using the *Lorentz reciprocity theorem* [1]. By using (7.14) and (7.15), we find the Lorentz reciprocity theorem:

$$\mathbf{V} \cdot (\mathbf{E}_1 \times \mathbf{H}_2^* + \mathbf{E}_2^* \times \mathbf{H}_1) = -\mathrm{i}\omega(\mathbf{E}_1 \cdot \Delta\mathbf{P}_2^* - \mathbf{E}_2^* \cdot \Delta\mathbf{P}_1), \tag{7.19}$$

where $(\mathbf{E}_1, \mathbf{H}_1)$ and $(\mathbf{E}_2, \mathbf{H}_2)$ are the solutions of (7.14) and (7.15) for a perturbation of $\Delta\mathbf{P}_1$ and $\Delta\mathbf{P}_2$, respectively. The reciprocity theorem holds true for any two arbitrary sets of fields, including those of the normal modes of the unperturbed waveguide with $\Delta\mathbf{P} = 0$.

A perturbation that makes the normal modes couple can be a spatial modulation on the waveguide structure, a temporal modulation, or an optical nonlinearity. It can be generally represented by a perturbation polarization $\Delta\mathbf{P}$. By applying the Lorentz reciprocity theorem while taking (7.16) and (7.17) of the perturbed waveguide with $\Delta\mathbf{P}_1 = \Delta\mathbf{P}$ for $(\mathbf{E}_1, \mathbf{H}_1)$, and the normal mode fields of the unperturbed waveguide with $\Delta\mathbf{P}_2 = 0$ for $(\mathbf{E}_2, \mathbf{H}_2)$, we find that

$$\sum_v \frac{\mathrm{d}}{\mathrm{d}z} A_v(z) e^{\mathrm{i}(\beta_v - \beta_\mu)z} \int_{-\infty}^{\infty}\int_{-\infty}^{\infty} (\hat{\boldsymbol{\mathcal{E}}}_v \times \hat{\boldsymbol{\mathcal{H}}}_\mu^* + \hat{\boldsymbol{\mathcal{E}}}_\mu^* \times \hat{\boldsymbol{\mathcal{H}}}_v) \cdot \hat{z}\,\mathrm{d}x\mathrm{d}y = \mathrm{i}\omega\, e^{-\mathrm{i}\beta_\mu z} \int_{-\infty}^{\infty}\int_{-\infty}^{\infty} \hat{\boldsymbol{\mathcal{E}}}_\mu^* \cdot \Delta\mathbf{P}\mathrm{d}x\mathrm{d}y. \tag{7.20}$$

By applying the orthonormality relation given in (7.9), we obtain the *general coupled-mode equation* [2]:

$$\pm\frac{\mathrm{d}A_v}{\mathrm{d}z} = \mathrm{i}\omega\, e^{-\mathrm{i}\beta_v z} \int_{-\infty}^{\infty}\int_{-\infty}^{\infty} \hat{\boldsymbol{\mathcal{E}}}_v^* \cdot \Delta\mathbf{P}\mathrm{d}x\mathrm{d}y, \tag{7.21}$$

where the plus sign is used for $\beta_v > 0$ when mode v is a forward-propagating mode that travels in the z direction, and the minus sign is used for $\beta_v < 0$ when mode v is a backward-propagating mode that travels in the $-z$ direction. The perturbation polarization can be generally expressed as

$$\Delta \mathbf{P} = \Delta \mathbf{P}_{\mathrm{L}} + \mathbf{P}_{\mathrm{NL}} = \Delta \boldsymbol{\epsilon} \cdot \mathbf{E} + \mathbf{P}_{\mathrm{NL}}, \tag{7.22}$$

where $\Delta \mathbf{P}_{\mathrm{L}} = \mathbf{P}^{(1)} = \Delta \boldsymbol{\epsilon} \cdot \mathbf{E}$ is a linear perturbation polarization with $\Delta \boldsymbol{\epsilon}$ being the linear perturbation permittivity tensor that is independent of the optical field $\mathbf{E}$ but only dependent on structural or temporal modulations on the waveguide, and $\mathbf{P}_{\mathrm{NL}}$ is a nonlinear perturbation polarization, which can be $\mathbf{P}^{(2)}$ or $\mathbf{P}^{(3)}$, for example.

For linear mode coupling in the absence of any nonlinear interaction, we have

$$\Delta \mathbf{P} = \Delta \mathbf{P}_{\mathrm{L}} = \Delta \boldsymbol{\epsilon} \cdot \mathbf{E} \text{ and } \mathbf{P}_{\mathrm{NL}} = 0. \tag{7.23}$$

Then, by using (7.16), we have

$$\Delta \mathbf{P} = \Delta \boldsymbol{\epsilon} \cdot \mathbf{E} = \Delta \boldsymbol{\epsilon} \cdot \sum_v A_v \hat{\boldsymbol{\mathcal{E}}}_v \, \mathrm{e}^{\mathrm{i}\beta_v z}. \tag{7.24}$$

By applying this relation to (7.21), we obtain the *coupled-mode equation for a linear waveguide* [3–5]:

$$\pm \frac{\mathrm{d}A_v}{\mathrm{d}z} = \sum_\mu \mathrm{i}\, \kappa_{v\mu} A_\mu \, \mathrm{e}^{\mathrm{i}(\beta_\mu - \beta_v)z}, \tag{7.25}$$

where

$$\kappa_{v\mu} = \omega \int_{-\infty}^{\infty} \int_{-\infty}^{\infty} \hat{\boldsymbol{\mathcal{E}}}_v^* \cdot \Delta \boldsymbol{\epsilon} \cdot \hat{\boldsymbol{\mathcal{E}}}_\mu \mathrm{d}x \mathrm{d}y. \tag{7.26}$$

In a *lossless waveguide*, $\Delta \epsilon_{ij} = \Delta \epsilon_{ji}^*$ so that

$$\kappa_{v\mu} = \kappa_{\mu v}^*. \tag{7.27}$$

In a lossy waveguide, (7.27) is not valid. The coupled-mode equation of the form given in (7.25) is generally applicable to mode coupling in linear waveguides, including the coupling among modes of multiple waveguides. However, (7.26) and (7.27) are not valid for the coupling among modes of different waveguides; they have to be properly modified for mode coupling of multiple linear waveguides [2, 6].

7.1.2 Coupled Waves and Modes

For a combined coupled-wave and coupled-mode formulation, the total field of the interacting waves is first expanded in terms of the fields of individual frequencies in a manner similar to (2.24) and (6.5):

$$\mathbf{E}(\mathbf{r}, t) = \sum_q \mathbf{E}_q(\mathbf{r}) \exp\left(-\mathrm{i}\,\omega_q t\right), \tag{7.28}$$

where $\mathbf{E}_q(\mathbf{r})$ is the spatially dependent total field amplitude for the frequency ω_q. Different from (2.24) and (6.5), where $\mathbf{E}_q(\mathbf{r})$ represents the spatial distribution of a plane-wave mode or a Gaussian mode of a spatially homogeneous medium, in (7.28) $\mathbf{E}_q(\mathbf{r})$ represents the spatial distribution of the field in an optical structure, which can be a superposition of multiple spatial modes of the structure. Therefore, instead of taking out a uniquely defined fast-varying spatial variation as done in (6.5) for the formulation of the coupled-wave theory, here we further expand each field $\mathbf{E}_q(\mathbf{r})$ at a given frequency ω_q in terms of the waveguide modes:

$$\mathbf{E}_q(\mathbf{r}) = \sum_\nu A_{q,\nu}(z)\hat{\boldsymbol{\mathcal{E}}}_{q,\nu}(x, y) \exp\left(\mathrm{i}\beta_{q,\nu}z\right), \tag{7.29}$$

where $\hat{\boldsymbol{\mathcal{E}}}_{q,\nu}$ is the normalized electric field distribution of the waveguide mode ν at the frequency ω_q, $\beta_{q,\nu}$ is the propagation constant of this mode at this frequency, and the expansion coefficient $A_{q,\nu}$ is the amplitude of the mode at the frequency.

Note that the propagation constant $\beta_{q,\nu}$ in (7.29) is a function of both the optical frequency ω_q and the waveguide mode index ν. Note also that the expansion of (7.29) is valid only when the nonlinear polarization $\mathbf{P}^{(n)}$ is small compared to the linear background polarization of $\boldsymbol{\epsilon} \cdot \mathbf{E}$ of the waveguide so that the waveguide modes defined by the linear optical properties of the medium remain a valid concept. From (7.18), the power contained in the waveguide mode ν at the frequency ω_q is simply given by the relation:

$$P_{q,\nu} = \left|A_{q,\nu}\right|^2 = A_{q,\nu} A_{q,\nu}^*. \tag{7.30}$$

By following the procedures used in formulating the coupled-mode equations as discussed in the preceding subsection, we find the general *coupled-wave-and-mode equation*:

$$\pm \frac{\mathrm{d}A_{q,\nu}}{\mathrm{d}z} = \mathrm{i}\,\omega_q\, e^{-\mathrm{i}\beta_{q,\nu}z} \int\limits_{-\infty}^{\infty} \int\limits_{-\infty}^{\infty} \hat{\boldsymbol{\mathcal{E}}}_{q,\nu}^* \cdot \Delta\mathbf{P}_q \,\mathrm{d}x\mathrm{d}y. \tag{7.31}$$

By allowing for any possible linear coupling besides nonlinear coupling, we can express the total perturbation polarization that includes an nth-order nonlinear interaction as

$$\Delta\mathbf{P}_q = \Delta\mathbf{P}_q^{\mathrm{L}} + \mathbf{P}_q^{\mathrm{NL}} = \Delta\boldsymbol{\epsilon} \cdot \mathbf{E}_q + \mathbf{P}_q^{(n)}, \tag{7.32}$$

where $\Delta\mathbf{P}_q^{\mathrm{L}} = \mathbf{P}_q^{(1)} = \Delta\boldsymbol{\epsilon} \cdot \mathbf{E}_q$ and $\mathbf{P}_q^{\mathrm{NL}} = \mathbf{P}_q^{(n)}$. We then find the coupled-wave-and-mode equation that accounts for an nth-order nonlinear interaction in a waveguide structure:[1]

---

[1] To be precise, just like that of the linear coupling coefficient $\kappa_{q,\nu\mu}$, the form of the nonlinear term on the right-hand side of (7.33) also has to be modified when $\hat{\boldsymbol{\mathcal{E}}}_{q,\nu}$ represents nonorthogonal modes of different individual waveguides in a structure that consists of multiple waveguides. Such modification is normally not significant and is ignored here. No such approximation is incurred in the use of (7.33), however, if $\hat{\boldsymbol{\mathcal{E}}}_{q,\nu}$ represents the modes of the entire structure, which are mutually orthogonal. This is always true in the case of a single waveguide. It is also true if the supermodes of a structure that consists of multiple waveguides are used in the analysis.

$$\pm \frac{\mathrm{d}A_{q,v}}{\mathrm{d}z} = \sum_{\mu} \mathrm{i}\,\kappa_{q,v\mu} A_{q,\mu}\,\mathrm{e}^{\mathrm{i}(\beta_{q,\mu}-\beta_{q,v})z} + \mathrm{i}\,\omega_q\,\mathrm{e}^{-\mathrm{i}\beta_{q,v}z} \int\limits_{-\infty}^{\infty}\int\limits_{-\infty}^{\infty} \hat{\boldsymbol{\mathcal{E}}}_{q,v}^{*} \cdot \mathbf{P}_q^{(n)}\,\mathrm{d}x\mathrm{d}y, \qquad (7.33)$$

where the plus sign is taken for a forward-propagating mode with $\beta_{q,v} > 0$, and the minus sign is for a backward-propagating mode with $\beta_{q,v} < 0$. In (7.33), the linear coupling coefficient couples only the modes of the same frequency:

$$\kappa_{q,v\mu} = \omega_q \int\limits_{-\infty}^{\infty}\int\limits_{-\infty}^{\infty} \hat{\boldsymbol{\mathcal{E}}}_{q,v}^{*} \cdot \Delta\boldsymbol{\epsilon} \cdot \hat{\boldsymbol{\mathcal{E}}}_{q,\mu}\,\mathrm{d}x\mathrm{d}y. \qquad (7.34)$$

In a lossless waveguide, $\kappa_{q,v\mu} = \kappa_{q,\mu v}^{*}$.

In summary, for nonlinear interactions in optical waveguides, the expansion given in (7.29) replaces that given in (6.6), and the coupled-wave-and-mode equation given in (7.33) replaces the coupled-wave equation given in (6.22). Besides the nonlinear effect that is characterized by $\mathbf{P}^{(n)}$, linear effects such as a periodic grating also modify the behavior of the waves in a waveguide and lead to coupling between different waveguide modes. Linear coupling between different waveguides in the presence of optical nonlinearity is also possible. The first term on the right-hand side of (7.33) accounts for the possibility of such linear coupling effects based on the coupled-mode formulation discussed in the preceding subsection. Therefore, (7.33) can be viewed as a generalization of (7.25) to include the nonlinear perturbation. In the case of coupling between different waveguides, $\kappa_{q,v\mu}$ still has to be properly evaluated due to the nonorthogonality between modes of different waveguides [2].

Many guided-wave nonlinear optical devices have direct bulk counterparts. The use of a waveguide for such a device offers the advantages of improved efficiency, phase matching, or miniaturization of the device, but it is not required for the device function. Guided-wave optical frequency converters typically fall into this category. Some nonlinear optical devices rely on the waveguide geometry for their functions and thus have no bulk counterparts. All-optical switches and modulators that use waveguide interferometers or waveguide couplers belong to this category. Sometimes, the use of an optical waveguide is necessary for the practical reason that only a waveguide can provide the long interaction length required for the operation of a device though the basic function of the device does not depend on the waveguide geometry. Many nonlinear optical devices that use optical fibers belong to this category.

7.2 PARAMETRIC INTERACTIONS IN WAVEGUIDES

All second-order and third-order nonlinear optical processes that were discussed in Chapter 5 can take place in a waveguide structure. However, some specific phenomena rarely occur in a waveguide, whereas others take place only in a waveguide. For example, though the optical Kerr effect can happen both in a spatially homogeneous medium and in a waveguide, self-focusing and self-defocusing do not easily occur in a waveguide because of the wave confinement by the structure of a waveguide. By contrast, nonlinear optical phenomena that take advantage of waveguide modes, such as nonlinear mode mixing and nonlinear mode switching, are unique to waveguides.

The basic principles and characteristics of the nonlinear processes in a waveguide that have bulk counterparts are the same as those of their bulk counterparts, except that the characteristics of the waveguide modes have to be considered. For example, though a guided-wave optical frequency converter generally takes the form of a single waveguide, there is often a possibility that multiple waveguide modes are involved in the frequency-conversion process. Each individual frequency component can consist of multiple waveguide modes, as expressed by (7.29). Even when each frequency component is represented by only one waveguide mode, it is still possible for the different interacting frequency components to be in different waveguide modes.

One significant difference in formulating an optical interaction in a waveguide structure versus that in a spatially homogeneous medium is that the coupled-wave-and-mode equations for the process in a waveguide is formulated in terms of the amplitude $A_{q,v}$ of the frequency ω_q and the mode v, as in (7.33), whereas the coupled-wave equations for the process in a spatially homogeneous medium is formulated in terms of only the field amplitude $\mathcal{E}_q$ of the frequency ω_q without referring to a mode, as in (6.22). The coupled-wave equations for a nonlinear interaction in a spatially homogeneous medium lead to a set of coupled equations in terms of the intensities of the interacting waves, such as (6.32)–(6.34) for the three-frequency parametric interaction and (6.56) and (6.57) for stimulated Raman scattering. By contrast, the coupled-wave-and-mode equations for a nonlinear interaction in a waveguide structure can be converted into a set of coupled equations in terms of the *power* of each participating waveguide mode at an interacting frequency but not in terms of an intensity. The reason is simple: The coupled-wave-and-mode equations are expressed in terms of the mode amplitude $A_{q,v}$, and the power of a mode of index v at a frequency ω_q is $P_{q,v} = \left| A_{q,v} \right|^2$, according to (7.30). It only makes sense to talk about the power of a waveguide mode, but not the intensity of a mode.

7.2.1 Three-Frequency Parametric Interaction

For a parametric second-order process in a waveguide that involves three different frequencies with $\omega_3 = \omega_1 + \omega_2$, the nonlinear polarizations have to account for all of the mode fields:

$$\mathbf{P}_3^{(2)} = 2\epsilon_0 \sum_{\mu,\xi} \boldsymbol{\chi}^{(2)}(\omega_3 = \omega_1 + \omega_2) : \hat{\mathcal{E}}_{1,\mu} \hat{\mathcal{E}}_{2,\xi} A_{1,\mu} A_{2,\xi}\, \mathrm{e}^{\mathrm{i}(\beta_{1,\mu} + \beta_{2,\xi})z}, \tag{7.35}$$

$$\mathbf{P}_1^{(2)} = 2\epsilon_0 \sum_{v,\xi} \boldsymbol{\chi}^{(2)}(\omega_1 = \omega_3 - \omega_2) : \hat{\mathcal{E}}_{3,v} \hat{\mathcal{E}}_{2,\xi}^* A_{3,v} A_{2,\xi}^*\, \mathrm{e}^{\mathrm{i}(\beta_{3,v} - \beta_{2,\xi})z}, \tag{7.36}$$

$$\mathbf{P}_2^{(3)} = 2\epsilon_0 \sum_{v,\mu} \boldsymbol{\chi}^{(2)}(\omega_2 = \omega_3 - \omega_1) : \hat{\mathcal{E}}_{3,v} \hat{\mathcal{E}}_{1,\mu}^* A_{3,v} A_{1,\mu}^*\, \mathrm{e}^{\mathrm{i}(\beta_{3,v} - \beta_{1,\mu})z}. \tag{7.37}$$

Therefore, we have

$$\hat{\mathcal{E}}_{3,v}^{*} \cdot \mathbf{P}_{3}^{(2)} = 2\epsilon_{0} \sum_{\mu,\xi} \hat{\mathcal{E}}_{3,v}^{*} \cdot \chi^{(2)}(\omega_{3} = \omega_{1} + \omega_{2}) : \hat{\mathcal{E}}_{1,\mu} \hat{\mathcal{E}}_{2,\xi} A_{1,\mu} A_{2,\xi} \, e^{i(\beta_{1,\mu} + \beta_{2,\xi})z}, \quad (7.38)$$

$$\hat{\mathcal{E}}_{1,\mu}^{*} \cdot \mathbf{P}_{1}^{(2)} = 2\epsilon_{0} \sum_{v,\xi} \hat{\mathcal{E}}_{1,\mu}^{*} \cdot \chi^{(2)}(\omega_{1} = \omega_{3} - \omega_{2}) : \hat{\mathcal{E}}_{3,v} \hat{\mathcal{E}}_{2,\xi}^{*} A_{3,v} A_{2,\xi}^{*} \, e^{i(\beta_{3,v} - \beta_{2,\xi})z}, \quad (7.39)$$

$$\hat{\mathcal{E}}_{2,\xi}^{*} \cdot \mathbf{P}_{2}^{(3)} = 2\epsilon_{0} \sum_{v,\mu} \hat{\mathcal{E}}_{2,\xi}^{*} \cdot \chi^{(2)}(\omega_{2} = \omega_{3} - \omega_{1}) : \hat{\mathcal{E}}_{3,v} \hat{\mathcal{E}}_{1,\mu}^{*} A_{3,v} A_{1,\mu}^{*} \, e^{i(\beta_{3,v} - \beta_{1,\mu})z}, \quad (7.40)$$

which replace (6.23), (6.24), and (6.25), respectively, for the similar interaction in a spatially homogeneous medium.

The interacting waves in an efficient frequency converter normally propagate in the same direction, though contradirectional geometry is also possible. Here we consider only codirectional geometry with all of the interacting waves propagating in the forward direction. For simplicity, we also assume that there is no linear perturbation to the waveguide so that the coupling coefficients $\kappa_{q,v\mu}$ in (7.33) that account for linear coupling between waveguide modes vanish for all frequencies (i.e., $\kappa_{q,v\mu} = 0$). In case this assumption is not valid, the linear coupling terms with nonzero coupling coefficients have to be restored. Then, by using (7.33), we have

$$\frac{dA_{3,v}}{dz} = i\omega_{3} \sum_{\mu,\xi} C_{v\mu\xi} A_{1,\mu} A_{2,\xi} \, e^{i\Delta\beta_{v\mu\xi}z}, \quad (7.41)$$

$$\frac{dA_{1,\mu}}{dz} = i\omega_{1} \sum_{v,\xi} C_{v\mu\xi}^{*} A_{3,v} A_{2,\xi}^{*} \, e^{-i\Delta\beta_{v\mu\xi}z}, \quad (7.42)$$

$$\frac{dA_{2,\xi}}{dz} = i\omega_{2} \sum_{v,\mu} C_{v\mu\xi}^{*} A_{3,v} A_{1,\mu}^{*} \, e^{-i\Delta\beta_{v\mu\xi}z}, \quad (7.43)$$

where

$$\begin{aligned} C_{v\mu\xi} &= 2\epsilon_{0} \int_{-\infty}^{\infty} \int_{-\infty}^{\infty} \hat{\mathcal{E}}_{3,v}^{*} \cdot \chi^{(2)}(\omega_{3} = \omega_{1} + \omega_{2}) : \hat{\mathcal{E}}_{1,\mu} \hat{\mathcal{E}}_{2,\xi} \mathrm{d}x\mathrm{d}y \\ &= 2\epsilon_{0} \int_{-\infty}^{\infty} \int_{-\infty}^{\infty} \hat{\mathcal{E}}_{1,\mu} \cdot \chi^{(2)*}(\omega_{1} = \omega_{3} - \omega_{2}) : \hat{\mathcal{E}}_{3,v}^{*} \hat{\mathcal{E}}_{2,\xi} \mathrm{d}x\mathrm{d}y \\ &= 2\epsilon_{0} \int_{-\infty}^{\infty} \int_{-\infty}^{\infty} \hat{\mathcal{E}}_{2,\xi} \cdot \chi^{(2)*}(\omega_{2} = \omega_{3} - \omega_{1}) : \hat{\mathcal{E}}_{3,v}^{*} \hat{\mathcal{E}}_{1,\mu} \mathrm{d}x\mathrm{d}y \end{aligned} \quad (7.44)$$

is the *effective nonlinear coefficient* that accounts for the overlapping of the field distribution patterns of different waveguide modes, as well as for any possible spatial variations in $\chi^{(2)}$ due to the waveguide structure, and

$$\Delta\beta_{v\mu\xi} = \beta_{1,\mu} + \beta_{2,\xi} - \beta_{3,v} \quad (7.45)$$

is the phase mismatch among the coupled modes of the three frequencies.

Note that in (7.44), $\chi^{(2)*} = \chi^{(2)}$ because $\chi^{(2)}$ is real for the parametric interaction. In comparison to the effective nonlinear susceptibility, χ_{eff}, defined in (6.26) for the interaction of plane waves, the effective nonlinear coefficient defined above for the interaction of waveguide modes has the relation:

$$\left|C_{\nu\mu\xi}\right|^2 = \frac{\left|\chi_{\text{eff}}\right|^2}{2c^3\epsilon_0 n_{3,\nu} n_{1,\mu} n_{2,\xi}} \frac{\Gamma_{\nu\mu\xi}}{\mathcal{A}} = \frac{2\left|d_{\text{eff}}\right|^2}{c^3\epsilon_0 n_{3,\nu} n_{1,\mu} n_{2,\xi}} \frac{\Gamma_{\nu\mu\xi}}{\mathcal{A}}, \tag{7.46}$$

where $n_{q,\nu} = c\beta_{q,\nu}/\omega_q$ is the effective refractive index of a waveguide mode, $\Gamma_{\nu\mu\xi}$ is the *overlap factor* for the interacting waveguide modes, and $\mathcal{A}$ is the cross-sectional area of the waveguide core. The overlap factor $\Gamma_{\nu\mu\xi}$ accounts for the differences in the mode field distributions among the interacting waves and any transverse spatial variations in the nonlinear susceptibility. An effective area for the interaction can be defined as $\mathcal{A}_{\text{eff}} = \mathcal{A}/\Gamma_{\nu\mu\xi}$.

All of the general concepts discussed in Subsection 6.2.1 for the three-frequency parametric interaction are equally valid for such interaction in an optical waveguide, except that the form of each relation has to be modified to account for the characteristics of the waveguide modes. For instance, Manley–Rowe relations still exist but such relations have to be expressed in terms of the optical powers of the waveguide modes. By using the relation $P_{q,\nu} = |A_{q,\nu}|^2$ given in (7.30), we can convert the coupled equations in terms of the mode amplitudes given in (7.41)–(7.43) into coupled equations in terms of the mode powers:

$$\frac{dP_{3,\nu}}{dz} = -2\omega_3 \sum_{\mu,\xi} \left|C_{\nu\mu\xi}\right| P_{3,\nu}^{1/2} P_{1,\mu}^{1/2} P_{2,\xi}^{1/2} \sin\varphi_{\nu\mu\xi}, \tag{7.47}$$

$$\frac{dP_{1,\mu}}{dz} = 2\omega_1 \sum_{\nu,\xi} \left|C_{\nu\mu\xi}\right| P_{3,\nu}^{1/2} P_{1,\mu}^{1/2} P_{2,\xi}^{1/2} \sin\varphi_{\nu\mu\xi}, \tag{7.48}$$

$$\frac{dP_{2,\xi}}{dz} = 2\omega_2 \sum_{\nu,\mu} \left|C_{\nu\mu\xi}\right| P_{3,\nu}^{1/2} P_{1,\mu}^{1/2} P_{2,\xi}^{1/2} \sin\varphi_{\nu\mu\xi}, \tag{7.49}$$

where

$$\varphi_{\nu\mu\xi} = \varphi_C + \varphi_{1,\mu} + \varphi_{2,\xi} - \varphi_{3,\nu} + \Delta\beta_{\nu\mu\xi}z, \tag{7.50}$$

φ_C is the phase of $C_{\nu\mu\xi}$, and $\varphi_{1,\mu}$, $\varphi_{2,\xi}$, and $\varphi_{3,\nu}$ are the phases of $A_{1,\mu}$, $A_{2,\xi}$, and $A_{3,\nu}$, respectively.

The total power in all frequencies and all modes is $P = \sum_\mu P_{1,\mu} + \sum_\xi P_{2,\xi} + \sum_\nu P_{3,\nu}$. By using (7.47)–(7.49) and the fact that $\omega_3 = \omega_1 + \omega_2$, we find the Manley–Rowe relations in terms of the mode powers:

$$\frac{dP}{dz} = \frac{d}{dz}\left(\sum_\mu P_{1,\mu} + \sum_\xi P_{2,\xi} + \sum_\nu P_{3,\nu}\right) = 0 \tag{7.51}$$

and

$$\frac{d}{dz}\left(\frac{\sum_\mu P_{1,\mu}}{\omega_1}\right) = \frac{d}{dz}\left(\frac{\sum_\xi P_{2,\xi}}{\omega_2}\right) = -\frac{d}{dz}\left(\frac{\sum_\nu P_{3,\nu}}{\omega_3}\right). \tag{7.52}$$

The relation in (7.51) is equivalent to (6.36) and has the meaning of conservation of optical energy, as expected for a parametric process. The relation in (7.52) is equivalent to (6.37); it indicates that the interaction takes place by combining one photon at ω_1 and one photon at ω_2 to generate one photon at ω_3 or, reversely, by splitting one photon at ω_3 into one photon at ω_1 and another photon at ω_2.

Phase matching is also most important for an efficient interaction in a waveguide, but it is now determined by the propagation constants of the interacting waveguide modes. Therefore, the coherence length for the coupling of the mode fields $\hat{\mathcal{E}}_{3,\nu}$, $\hat{\mathcal{E}}_{1,\mu}$, and $\hat{\mathcal{E}}_{2,\varsigma}$ is

$$l_{\nu\mu\varsigma}^{\text{coh}} = \frac{\pi}{|\Delta\beta_{\nu\mu\varsigma}|}. \tag{7.53}$$

Because the propagation constant $\beta_{q,\nu}$ for a given frequency ω_q is mode dependent due to *modal dispersion*, the phase mismatch and, consequently, the efficiency of an interaction depend on the specific combination of the modes among the interacting frequency components. In a multimode waveguide, there can be many different mode combinations, as is indicated by the summation over the mode indices on the right-hand side of the coupled equations in (7.41)–(7.43). However, it is unlikely and undesirable, though not impossible, for multiple mode combinations to be simultaneously phase matched in a particular interaction. In a practical optical frequency converter, normally only one waveguide mode for each frequency component is efficiently coupled to other frequency components in the interaction.

When an interaction involves only one waveguide mode in each frequency component, the coupled equations have the form of those for the corresponding interaction in a spatially homogeneous medium, though the mode amplitudes are used instead of the field amplitudes and the coefficients in the equations look different. Then, the characteristics of any guided-wave parametric interaction can be obtained by converting those of its bulk counterpart with the modifications:

1. The mode power $P_{q,\nu}$ is used in place of $I_q A_q$.
2. For phase mismatch, $\Delta\beta$ is used instead of Δk.
3. The nonlinear coefficient $C_{\nu\mu\varsigma}$ is used in place of χ_{eff} by replacing a compound coefficient of the form on the right-hand side of (7.46) in any relation for a bulk device with $|C_{\nu\mu\varsigma}|^2$ for a guided-wave device.

7.2.2 Second-Harmonic Generation

In the case of second-harmonic generation in a waveguide, the nonlinear polarizations are

$$\mathbf{P}_{2\omega}^{(2)} = \epsilon_0 \sum_{\mu,\varsigma} \boldsymbol{\chi}^{(2)}(2\omega = \omega + \omega) : \hat{\mathcal{E}}_{\omega,\mu}\hat{\mathcal{E}}_{\omega,\varsigma} A_{\omega,\mu} A_{\omega,\varsigma}\, \mathrm{e}^{\mathrm{i}(\beta_{\omega,\mu}+\beta_{\omega,\varsigma})z}, \tag{7.54}$$

$$\mathbf{P}_{\omega}^{(2)} = 2\epsilon_0 \sum_{\nu,\varsigma} \boldsymbol{\chi}^{(2)}(\omega = 2\omega - \omega) : \hat{\mathcal{E}}_{2\omega,\nu}\hat{\mathcal{E}}_{\omega,\varsigma}^* A_{2\omega,\nu} A_{\omega,\varsigma}^*\, \mathrm{e}^{\mathrm{i}(\beta_{2\omega,\nu}-\beta_{\omega,\varsigma})z}. \tag{7.55}$$

Therefore, we have

$$\hat{\mathcal{E}}^*_{2\omega,\nu} \cdot \mathbf{P}^{(2)}_{2\omega} = \epsilon_0 \sum_{\mu,\xi} \hat{\mathcal{E}}^*_{2\omega,\nu} \cdot \chi^{(2)}(2\omega = \omega + \omega) : \hat{\mathcal{E}}_{\omega,\mu}\hat{\mathcal{E}}_{\omega,\xi} A_{\omega,\mu} A_{\omega,\xi}\, e^{i(\beta_{\omega,\mu}+\beta_{\omega,\xi})z}, \quad (7.56)$$

$$\hat{\mathcal{E}}^*_{\omega,\mu} \cdot \mathbf{P}^{(2)}_{\omega} = 2\epsilon_0 \sum_{\nu,\xi} \hat{\mathcal{E}}^*_{\omega,\mu} \cdot \chi^{(2)}(\omega = 2\omega - \omega) : \hat{\mathcal{E}}_{2\omega,\nu}\hat{\mathcal{E}}^*_{\omega,\xi} A_{2\omega,\nu} A^*_{\omega,\xi}\, e^{i(\beta_{2\omega,\nu}-\beta_{\omega,\xi})z}. \quad (7.57)$$

By assuming that all waveguide modes for the fundamental and second-harmonic frequencies propagate in the same direction in a codirectional geometry and further that there is no linear perturbation to the waveguide so that $\kappa_{q,\nu\mu} = 0$, we can write the coupled equations for second-harmonic generation in a waveguide as

$$\frac{dA_{2\omega,\nu}}{dz} = i\omega \sum_{\mu,\xi} C_{\nu\mu\xi} A_{\omega,\mu} A_{\omega,\xi}\, e^{i\,\Delta\beta_{\nu\mu\xi}z}, \quad (7.58)$$

$$\frac{dA_{\omega,\mu}}{dz} = i\omega \sum_{\nu,\xi} C^*_{\nu\mu\xi} A_{2\omega,\nu} A^*_{\omega,\xi}\, e^{-i\,\Delta\beta_{\nu\mu\xi}z}, \quad (7.59)$$

where

$$C_{\nu\mu\xi} = 2\epsilon_0 \int_{-\infty}^{\infty}\int_{-\infty}^{\infty} \hat{\mathcal{E}}^*_{2\omega,\nu} \cdot \chi^{(2)}(2\omega = \omega + \omega) : \hat{\mathcal{E}}_{\omega,\mu}\hat{\mathcal{E}}_{\omega,\xi}\,dxdy$$

$$= 2\epsilon_0 \int_{-\infty}^{\infty}\int_{-\infty}^{\infty} \hat{\mathcal{E}}_{\omega,\mu} \cdot \chi^{(2)*}(\omega = 2\omega - \omega) : \hat{\mathcal{E}}^*_{2\omega,\nu}\hat{\mathcal{E}}_{\omega,\xi}\,dxdy \quad (7.60)$$

and

$$\Delta\beta_{\nu\mu\xi} = \beta_{\omega,\mu} + \beta_{\omega,\xi} - \beta_{2\omega,\nu}. \quad (7.61)$$

The effective nonlinear coefficient defined above for second-harmonic generation in a waveguide has the relation:

$$\left|C_{\nu\mu\xi}\right|^2 = \frac{|\chi_{\text{eff}}|^2}{2c^3\epsilon_0 n_{2\omega,\nu}n_{\omega,\mu}n_{\omega,\xi}}\frac{\Gamma_{\nu\mu\xi}}{\mathcal{A}} = \frac{2|d_{\text{eff}}|^2}{c^3\epsilon_0 n_{2\omega,\nu}n_{\omega,\mu}n_{\omega,\xi}}\frac{\Gamma_{\nu\mu\xi}}{\mathcal{A}}. \quad (7.62)$$

The coherence length for second-harmonic generation in a waveguide has the same form as that given in (7.53): $l^{\text{coh}}_{\nu\mu\xi} = \pi/|\Delta\beta_{\nu\mu\xi}|$ with the phase mismatch $\Delta\beta_{\nu\mu\xi}$ given in (7.61).

By using the relation $P_{q,\nu} = |A_{q,\nu}|^2$ given in (7.30), we can convert the coupled equations in terms of the mode amplitudes given in (7.58) and (7.59) into coupled equations in terms of the mode powers:

$$\frac{dP_{2\omega,\nu}}{dz} = -2\omega \sum_{\mu,\xi} \left|C_{\nu\mu\xi}\right| P^{1/2}_{2\omega,\nu} P^{1/2}_{\omega,\mu} P^{1/2}_{\omega,\xi} \sin\varphi_{\nu\mu\xi}, \quad (7.63)$$

$$\frac{dP_{\omega,\mu}}{dz} = 2\omega \sum_{\nu,\xi} \left|C_{\nu\mu\xi}\right| P^{1/2}_{2\omega,\nu} P^{1/2}_{\omega,\mu} P^{1/2}_{\omega,\xi} \sin\varphi_{\nu\mu\xi}, \quad (7.64)$$

where

$$\varphi_{\nu\mu\xi} = \varphi_C + \varphi_{\omega,\mu} + \varphi_{\omega,\xi} - \varphi_{2\omega,\nu} + \Delta\beta_{\nu\mu\xi}z; \tag{7.65}$$

φ_C is the phase of $C_{\nu\mu\xi}$; and $\varphi_{\omega,\mu}$, $\varphi_{\omega,\xi}$, and $\varphi_{2\omega,\nu}$ are the phases of $A_{\omega,\mu}$, $A_{\omega,\xi}$, and $A_{2\omega,\nu}$, respectively.

The total power in all frequencies and all modes is $P = \sum_\mu P_{\omega,\mu} + \sum_\nu P_{2\omega,\nu}$. By using (7.63) and (7.64), we find the Manley–Rowe relations for second-harmonic generation in terms of the mode powers:

$$\frac{\mathrm{d}P}{\mathrm{d}z} = \frac{\mathrm{d}}{\mathrm{d}z}\left(\sum_\mu P_{\omega,\mu} + \sum_\nu P_{2\omega,\nu}\right) = 0 \tag{7.66}$$

and

$$\frac{\mathrm{d}}{\mathrm{d}z}\left(\frac{\sum_\mu P_{\omega,\mu}}{\omega}\right) = -2\frac{\mathrm{d}}{\mathrm{d}z}\left(\frac{\sum_\nu P_{2\omega,\nu}}{2\omega}\right). \tag{7.67}$$

The relation in (7.66) is equivalent to (6.47) and has the meaning of conservation of optical energy, as expected for second-harmonic generation, which is a parametric process. The relation in (7.67) is equivalent to (6.48); it indicates that the interaction takes place by combining two photons at the fundamental frequency of ω to generate one photon at the second-harmonic frequency of 2ω or, reversely, by splitting one photon at 2ω into two photons at ω.

For the formulation discussed above, we generally consider the possibility that the wave of each frequency contains many different waveguide modes. The formulation can be simplified in the case that each wave contains only one waveguide mode. Note that in this simple case, different waves can still be in different modes. For example, consider a guided-wave second-harmonic generator in which the fundamental wave and the second-harmonic wave each contains only one mode, say, mode μ for the fundamental wave and mode ν for the second-harmonic wave, but the two modes are not necessarily the same. Then the coupled equations given in (7.58) and (7.59) are simplified as

$$\frac{\mathrm{d}A_{2\omega,\nu}}{\mathrm{d}z} = \mathrm{i}\,\omega C A_{\omega,\mu}^2\,\mathrm{e}^{\mathrm{i}\,\Delta\beta z}, \tag{7.68}$$

$$\frac{\mathrm{d}A_{\omega,\mu}}{\mathrm{d}z} = \mathrm{i}\,\omega C^* A_{2\omega,\nu} A_{\omega,\mu}^*\,\mathrm{e}^{-\mathrm{i}\,\Delta\beta z}, \tag{7.69}$$

where $C = C_{\nu\mu\mu}$ and $\Delta\beta = \Delta\beta_{\nu\mu\mu} = 2\beta_{\omega,\mu} - \beta_{2\omega,\nu}$. Meanwhile, (7.63) and (7.64) reduce to the forms:

$$\frac{\mathrm{d}P_{2\omega,\nu}}{\mathrm{d}z} = -2\omega|C|P_{2\omega,\nu}^{1/2}P_{\omega,\mu}\sin\varphi, \tag{7.70}$$

$$\frac{\mathrm{d}P_{\omega,\mu}}{\mathrm{d}z} = 2\omega|C|P_{2\omega,\nu}^{1/2}P_{\omega,\mu}\sin\varphi, \tag{7.71}$$

where $\varphi = \varphi_{\nu\mu\mu} = \varphi_C + 2\varphi_{\omega,\mu} - \varphi_{2\omega,\nu} + \Delta\beta z$. The total power is now simply $P = P_{\omega,\mu} + P_{2\omega,\nu}$, and the Manley–Rowe relations in (7.66) and (7.67) are simplified as

$$\frac{dP}{dz} = \frac{d}{dz}(P_{\omega,\mu} + P_{2\omega,\nu}) = 0 \tag{7.72}$$

and

$$\frac{d}{dz}\left(\frac{P_{\omega,\mu}}{\omega}\right) = -2\frac{d}{dz}\left(\frac{P_{2\omega,\nu}}{2\omega}\right). \tag{7.73}$$

7.2.3 Optical Kerr Effect

All-optical modulation is generally accomplished through a nonlinear optical process, most commonly a third-order process. An all-optical modulator can be either of the *refractive type*, which utilizes $\chi^{(3)\prime}$, or of the *absorptive type*, which utilizes $\chi^{(3)\prime\prime}$. For a guided-wave nonlinear optical device, however, any absorptive loss in the waveguide is detrimental to the device function because the primary advantage of using an optical waveguide for the device is the long interaction length that is made possible by the waveguiding effect. Therefore, all practical guided-wave all-optical modulators and switches are of the refractive type based on the optical Kerr effect. The majority of such devices require only one optical frequency for their operation, though some involve two or more frequencies at a time.

For a guided-wave all-optical modulator or switch that requires only one frequency at a time for its operation, we have

$$\mathbf{P}^{(3)} = 3\epsilon_0 \sum_{\mu,\xi,\zeta} \chi^{(3)}(\omega = \omega + \omega - \omega) \vdots \hat{\mathcal{E}}_\mu \hat{\mathcal{E}}_\xi \hat{\mathcal{E}}_\zeta^* A_\mu A_\xi A_\zeta^* e^{i(\beta_\mu + \beta_\xi - \beta_\zeta)z}, \tag{7.74}$$

so that

$$\hat{\mathcal{E}}_\nu^* \cdot \mathbf{P}^{(3)} = 3\epsilon_0 \sum_{\mu,\xi,\zeta} \hat{\mathcal{E}}_\nu^* \cdot \chi^{(3)}(\omega = \omega + \omega - \omega) \vdots \hat{\mathcal{E}}_\mu \hat{\mathcal{E}}_\xi \hat{\mathcal{E}}_\zeta^* A_\mu A_\xi A_\zeta^* e^{i(\beta_\mu + \beta_\xi - \beta_\zeta)z}. \tag{7.75}$$

Then, we find from (7.33) the general coupled-wave-and-mode equation for the single-frequency optical Kerr effect in a waveguide structure:

$$\pm\frac{dA_\nu}{dz} = \sum_\mu i\kappa_{\nu\mu} A_\mu e^{i(\beta_\mu - \beta_\nu)z} + i\omega \sum_{\mu,\xi,\zeta} C_{\nu\mu\xi\zeta} A_\mu A_\xi A_\zeta^* e^{i(\beta_\mu + \beta_\xi - \beta_\zeta - \beta_\nu)z}, \tag{7.76}$$

where $\kappa_{\nu\mu}$ is the linear coupling coefficient caused by any linear perturbations to the waveguide other than nonlinear optical perturbations, and $C_{\nu\mu\xi\zeta}$ is the effective nonlinear coefficient given as

$$C_{\nu\mu\xi\zeta} = 3\epsilon_0 \int\limits_{-\infty}^{\infty} \int\limits_{-\infty}^{\infty} \hat{\mathcal{E}}_\nu^* \cdot \chi^{(3)} \vdots \hat{\mathcal{E}}_\mu \hat{\mathcal{E}}_\xi \hat{\mathcal{E}}_\zeta^* dxdy. \tag{7.77}$$

As an example, we consider the coupling of two waveguides modes, a and b, which can be two modes at the same frequency ω of the same waveguide, or of two different waveguides. The total field of the two modes is $\mathbf{E}(\mathbf{r}, t) = \mathbf{E}(\mathbf{r}) \exp(-i \omega t)$, with

$$\mathbf{E}(\mathbf{r}) = A(z)\hat{\mathcal{E}}_a(x,y)e^{i\beta_a z} + B(z)\hat{\mathcal{E}}_b(x,y)e^{i\beta_b z}. \tag{7.78}$$

By using (7.76) for the amplitudes A and B of the two modes, we find the coupled equations:

$$\pm\frac{\mathrm{d}A}{\mathrm{d}z} = i\,\kappa_{aa}A + i\,\kappa_{ab}B\,e^{i(\beta_b-\beta_a)z} + i\,\sigma_{aaaa}|A|^2A + \text{nonlinear cross terms}, \tag{7.79}$$

$$\pm\frac{\mathrm{d}B}{\mathrm{d}z} = i\,\kappa_{bb}B + i\,\kappa_{ba}A\,e^{i(\beta_a-\beta_b)z} + i\,\sigma_{bbbb}|B|^2B + \text{nonlinear cross terms}, \tag{7.80}$$

where κ_{aa}, κ_{ab}, κ_{bb}, and κ_{ba} are the *linear coupling coefficients* of the two waveguide modes, and

$$\sigma_{aaaa} = \omega C_{aaaa} = 3\omega\epsilon_0 \int_{-\infty}^{\infty}\int_{-\infty}^{\infty} \hat{\mathcal{E}}_a^* \cdot \chi^{(3)} : \hat{\mathcal{E}}_a\hat{\mathcal{E}}_a\hat{\mathcal{E}}_a^*\mathrm{d}x\mathrm{d}y \tag{7.81}$$

and

$$\sigma_{bbbb} = \omega C_{bbbb} = 3\omega\epsilon_0 \int_{-\infty}^{\infty}\int_{-\infty}^{\infty} \hat{\mathcal{E}}_b^* \cdot \chi^{(3)} : \hat{\mathcal{E}}_b\hat{\mathcal{E}}_b\hat{\mathcal{E}}_b^*\mathrm{d}x\mathrm{d}y \tag{7.82}$$

are the *nonlinear coupling coefficients*. In general, $\sigma_{aaaa} \neq \sigma_{bbbb}$. The two terms $i\sigma_{aaaa}|A|^2A$ and $i\sigma_{bbbb}|B|^2B$ represent self-phase modulation of each mode due to the optical Kerr effect; they are always phase matched. The nonlinear cross terms represent nonlinear coupling through cross-phase modulation between the two modes, also due to the optical Kerr effect, with the effective nonlinear coefficients of C_{aaab}, C_{aaba}, C_{abaa}, C_{aabb}, C_{abab}, C_{abba}, and C_{abbb} for (7.79), and C_{bbba}, C_{bbab}, C_{babb}, C_{bbaa}, C_{baba}, C_{baab}, and C_{baaa} for (7.80). These nonlinear cross terms are generally much smaller than the self-modulation terms that are represented by σ_{aaaa} and σ_{bbbb} in (7.79) and (7.80), respectively. Furthermore, the cross terms are not necessarily phase matched because of the *modal dispersion* among the waveguide modes. Among them, the two terms with C_{aabb} and C_{abab} for (7.79) and the two terms with C_{bbaa} and C_{baba} for (7.80) are always phase matched. The term with C_{aaab} for (7.79) and the three terms with C_{bbab}, C_{babb}, and C_{baaa} for (7.80) have a phase mismatch of $\beta_a - \beta_b$; the three terms with C_{aaba}, C_{abaa}, and C_{abbb} for (7.79) and the term with C_{bbba} for (7.80) have a phase mismatch of $\beta_b - \beta_a$; the term with C_{abba} for (7.79) has a phase mismatch of $2(\beta_b - \beta_a)$; and the term with C_{baab} for (7.80) has a phase mismatch of $2(\beta_a - \beta_b)$. Therefore, five cross terms in each equation are not phase matched in the case that $\beta_a \neq \beta_b$. The effects of these terms are further reduced when they are phase mismatched.

7.3 NONPARAMETRIC INTERACTIONS IN WAVEGUIDES

The coupled-wave-and-mode equation given in (7.33) can also be used to formulate nonparametric interactions in waveguides in terms of the mode amplitude $A_{q,v}$ for a wave at the frequency of ω_q in mode v. In a nonparametric process, optical energy is usually expended to excite the medium through resonant transitions associated with the process because the medium is normally in its ground state before the interaction takes place. As stated above, an absorptive loss in a waveguide is detrimental to the functioning of a nonlinear optical device because the loss is substantially enhanced by the long interaction length that is made possible by the waveguide. However, this issue does not completely rule out the use of waveguides for nonparametric interactions, particularly when the optical loss is the purpose or it is the collateral loss for some gain. For example, in stimulated Raman scattering, the energy loss is accompanied by the amplification of the Stokes signal. While the long interaction length of a waveguide increases the loss, it simultaneously enhances the amplification of the Stokes signal.

A nonparametric interaction is automatically phase matched because any momentum difference among the participating photons can be absorbed by the medium through a momentum exchange that accompanies the necessary energy exchange between the optical fields and the medium. This is generally true, as we have seen in the preceding chapters. While this is also true for nonparametric interactions in waveguides, the optical fields in the waveguide modes that are involved in an interaction can still be phase mismatched because of the modal dispersion among the waveguide modes. This possibility was already seen and discussed in Subsection 7.2.3 for the single-frequency optical Kerr effect in a two-mode waveguide. The optical Kerr effect is automatically phase matched when it takes place in a spatially homogeneous medium where modal dispersion does not exist, but the modal dispersion of a waveguide can still cause phase mismatch between the interacting modes of a waveguide when the modes propagate with different propagation constants. Nonetheless, there are terms that are always phase matched, particularly the self-coupling terms. For this reason, the phase-mismatched terms are generally unimportant and can be ignored. This is also true for nonparametric interactions in waveguides.

7.3.1 Stimulated Raman Scattering

We consider stimulated Raman scattering in a waveguide. For generality, we take the waveguide to be a multimode waveguide. The nonlinear polarizations that account for the mode fields are

$$\mathbf{P}_{S}^{(3)} = 6\epsilon_0 \sum_{\mu,\xi,\zeta} \boldsymbol{\chi}^{(3)}(\omega_{S} = \omega_{S} + \omega_{p} - \omega_{p}) \vdots \hat{\boldsymbol{\mathcal{E}}}_{S,\mu} \hat{\boldsymbol{\mathcal{E}}}_{p,\xi} \hat{\boldsymbol{\mathcal{E}}}_{p,\zeta}^{*} A_{S,\mu} A_{p,\xi} A_{p,\zeta}^{*} e^{i(\beta_{S,\mu}+\beta_{p,\xi}-\beta_{p,\zeta})z}, \quad (7.83)$$

$$\mathbf{P}_{p}^{(3)} = 6\epsilon_0 \sum_{\zeta,v,\mu} \boldsymbol{\chi}^{(3)}(\omega_{p} = \omega_{p} + \omega_{S} - \omega_{S}) \vdots \hat{\boldsymbol{\mathcal{E}}}_{p,\zeta} \hat{\boldsymbol{\mathcal{E}}}_{S,v} \hat{\boldsymbol{\mathcal{E}}}_{S,\mu}^{*} A_{p,\zeta} A_{S,v} A_{S,\mu}^{*} e^{i(\beta_{p,\zeta}+\beta_{S,v}-\beta_{S,\mu})z}. \quad (7.84)$$

Therefore, we have

$$
\hat{\boldsymbol{\mathcal{E}}}_{S,\nu}^{*} \cdot \mathbf{P}_{S}^{(3)} = 6\epsilon_0 \sum_{\mu,\xi,\zeta} \hat{\boldsymbol{\mathcal{E}}}_{S,\nu}^{*} \cdot \boldsymbol{\chi}^{(3)}(\omega_S = \omega_S + \omega_p - \omega_p) \vdots \hat{\boldsymbol{\mathcal{E}}}_{S,\mu}\hat{\boldsymbol{\mathcal{E}}}_{p,\xi}\hat{\boldsymbol{\mathcal{E}}}_{p,\zeta}^{*} A_{S,\mu}A_{p,\xi}A_{p,\zeta}^{*} \, e^{i(\beta_{S,\mu}+\beta_{p,\xi}-\beta_{p,\zeta})z},
$$

(7.85)

$$
\hat{\boldsymbol{\mathcal{E}}}_{p,\xi}^{*} \cdot \mathbf{P}_{p}^{(3)} = 6\epsilon_0 \sum_{\zeta,\nu,\mu} \hat{\boldsymbol{\mathcal{E}}}_{p,\xi}^{*} \cdot \boldsymbol{\chi}^{(3)}(\omega_p = \omega_p + \omega_S - \omega_S) \vdots \hat{\boldsymbol{\mathcal{E}}}_{p,\zeta}\hat{\boldsymbol{\mathcal{E}}}_{S,\nu}\hat{\boldsymbol{\mathcal{E}}}_{S,\mu}^{*} A_{p,\zeta}A_{S,\nu}A_{S,\mu}^{*} \, e^{i(\beta_{p,\zeta}+\beta_{S,\nu}-\beta_{S,\mu})z}.
$$

(7.86)

By using (7.33), we find the general coupled-wave-and-mode equations for stimulated Raman scattering in a waveguide structure:

$$
\pm\frac{dA_{S,\nu}}{dz} = \sum_{\mu} i\,\kappa_{S,\nu\mu}A_{S,\mu}\, e^{i(\beta_{S,\mu}-\beta_{S,\nu})z} + i\,\omega_S \sum_{\mu,\xi,\zeta} C_{\nu\mu\xi\zeta}A_{S,\mu}A_{p,\xi}A_{p,\zeta}^{*}\, e^{i(\beta_{S,\mu}+\beta_{p,\xi}-\beta_{p,\zeta}-\beta_{S,\nu})z},
$$

(7.87)

$$
\pm\frac{dA_{p,\xi}}{dz} = \sum_{\zeta} i\,\kappa_{p,\xi\zeta}A_{p,\zeta}\, e^{i(\beta_{p,\zeta}-\beta_{p,\xi})z} + i\,\omega_p \sum_{\zeta,\nu,\mu} C_{\xi\zeta\nu\mu}A_{p,\zeta}A_{S,\nu}A_{S,\mu}^{*}\, e^{i(\beta_{p,\zeta}+\beta_{S,\nu}-\beta_{S,\mu}-\beta_{p,\xi})z},
$$

(7.88)

where $\kappa_{S,\nu\mu}$ and $\kappa_{p,\xi\zeta}$ are the linear coupling coefficients for the waveguide modes at the Stokes frequency ω_S and the pump frequency ω_p, respectively; $\beta_{S,\mu}$ and $\beta_{S,\nu}$ are the propagation constants of modes μ and ν at the Stokes frequency; $\beta_{p,\xi}$ and $\beta_{p,\zeta}$ are the propagation constants of modes ξ and ζ at the pump frequency; and $C_{\nu\mu\xi\zeta}$ and $C_{\xi\zeta\nu\mu}$ are the effective nonlinear coefficients given as

$$
C_{\nu\mu\xi\zeta} = 6\epsilon_0 \int_{-\infty}^{\infty}\int_{-\infty}^{\infty} \hat{\boldsymbol{\mathcal{E}}}_{S,\nu}^{*} \cdot \boldsymbol{\chi}^{(3)}(\omega_S = \omega_S + \omega_p - \omega_p) \vdots \hat{\boldsymbol{\mathcal{E}}}_{S,\mu}\hat{\boldsymbol{\mathcal{E}}}_{p,\xi}\hat{\boldsymbol{\mathcal{E}}}_{p,\zeta}^{*} \, dxdy
$$

$$
= 6\epsilon_0 \int_{-\infty}^{\infty}\int_{-\infty}^{\infty} \hat{\boldsymbol{\mathcal{E}}}_{p,\xi} \cdot \boldsymbol{\chi}^{(3)*}(\omega_p = \omega_p + \omega_S - \omega_S) \vdots \hat{\boldsymbol{\mathcal{E}}}_{p,\zeta}^{*}\hat{\boldsymbol{\mathcal{E}}}_{S,\nu}^{*}\hat{\boldsymbol{\mathcal{E}}}_{S,\mu} \, dxdy = C_{\xi\zeta\nu\mu}^{*},
$$

(7.89)

where the permutation property of the Raman susceptibility given in (6.49) is used.

By using (7.89), we can express (7.87) and (7.88) in the forms:

$$
\pm\frac{dA_{S,\nu}}{dz} = \sum_{\mu} i\,\kappa_{S,\nu\mu}A_{S,\mu}\, e^{i(\beta_{S,\mu}-\beta_{S,\nu})z} + i\,\omega_S \sum_{\mu,\xi,\zeta} C_{\nu\mu\xi\zeta}A_{S,\mu}A_{p,\xi}A_{p,\zeta}^{*}\, e^{i(\beta_{S,\mu}+\beta_{p,\xi}-\beta_{p,\zeta}-\beta_{S,\nu})z},
$$

(7.90)

$$
\pm\frac{dA_{p,\xi}}{dz} = \sum_{\zeta} i\,\kappa_{p,\xi\zeta}A_{p,\zeta}\, e^{i(\beta_{p,\zeta}-\beta_{p,\xi})z} + i\,\omega_p \sum_{\zeta,\nu,\mu} C_{\nu\mu\xi\zeta}^{*}A_{p,\zeta}A_{S,\nu}A_{S,\mu}^{*}\, e^{i(\beta_{p,\zeta}+\beta_{S,\nu}-\beta_{S,\mu}-\beta_{p,\xi})z}.
$$

(7.91)

These equations are similar to those given in (6.54) and (6.55) except that there are many terms accounting for the coupling among waveguide modes. It can be seen that many terms have phase mismatches when the propagation constants are different for different modes of the same frequency, as discussed above. It can also be seen that there are nonlinear terms that are always phase matched. For this reason, the phase-mismatched nonlinear terms are generally not important and can be ignored. However, the linear terms with the coupling coefficients of $\kappa_{\mathrm{s},\nu\mu}$ and $\kappa_{\mathrm{p},\zeta\zeta}$ cannot be ignored even when they are phase mismatched if these linear coupling coefficients do not vanish; they can be ignored only when linear coupling does not exist so that the linear coupling coefficients vanish.

To make a comparison with the coupled-wave formulation discussed in Subsection 6.3.1 for stimulated Raman scattering in a spatially homogeneous medium, we consider the simplest situation of stimulated Raman scattering in a single-mode waveguide. We also assume that both the Stokes and the pump waves propagate in the forward direction. In this situation, (7.90) and (7.91) reduce to the forms:

$$\frac{\mathrm{d}A_{\mathrm{S}}}{\mathrm{d}z} = \mathrm{i}\,\kappa_{\mathrm{S}}A_{\mathrm{S}} + \mathrm{i}\,\omega_{\mathrm{S}}C_{\mathrm{R}}A_{\mathrm{S}}|A_{\mathrm{p}}|^2, \tag{7.92}$$

$$\frac{\mathrm{d}A_{\mathrm{p}}}{\mathrm{d}z} = \mathrm{i}\,\kappa_{\mathrm{p}}A_{\mathrm{p}} + \mathrm{i}\,\omega_{\mathrm{p}}C_{\mathrm{R}}^*A_{\mathrm{p}}|A_{\mathrm{S}}|^2, \tag{7.93}$$

where κ_{S} and κ_{p} are self-coupling coefficients for the single waveguide mode at the Stokes and the pump frequencies, respectively, caused possibly by any linear perturbation to the waveguide. In the following, we assume that the waveguide has no linear loss so that $\kappa_{\mathrm{S}} = \kappa_{\mathrm{S}}^*$ and $\kappa_{\mathrm{p}} = \kappa_{\mathrm{p}}^*$. From (7.89) and (6.53), we see that the nonlinear coupling coefficient C_{R} is proportional to χ_{R}, but it is integrated over the profiles of the mode fields. By using the relation given in (7.30) that $P_{q,\nu} = |A_{q,\nu}|^2$ for the mode power, we can transform (7.92) and (7.93) into two coupled equations in terms of the mode power:

$$\frac{\mathrm{d}P_{\mathrm{S}}}{\mathrm{d}z} = -2\omega_{\mathrm{S}}C_{\mathrm{R}}''P_{\mathrm{S}}P_{\mathrm{p}}, \tag{7.94}$$

$$\frac{\mathrm{d}P_{\mathrm{p}}}{\mathrm{d}z} = 2\omega_{\mathrm{p}}C_{\mathrm{R}}''P_{\mathrm{S}}P_{\mathrm{p}}. \tag{7.95}$$

These two equations have the same forms as (6.56) and (6.57), respectively, except that they are expressed in terms of the mode powers whereas (6.56) and (6.57) are expressed in terms of the optical intensities.

We find that the total power, $P = P_{\mathrm{S}} + P_{\mathrm{p}}$, varies as

$$\frac{\mathrm{d}P}{\mathrm{d}z} = \frac{\mathrm{d}(P_{\mathrm{S}} + P_{\mathrm{p}})}{\mathrm{d}z} = 2(\omega_{\mathrm{p}} - \omega_{\mathrm{S}})C_{\mathrm{R}}''P_{\mathrm{S}}P_{\mathrm{p}}. \tag{7.96}$$

As discussed in Subsection 6.3.1, optical energy is not conserved in the nonparametric process of stimulated Raman scattering. The direction of energy flow is determined by the sign of C_{R}'', which depends on the state of the material. If the material is in the ground state of the Raman transition, the imaginary part of $\chi^{(3)}(\omega_{\mathrm{S}} = \omega_{\mathrm{S}} + \omega_{\mathrm{p}} - \omega_{\mathrm{p}})$ is negative, resulting in $C_{\mathrm{R}}'' < 0$ so that there is a net loss in the optical energy. As seen in (7.96), this optical energy loss is proportional to the frequency difference of $\Omega = \omega_{\mathrm{p}} - \omega_{\mathrm{S}}$ between the pump and the Stokes waves, which is the resonance frequency of the material excitation that takes the energy from the pump optical wave in the process of stimulated Raman scattering. Because one Stokes photon is created for every pump photon that is annihilated, we still have the Manley–Rowe relation:

$$\frac{d}{dz}\left(\frac{P_S}{\omega_S}\right) = -\frac{d}{dz}\left(\frac{P_p}{\omega_p}\right) = -2C_R'' P_S P_p. \tag{7.97}$$

7.3.2 Two-Photon Absorption

In this subsection, we consider the process of two-photon absorption as another example of a nonparametric process. The polarizations for two-photon absorption have the same forms as those for stimulated Raman scattering. However, as discussed in Subsection 6.3.2, the coupled-wave equations for this process are different from those for the process of stimulated Raman scattering because of the difference in the symmetry properties of their nonlinear susceptibilities. Consequently, the Manley–Rowe relations are also different for the two processes.

The nonlinear polarizations of two-photon absorption that account for the mode fields are

$$\mathbf{P}_1^{(3)} = 6\epsilon_0 \sum_{\mu,\xi,\zeta} \boldsymbol{\chi}^{(3)}(\omega_1 = \omega_1 + \omega_2 - \omega_2) \vdots \hat{\boldsymbol{\mathcal{E}}}_{1,\mu}\hat{\boldsymbol{\mathcal{E}}}_{2,\xi}\hat{\boldsymbol{\mathcal{E}}}_{2,\zeta}^* A_{1,\mu}A_{2,\xi}A_{2,\zeta}^* \, \mathrm{e}^{\mathrm{i}(\beta_{1,\mu}+\beta_{2,\xi}-\beta_{2,\zeta})z}, \tag{7.98}$$

$$\mathbf{P}_2^{(3)} = 6\epsilon_0 \sum_{\zeta,\nu,\mu} \boldsymbol{\chi}^{(3)}(\omega_2 = \omega_2 + \omega_1 - \omega_1) \vdots \hat{\boldsymbol{\mathcal{E}}}_{2,\zeta}\hat{\boldsymbol{\mathcal{E}}}_{1,\nu}\hat{\boldsymbol{\mathcal{E}}}_{1,\mu}^* A_{2,\zeta}A_{1,\nu}A_{1,\mu}^* \, \mathrm{e}^{\mathrm{i}(\beta_{2,\zeta}+\beta_{1,\nu}-\beta_{1,\mu})z}. \tag{7.99}$$

Then, we have

$$\hat{\boldsymbol{\mathcal{E}}}_{1,\nu}^* \cdot \mathbf{P}_1^{(3)} = 6\epsilon_0 \sum_{\mu,\xi,\zeta} \hat{\boldsymbol{\mathcal{E}}}_{1,\nu}^* \cdot \boldsymbol{\chi}^{(3)}(\omega_1 = \omega_1 + \omega_2 - \omega_2) \vdots \hat{\boldsymbol{\mathcal{E}}}_{1,\mu}\hat{\boldsymbol{\mathcal{E}}}_{2,\xi}\hat{\boldsymbol{\mathcal{E}}}_{2,\zeta}^* A_{1,\mu}A_{2,\xi}A_{2,\zeta}^* \, \mathrm{e}^{\mathrm{i}(\beta_{1,\mu}+\beta_{2,\xi}-\beta_{2,\zeta})z},$$

$$\tag{7.100}$$

$$\hat{\boldsymbol{\mathcal{E}}}_{2,\xi}^* \cdot \mathbf{P}_2^{(3)} = 6\epsilon_0 \sum_{\zeta,\nu,\mu} \hat{\boldsymbol{\mathcal{E}}}_{2,\xi}^* \cdot \boldsymbol{\chi}^{(3)}(\omega_2 = \omega_2 + \omega_1 - \omega_1) \vdots \hat{\boldsymbol{\mathcal{E}}}_{2,\zeta}\hat{\boldsymbol{\mathcal{E}}}_{1,\nu}\hat{\boldsymbol{\mathcal{E}}}_{1,\mu}^* A_{2,\zeta}A_{1,\nu}A_{1,\mu}^* \, \mathrm{e}^{\mathrm{i}(\beta_{2,\zeta}+\beta_{1,\nu}-\beta_{1,\mu})z}.$$

$$\tag{7.101}$$

The general coupled-wave-and-mode equations for two-photon absorption in a waveguide structure are found by using (7.33):

$$\pm\frac{dA_{1,\nu}}{dz} = \sum_{\mu} \mathrm{i}\kappa_{1,\nu\mu}A_{1,\mu}\, \mathrm{e}^{\mathrm{i}(\beta_{1,\mu}-\beta_{1,\nu})z} + \mathrm{i}\omega_1 \sum_{\mu,\xi,\zeta} C_{\nu\mu\xi\zeta}A_{1,\mu}A_{2,\xi}A_{2,\zeta}^* \, \mathrm{e}^{\mathrm{i}(\beta_{1,\mu}+\beta_{2,\xi}-\beta_{2,\zeta}-\beta_{1,\nu})z}, \tag{7.102}$$

$$\pm\frac{dA_{2,\xi}}{dz} = \sum_{\zeta} \mathrm{i}\kappa_{2,\xi\zeta}A_{2,\zeta}\, \mathrm{e}^{\mathrm{i}(\beta_{2,\zeta}-\beta_{2,\xi})z} + \mathrm{i}\omega_2 \sum_{\zeta,\nu,\mu} C_{\xi\zeta\nu\mu}A_{2,\zeta}A_{1,\nu}A_{1,\mu}^* \, \mathrm{e}^{\mathrm{i}(\beta_{2,\zeta}+\beta_{1,\nu}-\beta_{1,\mu}-\beta_{2,\xi})z}, \tag{7.103}$$

where $\kappa_{1,\nu\mu}$ and $\kappa_{2,\xi\zeta}$ are the linear coupling coefficients for the waveguide modes at the frequencies ω_1 and ω_2, respectively; $\beta_{1,\nu}$ and $\beta_{1,\mu}$ are the propagation constants of modes μ and ν at ω_1; and $\beta_{2,\xi}$ and $\beta_{2,\zeta}$ are the propagation constants of modes ξ and ζ at ω_2.

So far, the equations for two-photon absorption parallel those for stimulated Raman scattering, as can be seen by comparing (7.98)–(7.103) to each corresponding equation of (7.83)–(7.88). As discussed in Section 6.3, the difference between the two processes is in the different resonances of the processes, which result in different permutation properties of their nonlinear susceptibilities. This difference between the two processes in turn leads to a key difference in their effective nonlinear coefficients. For two-photon absorption, the effective nonlinear coefficients $C_{\nu\mu\xi\zeta}$ and $C_{\xi\zeta\nu\mu}$ given in (7.102) and (7.103) are

$$
\begin{aligned}
C_{\nu\mu\xi\zeta} &= 6\epsilon_0 \int_{-\infty}^{\infty}\int_{-\infty}^{\infty} \hat{\mathcal{E}}_{1,\nu}^* \cdot \boldsymbol{\chi}^{(3)}(\omega_1 = \omega_1 + \omega_2 - \omega_2) \vdots \hat{\mathcal{E}}_{1,\mu}\hat{\mathcal{E}}_{2,\xi}\hat{\mathcal{E}}_{2,\zeta}^* \mathrm{d}x\mathrm{d}y \\
&= 6\epsilon_0 \int_{-\infty}^{\infty}\int_{-\infty}^{\infty} \hat{\mathcal{E}}_{2,\xi}^* \cdot \boldsymbol{\chi}^{(3)}(\omega_2 = \omega_2 + \omega_1 - \omega_1) \vdots \hat{\mathcal{E}}_{2,\zeta}\hat{\mathcal{E}}_{1,\nu}\hat{\mathcal{E}}_{1,\mu}^* \mathrm{d}x\mathrm{d}y = C_{\xi\zeta\nu\mu},
\end{aligned}
\tag{7.104}
$$

where the permutation property of the susceptibility of two-photon absorption given in (6.50) is used.

By using (7.104), the coupled-wave-and-mode equations given in (7.102) and (7.103) can be expressed in the forms:

$$
\pm\frac{\mathrm{d}A_{1,\nu}}{\mathrm{d}z} = \sum_{\mu} \mathrm{i}\kappa_{1,\nu\mu}A_{1,\mu}\,\mathrm{e}^{\mathrm{i}(\beta_{1,\mu}-\beta_{1,\nu})z} + \mathrm{i}\omega_1 \sum_{\mu,\xi,\zeta} C_{\nu\mu\xi\zeta}A_{1,\mu}A_{2,\xi}A_{2,\zeta}^*\,\mathrm{e}^{\mathrm{i}(\beta_{1,\mu}+\beta_{2,\xi}-\beta_{2,\zeta}-\beta_{1,\nu})z}, \tag{7.105}
$$

$$
\pm\frac{\mathrm{d}A_{2,\xi}}{\mathrm{d}z} = \sum_{\zeta} \mathrm{i}\kappa_{2,\xi\zeta}A_{2,\zeta}\,\mathrm{e}^{\mathrm{i}(\beta_{2,\zeta}-\beta_{2,\xi})z} + \mathrm{i}\omega_2 \sum_{\zeta,\nu,\mu} C_{\nu\mu\xi\zeta}A_{2,\zeta}A_{1,\nu}A_{1,\mu}^*\,\mathrm{e}^{\mathrm{i}(\beta_{2,\zeta}+\beta_{1,\nu}-\beta_{1,\mu}-\beta_{2,\xi})z}. \tag{7.106}
$$

Note that (7.104) is different from (7.89) in that $C_{\nu\mu\xi\zeta} = C_{\xi\zeta\nu\mu}$ for two-photon absorption but $C_{\nu\mu\xi\zeta} = C_{\xi\zeta\nu\mu}^*$ for stimulated Raman scattering, which leads to the difference between (7.106) and (7.91) though (7.103) has the same form as (7.88). Similar to the general coupled equations for stimulated Raman scattering, many terms in (7.105) and (7.106) have phase mismatches when the propagation constants are different for different modes of the same frequency. It is also true that there are nonlinear terms that are always phase matched; therefore, the phase-mismatched nonlinear terms are generally not important and can be ignored. However, the linear terms with the linear coupling coefficients of $\kappa_{1,\nu\mu}$ and $\kappa_{2,\xi\zeta}$ cannot be ignored even when they are phase mismatched if the coupling coefficients do not vanish; they can be ignored only when linear coupling does not exist so that the linear coupling coefficients vanish.

We now consider the simplest situation of two-photon absorption in a single-mode wave-guide. We also assume that both waves at the frequencies ω_1 and ω_2 propagate in the forward direction. Then, (7.105) and (7.106) reduce to the forms:

$$
\frac{\mathrm{d}A_1}{\mathrm{d}z} = \mathrm{i}\kappa_1 A_1 + \mathrm{i}\omega_1 C_{\mathrm{TPA}}A_1|A_2|^2, \tag{7.107}
$$

$$
\frac{\mathrm{d}A_2}{\mathrm{d}z} = \mathrm{i}\kappa_2 A_2 + \mathrm{i}\omega_2 C_{\mathrm{TPA}}A_2|A_1|^2, \tag{7.108}
$$

where κ_1 and κ_2 are self-coupling coefficients for the single waveguide mode at the frequencies ω_1 and ω_2, respectively, caused possibly by any linear perturbation to the waveguide. In the above, we assume that the waveguide has no linear loss so that $\kappa_1 = \kappa_1^*$ and $\kappa_2 = \kappa_2^*$. From (7.104) and (6.62), we see that the nonlinear coupling coefficient C_{TPA} is proportional to χ_{TPA}, but it is integrated over the profiles of the mode fields. By using the relation given in (7.30) that $P_{q,v} = |A_{q,v}|^2$ for the mode power, we can transform (7.107) and (7.108) into two coupled equations in terms of the mode power:

$$\frac{\mathrm{d}P_1}{\mathrm{d}z} = -2\omega_1 C_{\mathrm{TPA}}'' P_1 P_2, \tag{7.109}$$

$$\frac{\mathrm{d}P_2}{\mathrm{d}z} = -2\omega_2 C_{\mathrm{TPA}}'' P_1 P_2. \tag{7.110}$$

These two equations have the same forms as (6.65) and (6.66), except that they are expressed in terms of the mode powers whereas (6.65) and (6.66) are expressed in terms of the optical intensities.

We find that the total power, $P = P_1 + P_2$, varies as

$$\frac{\mathrm{d}P}{\mathrm{d}z} = \frac{\mathrm{d}(P_1 + P_2)}{\mathrm{d}z} = -2(\omega_1 + \omega_2) C_{\mathrm{TPA}}'' P_1 P_2. \tag{7.111}$$

When the optical medium is in its normal state in thermal equilibrium, a lower energy level is more populated than a higher energy level, so that $C_{\mathrm{TPA}}'' > 0$. As expected, optical energy is not conserved in the nonparametric two-photon absorption process but is absorbed by the medium through the resonant two-photon transition at the frequency $\omega_1 + \omega_2$. In this process, two photons are simultaneously absorbed; therefore, we have the Manley–Rowe relation:

$$\frac{\mathrm{d}}{\mathrm{d}z}\left(\frac{P_1}{\omega_1}\right) = \frac{\mathrm{d}}{\mathrm{d}z}\left(\frac{P_2}{\omega_2}\right) = -2C_{\mathrm{TPA}}'' P_1 P_2. \tag{7.112}$$

This Manley–Rowe relation for two-photon absorption is different from that for stimulated Raman scattering because in two-photon absorption the two photons at the two frequencies are simultaneously absorbed, whereas in stimulated Raman scattering a Stokes photon is generated at the same time as a pump photon is absorbed.

Problem Set

7.1.1 Discuss the advantages of using waveguides for nonlinear optical devices.

7.1.2 Derive the general coupled-mode equation given in (7.21). Show that it takes the form of (7.25) with the coupling coefficients having the relationships given in (7.27) for the coupling of the modes of a lossless waveguide.

7.1.3 Derive the general coupled-wave-and-mode equation given in (7.31). Show that it takes the form of (7.33) for the total perturbation polarization given in (7.32).

7.2.1 Show that the coupled-wave-and-mode equations that are given in (7.41)–(7.43) in terms of the mode-field amplitudes for a three-frequency parametric interaction can be converted to those given in (7.47)–(7.49) in terms of the mode powers. Then show that the Manley–Rowe relations for this interaction take the forms of (7.51) and (7.52) in terms of the mode powers.

7.3.1 For nonparametric processes, the forms of the coupled-wave-and-mode equations in terms of the mode powers and those of the Manley–Rowe relations depend on the permutation symmetry of the nonlinear susceptibility. This difference can be seen by comparing the two nonparametric processes of stimulated Raman scattering and two-photon absorption, which have different permutation symmetries for their effective nonlinear coefficients as given in (7.89) and (7.104), respectively.

(a) Show that the coupled-wave-and-mode equations given in (7.92) and (7.93) in terms of the mode-field amplitudes for stimulated Raman scattering can be converted to those given in (7.94) and (7.95) in terms of the mode powers. Then show that the Manley–Rowe relations for this interaction take the forms of (7.96) and (7.97) in terms of the mode powers.

(b) Show that the coupled-wave-and-mode equations given in (7.107) and (7.108) in terms of the mode-field amplitudes for two-photon absorption can be converted to those given in (7.109) and (7.110) in terms of the mode powers. Then show that the Manley–Rowe relations for this interaction take the forms of (7.111) and (7.112) in terms of the mode powers.

References

[1] S. L. Chuang, "A coupled mode formulation by reciprocity and a variational principle," *Journal of Lightwave Technology*, vol. 5, pp. 5–15, 1987.

[2] J. M. Liu, *Photonic Devices*. Cambridge: Cambridge University Press, 2005.

[3] A. Yariv, "Coupled-mode theory for guided-wave optics," *IEEE Journal of Quantum Electronics*, vol. 9, pp. 919–933, 1973.

[4] H. Haus, W. Huang, S. Kawakami, and N. Whitaker, "Coupled-mode theory of optical waveguides," *Journal of Lightwave Technology*, vol. 5, pp. 16–23, 1987.

[5] H. A. Haus and W. Huang, "Coupled-mode theory," *Proceedings of the IEEE*, vol. 79, pp. 1505–1518, 1991.

[6] S. L. Chuang, "A coupled-mode theory for multiwaveguide systems satisfying the reciprocity theorem and power conservation," *Journal of Lightwave Technology*, vol. 5, pp. 174–183, 1987.

8 Nonlinear Propagation Equations

8.1 PROPAGATION IN A SPATIALLY HOMOGENEOUS MEDIUM

In the preceding two chapters, we discussed the formulations for the nonlinear interactions of discrete frequencies. The coupled-wave formulation presented in Chapter 6 is used for the nonlinear interactions of optical waves of different frequencies in a homogeneous medium. The formulation of coupled waves and modes presented in Chapter 7 is used for nonlinear interactions of optical modes of different frequencies in a waveguide. In this chapter, we discuss the general formulation for optical propagation of a temporally varying field, such as an optical pulse, in a nonlinear medium [1–5]. In this section, we consider the propagation in a homogeneous medium. In Section 8.2, we will consider the propagation in a waveguide.

The spatial and temporal evolutions of an optical field can be described by the wave equation, which can be expressed in a few alternative forms, as described in Subsection 1.1.5. Here we take the form that is given in (1.44) for the complex field:

$$\mathbf{\nabla} \times \mathbf{\nabla} \times \mathbf{E} + \mu_0 \frac{\partial^2 \mathbf{D}}{\partial t^2} = -\mu_0 \frac{\partial^2 \Delta \mathbf{P}}{\partial t^2}, \tag{8.1}$$

where $\mathbf{D}$ accounts for the linear and nonresonant optical properties of the background medium, and $\Delta \mathbf{P} = \Delta \mathbf{P}_L + \mathbf{P}_{res} + \mathbf{P}_{sp} + \mathbf{P}_{NL} + \mathbf{P}_{ext}$ in general. In this chapter, we ignore the spontaneous emission noise that is described by $\mathbf{P}_{sp}$ and the external perturbation that is described by $\mathbf{P}_{ext}$, but we keep $\mathbf{P}_{res}$ and $\mathbf{P}_{NL}$ for the possibility that the medium has resonant susceptibility and optical nonlinearity. For the propagation in a spatially homogeneous medium discussed in this section, the linear perturbation polarization $\Delta \mathbf{P}_L$ as expressed in (7.23) due to a waveguide structure does not exit. Therefore, $\Delta \mathbf{P} = \mathbf{P}_{res} + \mathbf{P}_{NL}$ for our purpose.

The electric displacement $\mathbf{D}$ is related to the electric field $\mathbf{E}$ in the real space and time domain through the relation defined in (1.43). Here, we assume that the background medium is isotropic and spatially homogeneous, and its response to the electric field is spatially local but not temporally instantaneous so that the background linear permittivity of the optical medium is $\epsilon_b(t)$, which is a scalar that is independent of direction and spatial location. Then, in the time domain,

$$\mathbf{D}(\mathbf{r}, t) = \int_{-\infty}^{t} \epsilon_b(t - t') \mathbf{E}(\mathbf{r}, t') dt'. \tag{8.2}$$

In the frequency domain,

$$D(\mathbf{r}, \omega) = \epsilon_b(\omega)\mathbf{E}(\mathbf{r}, \omega) = \epsilon_0 n_0^2(\omega)\mathbf{E}(\mathbf{r}, \omega), \tag{8.3}$$

where $n_0(\omega) = \sqrt{\epsilon_b(\omega)/\epsilon_0}$ is the frequency-dependent nonresonant linear refractive index of the background medium. The frequency-dependent linear propagation constant is

$$k(\omega) = \omega\sqrt{\mu_0 \epsilon_b(\omega)} = \frac{n_0(\omega)\omega}{c}. \tag{8.4}$$

There are a few ways to treat the possible background absorption. One possibility is to include it in the background permittivity, and thus in the linear propagation constant, by making $\epsilon_b(\omega)$ and $k(\omega)$ complex quantities. Another approach is to treat it as a part of $\mathbf{P}_{\text{res}}$ while excluding it from $\mathbf{D}$ by taking $\epsilon_b(\omega)$ and $k(\omega)$ to be real quantities. In this chapter, we take the latter approach; thus, the propagation constant $k(\omega)$ is a real quantity.

We consider in general unidirectional propagation of an optical pulse. By taking the direction of pulse propagation to be the positive z direction, we express $\mathbf{E}(\mathbf{r}, t)$, $\mathbf{P}_{\text{res}}(\mathbf{r}, t)$, and $\mathbf{P}_{\text{NL}}(\mathbf{r}, t)$ in terms of their *slowly varying envelopes* $\mathcal{E}(\mathbf{r}, t)$, $\mathcal{P}_{\text{res}}(\mathbf{r}, t)$, and $\mathcal{P}_{\text{NL}}(\mathbf{r}, t)$ as

$$\mathbf{E}(\mathbf{r}, t) = \mathcal{E}(\mathbf{r}, t)\exp(\mathrm{i}k_0 z - \mathrm{i}\omega_0 t), \tag{8.5}$$

$$\mathbf{P}_{\text{res}}(\mathbf{r}, t) = \mathcal{P}_{\text{res}}(\mathbf{r}, t)\exp(\mathrm{i}k_0 z - \mathrm{i}\omega_0 t), \tag{8.6}$$

$$\mathbf{P}_{\text{NL}}(\mathbf{r}, t) = \mathcal{P}_{\text{NL}}(\mathbf{r}, t)\exp(\mathrm{i}k_0 z - \mathrm{i}\omega_0 t), \tag{8.7}$$

where k_0 and ω_0 are, respectively, the propagation constant and the center frequency of the optical carrier that are related to each other as

$$k_0 = \frac{n_0(\omega_0)\omega_0}{c}. \tag{8.8}$$

From (8.6) and (8.7), we can express $\Delta\mathbf{P}$ as

$$\Delta\mathbf{P}(\mathbf{r}, t) = \Delta\mathcal{P}(\mathbf{r}, t)\exp(\mathrm{i}k_0 z - \mathrm{i}\omega_0 t), \tag{8.9}$$

where $\Delta\mathcal{P}(\mathbf{r}, t) = \mathcal{P}_{\text{res}}(\mathbf{r}, t) + \mathcal{P}_{\text{NL}}(\mathbf{r}, t)$ for $\Delta\mathbf{P}(\mathbf{r}, t) = \mathbf{P}_{\text{res}}(\mathbf{r}, t) + \mathbf{P}_{\text{NL}}(\mathbf{r}, t)$. Note that for the coupled-wave analysis discussed in Chapter 6, the nonlinear polarization $\mathbf{P}_q^{(n)}(\mathbf{r})$ at a specific frequency ω_q does not necessarily contain a spatial phase factor of $\exp(\mathrm{i}\mathbf{k}_q \cdot \mathbf{r})$ unless perfect phase matching is achieved for the nonlinear process. Therefore, we did not define a slowly varying amplitude $\mathcal{P}_q^{(n)}(\mathbf{r})$ in space for the general coupled-wave formulation presented in Chapter 6. In this chapter, however, we define the slowly varying envelope $\mathcal{P}_{\text{NL}}(\mathbf{r}, t)$ as in (8.7), thus $\Delta\mathcal{P}(\mathbf{r}, t)$ as in (8.9), for convenience. In the case of perfect phase matching, there is no spatially varying phase difference between $\mathcal{P}_{\text{NL}}(\mathbf{r}, t)$ and $\mathcal{E}(\mathbf{r}, t)$. In the case that phase matching is not perfect, there is a spatially varying phase difference between $\mathcal{P}_{\text{NL}}(\mathbf{r}, t)$ and $\mathcal{E}(\mathbf{r}, t)$.

For an isotropic and spatially homogeneous medium that is considered here, the relation given in (8.2) leads to the relation that $\nabla \cdot \mathbf{E}(\mathbf{r}, t) = 0$ because $\nabla \cdot \mathbf{D}(\mathbf{r}, t) = 0$. Therefore, $\nabla \times \nabla \times \mathbf{E} = \nabla\nabla \cdot \mathbf{E} - \nabla^2\mathbf{E} = -\nabla^2\mathbf{E}$. The wave equation given in (8.1) then reduces to the form:

$$\nabla^2 \mathbf{E} - \mu_0 \frac{\partial^2 \mathbf{D}}{\partial t^2} = \mu_0 \frac{\partial^2 \Delta \mathbf{P}}{\partial t^2}. \tag{8.10}$$

Every term in this equation is a second-order derivative of the total field, either in space or in time. Therefore, this wave equation contains both forward-propagating and backward-propagating components, in the positive and negative z directions, respectively. For the propagation of an optical pulse, we need a propagation equation of the field envelope [2, 5] that describes unidirectional propagation [6–9]. Furthermore, because ultrashort pulses as short as a few optical cycles have been generated, the pulse propagation equation has to be applicable to pulses as short as one carrier cycle [5, 10–12].

8.1.1 Frequency-Domain Equation

To derive the frequency-domain pulse propagation equation, we begin by transforming the wave equation of (8.10) into the frequency domain. By using (8.5) for $\mathbf{E}(\mathbf{r}, t)$, which represents unidirectional propagation in the positive z direction, we find that

$$\mathbf{E}(\mathbf{r}, \omega) = \mathcal{F}\Big[\mathbf{E}(\mathbf{r}, t)\Big] = \int_{-\infty}^{\infty} \mathbf{E}(\mathbf{r}, t) e^{i\omega t} dt = \int_{-\infty}^{\infty} \mathcal{E}(\mathbf{r}, t) e^{ik_0 z - i\omega_0 t} e^{i\omega t} dt = \mathcal{E}(\mathbf{r}, \omega - \omega_0) e^{ik_0 z},$$

$$\tag{8.11}$$

where $\mathcal{E}(\mathbf{r}, \omega) = \mathcal{F}\Big[\mathcal{E}(\mathbf{r}, t)\Big]$. By using (8.2) and (8.3), we find that

$$\mathcal{F}\left[-\mu_0 \frac{\partial^2 \mathbf{D}(\mathbf{r}, t)}{\partial t^2}\right] = \omega^2 \mu_0 \epsilon_b(\omega) \mathbf{E}(\mathbf{r}, \omega) = k^2(\omega) \mathcal{E}(\mathbf{r}, \omega - \omega_0) e^{ik_0 z}, \tag{8.12}$$

where

$$k(\omega) = \omega \sqrt{\mu_0 \epsilon_b(\omega)} = \frac{n_0(\omega)\omega}{c}, \tag{8.13}$$

as given in (8.4). By using (8.9), we find that

$$\mathcal{F}\left[\mu_0 \frac{\partial^2 \Delta \mathbf{P}(\mathbf{r}, t)}{\partial t^2}\right] = -\omega^2 \mu_0 \Delta \mathcal{P}(\mathbf{r}, \omega - \omega_0) e^{ik_0 z}, \tag{8.14}$$

where $\Delta \mathcal{P}(\mathbf{r}, \omega) = \mathcal{F}\Big[\Delta \mathcal{P}(\mathbf{r}, t)\Big]$.

By plugging (8.11), (8.12), and (8.14) into the Fourier transform of (8.10), we obtain

$$\nabla_\perp^2 \mathcal{E}(\mathbf{r}, \omega - \omega_0) + \frac{\partial^2}{\partial z^2} \mathcal{E}(\mathbf{r}, \omega - \omega_0) + 2ik_0 \frac{\partial}{\partial z} \mathcal{E}(\mathbf{r}, \omega - \omega_0) + (k^2 - k_0^2) \mathcal{E}(\mathbf{r}, \omega - \omega_0)$$
$$= -\omega^2 \mu_0 \Delta \mathcal{P}(\mathbf{r}, \omega - \omega_0),$$

$$\tag{8.15}$$

where $\nabla_\perp^2 = \partial^2/\partial x^2 + \partial^2/\partial y^2$ is the *transverse Laplacian*. The propagation constant can be expanded in a Taylor series as

$$k(\omega) = \sum_{n=0}^{\infty} \frac{1}{n!} \left(\frac{\mathrm{d}^n k}{\mathrm{d}\omega^n} \right)_{\omega_0} (\omega - \omega_0)^n = k_0 + k_1(\omega - \omega_0) + \sum_{n=2}^{\infty} \frac{k_n}{n!} (\omega - \omega_0)^n, \quad (8.16)$$

where

$$k_n = \left(\frac{\mathrm{d}^n k}{\mathrm{d}\omega^n} \right)_{\omega_0}, \quad (8.17)$$

and k_1 and k_2 are related to the group velocity, v_{g} defined in (4.81), and the group-velocity dispersion, D defined in (4.83), as

$$k_1 = \left(\frac{\mathrm{d}k}{\mathrm{d}\omega} \right)_{\omega_0} = \frac{1}{v_{\mathrm{g}}} \quad \text{and} \quad k_2 = \left(\frac{\mathrm{d}^2 k}{\mathrm{d}\omega^2} \right)_{\omega_0} = \frac{D}{c\omega_0}. \quad (8.18)$$

By using (8.16) to expand $k^2 - k_0^2$ in (8.15) and then rearranging (8.15), we obtain

$$\frac{\omega}{\omega_0} \left[\frac{\partial \mathcal{E}}{\partial z} - i \sum_{n=1}^{\infty} \frac{k_n}{n!} (\omega - \omega_0)^n \mathcal{E} + \frac{\omega}{2ic\epsilon_0 n_0} \Delta \mathcal{P} \right] + \frac{1}{2ik_0} \nabla_\perp^2 \mathcal{E}$$

$$= \left(\frac{\omega}{\omega_0} - 1 \right) \left[\frac{\partial}{\partial z} - i \sum_{n=1}^{\infty} \frac{k_n}{n!} (\omega - \omega_0)^n \right] \mathcal{E} - \frac{1}{2ik_0} \left[\frac{\partial^2}{\partial z^2} + \left(\sum_{n=1}^{\infty} \frac{k_n}{n!} (\omega - \omega_0)^n \right)^2 \right] \mathcal{E} \quad (8.19)$$

$$= \left(1 - \frac{\omega_0 k_1}{k_0} \right) \left(\frac{\omega}{\omega_0} - 1 \right) \left[\frac{\partial}{\partial z} - i \sum_{n=1}^{\infty} \frac{k_n}{n!} (\omega - \omega_0)^n \right] \mathcal{E}$$

$$- \frac{1}{2ik_0} \left[\left(\frac{\partial}{\partial z} - ik_1(\omega - \omega_0) \right)^2 + \left(\sum_{n=2}^{\infty} \frac{k_n}{n!} (\omega - \omega_0)^n \right)^2 \right] \mathcal{E}.$$

The right-hand side of this equation is small compared to the left-hand side. It can be ignored [5] under the condition that

$$\left| 1 - \frac{\omega_0 k_1}{k_0} \right| = \left| 1 - \frac{v_{\mathrm{p}}}{v_{\mathrm{g}}} \right| \ll 1 \quad (8.20)$$

and under the *slowly varying envelope approximation* that

$$\left| \left(\frac{\partial}{\partial z} - ik_1(\omega - \omega_0) \right) \mathcal{E} \right| \ll |k_0 \mathcal{E}|. \quad (8.21)$$

The combination of these two conditions is the *slowly evolving wave approximation*, which is valid for ultrashort optical pulses that have pulsewidths down to the single-cycle regime [5]. Therefore, by neglecting the right-hand side of (8.19), we obtain the *general pulse propagation equation in the frequency domain*:

$$\frac{\partial \mathcal{E}(\mathbf{r}, \omega - \omega_0)}{\partial z} - i \sum_{n=1}^{\infty} \frac{k_n}{n!} (\omega - \omega_0)^n \mathcal{E}(\mathbf{r}, \omega - \omega_0) - \frac{i\omega_0}{2k_0\omega} \nabla_\perp^2 \mathcal{E}(\mathbf{r}, \omega - \omega_0)$$

$$\approx \frac{i\omega}{2c\epsilon_0 n_0} \Delta \mathcal{P}(\mathbf{r}, \omega - \omega_0). \tag{8.22}$$

8.1.2 Time-Domain Equation

The pulse propagation equation in the time domain can be obtained by simply taking the inverse Fourier transform of (8.22). Because the field envelope in the frequency domain is a function of $\omega - \omega_0$, the inverse Fourier transform to the time domain is obtained by making the conversion:

$$\omega - \omega_0 \rightarrow i\frac{\partial}{\partial t}, \quad \text{thus} \quad \omega \rightarrow \omega_0 \left(1 + \frac{i}{\omega_0}\frac{\partial}{\partial t}\right). \tag{8.23}$$

By using (8.23) to transform (8.22) to the time domain, we obtain the *general pulse propagation equation in the time domain*:

$$\frac{\partial \mathcal{E}(\mathbf{r}, t)}{\partial z} + \frac{1}{v_g}\frac{\partial \mathcal{E}(\mathbf{r}, t)}{\partial t} + \sum_{n=2}^{\infty} \frac{i^{n-1} k_n}{n!} \frac{\partial^n \mathcal{E}(\mathbf{r}, t)}{\partial t^n} - \frac{i}{2k_0}\left(1 + \frac{i}{\omega_0}\frac{\partial}{\partial t}\right)^{-1} \nabla_\perp^2 \mathcal{E}(\mathbf{r}, t)$$

$$\approx \frac{i\omega_0}{2c\epsilon_0 n_0}\left(1 + \frac{i}{\omega_0}\frac{\partial}{\partial t}\right) \Delta \mathcal{P}(\mathbf{r}, t), \tag{8.24}$$

where $\mathcal{E}(\mathbf{r}, t)$ is related to the optical intensity as defined in (4.36):

$$I(\mathbf{r}, t) = 2c\epsilon_0 n_0 |\mathcal{E}(\mathbf{r}, t)|^2. \tag{8.25}$$

As discussed in the preceding subsection, (8.24) is valid under the slowly evolving wave approximation, which is the combination of the conditions given in (8.20) and (8.21) and takes the form in the time domain:

$$\left|1 - \frac{\omega_0 k_1}{k_0}\right| = \left|1 - \frac{v_p}{v_g}\right| \ll 1 \quad \text{and} \quad \left|\left(\frac{\partial}{\partial z} + k_1\frac{\partial}{\partial t}\right)\mathcal{E}\right| \ll |k_0\mathcal{E}|. \tag{8.26}$$

These conditions are valid for ultrashort optical pulses with pulsewidths down to the single-cycle regime [5].

On the left-hand side of the general pulse propagation equation given in (8.24), the terms of the time derivatives on $\mathcal{E}(\mathbf{r}, t)$ account for the effects of linear material dispersion on the propagation of the pulse field. The first-order time derivative term takes care of pulse propagation with the group velocity, the second-order time derivative term accounts for the group-velocity dispersion, and all high-order time derivative terms reflect high-order dispersion. The time derivative term $(i/\omega_0)\partial/\partial t$ in the denominator of the diffraction term of the transverse Laplacian accounts for an enhanced beam divergence of the low-frequency components [5]. On the right-hand side, the same time derivative term $(i/\omega_0)\partial/\partial t$ accounts for the *self-steeping effect* and *optical shock formation*, on a time scale of $1/\omega_0$, due to the dispersion of the nonlinear effects [2, 5], including self-phase modulation and cross-phase modulation due to the Kerr nonlinearity and stimulated Raman scattering. Inclusion of this term allows the general pulse propagation equation to be valid for pulses of pulsewidths down to the single-cycle regime. These two terms of $(\mathrm{i}/\omega_0)\partial/\partial t$ are neglected in the *paraxial approximation*.

8.2 PROPAGATION IN A WAVEGUIDE

For an optical pulse that propagates in a waveguide, $\Delta\mathbf{P}_\mathrm{L} \neq 0$ because of the waveguide structure. The primary effect of the waveguide structure on the propagation of the pulse is that the optical field of the pulse propagates in waveguide modes with the propagation constants of the modes. Thus, we expand the field of the optical pulse in terms of waveguide modes as

$$\mathbf{E}(\mathbf{r}, t) = \sum_v A_v(z, t)\hat{\mathcal{E}}_v(x, y)\exp\left(\mathrm{i}\beta_{v0}z - \mathrm{i}\omega_0 t\right), \tag{8.27}$$

where ω_0 is the center carrier frequency of the pulse field and β_{v0} is the propagation constant of mode v at the frequency ω_0 (i.e., $\beta_{v0} = \beta_v(\omega_0)$). As discussed in Chapter 7, the mode field $\hat{\mathcal{E}}_v$ is normalized through (7.9) such that the mode power of the pulse is related to the mode-field amplitude as

$$P_v(z, t) = \left|A_v(z, t)\right|^2 = A_v(z, t)A_v^*(z, t). \tag{8.28}$$

Our purpose in this section is to derive a general pulse propagation equation in terms of the pulse mode-field amplitude $A_v(z, t)$.

The general equation for pulse propagation in a waveguide can be obtained from the general coupled-wave-and-mode equation given in (7.31). For this purpose, we first note that the mode-field amplitude $A_{q,v}$ that appears in (7.31) is a discrete spectral amplitude at a discrete frequency of ω_q because nonlinear interactions among discrete frequencies were considered in Chapter 7. Because $A_{q,v}$ is a function of frequency, it can be identified with the spectral amplitude $A_v(z, \omega)$ of the pulse. The difference between $A_{q,v}$ and $A_v(z, \omega)$ is their units because $A_{q,v}$ is the spectral amplitude of a discrete spectrum, whereas $A_v(z, \omega)$ is that of a continuous spectrum. The relations for $A_{q,v}$ that were obtained in Chapter 7 can be converted to those for $A_v(z, \omega)$. From (7.29), we have

$$\mathbf{E}(\mathbf{r}, \omega) = \sum_v A_v(z, \omega)\hat{\mathcal{E}}_v(x, y)\exp\left[\mathrm{i}\beta_v(\omega)z\right], \tag{8.29}$$

where the frequency dependence of the mode-field profile $\hat{\mathcal{E}}_v$ is ignored as an approximation. The coupled-wave-and-mode equation given in (7.31) is converted to the equation for $A_v(z, \omega)$ of a continuous spectrum for an optical pulse that propagates unidirectionally in the positive z direction:

$$\frac{\partial A_v(z, \omega)}{\partial z} = \mathrm{i}\omega\, \mathrm{e}^{-\mathrm{i}\beta_v(\omega)z} \int_{-\infty}^{\infty} \int_{-\infty}^{\infty} \hat{\mathcal{E}}_v^* \cdot \Delta\mathbf{P}(\mathbf{r}, \omega)\mathrm{d}x\mathrm{d}y. \tag{8.30}$$

Therefore,

$$\frac{\partial A_v(z,\omega)}{\partial z}e^{i[\beta_v(\omega)-\beta_{v0}]z} = i\,\omega\,e^{-i\beta_{v0}z}\int\limits_{-\infty}^{\infty}\int\limits_{-\infty}^{\infty}\hat{\mathcal{E}}_v^* \cdot \Delta\mathbf{P}(\mathbf{r},\omega)dxdy. \tag{8.31}$$

Note that $A_v(z,\omega)$ as identified above from its relationship with $A_{q,v}$ is not the Fourier transform of $A_v(z,t)$. Its relationship with $A_v(z,t)$ can be found through the relationship between $\mathbf{E}(\mathbf{r},t)$ and its Fourier transform $\mathbf{E}(\mathbf{r},\omega)$ by using (8.29):

$$\mathbf{E}(\mathbf{r},t) = \frac{1}{2\pi}\int\limits_0^{\infty}\mathbf{E}(\mathbf{r},\omega)e^{-i\omega t}d\omega = \frac{1}{2\pi}\int\limits_0^{\infty}\sum_v A_v(z,\omega)\hat{\mathcal{E}}_v(x,y)e^{i\beta_v(\omega)z}\,e^{-i\omega t}d\omega. \tag{8.32}$$

By comparing (8.32) with (8.27), we find that

$$A_v(z,t) = \frac{1}{2\pi}\int\limits_0^{\infty}A_v(z,\omega)e^{i[\beta_v(\omega)-\beta_{v0}]z}\,e^{-i(\omega-\omega_0)t}d\omega. \tag{8.33}$$

The mode propagation constant can be expanded in a Taylor series:

$$\beta_v(\omega) - \beta_{v0} = \sum_{n=1}^{\infty}\frac{1}{n!}\left(\frac{\partial^n\beta_v}{\partial\omega^n}\right)_{\omega_0}(\omega-\omega_0)^n = \sum_{n=1}^{\infty}\frac{\beta_{vn}}{n!}(\omega-\omega_0)^n, \tag{8.34}$$

where

$$\beta_{vn} = \left(\frac{\partial^n\beta_v}{\partial\omega^n}\right)_{\omega_0}, \quad \beta_{v1} = \frac{1}{v_{gv}}, \quad \text{and} \quad \beta_{v2} = \frac{D_v}{c\omega_0}. \tag{8.35}$$

Then, by using (8.33) and (8.34), we find that

$$\begin{aligned}\frac{\partial A_v(z,t)}{\partial z} &= \frac{1}{2\pi}\int\limits_0^{\infty}\frac{\partial A_v(z,\omega)}{\partial z}e^{i[\beta_v(\omega)-\beta_{v0}]z}\,e^{-i(\omega-\omega_0)t}\,d\omega \\ &\quad + \frac{1}{2\pi}\int\limits_0^{\infty}i\left[\beta_v(\omega)-\beta_{v0}\right]A_v(z,\omega)e^{i[\beta_v(\omega)-\beta_{v0}]z}\,e^{-i(\omega-\omega_0)t}\,d\omega \\ &= \frac{1}{2\pi}\int\limits_0^{\infty}\frac{\partial A_v(z,\omega)}{\partial z}e^{i[\beta_v(\omega)-\beta_{v0}]z}\,e^{-i(\omega-\omega_0)t}\,d\omega \\ &\quad + \frac{1}{2\pi}\int\limits_0^{\infty}i\sum_{n=1}^{\infty}\frac{\beta_{vn}}{n!}(\omega-\omega_0)^n A_v(z,\omega)e^{i[\beta_v(\omega)-\beta_{v0}]z}\,e^{-i(\omega-\omega_0)t}\,d\omega \\ &= \frac{1}{2\pi}\int\limits_0^{\infty}\frac{\partial A_v(z,\omega)}{\partial z}e^{i[\beta_v(\omega)-\beta_{v0}]z}\,e^{-i(\omega-\omega_0)t}\,d\omega + i\sum_{n=1}^{\infty}\frac{i^n\beta_{vn}}{n!}\frac{\partial^n A_v(z,t)}{\partial t^n}.\end{aligned} \tag{8.36}$$

Therefore,

$$\frac{1}{2\pi}\int_0^\infty \frac{\partial A_v(z,\omega)}{\partial z}\,\mathrm{e}^{\mathrm{i}[\beta_v(\omega)-\beta_{v0}]z}\,\mathrm{e}^{-\mathrm{i}(\omega-\omega_0)t}\,\mathrm{d}\omega = \frac{\partial A_v(z,t)}{\partial z} - \mathrm{i}\sum_{n=1}^\infty \frac{\mathrm{i}^n\beta_{vn}}{n!}\frac{\partial^n A_v(z,t)}{\partial t^n}. \tag{8.37}$$

By using (8.31) and (8.37), we have

$$\frac{\partial A_v(z,t)}{\partial z} - \mathrm{i}\sum_{n=1}^\infty \frac{\mathrm{i}^n\beta_{vn}}{n!}\frac{\partial^n A_v(z,t)}{\partial t^n}$$

$$= \mathrm{i}\,\mathrm{e}^{-\mathrm{i}\beta_{v0}z}\frac{1}{2\pi}\int_0^\infty \omega \int_{-\infty}^\infty \int_{-\infty}^\infty \hat{\boldsymbol{\mathcal{E}}}_v^* \cdot \Delta\mathbf{P}(\mathbf{r},\omega)\mathrm{d}x\mathrm{d}y\,\mathrm{e}^{-\mathrm{i}(\omega-\omega_0)t}\mathrm{d}\omega \tag{8.38}$$

$$= -\mathrm{e}^{-\mathrm{i}\beta_{v0}z+\mathrm{i}\omega_0 t}\frac{\partial}{\partial t}\int_{-\infty}^\infty \int_{-\infty}^\infty \hat{\boldsymbol{\mathcal{E}}}_v^* \cdot \Delta\mathbf{P}(\mathbf{r},t)\mathrm{d}x\mathrm{d}y.$$

Thus, we find the *general pulse propagation equation for a waveguide mode:*

$$\frac{\partial A_v(z,t)}{\partial z} + \frac{1}{v_{gv}}\frac{\partial A_v(z,t)}{\partial t} + \sum_{n=2}^\infty \frac{\mathrm{i}^{n-1}\beta_{vn}}{n!}\frac{\partial^n A_v(z,t)}{\partial t^n} = -\mathrm{e}^{-\mathrm{i}\beta_{v0}z+\mathrm{i}\omega_0 t}\frac{\partial}{\partial t}\int_{-\infty}^\infty \int_{-\infty}^\infty \hat{\boldsymbol{\mathcal{E}}}_v^* \cdot \Delta\mathbf{P}(\mathbf{r},t)\mathrm{d}x\mathrm{d}y.$$

$$\tag{8.39}$$

Note that this is a coupled-mode pulse propagation equation because different waveguide modes can be coupled through $\Delta\mathbf{P}(\mathbf{r},t)$.

We can define a mode-specific envelope function for $\Delta\mathbf{P}(\mathbf{r},t)$ as

$$\Delta\boldsymbol{\mathcal{P}}_v(\mathbf{r},t) = \Delta\mathbf{P}(\mathbf{r},t)\mathrm{e}^{-\mathrm{i}\beta_{v0}z+\mathrm{i}\omega_0 t}. \tag{8.40}$$

In terms of $\Delta\boldsymbol{\mathcal{P}}_v(\mathbf{r},t)$, the general coupled-mode pulse propagation equation of (8.39) can be expressed as

$$\frac{\partial A_v(z,t)}{\partial z} + \frac{1}{v_{gv}}\frac{\partial A_v(z,t)}{\partial t} + \sum_{n=2}^\infty \frac{\mathrm{i}^{n-1}\beta_{vn}}{n!}\frac{\partial^n A_v(z,t)}{\partial t^n} = \mathrm{i}\omega_0\left(1 + \frac{\mathrm{i}}{\omega_0}\frac{\partial}{\partial t}\right)\int_{-\infty}^\infty \int_{-\infty}^\infty \hat{\boldsymbol{\mathcal{E}}}_v^* \cdot \Delta\boldsymbol{\mathcal{P}}_v(\mathbf{r},t)\mathrm{d}x\mathrm{d}y.$$

$$\tag{8.41}$$

This equation has a form that is similar to that of (8.24). Compared to (8.24), however, this equation does not have a transverse Laplacian term because we have assumed that the wave-guiding effect is sufficiently strong for the optical field to stay guided as it propagates. Note that $\Delta\boldsymbol{\mathcal{P}}_v(\mathbf{r},t)$ is mode specific because β_{v0} is mode specific, as defined in (8.40). Therefore, when the pulse consists of more than one mode, $\Delta\boldsymbol{\mathcal{P}}_v(\mathbf{r},t)$ has different values for the coupled pulse propagation equations of different modes.

8.3   PROPAGATION IN AN OPTICAL KERR MEDIUM

In this section, we consider the propagation in an optical Kerr medium. For simplicity, we consider a lossless medium that has no resonant susceptibility so that the only perturbation polarization is the nonlinear polarization from the optical Kerr effect such that

$$\Delta\mathbf{P}(\mathbf{r},t) = \mathbf{P}^{(3)}(\mathbf{r},t) = 3\epsilon_0\boldsymbol{\chi}^{(3)}(\omega_0 = \omega_0 + \omega_0 - \omega_0) \vdots \mathbf{E}(\mathbf{r},t)\mathbf{E}(\mathbf{r},t)\mathbf{E}^*(\mathbf{r},t). \quad (8.42)$$

For the propagation in a spatially homogeneous medium, we consider only linearly polarized or circularly polarized fields, but not an elliptically polarized field. Then, by using (8.5), we have

$$\Delta\mathbf{P}(\mathbf{r},t) = 3\epsilon_0\chi_{\text{eff}}\left|\mathbf{E}(\mathbf{r},t)\right|^2\mathbf{E}(\mathbf{r},t) = 4c\epsilon_0^2 n_0^2 n_2 \left|\mathcal{E}(\mathbf{r},t)\right|^2 \mathcal{E}(\mathbf{r},t)e^{ik_0 z - i\omega_0 t}, \quad (8.43)$$

where χ_{eff} and n_2 depend on the polarization of the electric field $\mathbf{E}(\mathbf{r},t)$. As discussed in Section 12.1.1, $\chi_{\text{eff}} = \chi_{1111}^{(3)\prime}$ with n_2 given in (12.3) for a linearly polarized field, and $\chi_{\text{eff}} = \chi_{1122}^{(3)\prime} + \chi_{1212}^{(3)\prime}$ with n_2 given in (12.6) for a circularly polarized field. Then, by using (8.9), we find that

$$\Delta\mathcal{P}(\mathbf{r},t) = 4c\epsilon_0^2 n_0^2 n_2 \left|\mathcal{E}(\mathbf{r},t)\right|^2 \mathcal{E}(\mathbf{r},t). \quad (8.44)$$

For the propagation in a waveguide, by using (8.27), we have

$$\Delta\mathbf{P}(\mathbf{r},t) = 3\epsilon_0\boldsymbol{\chi}^{(3)} \vdots \mathbf{E}(\mathbf{r},t)\mathbf{E}(\mathbf{r},t)\mathbf{E}^*(\mathbf{r},t) = 3\epsilon_0\sum_{\mu,\xi,\zeta}\boldsymbol{\chi}^{(3)} \vdots \hat{\boldsymbol{\mathcal{E}}}_\mu\hat{\boldsymbol{\mathcal{E}}}_\xi\hat{\boldsymbol{\mathcal{E}}}_\zeta^* A_\mu A_\xi A_\zeta^* e^{i(\beta_{\mu 0}+\beta_{\xi 0}-\beta_{\zeta 0})z - i\omega_0 t}. \quad (8.45)$$

By using (8.40), we find that

$$\Delta\mathcal{P}_v(\mathbf{r},t) = 3\epsilon_0\sum_{\mu,\xi,\zeta}\boldsymbol{\chi}^{(3)} \vdots \hat{\boldsymbol{\mathcal{E}}}_\mu\hat{\boldsymbol{\mathcal{E}}}_\xi\hat{\boldsymbol{\mathcal{E}}}_\zeta^* A_\mu A_\xi A_\zeta^* e^{i(\beta_{\mu 0}+\beta_{\xi 0}-\beta_{\zeta 0}-\beta_{v 0})z}. \quad (8.46)$$

Thus,

$$\omega_0 \int\limits_{-\infty}^{\infty} \int\limits_{-\infty}^{\infty} \hat{\boldsymbol{\mathcal{E}}}_v^* \cdot \Delta\mathcal{P}_v(\mathbf{r},t)\mathrm{d}x\mathrm{d}y = \sum_{\mu,\xi,\zeta}\sigma_{v\mu\xi\zeta}A_\mu A_\xi A_\zeta^* e^{i(\beta_{\mu 0}+\beta_{\xi 0}-\beta_{\zeta 0}-\beta_{v 0})z}, \quad (8.47)$$

where

$$\sigma_{v\mu\xi\zeta} = 3\omega_0\epsilon_0 \int\limits_{-\infty}^{\infty} \int\limits_{-\infty}^{\infty} \hat{\boldsymbol{\mathcal{E}}}_v^* \cdot \boldsymbol{\chi}^{(3)} \vdots \hat{\boldsymbol{\mathcal{E}}}_\mu\hat{\boldsymbol{\mathcal{E}}}_\xi\hat{\boldsymbol{\mathcal{E}}}_\zeta^*\mathrm{d}x\mathrm{d}y. \quad (8.48)$$

8.3.1 Nonlinear Equation with Spatial Diffraction

The propagation of an optical pulse in a spatially homogeneous medium is generally governed by the equation given in (8.24). By using (8.44), we can reduce (8.24) to a scalar equation for the propagation in an optical Kerr medium:

$$\frac{\partial \mathcal{E}(\mathbf{r},t)}{\partial z} + \frac{1}{v_g}\frac{\partial \mathcal{E}(\mathbf{r},t)}{\partial t} + \sum_{n=2}^{\infty}\frac{\mathrm{i}^{n-1}k_n}{n!}\frac{\partial^n \mathcal{E}(\mathbf{r},t)}{\partial t^n} - \frac{\mathrm{i}}{2k_0}\left(1 + \frac{\mathrm{i}}{\omega_0}\frac{\partial}{\partial t}\right)^{-1}\nabla_\perp^2 \mathcal{E}(\mathbf{r},t)$$
$$= \mathrm{i}s\left(1 + \frac{\mathrm{i}}{\omega_0}\frac{\partial}{\partial t}\right)\left|\mathcal{E}(\mathbf{r},t)\right|^2 \mathcal{E}(\mathbf{r},t), \tag{8.49}$$

where

$$s = 2k_0 c \epsilon_0 n_2 = 2\omega_0 \epsilon_0 n_0 n_2. \tag{8.50}$$

In the case of a continuous wave, or a very long pulse, the time derivative terms can be neglected. Then (8.49) reduces to the simple form:

$$\mathrm{i}\frac{\partial \mathcal{E}}{\partial z} + \frac{1}{2k_0}\nabla_\perp^2 \mathcal{E} + s\left|\mathcal{E}\right|^2 \mathcal{E} = 0. \tag{8.51}$$

In (8.49) and (8.51), the transverse Laplacian term accounts for the effect of beam divergence that is caused by spatial diffraction. Therefore, these two equations can describe the phenomenon of self-focusing or self-defocusing, discussed in Section 12.2, due to spatially dependent self-phase modulation caused by the optical Kerr effect.

For the propagation in a waveguide as discussed in the preceding section, no spatial diffraction can take place because we have assumed that the waveguiding effect is sufficiently strong for the optical field to stay guided as it propagates. In the case that the waveguiding effect is not sufficiently strong to overcome the spatial diffraction that results from the optical Kerr effect in a waveguide, the mode expansion that was carried out in the preceding section is not valid. Then, we ignore the waveguide and use (8.49) or (8.51).

8.3.2 Nonlinear Schrödinger Equation

In many cases of pulse propagation, the effects of spatial diffraction and high-order dispersion above the second order can both be neglected. Furthermore, the time derivative term $(\mathrm{i}/\omega_0)\partial/\partial t$ for the self-steeping effect is also ignored. Then, for pulse propagation in an isotropic and spatially homogeneous optical Kerr medium, (8.49) reduces to the form:

$$\frac{\partial \mathcal{E}(z,t)}{\partial z} + \frac{1}{v_g}\frac{\partial \mathcal{E}(z,t)}{\partial t} + \mathrm{i}\frac{k_2}{2}\frac{\partial^2 \mathcal{E}(z,t)}{\partial t^2} = \mathrm{i}s\left|\mathcal{E}(z,t)\right|^2 \mathcal{E}(z,t). \tag{8.52}$$

For pulse propagation in a waveguide that is made of an isotropic optical Kerr medium, (8.41) reduces to the form:

$$\frac{\partial A_\nu(z,t)}{\partial z} + \frac{1}{v_{g\nu}}\frac{\partial A_\nu(z,t)}{\partial t} + \mathrm{i}\frac{\beta_{\nu 2}}{2}\frac{\partial^2 A_\nu(z,t)}{\partial t^2} = \mathrm{i}\sum_{\mu,\xi,\zeta}\sigma_{\nu\mu\xi\zeta}A_\mu(z,t)A_\xi(z,t)A_\zeta^*(z,t)\mathrm{e}^{\mathrm{i}(\beta_{\mu 0}+\beta_{\xi 0}-\beta_{\zeta 0}-\beta_{\nu 0})z},$$

$$\tag{8.53}$$

where (8.47) is used for the right-hand side. In the following, we consider for simplicity pulse propagation in a single-mode waveguide. Then (8.53) reduces to the simple form:

$$\frac{\partial A(z,t)}{\partial z} + \frac{1}{v_g}\frac{\partial A(z,t)}{\partial t} + i\frac{\beta_2}{2}\frac{\partial^2 A(z,t)}{\partial t^2} = i\sigma|A(z,t)|^2 A(z,t), \tag{8.54}$$

where the mode index is dropped because there is only one mode.

It can be seen that (8.52) and (8.54) have the same form. The only difference between the two equations is the different units of the field amplitudes $\mathcal{E}$ and A, and consequently the different units of the coefficients s and σ. The form of these pulse propagation equations can be simplified by making the change of variables:

$$\zeta = z \quad \text{and} \quad \tau = t - \frac{z}{v_g}. \tag{8.55}$$

Then, (8.52) and (8.54) are converted to the form of the *nonlinear Schrödinger equation* [1, 13] as

$$i\frac{\partial \mathcal{E}(\zeta,\tau)}{\partial \zeta} = \frac{k_2}{2}\frac{\partial^2 \mathcal{E}(\zeta,\tau)}{\partial \tau^2} - s|\mathcal{E}(\zeta,\tau)|^2 \mathcal{E}(\zeta,\tau) \tag{8.56}$$

and

$$i\frac{\partial A(\zeta,\tau)}{\partial \zeta} = \frac{\beta_2}{2}\frac{\partial^2 A(\zeta,\tau)}{\partial \tau^2} - \sigma|A(\zeta,\tau)|^2 A(\zeta,\tau). \tag{8.57}$$

These equations are sufficient for describing many nonlinear propagation characteristics of optical pulses, including spectral broadening and soliton formation, both caused by self-phase modulation but in different regions of group-velocity dispersion, as discussed in Chapter 18. They are valid under the following conditions:

1. The spatial divergence of the beam is negligible so that the transverse Laplacian term can be ignored.
2. The medium is not highly dispersive and the spectrum of the pulse is not extremely broad so that the linear dispersion above the second order can be ignored.
3. The pulsewidth is much larger than the time scale of $1/\omega_0$ so that the shock term $(i/\omega_0)\partial/\partial t$ can be ignored.

The first of these three conditions is satisfied for pulse propagation in a waveguide, as discussed in the preceding subsection. The other two conditions are usually satisfied for optical pulses of pulsewidths down to a few hundred femtoseconds in the case that they do not create extremely large nonlinear spectral broadening. However, these two conditions can be violated in the case of extreme spectral broadening, such as supercontinuum generation discussed in Chapter 19, or extremely short pulses that have pulsewidths of a few optical cycles. Then, both the high-order dispersion terms and the shock term have to be included in the pulse propagation equation.

8.3.3 Generalized Nonlinear Schrödinger Equation

As discussed above, the pulse propagation equations in the form of the nonlinear Schrödinger equation are not sufficient for describing the nonlinear propagation characteristics of an optical pulse that has a pulsewidth down to a few optical cycles or that undergoes extreme spectral broadening. In this case, the high-order dispersion terms and the shock term have to be restored [1]. Then, we have

$$\frac{\partial \mathcal{E}(z,t)}{\partial z} + \frac{1}{v_g}\frac{\partial \mathcal{E}(z,t)}{\partial t} + \sum_{n=2}^{\infty}\frac{i^{n-1}k_n}{n!}\frac{\partial^n \mathcal{E}(z,t)}{\partial t^n} = is\left(1 + \frac{i}{\omega_0}\frac{\partial}{\partial t}\right)\left|\mathcal{E}(z,t)\right|^2\mathcal{E}(z,t) \tag{8.58}$$

for pulse propagation in an isotropic and spatially homogeneous optical Kerr medium, and

$$\frac{\partial A(z,t)}{\partial z} + \frac{1}{v_g}\frac{\partial A(z,t)}{\partial t} + \sum_{n=2}^{\infty}\frac{i^{n-1}\beta_n}{n!}\frac{\partial^n A(z,t)}{\partial t^n} = i\sigma\left(1 + \frac{i}{\omega_0}\frac{\partial}{\partial t}\right)\left|A(z,t)\right|^2 A(z,t) \tag{8.59}$$

for pulse propagation in a single-mode waveguide. By using the change of variables of (8.55), we can convert these equations into the form of a *generalized nonlinear Schrödinger equation*:

$$\frac{\partial \mathcal{E}(\zeta,\tau)}{\partial \zeta} + \sum_{n=2}^{\infty}\frac{i^{n-1}k_n}{n!}\frac{\partial^n \mathcal{E}(\zeta,\tau)}{\partial \tau^n} = is\left(1 + \frac{i}{\omega_0}\frac{\partial}{\partial \tau}\right)\left|\mathcal{E}(\zeta,\tau)\right|^2\mathcal{E}(\zeta,\tau) \tag{8.60}$$

and

$$\frac{\partial A(\zeta,\tau)}{\partial \zeta} + \sum_{n=2}^{\infty}\frac{i^{n-1}\beta_n}{n!}\frac{\partial^n A(\zeta,\tau)}{\partial \tau^n} = i\sigma\left(1 + \frac{i}{\omega_0}\frac{\partial}{\partial \tau}\right)\left|A(\zeta,\tau)\right|^2 A(\zeta,\tau). \tag{8.61}$$

As the optical pulse gets shorter, down to a few optical cycles, the response time of the optical nonlinearity has to be considered. The equations obtained above within this section are based on the nonlinear polarization $\mathbf{P}^{(3)}(\mathbf{r},t)$ that is expressed in (8.42). Though it is a time-domain polarization, it has the form of a frequency-domain polarization in terms of a direct tensor product rather than in terms of a temporal convolution integral, as it should be in the time domain. The condition for the validity of this expression as a direct tensor product is that the optical nonlinearity responds to the optical excitation *instantaneously*, which requires that the response time of the optical nonlinearity be much shorter than the pulsewidth of the optical pulse. The optical Kerr effect of most dielectric materials satisfies this condition for pulses with very short pulsewidths, down to the hundred-femtosecond regime. Nonetheless, the optical nonlinearity has a nonzero response time; consequently, the response cannot be considered as instantaneous when the pulsewidth is sufficiently short. Furthermore, the optical Kerr effect cannot always be isolated from other nonlinear effects. Indeed, it can occur together with stimulated Raman scattering [2, 14]. Stimulated Raman scattering, which is a resonant process, has a slower response than the optical Kerr effect, which is a nonresonant process. In the frequency domain, the optical Kerr effect is characterized by the real part of $\chi^{(3)}$, whereas stimulated Raman scattering is characterized by the imaginary part of $\chi^{(3)}$. Theoretically, they can be separately modeled, but experimentally they cannot be separated as the material responses to the optical field. The total nonlinear response function in the time domain thus has a response time that is defined by the slower process of stimulated Raman scattering. Further discussions of the theoretical and experimental consequences of these effects will be given in Chapter 19 on supercontinuum generation.

To account for the nonzero response time of the optical nonlinearity, the nonlinear response in the time-domain pulse propagation equation has to be expressed as a temporal convolution integral. Then, we have

$$\frac{\partial \mathcal{E}(z,t)}{\partial z} + \frac{1}{v_g}\frac{\partial \mathcal{E}(z,t)}{\partial t} + \sum_{n=2}^{\infty} \frac{i^{n-1}k_n}{n!}\frac{\partial^n \mathcal{E}(z,t)}{\partial t^n} = is\left(1 + \frac{i}{\omega_0}\frac{\partial}{\partial t}\right)\left[\mathcal{E}(z,t)\int_{-\infty}^{\infty} R(t')\big|\mathcal{E}(z,t-t')\big|^2 dt'\right]$$

(8.62)

for pulse propagation in an isotropic and spatially homogeneous optical Kerr medium, and

$$\frac{\partial A(z,t)}{\partial z} + \frac{1}{v_g}\frac{\partial A(z,t)}{\partial t} + \sum_{n=2}^{\infty} \frac{i^{n-1}\beta_n}{n!}\frac{\partial^n A(z,t)}{\partial t^n} = i\sigma\left(1 + \frac{i}{\omega_0}\frac{\partial}{\partial t}\right)\left[A(z,t)\int_{-\infty}^{\infty} R(t')\big|A(z,t-t')\big|^2 dt'\right]$$

(8.63)

for pulse propagation in a single-mode waveguide. In (8.62) and (8.63), $R(t)$ is the normalized temporal response function of the optical nonlinearity, which generally accounts for both the optical Kerr effect and the stimulated Raman process. By using the change of variables of (8.55), we can convert these equations into generalized nonlinear Schrödinger equations of the forms:

$$\frac{\partial \mathcal{E}(\zeta,\tau)}{\partial \zeta} + \sum_{n=2}^{\infty} \frac{i^{n-1}k_n}{n!}\frac{\partial^n \mathcal{E}(\zeta,\tau)}{\partial \tau^n} = is\left(1 + \frac{i}{\omega_0}\frac{\partial}{\partial \tau}\right)\left[\mathcal{E}(\zeta,\tau)\int_{-\infty}^{\infty} R(\tau')\big|\mathcal{E}(\zeta,\tau-\tau')\big|^2 d\tau'\right] \qquad (8.64)$$

and

$$\frac{\partial A(\zeta,\tau)}{\partial \zeta} + \sum_{n=2}^{\infty} \frac{i^{n-1}\beta_n}{n!}\frac{\partial^n A(\zeta,\tau)}{\partial \tau^n} = i\sigma\left(1 + \frac{i}{\omega_0}\frac{\partial}{\partial \tau}\right)\left[A(\zeta,\tau)\int_{-\infty}^{\infty} R(\tau')\big|A(\zeta,\tau-\tau')\big|^2 d\tau'\right]. \qquad (8.65)$$

Problem Set

8.1.1 The equation for pulse propagation in a spatially homogeneous medium is obtained from the general wave equation given in (8.10). It is expressed in terms of the slowly varying envelopes $\mathcal{E}(\mathbf{r},t)$ and $\Delta\mathcal{P}(\mathbf{r},t)$ of $\mathbf{E}(\mathbf{r},t)$ and $\Delta\mathbf{P}(\mathbf{r},t)$ in (8.5) and (8.9), respectively, on the optical carrier that has a propagation constant of k_0 and a center frequency of ω_0.

    (a) Show that the wave equation given in (8.10) can be converted to the equation given in (8.19) in the frequency domain. Then show that under the slowly evolving wave approximation it leads to the pulse propagation equation given in (8.22) in the frequency domain.

(b) Show that the pulse propagation equation given in (8.22) in the frequency domain can be converted to that given in (8.24) in the time domain.

8.2.1 Identify the key steps in the derivation of the general pulse propagation equation for a waveguide mode given in (8.39). Show that (8.39) can be converted to the form given in (8.41).

References

[1] Y. Kodama and A. Hasegawa, "Nonlinear pulse propagation in a monomode dielectric guide," *IEEE Journal of Quantum Electronics*, vol. 23, pp. 510–524, 1987.

[2] K. J. Blow and D. Wood, "Theoretical description of transient stimulated Raman scattering in optical fibers," *IEEE Journal of Quantum Electronics*, vol. 25, pp. 2665–2673, 1989.

[3] P. V. Mamyshev and S. V. Chernikov, "Ultrashort-pulse propagation in optical fibers," *Optics Letters*, vol. 15, pp. 1076–1078, 1990.

[4] P. L. François, "Nonlinear propagation of ultrashort pulses in optical fibers: total field formulation in the frequency domain," *Journal of the Optical Society of America B*, vol. 8, pp. 276–293, 1991.

[5] T. Brabec and F. Krausz, "Nonlinear optical pulse propagation in the single-cycle regime," *Physical Review Letters*, vol. 78, pp. 3282–3285, 1997.

[6] M. Kolesik, J. V. Moloney, and M. Mlejnek, "Unidirectional optical pulse propagation equation," *Physical Review Letters*, vol. 89, p. 283902, 2002.

[7] M. Kolesik and J. V. Moloney, "Nonlinear optical pulse propagation simulation: from Maxwell's to unidirectional equations," *Physical Review E*, vol. 70, p. 036604, 2004.

[8] P. Kinsler, S. B. P. Radnor, and G. H. C. New, "Theory of directional pulse propagation," *Physical Review A*, vol. 72, p. 063807, 2005.

[9] P. Kinsler, "Unidirectional optical pulse propagation equation for materials with both electric and magnetic responses," *Physical Review A*, vol. 81, p. 023808, 2010.

[10] T. Brabec and F. Krausz, "Intense few-cycle laser fields: frontiers of nonlinear optics," *Reviews of Modern Physics*, vol. 72, pp. 545–591, 2000.

[11] N. Karasawa, S. Nakamura, N. Nakagawa, M. Shibata, R. Morita, H. Shigekawa, *et al.*, "Comparison between theory and experiment of nonlinear propagation for a-few-cycle and ultrabroadband optical pulses in a fused-silica fiber," *IEEE Journal of Quantum Electronics*, vol. 37, pp. 398–404, 2001.

[12] G. Genty, P. Kinsler, B. Kibler, and J. M. Dudley, "Nonlinear envelope equation modeling of sub-cycle dynamics and harmonic generation in nonlinear waveguides," *Optics Express*, vol. 15, pp. 5382–5387, 2007.

[13] A. Hasegawa and F. Tappert, "Transmission of stationary nonlinear optical pulses in dispersive dielectric fibers: I. Anomalous dispersion," *Applied Physics Letters*, vol. 23, pp. 142–144, 1973.

[14] R. H. Stolen, J. P. Gordon, W. J. Tomlinson, and H. A. Haus, "Raman response function of silica-core fibers," *Journal of the Optical Society of America B*, vol. 6, pp. 1159–1166, 1989.

9 Phase Matching

9.1 PHASE-MATCHING GEOMETRY

The general conditions for phase matching have been discussed in Subsection 5.1.1. The nonlinear processes that require efforts on phase matching are listed in Tables 5.1 and 5.2 for the second-order and third-order processes, respectively. As discussed in Chapter 5, all nonparametric processes are automatically phase matched. Parametric processes generally are not automatically phase matched; the exceptions are such parametric processes for which the phase-matching conditions are mathematically true irrespective of the values and the directions of the wavevectors, such as optical rectification and the optical Kerr effect. Most parametric frequency-conversion processes, such as second-harmonic generation and third-harmonic generation, are not automatically phase matched, thus requiring arrangements to achieve phase matching.

In Chapters 6 and 7 we saw the importance of phase matching for those parametric nonlinear processes that are not automatically phase matched. If a parametric frequency-conversion process is perfectly phase matched, optical power can be efficiently converted from one frequency to another. Otherwise, the process is periodically reversed; thus, the optical power shuttles back and forth among the interacting waves, as Fig. 6.1(c) shows. No matter how long the crystal is, the best efficiency we can expect from a parametric interaction that is not perfectly phase matched is that contributed by the interaction over a coherence length. Therefore, phase matching is one of the most important technical issues that have to be addressed in designing any efficient nonlinear optical device that is based on a parametric frequency-conversion process. In this chapter, we discuss the techniques of phase matching for parametric frequency-conversion processes. Most of the discussions in this chapter focus on the second-order nonlinear optical processes because the third-order parametric frequency-conversion processes, such as third-harmonic generation, are relatively inefficient compared to the second-order processes and are thus not practically useful in most situations. Some exceptions do exist. One example is third-order parametric four-wave mixing of small frequency differences in an optical fiber, as discussed in Subsection 9.4.1 and in Section 10.7. Because of the small frequency differences, the dispersion among the interacting frequencies is small, which can be either uncompensated or compensated by the modal dispersion of the fiber. Because of the long interaction length that is made possible by an optical fiber, the efficiency can be significant despite the small third-order susceptibility.

The phase-matching condition of a nonlinear optical process is a relation among the wavevectors of the interacting waves. When more than two different wavevectors are involved, phase matching can be either *collinear* or *noncollinear*, as respectively illustrated in Figs. 9.1(a) and

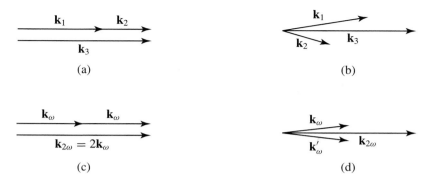

Figure 9.1 (a) Collinear and (b) noncollinear phase matching for a second-order process with the phase-matching condition $\mathbf{k}_3 = \mathbf{k}_1 + \mathbf{k}_2$. (c) Collinear and (d) noncollinear phase matching for second-harmonic generation with the phase-matching condition $\mathbf{k}_{2\omega} = \mathbf{k}_\omega + \mathbf{k}'_\omega$.

(b) for a second-order process. All of the wavevectors are parallel to one another in collinear phase matching, but they are not in noncollinear phase matching. In the case of second-harmonic generation with only one input fundamental wave, phase matching is always collinear because there are only two wavevectors, $\mathbf{k}_\omega$ and $\mathbf{k}_{2\omega}$, that are involved in the process, as shown in Fig. 9.1(c). However, noncollinear phase matching for second-harmonic generation is also possible if the input consists of two spatially separated fundamental waves, as shown in Fig. 9.1(d). In the latter situation, the wavevectors of the two distinct fundamental waves can have different magnitudes if the two waves are not polarized in the same direction inside a birefringent crystal.

For noncollinear phase matching, the interaction length is limited by the finite distance over which the beams overlap in space. Therefore, the collinear phase-matching arrangement is employed in most nonlinear optical devices, though noncollinear phase matching is also useful in some special applications, such as in an autocorrelator for measuring ultrashort laser pulses. In the following, we consider only collinear phase matching, mostly for the second-order nonlinear processes, but the general concepts can be easily extended to noncollinear phase matching.

With collinear beams, the phase-matching condition for a general second-order parametric process reduces to the simple scalar relation:

$$k_3 = k_1 + k_2, \quad \text{or} \quad n_3\omega_3 = n_1\omega_1 + n_2\omega_2. \tag{9.1}$$

Efficient parametric interactions are normally carried out in a spectral region away from the transition resonances of a medium to avoid the attenuation of the optical beams due to resonant absorption of the medium. As discussed in Subsection 4.4.1 and shown in Fig. 4.11, a material has normal dispersion in a spectral region away from resonances, meaning that $n(\omega_3) > n(\omega_1)$ and $n(\omega_3) > n(\omega_2)$ for $\omega_3 > \omega_1$ and $\omega_3 > \omega_2$. Clearly, within a spectral region of normal dispersion, collinear phase matching is not possible in a nonbirefringent medium, such as an isotropic material or a cubic crystal, nor is it possible in a birefringent crystal if all of the interacting waves have the same polarization. An isotropic material is of no practical use for second-order nonlinear interactions because $\chi^{(2)} = 0$ in the electric dipole approximation. A noncentrosymmetric cubic crystal, such as GaAs, has a decent $\chi^{(2)}$. However, such a crystal in the bulk, homogeneous form is also not useful for second-order frequency-conversion interactions because of its inability to support collinear phase matching in the normal dispersion region where the crystal is transparent to the interacting optical waves. It is useful for the second-order processes that are automatically phase matched, namely, the processes of optical rectification and Pockels effect.

Collinear phase matching can be achieved through the use of

1. the anomalous dispersion near the resonance frequency of a material;
2. the birefringence in a nonlinear crystal;
3. a periodic spatial modulation in the nonlinear coefficient of a medium for *quasi-phase matching*; or
4. the modal dispersion of an optical waveguide.

Among these possibilities, the use of anomalous dispersion is not very practical for device applications because of strong material absorption near a resonance frequency. The modal dispersion of a waveguide is usually not strong; thus, it is also of limited usefulness. Therefore, collinear phase matching for most practical applications is accomplished through birefringent phase matching or quasi-phase matching.

9.2 BIREFRINGENT PHASE MATCHING

Birefringent phase matching employs the birefringence of a uniaxial or biaxial crystal to accomplish phase matching of a nonlinear optical process. It is the most commonly used method of obtaining collinear phase matching for a second-order frequency-conversion process. Many useful nonlinear crystals, such as LiNbO$_3$, BBO, and KDP, are uniaxial, whereas other important nonlinear crystals, such as KTP, KTA, and LBO, are biaxial. The linear and nonlinear optical properties of some important nonlinear crystals for second-order applications are listed in Table 3.4. The index of refraction as a function of wavelength can be calculated by using the *Sellmeier equation*:

$$n^2 = A + \frac{B}{\lambda^2 - C} - D\lambda^2, \tag{9.2}$$

where λ is the wavelength measured in micrometers and A, B, C, and D are the *Sellmeier parameters* that are determined by fitting experimentally measured index data to the Sellmeier equation. Each index of refraction has a set of Sellmeier parameters. A uniaxial crystal has two sets for n_o and n_e, respectively. A biaxial crystal has three sets for n_x, n_y, and n_z, respectively. The Sellmeier parameters of the nonlinear crystals are also listed in Table 3.4.

9.2.1 Uniaxial Crystals

As discussed in Subsection 4.3.3 and illustrated in Fig. 4.5, there are two normal mode polarizations, $\hat{e}_o$ and $\hat{e}_e$, associated with each direction $\hat{k}$ of wave propagation in a uniaxial crystal. The ordinary wave of the polarization $\hat{e}_o$ has an ordinary refractive index of n_o that is independent of the direction of $\hat{k}$, whereas the extraordinary wave of the polarization $\hat{e}_e$ has an extraordinary refractive index of $n_e(\theta)$ that is given in (4.62) and is a function of the angle θ between the $\hat{k}$ vector and the optical axis $\hat{z}$. To satisfy the phase-matching condition given in (9.1) in a spectral region of normal dispersion, the wave at the highest frequency of ω_3 has to be associated with the smaller of the two indices. Consequently, in a positive uniaxial crystal the wave at ω_3, or that at 2ω in the case of second-harmonic generation, has to be an ordinary wave, whereas in a negative uniaxial crystal it has to be an extraordinary wave.

There are two different types of birefringent phase-matching methods. In *type I phase matching* the two low-frequency waves have the same polarization, whereas in *type II phase matching* they have orthogonal polarizations. Note that in collinear phase matching, the **k** vectors of the interacting waves are all parallel to one another. Therefore, their normal modes also have the same $\hat{e}_o$ and $\hat{e}_e$ vectors. Table 9.1 summarizes the characteristics of type I and type II phase-matching methods for uniaxial crystals. When $\omega_1 \neq \omega_2$, there are two different possibilities of type II phase matching in a positive uniaxial crystal and two possibilities in a negative uniaxial crystal, with the wave at ω_1 being the ordinary wave and that at ω_2 being the extraordinary wave, or the wave at ω_1 being the extraordinary wave and that at ω_2 being the ordinary wave. These possibilities are listed in Table 9.1. The angle θ_{PM} between the vector $\hat{k}$ and the optical axis that allows a particular phase-matching condition to be satisfied is known as the *phase-matching angle*.

For second-harmonic generation, $\omega_1 = \omega_2 = \omega$ and $\omega_3 = 2\omega$. For collinearly phase-matched second-harmonic generation, there is only one fundamental wave. In type I phase matching, the fundamental wave is completely polarized along one of the normal mode polarizations; thus, the phase-matching condition is simply

$$n_{2\omega}^o = n_\omega^e(\theta_{PM}) \tag{9.3}$$

for a positive uniaxial crystal, or

$$n_{2\omega}^e(\theta_{PM}) = n_\omega^o \tag{9.4}$$

Table 9.1 **Two types of birefringent phase matching for uniaxial crystals**

	Positive uniaxial ($n_e > n_o$)	Negative uniaxial ($n_e < n_o$)
Type I	$n_3^o\omega_3 = n_1^e(\theta_{PM})\omega_1 + n_2^e(\theta_{PM})\omega_2$	$n_3^e(\theta_{PM})\omega_3 = n_1^o\omega_1 + n_2^o\omega_2$
	$\chi_{eff} = \hat{e}_o \cdot \chi^{(2)}(\omega_3 = \omega_1 + \omega_2) : \hat{e}_e\hat{e}_e$	$\chi_{eff} = \hat{e}_e \cdot \chi^{(2)}(\omega_3 = \omega_1 + \omega_2) : \hat{e}_o\hat{e}_o$
Type II	$n_3^o\omega_3 = n_1^o\omega_1 + n_2^e(\theta_{PM})\omega_2$	$n_3^e(\theta_{PM})\omega_3 = n_1^o\omega_1 + n_2^e(\theta_{PM})\omega_2$
	$\chi_{eff} = \hat{e}_o \cdot \chi^{(2)}(\omega_3 = \omega_1 + \omega_2) : \hat{e}_o\hat{e}_e$	$\chi_{eff} = \hat{e}_e \cdot \chi^{(2)}(\omega_3 = \omega_1 + \omega_2) : \hat{e}_o\hat{e}_e$
	$n_3^o\omega_3 = n_1^e(\theta_{PM})\omega_1 + n_2^o\omega_2$	$n_3^e(\theta_{PM})\omega_3 = n_1^e(\theta_{PM})\omega_1 + n_2^o\omega_2$
	$\chi_{eff} = \hat{e}_o \cdot \chi^{(2)}(\omega_3 = \omega_1 + \omega_2) : \hat{e}_e\hat{e}_o$	$\chi_{eff} = \hat{e}_e \cdot \chi^{(2)}(\omega_3 = \omega_1 + \omega_2) : \hat{e}_e\hat{e}_o$

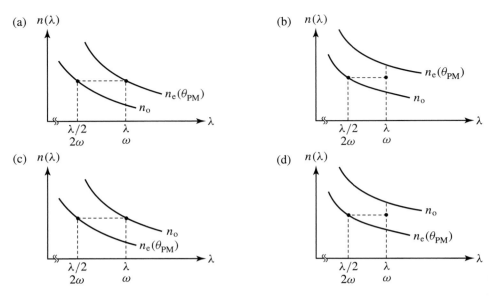

Figure 9.2 Different phase-matching methods in the region of normal dispersion for second-harmonic generation: (a) type I and (b) type II phase matching in a positive uniaxial crystal, and (c) type I and (d) type II phase matching in a negative uniaxial crystal.

for a negative uniaxial crystal. In type II phase matching, the fundamental wave consists of components in both normal mode polarizations. The two possibilities of type II phase matching listed in Table 9.1 for a parametric interaction with $\omega_1 \neq \omega_2$ become the same for second-harmonic generation with $\omega_1 = \omega_2 = \omega$. Therefore, the phase-matching condition becomes

$$n_{2\omega}^{o} = \frac{1}{2}\left[n_{\omega}^{o} + n_{\omega}^{e}(\theta_{PM})\right] \tag{9.5}$$

for a positive uniaxial crystal, or

$$n_{2\omega}^{e}(\theta_{PM}) = \frac{1}{2}\left[n_{\omega}^{o} + n_{\omega}^{e}(\theta_{PM})\right] \tag{9.6}$$

for a negative uniaxial crystal. These different phase-matching methods for second-harmonic generation are illustrated in Fig. 9.2.

Phase-Matching Angles for Three-Frequency Interactions

For parametric generation, sum-frequency generation, or difference-frequency generation with $\omega_3 = \omega_1 + \omega_2$ and $\omega_1 \neq \omega_2$, the phase-matching angles for type I phase matching in a negative uniaxial crystal and type II phase matching in a positive uniaxial crystal can be explicitly solved from the phase-matching conditions listed in Table 9.1. The phase-matching angles for type I phase matching in a positive uniaxial crystal and type II phase matching in a negative uniaxial crystal cannot be explicitly solved but have to be numerically found. Instead of the optical frequencies, we express these relations in terms of the optical wavelengths, λ_1, λ_2, and λ_3, which are commonly quoted in practical applications. Note that $\lambda_3^{-1} = \lambda_1^{-1} + \lambda_2^{-1}$ with $\omega_3 = \omega_1 + \omega_2$.

For type I phase matching in a positive uniaxial crystal, the phase-matching angle for a three-frequency parametric interaction has to be numerically found from the equation:

$$
\left\{ \left(\frac{n_1^o}{\lambda_1} \right)^{-2} + \left[\left(\frac{n_1^e}{\lambda_1} \right)^{-2} - \left(\frac{n_1^o}{\lambda_1} \right)^{-2} \right] \sin^2 \theta_{PM} \right\}^{-1/2}
$$
$$
+ \left\{ \left(\frac{n_2^o}{\lambda_2} \right)^{-2} + \left[\left(\frac{n_2^e}{\lambda_2} \right)^{-2} - \left(\frac{n_2^o}{\lambda_2} \right)^{-2} \right] \sin^2 \theta_{PM} \right\}^{-1/2} = \frac{n_3^o}{\lambda_3} .
\tag{9.7}
$$

For type I phase matching in a negative uniaxial crystal, the phase-matching angle for a three-frequency parametric interaction is

$$
\theta_{PM} = \sin^{-1} \left[\frac{(n_1^o/\lambda_1 + n_2^o/\lambda_2)^{-2} - (n_3^o/\lambda_3)^{-2}}{(n_3^e/\lambda_3)^{-2} - (n_3^o/\lambda_3)^{-2}} \right]^{1/2} .
\tag{9.8}
$$

For type II phase matching in a positive uniaxial crystal, there are two phase-matching angles for a three-frequency parametric interaction with $\lambda_1 \neq \lambda_2$:

$$
\theta_{PM} = \sin^{-1} \left[\frac{(n_3^o/\lambda_3 - n_1^o/\lambda_1)^{-2} - (n_2^o/\lambda_2)^{-2}}{(n_2^e/\lambda_2)^{-2} - (n_2^o/\lambda_2)^{-2}} \right]^{1/2}
\tag{9.9}
$$

and

$$
\theta_{PM} = \sin^{-1} \left[\frac{(n_3^o/\lambda_3 - n_2^o/\lambda_2)^{-2} - (n_1^o/\lambda_1)^{-2}}{(n_1^e/\lambda_1)^{-2} - (n_1^o/\lambda_1)^{-2}} \right]^{1/2} .
\tag{9.10}
$$

For type II phase matching in a negative uniaxial crystal, there are two phase-matching angles for a three-frequency parametric interaction with $\lambda_1 \neq \lambda_2$, which have to be numerically found from the equations:

$$
\left\{ \left(\frac{n_3^o}{\lambda_3} \right)^{-2} + \left[\left(\frac{n_3^e}{\lambda_3} \right)^{-2} - \left(\frac{n_3^o}{\lambda_3} \right)^{-2} \right] \sin^2 \theta_{PM} \right\}^{-1/2}
$$
$$
- \left\{ \left(\frac{n_2^o}{\lambda_2} \right)^{-2} + \left[\left(\frac{n_2^e}{\lambda_2} \right)^{-2} - \left(\frac{n_2^o}{\lambda_2} \right)^{-2} \right] \sin^2 \theta_{PM} \right\}^{-1/2} = \frac{n_1^o}{\lambda_1}
\tag{9.11}
$$

and

$$
\left\{ \left(\frac{n_3^o}{\lambda_3} \right)^{-2} + \left[\left(\frac{n_3^e}{\lambda_3} \right)^{-2} - \left(\frac{n_3^o}{\lambda_3} \right)^{-2} \right] \sin^2 \theta_{PM} \right\}^{-1/2}
$$
$$
- \left\{ \left(\frac{n_1^o}{\lambda_1} \right)^{-2} + \left[\left(\frac{n_1^e}{\lambda_1} \right)^{-2} - \left(\frac{n_1^o}{\lambda_1} \right)^{-2} \right] \sin^2 \theta_{PM} \right\}^{-1/2} = \frac{n_2^o}{\lambda_2} .
\tag{9.12}
$$

Phase-Matching Angles for Second-Harmonic Generation

The phase-matching angles of different phase-matching conditions for second-harmonic generation can be found from those of the different cases given above for a three-frequency parametric interaction by setting $\lambda_1 = \lambda_2 = \lambda$ for the fundamental wave and $\lambda_3 = \lambda/2$ for the second-harmonic wave. They can also be directly found by solving the specific phase-matching relation given in (9.3)–(9.6) for second-harmonic generation under the four different conditions. By using the relation that $n_e^{-2}(\theta) = n_o^{-2}\cos^2\theta + n_e^{-2}\sin^2\theta$ from (4.62), the phase-matching angles in (9.3)–(9.5) can be explicitly solved and expressed in a concise form, but that in (9.6) cannot be explicitly solved for a simple expression.

For type I phase matching in a positive uniaxial crystal, the phase-matching angle for second-harmonic generation is found from (9.3) to be

$$\theta_{PM} = \sin^{-1}\left[\frac{(n_{2\omega}^o)^{-2} - (n_\omega^o)^{-2}}{(n_\omega^e)^{-2} - (n_\omega^o)^{-2}}\right]^{1/2}. \tag{9.13}$$

For type I phase matching in a negative uniaxial crystal, the phase-matching angle for second-harmonic generation is found from (9.4) to be

$$\theta_{PM} = \sin^{-1}\left[\frac{(n_\omega^o)^{-2} - (n_{2\omega}^o)^{-2}}{(n_{2\omega}^e)^{-2} - (n_{2\omega}^o)^{-2}}\right]^{1/2}. \tag{9.14}$$

For type II phase matching in a positive uniaxial crystal, the phase-matching angle for second-harmonic generation is found from (9.5) to be

$$\theta_{PM} = \sin^{-1}\left[\frac{(2n_{2\omega}^o - n_\omega^o)^{-2} - (n_\omega^o)^{-2}}{(n_\omega^e)^{-2} - (n_\omega^o)^{-2}}\right]^{1/2}. \tag{9.15}$$

For type II phase matching in a negative uniaxial crystal, the phase-matching angle for second-harmonic generation cannot be expressed in a concise form. From (9.6), this phase-matching angle can be numerically found by solving the equation:

$$2\left\{\left(n_{2\omega}^o\right)^{-2} + \left[\left(n_{2\omega}^e\right)^{-2} - \left(n_{2\omega}^o\right)^{-2}\right]\sin^2\theta_{PM}\right\}^{-1/2}$$
$$-\left\{\left(n_\omega^o\right)^{-2} + \left[\left(n_\omega^e\right)^{-2} - \left(n_\omega^o\right)^{-2}\right]\sin^2\theta_{PM}\right\}^{-1/2} = n_\omega^o. \tag{9.16}$$

Angle for Maximizing Effective Nonlinear Susceptibility

We see from (4.59) that the $\hat{k}$ vector is a function of both angles θ and ϕ. In a phase-matched interaction, the value of θ is the phase-matching angle θ_{PM}, which is obtained by solving the condition for a specific phase-matching method. Phase matching in a uniaxial crystal is independent of the angle ϕ. Therefore, the value of θ_{PM} is determined without any requirement on the angle ϕ. The value of χ_{eff} is usually a function of both θ and ϕ, however. For example, KDP is a negative uniaxial crystal of the $\overline{4}2m$ point group with the only nonvanishing second-order nonlinear susceptibility elements being $\chi_{14}^{(2)} = \chi_{25}^{(2)}$ and $\chi_{36}^{(2)}$ under the conditions for index contraction. By using (4.60) and (4.61) for $\hat{e}_o$ and $\hat{e}_e$, respectively, we find that $\chi_{eff} = -\chi_{36}^{(2)} \sin\theta \sin 2\phi$, or $d_{eff} = -d_{36} \sin\theta \sin 2\phi$, for type I phase matching; and $\chi_{eff} = \left(\chi_{14}^{(2)} + \chi_{36}^{(2)}\right) \sin\theta \cos\theta \cos 2\phi$, or $d_{eff} = (d_{14} + d_{36}) \sin\theta \cos\theta \cos 2\phi$, for type II phase matching. Therefore, to maximize the value of $|\chi_{eff}|$ so that a second-order interaction in KDP is most efficient, ϕ has to be chosen to have one of the values among $\pi/4$, $-\pi/4$, $3\pi/4$, and $-3\pi/4$ in the case of type I phase matching and one of the values among 0, $\pi/2$, $-\pi/2$, and π in the case of type II phase matching. In some crystals, notably uniaxial crystals of the 4, 6, 422, 622, 4mm, and 6mm point groups, χ_{eff} is independent of the angle ϕ but is a function of θ only. Then the value of ϕ can be arbitrarily chosen, though that of θ is still determined by the phase-matching condition.

For a specific nonlinear interaction in a given crystal, the type I and type II phase-matching methods generally have different phase-matching angles and different effective nonlinear susceptibilities. In certain cases, only one type of phase matching is possible. Sometimes, both types are not possible in a particular crystal within a certain spectral range. In case it is possible to have both type I and type II phase matching, the choice between the two depends on many practical considerations, including efficiency, angular tolerance, temperature sensitivity, and beam walk-off. Usually the one with the larger value of $|\chi_{eff}|$ is chosen if it has no significant disadvantages from other considerations. Sometimes, χ_{eff} vanishes when phase matching is achieved. Clearly, such phase matching is of no practical usefulness. A simple example is type II phase matching in KDP with $\theta_{PM} = \pi/2$.

In summary, the condition for phase matching and the value of χ_{eff} have to be considered at the same time when designing a practical device. Phase matching by itself does not guarantee a desirable value of χ_{eff} and, in some special cases, can even lead to a vanishing χ_{eff}. For a collinearly phase-matched interaction in a uniaxial crystal, the value of θ is determined by phase matching to be $\theta = \theta_{PM}$, whereas that of ϕ is determined by maximizing the value of $|\chi_{eff}|$. When both θ and ϕ are determined, the propagation direction $\hat{k}$, which is common to all of the interacting waves in a collinear interaction, is fixed. In practice, a nonlinear crystal that is intended for a particular application is usually cut with the knowledge of the correct values of θ and ϕ for the application in a way that the $\hat{k}$ vector is normal to the input surface of the crystal, and the $\hat{e}_o$ and $\hat{e}_e$ polarizations are along certain convenient directions in the experimental setup.

EXAMPLE 9.1

Both type I and type II configurations of collinear birefringent phase matching are considered for second-harmonic generation in LiNbO₃ with a fundamental wave at the 1.064 μm wavelength. LiNbO₃ is a negative uniaxial crystal of the $3m$ point group. According to Table 3.4, the nonvanishing nonlinear susceptibility tensor elements of LiNbO₃ are $d_{31} = d_{32} = d_{24} = d_{15} = -4.4$ pm V⁻¹, $d_{22} = -d_{21} = -d_{16} = 2.4$ pm V⁻¹, and $d_{33} = -25.2$ pm V⁻¹. The refractive indices of LiNbO₃ at room temperature are $n_o = 2.2340$ and $n_e = 2.1554$ at the fundamental wavelength of 1.064 μm and $n_o = 2.3251$ and $n_e = 2.2330$ at the second-harmonic wavelength of 532 nm. Find the polarization directions of the interacting waves, the phase-matching angle, and the effective nonlinear susceptibility for each type.

Solution

The polarizations of the ordinary and extraordinary waves in a uniaxial crystal are $\hat{e}_o = \hat{x} \sin\phi - \hat{y} \cos\phi$ and $\hat{e}_e = -\hat{x} \cos\theta \cos\phi - \hat{y} \cos\theta \sin\phi + \hat{z} \sin\theta$, which are given in (4.60) and (4.61), respectively. The extraordinary index at an angle θ is given by (4.62) as $n_e^{-2}(\theta) = n_o^{-2} \cos^2\theta + n_e^{-2} \sin^2\theta$.

For type I phase matching with a negative uniaxial crystal such as LiNbO₃, we find from Table 9.1 that the fundamental wave is an ordinary wave with $\hat{e}_\omega = \hat{e}_o$ and the second-harmonic wave has to be an extraordinary wave with $\hat{e}_{2\omega} = \hat{e}_e$. With the given nonvanishing nonlinear susceptibility elements, we find that the effective nonlinear susceptibility is

$$\chi^I_{\text{eff}} = \hat{e}_e \cdot \chi^{(2)} : \hat{e}_o \hat{e}_o = \chi^{(2)}_{31} \sin\theta - \chi^{(2)}_{22} \cos\theta \sin 3\phi$$

or, equivalently, $d^I_{\text{eff}} = d_{31} \sin\theta - d_{22} \cos\theta \sin 3\phi$, where $\theta = \theta^I_{\text{PM}}$. The phase-matching angle θ^I_{PM} can be found by using the relation given in (9.4) for $n^e_{2\omega}(\theta_{\text{PM}}) = n^o_\omega$. By using the formula for $n_e(\theta)$, we find from (9.14) that

$$\theta^I_{\text{PM}} = \sin^{-1}\left[\frac{(n^o_\omega)^{-2} - (n^o_{2\omega})^{-2}}{(n^e_{2\omega})^{-2} - (n^o_{2\omega})^{-2}}\right]^{1/2} = \sin^{-1}\left[\frac{2.2340^{-2} - 3.2351^{-2}}{2.2330^{-2} - 3.2351^{-2}}\right]^{1/2} = 83.84°.$$

The angle ϕ is chosen so that $|d^I_{\text{eff}}|$ is maximized because ϕ is irrelevant to phase matching in a uniaxial crystal. Because $d_{31} < 0$, $d_{22} > 0$, and $0° < \theta < 90°$, we can maximize $|d^I_{\text{eff}}|$ by simply making $\sin 3\phi = 1$. Therefore, ϕ is chosen to be $-90°$, $30°$, or $150°$. We then find that $d^I_{\text{eff}} = -4.63$ pm V⁻¹ and $\chi^I_{\text{eff}} = -9.26$ pm V⁻¹ for $\theta = 83.84°$ and $\phi = -90°$, $30°$, or $150°$.

For type II phase matching with a negative uniaxial crystal such as LiNbO₃, we find from Table 9.1 that the fundamental wave is required to have both ordinary and extraordinary components, but the second-harmonic wave is an extraordinary wave with $\hat{e}_{2\omega} = \hat{e}_e$. With the given nonvanishing nonlinear susceptibility elements, we find that the effective nonlinear susceptibility is

$$\chi^{II}_{\text{eff}} = \hat{e}_e \cdot \chi^{(2)} : \hat{e}_o \hat{e}_e = \chi^{(2)}_{22} \cos^2\theta \cos 3\phi$$

or, equivalently, $d^{II}_{\text{eff}} = d_{22} \cos^2\theta \cos 3\phi$, where $\theta = \theta^{II}_{\text{PM}}$. The phase-matching angle θ^{II}_{PM} can be found by using the relation given in (9.6). Plugging the formula for $n_e(\theta)$ into (9.6) results in

(9.16), which can be solved either graphically or numerically to find that there is no solution for θ_{PM}^{II} in the range from $0°$ to $90°$ for the given values of n_ω^o, n_ω^e, $n_{2\omega}^o$, and $n_{2\omega}^e$ of LiNbO$_3$ at room temperature. Therefore, type II phase matching is not possible for second-harmonic generation in LiNbO$_3$ at $\lambda = 1.064$ μm at room temperature. The angle ϕ can still be chosen so that $|d_{eff}^{II}|$ is maximized because ϕ is irrelevant to phase matching in a uniaxial crystal. In type II phase matching, $|d_{eff}^{II}|$ can be maximized by making $\cos 3\phi = \pm 1$ so that $|\cos 3\phi| = 1$. For this purpose, ϕ can be chosen as one of these values: $0°$, $\pm 60°$, $\pm 120°$, or $180°$. Because phase matching is not possible, maximizing $|d_{eff}^{II}|$ in this situation serves no practical purpose.

From this example, we see that type I or type II phase matching is not always possible. It depends on the specific crystal, the specific wavelengths of the interacting waves, and temperature. For second-harmonic generation in LiNbO$_3$ at room temperature, type I phase matching is possible only for a fundamental wave that has a wavelength in the range that 1.057 μm $\leq \lambda \leq 3.755$ μm, whereas type II phase matching is possible only for a fundamental wave that has a wavelength in the range that 1.686 μm $\leq \lambda \leq 2.420$ μm. Therefore, both type I and type II phase matching are not possible at room temperature for a fundamental wave that has a wavelength shorter than 1.057 μm or longer than 3.755 μm. Only type I phase matching is possible at room temperature for a fundamental wave that has a wavelength in the range that 1.057 μm $\leq \lambda \leq 1.686$ μm or in the range that 2.420 μm $\leq \lambda \leq 3.755$ μm. Both type I and type II phase matching are possible at room temperature for a fundamental wave that has a wavelength in the range that 1.686 μm $\leq \lambda \leq 2.420$ μm. (See Problem 9.2.3.)

Note that it is still necessary to consider the value of χ_{eff} when a particular type of phase matching is possible. It can be seen from the forms of χ_{eff}^I and χ_{eff}^{II} found above that χ_{eff}^I does not vanish for any phase-matching angle θ_{PM}^I, whereas $\chi_{eff}^{II} = 0$ when $\theta_{PM}^{II} = 90°$. Therefore, type II phase matching at $\theta_{PM}^{II} = 90°$ is not useful even when it could be accomplished. Because $d_{31} = -4.4$ pm V^{-1} and $d_{22} = 2.4$ pm V^{-1}, we can compare χ_{eff}^I and χ_{eff}^{II} that are obtained above to see that for LiNbO$_3$ type I interaction is more efficient than type II interaction when phase matching is possible for both types.

9.2.2 Biaxial Crystals

Phase matching with a biaxial crystal [1–4] is conceptually the same as that with a uniaxial crystal. It is still accomplished by properly utilizing the birefringence and the dispersion of a crystal. The technical difference is that with a biaxial crystal, phase matching depends on both the angle θ, which is the angle between the $\hat{k}$ vector and the z principal axis that is designated as the one with the largest principal index of refraction, and the angle ϕ, which is the angle between the projection of $\hat{k}$ on the xy plane and the x principal axis that is designated as the one with the smallest principal index of refraction. Therefore, a given phase-matching condition specifies a set of two phase-matching angles, θ_{PM} and ϕ_{PM}, not just a phase-matching angle θ_{PM}. With a biaxial crystal, there is no freedom to choose the angle ϕ for a given θ_{PM} as can be done for phase matching with a uniaxial crystal.

Similar to the categorization for phase matching in a uniaxial crystal, phase matching in a biaxial crystal can also be categorized as type I and type II. Again, in type I phase matching,

Table 9.2 **Two types of birefringent phase matching for biaxial crystals**

Type I	$n_3^f(\theta_{PM}, \phi_{PM})\omega_3 = n_1^s(\theta_{PM}, \phi_{PM})\omega_1 + n_2^s(\theta_{PM}, \phi_{PM})\omega_2$
	$\chi_{eff} = \hat{e}_f \cdot \chi^{(2)}(\omega_3 = \omega_1 + \omega_2) : \hat{e}_s\hat{e}_s$
Type II	$n_3^f(\theta_{PM}, \phi_{PM})\omega_3 = n_1^f(\theta_{PM}, \phi_{PM})\omega_1 + n_2^s(\theta_{PM}, \phi_{PM})\omega_2$
	$\chi_{eff} = \hat{e}_f \cdot \chi^{(2)}(\omega_3 = \omega_1 + \omega_2) : \hat{e}_f\hat{e}_s$
	$n_3^f(\theta_{PM}, \phi_{PM})\omega_3 = n_1^s(\theta_{PM}, \phi_{PM})\omega_1 + n_2^f(\theta_{PM}, \phi_{PM})\omega_2$
	$\chi_{eff} = \hat{e}_f \cdot \chi^{(2)}(\omega_3 = \omega_1 + \omega_2) : \hat{e}_s\hat{e}_f$

the two low-frequency waves at ω_1 and ω_2 are in the same polarization mode. In type II phase matching, the two low-frequency waves at ω_1 and ω_2 are in mutually orthogonal polarization modes. Note that the polarization normal modes of a wave that propagates in a biaxial crystal are not classified as ordinary wave and extraordinary wave but as fast wave and slow wave. The reason is that, in general, neither of the two modes is an ordinary wave except when the propagation direction is along an optical axis. Therefore, for phase matching in the spectral region of normal dispersion, which is commonly true, it is only possible that the high-frequency wave at ω_3 is a fast wave. This is true irrespective of whether the biaxial crystal is positive biaxial or negative biaxial. Table 9.2 summarizes the characteristics of type I and type II phase matching for biaxial crystals. When $\omega_1 \neq \omega_2$, there are two different possibilities of type II phase matching, with the wave at ω_1 being the fast wave and that at ω_2 being the slow wave, or the wave at ω_1 being the slow wave and that at ω_2 being the fast wave. These two possibilities are listed in Table 9.2.

For second-harmonic generation, $\omega_1 = \omega_2 = \omega$ and $\omega_3 = 2\omega$. For collinearly phase-matched second-harmonic generation, there is only one fundamental wave. In type I phase matching, the fundamental wave is completely in the slow mode. Therefore, the phase-matching condition is simply

$$n_{2\omega}^f(\theta_{PM}, \phi_{PM}) = n_\omega^s(\theta_{PM}, \phi_{PM}). \tag{9.17}$$

In type II phase matching, the fundamental wave consists of components in both fast and slow modes. The two possibilities of type II phase matching listed in Table 9.2 for a parametric interaction with $\omega_1 \neq \omega_2$ become the same for second-harmonic generation with $\omega_1 = \omega_2 = \omega$. Therefore, the phase-matching condition becomes

$$n_{2\omega}^f(\theta_{PM}, \phi_{PM}) = \frac{1}{2}\left[n_\omega^f(\theta_{PM}, \phi_{PM}) + n_\omega^s(\theta_{PM}, \phi_{PM})\right]. \tag{9.18}$$

Though there is no freedom to arbitrarily choose the angle ϕ for a given θ_{PM}, or to arbitrarily choose the angle θ for a given ϕ_{PM}, while maintaining phase matching for a given interaction, it is possible to first fix the value of θ or ϕ and then find the value of the other angle that can accomplish phase matching. In this case, the value of the second angle for phase matching depends on the fixed value of the first angle. There are in principle an infinite number of choices in this approach. In practice, there are four common arrangements for phase matching in a biaxial crystal [5, 6]:

1. general phase matching by finding all possible combinations of θ_{PM} and ϕ_{PM} without fixing either θ or ϕ, followed by choosing a specific combination from the possible combinations [1, 2, 4];
2. phase matching on the xy principal plane by fixing the angle θ at $\theta = 90°$, for $\hat{k} = \hat{x} \cos\phi + \hat{y} \sin\phi$, while finding a phase-matching angle ϕ_{PM} for ϕ;
3. phase matching on the yz principal plane by fixing the angle ϕ at $\phi = 90°$, for $\hat{k} = \hat{y} \sin\theta + \hat{z} \cos\theta$, while finding a phase-matching angle θ_{PM} for θ; and
4. phase matching on the zx principal plane by fixing the angle ϕ at $\phi = 0°$, for $\hat{k} = \hat{x} \sin\theta + \hat{z} \cos\theta$, while finding a phase-matching angle θ_{PM} for θ.

Figure 9.3 shows the arrangements for phase matching on the three principal planes. Each arrangement can have type I and type II phase matching.

Angles for General Phase Matching

For general phase matching, the concept is straightforward, but numerical solution is generally required. We first find the indices of refraction, n_f and n_s, of the fast and slow waves as functions of θ and ϕ by using (4.64) or (4.65). Then we use the phase-matching relations that are listed in Table 9.2 to simultaneously find the possible combinations of phase-matching angles θ_{PM} and ϕ_{PM}.

The relation given in (4.65) can be express in the form [2]:

$$x^2 - Bx + C = 0 \qquad (9.19)$$

by letting $x = n^{-2}$, $B = k_x^2(b + c) + k_y^2(c + a) + k_z^2(a + b)$, and $C = k_x^2 bc + k_y^2 ca + k_z^2 ab$, where $a = n_x^{-2}$, $b = n_y^{-2}$, $c = n_z^{-2}$, $k_x = \sin\theta \cos\phi$, $k_y = \sin\theta \sin\phi$, and $k_z = \cos\theta$. By solving (9.19), we find two solutions for the indices of refraction, the smaller one for the fast wave and the larger one for the slow wave:

$$n_f = \frac{\sqrt{2}}{\sqrt{B + \sqrt{B^2 - 4C}}} \quad \text{and} \quad n_s = \frac{\sqrt{2}}{\sqrt{B - \sqrt{B^2 - 4C}}}. \qquad (9.20)$$

For type I phase matching, we then find the possible combinations of phase-matching angles θ_{PM} and ϕ_{PM} by solving the equation:

$$\frac{\sqrt{2}}{\lambda_3\sqrt{B_3 + \sqrt{B_3^2 - 4C_3}}} = \frac{\sqrt{2}}{\lambda_1\sqrt{B_1 - \sqrt{B_1^2 - 4C_1}}} + \frac{\sqrt{2}}{\lambda_2\sqrt{B_2 - \sqrt{B_2^2 - 4C_2}}}. \qquad (9.21)$$

For type II phase matching, we find two sets of possible combinations of phase-matching angles θ_{PM} and ϕ_{PM} by solving

$$\frac{\sqrt{2}}{\lambda_3\sqrt{B_3 + \sqrt{B_3^2 - 4C_3}}} = \frac{\sqrt{2}}{\lambda_1\sqrt{B_1 + \sqrt{B_1^2 - 4C_1}}} + \frac{\sqrt{2}}{\lambda_2\sqrt{B_2 - \sqrt{B_2^2 - 4C_2}}} \qquad (9.22)$$

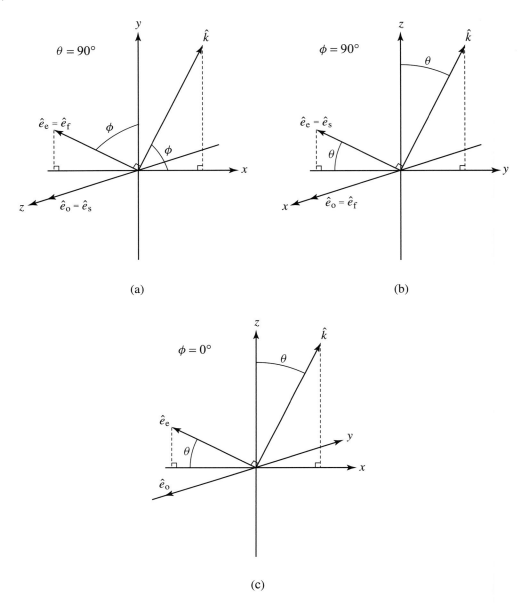

Figure 9.3 Propagation directions and polarization normal modes for phase matching in a biaxial crystal on three different principal planes: (a) on the xy principal plane, (b) on the yz principal plane, and (c) on the zx principal plane.

and

$$\frac{\sqrt{2}}{\lambda_3\sqrt{B_3 + \sqrt{B_3^2 - 4C_3}}} = \frac{\sqrt{2}}{\lambda_1\sqrt{B_1 - \sqrt{B_1^2 - 4C_1}}} + \frac{\sqrt{2}}{\lambda_2\sqrt{B_2 + \sqrt{B_2^2 - 4C_2}}}. \qquad (9.23)$$

Angles for Phase Matching on the *xy* Principal Plane

For phase matching on the xy principal plane with $\theta = 90°$ for $\hat{k} = \hat{x}\cos\phi + \hat{y}\sin\phi$, as shown in Fig. 9.3(a), the situation is similar to phase matching in a *negative* uniaxial crystal with the fast wave similar to the extraordinary wave, $\hat{e}_e = \hat{e}_f$, which has an index of refraction being a function of the angle ϕ, and the slow wave similar to the ordinary wave, $\hat{e}_o = \hat{e}_s$, which has an index of refraction being independent of the angle ϕ. From Fig. 9.3(a), we find that

$$\hat{e}_f = -\hat{x}\sin\phi + \hat{y}\cos\phi, \quad n_f(\phi) = \left(\frac{\sin^2\phi}{n_x^2} + \frac{\cos^2\phi}{n_y^2}\right)^{-1/2};$$

$$\hat{e}_s = \hat{z}, \qquad\qquad n_s = n_z. \tag{9.24}$$

For type I phase matching, we find from Table 9.2 the phase-matching angle:

$$\phi_{PM} = \sin^{-1}\left[\frac{(n_{z1}/\lambda_1 + n_{z2}/\lambda_2)^{-2} - (n_{y3}/\lambda_3)^{-2}}{(n_{x3}/\lambda_3)^{-2} - (n_{y3}/\lambda_3)^{-2}}\right]^{1/2}. \tag{9.25}$$

For type II phase matching, there are two phase-matching angles for $\lambda_1 \neq \lambda_2$, which can be numerically found from the equations:

$$\left\{\left(\frac{n_{y3}}{\lambda_3}\right)^{-2} + \left[\left(\frac{n_{x3}}{\lambda_3}\right)^{-2} - \left(\frac{n_{y3}}{\lambda_3}\right)^{-2}\right]\sin^2\phi_{PM}\right\}^{-1/2}$$

$$-\left\{\left(\frac{n_{y2}}{\lambda_2}\right)^{-2} + \left[\left(\frac{n_{x2}}{\lambda_2}\right)^{-2} - \left(\frac{n_{y2}}{\lambda_2}\right)^{-2}\right]\sin^2\phi_{PM}\right\}^{-1/2} = \frac{n_{z1}}{\lambda_1} \tag{9.26}$$

and

$$\left\{\left(\frac{n_{y3}}{\lambda_3}\right)^{-2} + \left[\left(\frac{n_{x3}}{\lambda_3}\right)^{-2} - \left(\frac{n_{y3}}{\lambda_3}\right)^{-2}\right]\sin^2\phi_{PM}\right\}^{-1/2}$$

$$-\left\{\left(\frac{n_{y1}}{\lambda_1}\right)^{-2} + \left[\left(\frac{n_{x1}}{\lambda_1}\right)^{-2} - \left(\frac{n_{y1}}{\lambda_1}\right)^{-2}\right]\sin^2\phi_{PM}\right\}^{-1/2} = \frac{n_{z2}}{\lambda_2}. \tag{9.27}$$

Angles for Phase Matching on the *yz* Principal Plane

For phase matching on the yz principal plane with $\phi = 90°$ for $\hat{k} = \hat{y}\sin\theta + \hat{z}\cos\theta$, as shown in Fig. 9.3(b), the situation is similar to phase matching in a *positive* uniaxial crystal with the fast wave similar to the ordinary wave, $\hat{e}_o = \hat{e}_f$, which has an index of refraction being independent of the angle θ, and the slow wave similar to the extraordinary wave, $\hat{e}_e = \hat{e}_s$, which has an index of refraction being a function of the angle θ. From Fig. 9.3(b), we find that

$$\hat{e}_f = \hat{x}, \qquad\qquad n_f = n_x;$$

$$\hat{e}_s = -\hat{y}\cos\theta + \hat{z}\sin\theta, \quad n_s(\theta) = \left(\frac{\cos^2\theta}{n_y^2} + \frac{\sin^2\theta}{n_z^2}\right)^{-1/2}. \tag{9.28}$$

For type I phase matching, the phase-matching angle can be numerically found from the equation:

$$\left\{\left(\frac{n_{y1}}{\lambda_1}\right)^{-2} + \left[\left(\frac{n_{z1}}{\lambda_1}\right)^{-2} - \left(\frac{n_{y1}}{\lambda_1}\right)^{-2}\right]\sin^2\theta_{PM}\right\}^{-1/2} \tag{9.29}$$

$$+\left\{\left(\frac{n_{y2}}{\lambda_2}\right)^{-2} + \left[\left(\frac{n_{z2}}{\lambda_2}\right)^{-2} - \left(\frac{n_{y2}}{\lambda_2}\right)^{-2}\right]\sin^2\theta_{PM}\right\}^{-1/2} = \frac{n_{x3}}{\lambda_3}.$$

For type II phase matching, there are two phase-matching angles for $\lambda_1 \neq \lambda_2$:

$$\theta_{PM} = \sin^{-1}\left[\frac{(n_{x3}/\lambda_3 - n_{x1}/\lambda_1)^{-2} - (n_{y2}/\lambda_2)^{-2}}{(n_{z2}/\lambda_2)^{-2} - (n_{y2}/\lambda_2)^{-2}}\right]^{1/2} \tag{9.30}$$

and

$$\theta_{PM} = \sin^{-1}\left[\frac{(n_{x3}/\lambda_3 - n_{x2}/\lambda_2)^{-2} - (n_{y1}/\lambda_1)^{-2}}{(n_{z1}/\lambda_1)^{-2} - (n_{y1}/\lambda_1)^{-2}}\right]^{1/2}. \tag{9.31}$$

Angles for Phase Matching on the *zx* Principal Plane

For phase matching on the *zx* principal plane with $\phi = 0°$ for $\hat{k} = \hat{x}\sin\theta + \hat{z}\cos\theta$, as shown in Fig. 9.3(c), the situation is more complicated than phase matching on the other principal planes because the optical axes lie on this plane at the angle θ_{OA} with respect to the *z* axis on the two sides of the axis. It is still similar to phase matching in a uniaxial crystal with $\hat{e}_o = -\hat{y}$ and $n_o = n_y$, which are independent of the angle θ, and $\hat{e}_e = -\hat{x}\cos\theta + \hat{z}\sin\theta$ and $n_e(\theta) = (\cos^2\theta/n_x^2 + \sin^2\theta/n_z^2)^{-1/2}$, which are functions of the angle θ. However, whether the extraordinary wave is fast or slow, and correspondingly the ordinary wave is slow or fast, depends on whether the angle θ is smaller or larger than θ_{OA}.

For $\theta < \theta_{OA}$, the situation is similar to phase matching in a *negative* uniaxial crystal for

$$\hat{e}_f = \hat{e}_e = -\hat{x}\cos\theta + \hat{z}\sin\theta, \quad n_f(\theta) = n_e(\theta) = \left(\frac{\cos^2\theta}{n_x^2} + \frac{\sin^2\theta}{n_z^2}\right)^{-1/2}; \tag{9.32}$$

$$\hat{e}_s = \hat{e}_o = -\hat{y}, \qquad\qquad n_s = n_o = n_y.$$

For type I phase matching, the phase-matching angle is

$$\theta_{PM} = \sin^{-1}\left[\frac{(n_{y1}/\lambda_1 + n_{y2}/\lambda_2)^{-2} - (n_{x3}/\lambda_3)^{-2}}{(n_{z3}/\lambda_3)^{-2} - (n_{x3}/\lambda_3)^{-2}}\right]^{1/2}. \tag{9.33}$$

For type II phase matching, there are two phase-matching angles for $\lambda_1 \neq \lambda_2$, which can be numerically found from the equations:

$$
\left\{ \left(\frac{n_{x3}}{\lambda_3} \right)^{-2} + \left[\left(\frac{n_{z3}}{\lambda_3} \right)^{-2} - \left(\frac{n_{x3}}{\lambda_3} \right)^{-2} \right] \sin^2 \theta_{PM} \right\}^{-1/2}
$$
$$
- \left\{ \left(\frac{n_{x2}}{\lambda_2} \right)^{-2} + \left[\left(\frac{n_{z2}}{\lambda_2} \right)^{-2} - \left(\frac{n_{x2}}{\lambda_2} \right)^{-2} \right] \sin^2 \theta_{PM} \right\}^{-1/2} = \frac{n_{y1}}{\lambda_1}
$$

(9.34)

and

$$
\left\{ \left(\frac{n_{x3}}{\lambda_3} \right)^{-2} + \left[\left(\frac{n_{z3}}{\lambda_3} \right)^{-2} - \left(\frac{n_{x3}}{\lambda_3} \right)^{-2} \right] \sin^2 \theta_{PM} \right\}^{-1/2}
$$
$$
- \left\{ \left(\frac{n_{x1}}{\lambda_1} \right)^{-2} + \left[\left(\frac{n_{z1}}{\lambda_1} \right)^{-2} - \left(\frac{n_{x1}}{\lambda_1} \right)^{-2} \right] \sin^2 \theta_{PM} \right\}^{-1/2} = \frac{n_{y2}}{\lambda_2}.
$$

(9.35)

For $\theta > \theta_{OA}$, the situation is similar to phase matching in a *positive* uniaxial crystal for

$$
\hat{e}_f = \hat{e}_o = -\hat{y}, \qquad n_f = n_o = n_y;
$$
$$
\hat{e}_s = \hat{e}_e = -\hat{x} \cos \theta + \hat{z} \sin \theta, \quad n_s(\theta) = n_e(\theta) = \left(\frac{\cos^2 \theta}{n_x^2} + \frac{\sin^2 \theta}{n_z^2} \right)^{-1/2}.
$$

(9.36)

For type I phase matching, the phase-matching angle can be numerically found from the equation:

$$
\left\{ \left(\frac{n_{x1}}{\lambda_1} \right)^{-2} + \left[\left(\frac{n_{z1}}{\lambda_1} \right)^{-2} - \left(\frac{n_{x1}}{\lambda_1} \right)^{-2} \right] \sin^2 \theta_{PM} \right\}^{-1/2}
$$
$$
+ \left\{ \left(\frac{n_{x2}}{\lambda_2} \right)^{-2} + \left[\left(\frac{n_{z2}}{\lambda_2} \right)^{-2} - \left(\frac{n_{x2}}{\lambda_2} \right)^{-2} \right] \sin^2 \theta_{PM} \right\}^{-1/2} = \frac{n_{y3}}{\lambda_3}.
$$

(9.37)

For type II phase matching, there are two phase-matching angles for $\lambda_1 \neq \lambda_2$:

$$
\theta_{PM} = \sin^{-1} \left[\frac{(n_{y3}/\lambda_3 - n_{y1}/\lambda_1)^{-2} - (n_{x2}/\lambda_2)^{-2}}{(n_{z2}/\lambda_2)^{-2} - (n_{x2}/\lambda_2)^{-2}} \right]^{1/2}
$$

(9.38)

and

$$
\theta_{PM} = \sin^{-1} \left[\frac{(n_{y3}/\lambda_3 - n_{y2}/\lambda_2)^{-2} - (n_{x1}/\lambda_1)^{-2}}{(n_{z1}/\lambda_1)^{-2} - (n_{x1}/\lambda_1)^{-2}} \right]^{1/2}.
$$

(9.39)

Phase-Matching Angles for Second-Harmonic Generation

Similar to second-harmonic generation in a uniaxial crystal discussed in Subsection 9.2.1, the phase-matching angles of different phase-matching conditions for second-harmonic generation in a biaxial crystal can be found from those of the different cases given above for a three-frequency parametric interaction by setting $\lambda_1 = \lambda_2 = \lambda$ for the fundamental wave and $\lambda_3 = \lambda/2$ for the second-harmonic wave. They can also be directly found by solving the specific phase-matching relation given in (9.17) and (9.18) for second-harmonic generation under various conditions. Because it is very lengthy to list the phase-matching angles for all cases of second-harmonic generation, we do not do so here.

9.2.3 Angle Tuning

The phase-matching angle for a specific interaction in a given nonlinear crystal is a function of the frequencies, or the wavelengths, of the interacting waves. When the frequencies of the interacting waves are varied, the angle θ, or the angle ϕ in the case of phase matching on the xy principal plane of a biaxial crystal, has to be varied accordingly for the interaction to remain phase matched. In practice, this angle tuning is normally carried out by rotating the crystal while maintaining the beam propagation direction, though it can also be achieved by varying the beam propagation direction while fixing the orientation of the crystal. One situation in which this tuning is necessary is in an application with a wavelength-tunable input wave, such as in the generation of a wavelength-tunable difference- or sum-frequency wave or in the frequency doubling of the output from a wavelength-tunable laser. In optical parametric generation where only the pump-wave frequency at ω_3 is fixed, the parametrically generated frequencies at ω_1 and ω_2 are varied when the parameters for phase matching are varied. Therefore, angle tuning of a nonlinear crystal is a convenient means for tuning the parametric wavelengths.

Figure 9.4 shows as an example the angle-tuning curves of the parametric wavelengths for type I and type II collinear phase matching in the uniaxial $LiNbO_3$ crystal with a fixed pump wavelength at 527 nm. $LiNbO_3$ is a negative uniaxial crystal of the $3m$ point group, in which the type I interaction is more efficient than the type II interaction. As another example, Fig. 9.5 shows the angle-tuning curves of the parametric wavelengths for type II collinear phase matching on the three different principal planes in the biaxial KTP crystal with a fixed pump wavelength at 527 nm. KTP is a positive biaxial crystal of the $mm2$ point group, in which type I phase matching is not useful because $\chi_{\text{eff}} = 0$ for type I phase matching at all of the possible phase-matching angles. (See Problem 9.2.8.) Therefore, we only show its tuning curves for type II phase matching.

When the frequencies of the interacting waves are fixed, any deviation of the wave propagation direction away from the phase-matched direction results in a phase mismatch. In the rest of this section, we consider phase matching in a uniaxial crystal so that the phase-matching angle is θ_{PM}. The same concept applies to phase matching in a biaxial crystal, except that either θ_{PM} or ϕ_{PM} is varied depending on the specific phase-matching arrangement used. The amount of this phase mismatch can be calculated by expanding $\Delta k = k_1 + k_2 - k_3$ around the phase-matching angle:

$$\Delta k = (k_1 + k_2 - k_3)_{\theta_{PM}} + \Delta\theta \left[\frac{d}{d\theta}(k_1 + k_2 - k_3) \right]_{\theta_{PM}} + \frac{(\Delta\theta)^2}{2} \left[\frac{d^2}{d\theta^2}(k_1 + k_2 - k_3) \right]_{\theta_{PM}} + \cdots$$

$$= \frac{\Delta\theta}{c} \left(\omega_1 \frac{dn_1}{d\theta} + \omega_2 \frac{dn_2}{d\theta} - \omega_3 \frac{dn_3}{d\theta} \right)_{\theta_{PM}} + \frac{(\Delta\theta)^2}{2c} \left(\omega_1 \frac{d^2 n_1}{d\theta^2} + \omega_2 \frac{d^2 n_2}{d\theta^2} - \omega_3 \frac{d^2 n_3}{d\theta^2} \right)_{\theta_{PM}} + \cdots .$$

$$(9.40)$$

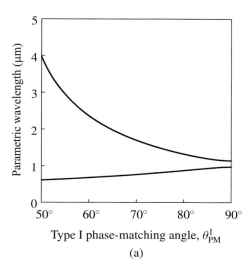

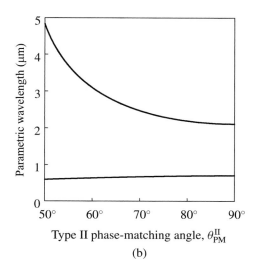

Figure 9.4 Angle-tuning curves showing parametric wavelengths as a function of the phase-matching angle for (a) type I and (b) type II collinear phase matching in LiNbO$_3$ with a fixed pump wavelength of 527 nm. LiNbO$_3$ is a negative uniaxial crystal. The data listed in Table 3.4 are used to generate these curves.

The acceptable angular tolerance in a nonlinear interaction is set by the amount of the acceptable phase mismatch. A common rule for setting this tolerance is $\Delta k l < \pi$, where l is the interaction length. In most applications of nonlinear optical devices, the interacting beams are focused to increase the efficiency. Because focusing increases the divergence, thus the angular spread, of a beam, an interaction that has a small angular tolerance requires the interacting beams to be well collimated and critically aligned.

Because $n_e(0°) = n_o$, a phase-matching angle in a uniaxial crystal cannot have the value of $0°$. Therefore, it can be shown by using (4.62) that $(dn_e(\theta)/d\theta)_{\theta_{PM}} \neq 0$, except when $\theta_{PM} = 90°$. For $\theta_{PM} \neq 90°$, the first-order term in (9.40) exists; thus, $\Delta k \propto \Delta\theta$ approximately. For phase matching with $\theta_{PM} = 90°$, known as *90° phase matching*, the first-order term in (9.40) vanishes; thus, $\Delta k \propto (\Delta\theta)^2$. Because $\Delta\theta \ll 1$, 90° phase matching has a smaller phase mismatch for a given angular deviation or, equivalently, a larger angular tolerance for a given acceptable phase mismatch than phase matching with $\theta_{PM} \neq 90°$. In 90° phase matching, an extraordinary wave is polarized along the optical axis; thus, $n_e(90°) = n_e$.

As can be seen from Table 9.1, for any method of birefringent phase matching in a uniaxial crystal, there is always at least one extraordinary wave involved in the interaction. For an extraordinary wave that is not polarized along a principal axis of the crystal, there is a walk-off angle α given by (4.73) between its direction of propagation, defined by $\hat{k}$, and its direction of power flow, defined by its Poynting vector **S**. In a collinear interaction, all of the interacting waves have the same direction of propagation, but not necessarily the same direction of power flow. When two interacting beams have different directions of power flow, there is a *walk-off angle ρ of power flow* between these two beams, which is defined as the angle between their Poynting vectors. Note the fine difference between the walk-off angle ρ of power flow between two waves and the walk-off angle α of an extraordinary wave. As is shown in Fig. 9.6(a), the

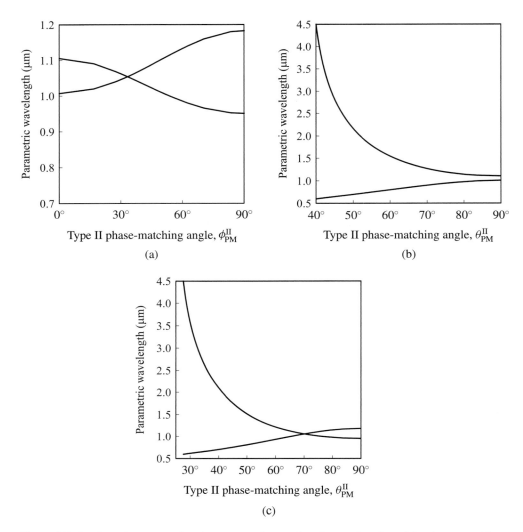

Figure 9.5 Angle-tuning curves showing parametric wavelengths as a function of the phase-matching angle for type II collinear phase matching on (a) the *xy* principal plane, (b) the *yz* principal plane, and (c) the *zx* principal plane in KTP with a fixed pump wavelength of 527 nm. KTP is a positive biaxial crystal. The data listed in Table 3.4 are used to generate these curves.

walk-off angle of power flow between an ordinary beam and an extraordinary beam is simply $\rho = |\alpha|$, which is determined only by the walk-off angle α of the extraordinary beam. However, as illustrated in Fig. 9.6(b), the walk-off angle of power flow between two collinear extraordinary beams is the difference of the walk-off angles of these two beams: $\rho = |\alpha_1 - \alpha_2|$, which exists between two extraordinary beams of different frequencies because of dispersion.

Because optical beams have finite transverse dimensions in reality, the existence of a walk-off angle ρ of power flow between two interacting beams limits the effective interaction length, as Fig. 9.6 shows. For the interaction of two Gaussian beams of the same radius of w_0 at the

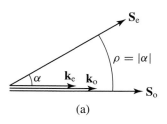
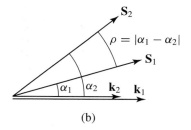

Figure 9.6 Walk-off between (a) an ordinary beam and an extraordinary beam and (b) two extraordinary beams when the beams propagate collinearly.

beam waist, the effective interaction length subject to the limitation of beam walk-off is the *aperture length*, also called the *aperture distance*:

$$l_a = \frac{\pi^{1/2} w_0}{\rho}. \tag{9.41}$$

Clearly, the aperture length decreases as the beams are increasingly focused. When $\theta_{PM} = 90°$, an extraordinary wave is polarized along the extraordinary principal axis and thus has no walk-off, as can be verified with (4.73). Consequently, there is no walk-off between any two interacting beams in the case of 90° phase matching. For this reason and for the reason discussed above that it has a larger angular tolerance than phase matching with $\theta_{PM} \neq 90°$ has, 90° phase matching is also called *noncritical phase matching*.

In the above discussion regarding noncritical phase matching by using $\theta_{PM} = 90°$, we consider phase matching in a uniaxial crystal. The same concept applies to biaxial crystal as well. With a biaxial crystal, 90° phase matching is accomplished when $\phi_{PM} = 90°$ for phase matching on the *xy* principal plane, and when $\theta_{PM} = 90°$ for phase matching on the *yz* or *zx* principal plane.

EXAMPLE 9.2

A Gaussian beam of the fundamental wave at 1.064 μm is used for second-harmonic generation with type I collinear angle phase matching in LiNbO$_3$, as discussed in Example 9.1. Find the walk-off angle ρ of power flow between the fundamental and the second-harmonic beams. If the fundamental beam is focused to a waist size of $w_0 = 50$ μm, what is the aperture length that is limited by the walk-off effect?

Solution

For type I phase matching, only the second-harmonic beam, which is an extraordinary wave, has walk-off with an angle α between $S_{2\omega}$ and $k_{2\omega}$. The fundamental beam is an ordinary wave with $S_\omega \| k_\omega$. Therefore, the walk-off between the two Poynting vectors S_ω and $S_{2\omega}$, which is what matters in this interaction, is $\rho = |\alpha|$ for collinear phase matching with $k_\omega \| k_{2\omega}$. By using (4.73) for α and taking the refractive indices to be $n_o = 2.3251$ and $n_e = 2.2330$ at the second-harmonic wavelength of 532 nm, we find, with $\theta = \theta^I_{PM} = 83.84°$, the walk-off angle:

$$\rho = |\alpha| = \left| \tan^{-1} \left(\frac{2.3251^2}{2.2330^2} \tan 83.84° \right) - 83.84° \right| = 0.475° = 8.29 \text{ mrad.}$$

For $w_0 = 50$ μm, the aperture length is

$$l_a = \frac{\pi^{1/2} w_0}{\rho} = \frac{\pi^{1/2} \times 50 \times 10^{-6}}{8.29 \times 10^{-3}} \text{ m} = 1.07 \text{ cm.}$$

The waist size of the second-harmonic beam is generally different from that of the fundamental beam. In the presence of walk-off, a second-harmonic beam generated by a circular Gaussian fundamental beam can have an elliptical spot shape. Such complications are ignored here.

9.2.4 Temperature Tuning

It is clear from the above discussions that 90° phase matching is most desirable for both type I and type II phase-matching methods. In general, the ordinary and the extraordinary indices of a uniaxial crystal have different temperature dependences, and the three principal indices of a biaxial crystal also change with temperature at different rates. In certain cases, it is possible to fix the angle θ at 90° while varying the temperature to achieve phase matching. The temperature, T_{PM}, at which 90° phase matching is achieved in a crystal is called the *phase-matching temperature*. Whether 90° phase matching by tuning the temperature is possible or not depends on the temperature dependence of the birefringence of a crystal, as well as on the wavelengths of the interacting waves in a given nonlinear process.

As discussed above, it is important to examine the value of χ_{eff} at $\theta_{PM} = 90°$ for a given phase-matching method. It turns out that $\chi_{eff} = 0$ for 90° phase matching in any uniaxial crystal if there are two extraordinary waves and one ordinary wave in the interaction, as well as in crystals of certain point groups when there are two ordinary waves and one extraordinary wave in the interaction. Specifically, *among all uniaxial crystals, 90° phase matching with $\chi_{eff} \neq 0$ is possible only for type I phase matching in a negative uniaxial crystal and for type II phase matching in a positive uniaxial crystal, but only if the crystal belongs to one of the point groups 3, 4, 6, 4mm, 6mm, 3m, $\overline{4}$, and $\overline{4}2m$ in either case.* For biaxial crystals, the situation is more complicated.

Because the phase-matching temperature for a specific nonlinear interaction in a given crystal is a function of the frequencies of the interacting waves, it has to be tuned when the wavelengths of the waves are varied. Alternatively, in optical parametric generation with a fixed pump wavelength, the parametrically generated wavelengths can be tuned by tuning the temperature while keeping both the propagation direction of the beams and the orientation of the crystal fixed. Figure 9.7 shows as an example the temperature-tuning curves of the parametric wavelengths for type I and type II collinear phase matching in LiNbO$_3$ with a fixed pump wavelength at 527 nm. The ordinary and extraordinary refractive indices of LiNbO$_3$ as functions of wavelength and temperature are given by the Sellmeier equations:

$$n_o^2 = 4.9130 + \frac{0.1173 + 1.65 \times 10^{-8}\, T^2}{\lambda^2 - (0.212 + 2.7 \times 10^{-8}\, T^2)^2} - 0.0278\, \lambda^2, \tag{9.42}$$

$$n_e^2 = 4.5567 + 2.605 \times 10^{-7}\, T^2 + \frac{0.0970 + 2.70 \times 10^{-8}\, T^2}{\lambda^2 - (0.201 + 5.4 \times 10^{-8}\, T^2)^2} - 0.0224\, \lambda^2, \tag{9.43}$$

where λ is the optical wavelength in micrometers and T is the temperature in kelvin. It can be seen from (9.42) and (9.43) that the ordinary and extraordinary indices of refraction have very different temperature dependences, which can be utilized for temperature tuning. As shown in Fig. 9.7(b), temperature tuning with type II collinear phase matching for parametric frequency generation with LiNbO$_3$ is possible, but it is not practically useful because χ_{eff} vanishes for type II 90° phase matching in LiNbO$_3$. Therefore, temperature tuning for LiNbO$_3$ is useful only with type I phase matching.

In temperature phase matching, deviation from the phase-matching temperature results in a phase mismatch given as

$$\begin{aligned}
\Delta k &= (k_1 + k_2 - k_3)_{T_{\mathrm{PM}}} + \Delta T \left[\frac{\mathrm{d}}{\mathrm{d}T}(k_1 + k_2 - k_3) \right]_{T_{\mathrm{PM}}} + \cdots \\
&= \frac{\Delta T}{c} \left(\omega_1 \frac{\mathrm{d}n_1}{\mathrm{d}T} + \omega_2 \frac{\mathrm{d}n_2}{\mathrm{d}T} - \omega_3 \frac{\mathrm{d}n_3}{\mathrm{d}T} \right)_{T_{\mathrm{PM}}} + \cdots .
\end{aligned} \tag{9.44}$$

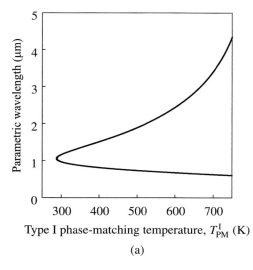

Type I phase-matching temperature, $T_{\mathrm{PM}}^{\mathrm{I}}$ (K)

(a)

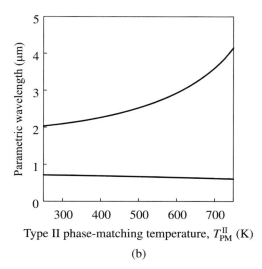

Type II phase-matching temperature, $T_{\mathrm{PM}}^{\mathrm{II}}$ (K)

(b)

Figure 9.7 Temperature-tuning curves showing parametric wavelengths as a function of the phase-matching temperature for (a) type I and (b) type II collinear phase matching in LiNbO$_3$ with a fixed pump wavelength at 527 nm. The temperature and wavelength dependences of the refractive indices given in (9.42) and (9.43) are used to generate these curves.

For a practical device that is temperature tuned, the first-order term in (9.44) normally does not vanish. Therefore, $\Delta k \propto \Delta T$ to first order.

In comparison with angle tuning, temperature tuning has all of the advantages of 90° phase matching discussed above. In addition, because the crystal orientation and the beam propagation direction are both fixed in a temperature-tuned device, temperature tuning also eliminates all of the troubles that come with angle tuning in mechanically changing the crystal orientation while trying to keep the optical beams aligned.

EXAMPLE 9.3

In this example, we consider 90° phase matching in both type I and type II configurations for second-harmonic generation in LiNbO$_3$ with a fundamental wave at the 1.064 μm wavelength. The ordinary and extraordinary refractive indices of LiNbO$_3$ as functions of wavelength and temperature are given by (9.42) and (9.43), respectively. Use the data given in Example 9.1 to find the polarization directions of the interacting waves, the phase-matching temperature, and the effective nonlinear susceptibility for each type.

Solution

For type I phase matching with LiNbO$_3$, which is negative uniaxial, the fundamental wave is an ordinary wave with $\hat{e}_\omega = \hat{e}_o$, and the second-harmonic wave has to be an extraordinary wave with $\hat{e}_{2\omega} = \hat{e}_e$. The temperature, T_{PM}^{I}, for 90° type I phase matching is found by solving the relation: $n_{2\omega}^e(T_{\text{PM}}^{\text{I}}) = n_\omega^o(T_{\text{PM}}^{\text{I}})$. By using the relations given in (9.42) and (9.43), we find the phase-matching temperature to be $T_{\text{PM}}^{\text{I}} = 316.1$ K (i.e., 43.1 °C). Because $\theta_{\text{PM}} = 90°$ for 90° phase matching, we find from Example 9.1 that $d_{\text{eff}}^{\text{I}} = d_{31} = -4.4$ pm V^{-1} and $\chi_{\text{eff}}^{\text{I}} = \chi_{31}^{(2)} = -8.8$ pm V^{-1}. Because $\chi_{\text{eff}}^{\text{I}}$ does not depend on the angle ϕ in this situation, the value of ϕ can be arbitrarily chosen.

For type II phase matching, the fundamental wave is required to have both ordinary and extraordinary components, and the second-harmonic wave is an extraordinary wave with $\hat{e}_{2\omega} = \hat{e}_e$. The temperature, $T_{\text{PM}}^{\text{II}}$, for 90° type II phase matching is found by solving the relation: $n_{2\omega}^e(T_{\text{PM}}^{\text{II}}) = \left[n_\omega^o(T_{\text{PM}}^{\text{II}}) + n_\omega^e(T_{\text{PM}}^{\text{II}})\right]/2$. By using the temperature dependences of the indices given in (9.42) and (9.43), we find that there is no solution for $T_{\text{PM}}^{\text{II}}$. Therefore, 90° type II phase matching by temperature tuning is not possible for second-harmonic generation in LiNbO$_3$ at a fundamental wavelength of 1.064 μm. Combining this finding and that in Example 9.1, we find that type II phase matching is simply not possible for second-harmonic generation in LiNbO$_3$ at the 1.064 μm fundamental wavelength.

As demonstrated above in Fig. 9.4, birefringent type II phase matching in LiNbO$_3$ is possible for parametric generation at certain wavelengths. However, 90° type II phase matching in LiNbO$_3$, and in any other crystal of the 3m point group alike, is useless anyway because $\chi_{\text{eff}}^{\text{II}} = 0$ for $\theta = 90°$, as discussed in the text above and can be seen in Example 9.1. For 90° collinear phase matching, there is no walk-off between $\mathbf{S}_\omega$ and $\mathbf{S}_{2\omega}$ because $\mathbf{S}_\omega \| \mathbf{k}_\omega \| \mathbf{k}_{2\omega} \| \mathbf{S}_{2\omega}$. In this situation, the interaction is not limited by an aperture length because $l_a = \infty$, effectively.

9.3 QUASI-PHASE MATCHING

A very different phase-matching technique involves the introduction of a periodic modulation in a nonlinear medium, which is actually the first proposed method for phase matching [7]. This approach is known as *quasi-phase matching* because phase mismatch is not eliminated within each modulation period but is periodically compensated. In principle, the periodic modulation can be on either the linear or the nonlinear susceptibility of the medium. In practice, however, modulating the linear susceptibility is less efficient than modulating the nonlinear susceptibility.

The principle of quasi-phase matching can be intuitively understood by following the discussions in Section 6.2 on the physical significance of the phase φ that is given in (6.35). The existence of a phase mismatch Δk leads to a change of the phase φ by an amount of π over a coherence length, resulting in a change of sign in $\sin \varphi$ and a reversal of the direction of energy flow in a parametric process, as illustrated in Fig. 6.1(c). If the nonlinear susceptibility is periodically modulated such that a phase of $\varphi_\chi = \pi$ is introduced over each coherence length, the total phase φ is reset to its initial value so that the reversal of the process is prevented. Then energy can continue to flow in the desired direction. The simplest and most effective approach to implementing such a periodic modulation is to periodically change the sign of $\chi^{(2)}$, as illustrated in Fig. 9.8(a). For a ferroelectric nonlinear crystal, such as $LiNbO_3$, $LiTaO_3$, or KTP, the periodic change of sign in $\chi^{(2)}$ can be achieved by periodic poling with an external electric field for periodic ferroelectric domain reversal. Periodically poled $LiNbO_3$ (PPLN) and periodically poled KTP (PPKTP) are of great interest.

Any periodic modulation can be viewed as a grating. Therefore, the effect of a periodically modulated nonlinear susceptibility can be formally analyzed with a procedure similar to that used in the analysis of a grating coupler. In the presence of a periodic spatial modulation, the effective susceptibility χ_{eff} that is defined in (6.26) for parametric frequency conversion, or in (6.41) for second-harmonic generation, becomes a periodic function of z. It can be expressed in terms of a Fourier series expansion as

$$\chi_{eff}(z) = \sum_q \chi_{eff}(q)\, e^{iqKz},\tag{9.45}$$

where $K = 2\pi/\Lambda$, Λ is the modulation period, and

$$\chi_{eff}(q) = \frac{1}{\Lambda}\int_0^\Lambda \chi_{eff}(z)\, e^{-iqKz}dz.\tag{9.46}$$

By substituting $\chi_{eff}(z)$ of (9.45) for χ_{eff} in (6.27), we have

$$\frac{d\mathcal{E}_3}{dz} = \frac{i\,\omega_3^2}{c^2 k_{3,z}}\mathcal{E}_1\mathcal{E}_2\sum_q \chi_{eff}(q)\, e^{i(\Delta k + qK)z} \approx \frac{i\,\omega_3^2}{c^2 k_{3,z}}\chi_Q\mathcal{E}_1\mathcal{E}_2\, e^{i\Delta k_Q z},\tag{9.47}$$

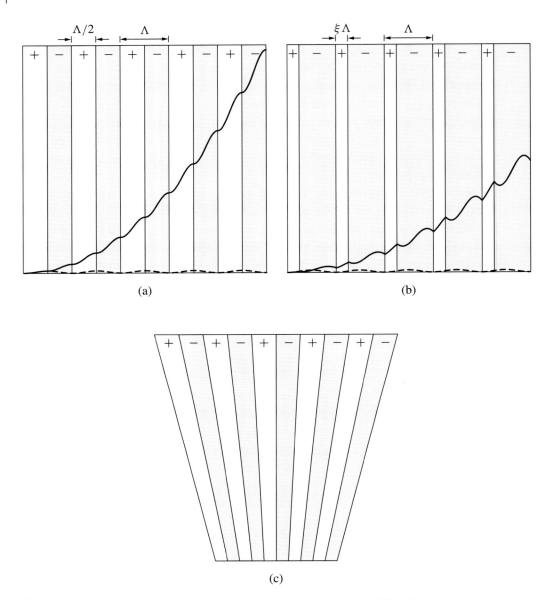

Figure 9.8 Structures with periodic sign reversal of the nonlinear susceptibility for quasi-phase matching: (a) first-order structure with a 50% duty factor, (b) general structure with a duty factor of ξ, and (c) fanned structure for wavelength tuning. Also shown in (a) and (b) is the growth of the second-harmonic intensity with quasi-phase matching, in solid curves, as a function of propagation distance when a first-order structure with $\Lambda = 2l_{coh}$ is used for second-harmonic generation. Shown for comparison, in dashed curves, is the second-harmonic intensity without quasi-phase matching. The plus and minus signs refer to the sign of χ_{eff} in each region.

where

$$\chi_Q = \chi_{eff}(q) \tag{9.48}$$

and

$$\Delta k_Q = \Delta k + qK \tag{9.49}$$

for an integer q that minimizes the value of $|\Delta k + qK|$. The other two coupled parametric equations in (6.28) and (6.29) can also be transformed in a similar manner. Therefore, all of the results obtained in Section 6.2 for parametric interactions are still valid in the case of quasi-phase matching after making the substitution of χ_Q and Δk_Q for χ_{eff} and Δk, respectively.

Perfect quasi-phase matching is achieved when $\Delta k_Q = 0$. This happens when the modulation period is chosen to be

$$\Lambda = -q\frac{2\pi}{\Delta k} = |q| \cdot 2l_{\text{coh}}. \tag{9.50}$$

Therefore, first-order quasi-phase matching occurs when $\Lambda = 2l_{\text{coh}}$ for $q = 1$ or -1. Quasi-phase matching at a high order is also possible.

When designing a periodic structure for quasi-phase matching, it is important to maximize the value of χ_Q to obtain the best efficiency for an interaction. In principle, any periodic structure is potentially useful. The simplest structure is one in which the sign of the nonlinear susceptibility is periodically reversed at abrupt boundaries. It is efficient and easy to fabricate. For such a structure that has a duty factor of ξ, as shown in Fig. 9.8(b), we find that

$$\chi_Q = \frac{1}{\Lambda}\left[\int_0^{\xi\Lambda} \chi_{\text{eff}}\, e^{-iqKz}dz - \int_{\xi\Lambda}^{\Lambda} \chi_{\text{eff}}\, e^{-iqKz}dz\right] = 2\chi_{\text{eff}}\frac{\sin \xi q\pi}{q\pi}e^{-i\xi q\pi}. \tag{9.51}$$

Note that the optimum value for the duty factor ξ depends on the order q of a structure. Clearly, a first-order structure with a 50% duty factor ($\xi = 1/2$), which is shown in Fig. 9.8(a), has the largest effective nonlinear susceptibility:

$$|\chi_Q| = \frac{2}{\pi}|\chi_{\text{eff}}|. \tag{9.52}$$

For a given interaction, $|\chi_Q|$ is always smaller than $|\chi_{\text{eff}}|$.

It seems that an interaction with birefringent phase matching is always more efficient than that with quasi-phase matching. This is not true, however. In an interaction with birefringent phase matching, χ_{eff} is subject to the constraints that are imposed by the phase-matching configuration on the propagation direction and the polarizations of the interacting waves. Therefore, other than choosing the proper angles θ and ϕ for the wave propagation direction and considering the difference between type I and type II phase-matching methods, there is little freedom in maximizing the effective nonlinear susceptibility when using birefringent phase matching. Quasi-phase matching is not subject to such constraints because it depends on an externally imposed structure, rather than on intrinsic material properties, for phase matching. Therefore, there is much freedom in seeking a high susceptibility in an interaction with quasi-phase matching. For example, for LiNbO$_3$, $|\chi_{22}^{(2)}| < |\chi_{31}^{(2)}| \approx |\chi_{33}^{(2)}|/6$. From Examples 9.1 and 9.3, we find that $|\chi_{\text{eff}}^{\text{I}}| < \left(\left|\chi_{31}^{(2)}\right|^2 + \left|\chi_{22}^{(2)}\right|^2\right)^{1/2}$ and $|\chi_{\text{eff}}^{\text{II}}| < |\chi_{31}^{(2)}|$ for type I and type II birefringent phase matching in LiNbO$_3$, respectively. With birefringent phase matching, it is not possible to exploit the largest nonlinear susceptibility element, $\chi_{33}^{(2)}$, of LiNbO$_3$ because $\chi_{33}^{(2)}$ can be used only when all of the interacting waves are polarized along the extraordinary principal axis. By contrast, with quasi-phase matching, all of the interacting waves can be polarized along the extraordinary axis so that $\chi_{\text{eff}} = \chi_{33}^{(2)}$. If a first-order periodic modulation with a 50% duty factor is used for quasi-phase matching to compensate for the phase mismatch among the extraordinary waves, we have $|\chi_Q| = 2|\chi_{\text{eff}}|/\pi = 2|\chi_{33}^{(2)}|/\pi$, which is about four times the value of $|\chi_{\text{eff}}|$ for the most efficient interaction with type I or type II birefringent phase matching.

From the above discussions, it is clear that *one important advantage of quasi-phase matching is that it makes possible efficient nonlinear interactions for which birefringent phase matching is not possible*. Nonlinear interactions in nonbirefringent nonlinear materials, such as III–V semiconductors, can also be phase matched with quasi-phase matching. The polarization directions of the interacting waves are not restricted in quasi-phase matching as they are in birefringent phase matching. This flexibility allows a collinear interaction within the transparency spectral region of a nonlinear material to be noncritically phase matched with no beam walk-off at any temperature. High efficiency is possible by arranging the polarization directions of the waves for an interaction to use the largest susceptibility element of a nonlinear crystal. For wavelength tuning, the modulation period Λ has to be varied. With a fanned structure such as that shown in Fig. 9.14(c), continuous wavelength tuning can be accomplished by transversely translating the crystal through the beam path.

EXAMPLE 9.4

A PPLN crystal is used for second-harmonic generation of a fundamental beam at the 1.064 μm wavelength. Find the required grating period for quasi-phase matching and the largest effective nonlinear susceptibility available for this interaction.

Solution

From Table 3.4, the largest nonlinear susceptibility element of LiNbO$_3$ is $d_{33} = -25.2$ pm V^{-1}, thus $\chi_{33}^{(2)} = 2d_{33} = -50.4$ pm V^{-1}. From the above discussions in the text, we know that both the fundamental and the second-harmonic waves have to be extraordinary waves polarized along the principal z axis of the crystal in order to obtain the largest value of $|\chi_Q|$ for a PPLN crystal. Therefore, we have to take $n_\omega^e = 2.1554$ for the fundamental wave at $\lambda = 1.064$ μm and $n_{2\omega}^e = 2.2330$ for the second harmonic at $\lambda/2 = 532$ nm to calculate the phase mismatch Δk and the coherence length l_{coh} as in case (b) of Example 6.2:

$$l_{\text{coh}} = \frac{\pi}{|\Delta k|} = \frac{\lambda}{4\left|n_\omega^e - n_{2\omega}^e\right|} = \frac{1.064 \times 10^{-6}}{4 \times |2.1554 - 2.2330|} \text{ m} = 3.43 \text{ μm}.$$

From the discussions following (9.51), we know that the largest value for $|\chi_Q|$ is that given in (9.52) obtained with a first-order structure with a 50% duty factor. Therefore, the required grating period is

$$\Lambda = 2l_{\text{coh}} = 6.86 \ \mu\text{m}$$

for $|q| = 1$, and the effective nonlinear susceptibility is

$$|\chi_Q| = \frac{2}{\pi} |\chi_{33}^{(2)}| = 32.08 \ \text{pm V}^{-1}$$

or, equivalently, $|d_Q| = 16.04 \ \text{pm V}^{-1}$.

In this scheme of quasi-phase matching, $\mathbf{S}_\omega \| \mathbf{k}_\omega \| \mathbf{k}_{2\omega} \| \mathbf{S}_{2\omega}$ because both the fundamental and the second-harmonic fields are polarized along the principal z axis. Therefore, there is no walk-off between $\mathbf{S}_\omega$ and $\mathbf{S}_{2\omega}$. This interaction is not limited by an aperture length, which is effectively $l_a = \infty$.

9.4 PHASE MATCHING IN WAVEGUIDES

The modal dispersion of a waveguide can be utilized for phase matching under suitable conditions. Phase matching in a waveguide is usually collinear, though two-dimensional noncollinear phase matching in a planar waveguide is also possible. For this reason, we consider only collinear phase matching in a waveguide, which can be either planar or nonplanar.

The concept of phase matching in a waveguide is straightforward. In a waveguide, a wave of a given frequency propagates as a linear superposition of waveguide modes, each with its characteristic propagation constant, as discussed in Subsection 7.1.1 and shown in (7.1) and (7.2). Therefore, the propagation constant $k(\omega)$ for a wave at a frequency ω that propagates in a spatially homogeneous medium is replaced by the propagation constant $\beta_\nu(\omega)$ for a waveguide mode of mode index ν at a frequency of ω that propagates in a waveguide. With collinear waveguide modes, the phase-matching condition for a general second-order parametric process in a waveguide with $\omega_3 = \omega_1 + \omega_2$ is obtained by replacing that given in (9.1) with the condition:

$$\beta_{3,\nu} = \beta_{1,\mu} + \beta_{2,\xi}, \quad \text{or} \quad n_{3,\nu}\omega_3 = n_{1,\mu}\omega_1 + n_{2,\xi}\omega_2, \tag{9.53}$$

where $n_{q,\nu} = c\beta_{q,\nu}/\omega_q$ is the effective refractive index of a waveguide mode. The collinear phase-matching condition for a general third-order parametric process in a waveguide with $\omega_4 = \omega_1 + \omega_2 + \omega_3$ is

$$\beta_{4,\nu} = \beta_{1,\mu} + \beta_{2,\xi} + \beta_{3,\zeta}, \quad \text{or} \quad n_{4,\nu}\omega_4 = n_{1,\mu}\omega_1 + n_{2,\xi}\omega_2 + n_{3,\zeta}\omega_3. \tag{9.54}$$

Note that in (9.53) and (9.54), the propagation constant $\beta_{q,\nu}$ of mode ν at a frequency of ω_q accounts for both the linear optical properties of the waveguide material and the optical confinement of the waveguide structure. In other words, both the material dispersion and the modal dispersion are included in $\beta_{q,\nu}$. When applying (9.53) or (9.54) for phase matching in a waveguide, it is generally not necessary to intentionally separate the two effects, but is only necessary to consider their combined effect by using $\beta_{q,\nu}$. Because of the modal dispersion caused by the waveguiding effect, it is in principle possible to accomplish collinear phase matching in a waveguide that is made of an optically isotropic material, such as a glass or a cubic crystal.

Though the concept of phase matching in a waveguide is straightforward, the implementation is usually not easy, particularly when the waveguide material is optically isotropic or when any two of the interacting frequencies are far apart. The reason is that to accomplish phase matching in a waveguide that is made of a nonbirefringent material, all of the phase mismatch caused by the material dispersion has to be compensated by the difference in the waveguiding effect on different waveguide modes. However, the contribution to the modal dispersion from the waveguiding effect is normally not sufficient to compensate for the phase mismatch caused by the material dispersion. When any two of the interacting frequencies are far apart, this problem is exacerbated by the large difference in the refractive indices at the two frequencies. This difficulty generally applies to a second-order parametric frequency-conversion process. By contrast, the technologies for birefringent phase matching and quasi-phase matching discussed in the preceding two sections are well developed and efficient for second-order parametric frequency-conversion processes, particularly for second-harmonic generation. For these reasons, phase matching for second-order parametric frequency conversion by using modal dispersion in a waveguide is not a competitive technology and is thus not practically useful.

The situation is different for third-order parametric frequency conversion, particularly when the differences among all participating frequencies are small, which is possible for third-order parametric frequency conversion but is not possible for second-order frequency conversion. Because $\chi^{(3)}$ exists in all materials, the waveguide can be made of any material, including glass such as a silica or germania optical fiber. Furthermore, even though the material is optically isotropic, birefringence can be incorporated into the waveguide through structural design. Indeed, *birefringent optical fibers*, also known as *polarization-maintaining optical fibers*, are well developed and commercially available. In a birefringent fiber, the cross-sectional xy plane that is normal to the longitudinal axis, z, of the fiber has two birefringent axes, the *fast axis* with an index of n_f and the *slow axis* with an index of n_s, where $n_f < n_s$. Any spatial normal mode of the fiber has two polarization normal modes, a *fast mode* polarized along the fast axis and a *slow mode* polarized along the slow axis, which have different phase velocities. This birefringence, together with the dispersion of the optical fiber, can be exploited to phase match for parametric frequency conversion in an optical fiber [8–10].

9.4.1 Phase Matching in a Birefringent Optical Fiber

In the following, we consider the phase-matched processes with a single pump frequency ω_p for *nearly degenerate four-wave mixing* in a birefringent optical fiber. This process is also called nearly degenerate four-photon mixing because four photons are involved in the process and two of them have the same frequency. Because we are interested in phase matching by utilizing the birefringence and the dispersion of the fiber, we restrict our discussion to only the fundamental HE_{11} spatial mode of the fiber, which consists of two nondegenerate fast and slow polarization modes in the birefringent fiber, as mentioned above. The nearly degenerate four-photon mixing process generates a *down-shifted frequency* at ω_S, commonly called the *Stokes frequency*, and an *up-shifted frequency* at ω_{AS}, commonly called the *anti-Stokes frequency*.[1] The frequency shift is

$$\Omega = \omega_p - \omega_S = \omega_{AS} - \omega_p, \tag{9.55}$$

which is measured in radians as expressed in (9.55), but is also commonly expressed in the unit of cm^{-1} as $\Omega_\lambda = \lambda_p^{-1} - \lambda_S^{-1} = \lambda_{AS}^{-1} - \lambda_p^{-1}$ or in hertz as $f = \nu_p - \nu_S = \nu_{AS} - \nu_p$.

The phase-matched parametric four-photon mixing processes with one pump frequency at ω_p that can possibly happen for the fundamental HE$_{11}$ spatial mode of a birefringent fiber are summarized in Table 9.3 [10]. In this table, D is the dimensionless group-velocity dispersion of the fiber, as defined in (4.83), B is the birefringence of the fiber defined as

$$B = n_s - n_f \approx N_s - N_f, \tag{9.56}$$

where N is the group index defined in (4.90), and D'' is the second-order derivative defined through the relations:

$$D(\lambda) = \lambda^2 \frac{d^2 n}{d\lambda^2}, \quad D'(\lambda) = \frac{d}{d\lambda}[\lambda D(\lambda)], \quad D''(\lambda) = \frac{1}{\lambda}\frac{d}{d\lambda}[\lambda^2 D'(\lambda)]. \tag{9.57}$$

These parameters can also be expressed in terms of the frequency ω as

$$B = \frac{c}{\omega}(\beta_0^s - \beta_0^f) \approx c(\beta_1^s - \beta_1^f), \tag{9.58}$$

and

$$N = c\beta_1, \quad D = c\omega\beta_2, \quad D' = -c\omega^2\beta_3, \quad D'' = c\omega^3\beta_4, \tag{9.59}$$

where $\beta_n = \partial^n \beta / \partial \omega^n$, as defined in (8.35).

From the symmetry of $\chi^{(3)}$ for an isotropic material like silica or germania glass of an optical fiber, the only nonvanishing elements of $\chi^{(3)}$ are those of the types $\chi_{1111}^{(3)}, \chi_{1122}^{(3)}, \chi_{1212}^{(3)}$, and $\chi_{1221}^{(3)}$, as discussed in Subsection 3.5.3 and listed in Table 3.3. Therefore, the polarizations of the four photons in a four-photon mixing process in a birefringent fiber have to appear in pairs. The six phase-matched processes listed in Table 9.3 are those that can possibly take place in a birefringent optical fiber. There are other four-photon mixing processes that are phase matched in a birefringent fiber [9], but they cannot happen because they use the nonlinear susceptibility elements of the types $\chi_{1112}^{(3)}, \chi_{1222}^{(3)}$, and $\chi_{1121}^{(3)}$, which vanish in a glass fiber.

There are two groups among the six processes listed in Table 9.3. Group I consists of those of a single-polarization pump field, either fast or slow, so that both the down-shifted Stokes and the up-shifted anti-Stokes fields have the same polarization along one of the birefringent axes.

---

[1] Unlike those in a Raman or Brillouin process, these Stokes and anti-Stokes frequencies are not in resonance with any material transition. They are similar to a Raman or Brillouin process only in that the Stokes frequency is down-shifted from the pump frequency and the anti-Stokes frequency is up-shifted.

Table 9.3 **Phase-matched parametric four-photon mixing processes of the fundamental mode fields in a birefringent optical fiber**

Group	Process	Polarizations $\omega_p\ \omega_p\ \omega_S\ \omega_{AS}$				Susceptibility	Frequency shift Ω	Condition
I	1	s	s	s	s	$\chi^{(3)}_{1111}$	$\omega_p\left.\left|\dfrac{12D}{D''}\right|^{1/2}\right|_{\omega_p}=\left.\left|\dfrac{12\beta_2}{\beta_4}\right|^{1/2}\right|_{\omega_p}$	$DD'' < 0$
I	2	f	f	f	f	$\chi^{(3)}_{1111}$	$\omega_p\left.\left|\dfrac{12D}{D''}\right|^{1/2}\right|_{\omega_p}=\left.\left|\dfrac{12\beta_2}{\beta_4}\right|^{1/2}\right|_{\omega_p}$	$DD'' < 0$
I	3	s	s	f	f	$\chi^{(3)}_{1122}$	$\omega_p\left.\left|\dfrac{2B}{D}\right|^{1/2}\right|_{\omega_p}=\left.\left|\dfrac{2\Delta\beta}{\beta_2}\right|^{1/2}\right|_{\omega_p}$	$D > 0$
I	4	f	f	s	s	$\chi^{(3)}_{1122}$	$\omega_p\left.\left|\dfrac{2B}{D}\right|^{1/2}\right|_{\omega_p}=\left.\left|\dfrac{2\Delta\beta}{\beta_2}\right|^{1/2}\right|_{\omega_p}$	$D < 0$
II	5	s	f	s	f	$\chi^{(3)}_{1212}$	$\omega_p\left.\left|\dfrac{B}{D}\right|\right|_{\omega_p}=\dfrac{1}{\omega_p}\left.\left|\dfrac{\Delta\beta}{\beta_2}\right|\right|_{\omega_p}$	$D > 0$
II	6	s	f	f	s	$\chi^{(3)}_{1221}$	$\omega_p\left.\left|\dfrac{B}{D}\right|\right|_{\omega_p}=\dfrac{1}{\omega_p}\left.\left|\dfrac{\Delta\beta}{\beta_2}\right|\right|_{\omega_p}$	$D < 0$

Group II consists of those of a divided-polarization pump field, which has both fast and slow polarization components, so that the down-shifted Stokes and the up-shifted anti-Stokes fields are orthogonally polarized. Processes 1 and 2 are high-order processes that depend on the group-velocity dispersion D and its second-order derivative D''; they can only happen near the zero-dispersion point where D has a small value. Processes 3–6 are low-order processes that depend on the birefringence B and the group-velocity dispersion D; their frequency shifts can be tuned by varying the fiber birefringence. For a given frequency shift, processes 3 and 4 require a smaller birefringence than processes 5 and 6.

Problem Set

9.2.1 Consider a second-order collinear frequency-conversion process, such as second-harmonic generation, sum-frequency generation, or difference-frequency generation, in a uniaxial crystal. The waves propagate in a direction of $\hat{k} = \hat{x}\sin\theta\cos\phi + \hat{y}\sin\theta\sin\phi + \hat{z}\cos\theta$, which makes an angle of θ with the optical axis $\hat{z}$ and an angle of ϕ with the principal axis $\hat{x}$.

(a) Show, by using Table 3.2 for the nonvanishing elements of the $\chi^{(2)}$ tensor, that the effective second-order susceptibility, χ_{eff}, is independent of the angle ϕ for the uniaxial crystals of the 4, 6, 422, 622, 4mm, and 6mm point groups for both type I and type II phase matching.

(b) Identify the cases that have a vanishing χ_{eff} (i.e., $\chi_{\text{eff}} = 0$).

(c) What are the forms of χ_{eff} for second-harmonic generation for which full permutation symmetry is valid for the $\chi^{(2)}$ elements?

9.2.2 Efficient second-order nonlinear optical frequency conversion using temperature tuning with 90° phase matching in a uniaxial crystal is not always possible. One reason is that the temperature for 90° phase matching does not exist or it is too low or too high to be practically applicable. In this problem, we consider the second reason that the effective nonlinear susceptibility is zero when phase matching is accomplished in this manner.

(a) Show that efficient 90° phase matching in a negative uniaxial crystal is not possible for type II phase matching but is possible for type I phase matching only in crystals of point groups 3, 4, 6, 4mm, 6mm, 3m, $\bar{4}$, and $\bar{4}2m$.

(b) Show that efficient 90° phase matching in a positive uniaxial crystal is not possible for type I phase matching but is possible for type II phase matching only in crystals of point groups 3, 4, 6, 4mm, 6mm, 3m, $\bar{4}$, and $\bar{4}2m$.

9.2.3 The ordinary and extraordinary indices of refraction of LiNbO$_3$ as functions of wavelength and temperature are given in (9.42) and (9.43), respectively. Consider second-harmonic generation with LiNbO$_3$ at room temperature.

(a) Find the fundamental wavelength or wavelengths for 90° type I phase matching with LiNbO$_3$ at room temperature. What is the wavelength range of the fundamental wave in which type I phase matching for second-harmonic generation is possible with LiNbO$_3$ at room temperature?

(b) Find the fundamental wavelength or wavelengths for 90° type II phase matching with LiNbO$_3$ at room temperature. What is the wavelength range of the fundamental wave in which type II phase matching for second-harmonic generation is possible with LiNbO$_3$ at room temperature?

9.2.4 LiNbO$_3$ is a negative uniaxial crystal of the 3m point group that has these nonvanishing elements of $\chi^{(2)}$: $\chi_{15}^{(2)} = \chi_{24}^{(2)}$, $\chi_{31}^{(2)} = \chi_{32}^{(2)}$, $\chi_{33}^{(2)}$, $\chi_{22}^{(2)} = -\chi_{21}^{(2)} = -\chi_{16}^{(2)}$. The ordinary and extraordinary refractive indices as functions of wavelength and temperature are given in (9.42) and (9.43), respectively. Its melting temperature is 1257 °C. The optical axis is the z axis. A laser beam at the fundamental wavelength of 2 µm propagates through the crystal at an angle of θ with respect to $\hat{z}$ in a plane at an angle of ϕ with respect to $\hat{x}$ to collinearly generate the second harmonic at the 1 µm wavelength.

(a) How should the polarizations of the fundamental and second-harmonic waves be chosen, respectively, for type I phase matching? Find the type I phase-matching angle at room temperature. Find χ_{eff} for type I phase matching as a function of θ, ϕ, and the nonvanishing elements of $\chi^{(2)}$.

(b) Answer the questions in (a) for type II phase matching.

(c) Find the phase-matching temperature for $90°$ type I phase matching. In this situation, how are the fundamental and second-harmonic fields polarized with respect to each other? What is χ_{eff} for this phase matching as a function of θ, ϕ, and the nonvanishing elements of $\chi^{(2)}$?

(d) Answer the questions in (c) for $90°$ type II phase matching.

(e) With $\chi_{31}^{(2)} < 0$ and $\chi_{22}^{(2)} > 0$, what are the best choices for the value of ϕ and the maximum value of $|\chi_{\text{eff}}|$ for each type of phase matching?

9.2.5 LiNbO$_3$ is a uniaxial crystal of the $3m$ point group. Its ordinary and extraordinary indices of refraction as functions of wavelength and temperature are given in (9.42) and (9.43). A laser beam at the fundamental wavelength propagates through the crystal at an angle of θ with respect to the optical axis $\hat{z}$ in a plane at an angle of ϕ with respect to the x axis to generate the second harmonic wave.

(a) Consider second-harmonic generation with a fundamental wave at a wavelength of 1.3 µm. Find the phase-matching angles for type I and type II phase matching, respectively, at room temperature. Find the phase-matching temperatures for type I and type II phase matching, respectively.

(b) Answer the questions in (a) for a fundamental wave at a wavelength of 800 nm.

9.2.6 In this problem, we consider the angular tolerance of collinear sum-frequency generation, with $\omega_3 = \omega_1 + \omega_2$, for $90°$ phase matching in uniaxial crystals.

(a) According to Problem 9.2.2, efficient $90°$ phase matching in a negative uniaxial crystal is possible only for type I phase matching in certain crystals. Find the angular tolerance for this case.

(b) Also according to Problem 9.2.2, efficient $90°$ phase matching in a positive uniaxial crystal is possible only for type II phase matching in certain crystals. Find the angular tolerance for this case.

(c) Because the efficiency of second-harmonic generation is proportional to the square of the fundamental intensity, the efficiency can usually be improved by simply focusing the beam. How much can the fundamental beam be focused while maintaining phase matching through the entire crystal length?

9.2.7 Phase matching with the BBO crystal is possible for second-harmonic and sum-frequency generation in the spectral range from the near-infrared to the near-ultraviolet. BBO is a negative uniaxial crystal. Its ordinary and extraordinary indices at room temperature as functions of optical wavelength are given by the Sellmeier equations:

$$n_{\text{o}}^2 = 2.7359 + \frac{0.01878}{\lambda^2 - 0.01822} - 0.01354\,\lambda^2, \tag{9.60}$$

$$n_{\text{e}}^2 = 2.3753 + \frac{0.01224}{\lambda^2 - 0.01667} - 0.01516\,\lambda^2, \tag{9.61}$$

where λ is in micrometers. It is desired to generate the second and third harmonics of the fundamental wavelength at 1.064 µm of a Nd:YAG laser by using BBO crystals. The second harmonic is generated by doubling the fundamental frequency, while the third harmonic is generated by summing the fundamental and the second-harmonic frequencies. We consider collinear phase matching with the waves propagating in a direction making an angle of θ with the z axis, which is the unique optical axis, and an angle of ϕ with the x axis.

(a) For the generation of a second-harmonic wave at the 532 nm wavelength through second-harmonic generation, write down the equations required to be solved for the phase-matching angles for type I and type II phase-matching conditions, respectively. Calculate the phase-matching angle for type I phase matching.

(b) Calculate the walk-off angle for the second-harmonic generation under type I phase matching in (a). Find the aperture distance by taking the fundamental beam to be a Gaussian beam that is focused to a beam waist size of $w_0 = 100$ μm. Under what conditions does the efficiency of second-harmonic generation increase quadratically with the length of the nonlinear crystal?

(c) For the generation of a third-harmonic wave at the 354.7 nm wavelength through sum-frequency generation by using 1.064 μm and 532 nm input waves, calculate the type I phase-matching angle.

(d) For the generation of a third-harmonic wave at the 354.7 nm wavelength through sum-frequency generation, what are the possibilities for type II phase matching? Find the phase-matching angle.

9.2.8 KTP is a positive uniaxial crystal of the $mm2$ point group. The only nonvanishing elements of its second-order nonlinear susceptibility are $\chi_{15}^{(2)}$, $\chi_{24}^{(2)}$, $\chi_{31}^{(2)}$, $\chi_{32}^{(2)}$, and $\chi_{33}^{(2)}$. Consider angle phase matching for a second-order parametric interaction on the three principal planes. The interacting waves propagate collinearly in a direction of $\hat{k} = \hat{x} \sin\theta \cos\phi + \hat{y} \sin\theta \sin\phi + \hat{z} \cos\theta$, where $\hat{x}$, $\hat{y}$, and $\hat{z}$ are the principal axes of the KTP crystal.

(a) Find the effective nonlinear susceptibilities for type I and type II phase matching on the xy principal plane in terms of the nonvanishing $\chi^{(2)}$ elements and the angles θ and ϕ.

(b) Find the effective nonlinear susceptibilities for type I and type II phase matching on the yz principal plane in terms of the nonvanishing $\chi^{(2)}$ elements and the angles θ and ϕ.

(c) Find the effective nonlinear susceptibilities for type I and type II phase matching on the zx principal plane in terms of the nonvanishing $\chi^{(2)}$ elements and the angles θ and ϕ.

(d) Use the results obtained in the above to find a conclusion on the possibility of each type of phase matching for KTP.

9.3.1 Find the grating period that is required for second-order quasi-phase matching. Find the duty factor that has to be chosen to maximize the value of $|\chi_Q|$ for such a second-order grating. What is this $|\chi_Q|$? How does it compare with that of the optimized first-order grating?

9.3.2 KTP is a positive biaxial crystal of the $mm2$ point group. The nonvanishing elements of its second-order nonlinear susceptibility have the values: $d_{15} = 3.7$ pm V^{-1}, $d_{24} = 1.9$ pm V^{-1}, $d_{31} = 3.7$ pm V^{-1}, $d_{32} = 2.2$ pm V^{-1}, and $d_{33} = 14.6$ pm V^{-1}. A PPKTP crystal is used for frequency doubling of an optical wave at the fundamental wavelength of $\lambda_\omega = 1.064$ μm to its second harmonic at $\lambda_{2\omega} = 532$ nm. The principal indices of refraction at these wavelengths are: $n_\omega^x = 1.7399$, $n_\omega^y = 1.7475$, and $n_\omega^z = 1.8296$; $n_{2\omega}^x = 1.7790$, $n_{2\omega}^y = 1.7900$, and $n_{2\omega}^z = 1.8868$. Find the polarization directions that have to be chosen for the fundamental and second-harmonic waves, respectively, in order to have the largest nonlinear susceptibility under quasi-phase matching. The grating period of this PPKTP is chosen for quasi-phase matching of this process. Find the required first-order grating period at room temperature. Find the effective nonlinear susceptibility $|\chi_Q|$ for this first-order PPKTP.

9.3.3 A first-order PPKTP crystal is pumped at $\lambda_3 = 532$ nm to generate a parametric signal at $\lambda_1 = 1.3$ μm. The properties of KTP are listed in Table 3.4.

(a) What is the idler wavelength λ_2?

(b) How should the pump wave be polarized for the largest nonlinear susceptibility under quasi-phase matching? What are the polarizations of the signal and idler waves, respectively?

(c) What is the grating period required for quasi-phase matching at room temperature?

(d) If the pump wavelength is at $\lambda_3 = 860$ nm instead, what is the idler wavelength λ_2? What is the required grating period? Compare this grating period with that obtained in (c).

9.4.1 Consider nearly degenerate four-photon mixing processes in a birefringent optical fiber with a pump frequency at ω_p to generate a Stokes signal at ω_S and an anti-Stokes signal at ω_{AS} subject to the condition that $\Omega = \omega_p - \omega_S = \omega_{AS} - \omega_p$. The refractive indices of the birefringent fiber for the slow and fast axes are n_s and n_f, respectively, for which $n_s > n_f$. The group index of refraction and the group-velocity dispersion are defined for each axis as

$$N_s = \frac{d}{d\omega}(\omega n_s) = n_s - \lambda \frac{dn_s}{d\lambda}, \quad N_f = \frac{d}{d\omega}(\omega n_f) = n_f - \lambda \frac{dn_f}{d\lambda};$$

and

$$D_s = \omega \frac{d^2}{d\omega^2}(\omega n_s) = \lambda^2 \frac{d^2 n_s}{d\lambda^2}, \quad D_f = \omega \frac{d^2}{d\omega^2}(\omega n_f) = \lambda^2 \frac{d^2 n_f}{d\lambda^2}.$$

It is a good assumption that $dn_s/d\lambda \approx dn_f/d\lambda$ and $d^2 n_s/d\lambda^2 \approx d^2 n_f/d\lambda^2$. Then, the birefringence of the fiber is

$$B = n_s - n_f = N_s - N_f,$$

and the group-velocity dispersion is

$$D = D_s = D_f.$$

(a) Show that the only processes that are phase matched and have nonvanishing effective third-order susceptibilities are the six processes listed in Table 9.3.

(b) Verify the frequency shift listed in Table 9.3 for each process.

References

[1] M. V. Hobden, "Phase-matched second-harmonic generation in biaxial crystals," *Journal of Applied Physics*, vol. 38, pp. 4365–4372, 1967.

[2] J. Q. Yao and T. S. Fahlen, "Calculations of optimum phase match parameters for the biaxial crystal KTiOPO$_4$," *Journal of Applied Physics*, vol. 55, pp. 65–68, 1984.

[3] H. Ito, H. Naito, and H. Inaba, "Generalized study on angular dependence of induced second-order nonlinear optical polarizations and phase matching in biaxial crystals," *Journal of Applied Physics*, vol. 46, pp. 3992–3998, 1975.

[4] W. Zhang, H. Yu, H. Wu, and P. S. Halasyamani, "Phase-matching in nonlinear optical compounds: a materials perspective," *Chemistry of Materials*, vol. 29, pp. 2655–2668, 2017.

[5] J. Chung and A. E. Siegman, "Singly resonant continuous-wave mode-locked KTiOPO$_4$ optical parametric oscillator pumped by a Nd:YAG laser," *Journal of the Optical Society of America B*, vol. 10, pp. 2201–2210, 1993.

[6] L. P. Chen, Y. Wang, and J. M. Liu, "Singly resonant optical parametric oscillator synchronously pumped by frequency-doubled additive-pulse mode-locked Nd:YLF laser pulses," *Journal of the Optical Society of America B*, vol. 12, pp. 2192–2198, 1995.

[7] J. A. Armstrong, N. Bloembergen, J. Ducuing, and P. S. Pershan, "Interactions between light waves in a nonlinear dielectric," *Physical Review*, vol. 127, pp. 1918–1939, 1962.

[8] R. H. Stolen, M. A. Bösch, and C. Lin, "Phase matching in birefringent fibers," *Optics Letters*, vol. 6, pp. 213–215, 1981.

[9] R. K. Jain and K. Stenersen, "Phase-matched four-photon mixing processes in birefringent fibers," *Applied Physics B*, vol. 35, pp. 49–57, 1984.

[10] P. N. Morgan and J. M. Liu, "Parametric four-photon mixing followed by stimulated Raman scattering with optical pulses in birefringent optical fibers," *IEEE Journal of Quantum Electronics*, vol. 27, pp. 1011–1021, 1991.

10 Optical Frequency Conversion

10.1 SECOND-HARMONIC GENERATION

A very important class of nonlinear optical devices is the optical frequency converters. Nonlinear optical frequency conversion is the only means of directly converting optical energy from one frequency to another. Indeed, the discipline of nonlinear optics was born out of the first observation of second-harmonic generation by Franken *et al.* in 1961 [1]. Besides the electro-optic devices based on the Pockels effect, the most widely used nonlinear optical devices so far are the second-harmonic generators.

There are basically two types of nonlinear optical frequency converters. The majority are based on parametric processes, particularly the parametric second-order processes, which require phase matching. Optical harmonic generators, sum-frequency generators, difference-frequency generators, and parametric amplifiers and oscillators belong to this type. Devices that use the nonparametric third-order process of stimulated Raman or Brillouin scattering to shift the optical frequency are the other type. In this chapter, we consider only those based on parametric processes. With a few exceptions, such as the parametric four-wave mixing of small frequency shifts discussed in Section 10.7 and the generation of high-order harmonics discussed in Section 10.8, almost all practical optical frequency converters are based on second-order parametric processes for two reasons:

1. The second-order nonlinear susceptibility, if it exists, is orders of magnitude larger than the third-order and high-order susceptibilities.
2. Collinear phase matching for efficient frequency conversion is easier to accomplish for a second-order process than for a third-order or high-order process.

The devices that are based on stimulated Raman or Brillouin scattering will be discussed in Chapter 13.

An optical harmonic generator produces an optical wave at a frequency that is an integral multiple of the fundamental frequency of the input wave. A second-harmonic generator produces a wave at twice the fundamental frequency of the input wave; thus, it is also called an *optical frequency doubler*. In the application of a second-harmonic generator, only two optical waves are involved in the interaction: one input wave at the fundamental frequency of ω and a nonlinearly generated wave at the second-harmonic frequency of 2ω, as schematically illustrated in Fig. 5.1.

The basic concept of second-harmonic generation was already discussed in Subsection 5.2.1, and the coupled-wave equations and the Manley–Rowe relations were covered in Subsection 6.2.2. In this section, we start with the general coupled-wave equations given in (6.42) and (6.43) for second-harmonic generation:

$$\frac{d\mathcal{E}_{2\omega}}{dz} = \frac{i(2\omega)^2}{2c^2 k_{2\omega,z}} \chi_{\text{eff}} \mathcal{E}_\omega^2\, e^{i\Delta kz} = \frac{i\omega}{cn_{2\omega,z}} \chi_{\text{eff}} \mathcal{E}_\omega^2\, e^{i\Delta kz}, \tag{10.1}$$

$$\frac{d\mathcal{E}_\omega}{dz} = \frac{i\omega^2}{c^2 k_{\omega,z}} \chi_{\text{eff}}^* \mathcal{E}_{2\omega} \mathcal{E}_\omega^*\, e^{-i\Delta kz} = \frac{i\omega}{cn_{\omega,z}} \chi_{\text{eff}}^* \mathcal{E}_{2\omega} \mathcal{E}_\omega^*\, e^{-i\Delta kz}, \tag{10.2}$$

where $\chi_{\text{eff}} = \hat{e}_{2\omega}^* \cdot \boldsymbol{\chi}^{(2)}(2\omega = \omega + \omega) : \hat{e}_\omega \hat{e}_\omega$ and $\Delta\mathbf{k} = 2\mathbf{k}_\omega - \mathbf{k}_{2\omega} = \Delta k\hat{z}$ are the effective nonlinear susceptibility and the phase mismatch, respectively, for second-harmonic generation. These two coupled equations can be solved for various operating conditions of second-harmonic generation.

10.1.1 Low-Efficiency Limit

In the low-efficiency limit, the depletion of the intensity of the fundamental beam can be neglected. Then, (10.1) can be directly integrated for $\mathcal{E}_{2\omega}(z)$ by taking $\mathcal{E}_\omega$ to be a constant that is independent of z. By using the relation in (6.31) for light intensity, we find that the second-harmonic intensity as a function of the interaction length, l, is

$$\begin{aligned}
I_{2\omega}(l) &= \frac{\omega^2 |\chi_{\text{eff}}|^2}{2c^3 \epsilon_0 n_{\omega,z}^2 n_{2\omega,z}} I_\omega^2 l^2 \frac{\sin^2(\Delta kl/2)}{(\Delta kl/2)^2} \\
&= \frac{8\pi^2 |d_{\text{eff}}|^2}{c\epsilon_0 n_{\omega,z}^2 n_{2\omega,z}\lambda^2} I_\omega^2 l^2 \frac{\sin^2(\Delta kl/2)}{(\Delta kl/2)^2},
\end{aligned} \tag{10.3}$$

where the intensity I_ω of the fundamental wave is assumed to remain constant at its input value such that $I_\omega = I_\omega(0)$, $d_{\text{eff}} = \chi_{\text{eff}}/2$, and $\lambda = 2\pi c/\omega$ is the wavelength of the fundamental wave in free space. In the case of quasi-phase matching, χ_{eff} and d_{eff} in (10.3) are replaced by χ_Q and d_Q, respectively.

The effect of phase mismatch is characterized by a function of the form:

$$\frac{I_{2\omega}}{I_{2\omega}^{\text{PM}}} = \frac{\sin^2(\Delta kl/2)}{(\Delta kl/2)^2}, \tag{10.4}$$

which is plotted in Fig. 10.1. When phase mismatch exists so that $\Delta k \neq 0$, it does not pay to have a crystal longer than the coherence length of the interaction, as discussed in Section 6.2 and illustrated in Fig. 6.1(c). When perfect phase matching is accomplished so that $\Delta k = 0$, the intensity of the second-harmonic wave grows quadratically with the interaction length as $I_{2\omega}(l) = I_{2\omega}^{\text{PM}}(l) \propto l^2/\lambda_\omega^2 \propto l^2/\lambda_{2\omega}^2$ in the low-efficiency limit.

We see from (10.3) that $I_{2\omega} \propto |d_{\text{eff}}|^2 I_\omega^2$ in the low-efficiency limit. Therefore, in this limit, the intensity of the second-harmonic wave grows *quadratically* with the intensity of the fundamental wave. The conversion efficiency increases linearly with the fundamental intensity.

In the above, we have assumed that the interacting waves are perfect plane waves. In reality, each optical beam has a finite cross-section and thus a nonuniform intensity distribution. This and other spatial effects have to be carefully considered in a detailed analysis of a second-harmonic generation process, as well as in that of any other nonlinear process to be discussed

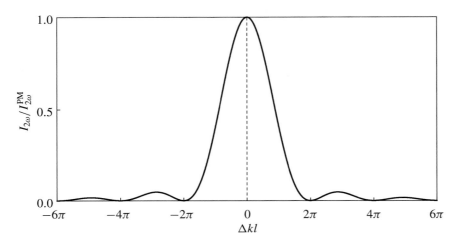

Figure 10.1 Effect of phase mismatch on the efficiency of second-harmonic generation in the low-efficiency limit.

later. Without carrying out such an analysis, we point out the important yet easily understood fact that the interaction between two or more optical beams takes place only in the area where these beams overlap spatially and, if the beams are optical pulses, also temporally. The spatial overlap of the fundamental and second-harmonic waves in the second-harmonic generation process is primarily limited by three factors:

1. the possible spatial walk-off caused by the birefringence of the nonlinear crystal;
2. the differences in the focal points and the focused spot sizes of the two beams caused by the dispersion of the crystal; and
3. the different spatial profiles of the two beams because of the nonlinear process.

In any event, the effective cross-sectional area, $\mathcal{A}_{2\omega}$, of the second-harmonic wave is never larger than the effective cross-sectional area, $\mathcal{A}_{\omega}$, of the fundamental wave – that is, $\mathcal{A}_{2\omega} \leq \mathcal{A}_{\omega}$. In the low-efficiency limit, $P_{2\omega} \propto |d_{\text{eff}}|^2 P_{\omega}^2 \mathcal{A}_{2\omega}/\mathcal{A}_{\omega}^2$.

10.1.2 Perfect Phase Matching

Perfect phase matching with $\Delta k = 0$ is required if a high efficiency for second-harmonic generation is desired. In addition, according to the discussions in Subsection 6.2.2, it is also necessary to have $\varphi = -\pi/2$. This condition is automatically satisfied if perfect phase matching is accomplished and if the input consists of only the fundamental wave because, without any coherent second-harmonic field at the input, only the second-harmonic field that has the most favorable phase is generated and subsequently amplified. The Manley–Rowe relation given in (6.47) states that the total intensity of the fundamental and second-harmonic waves remains constant throughout the interaction: $I = I_{\omega}(z) + I_{2\omega}(z) = I_{\omega}(0)$ for $I_{2\omega}(0) = 0$. Under these conditions, (6.45) can be expressed in the form:

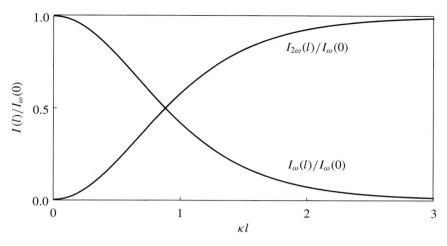

Figure 10.2 Intensities of the fundamental and second-harmonic waves, normalized to the total intensity, as a function of the interaction length in a second-harmonic generator with perfect phase matching.

$$\frac{dI_{2\omega}}{dz} = \frac{2\omega|\chi_{\text{eff}}|}{(2c^3\epsilon_0 n_{\omega,z}^3)^{1/2}} \left[I_\omega(0) - I_{2\omega}\right] I_{2\omega}^{1/2}. \tag{10.5}$$

Note that with perfect phase matching, $n_{2\omega,z} = n_{\omega,z}$, which is used to convert (6.45) to (10.5). By making the change of variable $u^2 = I_{2\omega}/I_\omega(0)$ and by using the fact that $u = \tanh \kappa z$ is the solution of the equation $du/dz = \kappa(1 - u^2)$, we solve (10.5) to obtain the general results for second-harmonic generation with perfect phase matching:

$$I_{2\omega}(l) = I_\omega(0)\tanh^2 \kappa l, \tag{10.6}$$

$$I_\omega(l) = I_\omega(0)\text{sech}^2 \kappa l, \tag{10.7}$$

where

$$\kappa = \left[\frac{\omega^2|\chi_{\text{eff}}|^2}{2c^3\epsilon_0 n_{\omega,z}^3} I_\omega(0)\right]^{1/2} = \left[\frac{8\pi^2|d_{\text{eff}}|^2}{c\epsilon_0 n_{\omega,z}^3 \lambda^2} I_\omega(0)\right]^{1/2}. \tag{10.8}$$

These results are plotted in Fig. 10.2. *With perfect phase matching, it is theoretically possible to convert nearly all of the fundamental power to the second harmonic if the interaction length is sufficiently long.* In the case of quasi-phase matching, χ_{eff} and d_{eff} in (10.8) are replaced by χ_Q and d_Q, respectively.

10.1.3 Conversion Efficiency

The conversion efficiency of a second-harmonic generator is commonly defined as

$$\eta_{SH} = \frac{P_{2\omega}(l)}{P_\omega(0)}. \tag{10.9}$$

By using (10.6) and (10.8) with (10.9), the conversion efficiency under the condition of perfect phase matching can be generally expressed as

$$\eta_{SH} = \frac{A_{2\omega}}{A_\omega} \tanh^2 \kappa l, \tag{10.10}$$

with κ expressed in terms of the power of the fundamental beam as

$$\kappa = \left[\frac{\omega^2 |\chi_{eff}|^2}{2c^3 \epsilon_0 n_{\omega,z}^3} \frac{P_\omega(0)}{A_\omega}\right]^{1/2} = \left[\frac{8\pi^2 |d_{eff}|^2}{c\epsilon_0 n_{\omega,z}^3 \lambda^2} \frac{P_\omega(0)}{A_\omega}\right]^{1/2}. \tag{10.11}$$

As can be seen from (10.10) and Fig. 10.2, the conversion efficiency initially increases linearly with the increase in the intensity of the fundamental wave. Then it grows slower than linear increase, and it eventually saturates as the fundamental intensity continues to increase. The conversion efficiency asymptotically approaches 100% for a very high power of the fundamental wave.

In the low-efficiency limit with perfect phase matching, the conversion efficiency can be approximated as

$$\eta_{SH} \approx \frac{A_{2\omega}}{A_\omega} (\kappa l)^2 = \frac{\omega^2 |\chi_{eff}|^2}{2c^3 \epsilon_0 n_{\omega,z}^3} \frac{A_{2\omega}}{A_\omega^2} P_\omega(0) l^2 = \frac{8\pi^2 |d_{eff}|^2}{c\epsilon_0 n_{\omega,z}^3 \lambda^2} \frac{A_{2\omega}}{A_\omega^2} P_\omega(0) l^2. \tag{10.12}$$

The relation in (10.12) indicates that the conversion efficiency of second-harmonic generation is linearly proportional to the power of the fundamental wave in the low-efficiency limit. This linear relationship is true only when the conversion efficiency is low, such that $\eta_{SH} \approx A_{2\omega}(\kappa l)^2 / A_\omega \ll 1$.

Because η_{SH} in the low-efficiency limit is linearly proportional to the fundamental power, it is convenient to define a normalized second-harmonic conversion efficiency as

$$\hat{\eta}_{SH} = \frac{\eta_{SH}}{P_\omega(0)} = \frac{\omega^2 |\chi_{eff}|^2}{2c^3 \epsilon_0 n_{\omega,z}^3} \frac{A_{2\omega}}{A_\omega^2} l^2 = \frac{8\pi^2 |d_{eff}|^2}{c\epsilon_0 n_{\omega,z}^3 \lambda^2} \frac{A_{2\omega}}{A_\omega^2} l^2. \tag{10.13}$$

There is a relation between $A_{2\omega}$ and A_ω that depends on the cross-sectional profile of the fundamental beam. For example, in the case that there is no spatial walk-off between the fundamental and the second-harmonic waves, $A_{2\omega} = A_\omega/2$ if the fundamental beam has a Gaussian profile, but $A_{2\omega} = A_\omega$ if the beam has a uniform profile. We see from (10.13) that the conversion efficiency can be raised by focusing the fundamental beam to reduce its cross-sectional area, provided that the beam remains well collimated. Focusing the beam too tightly increases the beam divergence, thus reducing its intensity outside the Rayleigh range from the beam waist. In addition, the conversion efficiency can be reduced by any walk-off between the interacting beams.

The second-harmonic generation efficiency of a focused Gaussian beam is a function of three characteristic lengths:

1. the crystal length l,
2. the confocal parameter $b = 2\pi n w_0^2/\lambda$, and
3. the aperture length $l_a = \pi^{1/2} w_0/\rho$, defined in (9.41).

For $b \gg l$ and $l_a \gg l$, the dependence of $\hat{\eta}_{SH}$ on l^2 seen in (10.13) is valid, and $\hat{\eta}_{SH}$ can be expressed in the form:

$$\hat{\eta}_{SH} = \frac{\eta_{SH}}{P_\omega(0)} = \frac{\omega^3 |\chi_{eff}|^2}{2\pi c^4 \epsilon_0 n_{\omega,z}^2} \frac{l^2}{b} = \frac{16\pi^2 |d_{eff}|^2}{c\epsilon_0 n_{\omega,z}^2 \lambda^3} \frac{l^2}{b}. \tag{10.14}$$

For $b < l < 10b$ or $l_a < l$, (10.13) and (10.14) are not valid, but the normalized conversion efficiency can be approximated as

$$\hat{\eta}_{SH} = \frac{\eta_{SH}}{P_\omega(0)} = \frac{\omega^3 |\chi_{eff}|^2}{2\pi c^4 \epsilon_0 n_{\omega,z}^2} \frac{1.068l}{1 + lb/l_a^2} = \frac{16\pi^2 |d_{eff}|^2}{c\epsilon_0 n_{\omega,z}^2 \lambda^3} \frac{1.068l}{1 + lb/l_a^2}. \tag{10.15}$$

We see that if $lb \gg l_a^2$, the normalized conversion efficiency is independent of the crystal length because $\hat{\eta}_{SH} \propto l_a^2/b$ in this situation. The best normalized efficiency that can be obtained with an optimally focused Gaussian beam is

$$\hat{\eta}_{SH} = \frac{\eta_{SH}}{P_\omega(0)} = \frac{\omega^3 |\chi_{eff}|^2}{2\pi c^4 \epsilon_0 n_{\omega,z}^2} (1.068l) = \frac{16\pi^2 |d_{eff}|^2}{c\epsilon_0 n_{\omega,z}^2 \lambda^3} (1.068l), \tag{10.16}$$

which occurs under the conditions of no walk-off so that $l_a = \infty$ and the crystal length is chosen to be $l = 2.84b$. We see that, with the fundamental beam optimally focused for the best efficiency, the conversion efficiency increases only linearly with the crystal length but the focused beam waist spot area has to vary linearly with the crystal length to maintain this optimum condition. Note that in the case of quasi-phase matching, χ_{eff} and d_{eff} that appear in the expressions of η_{SH} and $\hat{\eta}_{SH}$ given in (10.12)–(10.16) have to be replaced by χ_Q and d_Q, respectively.

We see that the values of the normalized efficiency $\hat{\eta}_{SH}$ expressed in (10.13)–(10.16) for various conditions are all independent of $P_\omega(0)$, meaning that the corresponding values of the conversion efficiency η_{SH} all increase linearly with an increase in the input power of the fundamental beam. This statement is true as long as we stay in the low-efficiency limit so that the depletion of the fundamental beam is negligible. Therefore, it should be noted that the relations in (10.13)–(10.16) are valid only in the low-efficiency limit. They are not valid at high power levels where the efficiency is sufficiently high for the power of the fundamental wave to be significantly depleted. In the high-efficiency regime, the conversion efficiency increases sublinearly with the input power of the fundamental beam. According to (10.6), it is theoretically possible to approach a conversion efficiency of 100% for second-harmonic generation if the input power is sufficiently high and the interaction

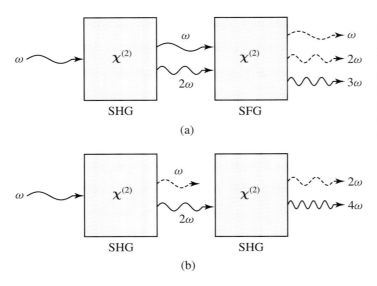

Figure 10.3 (a) A third-harmonic generator consisting of a second-harmonic generator (SHG) and a sum-frequency generator (SFG) in cascade. (b) A fourth-harmonic generator consisting of two second-harmonic generators in cascade.

length is sufficiently large. However, the conversion efficiency of a practical device is usually limited by the damage threshold of a nonlinear crystal, as well as by many complicated spatial and temporal effects.

For many practical applications, it is often necessary to generate the third harmonic or the fourth harmonic of a fundamental wave. As discussed in Subsection 5.3.1, the third harmonic can be generated with a parametric third-order nonlinear process characterized by $\chi^{(3)}(3\omega = \omega + \omega + \omega)$. However, a third-harmonic generator using a third-order nonlinear process is of little practical usefulness for two reasons:

1. The value of $\chi^{(3)}$, though always nonvanishing, is orders of magnitude smaller than the value of $\chi^{(2)}$ of any commonly used nonlinear crystal.
2. Phase matching is very difficult for a third-order parametric process.

In practice, efficient third-harmonic generation is normally carried out by following second-harmonic generation with sum-frequency generation for $\omega + 2\omega \rightarrow 3\omega$, as shown in Fig. 10.3(a), which is also shown in Fig. 5.7. Similarly, fourth-harmonic generation is accomplished by cascading two second-harmonic generators by first doubling ω and then doubling 2ω to obtain 4ω, as shown in Fig. 10.3(b). These possibilities are already demonstrated in Example 5.1.

EXAMPLE 10.1

The second-harmonic conversion efficiency is considered for a focused Gaussian beam at the fundamental wavelength of $\lambda = 1.064$ μm in LiNbO$_3$, under the phase-matching conditions that are discussed in Examples 9.1–9.4. Perfect phase matching is assumed for each case, with $\Delta k_Q = 0$ in the case of quasi-phase matching. The fundamental beam is focused to have its beam waist located at the center of a crystal that has a length of $l = 1.5$ cm.

(a) With angle phase matching as described in Example 9.1, find the normalized conversion efficiency $\hat{\eta}_{SH}$ if the Gaussian fundamental beam is focused to have a beam waist radius of $w_0 = 50$ μm. (b) With 90° phase matching by temperature tuning as described in Example 9.3, find $\hat{\eta}_{SH}$ for $w_0 = 50$ μm. (c) With 90° phase matching, the conversion efficiency can be increased by optimum focusing. Find the optimum beam waist radius for this purpose. What is the best conversion efficiency? (d) With quasi-phase matching in a PPLN crystal as described in Example 9.4, find the optimum beam waist radius and the highest attainable normalized conversion efficiency $\hat{\eta}_{SH}$.

Solution

(a) With type I angle phase matching, we have $|d_{eff}| = 4.63$ pm V^{-1} and $n_{\omega,z} = n_\omega^o = 2.2340$ from Example 9.1. With $w_0 = 50$ μm and $\lambda = 1.064$ μm, we find that $l_a = 1.07$ cm from Example 9.2 and

$$b = \frac{2\pi n_\omega^o w_0^2}{\lambda} = \frac{2\pi \times 2.2340 \times (50 \times 10^{-6})^2}{1.064 \times 10^{-6}} \text{ m} = 3.3 \text{ cm.}$$

Because $l_a < l$, (10.15) has to be used to find the normalized conversion efficiency:

$$\hat{\eta}_{SH} \approx \frac{16\pi^2 |d_{eff}|^2}{c\epsilon_0 n_{\omega,z}^2 \lambda^3} \frac{1.068l}{1 + lb/l_a^2}$$

$$= \frac{16\pi^2 \times (4.63 \times 10^{-12})^2}{3 \times 10^8 \times 8.85 \times 10^{-12} \times 2.2340^2 \times (1.064 \times 10^{-6})^3}$$

$$\times \frac{1.068 \times 1.5 \times 10^{-2}}{1 + 1.5 \times 10^{-2} \times 3.3 \times 10^{-2}/(1.07 \times 10^{-2})^2} \text{ W}^{-1}$$

$$= 0.064\% \text{ W}^{-1}.$$

(b) With 90° type I phase matching, we have $|d_{eff}| = 4.4$ pm V^{-1}, $n_{\omega,z} = n_\omega^o = 2.2341$ at $T = 316.1$ K, and $l_a = \infty$ from Example 9.3. With $w_0 = 50$ μm, $\lambda = 1.064$ μm, and $n_\omega^o = 2.2341$, we still have $b = 3.3$ cm. In this case, (10.14) is valid because $b > 2l$ and $l_a \gg l$. Therefore, we find that

$$\hat{\eta}_{SH} \approx \frac{16\pi^2 |d_{eff}|^2 l^2}{c\epsilon_0 n_{\omega,z}^2 \lambda^3 b}$$

$$= \frac{16\pi^2 \times (4.4 \times 10^{-12})^2 \times (1.5 \times 10^{-2})^2}{3 \times 10^8 \times 8.85 \times 10^{-12} \times 2.2341^2 \times (1.064 \times 10^{-6})^3 \times 3.3 \times 10^{-2}} \text{ W}^{-1}$$

$$= 0.13\% \text{ W}^{-1}.$$

We see that this normalized conversion efficiency is two times that found in (a) by using 90° phase matching to eliminate the walk-off effect.

(c) In the absence of walk-off for 90° phase matching, the best efficiency can be obtained by making $b = l/2.84 = 5.28$ mm, which can be accomplished by focusing the fundamental beam to the optimum beam waist radius:

$$w_0 = \left(\frac{\lambda b}{2\pi n_\omega^o}\right)^{1/2} = \left(\frac{1.064 \times 10^{-6} \times 5.28 \times 10^{-3}}{2\pi \times 2.2341}\right)^{1/2} \text{ m} = 20.0 \text{ μm.}$$

The normalized conversion efficiency is found by using (10.16) to be

$$\hat{\eta}_{SH} \approx \frac{16\pi^2 |d_{eff}|^2}{c\epsilon_0 n_{\omega,z}^2 \lambda^3}(1.068l)$$

$$= \frac{16\pi^2 \times (4.4 \times 10^{-12})^2 \times 1.068 \times 1.5 \times 10^{-2}}{3 \times 10^8 \times 8.85 \times 10^{-12} \times 2.2341^2 \times (1.064 \times 10^{-6})^3} \text{W}^{-1}$$

$$= 0.307\% \text{ W}^{-1}.$$

We see that this normalized conversion efficiency is more than two times that found in (b) by optimally focusing the beam in the absence of walk-off.

(d) Because there is no walk-off in the case of quasi-phase matching in a PPLN crystal as described in Example 9.4, we can still take $b = l/2.84 = 5.28$ mm. Because $n_{\omega,z} = n_\omega^e = 2.1554$ and $n_{2\omega}^e = 2.2330$ in this situation, we find the optimum beam waist radius:

$$w_0 = \left(\frac{\lambda b}{2\pi n_\omega^e}\right)^{1/2} = \left(\frac{1.064 \times 10^{-6} \times 5.28 \times 10^{-3}}{2\pi \times 2.1554}\right)^{1/2} \text{ m} = 20.4 \text{ μm,}$$

which is slightly larger than that found in (c). The highest normalized conversion efficiency is still found by using (10.16) but with $|d_{eff}|$ replaced by $|d_Q| = 16.04$ pm V^{-1}, found in Example 9.4. Therefore,

$$\hat{\eta}_{SH} \approx \frac{16\pi^2 |d_Q|^2}{c\epsilon_0 n_{\omega,z} n_{2\omega,z} \lambda^3}(1.068l)$$

$$= \frac{16\pi^2 \times (16.04 \times 10^{-12})^2 \times 1.068 \times 1.5 \times 10^{-2}}{3 \times 10^8 \times 8.85 \times 10^{-12} \times 2.1554 \times 2.2330 \times (1.064 \times 10^{-6})^3} \text{W}^{-1}$$

$$= 4.23\% \text{ W}^{-1}.$$

This normalized conversion efficiency is about 14 times that found in (c) because quasi-phase matching by using a PPLN crystal allows us to take advantage of the largest nonlinear susceptibility element, d_{33}, of LiNbO$_3$.

This example illustrates how the efficiency of a second-harmonic generator can be substantially increased by a combination of optimization procedures. Further increase in efficiency is possible by using a waveguide structure, as illustrated later in Example 10.2, or by using short optical pulses to increase the peak intensity at a given average power level. The same techniques can be generally applied to other nonlinear frequency converters that are discussed in this chapter to increase their conversion efficiencies.

10.1.4 Second-Harmonic Generation in a Waveguide

The general formulation for second-harmonic generation in a waveguide was discussed in Subsection 7.2.2. In general, multiple waveguide modes can be involved in the process. In the case of a guided-wave second-harmonic generator in which each of the fundamental and second-harmonic waves is in only one waveguide mode, the coupled equations take the forms of (7.68) and (7.69):

$$\frac{dA_{2\omega,v}}{dt} = i\,\omega C A^2_{\omega,\mu} e^{i\Delta\beta z}, \tag{10.17}$$

$$\frac{dA_{\omega,\mu}}{dt} = i\,\omega C^* A_{2\omega,v} A^*_{\omega,\mu} e^{-i\Delta\beta z}, \tag{10.18}$$

where $C = C_{v\mu\mu}$ is the effective nonlinear coefficient given in (7.62):

$$\left|C\right|^2 = \left|C_{v\mu\mu}\right|^2 = \frac{\left|\chi_{\mathrm{eff}}\right|^2}{2c^3\epsilon_0 n_{2\omega,v} n^2_{\omega,\mu}} \frac{\Gamma_{v\mu\mu}}{\mathcal{A}} = \frac{2\left|d_{\mathrm{eff}}\right|^2}{c^3\epsilon_0 n_{2\omega,v} n^2_{\omega,\mu}} \frac{\Gamma_{v\mu\mu}}{\mathcal{A}}, \tag{10.19}$$

and

$$\Delta\beta = \Delta\beta_{v\mu\mu} = 2\beta_{\omega,\mu} - \beta_{2\omega,v} \tag{10.20}$$

is the phase mismatch.

In the low-efficiency limit, the depletion of the power in the fundamental wave is negligible. In this case, we can obtain, by integrating (7.68) or by converting (10.3), the power of the second-harmonic wave for a waveguide that has a length of l:

$$P_{2\omega,v}(l) = \omega^2 \left|C\right|^2 P^2_{\omega,\mu} l^2 \frac{\sin^2(\Delta\beta l/2)}{(\Delta\beta l/2)^2} = \frac{4\pi^2 c^2}{\lambda^2}\left|C\right|^2 P^2_{\omega,\mu} l^2 \frac{\sin^2(\Delta\beta l/2)}{(\Delta\beta l/2)^2}. \tag{10.21}$$

In the high-efficiency limit with perfect phase matching, we have

$$P_{2\omega,v}(l) = P_{\omega,\mu}(0)\tanh^2 \kappa l, \tag{10.22}$$

$$P_{\omega,\mu}(l) = P_{\omega,\mu}(0)\mathrm{sech}^2 \kappa l, \tag{10.23}$$

with the coefficient κ given as

$$\kappa = \left[\omega^2\left|C\right|^2 P_{\omega,\mu}(0)\right]^{1/2} = \left[\frac{4\pi^2 c^2}{\lambda^2}\left|C\right|^2 P_{\omega,\mu}(0)\right]^{1/2}. \tag{10.24}$$

The techniques for phase matching that were discussed in Chapter 9 are also applicable to guided-wave devices. In principle, it is possible to phase match by using the modal dispersion of a waveguide if modes of different orders are involved in the interaction. However, because of the large frequency difference between the fundamental wave and the second-harmonic wave, the modal dispersion of the typical waveguide is usually insufficient to compensate for the material dispersion to achieve phase matching. Often, a combination of different techniques is employed. For example, a waveguide is fabricated along a certain direction in a crystal for collinear birefringent phase matching, but temperature is tuned for fine matching once the wave propagation direction is fixed by the waveguide structure. Quasi-phase matching is particularly useful for guided-wave devices because of its advantages discussed in Section 9.3 and because of its compatibility with the microfabrication technology. For a guided-wave device that is quasi-phase matched by using a periodic structure with a duty factor of ξ, the coupled equations can be transformed in a manner similar to the transformation shown in (9.47), resulting in a phase mismatch of

$$\Delta\beta_Q = \Delta\beta + qK \tag{10.25}$$

that is minimized with a particular integer q and a nonlinear coefficient of C_Q given as

$$C_Q = 2C \frac{\sin \xi q\pi}{q\pi} e^{-i\xi q\pi} \tag{10.26}$$

according to (9.51). With quasi-phase matching, we have to replace the phase mismatch $\Delta\beta$ in (10.21) with $\Delta\beta_Q$, and the nonlinear coefficient C in (10.21) and (10.24) with C_Q. For a first-order structure of a 50% duty factor, $\left|C_Q\right| = 2\left|C\right|/\pi$.

EXAMPLE 10.2

A PPLN waveguide is used for second-harmonic generation of a fundamental wave at the 1.064 μm wavelength. The waveguide is a diffused channel waveguide formed by Ti diffusion into a PPLN crystal that is similar to the one described in Example 9.4. It has a diffusion depth of $d = 4$ μm and a width of $w = 5$ μm, for an effective waveguide core area of $\mathcal{A} = wd = 20$ μm^2. It is a single-mode waveguide for the fundamental wavelength at 1.064 μm. The overlap factor for second-harmonic generation at this wavelength in this waveguide is $\Gamma = 0.6$. The grating period of the PPLN is properly chosen as a first-order grating of a 50% duty factor for quasi-phase matching of the interacting waveguide modes. For easy comparison to second-harmonic generation in the bulk PPLN crystal that is considered in Example 10.1(d), we take the waveguide length to be $l = 1.5$ cm. Find the normalized second-harmonic conversion efficiency for this device.

Solution

By replacing $\Delta\beta$ with $\Delta\beta_Q$ and C with C_Q in (10.21), we have the normalized efficiency in the low-efficiency limit for the PPLN waveguide:

$$\hat{\eta}_{SH} = \frac{P_{2\omega}(l)}{P_\omega^2} = \frac{4\pi^2 c^2}{\lambda^2} \left|C_Q\right|^2 l^2 \frac{\sin^2 (\Delta\beta_Q l/2)}{(\Delta\beta_Q l/2)^2}. \tag{10.27}$$

For perfect quasi-phase matching with a first-order grating that has a 50% duty factor, we have

$$\hat{\eta}_{SH} = \frac{4\pi^2 c^2}{\lambda^2} \left|C_Q\right|^2 l^2 = \frac{32 \left|d_{eff}\right|^2}{c\epsilon_0 n_\omega^2 n_{2\omega} \lambda^2} \frac{\Gamma}{\mathcal{A}} l^2, \tag{10.28}$$

where n_ω and $n_{2\omega}$ are the effective refractive index, n_β, of the waveguide modes at the fundamental and second-harmonic frequencies, respectively. Because the index change created by Ti diffusion is very small, typically around 0.5%, we can simply take the refractive index of the bulk PPLN as a very good approximation for the effective refractive index of a waveguide mode. From Example 9.4, we have $d_{\text{eff}} = d_{33} = -25.2 \text{ pm V}^{-1}$, $n_\omega = n_\omega^{\text{e}} = 2.1554$, $n_{2\omega} = n_{2\omega}^{\text{e}} = 2.2330$. We then find the normalized conversion efficiency:

$$\hat{\eta}_{\text{SH}} = \frac{32 \times (25.2 \times 10^{-12})^2 \times 0.6 \times (1.5 \times 10^{-2})^2}{3 \times 10^8 \times 8.85 \times 10^{-12} \times 2.1554^2 \times 2.2330 \times (1.064 \times 10^{-6})^2 \times 20 \times 10^{-12}} \text{ W}^{-1}$$

$$= 440\% \text{ W}^{-1}.$$

This normalized conversion efficiency for the PPLN waveguide is 104 times that obtained in Example 10.1(d) for the bulk PPLN.

Further increase in the efficiency is possible by increasing the length of the waveguide. In the waveguide device, the efficiency continues to increase quadratically with the waveguide length l because the optical waves remain confined as the waveguide length is increased. In a bulk device, the best efficiency only increases linearly with the length l, as seen in (10.16), because of the limitation imposed by diffraction. Note that (10.27) and (10.28) are valid only in the low-efficiency limit. Clearly, $\hat{\eta}_{\text{SH}} = 440\% \text{ W}^{-1}$ obtained in this example does not mean that it is possible to obtain an unphysical efficiency of 440% by launching a fundamental beam of 1 W power into the waveguide. Nor does it mean that a conversion efficiency of 100% is obtained by launching a fundamental beam of 227 mW into the waveguide. It only means that a very low input power of the fundamental wave is needed to obtain a decent conversion efficiency. For example, an input fundamental power of only $P_\omega = 22.7$ mW is required to have an output second-harmonic power of $P_{2\omega} = 2.27$ mW for a conversion efficiency of 10%. A conversion efficiency approaching 100% is theoretically possible, but with an input funda-

mental power found by using the relation in (10.22) , with $\kappa l = \left[\omega^2 \left| C_Q \right|^2 l^2 P_\omega(0) \right]^{1/2} = \left[\hat{\eta}_{\text{SH}} P_\omega(0) \right]^{1/2}$, for the high-efficiency limit. (See Problem 10.1.4.)

10.2 SUM-FREQUENCY GENERATION

The basic function of a sum-frequency generator is the generation of an optical wave at a high frequency, ω_3, by mixing two optical waves at low frequencies, ω_1 and ω_2, as schematically shown in Fig. 5.2. The general application of a sum-frequency generator is straightforward. It is most often used to obtain, through mixing available optical waves at long wavelengths, a coherent optical beam at a desired short wavelength that is not readily available from other sources. If one of the two input waves is tunable in wavelength, a wavelength-tunable sum-frequency output wave is obtained. For example, a wavelength-tunable optical beam in the ultraviolet spectral region can be obtained with a sum-frequency generator that mixes the output of a tunable laser in the visible spectral region with that of another laser at a fixed wavelength also in the visible spectral region.

The process of sum-frequency generation is generally described by the coupled equations that are given in (6.27)–(6.29) with the condition that $\mathcal{E}_1(0) \neq 0$ and $\mathcal{E}_2(0) \neq 0$ but $\mathcal{E}_3(0) = 0$ at the input surface, $z = 0$, of a nonlinear crystal:

$$\frac{d\mathcal{E}_3}{dz} = \frac{i\omega_3^2}{c^2 k_{3,z}} \chi_{\text{eff}} \mathcal{E}_1 \mathcal{E}_2 e^{i\Delta k z}, \tag{10.29}$$

$$\frac{d\mathcal{E}_1}{dz} = \frac{i\omega_1^2}{c^2 k_{1,z}} \chi_{\text{eff}}^* \mathcal{E}_3 \mathcal{E}_2^* e^{-i\Delta k z}, \tag{10.30}$$

$$\frac{d\mathcal{E}_2}{dz} = \frac{i\omega_2^2}{c^2 k_{2,z}} \chi_{\text{eff}}^* \mathcal{E}_3 \mathcal{E}_1^* e^{-i\Delta k z}, \tag{10.31}$$

where $\chi_{\text{eff}} = \hat{e}_3^* \cdot \boldsymbol{\chi}^{(2)}(\omega_3 = \omega_1 + \omega_2) : \hat{e}_1 \hat{e}_2$, from (6.26), is the effective nonlinear susceptibility and $\Delta \mathbf{k} = \mathbf{k}_1 + \mathbf{k}_2 - \mathbf{k}_3 = \Delta k \hat{z}$ is the phase mismatch for sum-frequency generation. The general solutions of these coupled equations can be found in terms of the Jacobi elliptic functions [2]. However, simpler, and often more useful, solutions can be found for specific experimental conditions of interest.

The simplest situation is the low-efficiency limit that the efficiency of a sum-frequency generator is low so that the intensities of both input waves at ω_1 and ω_2 are not significantly depleted throughout the interaction. We can then take $\mathcal{E}_1$ and $\mathcal{E}_2$ to be independent of z, ignore (10.30) and (10.31) in the coupled equations, and directly integrate (10.29) to find $\mathcal{E}_3(z)$. By using the relation given in (6.31) for light intensity, we find that, in the low-efficiency limit, the intensity of the wave at the sum frequency as a function of the interaction length l can be expressed as

$$\begin{aligned} I_3(l) &= \frac{\omega_3^2 |\chi_{\text{eff}}|^2}{2c^3 \epsilon_0 n_{1,z} n_{2,z} n_{3,z}} I_1 I_2 l^2 \frac{\sin^2(\Delta k l/2)}{(\Delta k l/2)^2} \\ &= \frac{8\pi^2 |d_{\text{eff}}|^2}{c \epsilon_0 n_{1,z} n_{2,z} n_{3,z} \lambda_3^2} I_1 I_2 l^2 \frac{\sin^2(\Delta k l/2)}{(\Delta k l/2)^2}, \end{aligned} \tag{10.32}$$

where the intensities I_1 and I_2 of the two low-frequency waves are assumed to remain constant at their input values such that $I_1 = I_1(0)$ and $I_2 = I_2(0)$, $d_{\text{eff}} = \chi_{\text{eff}}/2$, and $\lambda_3 = 2\pi c/\omega_3$ is the wavelength of the sum-frequency wave in free space. In the case of quasi-phase matching, χ_{eff} and d_{eff} in (10.32) are replaced by χ_Q and d_Q, respectively.

From (10.32), we see that the effect of phase mismatch for sum-frequency generation is characterized by a function of the same form as that given in (10.4) and plotted in Fig. 10.1. When perfect phase matching is not accomplished, such that $\Delta k \neq 0$, the largest useful length of a nonlinear crystal is the coherence length of the interaction. When perfect phase matching is accomplished such that $\Delta k = 0$, the intensity of the sum-frequency wave grows quadratically with the interaction length as $I_3(l) = I_3^{\text{PM}}(l) \propto l^2/\lambda_3^2$, as long as the interaction stays in the low-efficiency limit.

We also see from (10.32) that $I_3 \propto |d_{\text{eff}}|^2 I_1 I_2$ in the low-efficiency limit. Therefore, the output intensity of the sum-frequency wave increases *linearly* with the input intensity of each of the two low-frequency waves. Increasing the intensity of either input wave increases the output intensity of the sum-frequency wave but only linearly if the intensity of the other input wave is not simultaneously increased. If the purpose of an application is to produce a significant intensity for the sum-frequency wave, the two input waves need to have high and comparable intensities.

In the above discussion, the interacting waves are assumed to be perfect plane waves. In reality, each optical beam has a finite cross-section and a nonuniform intensity distribution. As mentioned in the preceding section for second-harmonic generation, this and other spatial effects also have to be carefully considered in a detailed analysis of a sum-frequency generation process. The key point is that the interaction among the optical beams takes place only in the area where they spatially overlap and during the time when they also temporally overlap if the beams are optical pulses. Therefore, in terms of optical power, I_1, I_2, and I_3 in (10.32) have to be replaced by $P_1/\mathcal{A}_1$, $P_2/\mathcal{A}_2$, and $P_3/\mathcal{A}_3$, respectively, where P_q is the power of the wave at the frequency ω_q and $\mathcal{A}_q$ is its effective cross-sectional area. In the low-efficiency limit, we then have $P_3 \propto \left|d_{\text{eff}}\right|^2 P_1 P_2 \mathcal{A}_3/\mathcal{A}_1\mathcal{A}_2$. It is important to realize that $\mathcal{A}_3 \leq \min(\mathcal{A}_1, \mathcal{A}_2)$ because the sum-frequency wave is generated only within the area where the two input waves overlap. We thus arrive at these conclusions:

1. To maximize the efficiency of sum-frequency generation with two input waves at given power levels, it is important to collimate these two beams to the same cross-sectional area and to have them overlap uniformly so that $\mathcal{A}_3$ is maximized.
2. It is possible to increase the conversion efficiency by focusing the input waves to simultaneously reduce $\mathcal{A}_1$ and $\mathcal{A}_2$ so long as the effective interaction length is not reduced due to the increased divergence of the focused beams.
3. It does not pay to just focus one input beam or to focus the two input beams unevenly because doing so results in a corresponding reduction in $\mathcal{A}_3$.

10.3 DIFFERENCE-FREQUENCY GENERATION

By mixing two optical waves, taken to be at the frequencies ω_3 and ω_1, respectively, a difference-frequency generator produces a third optical wave at the difference frequency $\omega_2 = \omega_3 - \omega_1$, as schematically shown in Fig. 5.3. Difference-frequency generators are the simplest devices for the generation of coherent infrared radiation, particularly the radiation in the mid- to far-infrared region where efficient laser materials are rare. For this purpose, both input waves can be in the visible region, or one in the visible and another in the near-infrared region, where many efficient laser sources are available. Wavelength-tunable infrared radiation can be obtained if one of the input waves is from a wavelength-tunable source.

The equations for the description of the difference-frequency generation process are also those given in (10.29)–(10.31), but the initial conditions are $\mathcal{E}_3(0) \neq 0$, $\mathcal{E}_1(0) \neq 0$, and $\mathcal{E}_2(0) = 0$ at the input surface of a nonlinear crystal. Similar to the case of sum-frequency generation, the general solutions of the coupled equations with the initial conditions for difference-frequency generation can be found in terms of elliptic functions. However, also similar to the case of sum-frequency generation, simple solutions under special situations are often more useful.

In the low-efficiency limit, the depletion of the intensities of the two input waves is negligible. By taking the two input fields, $\mathcal{E}_3$ and $\mathcal{E}_1$, to be independent of z, (10.31) can be directly integrated for the field $\mathcal{E}_2(z)$ at the difference frequency ω_2. The solution for the intensity of the difference-frequency wave is found:

$$
\begin{aligned}
I_2(l) &= \frac{\omega_2^2 \left|\chi_{\text{eff}}\right|^2}{2c^3 \epsilon_0 n_{1,z} n_{2,z} n_{3,z}} I_3 I_1 l^2 \frac{\sin^2(\Delta k l/2)}{(\Delta k l/2)^2} \\
&= \frac{8\pi^2 \left|d_{\text{eff}}\right|^2}{c\epsilon_0 n_{1,z} n_{2,z} n_{3,z} \lambda_2^2} I_3 I_1 l^2 \frac{\sin^2(\Delta k l/2)}{(\Delta k l/2)^2},
\end{aligned} \tag{10.33}
$$

where $\lambda_2 = 2\pi c/\omega_2$ is the wavelength of the difference-frequency wave in free space. In the case of quasi-phase matching, χ_{eff} and d_{eff} in (10.33) are replaced by χ_Q and d_Q, respectively.

The relation in (10.33) has the same form as that in (10.32). The effect of phase mismatch is also that shown in Fig. 10.1. With perfect phase matching such that $\Delta k = 0$, the intensity of the difference-frequency wave grows quadratically with the interaction length as $I_2(l) = I_2^{\text{PM}}(l) \propto l^2/\lambda_2^2$ as long as the interaction stays in the low-efficiency limit. To produce a significant intensity for the difference-frequency wave, it is also desirable to have two strong input waves with comparable intensities because $I_2 \propto \left|d_{\text{eff}}\right|^2 I_3 I_1$. In terms of optical power, we have $P_2 \propto \left|d_{\text{eff}}\right|^2 P_3 P_1 \mathcal{A}_2/\mathcal{A}_3 \mathcal{A}_1$, where $\mathcal{A}_2 \leq \min(\mathcal{A}_3, \mathcal{A}_1)$. Therefore, in the low-efficiency limit, the wave produced by a difference-frequency generator has the same general characteristics as those of the wave produced by a sum-frequency generator.

One word of caution in the application of a difference-frequency generator goes to the generation of far-infrared radiation. When the wavelength of the difference-frequency wave in the far-infrared region becomes comparable to, or even larger than, one of the cross-sectional beam diameters of the input waves, the diffraction effect of the long-wavelength difference-frequency wave becomes significant. As a result, the relation given in (10.33) is no longer valid. Instead, the spatially nonuniform distribution of the difference-frequency wave caused by this diffraction has to be considered, though the total power integrated over the entire cross-section of the difference-frequency wave is not changed by the diffraction effect.

10.4 OPTICAL PARAMETRIC FREQUENCY CONVERSION

The function of an optical parametric frequency converter is the conversion of a signal-carrying optical wave from one carrier frequency to another through *parametric up-conversion* or *parametric down-conversion*. Parametric up-conversion is a special case of sum-frequency generation with the objective of converting a signal-carrying optical wave at a low frequency, typically in the mid- or far-infrared region, where sensitive detectors are not available, to an optical wave that carries the same signal at a frequency in the visible region, where efficient detection can be easily made. Parametric down-conversion is a special case of difference-frequency generation in which a signal-carrying optical wave at a high frequency, often in the ultraviolet region, is converted to one at a low frequency in the visible or the infrared region. The signal-carrying input wave, which is called the *signal*, is generally very weak in comparison to the other input wave, which is called the *pump*. In the following analysis, the strong pump wave is taken to be at the frequency ω_2. The signal is taken to be at ω_1 for up-conversion, and it is taken to be at ω_3 for down-conversion. The relation $\omega_3 = \omega_1 + \omega_2$ applies to both cases.

Because the pump is much stronger than the signal, the intensity of the pump wave can be considered to be constant, though that of the signal wave is not. As a result, we have two coupled equations for parametric conversion processes:

$$\frac{d\mathcal{E}_3}{dz} = i\left(\frac{\omega_3^2}{c^2 k_{3,z}}\chi_{\text{eff}}\mathcal{E}_2\right)\mathcal{E}_1 e^{i\Delta kz} = i\kappa_{31}\mathcal{E}_1 e^{i\Delta kz}, \tag{10.34}$$

$$\frac{d\mathcal{E}_1}{dz} = i\left(\frac{\omega_1^2}{c^2 k_{1,z}}\chi_{\text{eff}}^*\mathcal{E}_2^*\right)\mathcal{E}_3 e^{-i\Delta kz} = i\kappa_{13}\mathcal{E}_3 e^{-i\Delta kz}, \tag{10.35}$$

where $\chi_{\text{eff}} = \hat{e}_3^* \cdot \chi^{(2)}(\omega_3 = \omega_1 + \omega_2) : \hat{e}_1\hat{e}_2$ and $\Delta \mathbf{k} = \mathbf{k}_1 + \mathbf{k}_2 - \mathbf{k}_3 = \Delta k\hat{z}$. In (10.34) and (10.35), $\mathcal{E}_2$ of the pump wave is taken to be a constant that is independent of z. Because the signal is normally weak in the application of a parametric converter, a high conversion efficiency is most desirable. Therefore, the device is normally used under the condition of perfect phase matching with $\Delta k = 0$.

10.4.1 Optical Parametric Up-Conversion

For up-conversion, the initial conditions for solving the coupled equations given in (10.34) and (10.35) are $\mathcal{E}_1(0) \neq 0$ and $\mathcal{E}_3(0) = 0$. The schematic diagram of an optical parametric up-converter is shown in Fig. 10.4(a). The solutions under the condition of perfect phase matching with $\Delta k = 0$ are

$$\mathcal{E}_3(l) = \frac{i\kappa_{31}}{\kappa}\mathcal{E}_1(0)\sin\kappa l, \tag{10.36}$$

$$\mathcal{E}_1(l) = \mathcal{E}_1(0)\cos\kappa l, \tag{10.37}$$

where

$$\kappa = (\kappa_{31}\kappa_{13})^{1/2} = \left(\frac{\omega_1\omega_3|\chi_{\text{eff}}|^2}{2c^3\epsilon_0 n_{1,z}n_{2,z}n_{3,z}}I_2\right)^{1/2} = \left(\frac{8\pi^2|d_{\text{eff}}|^2}{c\epsilon_0 n_{1,z}n_{2,z}n_{3,z}\lambda_1\lambda_3}I_2\right)^{1/2}. \tag{10.38}$$

In the case of quasi-phase matching, χ_{eff} and d_{eff} in (10.38) are replaced by χ_Q and d_Q, respectively.

For a parametric up-converter with perfect phase matching, the intensities of the three interacting beams vary with the interaction length as

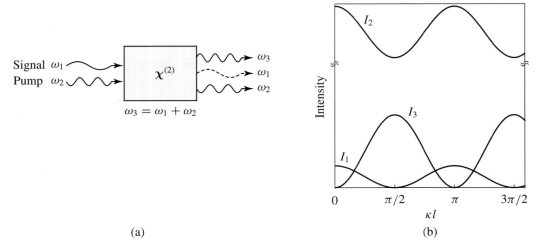

Figure 10.4 (a) Schematic of an optical parametric up-converter. (b) Intensity variations of the interacting optical waves as a function of the interaction length. The pump wave is at ω_2. The signal wave is at ω_1 for up-conversion to ω_3.

$$I_3(l) = \frac{\omega_3}{\omega_1} I_1(0) \sin^2 \kappa l, \tag{10.39}$$

$$I_1(l) = I_1(0) \cos^2 \kappa l, \tag{10.40}$$

$$I_2(l) = I_2(0) - \frac{\omega_2}{\omega_1} I_1(0) \sin^2 \kappa l \approx I_2(0). \tag{10.41}$$

Figure 10.4(b) illustrates these intensity variations. Complete up-conversion of the signal occurs at an interaction length of the *phase-matched coupling length* $l_c^{\mathrm{PM}} = \pi/2\kappa$, as well as a length of any odd multiple of l_c^{PM}, as expected of phase-matched codirectional coupling. The value of this length can be varied by varying the pump intensity because the value of κ depends on that of the strong pump intensity I_2. Note that when the signal intensity, $I_1(l)$, is completely depleted by up-conversion, the intensity of the up-converted sum-frequency wave, $I_3(l)$, reaches a maximum value of $I_3^{\mathrm{max}} = I_1(0)\omega_3/\omega_1 > I_1(0) = I_1^{\mathrm{max}}$ because $\omega_3 > \omega_1$ and the total number of the sum-frequency photons that are created at ω_3 is equal to the total number of the signal photons that are annihilated at ω_1.

10.4.2 Optical Parametric Down-Conversion

Parametric down-conversion is simply the reverse process of up-conversion, and vice versa. The same parametric converter can function as either an up-converter or a down-converter. The only difference is the initial conditions at the input. If the initial conditions are $\mathcal{E}_1(0) = 0$ and $\mathcal{E}_3(0) \neq 0$, then the device functions as a down-converter. The schematic diagram of an optical parametric down-converter is shown in Fig. 10.5(a). With these initial conditions, the solutions under the condition of perfect phase matching with $\Delta k = 0$ are

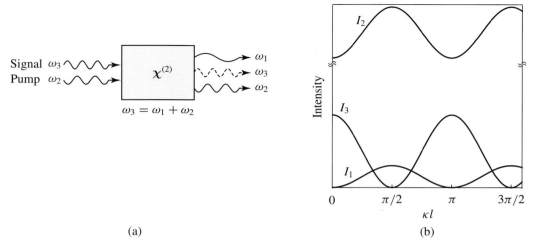

Figure 10.5 (a) Schematic of an optical parametric down-converter. (b) Intensity variations of the interacting optical waves as a function of the interaction length. The pump wave is at ω_2. The signal wave is at ω_3 for down-conversion to ω_1.

$$\mathcal{E}_1(l) = \frac{i\kappa_{13}}{\kappa}\mathcal{E}_3(0)\sin\kappa l, \tag{10.42}$$

$$\mathcal{E}_3(l) = \mathcal{E}_3(0)\cos\kappa l, \tag{10.43}$$

where κ is the same as that given in (10.38).

For a parametric down-converter with perfect phase matching, the intensities of the three interacting beams vary with the interaction length as

$$I_1(l) = \frac{\omega_1}{\omega_3}I_3(0)\sin^2\kappa l, \tag{10.44}$$

$$I_3(l) = I_3(0)\cos^2\kappa l, \tag{10.45}$$

$$I_2(l) = I_2(0) + \frac{\omega_2}{\omega_3}I_3(0)\sin^2\kappa l \approx I_2(0). \tag{10.46}$$

These intensity variations are illustrated in Fig. 10.5(b). Similar to the case of up-conversion, complete down-conversion of the signal occurs at an interaction length of the phase-matched coupling length, $l_{\mathrm{c}}^{\mathrm{PM}} = \pi/2\kappa$, or a length of any odd multiple of $l_{\mathrm{c}}^{\mathrm{PM}}$. The value of this length can be varied by varying the pump intensity because the value of κ depends on that of the strong pump intensity I_2. When the signal intensity, $I_3(l)$, is completely depleted by down-conversion, the intensity of the down-converted difference-frequency wave, $I_1(l)$, reaches a maximum value of $I_1^{\mathrm{max}} = I_3(0)\omega_1/\omega_3 < I_3(0) = I_3^{\mathrm{max}}$ because $\omega_1 < \omega_3$ and the total number of the difference-frequency photons that are created at ω_1 is equal to the total number of the signal photons that are annihilated at ω_3.

In Fig. 10.4(b), we see that when the intensity of the signal wave at ω_1 is completely depleted, for example, at a distance of $l = l_{\mathrm{c}}^{\mathrm{PM}}$, further interaction in the parameter up-converter leads to down-conversion from the wave at ω_3 back to the wave at ω_1. Similarly, in Fig. 10.5(b), when the intensity of the signal wave at ω_3 is completely depleted at a distance of any odd multiple of l_c^{PM}, further interaction in the parameter down-converter leads to up-conversion from the wave at ω_1 back to the wave at ω_3. It is clear from this observation that a parametric up-converter can be used as a parametric down-converter for the same three frequencies involved. With the same pump beam at the frequency ω_2, the only difference between these two devices is the initial conditions of the input intensities at the frequencies ω_1 and ω_3. If the input is a low-frequency signal at ω_1, the device functions as an up-converter to convert this low signal frequency to the higher frequency at ω_3. If the input is a high-frequency signal at ω_3, then the device functions as an down-converter to convert this high signal frequency to the lower frequency at ω_1.

10.5 OPTICAL PARAMETRIC AMPLIFICATION

The physical process involved in an *optical parametric amplifier*, commonly called an OPA, is basically the same as that in a difference-frequency generator. The only difference is in the usage of the device. In either case, there are two input waves at the frequencies of ω_1 and ω_3. While the usage of a difference-frequency generator is for the generation of a wave at the difference frequency of $\omega_2 = \omega_3 - \omega_1$, that of an OPA is for the amplification of the input wave at ω_1. The wave at the difference frequency of ω_2 is still generated in an OPA, though it is not the purpose of this application. Therefore, for an OPA, the high-frequency input wave at ω_3 is called the *pump* wave, the low-frequency input wave at ω_1 is called the *signal* wave, and the side product at ω_2 is called the *idler* wave, as shown in Fig. 10.6(a).

The general coupled-wave equations for an OPA are those for sum-frequency generation given in (10.29)–(10.31). Normally the pump wave, at ω_3, of an OPA is much stronger than the signal wave at ω_1 and the idler wave at ω_2 so that its intensity can be considered constant throughout the interaction with $\mathcal{E}_3(z) \approx \mathcal{E}_3(0)$. Therefore, only (10.30) and (10.31) have to be

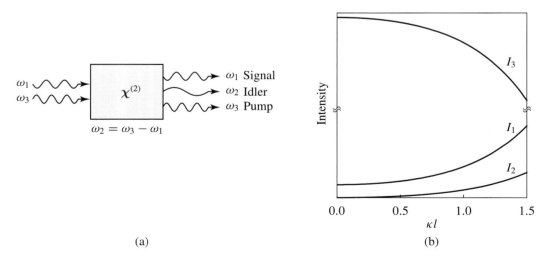

(a) (b)

Figure 10.6 (a) Schematic of an OPA. (b) Intensity variations of the pump, signal, and idler waves of an OPA with a strong pump as functions of the interaction length in the case of perfect phase matching.

considered for the spatial variations of $\mathcal{E}_1$ and $\mathcal{E}_2$, and the initial conditions are $\mathcal{E}_1(0) \neq 0$ and $\mathcal{E}_2(0) = 0$. The coupled equations for an OPA that is pumped with a strong pump wave at ω_3 are

$$\frac{d\mathcal{E}_1}{dz} = i\left(\frac{\omega_1^2}{c^2 k_{1,z}}\chi_{\text{eff}}^*\mathcal{E}_3\right)\mathcal{E}_2^* e^{-i\Delta kz} = i\kappa_{12}\mathcal{E}_2^* e^{-i\Delta kz}, \tag{10.47}$$

$$\frac{d\mathcal{E}_2^*}{dz} = i\left(-\frac{\omega_2^2}{c^2 k_{2,z}}\chi_{\text{eff}}\mathcal{E}_3^*\right)\mathcal{E}_1 e^{i\Delta kz} = i\kappa_{21}\mathcal{E}_1 e^{i\Delta kz}, \tag{10.48}$$

where (10.48) is obtained by taking the complex conjugate of (10.31). For efficient parametric amplification, phase matching is required. With $\Delta k = 0$ for perfect phase matching, and with the initial conditions $\mathcal{E}_1(0) \neq 0$ and $\mathcal{E}_2(0) = 0$, these two coupled equations have the solutions:

$$\mathcal{E}_1(z) = \mathcal{E}_1(0)\cosh \kappa z, \tag{10.49}$$

$$\mathcal{E}_2^*(z) = \frac{i\kappa_{21}}{\kappa}\mathcal{E}_1(0)\sinh \kappa z, \tag{10.50}$$

where

$$\kappa = (-\kappa_{12}\kappa_{21})^{1/2} = \left(\frac{\omega_1\omega_2\left|\chi_{\text{eff}}\right|^2}{2c^3\epsilon_0 n_{1,z}n_{2,z}n_{3,z}}I_3\right)^{1/2} = \left(\frac{8\pi^2\left|d_{\text{eff}}\right|^2}{c\epsilon_0 n_{1,z}n_{2,z}n_{3,z}\lambda_1\lambda_2}I_3\right)^{1/2}. \tag{10.51}$$

In the case of quasi-phase matching, χ_{eff} and d_{eff} in (10.51) are replaced by χ_Q and d_Q, respectively.

With perfect phase matching, the intensities of the signal, idler, and pump waves vary with the interaction length as

$$I_1(l) = I_1(0)\cosh^2 \kappa l, \tag{10.52}$$

$$I_2(l) = \frac{\omega_2}{\omega_1}I_1(0)\sinh^2 \kappa l, \tag{10.53}$$

$$I_3(l) = I_3(0) - \frac{\omega_3}{\omega_1}I_1(0)\sinh^2 \kappa l \approx I_3(0), \tag{10.54}$$

which are plotted in Fig. 10.6(b). We see that while the intensity of the signal wave grows as a result of parametric amplification, the intensity of the idler wave also increases because an idler photon is simultaneously generated with each signal photon that is generated in the parametric process.

With perfect phase matching, the *amplification factor*, or the *intensity gain*, of the signal wave for a single pass through an OPA is

$$G = \frac{P_1(l)}{P_1(0)} = \frac{I_1(l)}{I_1(0)} = \cosh^2 \kappa l \approx \begin{cases} 1 + \kappa^2 l^2, & \text{in the low-gain limit;} \\ \dfrac{e^{2\kappa l}}{4}, & \text{in the high-gain limit.} \end{cases} \tag{10.55}$$

The conversion efficiency from the pump to the signal of an OPA is

$$\eta_{OPA} = \frac{P_1(l)}{P_3(0)} = \frac{\mathcal{A}_1}{\mathcal{A}_3} \frac{I_1(l)}{I_3(0)} \approx \frac{\mathcal{A}_1}{\mathcal{A}_3} \frac{I_1(0)}{I_3(0)} G, \tag{10.56}$$

where $\mathcal{A}_1 \leq \mathcal{A}_3$ even though $\lambda_1 > \lambda_3$ because only the signal field that spatially overlaps with the pump field gets amplified. Note that the low-gain limit is not the same as the low-efficiency limit, and the high-gain limit is not the same as the high-efficiency limit. *A large gain factor does not necessarily imply a high conversion efficiency from the pump to the signal and idler* because the input signal can be extremely weak. Therefore, it is possible that the pump is not much depleted when the signal is amplified by a large gain factor but the conversion efficiency is low. When the input signal is strong, however, it is also possible that pump depletion is significant but the gain factor is small.

EXAMPLE 10.3

An OPA for a signal wavelength of $\lambda_1 = 1.55$ μm consists of a PPLN crystal that has a length of $l = 1$ cm. It is pumped with a Gaussian beam at $\lambda_3 = 527$ nm, which is focused to a beam waist radius of $w_0 = 50$ μm. The largest nonlinear susceptibility element of LiNbO$_3$ is $d_{33} = -25.2$ pm V^{-1}. (a) What is the idler wavelength? (b) What is the required first-order grating period for quasi-phase matching? (c) What is the amplification factor for the signal if the power of the pump beam is $P_3 = 1$ W? (d) What is the required pump power for an amplification factor of $G = 10^3$? Consider only the situation that the pump is not much depleted even when $G = 10^3$.

Solution

(a) With $\omega_1 = \omega_3 - \omega_2$ for the signal frequency of an OPA, the idler frequency is $\omega_2 = \omega_3 - \omega_1$. Therefore, the idler wavelength is

$$\lambda_2 = \left(\frac{1}{\lambda_3} - \frac{1}{\lambda_1} \right)^{-1} = \left(\frac{1}{527 \times 10^{-9}} - \frac{1}{1.55 \times 10^{-6}} \right)^{-1} m = 798 \text{ nm.}$$

(b) For the most efficient interaction in a PPLN crystal, all of the interacting waves have to be extraordinary waves polarized in the principal z direction because the largest $\chi^{(2)}$ element of LiNbO$_3$ is $\chi_{33}^{(2)}$. By using the data given in Table 3.4 for the Sellmeier equation of LiNbO$_3$, we find that $n_3^e = 2.2351$ at $\lambda_3 = 527$ nm, $n_1^e = 2.1373$ at $\lambda_1 = 1.55$ μm, and $n_2^e = 2.1755$ at $\lambda_2 = 798$ nm. The phase mismatch is $\Delta k = k_1 + k_2 - k_3 = 2\pi(n_1/\lambda_1 + n_2/\lambda_2 - n_3/\lambda_3)$ for collinear interaction. Therefore, according to (9.50), the required first-order grating period is

$$\Lambda = \frac{2\pi}{|\Delta k|} = \left| \frac{n_1}{\lambda_1} + \frac{n_2}{\lambda_2} - \frac{n_3}{\lambda_3} \right|^{-1} = \left| \frac{2.1373}{1.55 \times 10^{-6}} + \frac{2.1755}{798 \times 10^{-9}} - \frac{2.2351}{527 \times 10^{-9}} \right|^{-1} m = 7.35 \text{ μm.}$$

(c) For a Gaussian pump beam that is focused to a beam waist size of $w_0 = 50$ μm, we find that its confocal parameter is $b = 2\pi n_3^e w_0^2/\lambda_3 = 6.66$ cm. Because $b \gg l = 1$ cm, we can ignore the complicated effect of diffraction due to focusing and take $I_3 = P_3/\mathcal{A}_3 = 2P_3/\pi w_0^2$ over the entire length of the PPLN crystal. For this interaction, we have $|d_Q| = |2d_{\text{eff}}/\pi| = 16.04$ pm V^{-1}. Then, by using (10.51) with d_{eff} replaced by d_Q, we can express $\kappa^2 l^2$ as a function of the pump power as

$$\kappa^2 l^2 = \frac{16\pi|d_Q|^2 l^2}{c\epsilon_0 n_1^e n_2^e n_3^e \lambda_1 \lambda_2 w_0^2} P_3$$

$$= \frac{16\pi \times (16.04 \times 10^{-12})^2 \times (1 \times 10^{-2})^2}{3 \times 10^8 \times 8.85 \times 10^{-12} \times 2.1373 \times 2.1755 \times 2.2351 \times 1.55 \times 10^{-6} \times 798 \times 10^{-9} \times (50 \times 10^{-6})^2} P_3 \text{ W}^{-1}$$

$$= 0.015 P_3 \text{ W}^{-1}.$$

For $P_3 = 1$ W, the single-pass amplification factor is $G \approx 1 + \kappa^2 l^2 = 1.015$ in the low-gain limit, according to (10.55). The signal intensity grows only 1.5% in a single pass through the parametric amplifier.

(d) For an amplification factor of $G = 10^3$, we find by using the high-gain limit of (10.55) that $\kappa l = 4.147$ is required. From the dependence of $\kappa^2 l^2$ on P_3 found in (c), we find that the required pump power for $G = 10^3$ is

$$P_3 = \frac{\kappa^2 l^2}{0.015} \text{ W} = \frac{4.147^2}{0.015} \text{ W} = 1.15 \text{ kW}.$$

This pump power looks unrealistically high. It is indeed unrealistic if we consider only the possibility of a CW pump beam. It is not if we consider pulse pumping. For example, by using a Q-switched laser pulse that has a pulsewidth of $\Delta t_{\text{ps}} = 100$ ns, such a pump power requires a very common pump pulse energy of $U_{\text{ps}} = P_{\text{pk}}\Delta t_{\text{ps}} = 115$ μJ. As another example, if mode-locked pulses of $\Delta t_{\text{ps}} = 10$ ps pulsewidth at a repetition rate of $f_{\text{ps}} = 100$ MHz are used to pump the amplifier, the average power of the pulsed pump beam is again at a realistic level of $\overline{P} = P_{\text{pk}}\Delta t_{\text{ps}} f_{\text{ps}} = 1.15$ W.

10.6 OPTICAL PARAMETRIC GENERATION AND OSCILLATION

As discussed in Subsection 5.2.4, *spontaneous parametric fluorescence* is generated when a nonlinear crystal is pumped by an optical wave. The frequencies of the waves generated by this process are determined by the conservation of photon energy and the phase-matching condition. By pumping with a pump wave at a frequency of ω_3 that propagates with a wavevector of $\mathbf{k}_3$ in the crystal, the frequencies and, correspondingly, the wavelengths of the parametric fluorescence are subject to the conditions:

$$\omega_3 = \omega_1 + \omega_2 \quad \text{and} \quad \frac{1}{\lambda_3} = \frac{1}{\lambda_1} + \frac{1}{\lambda_2}, \tag{10.57}$$

which are required by conservation of energy because one photon at ω_3 splits into a pair of photons at ω_1 and ω_2. The exact frequencies to be generated, and the emission directions of the fluorescence, are further dictated by the phase-matching condition:

$$\mathbf{k}_3 = \mathbf{k}_1 + \mathbf{k}_2, \tag{10.58}$$

which is determined by the properties and the physical arrangement of the nonlinear crystal.

10.6.1 Optical Parametric Generation

A thorough analysis yields the *spectral power densities* of the spontaneous fluorescence of the two parametric waves [3]:

$$P_1(\omega_1)d\omega_1 = \frac{\hbar\omega_1^3\omega_2^2\left|\chi_{\mathrm{eff}}\right|^2}{4\pi c^3\epsilon_0 n_{3,z}^2\omega_3}P_3 l\, d\omega_1 = \frac{\hbar\omega_1^3\omega_2^2\left|d_{\mathrm{eff}}\right|^2}{\pi c^3\epsilon_0 n_{3,z}^2\omega_3}P_3 l\, d\omega_1, \tag{10.59}$$

$$P_2(\omega_2)d\omega_2 = \frac{\hbar\omega_1^2\omega_2^3\left|\chi_{\mathrm{eff}}\right|^2}{4\pi c^3\epsilon_0 n_{3,z}^2\omega_3}P_3 l\, d\omega_2 = \frac{\hbar\omega_1^2\omega_2^3\left|d_{\mathrm{eff}}\right|^2}{\pi c^3\epsilon_0 n_{3,z}^2\omega_3}P_3 l\, d\omega_2, \tag{10.60}$$

where χ_{eff} is the effective nonlinear susceptibility, $d_{\mathrm{eff}} = \chi_{\mathrm{eff}}/2$, P_3 is the pump power, and l is the interaction length. In the case of quasi-phase matching, χ_{eff} and d_{eff} in (10.59) and (10.60) are replaced by χ_{Q} and d_{Q}, respectively. By integrating (10.59) and (10.60) over the respective spectra, the total fluorescence powers of the two frequencies are, respectively,

$$P_1 = \frac{\hbar\omega_1^4\omega_2^2\left|\chi_{\mathrm{eff}}\right|^2}{16\pi c^3\epsilon_0 n_{3,z}^2\omega_3}P_3 l = \frac{\hbar\omega_1^4\omega_2^2\left|d_{\mathrm{eff}}\right|^2}{4\pi c^3\epsilon_0 n_{3,z}^2\omega_3}P_3 l = \frac{8\pi^4\hbar c^2\lambda_3\left|d_{\mathrm{eff}}\right|^2}{\epsilon_0 n_{3,z}^2\lambda_1^4\lambda_2^2}P_3 l, \tag{10.61}$$

$$P_2 = \frac{\hbar\omega_1^2\omega_2^4\left|\chi_{\mathrm{eff}}\right|^2}{16\pi c^3\epsilon_0 n_{3,z}^2\omega_3}P_3 l = \frac{\hbar\omega_1^2\omega_2^4\left|d_{\mathrm{eff}}\right|^2}{4\pi c^3\epsilon_0 n_{3,z}^2\omega_3}P_3 l = \frac{8\pi^4\hbar c^2\lambda_3\left|d_{\mathrm{eff}}\right|^2}{\epsilon_0 n_{3,z}^2\lambda_1^2\lambda_2^4}P_3 l. \tag{10.62}$$

From (10.59)–(10.62), we find that the spontaneous fluorescence power of each parametric frequency is *linearly proportional* to the power of the pump wave and to the interaction length. By comparison, in the low-efficiency limit, the power of the signal wave in every nonlinear frequency-conversion process discussed in the preceding sections is *quadratically proportional* to the pump power and to the interaction length. Because spontaneous fluorescence emission always takes place at a low efficiency, this difference clearly indicates that *spontaneous parametric fluorescence is not a nonlinear process but is a linear process*, as was already mentioned in Subsection 5.2.4. It is a linear process despite the fact that its efficiency is determined by the nonlinear susceptibility as a function of $|\chi_{\mathrm{eff}}|^2$. *Only the stimulated processes*, such as optical parametric conversion and optical parametric amplification, discussed in the preceding sections, as well as optical parametric oscillation, discussed below, *are second-order nonlinear processes*.

The spontaneous parametric fluorescence that is generated from a simple *optical parametric generator* (OPG), as shown in Fig. 10.7(a), is generally very weak, and thus of little practical usefulness as a light source. Nevertheless, it generates the seed parametric frequencies that can be amplified through an OPA. There are generally two approaches, which are schematically illustrated in Figs. 10.7(b) and (c). In one approach, shown in Fig. 10.7(b), a simple, low-efficiency OPG that generates the seed parametric fluorescence is followed by a separate OPA

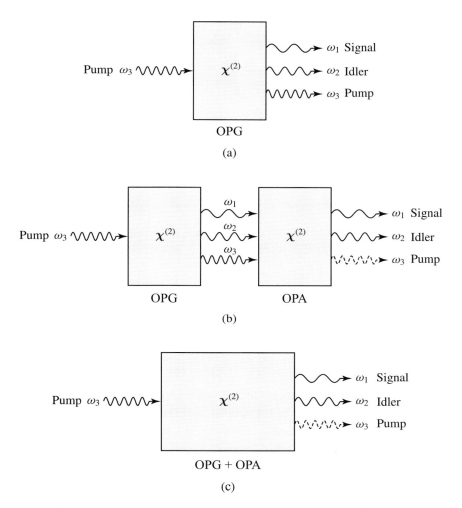

Figure 10.7 Schematics of optical parametric generation. (a) A simple OPG generating low-power parametric fluorescence. (b) An OPG followed by an OPA to increase the powers of the signal and idler waves. (c) An OPG combined with an OPA, pumped by a strong pump wave, to generate high-power signal and idler waves through amplified parametric fluorescence in one stage.

in cascade to amply the parametric fluorescence to a significant power. In the second approach, an OPG of a long interaction length is pumped with a strong pump beam so that the spontaneously generated parametric fluorescence is amplified within the OPG itself to a high power level. This process is similar to the *amplified spontaneous emission* (ASE) in a laser amplifier of a sufficiently long length and a high gain.

In practical applications, a sufficiently high gain for a useful OPG usually requires pumping with very intense laser pulses, such as the high-power nanosecond pulses from a Q-switched laser or the high-peak-power picosecond or femtosecond pulses from mode-locked lasers.

10.6.2 Optical Parametric Oscillation

The parametric gain can be utilized to construct an *optical parametric oscillator*, commonly called an OPO, by placing a parametric amplifier in a resonant optical cavity that provides

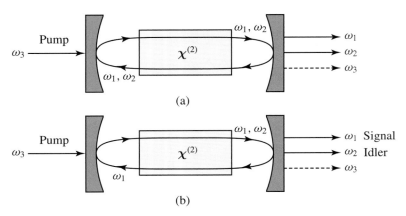

Figure 10.8 Schematic diagrams of (a) a doubly resonant OPO, in which both ω_1 and ω_2 are resonated, and (b) a singly resonant OPO, in which only ω_1 is resonated.

optical feedback to the parametric amplifier. In this manner, the spontaneously generated fluorescence is amplified through the feedback. At a sufficiently high pumping level, the OPO reaches its oscillation threshold and emits coherent output waves at the two parametric frequencies. This process is similar to the one that takes place in a laser oscillator, except that the gain medium of an OPO is a nonlinear crystal, and the gain is provided through a parametric nonlinear optical process.

There are basically two different types of OPOs. In a *doubly resonant* OPO, both waves at ω_1 and ω_2 are resonated because the mirrors of the optical cavity are highly reflective at both frequencies, as shown in Fig. 10.8(a). In a *singly resonant* OPO, the mirrors of the optical cavity are highly reflective at only one frequency, either ω_1 or ω_2, and only one wave is resonated, as shown in Fig. 10.8(b). The cavity mirrors are transparent to the pump wave and, in the singly resonant case, also to the nonresonant parametric wave. The input to an OPO consists of only the pump wave at ω_3 to pump the nonlinear crystal for a parametric gain. When the parametric gain is sufficiently high so that the round-trip loss in the optical resonator is fully compensated by the parametric gain, the oscillator reaches its threshold and parametric oscillation occurs. Because the parametric gain is a function of the pump intensity, the threshold parametric gain of an OPO translates into a *threshold pump intensity* that is required of the pump beam. Resonant oscillation builds up from the spontaneous emission noise of parametric fluorescence. No signal input is needed.

Because both low-frequency parametric waves at ω_1 and ω_2 are generated in the oscillator without an input signal, either of them can be called the signal or the idler. The designation of one particular wave to be called the signal is purely a matter of one's subjective interest. However, the choice of the resonating frequency in a singly resonant OPO is usually not arbitrary but is based on many practical considerations, such as the availability of high-quality cavity mirrors at either of the two parametric frequencies, the spectral characteristics of the transmittance of the nonlinear crystal, and other wavelength-dependent characteristics of the optical cavity.

The frequencies and, correspondingly, the wavelengths of a parametric oscillator are subject to the condition given in (10.57) required by conservation of energy. The exact frequencies to be generated by the oscillator are further dictated by the two conditions:

1. the phase-matching condition given in (10.58), which is determined by the properties and the physical arrangement of the nonlinear crystal; and
2. the resonance condition of the optical cavity, which depends on the physical parameters of the cavity and determines the resonance optical frequencies.

The peak parametric gain appears at the frequencies that exactly satisfy the phase-matching condition. The oscillation frequencies are those, subject to the condition given in (10.57) that $\omega_1 + \omega_2 = \omega_3$, that satisfy the resonance condition of the optical resonator with the least amount of phase mismatch. Therefore, the signal and idler frequencies, ω_1 and ω_2, of an OPO can be simultaneously tuned, though in opposite directions due to the constraint of (10.57), by varying the phase-matching condition in the crystal while the pump frequency, ω_3, is fixed. This wavelength tunability is one of the most important characteristics of OPOs. Another important characteristic is that the parametric gain is not tied to any resonant transition in the gain medium because the gain medium is a parametric nonlinear crystal. These two key characteristics make the OPOs unique devices for the generation of wavelength-tunable coherent optical waves in any spectral region where efficient laser materials do not exist, provided that an efficient nonlinear crystal and a commonly available laser source at a higher frequency, to serve as the pump, can be found.

A doubly resonant OPO generally has a lower oscillation threshold than a singly resonant one of comparable physical parameters. However, a doubly resonant OPO is difficult to operate because of its intrinsic instability. To resonate both the signal and the idler waves, both frequencies of ω_1 and ω_2 have to satisfy the resonance condition of the optical cavity. With the constraint of (10.57), this requirement cannot be met with an arbitrary cavity length but only with some specific values of the cavity length. This situation limits the tunability of the parametric oscillator. In addition, any variations in the cavity length due to mechanical or thermal fluctuations can lead to instability in the oscillation frequencies and the amplitudes of the optical fields. These problems do not exist in a singly resonant optical parametric resonator. Therefore, most OPOs that are designed for practical applications are of the singly resonant type.

EXAMPLE 10.4

The PPLN parametric amplifier described in Example 10.3 is placed in a properly designed optical cavity to make a singly resonant OPO. When the OPO is sufficiently pumped above its threshold with a pump beam at the 527 nm wavelength that has a pump power of $P_3 = 2$ W, it is found that 5% of the pump power is converted to the combined output power of the signal and the idler. Find the output powers of the signal and idler beams, respectively.

Solution

The total output power from this OPO is $P_{out} = 0.05P_3 = 100$ mW. In a parametric conversion process, an idler photon is simultaneously generated each time a signal photon is generated, while a pump photon is annihilated because of the relation $\omega_3 = \omega_1 + \omega_2$ required by (10.57). Therefore, the total number of signal photons has to be equal to that of idler photons because there are no input signal or idler photons to an OPO. If the signal and idler photons are subject to the same fractional loss, the power ratio between the signal and the idler is

$$\frac{P_1^{out}}{P_2^{out}} = \frac{\omega_1}{\omega_2} = \frac{\lambda_2}{\lambda_1}, \tag{10.63}$$

which leads to the power split:

$$P_1^{out} = \frac{\lambda_2}{\lambda_1 + \lambda_2} P_{out}, \qquad P_2^{out} = \frac{\lambda_1}{\lambda_1 + \lambda_2} P_{out}. \tag{10.64}$$

With $P_{out} = 100$ mW, we find that $P_1^{out} = 34$ mW for the signal at $\lambda_1 = 1.55$ μm and $P_2^{out} = 66$ mW for the idler at $\lambda_2 = 798$ nm.

For the split of output power expressed in (10.64), it is assumed that the signal and the idler suffer the same fractional loss in the OPO. In practice, this assumption may not be true, particularly when the wavelengths of the signal and the idler are far apart from each other. When the signal and the idler experience significantly disparate losses, the output power split can be very different from that described by (10.64). Even in this situation, it is still true that equal numbers of signal and idler photons are generated from converting the same number of pump photons when they interact in the nonlinear crystal.

Many lasers are available in the visible and near-infrared wavelength regions for pumping second-harmonic generators to generate short-wavelength optical waves well into the deep-ultraviolet region and for pumping OPGs and OPOs to generate wavelength-tunable optical waves in a broad infrared region. With advances in laser sources and crystal technology, the wavelengths that can be reached by nonlinear frequency conversion are basically only limited by the transmission windows of available nonlinear crystals. Figure 10.9 shows the transmission windows of various nonlinear optical crystals that can be chosen for frequency converters and the wavelengths of several lasers that can be used as pump sources. From this figure, we see that second-order nonlinear frequency conversion can cover a spectral range from about 200 nm in the deep-ultraviolet to about 18 μm in the mid-infrared. Depending on the pump lasers used, the coherent optical waves generated by these frequency converters in this wide spectral range cover the entire range of temporal characteristics, from CW beams through nanosecond Q-switched pulses to picosecond and femtosecond mode-locked pulses. Through these nonlinear optical devices, optical sources over a wide range of spectral and temporal characteristics are made available and flexible for many applications.

Compared with an OPG, a well-designed OPO can have a much lower pump threshold, and its output is less noisy than that of an OPG. An OPO that is pumped by nanosecond Q-switched

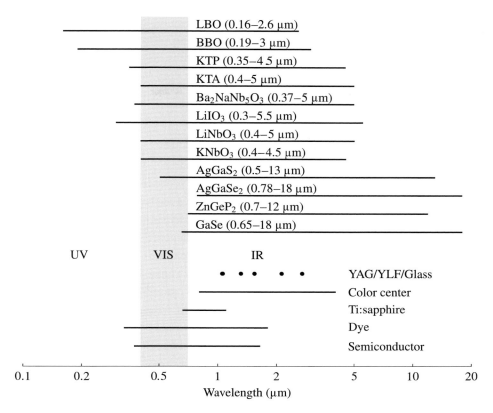

Figure 10.9 Transmission windows of various nonlinear optical crystals for frequency converters and wavelengths of several lasers that can be used as pump sources.

laser pulses can generate high-power nanosecond parametric pulses. An OPO that is synchronously pumped by a train of picosecond or femtosecond mode-locked laser pulses, known as a *synchronously pumped OPO*, can generate wavelength-tunable coherent picosecond or femtosecond parametric laser pulses [4–9]. The parametric pulses generated by an OPO that is pumped by laser pulses have these nice characteristics [4]:

1. Each pair of parametric signal and idler pulses are coherent to each other and are both coherent to the pump pulse.
2. The pulse durations of the signal and idler pulses are both shorter than that of the pump pulse because of the nonlinear generation process.
3. The wavelengths of the signal and idler pulses can be simultaneously tuned, in opposite directions, by tuning the phase-matching condition together with the resonance condition of the optical cavity.
4. An OPO that is synchronously pumped by a train of continuously mode-locked pulses has a low oscillation threshold.
5. A synchronously pumped OPO generates a pulse train of the same repetition rate as that of the pump pulses, and the signal and idler pulses are synchronized to the pump pulses.

With CW synchronously pumped OPOs, wavelength-tunable picosecond and femtosecond laser pulses covering the broad spectral range from the near-ultraviolet to the mid-infrared are readily available. The advantage of an OPG is that it is much simpler to set up and operate because it does not require an oscillator. However, the very high pump power required of an OPG often risks damaging the nonlinear crystal.

10.7 FOUR-WAVE MIXING

Four-wave mixing is a third-order nonlinear optical process that involves four interacting waves. For *nondegenerate four-wave mixing*, all four waves have either different wavevectors or different frequencies. For partially *degenerate four-wave mixing*, two or three of the four waves have the same wavevector and the same frequency. For fully degenerate four-wave mixing, all four waves are the same. In any event, four photons are involved. For simplicity, in this section, we consider the partially degenerate case, as considered in Section 9.4, with a single pump frequency of ω_p for four-photon mixing that two pump photons of the same frequency combine to generate a *Stokes* photon at a down-shifted frequency of ω_S and an *anti-Stokes* photon at an up-shifted frequency of ω_{AS}:

$$\omega_p = \omega_0, \quad \omega_S = \omega_0 - \Omega, \quad \omega_{AS} = \omega_0 + \Omega, \tag{10.65}$$

such that $2\omega_p = \omega_S + \omega_{AS}$, where Ω is the frequency shift from the pump frequency. In this section, we consider a small frequency shift such that $\Omega \ll \omega_0$. In this case, this process is *nearly degenerate four-wave mixing* because the frequencies of the four interacting photons are very close to one another.

Optical frequency conversion through four-wave mixing is a parametric process that requires phase matching. There are many possible phase-matching techniques for four-photon mixing [10]. One method is phase matching by using a birefringent optical fiber, as described in Section 9.4. Another possibility is to use different waveguide modes. In this section, we consider four-wave mixing in a nonbirefringent single-mode waveguide, such as a nonbirefringent single-mode fiber. In this case, the pump, Stokes, and anti-Stokes fields each consists of only one waveguide mode: $\hat{\mathcal{E}}_p$, $\hat{\mathcal{E}}_S$, and $\hat{\mathcal{E}}_{AS}$, respectively.

We consider the usual case of a strong and undepleted pump wave that has a much higher power than the Stokes and anti-Stokes waves throughout the interaction [10]. Then, the third-order nonlinear polarizations for the three frequencies are

$$\mathbf{P}_p^{(3)} \approx 3\epsilon_0 \chi^{(3)}(\omega_p = \omega_p + \omega_p - \omega_p) \vdots \hat{\mathcal{E}}_p \hat{\mathcal{E}}_p \hat{\mathcal{E}}_p^* A_p A_p A_p^* \, e^{i\beta_p z}, \tag{10.66}$$

$$\begin{aligned} \mathbf{P}_S^{(3)} &= 6\epsilon_0 \chi^{(3)}(\omega_S = \omega_p - \omega_p + \omega_S) \vdots \hat{\mathcal{E}}_p \hat{\mathcal{E}}_p^* \hat{\mathcal{E}}_S A_p A_p^* A_S \, e^{i\beta_S z} \\ &\quad + 3\epsilon_0 \chi^{(3)}(\omega_S = \omega_p + \omega_p - \omega_{AS}) \vdots \hat{\mathcal{E}}_p \hat{\mathcal{E}}_p \hat{\mathcal{E}}_{AS}^* A_p A_p A_{AS}^* \, e^{i(2\beta_p - \beta_{AS})z}, \end{aligned} \tag{10.67}$$

$$\mathbf{P}_{\mathrm{AS}}^{(3)} = 6\epsilon_0 \chi^{(3)}(\omega_{\mathrm{AS}} = \omega_{\mathrm{p}} - \omega_{\mathrm{p}} + \omega_{\mathrm{AS}}) \vdots \hat{\boldsymbol{\mathcal{E}}}_{\mathrm{p}} \hat{\boldsymbol{\mathcal{E}}}_{\mathrm{p}}^* \hat{\boldsymbol{\mathcal{E}}}_{\mathrm{AS}} A_{\mathrm{p}} A_{\mathrm{p}}^* A_{\mathrm{AS}} \, e^{i\beta_{\mathrm{AS}} z}$$

$$+ 3\epsilon_0 \chi^{(3)}(\omega_{\mathrm{AS}} = \omega_{\mathrm{p}} + \omega_{\mathrm{p}} - \omega_{\mathrm{S}}) \vdots \hat{\boldsymbol{\mathcal{E}}}_{\mathrm{p}} \hat{\boldsymbol{\mathcal{E}}}_{\mathrm{p}} \hat{\boldsymbol{\mathcal{E}}}_{\mathrm{S}}^* A_{\mathrm{p}} A_{\mathrm{p}} A_{\mathrm{S}}^* \, e^{i(2\beta_{\mathrm{p}} - \beta_{\mathrm{S}})z}. \tag{10.68}$$

In (10.66), the contributions to $\mathbf{P}_{\mathrm{p}}^{(3)}$ from the mixing with the Stokes and anti-Stokes frequencies are ignored because they are much smaller than the self-modulation of the strong pump. Similarly, in (10.67) and (10.68), the self-modulation and cross-modulation terms that involve only the Stokes and anti-Stokes frequencies are ignored; only the terms that involve the pump frequency are considered.

By using (7.33), we find the coupled equations:

$$\frac{\mathrm{d}A_{\mathrm{p}}}{\mathrm{d}z} \approx i\,\omega_{\mathrm{p}} C_{\mathrm{pppp}} |A_{\mathrm{p}}|^2 A_{\mathrm{p}}, \tag{10.69}$$

$$\frac{\mathrm{d}A_{\mathrm{S}}}{\mathrm{d}z} = i\,\omega_{\mathrm{S}} C_{\mathrm{SppS}} |A_{\mathrm{p}}|^2 A_{\mathrm{S}} + i\omega_{\mathrm{S}} C_{\mathrm{SppA}} A_{\mathrm{p}} A_{\mathrm{p}} A_{\mathrm{AS}}^* \, e^{i(2\beta_{\mathrm{p}} - \beta_{\mathrm{S}} - \beta_{\mathrm{AS}})z}, \tag{10.70}$$

$$\frac{\mathrm{d}A_{\mathrm{AS}}}{\mathrm{d}z} = i\,\omega_{\mathrm{AS}} C_{\mathrm{AppA}} |A_{\mathrm{p}}|^2 A_{\mathrm{AS}} + i\omega_{\mathrm{AS}} C_{\mathrm{AppS}} A_{\mathrm{p}} A_{\mathrm{p}} A_{\mathrm{S}}^* \, e^{i(2\beta_{\mathrm{p}} - \beta_{\mathrm{S}} - \beta_{\mathrm{AS}})z}, \tag{10.71}$$

where

$$C_{\mathrm{pppp}} = 3\epsilon_0 \int_{-\infty}^{\infty} \int_{-\infty}^{\infty} \hat{\boldsymbol{\mathcal{E}}}_{\mathrm{p}}^* \cdot \chi^{(3)} \vdots \hat{\boldsymbol{\mathcal{E}}}_{\mathrm{p}} \hat{\boldsymbol{\mathcal{E}}}_{\mathrm{p}} \hat{\boldsymbol{\mathcal{E}}}_{\mathrm{p}}^* \, \mathrm{d}x\mathrm{d}y, \tag{10.72}$$

$$C_{\mathrm{SppS}} = 6\epsilon_0 \int_{-\infty}^{\infty} \int_{-\infty}^{\infty} \hat{\boldsymbol{\mathcal{E}}}_{\mathrm{S}}^* \cdot \chi^{(3)} \vdots \hat{\boldsymbol{\mathcal{E}}}_{\mathrm{p}} \hat{\boldsymbol{\mathcal{E}}}_{\mathrm{p}}^* \hat{\boldsymbol{\mathcal{E}}}_{\mathrm{S}} \, \mathrm{d}x\mathrm{d}y, \tag{10.73}$$

$$C_{\mathrm{SppA}} = 3\epsilon_0 \int_{-\infty}^{\infty} \int_{-\infty}^{\infty} \hat{\boldsymbol{\mathcal{E}}}_{\mathrm{S}}^* \cdot \chi^{(3)} \vdots \hat{\boldsymbol{\mathcal{E}}}_{\mathrm{p}} \hat{\boldsymbol{\mathcal{E}}}_{\mathrm{p}} \hat{\boldsymbol{\mathcal{E}}}_{\mathrm{AS}}^* \, \mathrm{d}x\mathrm{d}y, \tag{10.74}$$

$$C_{\mathrm{AppA}} = 6\epsilon_0 \int_{-\infty}^{\infty} \int_{-\infty}^{\infty} \hat{\boldsymbol{\mathcal{E}}}_{\mathrm{AS}}^* \cdot \chi^{(3)} \vdots \hat{\boldsymbol{\mathcal{E}}}_{\mathrm{p}} \hat{\boldsymbol{\mathcal{E}}}_{\mathrm{p}}^* \hat{\boldsymbol{\mathcal{E}}}_{\mathrm{AS}} \, \mathrm{d}x\mathrm{d}y, \tag{10.75}$$

$$C_{\mathrm{AppS}} = 3\epsilon_0 \int_{-\infty}^{\infty} \int_{-\infty}^{\infty} \hat{\boldsymbol{\mathcal{E}}}_{\mathrm{AS}}^* \cdot \chi^{(3)} \vdots \hat{\boldsymbol{\mathcal{E}}}_{\mathrm{p}} \hat{\boldsymbol{\mathcal{E}}}_{\mathrm{p}} \hat{\boldsymbol{\mathcal{E}}}_{\mathrm{S}}^* \, \mathrm{d}x\mathrm{d}y. \tag{10.76}$$

By using the fact that $\Omega \ll \omega_0$, we can take the approximation that

$$\omega_{\mathrm{S}} \approx \omega_{\mathrm{AS}} \approx \omega_{\mathrm{p}} = \omega_0 \quad \text{and} \quad \hat{\boldsymbol{\mathcal{E}}}_{\mathrm{S}} \approx \hat{\boldsymbol{\mathcal{E}}}_{\mathrm{AS}} \approx \hat{\boldsymbol{\mathcal{E}}}_{\mathrm{p}}. \tag{10.77}$$

Then,

$$\sigma = \omega_{\mathrm{p}} C_{\mathrm{pppp}} = \frac{1}{2}\omega_{\mathrm{S}} C_{\mathrm{SppS}} = \omega_{\mathrm{S}} C_{\mathrm{SppA}} = \frac{1}{2}\omega_{\mathrm{AS}} C_{\mathrm{AppA}} = \omega_{\mathrm{AS}} C_{\mathrm{AppS}}. \tag{10.78}$$

According to (7.18), the pump power is

$$P_0 = |A_p|^2. \tag{10.79}$$

The coupled equations given in (10.69)–(10.71) can then be expressed as

$$\frac{dA_p}{dz} \approx i\sigma P_0 A_p = i\beta_p^{NL} A_p, \tag{10.80}$$

$$\frac{dA_S}{dz} = i2\sigma P_0 A_S + i\sigma A_p A_p A_{AS}^* e^{i(2\beta_p - \beta_S - \beta_{AS})z} = i\beta_S^{NL} A_S + i\sigma A_p A_p A_{AS}^* e^{i\Delta\beta z}, \tag{10.81}$$

$$\frac{dA_{AS}}{dz} = i2\sigma P_0 A_{AS} + i\sigma A_p A_p A_S^* e^{i(2\beta_p - \beta_S - \beta_{AS})z} = i\beta_{AS}^{NL} A_{AS} + i\sigma A_p A_p A_S^* e^{i\Delta\beta z}, \tag{10.82}$$

where

$$\beta_p^{NL} = \sigma P_0, \quad \beta_S^{NL} = \beta_{AS}^{NL} = 2\sigma P_0, \quad \text{and} \quad \Delta\beta = 2\beta_p - \beta_S - \beta_{AS}. \tag{10.83}$$

By integrating (10.80), we find that

$$A_p(z) = A_p(0)e^{i\beta_p^{NL}z}, \tag{10.84}$$

where $A_p(0)$ can be taken to be real without loss of generality. Therefore,

$$A_p(z)A_p(z) = A_p(0)A_p(0)e^{i2\beta_p^{NL}z} = P_0 e^{i2\beta_p^{NL}z}. \tag{10.85}$$

By using (10.85) and by defining $\widetilde{A}_S$ and $\widetilde{A}_{AS}$ through the relations that

$$A_S(z) = \widetilde{A}_S(z)e^{i\beta_S^{NL}z} \quad \text{and} \quad A_{AS}(z) = \widetilde{A}_{AS}(z)e^{i\beta_{AS}^{NL}z}, \tag{10.86}$$

we can express (10.81) and (10.82) as

$$\frac{d\widetilde{A}_S}{dz} = i\kappa \widetilde{A}_{AS}^* e^{i2\delta z}, \tag{10.87}$$

$$\frac{d\widetilde{A}_{AS}^*}{dz} = -i\kappa^* \widetilde{A}_S e^{-i2\delta z}, \tag{10.88}$$

where

$$\kappa = \sigma P_0, \tag{10.89}$$

and

$$2\delta = \Delta\beta + 2\beta_p^{NL} - \beta_S^{NL} - \beta_{AS}^{NL} \tag{10.90}$$

is the total phase mismatch including the linear contribution $\Delta\beta$ of the dispersion from the material and the waveguide, as well as the nonlinear contribution $2\beta_p^{NL} - \beta_S^{NL} - \beta_{AS}^{NL}$ from the power-dependent phase modulation caused by the strong pump.

The coupled equations given in (10.87) and (10.88) have the general solution:

$$
\begin{bmatrix} \widetilde{A}_{\mathrm{S}}(z) \\ \widetilde{A}_{\mathrm{p}}(z) \end{bmatrix} = \begin{bmatrix} \left(\cosh\dfrac{gz}{2} - \mathrm{i}\dfrac{2\delta}{g}\sinh\dfrac{gz}{2} \right)\mathrm{e}^{\mathrm{i}\delta z} & \mathrm{i}\dfrac{2\kappa}{g}\sinh\dfrac{gz}{2}\mathrm{e}^{\mathrm{i}\delta z} \\ -\mathrm{i}\dfrac{2\kappa^*}{g}\sinh\dfrac{gz}{2}\mathrm{e}^{-\mathrm{i}\delta z} & \left(\cosh\dfrac{gz}{2} + \mathrm{i}\dfrac{2\delta}{g}\sinh\dfrac{gz}{2} \right)\mathrm{e}^{-\mathrm{i}\delta z} \end{bmatrix} \begin{bmatrix} \widetilde{A}_{\mathrm{S}}(0) \\ \widetilde{A}_{\mathrm{p}}(0) \end{bmatrix},
$$

$$(10.91)$$

where

$$
g = 2\left(\kappa\kappa^* - \delta^2\right)^{1/2} = 2\left(|\kappa|^2 - \delta^2\right)^{1/2}. \tag{10.92}
$$

Note that g is the *power gain coefficient* as defined in (4.39) for the power, or the intensity, of an optical wave to grow as $\exp(gz)$, but for its field amplitude to grow as $\exp(gz/2)$. Note also that, according to (7.18) and (10.86), the powers of the Stokes and anti-Stokes waves are $P_{\mathrm{S}} = \left|A_{\mathrm{S}}\right|^2 = \left|\widetilde{A}_{\mathrm{S}}\right|^2$ and $P_{\mathrm{AS}} = \left|A_{\mathrm{AS}}\right|^2 = \left|\widetilde{A}_{\mathrm{AS}}\right|^2$, respectively.

10.7.1 Parametric Amplification

We see from (10.91) and (10.92) that a Stokes or anti-Stokes signal can be amplified by the four-wave mixing process if $|\kappa|^2 > \delta^2$ such that g has a real value. Because of the symmetry between the Stokes and the anti-Stokes waves seen in (10.91), similar results are obtained for the amplification of a Stokes or an anti-Stokes signal, except that the phases are different. Because a Stokes photon and an anti-Stokes photon are simultaneously generated in the four-wave mixing process, the anti-Stokes signal grows while the Stokes signal is amplified, and vice versa. Here we consider only the case that the input is a Stokes signal with $P_{\mathrm{S}}(0) \neq 0$ but $P_{\mathrm{AS}}(0) = 0$. Then, for an interaction length of l, we find from (10.91) the gain factors:

$$
G_{\mathrm{S,S}} = \frac{P_{\mathrm{S}}(l)}{P_{\mathrm{S}}(0)} = 1 + \frac{4|\kappa|^2}{g^2}\sinh^2\frac{gl}{2}, \tag{10.93}
$$

$$
G_{\mathrm{AS,S}} = \frac{P_{\mathrm{AS}}(l)}{P_{\mathrm{S}}(0)} = \frac{4|\kappa|^2}{g^2}\sinh^2\frac{gl}{2}. \tag{10.94}
$$

In the case that the input is an anti-Stokes signal with $P_{\mathrm{AS}}(0) \neq 0$ but $P_{\mathrm{S}}(0) = 0$, we have $G_{\mathrm{AS,AS}} = G_{\mathrm{S,S}}$ as given in (10.93), and $G_{\mathrm{S,AS}} = G_{\mathrm{AS,S}}$ as given in (10.94).

The peak gain occurs when the total phase mismatch is zero, such that $\delta = 0$ and $g = 2|\kappa|$. In the high-gain limit, $gl \gg 1$. Then, the gains for both Stokes and anti-Stokes signals grow exponentially as

$$
G_{\mathrm{S,S}} \approx G_{\mathrm{AS,S}} \approx G_{\mathrm{FWM}} \approx \sinh^2\frac{gl}{2} \approx \frac{\mathrm{e}^{gl}}{4}. \tag{10.95}
$$

In the other limit that the phase mismatch is large such that $\delta^2 \gg |\kappa|^2$, then $g \approx \mathrm{i}2\delta$. In this case, with $P_{\mathrm{S}}(0) \neq 0$ but $P_{\mathrm{AS}}(0) = 0$, we have

$$G_{\mathrm{S,S}} \approx 1 + |\kappa|^2 l^2 \frac{\sin^2 \delta l}{(\delta l)^2} \approx 1 \quad \text{and} \quad G_{\mathrm{AS,S}} \approx |\kappa|^2 l^2 \frac{\sin^2 \delta l}{(\delta l)^2}. \tag{10.96}$$

10.7.2 Phase Matching

As mentioned above, here we consider the case that the pump, Stokes, and anti-Stokes waves are in the same mode. The linear propagation constant of each wave includes linear material dispersion and waveguide dispersion. With $\omega_{\mathrm{p}} = \omega_0$, $\omega_{\mathrm{S}} = \omega_0 - \Omega$, and $\omega_{\mathrm{AS}} = \omega_0 + \Omega$, as given in (10.65), the propagation constants of the Stokes and anti-Stokes waves can be expanded as

$$\beta(\omega_0 \pm \Omega) = \beta(\omega_0) + \sum_{n=1}^{\infty} \frac{\beta_n}{n!} (\pm\Omega)^n, \tag{10.97}$$

where

$$\beta_n = \left(\frac{\partial^n \beta}{\partial \omega^n} \right)_{\omega_0}. \tag{10.98}$$

We then find from (10.83) that the linear phase mismatch is

$$\Delta\beta = 2\beta_{\mathrm{p}} - \beta_{\mathrm{S}} - \beta_{\mathrm{AS}} = 2\beta(\omega_0) - \beta(\omega_0 - \Omega) - \beta(\omega_0 + \Omega) = -2\sum_{m=1}^{\infty} \frac{\beta_{2m}}{(2m)!} \Omega^{2m}, \tag{10.99}$$

and the total phase mismatch is

$$2\delta = \Delta\beta + 2\beta_{\mathrm{p}}^{\mathrm{NL}} - \beta_{\mathrm{S}}^{\mathrm{NL}} - \beta_{\mathrm{AS}}^{\mathrm{NL}} = -2\sum_{m=1}^{\infty} \frac{\beta_{2m}}{(2m)!} \Omega^{2m} - 2\sigma P_0. \tag{10.100}$$

Note that only the even orders, β_{2m}, of the linear dispersion appear in the phase mismatch.

For most materials, such as a silica optical fiber, $\sigma \propto n_2 > 0$, where n_2 is the nonlinear refractive index. In this case, from (10.100), it is seen that phase matching is possible if $\Delta\beta > 0$ because $2\sigma P_0 > 0$. The leading term of $\Delta\beta$ is

$$\Delta\beta \approx -\beta_2 \Omega^2 = -\frac{D}{c\omega_0} \Omega^2, \tag{10.101}$$

where D is the group-velocity dispersion defined in (4.83); here, $D = c\omega_0\beta_2$, which includes the dispersion from both the material and the waveguide structure. Then,

$$2\delta = -\beta_2\Omega^2 - 2\sigma P_0 = -\frac{D}{c\omega_0} \Omega^2 - 2\sigma P_0. \tag{10.102}$$

We find from (10.102) that phase matching for four-wave mixing in a single-mode nonbirefringent waveguide is possible if β_2 and σ have opposite signs – that is, if D and n_2 have opposite signs. This means that for the common material that has a positive n_2, *phase matching can be accomplished in the spectral region of negative group-velocity dispersion.*

Note that the above conclusion does not preclude the possibility for phase matching in the region of positive group-velocity dispersion where $D > 0$ so that $\beta_2 > 0$ even when $n_2 > 0$, so that $\sigma > 0$. The reason is that β_2 is only the leading term of $\Delta\beta$, as seen in (10.99). When $\beta_2 > 0$, phase matching with $2\delta = 0$ is still possible if the sum of the high-order terms are sufficiently negative such that $\Delta\beta > 0$ for (10.99). This is possible in a region of small positive group-velocity dispersion near the *zero-dispersion wavelength*, where the value of β_2 is small but positive.

10.7.3 Gain Bandwidth

The gain bandwidth of four-wave mixing is determined by the condition that the power gain coefficient be real and positive. Here we consider the common case that $n_2 > 0$ so that $\sigma > 0$, and we consider only the leading term of the linear dispersion with a negative group-velocity dispersion that $D < 0$ so that $\beta_2 < 0$. Then, with $\kappa = \kappa^* = \sigma P_0 > 0$ and $2\delta = -\beta_2\Omega^2 - 2\sigma P_0$ as given in (10.102), we find the power gain coefficient as

$$g(\pm\Omega) = 2\left(|\kappa|^2 - \delta^2\right)^{1/2} = |\beta_2|\Omega\left(\Omega_c^2 - \Omega^2\right)^{1/2}, \tag{10.103}$$

where the *critical frequency* is

$$\Omega_c = 2\left(\frac{\sigma}{|\beta_2|}\right)^{1/2} P_0^{1/2}. \tag{10.104}$$

In terms of the pump power,

$$g = 4\sigma P_c^{1/2}(P_0 - P_c)^{1/2}, \tag{10.105}$$

where the *critical power* is

$$P_c = \frac{|\beta_2|}{4\sigma}\Omega^2. \tag{10.106}$$

For a given pump power of P_0, the power gain coefficient as a function of the frequency shift as given in (10.103) has a maximum value of

$$g_{max} = 2\sigma P_0 \tag{10.107}$$

at the frequency shift of

$$\Omega_{max} = \frac{\Omega_c}{\sqrt{2}}. \tag{10.108}$$

In terms of g_{max} and Ω_c, the four-wave mixing power gain spectrum, $g(\pm\Omega)$, can be expressed in the normalized form:

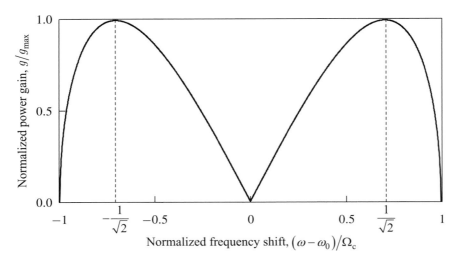

Figure 10.10 Four-wave mixing gain spectrum of the Stokes and anti-Stokes waves as a function of the frequency shift from the pump frequency ω_0.

$$g(\pm\Omega) = 2g_{max} \frac{\Omega}{\Omega_c} \left(1 - \frac{\Omega^2}{\Omega_c^2}\right)^{1/2}. \tag{10.109}$$

The normalized four-wave mixing power gain spectrum of the Stokes and anti-Stokes frequencies as a function of the frequency shift from the pump frequency, normalized to the critical frequency, is plotted in Fig. 10.10.

10.8 HIGH-ORDER HARMONIC GENERATION

Harmonic generation beyond the second order took the first step when Maker and Terhune reported the first third-harmonic generation in solid crystals in 1965 [11], followed by the observation of third-harmonic generation in inert gases by New and Ward in 1967 [12], both by using a Q-switched ruby laser. *High-order harmonic generation*, up to the 11th order, was first reported by Burnett *et al.* in 1977 [13] by illuminating flat solid aluminum targets at intensities higher than 10^{14} W cm^{-2} with short CO_2 laser pulses of a pulse energy in the range of 20–50 J and a pulse duration of 1.8–2.0 ns. In 1981, Carman *et al.* reported high-order harmonic generation up to the 29th order from various plane solid targets with CO_2 laser pulses of a pulse energy in the range of 100–150 J and a pulse duration of 1.0–1.2 ns [14]. A nearly constant efficiency for high harmonic orders and a sharp cutoff were observed. High-order harmonic generation in gaseous targets, up to the 17th order in Ne, was first reported by McPherson *et al.* in 1987 [15]. Inert gases of Ar, Kr, and Xe were irradiated with a focused KrF* laser pulse of a 15 ps pulsewidth at a wavelength of 248 nm and an intensity of the order of 10^{15}–10^{16} W cm^{-2}. This work was followed by Ferray *et al.* in 1988 [16] by using tightly focused high-power Nd:YAG laser pulses of a pulsewidth of 30 ps and a peak power of 1 GW to excite a gaseous target of Ar, Kr, or Xe at a wavelength of 1064 nm and an intensity of the order of 10^{13}–10^{14} W cm^{-2}. Harmonics up to the 33rd order were generated with an Ar target for a harmonic wavelength as short as 32.2 nm in the extreme-ultraviolet region. Following this initial work, intensive research activities took off on high-order harmonic generation in gaseous targets, mostly inert gases, including He and Ne, but also other gases. In 1997, by using ultrashort laser pulses of a pulsewidth of 26 fs from a Ti:sapphire laser centered at the wavelength of 800 nm, Chang *et al.* [17] generated coherent soft-X-ray harmonics in He with discrete peaks up to order 221 and unresolved harmonic emission up to order 297 at wavelengths as short as 2.7 nm (a photon energy of 460 eV), and in Ne with discrete peaks up to order 155 at a wavelength of 5.2 nm (a photon energy of 239 eV). Though high-order harmonic generation from the surface of a solid target was observed before the work on inert gases, high-order harmonic generation in a bulk crystal was first reported by Ghimire *et al.* in 2011 [18]. This work opened up a new field of theoretical and experimental research on high-order harmonic generation in solids [19].

High-order harmonic generation by using femtosecond Ti:sapphire laser pulses as the excitation source has generated extremely short pulses of pulsewidths in the attosecond range at wavelengths as short as a few nanometers in the soft-X-ray region [20]. It is one of the best methods to produce ultrashort coherent light pulses covering a wavelength range from the vacuum-ultraviolet to the soft-X-ray region. Thus, high-order harmonic generation forms the basis of table-top sources for attosecond science and ultrafast X-ray imaging [21, 22].

Several outstanding characteristics have been identified in the experiments of high-order harmonics with gases and solids:

1. Harmonics at *both even and odd orders* are observed in the experiments with solid targets [13, 14, 19, 20].
2. Only harmonics of *odd orders* are generated in the experiments with gases [15–17].
3. The strengths of the signals decrease very slowly for the high harmonic orders, thus forming a *plateau* of a nearly constant efficiency for high-order harmonics [14–20].
4. The efficiency of high-order harmonic generation depends on the polarization of the laser field [19, 20]. In the case of a gaseous target, the efficiency declines rapidly when the ellipticity of the field polarization increases [23].
5. The harmonics terminate with a sharp *cutoff* at a high harmonic order [23–25].
6. The *cutoff energy* scales linearly with the peak intensity, thus quadratically with the peak field, of the excitation pulse for a gaseous target [23–25], but it scales linearly with the peak field for a solid target [19].

The characteristics described above cannot be understood through the perturbative treatment of optical nonlinearity that was described in Chapter 2. High-order harmonic generation is a representative example of nonperturbative nonlinear optical processes. The field strength of the high laser intensity that is involved in the process is of the order of the ionization field strength of the target atom. For this reason, the assumption for the perturbation on the target atom by the laser field is not valid. Instead, a nonperturbative approach that accounts for the

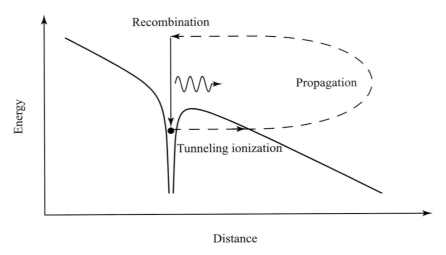

Figure 10.11 Three-step recollision model of high-order harmonic generation.

complete interaction of the laser field and the target has to be taken. For a gaseous atomic target, a semiclassical *three-step recollision* model described by Corkum in 1993 works surprisingly well [23]. For a solid target, a more sophisticated model that takes into consideration the band structure of the solid material is required [19]. In the following, we consider only high-order harmonic generation with gaseous atoms.

Figure 10.11 shows the three-step recollision model for high-order harmonic generation with a gaseous atom. The first step is *ionization* by the strong field at or near its peak value to create a free electron and an ion from the atom. In this step, an electron is lifted to the continuum at the position of the nucleus through tunneling to gain a potential energy of the ionization potential I_p but no kinetic energy. The second step is *propagation* of the free electron, which is described classically by the acceleration of the oscillating laser field. In this step, the electron gains a kinetic energy of K. Because of the oscillating nature of the electric field, the electron can return to the nuclear position to *recollide* with the ion under the right conditions. Thus, the third step is the *recombination* of the electron with the ion. In this final step, the sum of the kinetic and potential energies of the electron is emitted as a harmonic photon. This model is semiclassical because the processes of ionization and recombination are inherently quantum mechanical but the propagation process is treated classically. All of the three steps are completed within one cycle of the excitation laser field. Thus, a high-order harmonic field can be emitted as a femtosecond or attosecond pulse if the wavelength of the excitation field is sufficiently short.

The propagation step can be analyzed by considering the behavior of an electron in a laser field. We consider a laser field that is *linearly polarized* in the x direction. The *real field* has the form:

$$\boldsymbol{E}(t) = \hat{x}E_0 \cos \omega t = \hat{x}E_0 \cos \varphi, \qquad (10.110)$$

where ω is the center frequency of the excitation laser field and $\varphi = \omega t$ is the phase of the excitation field. The time of ionization is t_i, which marks the beginning of the electron propagation, and the time of recombination upon recollision with the ion is t_r, which marks the end of the electron propagation. The phases of the excitation laser field at these two points of time are, respectively,

$$\varphi_i = \omega t_i \quad \text{and} \quad \varphi_r = \omega t_r. \tag{10.111}$$

The propagation of the electron is governed by the equation for the acceleration:

$$\frac{d^2 \mathbf{r}}{dt^2} = \mathbf{a} = \frac{\mathbf{F}}{m_0} = \frac{-e\mathbf{E}}{m_0} = -\hat{x}\frac{eE_0}{m_0}\cos\omega t, \tag{10.112}$$

where $\mathbf{r}$ is the displacement vector, $\mathbf{a}$ is the acceleration vector, $\mathbf{F}$ is the force vector, e is the elementary charge, and m_0 is the rest mass of the electron. The origin of the displacement $\mathbf{r}$ is taken to be the location of the nucleus, and the initial condition is that the electron starts from this location with a zero velocity such that

$$\mathbf{r}(t_i) = \mathbf{0} \quad \text{and} \quad \frac{d\mathbf{r}}{dt}\bigg|_{t_i} = \mathbf{v}(t_i) = \mathbf{0}. \tag{10.113}$$

By using (10.113) to integrate (10.112), the velocity and displacement of the electron can be respectively found as

$$\mathbf{v}(t) = -\hat{x}\frac{eE_0}{m_0\omega}(\sin\omega t - \sin\omega t_i) = -\hat{x}\frac{eE_0}{m_0\omega}(\sin\varphi - \sin\varphi_i), \tag{10.114}$$

and

$$\mathbf{r}(t) = \hat{x}\frac{eE_0}{m_0\omega^2}[\cos\omega t - \cos\omega t_i + (\omega t - \omega t_i)\sin\omega t_i]$$
$$= \hat{x}\frac{eE_0}{m_0\omega^2}[\cos\varphi - \cos\varphi_i + (\varphi - \varphi_i)\sin\varphi_i]. \tag{10.115}$$

The condition for the recombination of the electron and the ion to take place is that the electron returns to the location of the nucleus to recollide with the ion at a time t_r that is later than t_i but within one cycle of the field, such that

$$\cos\varphi_r - \cos\varphi_i + (\varphi_r - \varphi_i)\sin\varphi_i = 0, \quad \text{with} \quad \varphi_i < \varphi_r < \varphi_i + 2\pi. \tag{10.116}$$

Then, the kinetic energy of the returning electron is

$$K = \frac{1}{2}m_0 v^2 = \frac{e^2 E_0^2}{2m_0\omega^2}(\sin\varphi_r - \sin\varphi_i)^2. \tag{10.117}$$

Each pair of the solution (φ_i, φ_r) found for (10.116) defines a trajectory of the electron that is ionized at t_i and recollides with the ion at t_r. For $0 < \varphi_i < \pi$, there are two sets of solutions, which represent the long and short trajectories:

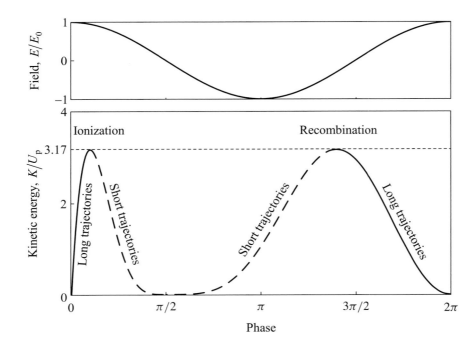

Figure 10.12 Laser field (top solid curve) and the kinetic energy of the electron as a function of the phases φ_i and φ_r of the laser field at the times of ionization and recombination. Only the solutions for $0 < \varphi_i < \pi$ are plotted. The same pattern for the kinetic energy exists for $\pi < \varphi_i < 2\pi$, which is not plotted.

$$0 < \varphi_i < 0.1\pi \quad \text{and} \quad 1.4\pi < \varphi_r < 2\pi \tag{10.118}$$

for a long trajectory, and

$$0.1\pi < \varphi_i < 0.5\pi \quad \text{and} \quad 0.5\pi < \varphi_r < 1.4\pi \tag{10.119}$$

for a short trajectory. The kinetic energy of the electron that propagates through a given trajectory is found from (10.117) for a given pair of (φ_i, φ_r) found for (10.116). Figure 10.12 shows the kinetic energy of the electron as a function of the phases φ_i and φ_r of the laser field at the times of ionization and recombination.

The maximum kinetic energy is found by maximizing (10.117) subject to the condition in (10.116). It is found at $\varphi_i = 0.1\pi$ and $\varphi_r = 1.4\pi$, where the regions for the long and short trajectories meet, as seen in Fig. 10.12. The maximum kinetic energy is

$$K_{max} = 3.17U_p, \tag{10.120}$$

where

$$U_p = \frac{e^2 E_0^2}{4m_0\omega^2} = \frac{e^2 I_{pk}}{8m_0 c\epsilon_0\omega^2} \tag{10.121}$$

is the *ponderomotive energy* of an electron in the laser field – that is, the average kinetic energy of an electron in a cycle of the laser field. In (10.121), $I_{pk} = 2c\epsilon_0 E_0^2$ is the peak intensity of the excitation laser pulse. Therefore, the *cutoff energy* for high-order harmonic generation is

$$E_c = \hbar\omega_c = 3.17U_p + I_p \approx n_{max}\hbar\omega, \tag{10.122}$$

where n_{max} is the highest order of the harmonics that can be generated. The relation given in (10.122) is found through the semiclassical analysis as described above. It is slightly modified through a quantum-mechanical analysis as described by Lewenstein *et al.* in 1994 [26]:

$$E_c = \hbar\omega_c = 3.17U_p + 1.3I_p \approx n_{max}\hbar\omega. \tag{10.123}$$

It can be clearly seen from (10.121)–(10.123) that the cutoff energy scales quadratically with the excitation laser field strength, thus linearly with the peak excitation laser intensity.

Note that the trajectories defined by (10.118) and (10.119) are those subject to the condition that $0 < \varphi_i < \pi$ – that is, within the first half-cycle of the laser field. From (10.118) and (10.119), it can be seen that the solutions exist only for $0 < \varphi_i < 0.5\pi$ but not for $0.5\pi < \varphi_i < \pi$. A trajectory with $0 < \varphi_i < 0.5\pi$ is bounded so that the electron returns to recollide with the ion; by contrast, a trajectory with $0.5\pi < \varphi_i < \pi$ is unbounded so that the electron is driven away by the laser field and is not able to recollide with the ion.

The trajectories defined by (10.118) and (10.119) do not cover the second half-cycle of the laser field, for which $\pi < \varphi_i < 2\pi$. It can be seen from (10.116) that if the pair (φ_i, φ_r) is a solution, the pair $(\varphi_i + \pi, \varphi_r + \pi)$ is also a solution. Furthermore, we find that the kinetic energy for the trajectory defined by (φ_i, φ_r) is the same as that for the trajectory defined by $(\varphi_i + \pi, \varphi_r + \pi)$. However, we also find from (10.112), (10.114), and (10.115) that electrons in these two trajectories are accelerated by fields of opposite signs and thus propagate in opposite directions. The harmonic fields that are radiated by these two sets of electrons have opposite signs. In the interaction of the laser field with an ensemble of atoms, the probability for an electron to be ionized in the first half-cycle of the field is equal to the probability for an electron to be ionized in the second half-cycle. Consequently, the time-domain field for the high-order harmonics has the symmetry of changing sign in half a cycle of the excitation field:

$$E_{HH}(t) = -E_{HH}\left(t - \frac{T}{2}\right) = -E_{HH}\left(t - \frac{\pi}{\omega}\right), \tag{10.124}$$

where T is the period of the excitation field. This function can be expanded as a Fourier series that contains only terms of frequencies that are *odd* multiples of ω. Therefore, in the frequency domain, only odd high-order harmonics are observed.

The excitation laser field is taken to be linearly polarized in the above analysis. In case it is elliptically polarized, the real field has the form:

$$\boldsymbol{E}(t) = E_0(\hat{x}\cos\varepsilon\cos\omega t + \hat{y}\sin\varepsilon\sin\omega t) = E_0(\hat{x}\cos\varepsilon\cos\varphi + \hat{y}\sin\varepsilon\sin\varphi), \tag{10.125}$$

where ε is the ellipticity of the field polarization, and it has a value in the range $-\pi/4 \leq \varepsilon \leq \pi/4$. The field is linearly polarized for $\varepsilon = 0$, and it is circularly polarized for $\varepsilon = -\pi/4$ or $\varepsilon = \pi/4$. By following through the analysis given above, it can be shown that the range of φ_i for the bound trajectories is quickly reduced as the ellipticity increases because it is more and more difficult to find the solution of φ_r for a given value of φ_i such that $\mathbf{r}(t_r) = \mathbf{r}(t_i) = 0$ for the electron to recollide with the ion as it is moving around in both the x and y directions. When the excitation field is circularly polarized, the electron never comes back to the location of the nucleus to recollide with the ion. Therefore, the efficiency of high-order harmonics quickly diminishes with an increasing ellipticity of the excitation field [23].

Though the semiclassical three-step recollision model works well in describing the key features of high-order harmonic generation with a gaseous target, it should be noted that high-order harmonic generation is inherently a quantum process that generates photons of energies that are integral multiples of the excitation photon energy. It is important to note that high-order harmonic generation is a parametric process, which does not involve net energy exchange between the optical field and the medium. As a parametric process, it is not automatically phase matched, and its efficiency decreases with the amount of phase mismatch. For the excitation of a target with a highly focused excitation beam, phase matching is facilitated by the large range of wavevectors provided by the divergence of the highly focused beam. Nevertheless, various phase-matching schemes are developed to improve the efficiency of high-order harmonic generation [20].

The analysis described above and the conclusions obtained from it are valid only for high-order harmonic generation with gaseous atoms. They can also be extended to gaseous ions. However, they are not valid for solid targets. For high-order harmonic generation with a solid target, both the interband nonlinear polarization and the intraband nonlinear current have to be considered in the analysis [19]. Both of these effects depend strongly on the band structure of the target material [27].

Problem Set

10.1.1 A KTP crystal is used for frequency doubling of a train of laser pulses at the fundamental wavelength of $\lambda_\omega = 1.064$ μm to its second harmonic at $\lambda_{2\omega} = 532$ nm. KTP is a positive biaxial crystal of the $mm2$ point group. The nonvanishing elements of its second-order nonlinear susceptibility have the values: $d_{15} = 3.7$ pm V^{-1}, $d_{24} = 1.9$ pm V^{-1}, $d_{31} = 3.7$ pm V^{-1}, $d_{32} = 2.2$ pm V^{-1}, and $d_{33} = 14.6$ pm V^{-1}. The principal indices of refraction at these wavelengths are: $n_\omega^x = 1.7399$, $n_\omega^y = 1.7475$, and $n_\omega^z = 1.8296$; $n_{2\omega}^x = 1.7790$, $n_{2\omega}^y = 1.7900$, and $n_{2\omega}^z = 1.8868$. The fundamental and second-harmonic waves propagate collinearly in a direction of $\hat{k} = \hat{x} \sin\theta \cos\phi + \hat{y} \sin\theta \sin\phi + \hat{z} \cos\theta$, where $\hat{x}$, $\hat{y}$, and $\hat{z}$ are the principal axes of the KTP crystal. Consider only type II phase matching on the xy plane in this problem.

(a) Find the phase-matching angle ϕ_{PM}^{II} and the walk-off angle ρ. Find the value of χ_{eff}^{II}.

(b) The KTP crystal has a length of $l = 1$ cm. The fundamental wave is a Gaussian beam that is focused to a beam waist radius of $w_0 = 75$ μm at the center of the crystal. Find the aperture length l_a and the confocal parameter b. Then find the normalized conversion efficiency $\hat{\eta}_{SH}$.

(c) The fundamental beam is a train of mode-locked laser pulses that have a pulsewidth of $\Delta t_{ps} = 30$ ps and a repetition rate of $f_{ps} = 100$ MHz. If a 10% conversion efficiency is desired, what are the required peak power of the pulse and the average power of the fundamental beam? What is the average power of the second-harmonic beam thus generated?

10.1.2 A train of mode-locked short optical pulses usually contains many equally spaced longitudinal mode frequencies. Consider a train of transform-limited Gaussian pulses that contains 100 longitudinal modes with the central mode at the frequency of ω_0 and the spacing between two adjacent modes being $\Delta\omega_L$. The Gaussian pulses have a full width at half-maximum pulsewidth of Δt_{ps} in the time domain and a spectral width of $\Delta\omega_{ps}$ in the frequency domain so that the time-domain intensity $I(t)$ and the frequency-domain spectral intensity $I(\omega)$ are, respectively,

$$I(t) = I_0 \exp\left[-4\ln 2\left(\frac{t}{\Delta t_{ps}}\right)^2\right] \quad \text{and} \quad I(\omega) = I(\omega_0)\exp\left[-4\ln 2\left(\frac{\omega - \omega_0}{\Delta\omega_{ps}}\right)^2\right]. \quad (10.126)$$

(a) In a phase-matched second-harmonic generation process with this train of Gaussian pulses, how many longitudinal mode frequencies are found for the second-harmonic pulses? What is the mode spacing? Sketch the spectrum and identify the frequencies of these modes.

(b) If the phase-matching spectral range is much larger than $\Delta\omega_{ps}$, what are the spectral width and the pulsewidth of the second-harmonic pulses?

(c) If the phase-matching spectral range were smaller than $\Delta\omega_{ps}$, what would happen to the spectral width and the pulsewidth of the second-harmonic pulses?

10.1.3 Two nonlinear crystals are often used in tandem, as shown in Fig. 10.13, to increase the overall nonlinear conversion efficiency because of the limitation on the length of available crystals. Assume that both crystals are cut and oriented for perfect phase matching. Consider for simplicity the case of second-harmonic generation, although the answers to the following questions can be easily extended to other second-order frequency-generation processes. Consider also only the low-efficiency limit, which is usually the case when there is a need to use two crystals.

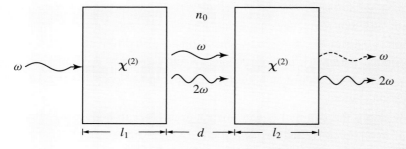

Figure 10.13 Second-harmonic generation with two nonlinear crystals in tandem.

(a) If the medium between the crystals is isotropic and nondispersive, what is the effect of the distance d between the two crystals on the overall nonlinear conversion efficiency? Explain.

(b) What is the dependence of the overall conversion efficiency on the lengths, l_1 and l_2, of the crystals and the distance d between them?

(c) Discuss what can happen if the medium between the crystals is birefringent but nondispersive. What has to be done in this case to ensure maximum conversion efficiency?

(d) What can happen if the medium between the crystals is isotropic but is dispersive? What has to be done to ensure maximum conversion efficiency?

(e) Answer the questions in (d) if the medium is nondispersive but is nonlinear with an intensity-dependent refractive index.

10.1.4 The PPLN waveguide second-harmonic generator described in Example 10.2 is considered. Perfect phase matching is accomplished.

(a) Find the required input fundamental power P_ω and the conversion efficiency η_{SH} for an output second-harmonic power of $P_{2\omega} = 1$ mW.

(b) Find the input power of the fundamental wave that is required for the PPLN second-harmonic generator to have a conversion efficiency of $\eta_{SH} = 99\%$.

(c) Answer the question in (a) if the length of the waveguide is doubled to $l = 3$ cm.

(d) Answer the question in (b) if the length of the waveguide is doubled to $l = 3$ cm.

10.2.1 A crystal that has a large $\chi^{(2)}$ is used for nonlinear frequency conversion.

(a) When two laser beams at the frequencies of ω_1 and ω_2 are sent together into the crystal, what determines whether you will see $\omega_1 + \omega_2$, $\omega_1 - \omega_2$, $2\omega_1$, or $2\omega_2$ at the output?

(b) When three beams at three the different frequencies of ω_1, ω_2, and ω_3, with $\omega_3 = \omega_1 + \omega_2$, simultaneously propagate in the crystal, what determines whether the sum-frequency process of $\omega_1 + \omega_2 \rightarrow \omega_3$ or the difference-frequency process of $\omega_3 - \omega_2 \rightarrow \omega_1$ will occur?

10.4.1 Parametric frequency up-conversion is similar to sum-frequency generation in that in both processes two low-frequency photons, at ω_1 and ω_2, are combined to a high-frequency photon at ω_3, whereas parametric frequency down-conversion is similar to difference-frequency generation in that a high-frequency photon at ω_3 is split into two low-frequency photons at ω_1 and ω_2. However, the characteristics of parametric up-conversion and sum-frequency generation are different, and those of parametric down-conversion and difference-frequency generation are also different. In the following, we compare optical parametric up-conversion and sum-frequency generation. Similar comparison can be made between optical parametric down-conversion and difference-frequency generation.

(a) Write down the coupled equation for parametric up-conversion from a signal at ω_1 to an up-converted frequency at ω_3 with a strong pump at ω_2. Examine the effect of phase mismatch on the output intensity, I_3, of the optical parametric up-conversion.

(b) Write down the coupled equation for sum-frequency generation from two input waves at ω_1 and ω_2 to the sum-frequency frequency at ω_3. Examine the effect of phase mismatch on the output intensity, I_3, of the sum-frequency generation.

(c) Compare the effects of phase mismatch for an optical parametric up-converter and a sum-frequency generator. What is the reason for the differences?

10.5.1 Find the intensity gain factor of an optical parametric amplifier with a phase mismatch of Δk in the case that $\Delta k/2 > \kappa$, where κ is that given in (10.51). Show that in the case that $\Delta k/2 \gg \kappa$, the parametric gain is approximately

$$G = 1 + \kappa^2 l^2 \frac{\sin^2(\Delta kl/2)}{(\Delta kl/2)^2}. \tag{10.127}$$

Also find the expression of the parametric gain as a function of Δk in the case that $\Delta k/2 < \kappa$, and the approximate expression when $\Delta k/2 \ll \kappa$.

10.5.2 A wavelength-tunable OPA or OPO is normally operated with a pump at a fixed wavelength of λ_3 and a fixed input intensity of $I_3(0)$ while the signal and idler wavelengths, λ_1 and λ_2, respectively, are tuned together in opposite directions. In this process, other characteristic parameters, such as χ_{eff} and the refractive indices of the crystal at the signal and idler wavelengths, of the device also vary accordingly. If such variations can be ignored, at what signal and idler wavelengths does a given wavelength-tunable OPA have the largest gain? What is the implication of this wavelength dependence of the parametric gain on the practical design and applications of OPAs and OPOs?

10.6.1 AgGaS$_2$ is a negative uniaxial crystal of the $\overline{4}2m$ point group. Its refractive indices can be described by the Sellmeier equations:

$$n_o^2 = 5.728 + \frac{0.24107}{\lambda^2 - 0.08703} - 0.00210\lambda^2, \tag{10.128}$$

$$n_e^2 = 5.497 + \frac{0.20259}{\lambda^2 - 0.13070} - 0.00233\lambda^2, \tag{10.129}$$

where λ is in micrometers. It is a good nonlinear crystal for parametric generation of infrared frequencies because it is transparent in the infrared spectral region. Its non-vanishing $\chi^{(2)}$ elements are $\chi_{xyz}^{(2)} = \chi_{yxz}^{(2)}$, $\chi_{xzy}^{(2)} = \chi_{yzx}^{(2)}$, and $\chi_{zxy}^{(2)} = \chi_{zyx}^{(2)}$. We can assume Kleinman's symmetry so that all of its nonvanishing $\chi^{(2)}$ elements have equal magnitude: $\chi_{14}^{(2)} = \chi_{25}^{(2)} = \chi_{36}^{(2)} = 57.4 \text{ pm V}^{-1}$. An OPO is constructed by using the uniaxial nonlinear crystal AgGaS$_2$ and pumped with a laser beam at 1.064 μm to generate wavelength-tunable infrared light in the wavelength range between 3 and 5 μm. A linearly polarized pump beam propagates through the crystal at an angle of θ with respect to the z principal axis in a plane at an angle of ϕ with respect to the x axis. The signal and idler waves propagate collinearly with the pump wave. Consider type I phase matching.

(a) What is the wavelength range of the idler?

(b) Find $\chi_{\text{eff}}^{\text{I}}$ for type I phase matching as a function of θ and ϕ. Choose an optimum value of ϕ that maximizes the efficiency of the OPO.

(c) It is desired that only one crystal be used for the entire signal wavelength range of 3–5 μm. Wavelength tuning in this range is to be accomplished by tuning the angle of the crystal. How should the crystal be cut so that the length of the crystal needed is minimized for a given pump beam cross-section? Sketch for clarity.

(d) Following (c), what is the range of the tuning angle for the desired wavelength-tuning range?

(e) Find the maximum value of the effective nonlinear susceptibility $|\chi_{\text{eff}}^{\text{I}}|$ by using the phase-matching angle and the optimum value of the angle ϕ.

(f) How should the direction of polarization of the pump wave be chosen for the maximum efficiency? What are the polarization directions of the signal and idler waves, respectively?

(g) What is the change in the nonlinear conversion efficiency across the signal wavelength range?

10.6.2 An AgGaS$_2$ OPO that is pumped at the $\lambda_3 = 1.053$ μm wavelength can generate signal and idler wavelengths that cover a range from 1 to 12 μm by angle tuning. The internal small-signal gain of the OPO is determined by $\kappa^2 l^2$ for a crystal of a length l, where κ is given in (10.51). The effective nonlinear susceptibility d_{eff} depends on the signal and idler wavelengths through its dependence on the tuning angle. For a given crystal length and a given pump intensity, the gain of the OPO thus varies with the signal and idler wavelengths through its dependence on d_{eff} and on the refractive indices in addition to its explicit dependence on λ_1 and λ_2 as seen in (10.51). Collinear phase matching is considered. From the data given in Table 3.4 for AgGaS$_2$, the ordinary and extraordinary refractive indices are given in (10.128) and (10.129), and the only nonvanishing d coefficients are $d_{14} = d_{25} = d_{36} = 28.7$ pm V^{-1}.

(a) Plot the angle-tuning curves in the form of the parametric wavelengths versus the phase-matching angle for both type I and type II phase matching.

(b) Find d_{eff} as a function of the angles θ and ϕ for both types of phase matching, where θ is the angle between the propagation vector $\mathbf{k}$ and the optical axis $\hat{z}$ and ϕ is that between $\mathbf{k}$ and the principal axis $\hat{x}$. For each type of phase matching, maximize the value of $|d_{\text{eff}}|$ by properly choosing the value for ϕ.

(c) Plot the maximum value of $|d_{\text{eff}}|$ as a function of the signal and idler wavelengths for both types of phase matching.

(d) Plot the small-signal gain $\kappa^2 l^2$, normalized to its peak value, for the OPO as a function of the signal and idler wavelengths for both types of phase matching. Compare these curves with those obtained in (c).

(e) Compare type I and type II phase matching in terms of the wavelength coverage, the tunability, and the efficiency.

10.6.3 The angle-tuning curves for a LiNbO$_3$ OPO that is pumped at $\lambda_3 = 527$ nm are shown in Fig. 9.4. The nonvanishing nonlinear susceptibility elements of LiNbO$_3$ are $d_{31} = d_{32} = d_{24} = d_{15} = -4.4$ pm V^{-1}, $d_{22} = -d_{21} = -d_{16} = 2.4$ pm V^{-1}, and $d_{33} = -25.2$ pm V^{-1}. The effective nonlinear susceptibilities are $d_{\text{eff}}^{\text{I}} = d_{31} \sin \theta - d_{22} \cos \theta \sin 3\phi$ for type I phase matching and $d_{\text{eff}}^{\text{II}} = d_{22} \cos^2 \theta \cos 3\phi$ for type II phase matching, as found in Example 9.1. The internal small-signal gain of the OPO is $\kappa^2 l^2$ for a crystal of a length l, where κ is given in (10.45). Answer the following questions for an angle-tuned LiNbO$_3$ OPO pumped at 527 nm with collinear phase matching. Use the data given in Table 3.4 for LiNbO$_3$.

(a) Plot the maximum value of $|d_{\text{eff}}|$ as a function of the signal and idler wavelengths for both types of phase matching.

(b) Plot the small-signal gain $\kappa^2 l^2$, normalized to its peak value, for the OPO as a function of the signal and idler wavelengths for both types of phase matching. Compare these curves with those obtained in (a).

(c) Compare type I and type II phase matching in terms of the wavelength coverage, the tunability, and the efficiency.

10.6.4 Answer the questions in Problem 10.6.3 for a temperature-tuned LiNbO$_3$ OPO that is pumped at $\lambda_3 = 527$ nm. The temperature-tuning curves are shown in Fig. 9.7.

10.6.5 In this problem, we consider a LiNbO$_3$ OPO pumped at 527 nm such as those considered in Problems 10.6.3 and 10.6.4, but with quasi-phase matching with a PPLN crystal that has a first-order grating of a 50% duty factor. Tuning of the parametric wavelengths is accomplished by varying the grating period in a fanned structure, as shown in Fig. 9.8(c). Use the data given in Table 3.4 for LiNbO$_3$ to answer the following questions.

(a) Plot the tuning curve in the form of the parametric wavelengths versus the phase-matching grating period.

(b) Plot the maximum value of $|d_{\text{eff}}|$ as a function of the signal and idler wavelengths.

(c) Plot the small-signal gain $\kappa^2 l^2$, normalized to its peak value, for the OPO as a function of the signal and idler wavelengths. Compare these curves with those obtained in (b).

(d) Compare this OPO with quasi-phase matching to the OPOs with angle tuning and temperature tuning that are considered in Problems 10.6.3 and 10.6.4, respectively, in terms of the wavelength coverage, the tunability, and the efficiency.

10.7.1 The coupled equations for the Stokes and anti-Stokes waves in the nearly degenerate photon-wave mixing process are those given in (10.87) and (10.88) with the coupling coefficient of $\kappa = \sigma P_0$, as given in (10.89).

(a) With an initial condition of $\widetilde{A}_{\text{S}}(0)$ and $\widetilde{A}_{\text{AS}}(0)$ for the mode amplitudes of the input Stokes and anti-Stokes waves, respectively, show that the coupled equations have the solution as given in (10.91) with the power gain coefficient of $g = 2(\kappa\kappa^* - \delta^2)^{1/2}$ as given in (10.91).

(b) When the input is a Stokes signal with $P_S(0) \neq 0$ but $P_{AS}(0) = 0$, show that the power gain factors, $G_{S,S}$ and $G_{AS,S}$, of the Stokes and anti-Stokes signals are those given in (10.93) and (10.94), respectively.

(c) Show that the power gain factors take the forms of (10.95) in the high-gain limit when phase mismatch vanishes, and the forms of (10.96) in the low-gain limit when phase mismatch is large.

10.7.2 Consider the power gain coefficient $g = 2(|\kappa|^2 - \delta^2)^{1/2}$ of nearly degenerate four-wave mixing in the case that the nonlinear index of refraction of the medium is positive, $n_2 > 0$, and the group-velocity dispersion of the medium is negative, $D < 0$.

(a) Show that the gain coefficient as a function of the frequency shift is that given in (10.103).

(b) Show that the gain coefficient as a function of the pump power is that given in (10.105).

(c) Discuss the possibilities of the four-wave mixing gain in the case that the group-velocity dispersion is positive, $D > 0$.

10.8.1 The three-step recollision model is used to describe the physical process that takes place for high-order harmonic generation. In this model, the propagation of an electron that is driven by a linearly polarized laser field is treated classically. The electron is ionized by the laser field at the time t_i, which marks the beginning of the electron propagation, and it recombines with the ion at the time t_r, which marks the end of the electron propagation. The phases of the laser field at these two times are $\varphi_i = \omega t_i$ and $\varphi_r = \omega t_r$, respectively. The condition for this electron to recombine with the ion within one cycle of the laser field is for φ_i and φ_r to satisfy the condition given in (10.116).

(a) Show, by finding the values of φ_r for $\varphi_i = 0$, 0.05π, 0.1π, 0.2π, 0.3π, 0.4π, 0.5π, 0.6π, 0.7π, 0.8π, 0.9π, and π that for the first half-cycle of the laser field, $0 < \varphi_i < \pi$, there are two sets of solutions as given in (10.118) and (10.119) for the trajectories of the electron. Show also that for the second half-cycle of the laser field, the pair $(\varphi_i + \pi, \varphi_r + \pi)$ is also a solution if the pair (φ_i, φ_r) is a solution.

(b) Show that the maximum kinetic energy of the electron is $K_{max} = 3.17U_p$, as given in (10.120), which occurs at $\varphi_i = 0.1\pi$ and $\varphi_r = 1.4\pi$.

References

[1] P. A. Franken, A. E. Hill, C. W. Peters, and G. Weinreich, "Generation of optical harmonics," *Physical Review Letters*, vol. 7, pp. 118–119, 1961.

[2] J. A. Armstrong, N. Bloembergen, J. Ducuing, and P. S. Pershan, "Interactions between light waves in a nonlinear dielectric," *Physical Review*, vol. 127, pp. 1918–1939, 1962.

[3] K. Koch, E. C. Cheung, G. T. Moore, S. H. Chakmakjian, and J. M. Liu, "Hot spots in parametric fluorescence with a pump beam of finite cross section," *IEEE Journal of Quantum Electronics*, vol. 31, pp. 769–781, 1995.

[4] E. C. Cheung and J. M. Liu, "Theory of a synchronously pumped optical parametric oscillator in steady-state operation," *Journal of the Optical Society of America B*, vol. 7, pp. 1385–1401, 1990.

[5] E. C. Cheung and J. M. Liu, "Efficient generation of ultrashort, wavelength-tunable infrared pulses," *Journal of the Optical Society of America B*, vol. 8, pp. 1491–1506, 1991.

[6] A. Piskarskas, V. Smil'gyavichyus, and A. Umbrasas, "Continuous parametric generation of picosecond light pulses," *Soviet Journal of Quantum Electronics*, vol. 18, pp. 155–156, 1988.

[7] D. C. Edelstein, E. S. Wachman, and C. L. Tang, "Broadly tunable high repetition rate femtosecond optical parametric oscillator," *Applied Physics Letters*, vol. 54, pp. 1728–1730, 1989.

[8] J. Chung and A. E. Siegman, "Singly resonant continuous-wave mode-locked $KTiOPO_4$ optical parametric oscillator pumped by a Nd:YAG laser," *Journal of the Optical Society of America B*, vol. 10, pp. 2201–2210, 1993.

[9] L. P. Chen, Y. Wang, and J. M. Liu, "Singly resonant optical parametric oscillator synchronously pumped by frequency-doubled additive-pulse mode-locked Nd:YLF laser pulses," *Journal of the Optical Society of America B*, vol. 12, pp. 2192–2198, 1995.

[10] R. Stolen and J. Bjorkholm, "Parametric amplification and frequency conversion in optical fibers," *IEEE Journal of Quantum Electronics*, vol. 18, pp. 1062–1072, 1982.

[11] P. D. Maker and R. W. Terhune, "Study of optical effects due to an induced polarization third order in the electric field strength," *Physical Review*, vol. 137, pp. A801–A818, 1965.

[12] G. H. C. New and J. F. Ward, "Optical third-harmonic generation in gases," *Physical Review Letters*, vol. 19, pp. 556–559, 1967.

[13] N. H. Burnett, H. A. Baldis, M. C. Richardson, and G. D. Enright, "Harmonic generation in CO_2 laser target interaction," *Applied Physics Letters*, vol. 31, pp. 172–174, 1977.

[14] R. L. Carman, D. W. Forslund, and J. M. Kindel, "Visible harmonic emission as a way of measuring profile steepening," *Physical Review Letters*, vol. 46, pp. 29–32, 1981.

[15] A. McPherson, G. Gibson, H. Jara, U. Johann, T. S. Luk, I. A. McIntyre, *et al.*, "Studies of multiphoton production of vacuum-ultraviolet radiation in the rare gases," *Journal of the Optical Society of America B*, vol. 4, pp. 595–601, 1987.

[16] M. Ferray, A. L'Huillier, X. F. Li, L. A. Lompre, G. Mainfray, and C. Manus, "Multiple-harmonic conversion of 1064 nm radiation in rare gases," *Journal of Physics B: Atomic, Molecular and Optical Physics*, vol. 21, pp. L31–L35, 1988.

[17] Z. Chang, A. Rundquist, H. Wang, M. M. Murnane, and H. C. Kapteyn, "Generation of coherent soft X rays at 2.7 nm using high harmonics," *Physical Review Letters*, vol. 79, pp. 2967–2970, 1997.

[18] S. Ghimire, A. D. DiChiara, E. Sistrunk, P. Agostini, L. F. DiMauro, and D. A. Reis, "Observation of high-order harmonic generation in a bulk crystal," *Nature Physics*, vol. 7, pp. 138–141, 2011.

[19] S. Ghimire and D. A. Reis, "High-harmonic generation from solids," *Nature Physics*, vol. 15, pp. 10–16, 2019.

[20] J. G. Eden, "High-order harmonic generation and other intense optical field–matter interactions: review of recent experimental and theoretical advances," *Progress in Quantum Electronics*, vol. 28, pp. 197–246, 2004.

[21] K. Midorikawa, "Ultrafast dynamic imaging," *Nature Photonics*, vol. 5, pp. 640–641, 2011.

[22] J. Li, J. Lu, A. Chew, S. Han, J. Li, Y. Wu, *et al.*, "Attosecond science based on high harmonic generation from gases and solids," *Nature Communications*, vol. 11, p. 2748, 2020.

[23] P. B. Corkum, "Plasma perspective on strong field multiphoton ionization," *Physical Review Letters*, vol. 71, pp. 1994–1997, 1993.

[24] J. L. Krause, K. J. Schafer, and K. C. Kulander, "High-order harmonic generation from atoms and ions in the high intensity regime," *Physical Review Letters*, vol. 68, pp. 3535–3538, 1992.

[25] K. J. Schafer, B. Yang, L. F. DiMauro, and K. C. Kulander, "Above threshold ionization beyond the high harmonic cutoff," *Physical Review Letters*, vol. 70, pp. 1599–1602, 1993.

[26] M. Lewenstein, Ph. Balcou, M. Yu Ivanov, A. L. Huillier, and P. B. Corkum, "Theory of high-harmonic generation by low-frequency laser fields," *Physical Review A*, vol. 49, pp. 2117–2132, 1994.

[27] C. Yu, S. Jiang, and R. Lu, "High order harmonic generation in solids: a review on recent numerical methods," *Advances in Physics: X*, vol. 4, p. 1562982, 2019.

11 Electro-optic Modulation

11.1 ELECTRO-OPTIC EFFECTS

The optical property of a dielectric material can be changed by an applied static or low-frequency electric field E_0 through an *electro-optic effect*. There are two well-known electro-optic effects, the *Pockels effect*, discussed in Subsection 5.2.7, and the *electro-optic Kerr effect*, discussed in Subsection 5.3.7. The electro-optic Kerr effect was discovered by Kerr in 1875 [1, 2], and the Pockels effect was discovered by Pockels in 1893. The electro-optic Kerr effect is also called the *DC Kerr effect*, in contrast to the *AC Kerr effect*, which refers to the *optical Kerr effect* discussed in Subsection 5.3.4 and in Section 12.1. A device that consists of a medium sandwiched between two electrodes for observing the electro-optic Kerr effect is called the *Kerr cell*; one that is constructed for observing the Pockels effect is called the *Pockels cell*. The Pockels effect and the electro-optic Kerr effect are respectively related to the second- and third-order nonlinear polarizations in the perturbation expansion of the polarization that results from the interaction between an electrostatic field and an optical field. However, they were traditionally not described in the context of nonlinear optics because they were established before the field of nonlinear optics was born in 1961. High-order nonlinear polarizations above the third order are very weak and generally ignored.

The Pockels effect is a second-order nonlinear optical process, in which the interaction between the electrostatic field and an optical field at a frequency of ω results in a second-order nonlinear polarization at the optical frequency ω:

$$
\begin{aligned}
\mathbf{P}^{(2)}(\omega) &= 2\epsilon_0\boldsymbol{\chi}^{(2)}(\omega = \omega + 0) : \mathbf{E}(\omega)\mathbf{E}(0) \\
&= 2\epsilon_0\boldsymbol{\chi}^{(2)}(\omega = \omega + 0) : \mathbf{E}(\omega)\mathbf{E}_0,
\end{aligned}
\tag{11.1}
$$

as is given in (5.17), where $\mathbf{E}(0) = \mathbf{E}_0$ for an electrostatic field.[1] In terms of the tensor elements of $\boldsymbol{\chi}^{(2)}$, (11.1) can be expressed as

$$
P_i^{(2)}(\omega) = 2\epsilon_0\sum_{j,k}\chi_{ijk}^{(2)}(\omega = \omega + 0)E_j(\omega)E_{0k}.
\tag{11.2}
$$

The electro-optic Kerr effect is a third-order nonlinear optical process, in which the interaction between the electrostatic field and an optical field at a frequency of ω results in a third-order nonlinear polarization at the optical frequency ω:

---

[1] A static field is not a harmonic field. Therefore, $\mathbf{E}(0)$ is not a complex field as defined in (1.37) for a harmonic field. Both $\mathbf{E}_0$ and $\mathbf{E}(0)$ are real fields so that we define $\mathbf{E}(0) = \mathbf{E}_0$.

$$\mathbf{P}^{(3)}(\omega) = 3\epsilon_0\chi^{(3)}(\omega = \omega + 0 + 0) \,\vdots\, \mathbf{E}(\omega)\mathbf{E}(0)\mathbf{E}(0) \tag{11.3}$$
$$= 3\epsilon_0\chi^{(3)}(\omega = \omega + 0 + 0) \,\vdots\, \mathbf{E}(\omega)\mathbf{E}_0\mathbf{E}_0,$$

as is given in (5.46). In terms of the tensor elements of $\chi^{(3)}$, (11.3) can be expressed as

$$P_i^{(3)}(\omega) = 3\epsilon_0\sum_{j,k,l}\chi_{ijkl}^{(3)}(\omega = \omega + 0 + 0)E_j(\omega)E_{0k}E_{0l}. \tag{11.4}$$

As discussed in Section 3.5, all elements of the second-order susceptibility $\chi^{(2)}$ vanish for a centrosymmetric material, whereas at least some elements of the third-order susceptibility $\chi^{(3)}$ do not vanish in any material. Therefore, the Pockels effect does not exist in a centrosymmetric material, as mentioned in Subsection 5.2.7, but the electro-optic Kerr effect exists in any material, as mentioned in Subsection 5.3.7.

11.1.1 Permittivity Tensor Representation

By accounting for the electro-optic effects, the total polarization for an optical field $\mathbf{E}(\omega)$ that propagates in a material subject to an electrostatic field $\mathbf{E}_0$ can be expressed as

$$\mathbf{P}(\omega, \mathbf{E}_0) = \mathbf{P}_{\mathrm{L}}(\omega) + \mathbf{P}_{\mathrm{NL}}(\omega, \mathbf{E}_0) = \mathbf{P}^{(1)}(\omega) + \mathbf{P}^{(2)}(\omega, \mathbf{E}_0) + \mathbf{P}^{(3)}(\omega, \mathbf{E}_0) + \cdots, \tag{11.5}$$

where

$$\mathbf{P}_{\mathrm{L}}(\omega) = \mathbf{P}^{(1)}(\omega) = \epsilon_0\chi^{(1)}(\omega) \cdot \mathbf{E}(\omega) \tag{11.6}$$

is the linear polarization that is independent of the electrostatic field $\mathbf{E}_0$, and

$$\mathbf{P}_{\mathrm{NL}}(\omega) = \mathbf{P}^{(2)}(\omega, \mathbf{E}_0) + \mathbf{P}^{(3)}(\omega, \mathbf{E}_0) + \cdots = \epsilon_0\chi_{\mathrm{NL}}(\omega, \mathbf{E}_0) \cdot \mathbf{E}(\omega) \tag{11.7}$$

is the nonlinear polarization that includes the second- and third-order nonlinear optical polarizations of the Pockels effect and the electro-optic Kerr effect. By defining a field-dependent total susceptibility as

$$\chi(\omega, \mathbf{E}_0) = \chi^{(1)}(\omega) + \chi_{\mathrm{NL}}(\omega, \mathbf{E}_0), \tag{11.8}$$

the total polarization can be expressed as

$$\mathbf{P}(\omega, \mathbf{E}_0) = \epsilon_0\chi(\omega, \mathbf{E}_0) \cdot \mathbf{E}(\omega) = \epsilon_0\chi^{(1)}(\omega) \cdot \mathbf{E}(\omega) + \epsilon_0\chi_{\mathrm{NL}}(\omega, \mathbf{E}_0) \cdot \mathbf{E}(\omega). \tag{11.9}$$

Note that in (11.8) and (11.9), the three susceptibility tensors, $\chi(\omega, \mathbf{E}_0)$, $\chi^{(1)}(\omega)$, and $\chi_{\mathrm{NL}}(\omega, \mathbf{E}_0)$, are all second-order tensors. By using (11.1) and (11.3) for (11.7), the tensor $\chi_{\mathrm{NL}}(\omega, \mathbf{E}_0)$ can be formally expressed as

$$\chi_{\mathrm{NL}}(\omega, \mathbf{E}_0) = 2\chi^{(2)}(\omega = \omega + 0) \cdot \mathbf{E}_0 + 3\chi^{(3)}(\omega = \omega + 0 + 0) : \mathbf{E}_0\mathbf{E}_0 + \cdots. \tag{11.10}$$

The tensor products in (11.10) have to be carefully carried out. To be clear, the tensor elements of $\chi_{\mathrm{NL}}(\omega, \mathbf{E}_0)$ can be found by using (11.2) and (11.4) for (11.7) as

$$\chi_{ij}^{NL}(\omega, \boldsymbol{E}_0) = 2\sum_k \chi_{ijk}^{(2)}(\omega = \omega + 0)E_{0k} + 3\sum_{k,l} \chi_{ijkl}^{(2)}(\omega = \omega + 0 + 0)E_{0k}E_{0l} + \cdots. \quad (11.11)$$

By using the general relation that $\boldsymbol{D}(\omega) = \epsilon_0\boldsymbol{E}(\omega) + \boldsymbol{P}(\omega) = \boldsymbol{\epsilon}(\omega) \cdot \boldsymbol{E}(\omega)$ for a field-dependent polarization $\boldsymbol{P}(\omega, \boldsymbol{E}_0)$ as given in (11.9), we can express a field-dependent electric displacement as

$$\boldsymbol{D}(\omega, \boldsymbol{E}_0) = \boldsymbol{\epsilon}(\omega, \boldsymbol{E}_0) \cdot \boldsymbol{E}(\omega) = \boldsymbol{\epsilon}(\omega) \cdot \boldsymbol{E}(\omega) + \Delta\boldsymbol{\epsilon}(\omega, \boldsymbol{E}_0) \cdot \boldsymbol{E}(\omega), \quad (11.12)$$

where

$$\Delta\boldsymbol{\epsilon}(\omega, \boldsymbol{E}_0) = \epsilon_0\boldsymbol{\chi}_{NL}(\omega, \boldsymbol{E}_0) \quad (11.13)$$

is the nonlinear permittivity that accounts for the electro-optic effects, and

$$\boldsymbol{\epsilon}(\omega, \boldsymbol{E}_0) = \boldsymbol{\epsilon}(\omega) + \Delta\boldsymbol{\epsilon}(\omega, \boldsymbol{E}_0) = \boldsymbol{\epsilon}(\omega) + \epsilon_0\boldsymbol{\chi}_{NL}(\omega, \boldsymbol{E}_0) \quad (11.14)$$

is the total permittivity in the presence of the electrostatic field $\boldsymbol{E}_0$.

The linear dielectric permittivity tensor $\boldsymbol{\epsilon}(\omega)$ in the absence of an electrostatic field is diagonal in the coordinate system defined by the intrinsic principal dielectric axes, $\hat{x}$, $\hat{y}$, and $\hat{z}$, of the dielectric material, as discussed in Subsection 3.5.1:

$$\boldsymbol{\epsilon}(\omega) = \begin{bmatrix} \epsilon_x & 0 & 0 \\ 0 & \epsilon_y & 0 \\ 0 & 0 & \epsilon_z \end{bmatrix}. \quad (11.15)$$

In the coordinate systems defined by the intrinsic x, y, and z principal axes of the material, the electro-optically induced changes usually generate off-diagonal elements in addition to changing the diagonal elements:

$$\boldsymbol{\epsilon}(\omega, \boldsymbol{E}_0) = \begin{bmatrix} \epsilon_x + \Delta\epsilon_{xx} & \Delta\epsilon_{xy} & \Delta\epsilon_{xz} \\ \Delta\epsilon_{yx} & \epsilon_y + \Delta\epsilon_{yy} & \Delta\epsilon_{yz} \\ \Delta\epsilon_{zx} & \Delta\epsilon_{zy} & \epsilon_z + \Delta\epsilon_{zz} \end{bmatrix}. \quad (11.16)$$

As discussed in Subsection 3.5.1, $\boldsymbol{\epsilon}(\omega)$ for a dielectric material is a symmetric tensor that can be diagonalized in a rectilinear coordinate system with real eigenvectors defining the principal axes for the coordinate system. This remains true for an electro-optic material subject to an electrostatic field because a nonmagnetic material remains nonmagnetic under an electrostatic field. Therefore, for the field-dependent permittivity tensors $\boldsymbol{\epsilon}(\omega, \boldsymbol{E}_0)$ in (11.16),

$$\epsilon_{ij}(\omega, \boldsymbol{E}_0) = \epsilon_{ji}(\omega, \boldsymbol{E}_0) \quad \text{and} \quad \Delta\epsilon_{ij}(\omega, \boldsymbol{E}_0) = \Delta\epsilon_{ji}(\omega, \boldsymbol{E}_0). \quad (11.17)$$

Being a symmetric tensor, the electro-optically induced permittivity tensor $\boldsymbol{\epsilon}(\omega, \boldsymbol{E}_0)$ given in (11.16) can again be diagonalized in a rectilinear coordinate system. Its orthonormalized eigenvectors, $\hat{X}$, $\hat{Y}$, and $\hat{Z}$, are the new principal dielectric axes of the material in the presence of the electrostatic field $\boldsymbol{E}_0$. The three principal axes generally depend on the direction of $\boldsymbol{E}_0$, and often also on the magnitude of E_0. If the unit vectors $\hat{X}$, $\hat{Y}$, and $\hat{Z}$ are expressed in terms of $\hat{x}$, $\hat{y}$, and $\hat{z}$ as

$$\hat{X} = a_1\hat{x} + b_1\hat{y} + c_1\hat{z}, \quad \hat{Y} = a_2\hat{x} + b_2\hat{y} + c_2\hat{z}, \quad \hat{Z} = a_3\hat{x} + b_3\hat{y} + c_3\hat{z}, \quad (11.18)$$

then the transformation between the old coordinate system defined by $\hat{x}$, $\hat{y}$, and $\hat{z}$ and the new coordinate system defined by $\hat{X}$, $\hat{Y}$, and $\hat{Z}$ can be carried out by using the transformation matrix:

$$\mathbf{T} = \begin{bmatrix} a_1 & b_1 & c_1 \\ a_2 & b_2 & c_2 \\ a_3 & b_3 & c_3 \end{bmatrix}. \quad (11.19)$$

Because both sets of vectors, $\{\hat{x},\hat{y},\hat{z}\}$ and $\{\hat{X},\hat{Y},\hat{Z}\}$, that define the transformation matrix $\mathbf{T}$ are orthonormal unit vectors, the transformation characterized by the matrix $\mathbf{T}$ is an orthogonal transformation that has the convenient characteristic:

$$\mathbf{T}^{-1} = \tilde{\mathbf{T}} = \begin{bmatrix} a_1 & a_2 & a_3 \\ b_1 & b_2 & b_3 \\ c_1 & c_2 & c_3 \end{bmatrix}, \quad (11.20)$$

where $\tilde{\mathbf{T}}$ is the transpose of $\mathbf{T}$.

The relation given in (11.18) between the old and the new principal axes can be expressed as

$$\begin{bmatrix} \hat{X} \\ \hat{Y} \\ \hat{Z} \end{bmatrix} = \mathbf{T} \begin{bmatrix} \hat{x} \\ \hat{y} \\ \hat{z} \end{bmatrix} \quad \text{and} \quad \begin{bmatrix} \hat{x} \\ \hat{y} \\ \hat{z} \end{bmatrix} = \tilde{\mathbf{T}} \begin{bmatrix} \hat{X} \\ \hat{Y} \\ \hat{Z} \end{bmatrix}. \quad (11.21)$$

The transformation of the coordinates of any vector $\mathbf{r} = x\hat{x} + y\hat{y} + z\hat{z} = X\hat{X} + Y\hat{Y} + Z\hat{Z}$ in space is given by the relations:

$$\begin{bmatrix} X \\ Y \\ Z \end{bmatrix} = \mathbf{T} \begin{bmatrix} x \\ y \\ z \end{bmatrix} = \begin{bmatrix} a_1x + b_1y + c_1z \\ a_2x + b_2y + c_2z \\ a_3x + b_3y + c_3z \end{bmatrix} \quad (11.22)$$

and

$$\begin{bmatrix} x \\ y \\ z \end{bmatrix} = \tilde{\mathbf{T}} \begin{bmatrix} X \\ Y \\ Z \end{bmatrix} = \begin{bmatrix} a_1X + a_2Y + a_3Z \\ b_1X + b_2Y + b_3Z \\ c_1X + c_2Y + c_3Z \end{bmatrix}. \quad (11.23)$$

Accordingly, the field components in the two coordinate systems are related through the relations:

$$\begin{bmatrix} E_X \\ E_Y \\ E_Z \end{bmatrix} = \mathbf{T} \begin{bmatrix} E_x \\ E_y \\ E_z \end{bmatrix}, \quad \begin{bmatrix} D_X \\ D_Y \\ D_Z \end{bmatrix} = \mathbf{T} \begin{bmatrix} D_x \\ D_y \\ D_z \end{bmatrix}, \quad (11.24)$$

and so on.

Diagonalization of $\epsilon(\omega, E_0)$ to obtain its eigenvalues can be carried out by using $\mathbf{T}$ as

$$\mathbf{T}\boldsymbol{\epsilon}(\omega, \boldsymbol{E}_0)\mathbf{T}^{-1} = \mathbf{T}\boldsymbol{\epsilon}(\omega, \boldsymbol{E}_0)\widetilde{\mathbf{T}} = \begin{bmatrix} \epsilon_X & 0 & 0 \\ 0 & \epsilon_Y & 0 \\ 0 & 0 & \epsilon_Z \end{bmatrix}. \tag{11.25}$$

The propagation characteristics of an optical wave in the presence of an electro-optic effect are then determined by ϵ_X, ϵ_Y, and ϵ_Z with the new principal indices of refraction:

$$n_X = \sqrt{\frac{\epsilon_X}{\epsilon_0}}, \quad n_Y = \sqrt{\frac{\epsilon_Y}{\epsilon_0}}, \quad n_Z = \sqrt{\frac{\epsilon_Z}{\epsilon_0}}. \tag{11.26}$$

As long as the electrostatic field $\boldsymbol{E}_0$ stays on and fixed in direction and magnitude, the new principal axes, together with the corresponding principal permittivities and refractive indices, have to be used to describe the optical properties of the material and the characteristics of all optical waves that propagate through it.

11.1.2 Impermeability Tensor Representation

The discussions in the preceding subsection described a formal and systematic approach to treating an electro-optic effect in terms of the changes in the permittivity tensor. However, electro-optic effects are traditionally defined in terms of the changes in the elements of the relative impermeability tensor as

$$\boldsymbol{\eta}(\omega, \boldsymbol{E}_0) = \boldsymbol{\eta}(\omega) + \Delta\boldsymbol{\eta}(\omega, \boldsymbol{E}_0), \tag{11.27}$$

which is expanded in the form:

$$\eta_{ij}(\omega, \boldsymbol{E}_0) = \eta_{ij}(\omega) + \Delta\eta_{ij}(\omega, \boldsymbol{E}_0) = \eta_{ij}(\omega) + \sum_k r_{ijk}(\omega)E_{0k} + \sum_{k,l} s_{ijkl}(\omega)E_{0k}E_{0l} + \cdots, \tag{11.28}$$

where the first term, $\eta_{ij}(\omega)$, is the field-independent component; the elements of the $r_{ijk}(\omega)$ tensor are the *linear electro-optic coefficients*, known as the *Pockels coefficients*; and those of the $s_{ijkl}(\omega)$ tensor are the *quadratic electro-optic coefficients*, known as the *electro-optic Kerr coefficients*. The *first-order electro-optic effect* that is characterized by the linear dependence of $\eta_{ij}(\omega, \boldsymbol{E}_0)$ on $\boldsymbol{E}_0$ through the coefficients $r_{ijk}(\omega)$ is called the *linear electro-optic effect*, which is also known as the *Pockels effect* and is the same as that described in Subsection 5.2.7 and in the beginning of this section. The *second-order electro-optic effect* that is characterized by the quadratic field dependence through the coefficients $s_{ijkl}(\omega)$ is called the *quadratic electro-optic effect*, which is also known as the *electro-optic Kerr effect* and is the same as that described in Subsection 5.3.7 and in the beginning of this section.

   Note that both the Pockels effect and the electro-optic Kerr effect are nonlinear optical effects. As discussed above, the Pockels effect is a second-order nonlinear optical effect that is associated with the second-order nonlinear susceptibility $\chi^{(2)}$, and the electro-optic Kerr effect is a third-order nonlinear optical effect that is associated with the third-order nonlinear susceptibility $\chi^{(3)}$. However, the Pockels effect is known as a *linear* electro-optic effect because it causes a change in the impermeability of $\Delta\eta_{ij}(\omega, \boldsymbol{E}_0) = \sum_k r_{ijk}(\omega)E_{0k}$, which is linearly proportional to the electrostatic field $\boldsymbol{E}_0$. And the electro-optic Kerr effect is known as a *quadratic electro-optic effect* because it causes a change in the impermeability of $\Delta\eta_{ij}(\omega, \boldsymbol{E}_0) = \sum_{k,l} s_{ijkl}(\omega)E_{0k}E_{0l}$, which quadratically depends on the electrostatic field $\boldsymbol{E}_0$. In any event, both linear and quadratic electro-optic effects are nonlinear optical effects if both the optical field and the electrostatic field are considered in the perturbation expansion of the total polarization in terms of linear and nonlinear polarizations, as is done in the formulation of nonlinear optics.

As discussed above and mentioned in Section 5.2.7, the Pockels effect does not exist in a centrosymmetric material, which is a material that possesses inversion symmetry. The structure and properties of such a material remain unchanged under the transformation of space inversion, which changes the signs of all rectilinear spatial coordinates from (x, y, z) to $(-x, -y, -z)$ (i.e., from $\mathbf{r}$ to $-\mathbf{r}$) and those of all polar vectors. As discussed in Subsection 1.1.3 and expressed in (1.22), an electric field vector is a polar vector that changes sign under the transformation of space inversion. By simply considering the effect of space inversion, it is clear that the electro-optically induced changes in the optical property of a centrosymmetric material are not affected by the sign change in the applied field from $\boldsymbol{E}_0$ to $-\boldsymbol{E}_0$, meaning that $\eta_{ij}(\omega, \boldsymbol{E}_0) = \eta_{ij}(\omega, -\boldsymbol{E}_0)$. As can be seen from (11.28), this condition requires that the Pockels coefficients $r_{ijk}(\omega)$ vanish. It can also be seen that the condition does not require the electro-optic Kerr coefficients $s_{ijkl}(\omega)$ to vanish. Consequently, the Pockels effect only exists in noncentrosymmetric materials, whereas the electro-optic Kerr effect exists in all materials, including centrosymmetric ones. This conclusion is consistent with that obtained in Subsections 5.2.7 and 5.3.7 and mentioned in the beginning of this section from the argument that $\chi^{(2)}$ vanishes in a centrosymmetric material, whereas $\chi^{(3)}$ does not vanish in any material.

In (11.28), indices i and j are associated with the optical field at a frequency of ω, while indices k and l are associated with the electrostatic field $\boldsymbol{E}_0$. Because $\eta_{ij} = \eta_{ji}$ and $\Delta\eta_{ij} = \Delta\eta_{ji}$, indices i and j can be contracted by using the index contraction rule of (3.32), thus reducing (11.28) to the form:

$$\eta_\alpha(\omega, \boldsymbol{E}_0) = \eta_\alpha(\omega) + \Delta\eta_\alpha(\omega, \boldsymbol{E}_0) = \eta_\alpha(\omega) + \sum_k r_{\alpha k}(\omega)E_{0k} + \sum_{k,l} s_{\alpha kl}(\omega)E_{0k}E_{0l} + \cdots,$$

$$(11.29)$$

where $\alpha = 1, 2, \ldots, 6$, with the meaning defined in (3.32).

As discussed in Section 4.3, the impermeability tensor defines an index ellipsoid according to (4.47), which is explicitly expressed as (4.48), or (4.49) by using index contraction, in a general rectilinear coordinate system (x_1, x_2, x_3). In general, the coefficients $2\eta_{23}$, $2\eta_{31}$, and $2\eta_{12}$ of the cross terms x_2x_3, x_3x_1, and x_1x_2 do not all vanish when the general coordinate system is arbitrarily chosen.

From the relation that $\boldsymbol{\eta} = (\epsilon/\epsilon_0)^{-1}$ defined in (4.46), it can be seen that $\boldsymbol{\eta}$ is diagonalized when $\boldsymbol{\epsilon}$ is diagonalized, and they have the same set of eigenvectors. In the absence of the electrostatic field $\boldsymbol{E}_0$, $\boldsymbol{\eta}$ is a diagonal tensor in the coordinate system that is defined by the principal axes $\hat{x}$, $\hat{y}$, and $\hat{z}$ with the eigenvalues given in (4.50): $\eta_x = \epsilon_0/\epsilon_x = 1/n_x^2$, $\eta_y = \epsilon_0/\epsilon_y = 1/n_y^2$, and $\eta_z = \epsilon_0/\epsilon_z = 1/n_z^2$, where n_x, n_y, and n_z are the principal refractive indices of the material in the absence of E_0. The three axes of the index ellipsoid are aligned with the three principal axes, with n_x, n_y, and n_z being the three semiaxes, as shown in Fig. 11.1. When expressed in the principal coordinate system of (x, y, z), the index ellipsoid has no cross-product terms:

$$\frac{x^2}{n_x^2} + \frac{y^2}{n_y^2} + \frac{z^2}{n_z^2} = 1. \tag{11.30}$$

The presence of an electrostatic field causes changes in the impermeability tensor through electro-optic effects, as expressed in (11.28). In general, $\boldsymbol{\eta}(\omega, E_0)$ in the presence of an electrostatic field E_0 is not diagonal in the original principal coordinate system (x, y, z) that is defined in the absence of the electrostatic field. The changes in the optical property of the material caused by an electro-optic effect deform the index ellipsoid into a new one described by the relation:

$$(\eta_1 + \Delta\eta_1)x^2 + (\eta_2 + \Delta\eta_2)y^2 + (\eta_3 + \Delta\eta_3)z^2 + 2\Delta\eta_4\,yz + 2\Delta\eta_5 zx + 2\Delta\eta_6 xy = 1. \tag{11.31}$$

The principal axes of this electro-optically deformed ellipsoid do not line up with $\hat{x}$, $\hat{y}$, and $\hat{z}$ unless $\Delta\eta_4 = \Delta\eta_5 = \Delta\eta_6 = 0$, which does happen but is not the general case. In general, at least one of the principal indices is changed. To find the principal axes of this new ellipsoid and their corresponding principal indices of refraction, we can perform a coordinate rotation in space to eliminate the cross-product terms containing yz, zx, or xy. From the above discussions, it can be seen that this procedure is the same as the coordinate rotation through the transformation matrix $\mathbf{T}$ to diagonalize $\boldsymbol{\epsilon}(\omega, E_0)$, as described in Subsection 11.1.1. Thus, we can use (11.23) to transform (11.31) into

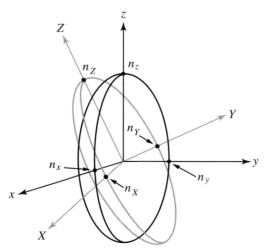

Figure 11.1 Transformation of index ellipsoid by an electro-optic effect. An electro-optic effect transforms an index ellipsoid that is originally aligned with the x, y, and z coordinates defined by the original principal axes $\hat{x}$, $\hat{y}$, and $\hat{z}$ into a new one that is aligned with the X, Y, and Z coordinates defined by the new principal axes $\hat{X}$, $\hat{Y}$, and $\hat{Z}$. Meanwhile, the principal indices of refraction are changed from n_x, n_y, and n_z to n_X, n_Y, and n_Z.

$$\frac{X^2}{n_X^2} + \frac{Y^2}{n_Y^2} + \frac{Z^2}{n_Z^2} = 1, \tag{11.32}$$

where n_X, n_Y, and n_Z are the same as those given in (11.26). The principal axes of this ellipsoid are simply the same $\hat{X}$, $\hat{Y}$, and $\hat{Z}$ as those found from the eigenvectors of $\boldsymbol{\epsilon}(\omega, \boldsymbol{E}_0)$ and given in (11.18), which are also the eigenvectors of $\boldsymbol{\eta}(\omega, \boldsymbol{E}_0)$. The concept described here is illustrated in Fig. 11.1.

11.1.3 Relation between Permittivity and Impermeability Representations

It is generally true that the changes caused by electro-optic effects are small, such that $|\Delta\epsilon_{ij}| \ll \epsilon_i, \epsilon_j$ and $|\Delta\eta_{ij}| \ll \eta_i, \eta_j$. By using the relations $\boldsymbol{\eta}(\omega, \boldsymbol{E}_0) \cdot \boldsymbol{\epsilon}(\omega, \boldsymbol{E}_0)/\epsilon_0 = 1$ and $\boldsymbol{\eta}(\omega) \cdot \boldsymbol{\epsilon}(\omega)/\epsilon_0 = 1$ under this condition, the relations between $\Delta\boldsymbol{\epsilon}$ and $\Delta\boldsymbol{\eta}$ can be found:

$$\Delta\boldsymbol{\epsilon} = -\frac{1}{\epsilon_0}\boldsymbol{\epsilon} \cdot \Delta\boldsymbol{\eta} \cdot \boldsymbol{\epsilon} \quad \text{and} \quad \Delta\boldsymbol{\eta} = -\frac{1}{\epsilon_0}\boldsymbol{\eta} \cdot \Delta\boldsymbol{\epsilon} \cdot \boldsymbol{\eta}. \tag{11.33}$$

In (11.33), $\boldsymbol{\epsilon} = \boldsymbol{\epsilon}(\omega)$ and $\boldsymbol{\eta} = \boldsymbol{\eta}(\omega)$, which are diagonalized in terms of the eigenvalues in the intrinsic principal coordinates defined for the material in the absence of $\boldsymbol{E}_0$. Therefore, the relations in (11.33) for the changes $\Delta\boldsymbol{\epsilon}$ and $\Delta\boldsymbol{\eta}$ that are caused by $\boldsymbol{E}_0$ can be explicitly written as

$$\Delta\epsilon_{ij} = -\epsilon_0\frac{\Delta\eta_{ij}}{\eta_i\eta_j} = -\epsilon_0 n_i^2 n_j^2 \Delta\eta_{ij} \quad \text{and} \quad \Delta\eta_{ij} = -\epsilon_0\frac{\Delta\epsilon_{ij}}{\epsilon_i\epsilon_j} = -\frac{\Delta\epsilon_{ij}}{\epsilon_0 n_i^2 n_j^2}. \tag{11.34}$$

Because the Pockels effect and the electro-optic Kerr effect were discovered and established before nonlinear optics was made possible by the invention of the laser in 1960, they were traditionally described in terms of the impermeability through the Pockels coefficients and the electro-optic Kerr coefficients, as defined in (11.28). In practice, the Pockels coefficients and the electro-optic Kerr coefficients are measured or given, which are used to find the change, $\Delta\boldsymbol{\eta}(\omega, \boldsymbol{E}_0)$, in the impermeability. Therefore, the analysis of an electro-optic effect is commonly carried out through finding the axes of the index ellipsoid in the presence of $\Delta\boldsymbol{\eta}(\omega, \boldsymbol{E}_0)$, as described in Subsection 11.1.2, rather than through the diagonalization of the $\Delta\boldsymbol{\epsilon}(\omega, \boldsymbol{E}_0)$ tensor, as described in Subsection 11.1.1. However, it is more systematic, and more consistent with the general analysis in nonlinear optics, to treat $\boldsymbol{\epsilon}(\omega, \boldsymbol{E}_0)$ rather than $\boldsymbol{\eta}(\omega, \boldsymbol{E}_0)$. To do so, we can simply find $\Delta\boldsymbol{\epsilon}(\omega, \boldsymbol{E}_0)$ from $\Delta\boldsymbol{\eta}(\omega, \boldsymbol{E}_0)$ by using (11.34). Then the procedure described in Subsection 11.1.1 can be carried out by diagonalizing the tensor $\boldsymbol{\epsilon}(\omega, \boldsymbol{E}_0)$ to find its eigenvalues and eigenvectors that represent the principal permittivities and principal axes of the optical material in the presence of the electrostatic field $\boldsymbol{E}_0$.

11.1.4 Electro-optic Coefficients and Nonlinear Susceptibilities

Because the Pockels effect is a second-order nonlinear optical process, the Pockels coefficients are related to the elements of the $\chi^{(2)}$ tensor. Similarly, the electro-optic Kerr coefficients are related to the elements of the $\chi^{(3)}$ tensor because the electro-optic Kerr effect is a third-order nonlinear optical process.

For the Pockels effect,

$$\Delta\epsilon_{ij}(\omega, \boldsymbol{E}_0) = 2\epsilon_0 \sum_k \chi_{ijk}^{(2)}(\omega = \omega + 0)E_{0k} \quad \text{and} \quad \Delta\eta_{ij}(\omega, \boldsymbol{E}_0) = \sum_k r_{ijk}(\omega)E_{0k}, \quad (11.35)$$

according to (11.11) and (11.28), respectively. Then, by using (11.34), we find the relation:

$$r_{ijk}(\omega) = -\frac{2}{n_i^2 n_j^2}\chi_{ijk}^{(2)}(\omega = \omega + 0) = -\frac{2}{n_i^2 n_j^2}\chi_{kij}^{(2)}(0 = \omega - \omega), \quad (11.36)$$

where the full permutation symmetry is used for $\chi_{ijk}^{(2)}(\omega = \omega + 0) = \chi_{kij}^{(2)}(0 = \omega - \omega)$ to move the zero frequency, together with its coordinate index, to the left-hand side of the equals sign. Then index contraction can be applied to $r_{ijk}(\omega)$ and $\chi_{kij}^{(2)}(0 = \omega - \omega)$ on the indices i and j, but not on the index k:

$$r_{\alpha k}(\omega) = -\frac{2}{n_i^2 n_j^2}\chi_{k\alpha}^{(2)}(0 = \omega - \omega). \quad (11.37)$$

For the electro-optic Kerr effect,

$$\Delta\epsilon_{ij}(\omega, \boldsymbol{E}_0) = 3\epsilon_0 \sum_{k,l} \chi_{ijkl}^{(2)}(\omega = \omega + 0 + 0)E_{0k}E_{0l} \quad \text{and} \quad \Delta\eta_{ij}(\omega, \boldsymbol{E}_0) = \sum_{k,l} s_{ijkl}(\omega)E_{0k}E_{0l}, \quad (11.38)$$

according to (11.11) and (11.28), respectively. By using (11.34), we find the relation:

$$s_{ijkl}(\omega) = -\frac{3}{n_i^2 n_j^2}\chi_{ijkl}^{(3)}(\omega = \omega + 0 + 0). \quad (11.39)$$

11.2 POCKELS EFFECT

As discussed above, the electro-optic Kerr effect exists in all materials, but the Pockels effect only exists in noncentrosymmetric crystals. Nevertheless, the majority of electro-optic devices are based on the Pockels effect. In a noncentrosymmetric crystal where the Pockels effect shows up, the electro-optic Kerr effect is overshadowed by the Pockels effect, and is thus negligible. In a centrosymmetric material where the Pockels effect disappears, the electro-optic Kerr effect is sometimes useful but is usually too weak for practical applications. For this reason, we only consider the Pockels effect in the rest of this chapter.

11.2.1 Properties of Electro-optic Crystals

Structurally isotropic materials, including all gases, liquids, and amorphous solids such as glass, show no Pockels effect because they are centrosymmetric. As discussed in Section 3.5 and shown in Table 3.1, the *linear optical properties* of a crystal are only determined by its crystal system. As also discussed in Section 3.5, the *nonlinear optical properties* of a crystal, including its Pockels coefficients, further depend on its point group. Among the 32 point groups in the 7 crystal systems, 11 groups are centrosymmetric, and the remaining 21 groups are noncentrosymmetric; these are listed in Table 3.2 together with the nonvanishing elements of their second-order nonlinear susceptibilities. As is expressed in (11.36), the Pockels coefficient $r_{ijk}(\omega)$ is directly related to $\chi_{ijk}^{(2)}(\omega = \omega + 0)$ and $\chi_{kij}^{(2)}(0 = \omega - \omega)$. Note the difference between $\chi_{ijk}^{(2)}(\omega = \omega + 0)$ and $\chi_{kij}^{(2)}(0 = \omega - \omega)$ in the sequences of the frequencies and the corresponding coordinate indices. In terms of index contraction, (11.37) clearly indicates that $r_{ak}(\omega)$ is directly related to $\chi_{ka}^{(2)}(0 = \omega - \omega)$. Therefore, all second-order nonlinear crystals, such as those listed in Table 3.4, are useful electro-optic crystals, and the symmetry properties of the Pockels coefficients $r_{ak}(\omega)$ can be translated to those of the second-order nonlinear susceptibilities $\chi_{ka}^{(2)}(0 = \omega - \omega)$ for optical rectification. For this reason, a material that does not show the Pockels effect is not useful for a second-order nonlinear process. However, a crystal that is not useful for second-order parametric frequency conversion might still be useful for electro-optic applications based on the Pockels effect because a parametric frequency-conversion process is not automatically phase matched but electro-optic modulation is automatically phase matched.

An instructive example is that all cubic crystals have isotropic linear optical properties but not isotropic crystal structures. Two cubic crystals belonging to different point groups can have very different nonlinear optical properties. Among the cubic crystals, C in the diamond crystal form, Si, and Ge are centrosymmetric materials of the diamond structure that show no Pockels effect, whereas GaAs, InP, and other III–V semiconductors are noncentrosymmetric materials that have nonvanishing Pockels coefficients.

For the Pockels effect,

$$\Delta\eta_\alpha = \sum_k r_{\alpha k} E_{0k}, \tag{11.40}$$

which can be explicitly written in the matrix form:

$$\begin{bmatrix} \Delta\eta_1 \\ \Delta\eta_2 \\ \Delta\eta_3 \\ \Delta\eta_4 \\ \Delta\eta_5 \\ \Delta\eta_6 \end{bmatrix} = \begin{bmatrix} r_{11} & r_{12} & r_{13} \\ r_{21} & r_{22} & r_{23} \\ r_{31} & r_{32} & r_{33} \\ r_{41} & r_{42} & r_{43} \\ r_{51} & r_{52} & r_{53} \\ r_{61} & r_{62} & r_{63} \end{bmatrix} \begin{bmatrix} E_{0x} \\ E_{0y} \\ E_{0z} \end{bmatrix}. \tag{11.41}$$

Even for a noncentrosymmetric crystal, the number of the nonvanishing independent elements in its r_{ak} matrix is generally reduced by its symmetry. Table 11.1 shows the matrix forms of the Pockels coefficients for the 21 noncentrosymmetric point groups.

A secondary effect due to the existence of *piezoelectricity* causes complexity in the determination of the Pockels coefficients of a crystal. A stress that is applied to a noncentrosymmetric polar crystal can induce an electric polarization in the crystal. This effect is called the *direct piezoelectric effect*. In the *converse piezoelectric effect*, an electric field that is applied to

Table 11.1 **Matrix form of Pockels coefficients for noncentrosymmetric point groups**[a]

Triclinic

$$1 \quad \begin{bmatrix} r_{11} & r_{12} & r_{13} \\ r_{21} & r_{22} & r_{23} \\ r_{31} & r_{32} & r_{33} \\ r_{41} & r_{42} & r_{43} \\ r_{51} & r_{52} & r_{53} \\ r_{61} & r_{62} & r_{63} \end{bmatrix} .$$

Monoclinic

$$\begin{matrix} 2 \\ (2 \parallel \hat{y}) \end{matrix} \begin{bmatrix} 0 & r_{21} & 0 \\ 0 & r_{22} & 0 \\ 0 & r_{23} & 0 \\ r_{41} & 0 & r_{43} \\ 0 & r_{52} & 0 \\ r_{61} & 0 & r_{63} \end{bmatrix} \qquad \begin{matrix} m \\ (m \perp \hat{y}) \end{matrix} \begin{bmatrix} r_{11} & 0 & r_{13} \\ r_{21} & 0 & r_{23} \\ r_{31} & 0 & r_{33} \\ 0 & r_{42} & 0 \\ r_{51} & 0 & r_{53} \\ 0 & r_{62} & 0 \end{bmatrix}$$

Orthorhombic

$$222 \begin{bmatrix} 0 & 0 & 0 \\ 0 & 0 & 0 \\ 0 & 0 & 0 \\ r_{41} & 0 & 0 \\ 0 & r_{52} & 0 \\ 0 & 0 & r_{63} \end{bmatrix} \qquad mm2 \begin{bmatrix} 0 & 0 & r_{13} \\ 0 & 0 & r_{23} \\ 0 & 0 & r_{33} \\ 0 & r_{42} & 0 \\ r_{51} & 0 & 0 \\ 0 & 0 & 0 \end{bmatrix}$$

Tetragonal

$$4 \begin{bmatrix} 0 & 0 & r_{13} \\ 0 & 0 & r_{13} \\ 0 & 0 & r_{33} \\ r_{41} & r_{42} & 0 \\ r_{42} & -r_{41} & 0 \\ 0 & 0 & 0 \end{bmatrix} \qquad \bar{4} \begin{bmatrix} 0 & 0 & r_{13} \\ 0 & 0 & -r_{13} \\ 0 & 0 & 0 \\ r_{41} & r_{42} & 0 \\ -r_{42} & r_{41} & 0 \\ 0 & 0 & r_{63} \end{bmatrix}$$

$$422 \begin{bmatrix} 0 & 0 & 0 \\ 0 & 0 & 0 \\ 0 & 0 & 0 \\ r_{41} & 0 & 0 \\ 0 & -r_{41} & 0 \\ 0 & 0 & 0 \end{bmatrix} \quad 4mm \begin{bmatrix} 0 & 0 & r_{13} \\ 0 & 0 & r_{13} \\ 0 & 0 & r_{33} \\ 0 & r_{42} & 0 \\ r_{42} & 0 & 0 \\ 0 & 0 & 0 \end{bmatrix} \quad \bar{4}2m \begin{bmatrix} 0 & 0 & 0 \\ 0 & 0 & 0 \\ 0 & 0 & 0 \\ r_{41} & 0 & 0 \\ 0 & r_{41} & 0 \\ 0 & 0 & r_{63} \end{bmatrix}$$

Trigonal

$$3 \begin{bmatrix} r_{11} & -r_{22} & r_{13} \\ -r_{11} & r_{22} & r_{13} \\ 0 & 0 & r_{33} \\ r_{41} & r_{42} & 0 \\ r_{42} & -r_{41} & 0 \\ -r_{22} & -r_{11} & 0 \end{bmatrix}$$

$$32 \begin{bmatrix} r_{11} & 0 & 0 \\ -r_{11} & 0 & 0 \\ 0 & 0 & 0 \\ r_{41} & 0 & 0 \\ 0 & -r_{41} & 0 \\ 0 & -r_{11} & 0 \end{bmatrix} \qquad 3m \begin{bmatrix} 0 & -r_{22} & r_{13} \\ 0 & r_{22} & r_{13} \\ 0 & 0 & r_{33} \\ 0 & r_{42} & 0 \\ r_{42} & 0 & 0 \\ -r_{22} & 0 & 0 \end{bmatrix}$$

Table 11.1 (*cont.*)

Hexagonal	6 $\begin{bmatrix} 0 & 0 & r_{13} \\ 0 & 0 & r_{13} \\ 0 & 0 & r_{33} \\ r_{41} & r_{42} & 0 \\ r_{42} & -r_{41} & 0 \\ 0 & 0 & 0 \end{bmatrix}$	$\bar{6}$ $\begin{bmatrix} r_{11} & -r_{22} & 0 \\ -r_{11} & r_{22} & 0 \\ 0 & 0 & 0 \\ 0 & 0 & 0 \\ 0 & 0 & 0 \\ -r_{22} & -r_{11} & 0 \end{bmatrix}$
622 $\begin{bmatrix} 0 & 0 & 0 \\ 0 & 0 & 0 \\ 0 & 0 & 0 \\ r_{41} & 0 & 0 \\ 0 & -r_{41} & 0 \\ 0 & 0 & 0 \end{bmatrix}$	$6mm$ $\begin{bmatrix} 0 & 0 & r_{13} \\ 0 & 0 & r_{13} \\ 0 & 0 & r_{33} \\ 0 & r_{42} & 0 \\ r_{42} & 0 & 0 \\ 0 & 0 & 0 \end{bmatrix}$	$\bar{6}m2$ $(m \perp \hat{x})$ $\begin{bmatrix} 0 & -r_{22} & 0 \\ 0 & r_{22} & 0 \\ 0 & 0 & 0 \\ 0 & 0 & 0 \\ 0 & 0 & 0 \\ -r_{22} & 0 & 0 \end{bmatrix}$
Cubic	432 $\begin{bmatrix} 0 & 0 & 0 \\ 0 & 0 & 0 \\ 0 & 0 & 0 \\ 0 & 0 & 0 \\ 0 & 0 & 0 \\ 0 & 0 & 0 \end{bmatrix}$	23 and $\bar{4}3m$ $\begin{bmatrix} 0 & 0 & 0 \\ 0 & 0 & 0 \\ 0 & 0 & 0 \\ r_{41} & 0 & 0 \\ 0 & r_{41} & 0 \\ 0 & 0 & r_{41} \end{bmatrix}$

[a] From I. P. Kaminow, *An Introduction to Electrooptic Devices*. Orlando, FL: Academic Press, 1974, pp. 110–111.

the same crystal can induce a strain in the crystal. The piezoelectric effect and the Pockels effect have similar symmetry properties: Both vanish in centrosymmetric materials, and both are restricted by crystal symmetry in similar manners. Consequently, the piezoelectric effect exists in a crystal that shows the Pockels effect. The strain generated in a crystal by an applied electric field can induce index changes through the *photoelastic effect*. In a free crystal, which is allowed to strain in response to the applied electric field, this secondary effect is comparable in magnitude to the primary effect of the index changes that are directly caused by the applied electric field. Pockels coefficients measured at constant strain (indicated by S) with a crystal that is clamped reflect only the primary effect, whereas those measured at constant stress (indicated by T) with a crystal that is free and unclamped reflect the sum of the primary and secondary effects.

The properties of some representative electro-optic crystals are listed in Table 11.2. Note that all of the crystals listed in Table 11.2, except for the cubic semiconductors GaAs and ZnTe of the $\bar{4}3m$ point group, are also commonly used nonlinear optical crystals. The III–V semiconductors, such as GaAs and InP, and the II–VI semiconductors, such as ZnTe and HgSe, are not useful as nonlinear optical crystals for parametric frequency conversion because they absorb light in the visible and near-infrared spectral regions and because their nonbirefringent linear optical properties do not allow birefringent phase matching. Nonetheless, they are important electro-optic crystals for the infrared.

In practical device applications, an electro-optic crystal is not clamped, but its electro-optic coefficient is a function of the modulation frequency. At low modulation frequencies, the

Table 11.2 **Properties of representative electro-optic crystals[a]**

Point group	Material	Pockels coefficients (pm V^{-1})	Refractive index (at 1 μm)
$\bar{4}3m$	GaAs	(S)$r_{41} = 1.2$	$n_o = 3.50$
	ZnTe	(S)$r_{41} = 4.3$	$n_o = 2.76$
$\bar{4}2m$	KDP[b]	(T)$r_{41} = 8.8$, (T)$r_{63} = 10.5$, (S)$r_{63} = 9.7$	$n_o = 1.51$, $n_e = 1.47$
	ADP	(T)$r_{41} = 24.5$, (T)$r_{63} = 8.5$, (S)$r_{63} = 5.5$	$n_o = 1.52$, $n_e = 1.48$
$3m$	LiNbO$_3$	(S)$r_{13} = 8.6$, (S)$r_{22} = 3.4$, (S)$r_{33} = 30.8$, (S)$r_{42} = 28$	$n_o = 2.238$, $n_e = 2.159$
	LiTaO$_3$	(S)$r_{13} = 8.5$, (S)$r_{22} = 1$, (S)$r_{33} = 30.5$, (S)$r_{42} = 20$	$n_o = 2.131$, $n_e = 2.134$
$mm2$	KTP	(S)$r_{13} = 8.8$, (S)$r_{23} = 13.8$, (S)$r_{33} = 35$, (S)$r_{42} = 8.8$, (S)$r_{51} = 6.9$	$n_x = 1.742$, $n_y = 1.750$, $n_z = 1.832$
	KTA[c]	(S)$r_{13} = 15$, (S)$r_{23} = 21$, (S)$r_{33} = 40$	$n_x = 1.783$, $n_y = 1.789$, $n_z = 1.870$

[a] Data are collected from various sources in the literature.
[b] KTP properties from J. D. Bierlein and H. Vanherzeele, "Potassium titanyl phosphate: properties and new applications," *Journal of the Optical Society of America B*, vol. 6, pp. 622–633, 1989.
[c] KTA properties from J. D. Bierlein and H. Vanherzeele, "Linear and nonlinear optical properties of flux-grown KTiOAsO$_4$," *Applied Physics Letters*, vol. 54, pp. 783–785, 1989.

electro-optic response of the crystal is that of a free crystal at constant stress because the photoelastic response can follow the low-frequency modulation signal. At high modulation frequencies, however, the photoelastic effect vanishes because the strain in the crystal cannot respond quickly enough to follow the modulation signal. Consequently, the Pockels coefficients measured at constant stress have to be used for low-frequency modulation, but those measured at constant strain have to be used for high-frequency modulation. Besides their dependence on the modulation frequency, the Pockels coefficients also vary with temperature and optical wavelength. Because of these complications, only the typical values of the Pockels coefficients measured at constant strain are listed in Table 11.2, except for those of KDP and ADP crystals, for which the r_{41} coefficient at constant strain is not available.

The technologically most important electro-optic crystals are the cubic $\bar{4}3m$ III–V semiconductors, particularly GaAs and InP and related compounds, and the uniaxial $3m$ crystals, such as LiNbO$_3$ and LiTaO$_3$. The biaxial $mm2$ crystals, such as KTP and KTA, are also popular and important as electro-optic crystals and as nonlinear optical crystals because of their large Pockels coefficients, and correspondingly large nonlinear susceptibilities. Electro-optic devices, and second-order frequency converters, that are based on $LiNbO_3$ are the most extensively studied and most well developed. Electro-optic devices based on the III–V semiconductors are also well developed because they can be monolithically integrated with other optoelectronic devices, including semiconductor lasers, amplifiers, and detectors. It can be seen from Table 11.2 that the Pockels coefficients of the III–V semiconductors are relatively small compared with those of other important electro-optic crystals. However, this disadvantage is generally compensated by using advanced semiconductor-processing technologies. For example, small waveguide structures with optimized overlap of the electrostatic field and the optical field can be made in a III–V semiconductor to maximize the electro-optic modulation efficiency. The intrinsic electro-optic effect in a III–V semiconductor can also be substantially enhanced by incorporating artificially tailored structures, such as quantum-well structures, in the semiconductor.

11.2.2 Induced Birefringence and Rotation of Principal Axes

The Pockels effect always causes a change in the birefringence of a crystal, either by inducing a birefringence that originally does not exist, such as in a cubic $\overline{4}3m$ crystal, or by changing the magnitude of the birefringence that originally exists, such as in a uniaxial $3m$ crystal or in a biaxial $mm2$ crystal. Depending on the symmetry of a specific crystal being used and the direction of the electrostatic field being applied to the crystal, the refractive-index changes induced by the Pockels effect may or may not be accompanied by a rotation of principal axes, and the axis rotation may or may not depend on the magnitude of the electrostatic field.

In this subsection, we illustrate four different scenarios through real examples on the popular electro-optic crystals of $LiNbO_3$ and GaAs. $LiNbO_3$ is a negative uniaxial crystal of the $3m$ point group; the analysis illustrated below for $LiNbO_3$ applies equally to other $3m$ crystals, irrespective of whether they are positive or negative uniaxial. The analysis illustrated below for GaAs applies equally to all other cubic $\overline{4}3m$ crystals, including other cubic III–V semiconductors and cubic II–VI semiconductors.

In the examples illustrated below, the electrostatic field is applied along one of the principal axes of the crystal. In practical applications, the field might be applied in a direction that does not line up with any of the principal axes. The problem can be solved by following the same general procedure illustrated below, but the mathematics might be more complicated.

Induced Birefringence without Axis Rotation

The electrostatic field is applied along the optical axis of $LiNbO_3$ such that $E_{0x} = E_{0y} = 0$ but $E_{0z} \neq 0$. In this case, the only changes in the impermeability tensor elements that are caused by the Pockels effect are $\Delta\eta_1 = r_{13}E_{0z}$, $\Delta\eta_2 = r_{23}E_{0z} = r_{13}E_{0z}$, and $\Delta\eta_3 = r_{33}E_{0z}$, where the fact that $r_{23} = r_{13}$ for a $3m$ crystal from Table 11.1 is used. With these changes, the index ellipsoid expressed in the original principal coordinates becomes

$$\left(\frac{1}{n_o^2} + r_{13}E_{0z}\right)x^2 + \left(\frac{1}{n_o^2} + r_{23}E_{0z}\right)y^2 + \left(\frac{1}{n_e^2} + r_{33}E_{0z}\right)z^2 = 1. \tag{11.42}$$

By using (11.34) to find $\Delta\epsilon_{ij}$ from $\Delta\eta_{ij}$, followed by using (11.16), the field-dependent dielectric permittivity tensor can be found:

$$\boldsymbol{\epsilon}(\omega, \boldsymbol{E}_0) = \epsilon_0 \begin{bmatrix} n_o^2 - n_o^4 r_{13}E_{0z} & 0 & 0 \\ 0 & n_o^2 - n_o^4 r_{13}E_{0z} & 0 \\ 0 & 0 & n_e^2 - n_e^4 r_{33}E_{0z} \end{bmatrix}. \tag{11.43}$$

It can be clearly seen from (11.42) that the equation for the field-dependent index ellipsoid contains no cross-product terms and, correspondingly, the field-dependent permittivity tensor given in (11.43) remains diagonal. Therefore, the principal axes are not rotated:

$$\hat{X} = \hat{x}, \quad \hat{Y} = \hat{y}, \quad \text{and} \quad \hat{Z} = \hat{z}. \tag{11.44}$$

The crystal remains uniaxial with the same optical axis, but all three indices of refraction are changed. Because the induced changes of the refractive indices are generally so small that $|r_{13}E_{0z}| \ll n_o^{-2}$ and $|r_{33}E_{0z}| \ll n_e^{-2}$, the field-dependent principal indices of refraction are

$$n_X = n_Y \approx n_o - \frac{n_o^3}{2} r_{13}E_{0z} \quad \text{and} \quad n_Z \approx n_e - \frac{n_e^3}{2} r_{33}E_{0z}. \tag{11.45}$$

In this scenario, the Pockels effect induces a field-dependent change in the birefringence of the crystal without changing its principal axes. The crystal also remains uniaxial, though the magnitude of its birefringence is changed.

Induced Birefringence with Field-Strength-Dependent Rotation of Two Axes
The electrostatic field is applied along the y axis of LiNbO$_3$ such that $E_{0x} = E_{0z} = 0$ but $E_{0y} \neq 0$. In this case, the only changes in the impermeability tensor elements that are caused by the Pockels effect are $\Delta\eta_1 = r_{12}E_{0y} = -r_{22}E_{0y}$, $\Delta\eta_2 = r_{22}E_{0y}$, and $\Delta\eta_4 = r_{42}E_{0y}$, where the fact that $r_{12} = -r_{22}$ for a 3m crystal from Table 11.1 is used. With these changes, the index ellipsoid expressed in the original principal coordinates becomes

$$\left(\frac{1}{n_o^2} - r_{22}E_{0y}\right)x^2 + \left(\frac{1}{n_o^2} + r_{22}E_{0y}\right)y^2 + \frac{z^2}{n_e^2} + 2r_{42}E_{0y}yz = 1. \tag{11.46}$$

By using (11.34) to find $\Delta\epsilon_{ij}$ from $\Delta\eta_{ij}$, followed by using (11.16), the field-dependent dielectric permittivity tensor can be found:

$$\boldsymbol{\epsilon}(\omega, \boldsymbol{E}_0) = \epsilon_0 \begin{bmatrix} n_o^2 + n_o^4 r_{22}E_{0y} & 0 & 0 \\ 0 & n_o^2 - n_o^4 r_{22}E_{0y} & -n_o^2 n_e^2 r_{42}E_{0y} \\ 0 & -n_o^2 n_e^2 r_{42}E_{0y} & n_e^2 \end{bmatrix}. \tag{11.47}$$

It can be seen from (11.46) that the equation for the field-dependent index ellipsoid contains one cross-product term in yz and, correspondingly, the field-dependent permittivity tensor given in (11.47) has two nonvanishing off-diagonal elements of $\Delta\epsilon_{yz} = \Delta\epsilon_{zy}$. The new principal axes and the field-dependent principal indices of refraction can be found by eliminating the cross-product yz term in the index ellipsoid equation given in (11.46), or by diagonalizing the field-dependent permittivity $\boldsymbol{\epsilon}(\omega, \boldsymbol{E}_0)$ given in (11.47). The new field-dependent principal axes $\hat{Y}$ and $\hat{Z}$ are rotated away from the original principal axes $\hat{y}$ and $\hat{z}$, though the principal axis $\hat{X}$ remains the same as $\hat{x}$:

$$
\begin{aligned}
\hat{X} &= \hat{x}, \\
\hat{Y} &= \hat{y}\cos\theta - \hat{z}\sin\theta \approx \frac{1}{\sqrt{1+\alpha^2}}\hat{y} - \frac{\alpha}{\sqrt{1+\alpha^2}}\hat{z}, \\
\hat{Z} &= \hat{y}\sin\theta + \hat{z}\cos\theta \approx \frac{\alpha}{\sqrt{1+\alpha^2}}\hat{y} + \frac{1}{\sqrt{1+\alpha^2}}\hat{z},
\end{aligned}
\tag{11.48}
$$

where $\alpha = n_\text{o}^2 n_\text{e}^2 r_{42}E_{0y}/(n_\text{o}^2 - n_\text{e}^2)$. For the negative uniaxial crystal LiNbO$_3$, $n_\text{o} > n_\text{e}$ and $n_\text{o}^2 - n_\text{e}^2 \gg |n_\text{o}^2 n_\text{e}^2 r_{42}E_{0y}| > |n_\text{o}^4 r_{22}E_{0y}|$ for any electrostatic field E_{0y} of a magnitude that is below the crystal breakdown field of the order of $100\,\text{MV m}^{-1}$. Therefore, $1 \gg |\alpha| > |\beta|$ where $\beta = n_\text{o}^4 r_{22}E_{0y}/(n_\text{o}^2 - n_\text{e}^2)$. Because $1 \gg |\alpha|$, the field-dependent principal axis $\hat{Y}$ is very close to $\hat{y}$ with a very small component in $\hat{z}$, and $\hat{Z}$ is very close to $\hat{z}$ with a very small component in $\hat{y}$. The angle θ of the axis rotation is very small:

$$
\theta \approx \tan^{-1}\alpha = \tan^{-1}\frac{n_\text{o}^2 n_\text{e}^2 r_{42}E_{0y}}{n_\text{o}^2 - n_\text{e}^2}.
\tag{11.49}
$$

We also find that

$$
\begin{aligned}
n_X &\approx n_\text{o} + \frac{n_\text{o}^3}{2}r_{22}E_{0y}, \\
n_Y &\approx n_\text{o} - \frac{n_\text{o}^3}{2}r_{22}E_{0y} + \frac{1}{2}\frac{n_\text{o}^3 n_\text{e}^4}{n_\text{o}^2 - n_\text{e}^2}(r_{42}E_{0y})^2, \\
n_Z &\approx n_\text{e} - \frac{1}{2}\frac{n_\text{e}^3 n_\text{o}^4}{n_\text{o}^2 - n_\text{e}^2}(r_{42}E_{0y})^2.
\end{aligned}
\tag{11.50}
$$

In this scenario, the Pockels effect induces a field-dependent change in the birefringence of the crystal while rotating two of its principal axes. The induced birefringence changes the crystal from uniaxial to biaxial because the three field-dependent principal refractive indices are all different, as seen in (11.50). The angle of rotation of the two rotated principal axes varies with the strength of the electrostatic field, as seen in (11.49).

Induced Birefringence with Field-Strength-Dependent Rotation of All Three Axes

The electrostatic field is applied along the x axis of LiNbO$_3$, such that $E_{0y} = E_{0z} = 0$ but $E_{0x} \neq 0$. In this case, the only changes in the impermeability tensor elements that are caused by the Pockels effect are $\Delta\eta_5 = r_{51}E_{0x} = r_{42}E_{0x}$ and $\Delta\eta_6 = r_{61}E_{0x} = -r_{22}E_{0x}$, where the fact that $r_{51} = r_{42}$ and $r_{61} = -r_{22}$ for a $3m$ crystal from Table 11.1 is used. With these changes, the index ellipsoid expressed in the original principal coordinates becomes

$$\frac{x^2}{n_o^2} + \frac{y^2}{n_o^2} + \frac{z^2}{n_e^2} + 2r_{42}E_{0x}zx - 2r_{22}E_{0x}xy = 1. \tag{11.51}$$

By using (11.34) to find $\Delta\epsilon_{ij}$ from $\Delta\eta_{ij}$, followed by using (11.16), the field-dependent dielectric permittivity tensor can be found:

$$\epsilon(\omega, \mathbf{E}_0) = \epsilon_0 \begin{bmatrix} n_o^2 & n_o^4 r_{22}E_{0x} & -n_o^2 n_e^2 r_{42}E_{0x} \\ n_o^4 r_{22}E_{0x} & n_o^2 & 0 \\ -n_o^2 n_e^2 r_{42}E_{0x} & 0 & n_e^2 \end{bmatrix}. \tag{11.52}$$

It can be seen from (11.51) that the equation for the field-dependent index ellipsoid contains two cross-product terms in zx and xy, and, correspondingly, the field-dependent permittivity tensor given in (11.52) has four nonvanishing off-diagonal elements of $\Delta\epsilon_{xy} = \Delta\epsilon_{yx}$ and $\Delta\epsilon_{zx} = \Delta\epsilon_{xz}$. By diagonalizing the field-dependent permittivity $\epsilon(\omega, \mathbf{E}_0)$ given in (11.52) we find that all three field-dependent principal axes $\hat{X}$, $\hat{Y}$, and $\hat{Z}$ are rotated away from the original principal axes $\hat{x}$, $\hat{y}$, and $\hat{z}$:

$$\hat{X} \approx \frac{1}{\sqrt{2 + \alpha^2}}\hat{x} + \frac{1}{\sqrt{2 + \alpha^2}}\hat{y} - \frac{\alpha}{\sqrt{2 + \alpha^2}}\hat{z},$$

$$\hat{Y} \approx -\frac{1}{\sqrt{2 + \alpha^2}}\hat{x} + \frac{1}{\sqrt{2 + \alpha^2}}\hat{y} + \frac{\alpha}{\sqrt{2 + \alpha^2}}\hat{z}, \tag{11.53}$$

$$\hat{Z} \approx \frac{\alpha}{\sqrt{1 + \alpha^2 + \alpha^2\beta^2}}\hat{x} - \frac{\alpha\beta}{\sqrt{1 + \alpha^2 + \alpha^2\beta^2}}\hat{y} + \frac{1}{\sqrt{1 + \alpha^2 + \alpha^2\beta^2}}\hat{z},$$

where $\alpha = n_o^2 n_e^2 r_{42}E_{0x}/(n_o^2 - n_e^2)$ and $\beta = n_o^4 r_{22}E_{0x}/(n_o^2 - n_e^2)$, with $1 \gg |\alpha| > |\beta|$ for LiNbO$_3$ for any electrostatic field E_{0x} of a magnitude that is below the crystal breakdown field of the order of $100\,\mathrm{MV\,m^{-1}}$. Note that the three principal axes $\hat{X}$, $\hat{Y}$, and $\hat{Z}$ are unit vectors, as required, but they do not exactly satisfy the requirement that $\hat{X} \times \hat{Y} = \hat{Z}$ because they are simplified under the condition that $1 \gg |\alpha| > |\beta|$. Because $1 \gg |\alpha| > |\beta|$, it is seen from (11.53) that the field-dependent principal axis $\hat{X}$ is very close to $(\hat{x} + \hat{y})/\sqrt{2}$ with a very small component in $\hat{z}$, $\hat{Y}$ is very close to $(-\hat{x} + \hat{y})/\sqrt{2}$ with a very small component in $\hat{z}$, and $\hat{Z}$ is very close to $\hat{z}$ with very small components in $\hat{x}$ and $\hat{y}$. We also find that

$$n_X \approx n_o + \frac{n_o^3}{2}r_{22}E_{0x} + \frac{n_o^3 n_e^4}{4(n_o^2 - n_e^2)}(r_{42}E_{0x})^2,$$

$$n_Y \approx n_o - \frac{n_o^3}{2}r_{22}E_{0x} + \frac{n_o^3 n_e^4}{4(n_o^2 - n_e^2)}(r_{42}E_{0x})^2, \tag{11.54}$$

$$n_Z \approx n_e - \frac{n_e^3 n_o^4}{2(n_o^2 - n_e^2)}(r_{42}E_{0x})^2.$$

In this scenario, the Pockels effect induces a field-dependent change in the birefringence of the crystal while rotating all three of its principal axes. The induced birefringence makes the crystal biaxial because the three field-dependent principal refractive indices are all different, as seen in (11.54). The amount of rotation for each of the three rotated principal axes varies with the strength of the electrostatic field, as seen in (11.53).

Induced Birefringence with Field-Strength-Independent Rotation of Two Axes

An important example is the Pockels effect in a III–V semiconductor of the cubic $\overline{4}3m$ point group, such as GaAs or InP. A similar effect is seen when an electrostatic field is applied along any of the original principal axes because $n_x = n_y = n_z = n_o$ and the only nonvanishing Pockels coefficients are $r_{41} = r_{52} = r_{63}$ for such a crystal. We therefore consider only the case that the field is applied along the z axis such that $E_{0x} = E_{0y} = 0$ but $E_{0z} \neq 0$. In this case, the only change in the impermeability tensor elements caused by the Pockels effect is $\Delta\eta_6 = r_{41}E_{0z}$. The index ellipsoid expressed in the original principal coordinates becomes

$$\frac{x^2}{n_o^2} + \frac{y^2}{n_o^2} + \frac{z^2}{n_o^2} + 2r_{41}E_{0z}xy = 1. \tag{11.55}$$

By using (11.34) to find $\Delta\epsilon_{ij}$ from $\Delta\eta_{ij}$, followed by using (11.16), the field-dependent dielectric permittivity tensor can be found:

$$\epsilon(\omega, E_0) = \epsilon_0 \begin{bmatrix} n_o^2 & -n_o^4 r_{41}E_{0z} & 0 \\ -n_o^4 r_{41}E_{0z} & n_o^2 & 0 \\ 0 & 0 & n_o^2 \end{bmatrix}. \tag{11.56}$$

The equation in (11.55) for the field-dependent index ellipsoid contains one cross-product term in xy and, correspondingly, the field-dependent permittivity tensor given in (11.56) has two nonvanishing off-diagonal elements of $\Delta\epsilon_{xy} = \Delta\epsilon_{yx}$. Therefore, the field-dependent principal axes $\hat{X}$ and $\hat{Y}$ are rotated away from the original principal axes $\hat{x}$ and $\hat{y}$:

$$\hat{X} = \frac{1}{\sqrt{2}}(\hat{x} + \hat{y}), \quad \hat{Y} = \frac{1}{\sqrt{2}}(-\hat{x} + \hat{y}), \quad \text{and} \quad \hat{Z} = \hat{z}. \tag{11.57}$$

The field-dependent principal indices are

$$n_X \approx n_o - \frac{n_o^3}{2}r_{41}E_{0z}, \quad n_Y \approx n_o + \frac{n_o^3}{2}r_{41}E_{0z}, \quad \text{and} \quad n_Z = n_o. \tag{11.58}$$

We find from (11.57) that though the application of an electrostatic field to the crystal causes the new principal axes $\hat{X}$ and $\hat{Y}$ to rotate by $45°$ from the original principal axes $\hat{x}$ and $\hat{y}$, these new principal axes are independent of the field strength because $n_x = n_y = n_o$. Nonetheless, the originally nonbirefringent cubic crystal becomes biaxial, and the birefringence varies with the field strength, as is seen in (11.58).

EXAMPLE 11.1

Find the changes in the refractive indices and in the birefringence at $\lambda = 1\,\mu\text{m}$ caused by an electrostatic field of $E_0 = 1\,\text{MV m}^{-1}$ that is applied to LiNbO_3 and GaAs, respectively, in the direction along the z principal axis of the crystal.

Solution

The values of the Pockels coefficients and the refractive indices at $\lambda = 1\,\mu\text{m}$ for both LiNbO_3 and GaAs are listed in Table 11.2. For LiNbO_3, an electrostatic field applied along its z axis does not rotate its principal axes but only causes changes in its refractive indices. The crystal remains uniaxial. From (11.45), we find that the change in the ordinary index is

$$\Delta n_{\text{o}} = -\frac{n_{\text{o}}^3}{2} r_{13} E_{0z} = -\frac{2.238^3}{2} \times 8.6 \times 10^{-12} \times 1 \times 10^6 = -4.82 \times 10^{-5},$$

whereas the change in the extraordinary index is

$$\Delta n_{\text{e}} = -\frac{n_{\text{e}}^3}{2} r_{33} E_{0z} = -\frac{2.159^3}{2} \times 30.8 \times 10^{-12} \times 1 \times 10^6 = -1.55 \times 10^{-4}.$$

The electro-optically induced change in the birefringence is $\Delta B = \Delta n_{\text{e}} - \Delta n_{\text{o}} \approx -1 \times 10^{-4}$, which is almost three orders of magnitude smaller than the intrinsic birefringence of $B = n_{\text{e}} - n_{\text{o}} = -0.08$ for LiNbO_3. In normal device applications, the applied electrostatic field typically falls in the range between 0.1 and $10\,\text{MV m}^{-1}$. Because the index changes are linearly proportional to the applied electrostatic field, the electro-optically induced birefringence is typically two to three orders of magnitude smaller than the intrinsic birefringence of LiNbO_3.

For GaAs, which is originally nonbirefringent, an electrostatic field that is applied along its z axis causes a rotation of its x and y principal axes and a birefringence between them. From (11.58), we find that the electro-optically induced index changes are

$$\Delta n_Y = -\Delta n_X \approx \frac{n_{\text{o}}^3}{2} r_{41} E_{0z} = \frac{3.5^3}{2} \times 1.2 \times 10^{-12} \times 1 \times 10^6 = 2.57 \times 10^{-5}.$$

The electro-optically induced birefringence is $B = n_Y - n_X = \Delta B = \Delta n_Y - \Delta n_X = 5.14 \times 10^{-5}$. Although this birefringence is smaller than that in the case of LiNbO_3, it is significant because GaAs is originally nonbirefringent.

11.3 ELECTRO-OPTIC MODULATORS

The changes caused by the Pockels effect in the principal refractive indices and in the birefringence of an electro-optic crystal can be utilized to construct a variety of electro-optic modulators, either in the bulk form or in waveguide structures. An electro-optically induced rotation of the principal axes is not required for the functioning of an electro-optic modulator, though it often accompanies the index changes. Nevertheless, the directions of the principal axes in the presence of an applied electrostatic field, whether rotated or not, have to be taken into consideration in the design and operation of an electro-optic modulator. An electro-optic modulator in the bulk form that is based on the Pockels effect is also called a *Pockels cell*. Electro-optic modulators in the waveguide form are also well developed. In this section, we consider the operational principles of basic electro-optic modulators.

11.3.1 Phase Modulators

As we shall see in this section and in the following chapter, *the most basic of all types of modulation on an optical wave is phase modulation*. If the phase of an optical wave can be modulated, it is generally possible to translate this phase modulation into a modulation on other features of the optical wave, including its polarization and its amplitude. However, the converse of this fact is not true. For example, a modulation on the intensity of an optical beam does not necessarily translate into a modulation on its phase. For this reason, a physical mechanism that can be utilized to perform phase modulation of an optical wave is generally useful.

The phase of an optical wave can be electro-optically modulated through an index change caused by the Pockels effect. For this type of application, the simplest arrangement is to make the optical wave linearly polarized in a direction that is parallel to a principal axis, $\hat{X}$, $\hat{Y}$, or $\hat{Z}$, that has a field-dependent principal index of refraction. The preferred choice is a principal axis that has a large electro-optically induced index change but remains in a fixed direction as the magnitude of the modulation electric field varies. For LiNbO$_3$, this possibility can be accomplished by applying the electric field along the z principal axis, as illustrated in the first scenario discussed in Subsection 11.2.2 and shown in Fig. 11.2. In this case, all three principal axes remained unchanged by the modulation electric field: $\hat{X} = \hat{x}$, $\hat{Y} = \hat{y}$, and $\hat{Z} = \hat{z}$.

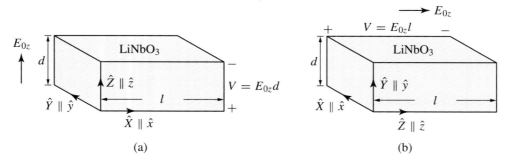

Figure 11.2 (a) LiNbO$_3$ transverse electro-optic phase modulator. (b) LiNbO$_3$ longitudinal electro-optic phase modulator. The $\hat{x}$, $\hat{y}$, and $\hat{z}$ unit vectors represent the original principal axes of the crystal in the absence of the applied electric field, and $\hat{X}$, $\hat{Y}$, and $\hat{Z}$ represent the new principal axes in the presence of the applied field. In this example, the principal axes are not rotated by the applied field because the modulation field is applied in the direction along the principal z axis.

There are two possible arrangements: *transverse modulation*, for which the optical wave propagates in a direction perpendicular to the modulation field, as shown in Fig. 11.2(a), and *longitudinal modulation*, for which the modulation field is parallel to the direction of optical wave propagation, as shown in Fig. 11.2(b).

Transverse Phase Modulators

We first consider the situation of the *transverse phase modulator*, for which the modulation electric field is applied in a direction that is transverse to the propagation direction of the optical wave. For the LiNbO$_3$ transverse phase modulator shown in Fig. 11.2(a), the modulation field is applied in the Z direction, whereas the optical wave propagates in the X direction. In this case, the optical wave can be polarized in either the Y or Z direction. If it is linearly polarized in the Z direction, its space and time dependence can be expressed as

$$\mathbf{E}(X,t) = \hat{Z}\mathcal{E}\exp\left(\mathrm{i}k^Z X - \mathrm{i}\omega t\right) = \hat{Z}\mathcal{E}\exp\left(\mathrm{i}\varphi_Z - \mathrm{i}\omega t\right), \tag{11.59}$$

where $k^Z = n_Z\omega/c$ is the propagation constant of an optical wave that is polarized in the Z direction, not the Z component of its wavevector. For the propagation over a crystal of a length l, the total phase shift is

$$\varphi_Z = k^Z l = \frac{\omega}{c}n_Z l = \frac{\omega}{c}\left(n_\mathrm{e}l - \frac{n_\mathrm{e}^3}{2}r_{33}E_{0z}l\right) = \frac{\omega}{c}\left(n_\mathrm{e}l - \frac{n_\mathrm{e}^3}{2}r_{33}V\frac{l}{d}\right), \tag{11.60}$$

where $V = E_{0z}d$ is the voltage that is transversely applied to the modulator in the Z direction, which is also the z direction, as shown in Fig. 11.2(a).

For a sinusoidal modulation at a frequency of $f = \Omega/2\pi$, the modulation voltage can be expressed as

$$V(t) = V_\mathrm{pk}\sin\Omega t, \tag{11.61}$$

which has a peak voltage of V_pk. The optical field at the output plane, $X = l$, of the crystal is

$$\mathbf{E}(l,t) = \hat{Z}\mathcal{E}\mathrm{e}^{\mathrm{i}\omega n_\mathrm{e}l/c}\exp[-\mathrm{i}\left(\omega t + \varphi_\mathrm{pk}\sin\Omega t\right)], \tag{11.62}$$

where

$$\varphi_\mathrm{pk} = \frac{\omega}{c}\frac{n_\mathrm{e}^3}{2}r_{33}V_\mathrm{pk}\frac{l}{d} = \frac{\pi}{\lambda}n_\mathrm{e}^3 r_{33}V_\mathrm{pk}\frac{l}{d} \tag{11.63}$$

is the peak phase shift, known as the *phase modulation depth*, for the Z-polarized optical wave.

If the optical field is instead linearly polarized in the Y direction, the propagation constant is $k^Y = n_Y\omega/c$ instead of k^Z. Then, the phase shift after the propagation through the crystal is

$$\varphi_Y = k^Y l = \frac{\omega}{c}n_Y l = \frac{\omega}{c}\left(n_\mathrm{o}l - \frac{n_\mathrm{o}^3}{2}r_{13}E_{0z}l\right) = \frac{\omega}{c}\left(n_\mathrm{o}l - \frac{n_\mathrm{o}^3}{2}r_{13}V\frac{l}{d}\right), \tag{11.64}$$

where $V = E_{0z}d$ is still valid because the modulation electric field is not changed with the polarization of the optical wave. The phase modulation depth for the Y-polarized optical wave is

$$\varphi_{pk} = \frac{\omega}{c} \frac{n_o^3}{2} r_{13} V_{pk} \frac{l}{d} = \frac{\pi}{\lambda} n_o^3 r_{13} V_{pk} \frac{l}{d}. \tag{11.65}$$

Because $n_o \approx n_e$ but $r_{33} \approx 3.6 r_{13}$, it can be seen by comparing (11.63) with (11.65) that for a desired modulation depth, the modulation voltage required for a Y-polarized optical wave is about 3.6 times that for a Z-polarized wave.

For practical applications, it is desired that the peak modulation depth reach π. Therefore, it is usually necessary that the peak modulation depth reach $\pi/2$ so that the phase can be modulated over a range of π when the modulation voltage varies from $-V_{pk}$ to V_{pk}. The voltage that is required for a phase modulation of π is known as the *half-wave voltage* and is denoted as V_π, also denoted as $V_{\lambda/2}$. The voltage for a phase modulation of $\pi/2$ is the *quarter-wave voltage*, denoted as $V_{\pi/2}$ or $V_{\lambda/4}$. For the LiNbO$_3$ transverse phase modulator discussed above, the half-wave and quarter-wave voltages are, respectively,

$$V_\pi = \frac{\lambda}{n_e^3 r_{33}} \frac{d}{l} \quad \text{and} \quad V_{\pi/2} = \frac{\lambda}{2 n_e^3 r_{33}} \frac{d}{l} \tag{11.66}$$

for the Z-polarized wave, and

$$V_\pi = \frac{\lambda}{n_o^3 r_{13}} \frac{d}{l} \quad \text{and} \quad V_{\pi/2} = \frac{\lambda}{2 n_o^3 r_{13}} \frac{d}{l} \tag{11.67}$$

for the Y-polarized wave.

Longitudinal Phase Modulators

We now consider the situation of the *longitudinal phase modulator*, for which the modulation electric field is applied along the propagation direction of the optical wave. For the LiNbO$_3$ longitudinal phase modulator shown in Fig. 11.2(b), both the modulation field and the optical wave propagation are in the Z direction. In this case, an optical wave of any polarization in the XY plane experiences the same amount of phase shift because $n_X = n_Y$ so that $k^X = k^Y$. For a crystal of a length l, as shown in Fig. 11.2(b), we have

$$\varphi_X = \varphi_Y = \frac{\omega}{c} \left(n_o l - \frac{n_o^3}{2} r_{13} E_{0z} l \right) = \frac{\omega}{c} \left(n_o l - \frac{n_o^3}{2} r_{13} V \right), \tag{11.68}$$

where $V = E_{0z}l$ for the longitudinal modulator. Therefore, with a sinusoidal modulation voltage as given in (11.61), the modulation depth of the longitudinal phase modulator is

$$\varphi_{pk} = \frac{\omega}{c} \frac{n_o^3}{2} r_{13} V_{pk} = \frac{\pi}{\lambda} n_o^3 r_{13} V_{pk}, \tag{11.69}$$

which is independent of the crystal length l. The half-wave and quarter-wave voltages for any polarization in the XY plane are, respectively,

$$V_\pi = \frac{\lambda}{n_0^3 r_{13}} \quad \text{and} \quad V_{\pi/2} = \frac{\lambda}{2n_0^3 r_{13}}. \tag{11.70}$$

It is seen that the voltage required for a given modulation depth is independent of the physical dimensions of the modulator in the case of longitudinal modulation, whereas it is proportional to d/l in the case of transverse modulation. One advantage of transverse modulation is that the required modulation voltage can be substantially lowered by reducing the d/l dimensional ratio of a transverse modulator. Furthermore, the geometry of transverse modulation is compatible with waveguide geometry, and the d/l dimensional ratio can be made very small in a guided-wave modulator. Another advantage is that the electrodes of a transverse modulator can be made with standard fabrication technology and can be patterned if desired, whereas those of a longitudinal modulator have to be made of transparent conductors, which can be very difficult, if not impossible, to fabricate in the dimensions of the typical optical waveguide. However, if a large input and output aperture is desired such that $d/l > 1$, it becomes advantageous to use longitudinal modulation rather than transverse modulation. The relative advantages and disadvantages of transverse versus longitudinal modulation discussed above also hold true for the polarization and intensity modulators discussed in the following.

EXAMPLE 11.2

As a practical example, consider the LiNbO$_3$ transverse and longitudinal phase modulators shown in Figs. 11.2(a) and (b), respectively, where the modulation voltage is applied along the z axis of the crystal. Find the half-wave voltage V_π that is required for a phase modulation depth of $\varphi_{\text{pk}} = \pi$ at the wavelength of $\lambda = 1$ μm for optical waves of different polarizations.

Solution

We first consider the transverse modulator shown in Fig. 11.2(a). By using (11.66) and (11.67) for the half-wave voltage and the parameters of $r_{13} = 8.6$ pm V^{-1}, $r_{33} = 30.8$ pm V^{-1}, $n_\text{e} = 2.159$, and $n_\text{e} = 2.238$ that are given in Table 11.2 for LiNbO$_3$ at $\lambda = 1$ μm, the half-wave voltage for a Z-polarized optical wave is found to be

$$V_\pi = \frac{\lambda}{n_\text{e}^3 r_{33}} \frac{d}{l} = \frac{1 \times 10^{-6}}{2.159^3 \times 30.8 \times 10^{-12}} \frac{d}{l} \, \text{V} = 3.23 \frac{d}{l} \, \text{kV},$$

and the half-wave voltage for a Y-polarized optical wave is

$$V_\pi = \frac{\lambda}{n_0^3 r_{13}} \frac{d}{l} = \frac{1 \times 10^{-6}}{2.238^3 \times 8.6 \times 10^{-12}} \frac{d}{l} \, \text{V} = 10.4 \frac{d}{l} \, \text{kV}.$$

For a bulk modulator, d and l are generally of the same order of magnitude, and the half-wave voltage is of the order of kilovolts. However, for a guided-wave modulator of the typical waveguide dimensions, the ratio d/l is of the order of 10^{-3}. For example, for a transverse waveguide modulator that has the dimensions of $d = 5$ μm and $l = 5$ mm, the half-wave voltages are reduced to 3.23 V and 10.4 V for Z-polarized and Y-polarized waves, respectively.

For the longitudinal modulator shown in Fig. 11.2(b), the optical wave is polarized in the XY plane. By using (11.70) for the half-wave voltage, we find that the half-wave voltage is always

$$V_\pi = \frac{\lambda}{n_o^3 r_{13}} = \frac{1 \times 10^{-6}}{2.238^3 \times 8.6 \times 10^{-12}} \text{ V} = 10.4 \text{ kV},$$

irrespective of the dimensions of the longitudinal modulator or the polarization of the optical wave.

11.3.2 Polarization Modulators

In the operation of an electro-optic polarization modulator, the optical wave is not linearly polarized in a direction that is parallel to any of the principal axes in the presence of the modulation field. The optical field can be decomposed into two linearly polarized normal modes. If the two normal modes see different field-induced indices of refraction, there is a field-dependent *phase retardation* between the two modes. The polarization of the optical wave at the output of the crystal can then be controlled by the modulation field.

The LiNbO$_3$ transverse modulator discussed in the preceding subsection becomes a polarization modulator if the input optical field is not linearly polarized along the Y or Z axis but is polarized in the YZ plane:

$$\mathbf{E}(0,t) = (\hat{Y}\mathcal{E}_Y + \hat{Z}\mathcal{E}_Z)e^{-i\omega t}, \tag{11.71}$$

where $\mathcal{E}_Y \neq 0$ and $\mathcal{E}_Z \neq 0$, as is shown in Fig. 11.3(a). At the output, we have

$$\mathbf{E}(l,t) = (\hat{Y}\mathcal{E}_Y e^{ik^Y l} + \hat{Z}\mathcal{E}_Z e^{ik^Z l})e^{-i\omega t} = (\hat{Y}\mathcal{E}_Y e^{i\Delta\varphi} + \hat{Z}\mathcal{E}_Z)e^{ik^Z l - i\omega t}, \tag{11.72}$$

where

$$\Delta\varphi = (k^Y - k^Z)l \tag{11.73}$$

is the phase retardation of the Z component with respect to the Y component. By using (11.45), or (11.60) and (11.64), we have

$$\Delta\varphi = \frac{\pi}{\lambda}\left[2(n_o - n_e)l + \left(n_e^3 r_{33} - n_o^3 r_{13}\right)V\frac{l}{d}\right] = \Delta\varphi_0 + \frac{\pi}{\lambda}\left(n_e^3 r_{33} - n_o^3 r_{13}\right)V\frac{l}{d}. \tag{11.74}$$

The intrinsic birefringence of a uniaxial LiNbO$_3$ crystal causes a voltage-independent background phase retardation of $\Delta\varphi_0 = 2\pi(n_o - n_e)l/\lambda$ at $V = 0$ in the absence of an applied field. For a given crystal length, this background phase retardation is fixed, and the function of the device cannot be varied if no modulation field is applied. By properly choosing the length l, the device can function as a quarter-wave or half-wave plate, as discussed in Subsection 4.3.2.

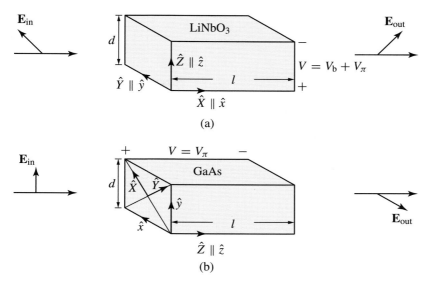

Figure 11.3 (a) LiNbO$_3$ transverse electro-optic polarization modulator. (b) GaAs longitudinal electro-optic polarization modulator. The $\hat{x}$, $\hat{y}$, and $\hat{z}$ unit vectors represent the original principal axes of the crystal, and $\hat{X}$, $\hat{Y}$, and $\hat{Z}$ represent the new principal axes.

An applied field causes an additional voltage-dependent phase retardation. The output polarization state of a given input optical wave that has nonvanishing Y and Z components can be varied by varying the modulation voltage. Thus, the device functions as a voltage-controlled polarization modulator. For proper operation of the device as a voltage-controlled polarization modulator, the background phase retardation has to be compensated with a fixed bias voltage, V_b, as shown in Fig. 11.3(a).

The background phase retardation contributed by the intrinsic birefringence is the major drawback of the LiNbO$_3$ transverse polarization modulator discussed here. Although it can be compensated with a bias voltage that falls within the range of $\pm V_\pi$, the requirement of such a DC bias voltage complicates the operation of the device, particularly when it is modulated at a high frequency. Because the bias voltage depends on the length and the refractive indices of the device, it is susceptible to changes in the operating condition, such as temperature variations caused by the operation of the device. Furthermore, the bias voltage is also a function of the optical wavelength because $\Delta\varphi_0$ varies with the optical wavelength. In practice, the bias voltage has to be carefully adjusted for each individual device in a given operating condition due to small variations in the length and the refractive indices of the device.

The LiNbO$_3$ longitudinal modulator shown in Fig. 11.2(b) cannot function as a polarization modulator because $n_X = n_Y$. Instead, we consider the GaAs longitudinal modulator shown in Fig. 11.3(b). In this case, the principal axes and their corresponding indices of refraction in the presence of a modulation field are those given in (11.57) and (11.58), respectively. The optical wave to be modulated propagates in the Z direction and has both X and Y field components. At the input end, it can be expressed as

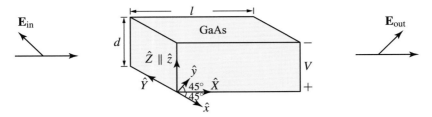

Figure 11.4 GaAs transverse electro-optic polarization modulator. The $\hat{x}$, $\hat{y}$, and $\hat{z}$ unit vectors represent the original principal axes of the crystal, and $\hat{X}$, $\hat{Y}$, and $\hat{Z}$ represent the new principal axes. The voltage is applied in the z direction, which rotates the principal axes $\hat{x}$ and $\hat{y}$ by $45°$ to $\hat{X}$ and $\hat{Y}$. This axis rotation is independent of the magnitude of the applied voltage. The input optical field contains both Y and Z components.

$$\mathbf{E}(0, t) = (\hat{X}\mathcal{E}_X + \hat{Y}\mathcal{E}_Y)e^{-i\omega t}, \tag{11.75}$$

where $\mathcal{E}_X \neq 0$ and $\mathcal{E}_Y \neq 0$, as is shown in Fig. 11.3(b). At the output, the optical field is

$$\mathbf{E}(l, t) = (\hat{X}\mathcal{E}_X e^{ik^X l} + \hat{Y}\mathcal{E}_Y e^{ik^Y l})e^{-i\omega t} = (\hat{X}\mathcal{E}_X + \hat{Y}\mathcal{E}_Y e^{i\Delta\varphi})e^{ik^X l - i\omega t}, \tag{11.76}$$

where

$$\Delta\varphi = (k^Y - k^X)l \tag{11.77}$$

is the phase retardation of the X component with respect to the Y component. By using (11.58) and the fact that $V = E_{0z}l$ for the longitudinal modulator, we have

$$\Delta\varphi = \frac{2\pi}{\lambda} n_o^3 r_{41} V = \frac{V}{V_\pi}\pi, \tag{11.78}$$

where the half-wave voltage of the GaAs longitudinal modulator is

$$V_\pi = \frac{\lambda}{2n_o^3 r_{41}}. \tag{11.79}$$

It can be seen from (11.78) that no bias voltage is needed for this GaAs modulator because both the x and y axes are ordinary axes in the absence of an applied field. However, because of the longitudinal modulation scheme, V_π is a constant that is independent of both dimensions l and d. Therefore, in comparison with the LiNbO$_3$ transverse modulator, the advantage of this modulator in requiring no bias voltage is offset by the disadvantage due to its longitudinal modulation scheme. In a GaAs transverse polarization modulator, as shown in Fig. 11.4, both problems can be eliminated.

EXAMPLE 11.3

For $\lambda = 1$ μm, the parameters of GaAs given in Table 11.2 yield $V_\pi \approx 9.72$ kV, from (11.79), for the GaAs longitudinal modulator shown in Fig. 11.3(b). This half-wave voltage is independent of the dimensions of the modulator because of longitudinal modulation. Though no bias voltage is needed because GaAs has no intrinsic birefringence, this half-wave voltage cannot be reduced by varying the dimensions of the modulator. For $\Delta\varphi$ to vary in the range between 0 and π, the modulation voltage has to be varied between 0 and 9.72 kV. The required half-wave voltage can be reduced by using the transverse modulation scheme shown in Fig. 11.4 while choosing a dimensional ratio of $d/l < 1$.

11.3.3 Amplitude Modulators

An *amplitude modulator* is also commonly called an *intensity modulator* because the intensity of an optical wave is modulated when the field amplitude is modulated. An electro-optic amplitude modulator can be constructed by simply placing a polarization modulator between a polarizer at the input end and another, often referred to as an *analyzer*, at the output end. Usually, the axis of the polarizer and that of the analyzer are arranged to be orthogonally crossed, although other arrangements are possible.

Figure 11.5 shows the typical setup of a GaAs longitudinal amplitude modulator. In this arrangement, the polarizer ensures that the input optical wave is linearly polarized in the y direction while the analyzer passes only the x component of the optical wave at the output end. The input field is $\mathbf{E}(0,t) = \hat{y}\mathcal{E}\mathrm{e}^{-\mathrm{i}\omega t}$, which can be written in the form of (11.75) with $\mathcal{E}_X = \mathcal{E}_Y = \mathcal{E}/\sqrt{2}$. Then, from (11.76), the field at the output end of the crystal is

$$\mathbf{E}(l,t) = \frac{\mathcal{E}}{\sqrt{2}}(\hat{X} + \hat{Y}\mathrm{e}^{\mathrm{i}\Delta\varphi})\mathrm{e}^{\mathrm{i}k^X l - \mathrm{i}\omega t} = \frac{\mathcal{E}}{2}\left[\hat{x}(1 - \mathrm{e}^{\mathrm{i}\Delta\varphi}) + \hat{y}(1 + \mathrm{e}^{\mathrm{i}\Delta\varphi})\right]\mathrm{e}^{\mathrm{i}k^X l - \mathrm{i}\omega t}, \quad (11.80)$$

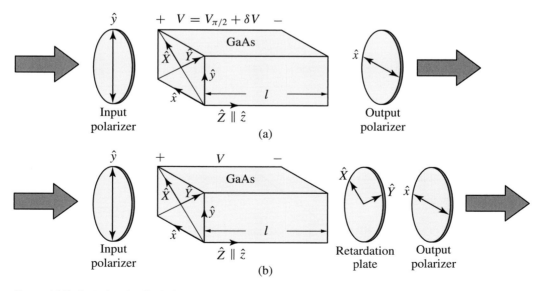

Figure 11.5 GaAs longitudinal electro-optic amplitude modulator. A bias phase retardation of $\pi/2$ can be introduced with (a) a bias voltage of $V_b = V_{\pi/2}$ or (b) a properly oriented quarter-wave plate. The $\hat{x}, \hat{y}$, and $\hat{z}$ unit vectors represent the original principal axes of the crystal, and $\hat{X}, \hat{Y}$, and $\hat{Z}$ represent the new principal axes in the presence of the modulation field.

where $\Delta\varphi$ is that given in (11.77) and (11.57) is used to transform the $\hat{X}$ and $\hat{Y}$ axes to the $\hat{x}$ and $\hat{y}$ axes. Because the analyzer passes only the x component of the optical field, the transmittance of the amplitude modulator is

$$T = \frac{I_{\text{out}}}{I_{\text{in}}} = \sin^2 \frac{\Delta\varphi}{2} = \frac{1}{2}(1 - \cos\Delta\varphi). \tag{11.81}$$

A similar result is obtained for an amplitude modulator constructed by placing any other polarization modulator, such as the LiNbO$_3$ transverse polarization modulator shown in Fig. 11.3(a), between a pair of properly oriented polarizer and analyzer.

Because $\Delta\varphi$ varies linearly with the applied voltage V, the amplitude modulator would have a linear response if its transmittance T varied linearly with $\Delta\varphi$. It can be seen from (11.81) that this is generally not true. However, for small variations of $\Delta\varphi$, it is approximately true near the point at $\Delta\varphi = \pi/2$, as can be seen by substituting

$$\Delta\varphi = \frac{\pi}{2} + \delta\varphi \tag{11.82}$$

in (11.81) to get

$$T = \frac{1}{2}(1 + \sin\delta\varphi) \approx \frac{1}{2}(1 + \delta\varphi) \tag{11.83}$$

for $|\delta\varphi| \ll 1$. By setting the operating point at a bias phase retardation of $\Delta\varphi_b = \pi/2$, the device has a linear small-signal response, as shown in Fig. 11.6. Then, with a sinusoidal modulation voltage such as that given in (11.61), the output intensity is sinusoidally modulated. This bias phase retardation can be implemented either by operating the device with a fixed bias voltage of $V_b = V_{\pi/2}$, as shown in Fig. 11.5(a), or by inserting a properly oriented quarter-wave plate

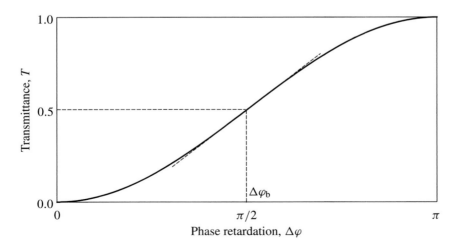

Figure 11.6 Transmission characteristics of the electro-optic amplitude modulator shown in Fig. 11.5. The response is nearly linear for small variations near the operating point at $\Delta\varphi_b = \pi/2$.

between the modulator crystal and the analyzer to introduce an extra fixed phase retardation of $\pi/2$, as shown in Fig. 11.5(b).

For large-signal applications, the response of the amplitude modulator is nonlinear. The device is then often used as an electro-optically controlled on–off modulator. In this type of application, a bias is neither useful nor necessary.

11.4 GUIDED-WAVE ELECTRO-OPTIC MODULATORS

Optical waveguides possess many unique characteristics that do not exist in bulk optics. An important one is their ability to guide optical waves within a small cross-sectional area over a long distance. This feature allows for the possibility of using the transverse modulation scheme to realize very efficient modulators at very low modulation voltages. In bulk optics, the ratio of the length to the cross-sectional dimensions is limited by the diffraction effect, thus limiting the advantage that can be realized by using transverse modulation. This limitation does not exist in waveguide optics. Another unique characteristic is the existence of waveguide modes. This results in many phenomena that have no counterpart in bulk optics, such as mode coupling, mode conversion, and modal dispersion. These unique features are the basis of many devices that take advantage of the waveguide configuration. In addition, guided-wave electro-optic devices are important building-block components of integrated optical and integrated optoelectronic systems.

The modulation electric field in a waveguide is usually the fringe field around surface electrodes or, in some cases of semiconductor waveguides, the field resulting from a junction voltage drop. Figure 11.7 shows two commonly used approaches for buried waveguides using surface-loading electrodes. In the configuration shown in Fig. 11.7(a), the electrodes are placed on the two sides of the waveguide, and the horizontal electric field $E_{0\parallel}$ is applied. In the configuration shown in Fig. 11.7(b), one of the electrodes is placed directly over the waveguide, and the applied electric field is the vertical electric field $E_{0\perp}$. The buried waveguide shown in Fig. 11.7 is a channel waveguide, but the index step at the air–crystal interface in the vertical direction is much higher than those at the other waveguide boundaries. Therefore, modes with electric fields polarized mainly parallel to the air–crystal interface are called the *TE-like modes*, whereas those with electric fields polarized mainly perpendicular to this interface are called the *TM-like modes*. When an electrode is placed directly over a waveguide, an insulating buffer layer, such as SiO_2 or Al_2O_3, between the electrode and the substrate crystal is needed to ensure low loss for the TM-like modes, as also shown in Fig. 11.7(b).

In a waveguide, the modulation electric field that is sensed by a particular waveguide mode depends on a number of parameters, including the geometric dimensions of the waveguide structure and the optical field distribution of the mode. In general, the modulation field is not uniformly distributed across the mode field distribution. The effect of electro-optic modulation in a waveguide can be calculated by using the coupled-mode theory discussed in Subsection 7.1.1. For modulation on a single mode, the effect is to introduce a change in the propagation constant of the mode. This change is equal to the self-coupling coefficient of the mode given by the relation:

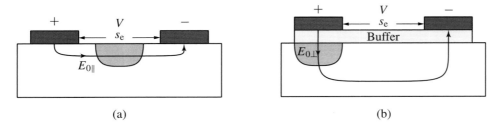

Figure 11.7 Configurations for applying a modulation field to a buried waveguide through surface-loading electrodes. In (a), $E_{0\parallel}$ is applied to the waveguide. In (b), $E_{0\perp}$ is applied. The buffer layer in (b) is required to reduce loss to the TM-like modes.

$$\Delta\beta_v = \kappa_{vv} = \omega \int_{-\infty}^{\infty} \hat{\boldsymbol{\mathcal{E}}}_v^* \cdot \Delta\boldsymbol{\epsilon} \cdot \hat{\boldsymbol{\mathcal{E}}}_v \mathrm{d}\boldsymbol{\rho}, \tag{11.84}$$

where $\Delta\boldsymbol{\epsilon}(\omega, \boldsymbol{E}_0)$ is the electro-optically induced change in the dielectric permittivity tensor and $\boldsymbol{\rho}$ is the two-dimensional vector in the cross-sectional plane of the waveguide. Note that in analyzing a guided-wave electro-optic modulator, we generally work with the intrinsic principal axes $\hat{x}$, $\hat{y}$, and $\hat{z}$ without finding the rotated principal axes. Therefore, $\Delta\boldsymbol{\epsilon}(\omega, \boldsymbol{E}_0)$ is expressed in terms of x, y, and z. One reason for using the intrinsic principal axes is that the geometry of a waveguide defines a coordinate system, which does not change with the applied field. Another reason is that the electric field that is sensed by the waveguide mode might not be in a fixed direction throughout the waveguide structure.

As an example, we consider phase modulation of a TE-like mode in a waveguide modulator that is fabricated in a LiNbO$_3$ crystal with the crystal surface perpendicular to the x principal axis and the longitudinal direction of the waveguide parallel to the y principal axis. This arrangement is shown in Fig. 11.8(a) and is referred to as *y propagating in an x-cut crystal*. The modulation field that appears in the waveguide area is $E_{0\parallel}$, which is E_{0z} in this configuration. Because a TE-like mode of this waveguide is predominantly polarized in the z direction and $\Delta\epsilon_{zz} = -\epsilon_0 n_e^4 r_{33} E_{0z}$ from (11.43), we have

$$\begin{aligned}
\Delta\beta_{\text{TE}} &= -\omega\epsilon_0 n_e^4 r_{33} \int_{-\infty}^{\infty}\int_{-\infty}^{\infty} E_{0z}(x,z)\left|\hat{\boldsymbol{\mathcal{E}}}_{\text{TE}}(x,z)\right|^2 \mathrm{d}x\mathrm{d}z \\
&= -\frac{\omega^2\mu_0\epsilon_0}{2\beta_{\text{TE}}} n_e^4 r_{33} \frac{V}{s_e}\Gamma_{\text{TE}} \\
&\approx -\frac{\pi}{\lambda} n_e^3 r_{33} \frac{V}{s_e}\Gamma_{\text{TE}},
\end{aligned} \tag{11.85}$$

where V is the applied voltage, s_e is the separation between the electrodes, β_{TE} is approximated by $n_e\omega/c$, (7.10) is used to normalize the mode field, and

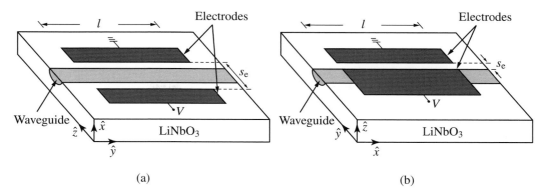

Figure 11.8 Waveguide phase modulators in (a) an x-cut, y-propagating LiNbO$_3$ crystal and (b) a z-cut, x-propagating LiNbO$_3$ crystal. The $\hat{x}$, $\hat{y}$, and $\hat{z}$ unit vectors represent the intrinsic principal axes of the crystal.

$$\Gamma_{\text{TE}} = \frac{s_e}{V} \frac{2\beta_{\text{TE}}}{\omega\mu_0} \int\limits_{-\infty}^{\infty} \int\limits_{-\infty}^{\infty} E_{0z}(x,z)\left|\hat{\boldsymbol{\mathcal{E}}}_{\text{TE}}(x,z)\right|^2 dxdz = \frac{s_e}{V} \frac{\displaystyle\int\limits_{-\infty}^{\infty}\int\limits_{-\infty}^{\infty} E_{0z}(x,z)\left|\hat{\boldsymbol{\mathcal{E}}}_{\text{TE}}(x,z)\right|^2 dxdz}{\displaystyle\int\limits_{-\infty}^{\infty}\int\limits_{-\infty}^{\infty}\left|\hat{\boldsymbol{\mathcal{E}}}_{\text{TE}}(x,z)\right|^2 dxdz} \qquad (11.86)$$

is the *overlap factor*, which accounts for the spatial overlap between the modulation electric field and the modulated optical mode. The overlap factor has a value between 0 and 1. The total electro-optically induced phase shift of this TE-like mode over the length l of the modulator is simply $\Delta\varphi = \Delta\beta_{\text{TE}}l$.

By comparing the result obtained above for the guided-wave phase modulator with that in (11.63) for the bulk phase modulator, we find that the net effect for a waveguide mode v can be approximated by using a single uniform effective modulation electric field given as

$$E_{\text{eff}} = \frac{V}{s_e}\Gamma_v, \qquad (11.87)$$

where the overlap factor Γ_v is evaluated by using the appropriate modulation field component for the device configuration under consideration. For example, $E_{0\parallel}$ is used for the electrode configuration in Fig. 11.7(a), whereas $E_{0\perp}$ is used for the configuration in Fig. 11.7(b). The value of Γ_v depends on the electrode configuration and is different for different waveguide modes in the same structure. For a given configuration and a given waveguide mode, it increases monotonically as the ratio of the electrode separation to the horizontal waveguide width increases.

EXAMPLE 11.4

An x-cut, y-propagating LiNbO$_3$ single-mode waveguide phase modulator for $\lambda = 1.3$ μm, as shown in Fig. 11.8(a), has a gap separation of $s_e = 20$ μm between its electrodes and an overlap factor of $\Gamma_{TE} = 0.57$ for its TE-like mode. It is modulated with an applied voltage of $V = 12$ V. At $\lambda = 1.3$ μm, $n_e = 2.145$ for LiNbO$_3$. What is the effective modulation electric field strength? Find the electro-optically induced change in the propagation constant of the TE-like mode. If an electro-optically controlled phase shift of π is desired, what is the required length of the device?

Solution
According to (11.87), the effective modulation electric field is

$$E_{eff} = \frac{V}{s_e}\Gamma_{TE} = \frac{12}{20 \times 10^{-6}} \times 0.57 \, \text{V m}^{-1} = 342 \, \text{kV m}^{-1}.$$

We then find, by using (11.85), that

$$\Delta\beta_{TE} = -\frac{\pi}{1.3 \times 10^{-6}} \times 2.145^3 \times 30.8 \times 10^{-12} \times \frac{12}{20 \times 10^{-6}} \times 0.57 \, \text{m}^{-1} = -251.23 \, \text{m}^{-1}.$$

For $\Delta\varphi = \pi$, the required length of the device is

$$l = \frac{\pi}{|\Delta\beta_{TE}|} = 12.5 \, \text{mm}.$$

As mentioned in the beginning of Subsection 11.3.1, phase modulation can be translated into all forms of modulation. This is also generally true for guided-wave modulators. Therefore, the general concept of the guided-wave electro-optic phase modulator discussed above can be utilized to design and implement various guided-wave electro-optic modulators, including those that have the bulk counterparts, such as polarization modulators and amplitude modulators, and those that have no bulk counterparts, such as mode converters and directional coupler switches. In the following, we show the forms of some well-established structures without going into the details of their characteristics.

11.4.1 Mach–Zehnder Waveguide Interferometers

An optical interferometer can convert the phase difference between two wave components into a change in the amplitude of the wave that combines the two components. Guided-wave electro-optic phase modulators can be used to construct waveguide interferometers for effective amplitude modulation of guided optical waves. A *Mach–Zehnder waveguide interferometer* consists of two parallel waveguides that are connected at the input and the output ends, respectively, by beam-splitting and beam-combining optical couplers. These couplers can be *Y-junction waveguides*, as in the devices shown in Fig. 11.9, or *directional couplers*, as shown in Fig. 11.10.

The Mach–Zehnder waveguide interferometer shown in Fig. 11.9(a) uses Y-junction couplers and is fabricated in an x-cut, y-propagating LiNbO$_3$ crystal. In this configuration, the electrodes are placed on the two sides of each waveguide to utilize r_{33}, which is the largest

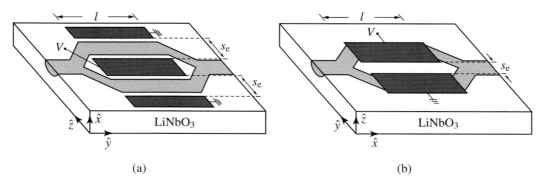

Figure 11.9 Mach–Zehnder waveguide interferometric modulator using Y junctions fabricated on (a) an x-cut, y-propagating LiNbO$_3$ substrate and (b) a z-cut, x-propagating LiNbO$_3$ substrate. The $\hat{x}$, $\hat{y}$, and $\hat{z}$ unit vectors represent the intrinsic principal axes of the crystal.

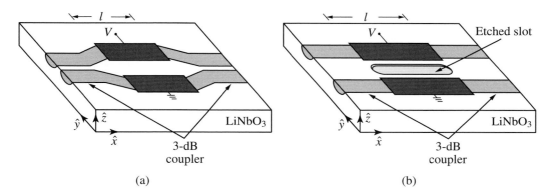

Figure 11.10 Balanced-bridge interferometers fabricated on z-cut, x-propagating LiNbO$_3$ substrates using (a) bent waveguides and (b) straight waveguides. Isolation between the two arms of the interferometer is accomplished with a large separation in (a) and with an etched slot in (b). The $\hat{x}$, $\hat{y}$, and $\hat{z}$ unit vectors represent the intrinsic principal axes of the crystal.

Pockels coefficient for LiNbO$_3$. By comparison, Fig. 11.9(b) shows a Mach–Zehnder interferometer that is fabricated on a z-cut, x-propagating LiNbO$_3$ substrate. In this configuration, the electrodes have to be placed directly over the waveguides in order to use r_{33}. Instead of Y junctions, *3-dB directional couplers* can be used at both the input and the output ends of a Mach–Zehnder waveguide interferometer. This type of Mach–Zehnder interferometer is called the *balanced-bridge interferometer*. Figure 11.10 shows two examples. It is also possible to use different types of couplers at the input and the output ends. For example, a Y-junction coupler is used at the input end while a directional coupler is used at the output end, or vice versa.

For each of the Mach–Zehnder waveguide interferometers shown in Figs. 11.9 and 11.10, the modulation electric fields are applied to two parallel waveguides, which form the two arms of the interferometer and are sufficiently separated to avoid direct coupling between them. Each waveguide by itself functions as an electro-optic phase modulator. The modulation electric fields that appear in the two arms point in opposite directions, resulting in a *push–pull operation* with equal but opposite phase shifts in the optical waves that propagate through the two arms. Constructive or destructive interference occurs at the output coupler if the phase difference between the two arms is, respectively, an even or odd multiple of π. By electro-optically controlling this phase difference through the applied voltage, the amplitude of the guided optical field at the output can be modulated.

11.4.2 Directional Coupler Switches

Two closely coupled, parallel waveguides form a *directional coupler*. A very important practical application of a directional coupler is to use it as an optical switch, which can be switched between cross and parallel states. In the *cross state*, an optical wave that is coupled to one waveguide exits the coupler from the other waveguide for a *coupling efficiency* of $\eta = 1$. In the *parallel state*, an optical wave that is coupled to one waveguide exits the coupler from the same waveguide with a coupling efficiency of $\eta = 0$. For a fabricated device with fixed geometric parameters, this switching can be accomplished by varying the phase mismatch or the coupling coefficient between the two waveguides through electro-optically induced changes in the refractive index of the waveguide material. Figure 11.11(a) shows the basic structure of an electro-optic directional coupler switch. This structure has the *uniform-Δβ configuration*, which has a two-electrode configuration similar to that of the two-electrode Mach–Zehnder interferometers. The only difference is that the two waveguides in a directional coupler are coupled, whereas those in an interferometer are isolated. Figure 11.11(b) shows an electro-optic directional coupler switch in the *reversed-Δβ configuration*. In this configuration, voltages of equal magnitude but opposite polarities are applied to the split electrodes. Configurations that are more sophisticated are possible.

11.4.3 Polarization-Mode Converters

The operation of a waveguide polarization modulator depends on the coupling between modes of different polarizations. As an example, we consider the coupling between fundamental TE-like and

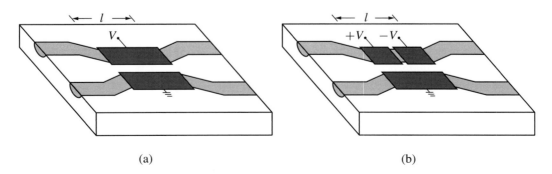

(a)　　　　　　　　　　　　　　　　(b)

Figure 11.11 (a) Schematic structures of (a) an electro-optic uniform-Δβ directional coupler switch and (b) an electro-optic reversed-Δβ directional coupler switch.

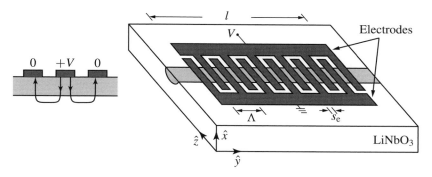

Figure 11.12 Waveguide polarization modulator fabricated on an x-cut, y-propagating LiNbO$_3$ substrate using a periodic electrode for phase matching between fundamental TE-like and TM-like modes. The $\hat{x}$, $\hat{y}$, and $\hat{z}$ unit vectors represent the intrinsic principal axes of the crystal.

TM-like modes in an x-cut, y-propagating LiNbO$_3$ waveguide shown in Fig. 11.12. With an input optical wave that is coupled into one of the two polarization modes, say the TE-like mode, the polarization of the output wave can be modulated by varying the coupling efficiency η from the TE-like mode to the TM-like mode. This coupling efficiency is determined by two parameters: (a) the coupling coefficients, $\kappa_{\text{EM}} = \kappa_{\text{ME}}^*$, and (b) the phase mismatch, $\Delta\beta = \beta_{\text{TM}} - \beta_{\text{TE}}$, between the two polarization modes. In the structure shown in Fig. 11.12, the phase mismatch is compensated by properly choosing the period Λ of the periodic *interdigital electrodes* to be $\Lambda = 2\pi q / |\Delta\beta|$, which is similar to (9.50) for quasi-phase matching discussed in Section 9.3. The cross-coupling coefficient κ_{EM} varies linearly with the applied voltage in a manner similar to the self-coupling coefficient given in (11.85). Then, under the condition of perfect phase matching by properly choosing the period Λ, a coupling efficiency of $\eta_{\text{PM}} = 1$ can be reached at a mode-conversion voltage V_{EM} that makes the coupling coefficient satisfy the condition that $2|\kappa_{\text{EM}}|l = \pi$ for a given device length of l. By switching the applied voltage off and on between 0 and V_{EM}, the output polarization can be switched between TE and TM.

11.4.4 Traveling-Wave Modulators

The Pockels effect has a very high response speed, thus allowing high-frequency modulation. In many applications, the modulation frequency is in the microwave or millimeter-wave range to take advantage of the high-bandwidth capacity of the optical carrier wave. The modulation efficiency of a lumped modulator drops drastically at high modulation frequencies because of its RC-limited frequency response. This problem can be overcome by using a traveling-wave configuration for the modulator and by matching the phase velocity of the microwave modulation field to that of the optical wave in the traveling-wave modulator. The electrodes of a traveling-wave modulator are made of strip transmission lines, as shown in Fig. 11.13. The electrodes are specifically designed for traveling-wave interactions. The high-frequency modulation signal is injected at one end, propagates along the same direction as the optical wave, and terminates at the other end of the electrode transmission lines. The traveling-wave configuration inherently requires the use of the transverse modulation scheme. This scheme is consistent with

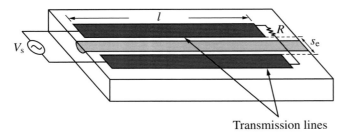

Figure 11.13 Traveling-wave electro-optic phase modulator.

the configurations of most guided-wave devices. Therefore, traveling-wave modulation can be applied to a large variety of guided-wave devices, including the Mach–Zehnder waveguide interferometers and the directional coupler switches that were respectively discussed in Subsections 11.4.1 and 11.4.2.

Problem Set

11.2.1 The only nonvanishing Pockels coefficients of a uniaxial crystal that has the symmetry of the 422 or 622 point group are $r_{52} = -r_{41}$. The uniaxial optical axis is taken to be the z axis. The ordinary index of refraction is n_o, and the extraordinary index is n_e. A general electrostatic field, $\boldsymbol{E}_0 = E_{0x}\hat{x} + E_{0y}\hat{y} + E_{0z}\hat{z}$, is applied to the crystal.

(a) Write down the index ellipsoid equation and the dielectric permittivity tensor in terms of the components of the applied electrostatic field.

(b) Find the new principal indices of refraction as functions of the components of the applied electrostatic field. Under what condition does the electrostatic field make the crystal biaxial? Under what condition does the crystal remain uniaxial?

(c) Find the new principal axes, corresponding to the new principal indices of refraction, as functions of the components of the applied electrostatic field.

11.2.2 KTP is a biaxial crystal of the $mm2$ point group, which has five nonvanishing Pockels coefficients: $r_{13} = 8.8\,\text{pm V}^{-1}$, $r_{23} = 13.8\,\text{pm V}^{-1}$, $r_{33} = 35\,\text{pm V}^{-1}$, $r_{42} = 8.8\,\text{pm V}^{-1}$, and $r_{51} = 6.9\,\text{pm V}^{-1}$. An electrostatic field, $\boldsymbol{E}_0 = E_0\hat{z}$, is applied along its z principal axis.

(a) Write down the index ellipsoid equation and the dielectric permittivity tensor as functions of the applied electrostatic field.

(b) Find the principal indices and their corresponding principal axes as functions of the applied electrostatic field.

(c) At $\lambda = 1\,\text{μm}$ wavelength, the principal indices of refraction of KTP are $n_x = 1.742$, $n_y = 1.750$, and $n_z = 1.832$. Find the changes in the principal indices and in the birefringence of the crystal caused by an applied electrostatic field of $E_0 = 1\,\text{MV m}^{-1}$.

11.2.3 Answer the questions in Problem 11.2.2 for the KTP crystal in the case that the electrostatic field $E_0 = E_0\hat{y}$ is applied along the y principal axis.

11.2.4 Answer the questions in Problem 11.2.2 for the KTP crystal in the case that the electrostatic field $E_0 = E_0\hat{x}$ is applied along the x principal axis.

11.2.5 A $\bar{4}3m$ crystal, such as GaAs or InP, is nonbirefringent so that $n_x = n_y = n_z = n_o$. It can become birefringent due to the Pockels effect when an electrostatic field is applied. The only nonvanishing Pockels coefficients are $r_{41} = r_{52} = r_{63}$.

 (a) An electrostatic field of $E_0 = E_0(\hat{x}\cos\phi + \hat{y}\sin\phi)$ is applied to the crystal in the xy plane, where ϕ is the angle between the direction of the electrostatic field and the x principal axis. Find the new principal axes and their corresponding new principal refractive indices.

 (b) A linearly polarized optical beam propagates through the crystal along the z principal axis. With the electrostatic field applied to the crystal at a given angle ϕ, how should the polarization of a linearly polarized optical beam be arranged so that it remains linearly polarized along its entire path through the crystal?

 (c) If the linearly polarized optical beam propagates along the y principal axis, how should the polarization directions of the electrostatic and optical fields be arranged so that the beam remains linearly polarized along its entire path through the crystal?

11.2.6 A hexagonal crystal of the 6 point group, such as $LiIO_3$, is a uniaxial crystal that has $n_x = n_y = n_o$ and $n_z = n_e$. Its only nonvanishing Pockels coefficients are $r_{13} = r_{23}$, r_{33}, $r_{41} = -r_{52}$, and $r_{42} = r_{51}$. A general electrostatic field, $E_0 = E_{0x}\hat{x} + E_{0y}\hat{y} + E_{0z}\hat{z}$, is applied to the crystal. Write down the index ellipsoid equation and the dielectric permittivity tensor in terms of the components of the applied electrostatic field. Is it possible to induce changes in the principal indices without changing the directions of the principal axes by properly choosing the applied field?

11.3.1 A KTP transverse phase modulator of the configuration shown in Fig. 11.2(a) and a KTP longitudinal phase modulator of the configuration shown in Fig. 11.2(b) for the $\lambda = 1$ μm wavelength are both modulated with the modulation voltage applied along its principal z axis. Use the results obtained in Problem 11.2.2(b) to answer each of the following questions for optical waves polarized in the x, y, and z directions, respectively, under different possible arrangements.

 (a) Find the phase modulation depth φ_{pk} as a function of the parameters of KTP, the dimensions of the modulator, and the peak modulation voltage V_{pk}. Find also the half-wave voltage V_π that is required for a phase modulation depth of $\varphi_{pk} = \pi$.

 (b) Use the parameters of KTP given in Problem 11.2.2 to find the half-wave voltage for transverse and longitudinal bulk modulators that have the dimensions of $d = 3$ mm and $l = 9$ mm. Compare the half-wave voltages for the two modulators.

 (c) Find the half-wave voltage for transverse and longitudinal waveguide modulators that have the dimensions of $d = 3$ μm and $l = 9$ mm.

11.3.2 A III–V semiconductor is an electro-optic crystal that has the $\bar{4}3m$ symmetry and a nonbirefringent index of refraction of n_o. A DC electric field E_0 is applied to the crystal in the [001] crystallographic direction. A light beam at a wavelength of λ travels in the [110] direction. Its polarization makes an angle of 45° with respect to the [001] axis.

(a) Show that the plane of polarization is rotated by 90° after the light has traveled a distance of

$$l = \frac{\lambda}{n_o^3 r_{41} E_0}.$$

(b) What is changed if the DC electric field is turned by 90° but is still transverse to the direction of wave propagation – that is, if the DC field is now parallel to the [1$\bar{1}$0] direction?

11.3.3 KDP is a uniaxial crystal of the $\bar{4}2m$ point group. Its nonvanishing Pockels coefficients are $r_{41} = r_{52} = 8.8$ pm V^{-1} and $r_{63} = 10.5$ pm V^{-1}. Its optical axis is the z axis. At $\lambda = 1$ μm, its refractive indices are $n_x = n_y = n_o = 1.51$ and $n_z = n_e = 1.47$.

(a) In an application of KDP for electro-optic phase modulation, describe how you would arrange the principal axes of the crystal with respect to the polarization directions of the electrostatic field and the optical field in the longitudinal and transverse modulation schemes, respectively. Show the arrangements with sketches.

(b) Find the half-wave voltage for an optical beam at $\lambda = 1$ μm wavelength in the longitudinal modulation scheme.

(c) What arrangement is necessary so that the half-wave voltage in the transverse modulation scheme is half that of the longitudinal modulation scheme found in (b)?

References

[1] J. Kerr, "XL. A new relation between electricity and light: dielectrified media birefringent," *The London, Edinburgh, and Dublin Philosophical Magazine and Journal of Science*, vol. 50, pp. 337–348, 1875.

[2] J. Kerr, "LIV. A new relation between electricity and light: dielectrified media birefringent (second paper)," *The London, Edinburgh, and Dublin Philosophical Magazine and Journal of Science*, vol. 50, pp. 446–458, 1875.

12 All-Optical Modulation

12.1 OPTICAL KERR EFFECT

All-optical modulation of an optical wave is accomplished through a nonlinear optical process that involves one or multiple optical waves. A nonlinear optical modulator can be based on either *self-modulation* or *cross-modulation*. In the case of self-modulation, only one optical beam is present, and the modulation on the beam is a function of the characteristics of the beam itself. In the case of cross-modulation, two or more optical beams are present, and one or more other beams that carry the modulation signals modulate the beam of interest. In either case, the modulation field is an optical field; no electrostatic, magnetostatic, or acoustic field is needed. Therefore, such nonlinear optical modulators and switches are also known as *all-optical modulators* and *all-optical switches*, respectively.

Most of all-optical modulators and switches are based on third-order nonlinear optical processes, but some rely on the high-order process of optical saturation, either absorption saturation or gain saturation. There are two fundamentally different types of all-optical modulators and switches. One is the *dispersive*, or *refractive*, type, which is based on the optical Kerr effect due to optical-field-induced changes in the real part of the permittivity of a material. Another is the *absorptive* type, which relies on an intensity-dependent absorption coefficient associated with the nonlinear characteristics of the imaginary part of the permittivity of a material. In this chapter, we only discuss all-optical modulation of the dispersive type based on the optical Kerr effect. All-optical modulation of the absorptive type can be based on multiphoton absorption, which will be discussed in Chapter 14, or optical saturation, which will be discussed in Chapter 15.

Kerr did not discover the optical Kerr effect because at his time there was no light that was sufficiently intense to make the optical Kerr effect observable. It is called the optical Kerr effect because it is similar to the electro-optic Kerr effect in that it is a third-order nonlinear optical process. The difference between the two is only that the electro-optic Kerr effect is caused by an electrostatic field, whereas the optical Kerr effect is caused by an optical field. Therefore, the optical Kerr effect is also known as the *AC Kerr effect*, whereas the electro-optic Kerr effect is also known as the *DC Kerr effect*. They share many common characteristics. One important fact is that they both exist in any material, including a centrosymmetric material. Being third-order effects, they are relatively weak and are usually overshadowed by second-order nonlinear optical processes in a noncentrosymmetric crystal. For these reasons and for simplifying the mathematics, we shall consider only isotropic materials in the discussions throughout this chapter. With proper modifications, most of the concepts and results also apply to cubic crystals, such as Si and GaAs.

The basic concept of the optical Kerr effect has been briefly discussed in Subsection 5.3.4, and the phenomena of self-phase modulation and cross-phase modulation have been discussed in Subsections 5.3.5 and 5.3.6. These processes have simple concepts, but they have broad applications. In this chapter, we discuss the applications of these processes.

12.1.1 Self-Phase Modulation

In Subsection 5.3.5, only self-phase modulation of a linearly polarized optical wave was considered. In fact, *the consequence of self-phase modulation depends on the polarization state of the input optical wave*. Here we examine the different features of self-phase modulation for different input polarization states. The outcome of self-phase modulation also depends on the symmetry property of the medium. For simplicity, we consider only isotropic materials such that the linear refractive index is n_0 for all polarization states. The only nonvanishing elements of the $\chi^{(3)}$ tensor are those of the types of $\chi_{1111}^{(3)}, \chi_{1122}^{(3)}, \chi_{1212}^{(3)}$, and $\chi_{1221}^{(3)}$, which have the relation that $\chi_{1111}^{(3)} = \chi_{1122}^{(3)} + \chi_{1212}^{(3)} + \chi_{1221}^{(3)}$. If Kleinman's symmetry condition is valid, then $\chi_{1122}^{(3)} = \chi_{1212}^{(3)} = \chi_{1221}^{(3)} = \chi_{1111}^{(3)}/3$. Without loss of generality, we assume that the input optical wave propagates in the z direction and is polarized in the xy plane.

Linearly Polarized Wave
For a linearly polarized optical wave, $\mathbf{E}(\omega) = \hat{e}E(\omega)$ with $\hat{e}$ being a real unit vector. For the propagation of this wave in an isotropic medium, we can take $\hat{e} = \hat{x}$ or $\hat{e} = \hat{y}$ without loss of generality to find that

$$
\begin{aligned}
\mathbf{P}^{(3)}(\omega) &= 3\epsilon_0 \chi^{(3)}(\omega = \omega + \omega - \omega) \mathbin{\vdots} \mathbf{E}(\omega)\mathbf{E}(\omega)\mathbf{E}^*(\omega) \\
&= 3\epsilon_0 \chi^{(3)} \mathbin{\vdots} \hat{e}\hat{e}\hat{e}E|E|^2 \\
&= \hat{e}3\epsilon_0 \chi_{1111}^{(3)}|E|^2 E.
\end{aligned}
\tag{12.1}
$$

Then, by following through the procedure from (5.37) to (5.39), we obtain

$$
n = n_0 + n_2 I,
\tag{12.2}
$$

as given in (5.38), with n_2 given in (5.39) for a linearly polarized optical field:

$$
n_2 = \frac{3\chi_{1111}^{(3)\prime}}{4c\epsilon_0 n_0^2}.
\tag{12.3}
$$

According to (5.42), the Kerr phase change caused by self-phase modulation of a linearly polarized optical wave that propagates in an isotropic medium over a distance of l is

$$
\varphi_K(\mathbf{r}, t) = \frac{\omega}{c} n_2 l I(\mathbf{r}, t) = \frac{3\omega \chi_{1111}^{(3)\prime}}{4c^2 \epsilon_0 n_0^2} l I(\mathbf{r}, t) = \frac{3\pi \chi_{1111}^{(3)\prime}}{2c\epsilon_0 n_0^2 \lambda} l I(\mathbf{r}, t).
\tag{12.4}
$$

Circularly Polarized Wave

For a circularly polarized optical wave, either left-circularly polarized[1] with a unit vector of $\hat{e}_+ = (\hat{x} + i\hat{y})/\sqrt{2}$ or right-circularly polarized with a unit vector of $\hat{e}_- = (\hat{x} - i\hat{y})/\sqrt{2}$, we have $\mathbf{E}(\omega) = \hat{e}_{\pm}E(\omega)$, meaning $\mathbf{E}(\omega) = \hat{e}_+E(\omega)$ or $\mathbf{E}(\omega) = \hat{e}_-E(\omega)$. Then, we find that

$$\mathbf{P}^{(3)}(\omega) = 3\epsilon_0\boldsymbol{\chi}^{(3)}(\omega = \omega + \omega - \omega) \vdots \mathbf{E}(\omega)\mathbf{E}(\omega)\mathbf{E}^*(\omega)$$

$$= 3\epsilon_0\boldsymbol{\chi}^{(3)} \vdots \hat{e}_{\pm}\hat{e}_{\pm}\hat{e}_{\pm}^* E|E|^2 \tag{12.5}$$

$$= \hat{e}_{\pm}3\epsilon_0\left(\chi_{1122}^{(3)} + \chi_{1212}^{(3)}\right)|E|^2E.$$

By following through the procedure from (5.37) to (5.39), we obtain n_2 for a circularly polarized optical field:

$$n_2 = \frac{3\left(\chi_{1122}^{(3)\prime} + \chi_{1212}^{(3)\prime}\right)}{4c\epsilon_0 n_0^2}. \tag{12.6}$$

According to (5.42), the Kerr phase change caused by self-phase modulation of a circularly polarized optical wave that propagates in an isotropic medium over a distance of l is

$$\varphi_K(\mathbf{r}, t) = \frac{\omega}{c}n_2 lI(\mathbf{r}, t) = \frac{3\omega\left(\chi_{1122}^{(3)\prime} + \chi_{1212}^{(3)\prime}\right)}{4c^2\epsilon_0 n_0^2}lI(\mathbf{r}, t) = \frac{3\pi\left(\chi_{1122}^{(3)\prime} + \chi_{1212}^{(3)\prime}\right)}{2c\epsilon_0 n_0^2\lambda}lI(\mathbf{r}, t). \tag{12.7}$$

We see from (12.6) that n_2 is the same for left- and right-circularly polarized waves, but it is different from that for a linearly polarized wave. Therefore, the Kerr phase change given in (12.7) due to self-phase modulation is the same for both circularly polarized waves, but it is different from that for a linearly polarized wave. Because $\left|\chi_{1122}^{(3)\prime} + \chi_{1212}^{(3)\prime}\right|$ is generally smaller than $\left|\chi_{1111}^{(3)\prime}\right|$, a linearly polarized optical wave causes a larger optical Kerr effect, and thus stronger self-phase modulation, than a circularly polarized wave.

Elliptically Polarized Wave

An elliptically polarized optical wave that propagates in an isotropic medium in the z direction can be expressed as

$$\mathbf{E}(\omega) = \hat{e}_+E_+(\omega) + \hat{e}_-E_-(\omega), \tag{12.8}$$

where the amplitude of the left-circularly polarized component is $E_+(\omega)$, and that of the right-circularly polarized component is $E_-(\omega)$. In general, $|E_+| \neq |E_-|$ for an elliptically polarized wave. This elliptically polarized field can also be expressed in terms of its x and y components as

---

[1] Unfortunately, there are two opposite conventions for defining the left and right circular polarizations. In this book, we define $\hat{e}_+$ to be the left circular polarization and $\hat{e}_-$ to be the right circular polarization.

$$\mathbf{E}(\omega) = \hat{x}E_x(\omega) + \hat{y}E_y(\omega), \tag{12.9}$$

with the relations that

$$E_x = \frac{E_+ + E_-}{\sqrt{2}} \quad \text{and} \quad E_y = \mathrm{i}\frac{E_+ - E_-}{\sqrt{2}}, \tag{12.10}$$

and

$$\hat{x} = \frac{\hat{e}_+ + \hat{e}_-}{\sqrt{2}} \quad \text{and} \quad \hat{y} = -\mathrm{i}\frac{\hat{e}_+ - \hat{e}_-}{\sqrt{2}}. \tag{12.11}$$

Thus, we have

$$
\begin{aligned}
\mathbf{P}^{(3)}(\omega) &= 3\epsilon_0 \boldsymbol{\chi}^{(3)}(\omega = \omega + \omega - \omega) \vdots \mathbf{E}(\omega)\mathbf{E}(\omega)\mathbf{E}^*(\omega) \\
&= 3\epsilon_0 \boldsymbol{\chi}^{(3)} \vdots (\hat{e}_+ E_+ + \hat{e}_- E_-)(\hat{e}_+ E_+ + \hat{e}_- E_-)(\hat{e}_+ E_+ + \hat{e}_- E_-)^* \\
&= 3\epsilon_0 \boldsymbol{\chi}^{(3)} \vdots (\hat{x}E_x + \hat{y}E_y)(\hat{x}E_x + \hat{y}E_y)(\hat{x}E_x + \hat{y}E_y)^*.
\end{aligned} \tag{12.12}
$$

By carrying out the tensor multiplication for (12.12), we find that $\mathbf{P}^{(3)}(\omega)$ can be expressed as

$$\mathbf{P}^{(3)}(\omega) = \hat{e}_+ P_+ + \hat{e}_- P_-, \tag{12.13}$$

with

$$
\begin{aligned}
P_+^{(3)} &= 3\epsilon_0 \left[\left(\chi_{1122}^{(3)} + \chi_{1212}^{(3)} \right) |E_+|^2 + \left(\chi_{1122}^{(3)} + \chi_{1212}^{(3)} + 2\chi_{1221}^{(3)} \right) |E_-|^2 \right] E_+, \\
P_-^{(3)} &= 3\epsilon_0 \left[\left(\chi_{1122}^{(3)} + \chi_{1212}^{(3)} \right) |E_-|^2 + \left(\chi_{1122}^{(3)} + \chi_{1212}^{(3)} + 2\chi_{1221}^{(3)} \right) |E_+|^2 \right] E_-.
\end{aligned} \tag{12.14}
$$

The field-dependent dielectric permittivities seen by the two circularly polarized components are, respectively,

$$
\begin{aligned}
\epsilon_+(\omega, \mathbf{E}) &= \epsilon(\omega) + 3\epsilon_0 \left[\left(\chi_{1122}^{(3)} + \chi_{1212}^{(3)} \right) |E_+|^2 + \left(\chi_{1122}^{(3)} + \chi_{1212}^{(3)} + 2\chi_{1221}^{(3)} \right) |E_-|^2 \right], \\
\epsilon_-(\omega, \mathbf{E}) &= \epsilon(\omega) + 3\epsilon_0 \left[\left(\chi_{1122}^{(3)} + \chi_{1212}^{(3)} \right) |E_-|^2 + \left(\chi_{1122}^{(3)} + \chi_{1212}^{(3)} + 2\chi_{1221}^{(3)} \right) |E_+|^2 \right].
\end{aligned} \tag{12.15}
$$

The refractive indices seen by the two circularly polarized components are, respectively,

$$
\begin{aligned}
n_+ &= n_0 + \Delta n_+ = n_0 + \frac{3}{2n_0} \left[\left(\chi_{1122}^{(3)\prime} + \chi_{1212}^{(3)\prime} \right) |E_+|^2 + \left(\chi_{1122}^{(3)\prime} + \chi_{1212}^{(3)\prime} + 2\chi_{1221}^{(3)\prime} \right) |E_-|^2 \right], \\
n_- &= n_0 + \Delta n_- = n_0 + \frac{3}{2n_0} \left[\left(\chi_{1122}^{(3)\prime} + \chi_{1212}^{(3)\prime} \right) |E_-|^2 + \left(\chi_{1122}^{(3)\prime} + \chi_{1212}^{(3)\prime} + 2\chi_{1221}^{(3)\prime} \right) |E_+|^2 \right].
\end{aligned}
$$
$$\tag{12.16}$$

For $|E_+| \neq |E_-|$, the two circularly polarized components experience different field-dependent index changes, thus different phase modulation. Therefore, an elliptically polarized wave causes a *field-dependent circular birefringence*:

$$B_{\rm c} = \Delta n_{\rm c} = \Delta n_+ - \Delta n_- = -\frac{3\chi_{1221}^{(3)\prime}}{n_0}\left(|E_+|^2 - |E_-|^2\right) = -\frac{3\chi_{1221}^{(3)\prime}}{2c\epsilon_0 n_0^2}(I_+ - I_-), \qquad (12.17)$$

where $I_+ = 2c\epsilon_0 n_0 |E_+|^2$ and $I_- = 2c\epsilon_0 n_0 |E_-|^2$ are the intensities of the two circularly polarized field components.

Because of the field-dependent circular birefringence, the two circularly polarized components experience different amounts of phase modulation, with a difference of

$$\Delta\varphi_{\rm K} = (k^+ - k^-)l = \frac{\omega}{c}\Delta n_{\rm c}l = -\frac{3\omega\chi_{1221}^{(3)\prime}}{2c^2\epsilon_0 n_0^2}l(I_+ - I_-) = -\frac{3\pi\chi_{1221}^{(3)\prime}}{c\epsilon_0 n_0^2 \lambda}l(I_+ - I_-), \qquad (12.18)$$

where λ is the optical wavelength of the field and l is the distance of propagation through the medium. The circular birefringence causes the ellipse of the elliptical polarization to rotate as the wave propagates through the medium. The *ellipse rotation angle* is

$$\Delta\alpha \approx -\frac{\Delta\varphi_{\rm K}}{2}, \qquad (12.19)$$

where the angle α for an elliptically polarized field is defined as $\alpha = \tan^{-1}|\mathcal{E}_y|/|\mathcal{E}_x|$. We find from (12.18) that the ellipse rotation angle due to the circular birefringence caused by self-phase modulation of an elliptically polarized wave depends on the intensity of the wave and the ellipticity of its polarization.

12.1.2 Cross-Phase Modulation

From the above discussions, we find that the characteristics and the magnitude of self-phase modulation depend not only on the properties of the medium, but also on the polarization state of the optical wave even though the process involves only one wave. The characteristics and the strength of cross-phase modulation depend not only on the polarization state of each optical wave, but also on the relationship between the polarization states of the two interacting waves. For simplicity, we consider only isotropic materials such that the linear refractive index is n_0 for all polarization states. The only nonvanishing elements of the $\chi^{(3)}$ tensor are those of the types of $\chi_{1111}^{(3)}$, $\chi_{1122}^{(3)}$, $\chi_{1212}^{(3)}$, and $\chi_{1221}^{(3)}$, which have the relation that $\chi_{1111}^{(3)} = \chi_{1122}^{(3)} + \chi_{1212}^{(3)} + \chi_{1221}^{(3)}$. If Kleinman's symmetry condition is valid, then $\chi_{1122}^{(3)} = \chi_{1212}^{(3)} = \chi_{1221}^{(3)} = \chi_{1111}^{(3)}/3$. Without loss of generality, we also assume that all of the optical waves propagate in the z direction and are polarized in the xy plane. The optical wave being modulated is called the *probe* wave or the *signal* wave, which is taken to be at the frequency ω, and that creating the modulation is called the *pump* wave, which has a frequency at ω'.

As expressed in (5.32), the nonlinear polarization $\mathbf{P}^{(3)}(\omega)$ of the probe wave in the interaction of two waves at the frequencies of ω and ω' consists of two terms: the first term accounts for self-modulation and the second for cross-modulation. In the discussions within this subsection, we focus our attention on cross-modulation. For this purpose, we shall only consider the second term of (5.32) for the modulation on the probe wave at the frequency of ω by the pump wave at the frequency of ω'. Therefore, we shall only express the relevant nonlinear polarization as

$$\mathbf{P}^{(3)}(\omega) = 6\epsilon_0 \boldsymbol{\chi}^{(3)}(\omega = \omega + \omega' - \omega') \vdots \mathbf{E}(\omega)\mathbf{E}(\omega')\mathbf{E}^*(\omega'). \tag{12.20}$$

Correspondingly, the field-dependent change in the permittivity caused by the cross-modulation is the second term in (5.36):

$$\Delta\epsilon(\omega, \mathbf{E}) = 6\epsilon_0 \boldsymbol{\chi}^{(3)}(\omega = \omega + \omega' - \omega') : \mathbf{E}(\omega')\mathbf{E}^*(\omega'). \tag{12.21}$$

Linearly Polarized Pump Wave

We first consider the case of a linearly polarized pump wave at the frequency of ω' and a probe wave at the frequency of ω that can be in any polarization state. The pump wave generates field-dependent changes in the optical properties of the medium, and the probe wave senses the changes. In an isotropic medium, we take the pump wave to be polarized in the x direction without loss of generality. Then,

$$\mathbf{E}(\omega') = \hat{x}E(\omega') \quad \text{and} \quad \mathbf{E}(\omega) = \hat{x}E_x(\omega) + \hat{y}E_y(\omega). \tag{12.22}$$

By using (12.20), we find that

$$\begin{aligned} P_x^{(3)}(\omega) &= 6\epsilon_0 \chi_{1111}^{(3)}(\omega = \omega + \omega' - \omega')|E(\omega')|^2 E_x(\omega), \\ P_y^{(3)}(\omega) &= 6\epsilon_0 \chi_{1122}^{(3)}(\omega = \omega + \omega' - \omega')|E(\omega')|^2 E_y(\omega). \end{aligned} \tag{12.23}$$

Therefore, we have

$$\begin{aligned} \Delta\epsilon_\parallel(\omega) &= \Delta\epsilon_{xx}(\omega) = 6\epsilon_0 \chi_{1111}^{(3)}(\omega = \omega + \omega' - \omega')|E(\omega')|^2 = \frac{3\chi_{1111}^{(3)}}{cn_0}I(\omega'), \\ \Delta\epsilon_\perp(\omega) &= \Delta\epsilon_{yy}(\omega) = 6\epsilon_0 \chi_{1122}^{(3)}(\omega = \omega + \omega' - \omega')|E(\omega')|^2 = \frac{3\chi_{1122}^{(3)}}{cn_0}I(\omega'). \end{aligned} \tag{12.24}$$

where $\Delta\epsilon_\parallel(\omega)$ and $\Delta\epsilon_\perp(\omega)$ are the permittivity changes sensed by the components of the probe wave that are, respectively, parallel and perpendicular to the polarization of the pump wave. The field-dependent refractive indices for the two polarization directions are, respectively,

$$\begin{aligned} n_\parallel(\omega) &= n_0 + n_{2\parallel}I(\omega') = n_0 + \frac{3\chi_{1111}^{(3)\prime}}{2c\epsilon_0 n_0^2}I(\omega'), \\ n_\perp(\omega) &= n_0 + n_{2\perp}I(\omega') = n_0 + \frac{3\chi_{1122}^{(3)\prime}}{2c\epsilon_0 n_0^2}I(\omega'). \end{aligned} \tag{12.25}$$

From (12.25), we see that the linearly polarized pump wave generates a *field-dependent linear birefringence*:

$$B = \Delta n = n_{\parallel} - n_{\perp} = \frac{3}{2c\epsilon_0 n_0^2}\left(\chi_{1111}^{(3)\prime} - \chi_{1122}^{(3)\prime}\right)I(\omega') = \frac{3}{2c\epsilon_0 n_0^2}\left(\chi_{1212}^{(3)\prime} + \chi_{1221}^{(3)\prime}\right)I(\omega'). \quad (12.26)$$

Because of this field-dependent linear birefringence, besides the field-dependent phase change caused by cross-phase modulation, the polarization state of the probe wave is also dependent on the intensity of the pump wave. The exception is when the probe wave is linearly polarized in a direction that is either parallel or perpendicular to the polarization direction of the pump wave. When the probe wave is linearly polarized in the direction that is parallel to the polarization direction of the pump wave, it only senses $n_{\parallel}(\omega)$. When the probe wave is linearly polarized in the direction that is perpendicular to the polarization direction of the pump wave, it only senses $n_{\perp}(\omega)$. When the probe wave is linearly polarized in a direction that is neither parallel nor perpendicular to the polarization direction of the pump wave, the field-dependent linear birefringence changes the polarization state of the probe wave as it propagates. For a probe wave that is circularly or elliptically polarized, both its phase and polarization state are changed by the index changes and the linear birefringence caused by the pump wave.

Circularly Polarized Pump Wave

We now consider the case of a circularly polarized pump wave at the frequency of ω' and a probe wave at the frequency of ω that can be in any polarization state. The pump wave can be either left- or right-circularly polarized:

$$\mathbf{E}(\omega') = \hat{e}_{\pm}E(\omega'). \quad (12.27)$$

The probe wave can be generally expressed as

$$\mathbf{E}(\omega) = \hat{e}_+ E_+(\omega) + \hat{e}_- E_-(\omega) = \hat{x}E_x(\omega) + \hat{y}E_y(\omega), \quad (12.28)$$

where E_+ and E_- are related to E_x and E_y as given in (12.10), and $\hat{e}_+$ and $\hat{e}_-$ are related to $\hat{x}$ and $\hat{y}$ as given in (12.11). By using (12.20), we find that

$$P_+^{(3)}(\omega) = 3\epsilon_0\left[\chi_{1111}^{(3)} + \chi_{1122}^{(3)} \pm \left(\chi_{1212}^{(3)} - \chi_{1221}^{(3)}\right)\right]\left|E(\omega')\right|^2 E_+(\omega),$$

$$P_-^{(3)}(\omega) = 3\epsilon_0\left[\chi_{1111}^{(3)} + \chi_{1122}^{(3)} \pm \left(\chi_{1221}^{(3)} - \chi_{1212}^{(3)}\right)\right]\left|E(\omega')\right|^2 E_-(\omega),$$

$$\quad (12.29)$$

where the $\pm$ sign refers to the $\hat{e}_\pm$ polarization state of the pump wave as expressed in (12.27), whereas the subscripts $+$ and $-$ of $P^{(3)}(\omega)$ and $E(\omega)$ refer to the circular-polarization components of the probe wave. From (12.29), we find the field-dependent permittivity changes that are respectively sensed by the left and right circular-polarization components of the probe wave as

$$\Delta\epsilon_+(\omega) = 3\epsilon_0\left[\chi_{1111}^{(3)} + \chi_{1122}^{(3)} \pm \left(\chi_{1212}^{(3)} - \chi_{1221}^{(3)}\right)\right]\left|E(\omega')\right|^2,$$

$$\Delta\epsilon_-(\omega) = 3\epsilon_0\left[\chi_{1111}^{(3)} + \chi_{1122}^{(3)} \pm \left(\chi_{1221}^{(3)} - \chi_{1212}^{(3)}\right)\right]\left|E(\omega')\right|^2.$$

$$\quad (12.30)$$

The field-dependent refractive indices seen by the two circular-polarization components are

$$n_+(\omega) = n_0 + n_{2+}I(\omega') = n_0 + \frac{3}{4c\epsilon_0 n_0^2}\left[\chi_{1111}^{(3)\prime} + \chi_{1122}^{(3)\prime} \pm \left(\chi_{1212}^{(3)\prime} - \chi_{1221}^{(3)\prime}\right)\right]I(\omega'),$$

$$n_-(\omega) = n_0 + n_{2-}I(\omega') = n_0 + \frac{3}{4c\epsilon_0 n_0^2}\left[\chi_{1111}^{(3)\prime} + \chi_{1122}^{(3)\prime} \pm \left(\chi_{1221}^{(3)\prime} - \chi_{1212}^{(3)\prime}\right)\right]I(\omega'). \tag{12.31}$$

From (12.31), we see that the circularly polarized pump wave generates a *field-dependent circular birefringence*:

$$B_c = \Delta n_c = n_+ - n_- = \pm\frac{3}{2c\epsilon_0 n_0^2}\left(\chi_{1212}^{(3)\prime} - \chi_{1221}^{(3)\prime}\right)I(\omega'), \tag{12.32}$$

where n_+ is sensed by the left-circularly polarized probe wave and n_- is sensed by the right-circularly polarized probe wave, whereas the $\pm$ sign refers to the polarization state of the pump wave.

The field-dependent circular birefringence is nonzero when $\chi_{1212}^{(3)\prime} \neq \chi_{1221}^{(3)\prime}$. This is possible for an isotropic material when Kleinman's symmetry condition is not valid. In this situation, a circularly polarized pump wave causes only an intensity-dependent phase change on a circularly polarized probe wave without changing its polarization state. However, it changes the polarization state of a linearly polarized or elliptically polarized probe wave as well as changing its phase. When Kleinman's symmetry condition is valid, $\chi_{1212}^{(3)\prime} = \chi_{1221}^{(3)\prime}$ and $B_c = 0$; then, a circularly polarized pump wave does not induce a birefringence through cross-phase modulation, neither linear birefringence nor circular birefringence. In this situation, we have

$$n(\omega) = n_+(\omega) = n_-(\omega) = n_0 + n_2 I(\omega') = n_0 + \frac{3}{4c\epsilon_0 n_0^2}\left(\chi_{1111}^{(3)\prime} + \chi_{1122}^{(3)\prime}\right)I(\omega') \tag{12.33}$$

for a probe wave of any polarization state.

Elliptically Polarized Pump Wave
An elliptically polarized pump wave at the frequency of ω' can be generally expressed as

$$\mathbf{E}(\omega') = \hat{e}_+ E_+(\omega') + \hat{e}_- E_-(\omega') = \hat{x}E_x(\omega') + \hat{y}E_y(\omega'). \tag{12.34}$$

The probe wave can also be generally expressed as

$$\mathbf{E}(\omega) = \hat{e}_+ E_+(\omega) + \hat{e}_- E_-(\omega) = \hat{x}E_x(\omega) + \hat{y}E_y(\omega). \tag{12.35}$$

By using (12.20), we find that, in terms of the linear-polarization field components,

$$\begin{aligned}
P_x^{(3)}(\omega) &= 6\epsilon_0\left[\chi_{1111}^{(3)}|E_x(\omega')|^2 + \chi_{1122}^{(3)}|E_y(\omega')|^2\right]E_x(\omega) \\
&\quad + 6\epsilon_0\left[\chi_{1212}^{(3)}E_x(\omega')E_y^*(\omega') + \chi_{1221}^{(3)}E_y(\omega')E_x^*(\omega')\right]E_y(\omega), \\
P_y^{(3)}(\omega) &= 6\epsilon_0\left[\chi_{1111}^{(3)}|E_y(\omega')|^2 + \chi_{1122}^{(3)}|E_x(\omega')|^2\right]E_y(\omega) \\
&\quad + 6\epsilon_0\left[\chi_{1212}^{(3)}E_y(\omega')E_x^*(\omega') + \chi_{1221}^{(3)}E_x(\omega')E_y^*(\omega')\right]E_x(\omega).
\end{aligned} \tag{12.36}$$

In terms of the circular-polarization field components, we have

$$
\begin{aligned}
P_+^{(3)}(\omega) = 6\epsilon_0 & \left[\left(\chi_{1122}^{(3)} + \chi_{1212}^{(3)} \right) \left| E_+(\omega') \right|^2 + \left(\chi_{1122}^{(3)} + \chi_{1221}^{(3)} \right) |E_-(\omega')|^2 \right] E_+(\omega) \\
& + 6\epsilon_0 \left(\chi_{1212}^{(3)} + \chi_{1221}^{(3)} \right) E_+(\omega') E_-^*(\omega') E_-(\omega), \\
P_-^{(3)}(\omega) = 6\epsilon_0 & \left[\left(\chi_{1122}^{(3)} + \chi_{1212}^{(3)} \right) \left| E_-(\omega') \right|^2 + \left(\chi_{1122}^{(3)} + \chi_{1221}^{(3)} \right) \left| E_+(\omega') \right|^2 \right] E_-(\omega) \\
& + 6\epsilon_0 \left(\chi_{1212}^{(3)} + \chi_{1221}^{(3)} \right) E_-(\omega') E_+^*(\omega') E_+(\omega).
\end{aligned}
\tag{12.37}
$$

From (12.36) we find that both $P_x^{(3)}(\omega)$ and $P_y^{(3)}(\omega)$ are functions of both $E_x(\omega)$ and $E_y(\omega)$. From (12.37), we also find that both $P_+^{(3)}(\omega)$ and $P_-^{(3)}(\omega)$ are functions of both $E_+(\omega)$ and $E_-(\omega)$. If $E_x(\omega') \neq 0$ and $E_y(\omega') \neq 0$, $\Delta\epsilon$ is not diagonal and has unequal diagonal elements when represented by using the basis of $\hat{x}$ and $\hat{y}$; thus, it causes a field-dependent linear birefringence. Similarly, if $E_+(\omega') \neq 0$ and $E_-(\omega') \neq 0$, $\Delta\epsilon$ is not diagonal and has unequal diagonal elements when represented by using the basis of $\hat{e}_+$ and $\hat{e}_-$; thus, it causes a field-dependent circular birefringence. Therefore, an elliptically polarized pump wave generally induces both linear and circular birefringence. It causes field-dependent changes in both the phase and the polarization state of the probe wave.

12.2 SELF-FOCUSING AND SELF-DEFOCUSING

In this section, we consider the effect of self-phase modulation caused by a linearly or circularly polarized optical wave that has a spatially varying intensity distribution. For a plane optical wave, which has a spatially uniform intensity distribution, the optical Kerr effect merely causes a uniform intensity-dependent phase shift across the wavefront. Thus, the beam remains a plane wave without any change in its spatial intensity distribution. If an optical beam has a nonuniform intensity distribution, the intensity-dependent index of refraction due to self-phase modulation leads to a nonuniform phase shift across the wavefront as the beam propagates through the nonlinear medium. This beam will then be focused or defocused as a result of the distortion in its phase front, known as *self-focusing* and *self-defocusing*, respectively. If the intensity distribution only varies with space but not with time, self-focusing or self-defocusing of the beam does not vary with time. If the intensity distribution also varies with time, such as in the case of an optical pulse, self-focusing or self-defocusing also evolves with time. The consequence of the optical Kerr effect on a spatially nonuniform intensity profile is discussed in this section.

12.2.1 Kerr Lens

We first consider the effects and the applications of self-focusing and self-defocusing that are caused by self-phase modulation in a *thin optical Kerr medium*, which acts as an intensity-dependent lens, known as a *Kerr lens*. As discussed in Subsection 12.1.1, the optical Kerr effect results in an intensity-dependent index of refraction of $n = n_0 + n_2 I$ from self-phase modulation of an optical wave that has an intensity of I. The phase shift of the wave at a frequency of ω after it propagates in the medium over a distance of l is

$$\varphi(\mathbf{r}, t) = kl - \omega t = \frac{n\omega}{c} l - \omega t. \tag{12.38}$$

For simplicity, we consider the propagation of a beam of circular cross-section, which has a transverse spatial intensity distribution of $I(r)$, where $r = (x^2 + y^2)^{1/2}$, assuming that the longitudinal direction of beam propagation is the z direction. After such a beam propagates through a thin nonlinear medium that has a thickness of l, the total intensity-dependent phase shift is

$$\varphi(r) = \frac{n\omega}{c} l = \frac{\omega}{c} \left[n_0 + n_2 I(r) \right] l = \varphi_0 + \varphi_K(r), \tag{12.39}$$

where $\varphi_0 = n_0 \omega l / c$ is the intensity-independent linear phase shift and

$$\varphi_K(r) = \frac{\omega}{c} n_2 l I(r) \tag{12.40}$$

is the intensity-dependent Kerr phase change caused by self-phase modulation. As discussed in Subsection 12.1.1, the expression of n_2 depends on the polarization of the optical beam. It is given in (12.3) for a linearly polarized wave, and in (12.6) for a circularly polarized wave.

The effect of a thin spherical lens that has a focal length of f is to cause a spatially varying phase shift of

$$\varphi(r) = -k \frac{r^2}{2f} = -\frac{\omega}{c} \frac{r^2}{2f} \tag{12.41}$$

in an optical wave that passes through the lens, where r is the transverse radial distance from the center of the lens. Therefore, if the intensity-dependent Kerr phase shift given in (12.40) has a quadratic dependence on the transverse radial coordinate, the optical Kerr effect in the thin nonlinear medium would be equivalent to the effect of a *thin lens*. A thin nonlinear medium with such a function is a Kerr lens.

In reality, no optical beam has an ideal quadratic spatial intensity distribution. However, if the intensity distribution near the beam center of a circular beam is approximately quadratic in r, the *effective focal length* f_K of the Kerr lens is given by the relation:

$$\frac{1}{f_K} = -a \frac{c}{\omega} \frac{\mathrm{d}^2 \varphi}{\mathrm{d} r^2} \bigg|_{r=0} = -a \frac{c}{\omega} \frac{\mathrm{d}^2 \varphi_K}{\mathrm{d} r^2} \bigg|_{r=0} = -a n_2 l \frac{\mathrm{d}^2 I(r)}{\mathrm{d} r^2} \bigg|_{r=0}, \tag{12.42}$$

where a is a correction factor to account for the difference between the true beam profile and the ideal quadratic profile. By using this relation, we find that the effective focal length of the Kerr lens for a circular Gaussian beam with an intensity distribution of $I(r) = I_0 \exp(-2r^2/w^2)$ is

$$f_K = \frac{w^2}{4an_2lI_0} = \frac{\pi w^4}{8an_2lP},$$

(12.43)

where w is the beam radius at the location of the optical Kerr medium, I_0 is the intensity at the beam center, and $P = I_0(\pi w_0^2/2)$ is the power of the beam. For a circular Gaussian beam, $a = 1.723$, and the *thin-lens condition* for (12.43) to be valid is that the thickness of the optical Kerr medium be small compared to the Rayleigh range, z_R, of the Gaussian beam:

$$l < z_R = \pi n_0 w_0^2/\lambda.$$

(12.44)

Note that n_2 can be either positive or negative because the $\chi^{(3)\prime}$ element that is responsible for the optical Kerr effect can be either positive or negative. Therefore, a Kerr lens can either focus or defocus a beam, depending on the sign of its effective focal length.

Most applications of Kerr lenses are based on the fact that the effective focal length f_K of a thin Kerr lens is inversely proportional to the peak intensity I_0 of an optical beam. As a result of this characteristic, the divergence of a beam that passes through a Kerr lens is a function of the intensity of the beam. In addition, the beam divergence also depends on the sign of n_2 and the location of the Kerr lens with respect to the beam waist, as illustrated in Fig. 12.1.

A Kerr lens is often used as an *optical power limiter* for the protection of a sensitive optical detector. In this application, the action of the Kerr lens is to increase the beam divergence as the input intensity of a beam is increased, thereby increasing the spread of the beam to reduce its intensity at the surface of the detector. As demonstrated in Figs. 12.1(a), (b), (e), and (f), with proper arrangement, either a Kerr lens of a positive effective focal length, $f_K > 0$, or one with a negative effective focal length, $f_K < 0$, can be used for this purpose. When a Kerr lens in such an arrangement is used as an optical power limiter, the light reaching the detector is only a fraction of the diverging optical beam within a finite central cross-sectional area that is defined either by the input surface area of a small detector or by a hole in a beam block. Because the divergence of the beam increases with its intensity, the optical power passing through the finite area to be received by the detector saturates at a certain level as the input power of the beam continues to increase. Without the Kerr lens, the beam divergence does not change with its intensity. Then, the optical power received by the detector increases linearly with the input power of the beam without a limit until the detector is damaged even if the detector has a very small area to intercept only a tiny fraction of the beam.

A Kerr lens can also be used as a passive optical switch or an *optical thresholding device*. For this purpose, an arrangement such as that shown in Fig. 12.1(c) or (d) is used, which leads to a reduction in beam divergence with an increase in input beam intensity. Similar to the setup of a power limiter, only a portion of the beam within a finite central area of the beam cross-section is allowed to pass. However, instead of a saturation, the optical power passing through this area increases nonlinearly with the input power of the beam. The transmittance is low at low power levels, but it is switched up at a threshold power level. This behavior can be used to provide a nonlinear feedback to an optical system or to switch on an optical device at a certain threshold. It has been used as the *passive mode locker* in a technique known as *Kerr-lens mode locking* for the generation of ultrashort laser pulses, which will be discussed in Subsection 17.4.2 and shown in Fig. 17.13(b).

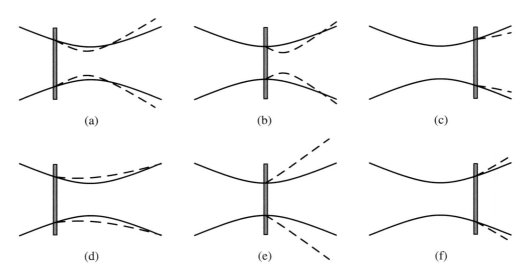

Figure 12.1 Nonlinear refraction caused by a Kerr lens as a function of beam intensity and the location of the Kerr lens with respect to the beam waist for (a), (b), and (c) $f_K > 0$, and for (d), (e), and (f) $f_K < 0$. The solid curves are the propagation lines of a Gaussian beam at a low intensity for which the effect of the Kerr lens is negligible, and the dashed curves are those at a high intensity for which the effect of the Kerr lens is significant.

EXAMPLE 12.1

A Ti:sapphire laser generates a train of laser pulses of $\lambda = 780$ nm wavelength and $\Delta t_{ps} = 100$ fs pulsewidth at a repetition rate of $f_{ps} = 100$ MHz. A beam of such pulses at an average power of $\overline{P} = 50$ mW is focused tightly on a thin silica plate of $l = 1$ mm thickness. The nonlinear response time of silica is shorter than 100 fs, so the response of the optical Kerr effect to the temporal variation of each pulse can be considered instantaneous. Silica has a linear refractive index of $n_0 = 1.4537$ at $\lambda = 780$ nm and, according to Example 5.2, a nonlinear refractive index of $n_2 = 2.4 \times 10^{-20}$ m^2 W^{-1}. If the laser beam is focused with its waist on the silica plate as tightly as allowed by the thin-lens condition, what is the effective focal length of the Kerr lens caused by self-phase modulation at the peaks of the optical pulses?

Solution

The peak power of the pulses is

$$P_{pk} = \frac{\overline{P}}{f_{ps}\Delta t_{ps}} = \frac{50 \times 10^{-3}}{100 \times 10^6 \times 100 \times 10^{-15}} \text{ W} = 5 \text{ kW}.$$

The thin-lens condition, $l < z_R = \pi n_0 w_0^2/\lambda$, requires that

$$w_0 > \left(\frac{l\lambda}{\pi n_0}\right)^{1/2} = \left(\frac{1 \times 10^{-3} \times 780 \times 10^{-9}}{\pi \times 1.4537}\right)^{1/2} \text{ m} = 13 \text{ μm}.$$

By focusing the beam to the limit of $w_0 = 13$ μm that is allowed by the thin-lens condition and by placing the beam waist on the silica plate, we have the following Kerr focal length at the peak of each pulse:

$$f_K = \frac{\pi w_0^4}{8an_2 l P_{pk}} = \frac{\pi \times (13 \times 10^{-6})^4}{8 \times 1.723 \times 2.4 \times 10^{-20} \times 1 \times 10^{-3} \times 5 \times 10^3} \text{ m} = 5.42 \text{ cm}.$$

Note that this is the Kerr focal length only at the temporal peak of each pulse. Because f_K is inversely proportional to the optical power and because the nonlinear refractive response of silica is faster than the 100 fs duration of each pulse, we can easily see that the value of f_K varies with time through the duration of a pulse. As a consequence of this temporally varying f_K, the divergence of the pulse after it passes through the silica plate is a function of time over the pulse duration. Kerr-lens mode locking of lasers takes advantage of this interesting phenomenon.

12.2.2 Self-Trapping and Catastrophic Self-Focusing

In the preceding subsection, we considered self-focusing and self-defocusing of an optical beam that propagates through a thin optical Kerr material that satisfies the condition given in (12.44). In this subsection, we consider the dramatic effects of self-focusing in a continuous bulk medium.

The possibility of self-focusing was first discussed in 1962 by Askaryan [1]. Self-focusing occurs in an optical Kerr medium that has a positive nonlinear refractive index, $n_2 > 0$. When an optical beam propagates in a continuous optical Kerr medium, the Kerr-lensing effect causes the beam diameter to decrease, thus increasing the intensity at the beam center. The increased intensity at the beam center increases the refractive index due to the optical Kerr effect, leading to further focusing of the beam to further reduce the beam diameter. Meanwhile, the effect of diffraction also increases as the beam diameter is reduced. Thus, the self-focusing process is counteracted by the diffraction of the beam. Depending on the power of the beam, a few possible scenarios can take place [2], as illustrated in Fig. 12.2.

If the power of the laser beam is below a characteristic *critical power*, P_c, the self-focusing effect is always weaker than diffraction. In this case, the optical Kerr effect causes a change in the divergence of the beam, but the change is not sufficiently strong to overcome diffraction. The beam still diverges due to diffraction as it continues to propagate after passing its beam waist, as shown in Fig. 12.2(a). At the critical power, P_c, the effect of diffraction exactly balances that of self-focusing when the beam is reduced to its beam waist size, w_0. In this case, the beam produces its own waveguide through the optical Kerr effect. It then propagates with a constant beam size as a nonlinearly guided wave without further focusing or diverging. This phenomenon is shown in Fig. 12.2(b) and is known as *self-trapping* [3, 4]. For a laser beam that has a power higher than the critical power, $P > P_c$, the effect of self-focusing is stronger than the effect of diffraction, such that the beam continues to self-focus into a very small focal spot. As shown in Fig. 12.2(c), the consequence is *catastrophic self-focusing* [5]. At an even higher power, $P \gg P_c$,

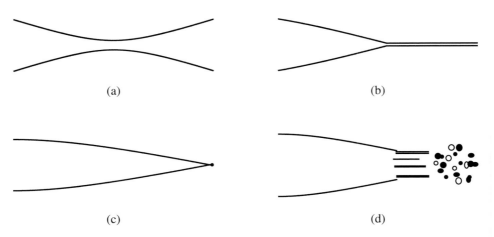

Figure 12.2 Scenarios of the propagation of an optical beam in a continuous optical Kerr medium that has a positive nonlinear refractive index. (a) Insufficient optical Kerr effect to overcome diffraction in the case that $P < P_c$. (b) Self-trapping in the case that $P = P_c$. (c) Catastrophic self-focusing in the case that $P > P_c$. (d) Beam breakup and filamentation in the case that $P \gg P_c$.

the laser beam breaks up into many self-trapped filaments, each of a few micrometers in diameter, containing a power of approximately P_c. This scenario is shown in Fig. 12.2(d).

For self-focusing, both the longitudinal and the transverse variations of the optical field have to be simultaneously considered. We assume that the optical field propagates in a homogeneous, isotropic optical Kerr medium that has $n_2 > 0$. Its propagation direction is taken to be the z direction. For simplicity, we consider a CW optical beam of a circular cross-section such that the wave propagation has cylindrical symmetry. Then the optical field can be expressed as

$$\mathbf{E}(r, z, t) = \hat{e}\mathcal{E}(r, z) \exp{(ik_0 z - i\omega_0 t)}, \tag{12.45}$$

where $r = (x^2 + y^2)^{1/2}$ and

$$k_0 = \frac{n_0 \omega_0}{c} = \frac{2\pi n_0}{\lambda}. \tag{12.46}$$

The refractive index of the optical Kerr medium is

$$n(r, z) = n_0 + n_2 I(r, z) = n_0 + 2c\epsilon_0 n_0 n_2 \left|\mathcal{E}(r, z)\right|^2, \tag{12.47}$$

where the optical intensity is $I = 2c\epsilon_0 n_0 \left|\mathcal{E}\right|^2$.

To estimate the critical power, we consider a fundamental Gaussian mode that propagates in the optical Kerr medium, as shown in Fig. 12.3. The beam is incident on the surface of the medium at the location $z = -d$. It converges to its beam spot size w_0 at its waist location at $z = 0$. The linear propagation of the Gaussian beam can be described as

$$\mathcal{E}_G(r, z) = \mathcal{E}_0 \frac{w_0}{w(z)} \exp\left[-\frac{r^2}{w^2(z)}\right] \exp\left[i\frac{k_0}{2}\frac{r^2}{\mathcal{R}(z)} + i\zeta(z)\right], \tag{12.48}$$

where $w(z)$ is the spot size and $\mathcal{R}(z)$ is the radius of curvature of the Gaussian beam at the location z:

$$w(z) = w_0 \left(1 + \frac{z^2}{z_R^2} \right)^{1/2} \quad \text{and} \quad \mathcal{R}(z) = z \left(1 + \frac{z_R^2}{z^2} \right), \tag{12.49}$$

z_R is the Rayleigh range:

$$z_R = \frac{k_0 w_0^2}{2} = \frac{\pi n_0 w_0^2}{\lambda}, \tag{12.50}$$

and $\zeta(z)$ is an on-axis phase variation along the z axis.

For the Gaussian field expressed in (12.48), the spatially varying phase factor

$$\varphi_d(r, z) = \frac{k_0}{2} \frac{r^2}{\mathcal{R}(z)} \tag{12.51}$$

accounts for the effect of linear diffraction. The difference of this linear diffraction phase between the two Gaussian-field paths at $r = 0$ and $r = w(z)$ as the beam propagates from the surface of the medium at $z = -d$ to the waist location at $z = 0$ is

$$\Delta\varphi_d = \left[\varphi_d(0,0) - \varphi_d(0,-d) \right] - \left[\varphi_d(w_0, 0) - \varphi_d\left(w(-d), -d \right) \right] = -\frac{d}{z_R}, \tag{12.52}$$

which is found by using the relations given in (12.49)–(12.51). From (12.47), the nonlinear phase that accounts for the optical Kerr effect as a function of space is

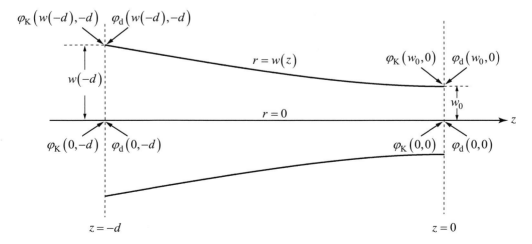

Figure 12.3 Propagation of a fundamental Gaussian beam. The linear and nonlinear phase changes, due to diffraction and optical Kerr effect, respectively, are considered between the two Gaussian-field paths at $r = 0$ and $r = w(z)$ as the beam propagates from the surface of the medium at $z = -d$ to the waist location at $z = 0$.

$$\varphi_K(r, z) = k_0 n_2 I(r, z) z. \qquad (12.53)$$

The difference of this nonlinear Kerr phase between the two Gaussian-field paths at $r = 0$ and $r = w(z)$ as the beam propagates from the surface of the medium at $z = -d$ to the waist location at $z = 0$ is

$$\Delta\varphi_K = \left[\varphi_K(0,0) - \varphi_K(0,-d)\right] - \left[\varphi_K(w_0,0) - \varphi_K\left(w(-d),-d\right)\right]$$

$$\approx \frac{k_0 n_2 I_0}{\alpha}\left[0 - (-d)\right] \qquad (12.54)$$

$$= \frac{k_0 n_2 I_0}{\alpha} d.$$

In (12.54), $I_0 = I(0,0)$ is the peak intensity of the beam at the beam focal point such that the power of the beam is

$$P = \frac{\pi w_0^2}{2} I_0. \qquad (12.55)$$

The factor α is an averaging factor that accounts for the varying beam intensity from $z = -d$ to $z = 0$ and between the two optical paths at $r = 0$ and $r = w(z)$. Because of the waveguiding effect caused by the optical Kerr effect, α is expected to have a value larger than unity because the propagation constant of a guided wave is smaller than that of an unguided wave that propagates in the same medium. Because the beam becomes collimated when self-trapping occurs, the value of α is not expected to be large. Thus, we can expect that $1 < \alpha < 2$.

The critical power P_c is found when the linear diffraction phase change is exactly balanced by the nonlinear Kerr phase change such that the total phase change is

$$\Delta\varphi_{\text{total}} = \Delta\varphi_d + \Delta\varphi_K = 0. \qquad (12.56)$$

By using (12.52) and (12.54) for (12.56) while setting the power P given in (12.55) to the critical power P_c, we find that

$$P_c = \alpha \frac{\lambda^2}{4\pi n_0 n_2}. \qquad (12.57)$$

Though (12.57) is obtained by considering the propagation of a Gaussian beam, its form is generally valid for any beam profile. The difference among different beam profiles is the value of the factor α, which depends on the transverse profile of the beam but not on other details of the beam, such as the initial beam width or the initial beam divergence [6, 7].

To find the exact value of α, it is necessary to analyze the nonlinear propagation of the field of (12.45) in an optical Kerr medium. The field amplitude $\mathcal{E}(r, z)$ is governed by (8.51):

$$i\frac{\partial \mathcal{E}}{\partial z} + \frac{1}{2k_0}\nabla_\perp^2 \mathcal{E} + s|\mathcal{E}|^2 \mathcal{E} = 0, \qquad (12.58)$$

where $\nabla_\perp^2 = \partial^2/\partial x^2 + \partial^2/\partial y^2$ and $s = 2k_0 c\epsilon_0 n_2 = 2\omega_0\epsilon_0 n_0 n_2$. This three-dimensional equation has no analytical solutions. It has to be numerically solved. However, it has been shown by

Weinstein [6] that the behavior of the optical field governed by this equation depends on its initial profile and power, but not on its initial width or divergence. Furthermore, the critical power has a lower bound such that [6, 7]

$$\alpha \geq N_c = 1.86225 \tag{12.59}$$

for all beam profiles. No self-trapping or catastrophic collapse can happen for any beam, irrespective of its profile, that has an initial power below the lower bound such that

$$P < N_c \frac{\lambda^2}{4\pi n_0 n_2}. \tag{12.60}$$

Above this lower bound, the value of α for the critical power depends on the beam profile; it has been numerically calculated for various profiles by Fibich and Gaeta [7]. The nonlinear Schrödinger equation given in (12.58) has a solution of self-trapped profile that is known as the *Townes soliton* [3], which has a value of $\alpha = N_c = 1.86225$ for its critical power. For a Gaussian beam that propagates in a homogeneous bulk medium, it is numerically found that $\alpha = 1.8962$, which is slightly larger than N_c. One significant result is that the critical power for the propagation of all profiles in a hollow waveguide is equal to that of the Townes soliton, with $\alpha = N_c = 1.86225$.

From the above discussion, a beam that has exactly the critical power can propagate in a homogeneous medium as a guided wave over a long distance due to self-trapping in a self-generated nonlinear waveguide. This state is unstable because it can be easily perturbed by many factors, such as a small deviation from the critical power or other nonlinear optical effects. A deviation in the power below the critical power causes the divergence of the beam, as discussed above. By contrast, a small increase in the power leads to the collapse of the beam through catastrophic self-focusing. This run-away process results in a very high optical intensity at the focal point, which can easily destroy the optical medium due to optical damage. This mechanism sets the upper limit of the power of a laser beam that can safely propagate through an optical Kerr medium to be at the critical power level. The distance that a beam with a power above the critical level, $P > P_c$, can propagate before catastrophic self-focusing takes place is approximated by the relation [5]:

$$z_{SF} = \frac{k_0 w_0^2}{(P/P_c - 1)^{1/2}} = \frac{2\pi w_0^2}{\lambda (P/P_c - 1)^{1/2}}. \tag{12.61}$$

As discussed above, the critical power does not depend on the original beam area, nor on the initial beam divergence. However, for a given optical power, a beam that has a larger initial cross-sectional area or a larger divergence gets through a larger propagation distance before it collapses. This effect is not represented by the relation given in (12.61). Furthermore, catastrophic self-focusing is a highly nonlinear effect that can lead to other nonlinear effects, such as thermal lensing, which can result in nonlinear feedback to accelerate or delay the self-focusing process. At an even higher power, the beam breaks up into filaments. Therefore, the relation given in (12.61) can only be considered as an estimate. Indeed, even the power dependence of $(P/P_c - 1)^{1/2}$ is not accurate for a beam power that is much higher than the critical power.

Besides optical damage to the medium, self-focusing can lead to other nonlinear optical effects. Among the most significant are stimulated Raman scattering and self-phase modulation, both of which are strongly enhanced by self-focusing because they both depend on the light intensity. For the same reason, multiphoton absorption can also be enhanced. Therefore, self-focusing is often not an independent process that can be analyzed without taking into consideration the accompanying nonlinear optical effects.

In this subsection, only the self-focusing of a CW optical beam is considered. If the optical beam consists of optical pulses, different temporal parts of a pulse self-focus at different locations due to their different power levels. The *transient self-focusing* dynamics strongly depend on the pulse duration, particularly for an ultrashort pulse that has a pulsewidth close to the characteristic response time of the nonlinear optical processes [2].

EXAMPLE 12.2

A high-power Gaussian laser beam at the wavelength of $\lambda = 780$ nm is sent through a spatially homogeneous block of silica glass. According to Example 12.1, silica has a linear refractive index of $n_0 = 1.4537$ at $\lambda = 780$ nm and a nonlinear refractive index of $n_2 = 2.4 \times 10^{-20} \, \text{m}^2 \, \text{W}^{-1}$. (a) Find the critical power for catastrophic self-focusing of this Gaussian laser beam in this silica glass block. (b) If the laser beam is a pulse that has a duration of 10 ns, what is the critical energy of the pulse? (c) If the power of the laser beam is $P = 5$ MW, find the distances that the laser beam can propagate if the beam is focused to a beam waist size of 1 mm, 100 μm, and 20 μm, respectively.

Solution

(a) For a Gaussian beam, the critical power for catastrophic self-focusing is given by (12.57) with the parameter $\alpha = 1.8962$. Therefore, with $\lambda = 780$ nm, $n_0 = 1.4537$, and $n_2 = 2.4 \times 10^{-20} \, \text{m}^2 \, \text{W}^{-1}$, we find that the critical power for catastrophic self-focusing of the laser beam at $\lambda = 780$ nm in the silica glass block is

$$P_c = \alpha \frac{\lambda^2}{4\pi n_0 n_2} = 1.8962 \times \frac{\left(780 \times 10^{-9}\right)^2}{4\pi \times 1.4537 \times 2.4 \times 10^{-20}} \, \text{W} = 2.63 \, \text{MW}.$$

This power level is quite high, but it is within reach of a high-power pulsed laser.

(b) If the laser beam is a pulse that has a duration of 10 ns, the critical energy for the pulse to undergo catastrophic self-focusing can be estimated as

$$U_c = P_c \Delta t_{ps} = 2.63 \times 10^6 \times 10 \times 10^{-9} \, \text{J} = 26.3 \, \text{mJ}.$$

This pulse energy is high, but it is quite reachable for a giant Q-switched pulse that has a pulsewidth of 10 ns.

(c) The distance that a laser beam of a power above the critical level can propagate before catastrophic self-focusing occurs can be estimated by using (10.61). Though the critical power does not depend on the beam spot size, this distance does depend on the spot size. By using (10.61) with $P = 5\,\text{MW}$, we find for the three spot sizes that

$$z_{\text{SF}} = \frac{2\pi w_0^2}{\lambda (P/P_c - 1)^{1/2}} = \frac{2\pi \times (1 \times 10^{-3})^2}{780 \times 10^{-9} \times (5/2.63 - 1)^{1/2}}\,\text{m} = 8.5\,\text{m} \quad \text{for} \quad w_0 = 1\,\text{mm},$$

$$z_{\text{SF}} = \frac{2\pi w_0^2}{\lambda (P/P_c - 1)^{1/2}} = \frac{2\pi \times (100 \times 10^{-6})^2}{780 \times 10^{-9} \times (5/2.63 - 1)^{1/2}}\,\text{m} = 8.5\,\text{cm} \quad \text{for} \quad w_0 = 100\,\mu\text{m},$$

$$z_{\text{SF}} = \frac{2\pi w_0^2}{\lambda (P/P_c - 1)^{1/2}} = \frac{2\pi \times (20 \times 10^{-6})^2}{780 \times 10^{-9} \times (5/2.63 - 1)^{1/2}}\,\text{m} = 3.4\,\text{mm} \quad \text{for} \quad w_0 = 20\,\mu\text{m}.$$

From the above results, we find that catastrophic self-focusing is not likely to occur for a beam that is not tightly focused such that $w_0 = 1\,\text{mm}$. However, it will take place when the beam is tightly focused to have a spot size of $w_0 = 20\,\mu\text{m}$ as it enters a silica block that has a thickness larger than 3.4 mm.

12.3 Z SCAN

One useful application of the Kerr lens is the Z-scan measurement technique, which was first analyzed and demonstrated in 1989 by Sheik-Bahae *et al.* [8] for measuring the sign and the strength of the Kerr nonlinearity of an optical material. The Z scan was extended in 1990, also by Sheik-Bahae *et al.* [9], to measure the nonlinear absorption coefficient. There are two basic types of Z scan: the *closed-aperture Z scan* [8, 9] and the *open-aperture Z scan* [9]. Because of its sensitivity and simplicity, this technique has become the standard method for measuring both the real and the imaginary parts of the nonlinear susceptibility. Many variants of this method, such as *eclipsing Z scan* [10], *two-color Z scan* [11], and *time-resolved pump-probe Z scan* [12], have been developed.

In the following, the principles and characteristics of the basic closed-aperture Z scan and open-aperture Z scan are summarized. The nonlinear parameters that are measured are the *nonlinear refractive index*, which represents the real part of a nonlinear susceptibility and causes an intensity-dependent phase change of an optical beam, and the *nonlinear absorption coefficient*, which represents the imaginary part of a nonlinear susceptibility and causes an intensity-dependent absorption of an optical beam. The advantages of the Z-scan technique are its simplicity, its sensitivity, and its ability to separate the real and the imaginary parts of the nonlinear susceptibility. One major limitation of the Z-scan technique, however, is that it does not differentiate physical mechanisms that contribute to the measured nonlinear parameters. Separate measurements or theoretical analyses have to be performed to identify the mechanisms that contribute to a measured nonlinear parameter.

In the following, we assume that the physical mechanisms that contribute to the measured nonlinear refractive index and nonlinear absorption coefficient are both from the third-order nonlinearity. In this case, the nonlinear refractive index n_2 for a linearly polarize optical wave is that given in (12.3):

$$n_2 = \frac{3\chi_{1111}^{(3)\prime}}{4c\epsilon_0 n_0^2}, \tag{12.62}$$

and the nonlinear absorption coefficient is the two-photon absorption coefficient β that is given in (14.5):

$$\beta = \frac{3\omega}{2c^2\epsilon_0 n_0^2}\chi_{\text{TPA}}^{\prime\prime}. \tag{12.63}$$

The linear absorption coefficient of the sample is α, and the thickness of the sample is l. Thus, the *effective length* of the sample for a laser beam that propagates through its thickness is

$$l_{\text{eff}} = \frac{1 - e^{-\alpha l}}{\alpha}. \tag{12.64}$$

The characteristics described below are valid for Z scan using a *thin sample* that satisfies the thin-lens condition given in (12.44) (i.e., $l < z_R$) for the Kerr lens, where z_R is the Rayleigh range of the Gaussian beam that is used for the Z-scan measurement.

12.3.1 Closed-Aperture Z Scan

We first consider the simple case that the nonlinearity of the material to be measured is purely refractive without nonlinear absorption such that $n_2 \neq 0$ but $\beta = 0$. The closed-aperture Z scan is used to measure the nonlinear refractive index n_2 through the Kerr-lensing effect by using a single Gaussian laser beam. Figure 12.4(a) shows the schematic setup. It essentially consists of a focused Gaussian beam, with its linear focal point taken to be $z = 0$ in the longitudinal direction, and an aperture before a photodetector, which are placed at a far-field distance, typically $20z_R$ to $100z_R$ from the focal point.

A thin sample of the material being measured is translated along the z axis through the focal point. The measurement is usually performed by using short laser pulses to take advantage of the high peak intensity of the pulse. Therefore, the focused laser beam undergoes a spatially and temporally varying Kerr phase shift in the sample:

$$\varphi_K(z, r, t) = \frac{\omega}{c} n_2 I(z, r, t) l_{\text{eff}}, \tag{12.65}$$

where z is the location of the sample, r is the transverse radial coordinate, and t is the temporal variable for the laser pulse. Because of the r dependence of the Kerr phase shift, the thin sample acts as a thin Kerr lens that changes the divergence of the focused Gaussian beam. This change of divergence varies with the location of the sample because of the z dependence of the Kerr phase shift. The transmittance through the aperture is recorded as a function of the z location of the sample; it is normalized to the linear transmittance that is measured with a negligible Kerr-lensing effect by placing the sample far away from the focal point – for example, at the location right in front of the aperture. The linear transmittance of a circular aperture that is aligned with the Gaussian beam center is

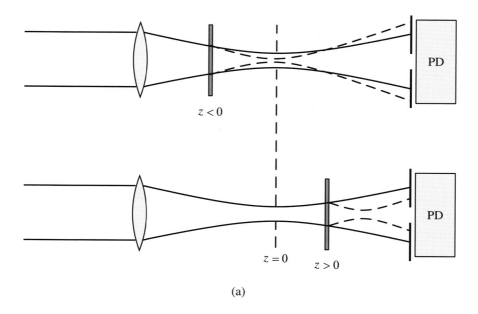

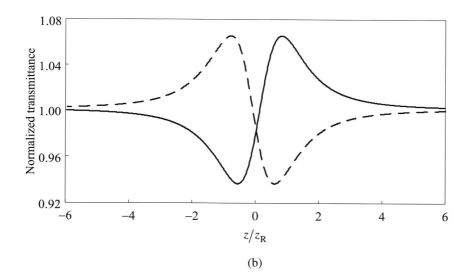

Figure 12.4 (a) Schematic setup for the closed-aperture Z-scan measurement. (b) Normalized transmittance through the aperture measured as a function of the sample position with respect to the linear focal point, which is taken to be $z = 0$. The solid curve is for $n_2 > 0$, and the dashed curve is for $n_2 < 0$.

$$S = 1 - \exp\left(-\frac{2r_a^2}{w_a^2}\right), \tag{12.66}$$

where r_a is the radius of the aperture and w_a is the e^{-2} radius of the linear intensity profile of the Gaussian beam at the location of the aperture. In most experiments, the transmittance of the aperture is chosen in the range of $0.1 < S < 0.5$.

The typical features of the *normalized transmittance* for the closed-aperture Z scan as a function of z are plotted in Fig. 12.4(b). For a positive nonlinear refractive index, $n_2 > 0$, the optical Kerr effect causes self-focusing. When the thin sample is placed at a location before the linear focal point, this self-focusing effect leads to an increased divergence of the Gaussian beam at the far-field location of the aperture, thus reducing the transmittance for $z < 0$. When it is placed at a location after the linear focal point, the self-focusing effect results in a decreased divergence of the Gaussian beam at the location of the aperture, thus increasing the transmittance for $z > 0$. These features are seen in the solid curve shown in Fig. 12.4(b) for $n_2 > 0$. For a negative nonlinear refractive index, $n_2 < 0$, the optical Kerr effect causes self-defocusing. It results in an increased transmittance for $z < 0$ and a decreased transmittance for $z > 0$, as seen in the dashed curve shown in Fig. 12.4(b). Note that the normalized transmittance has a peak value larger than unity, as seen in Fig. 12.4(b), because it is normalized to a reference value of the linear transmittance.

The normalized transmittance is analyzed to find the Kerr phase shift, through which the nonlinear refractive index is found. In general, it is sufficient to find the on-axis Kerr phase shift when the sample is placed at the focal point:

$$\Phi_K(t) = \varphi_K(0, 0, t) = \frac{\omega}{c} n_2 I(0, 0, t) l_{\text{eff}}, \tag{12.67}$$

where $I(0, 0, t)$ is the on-axis intensity of the laser pulse at the focal point. When the measurement is carried out by using laser pulses, particularly ultrashort laser pulses, what is measured is the time-averaged Kerr phase shift:

$$\Phi_{K0} = \overline{\Phi_K(t)} = A_\tau \frac{\omega}{c} n_2 I_0 l_{\text{eff}} = A_\tau \frac{2\pi}{\lambda} n_2 I_0 l_{\text{eff}}, \tag{12.68}$$

where $\overline{\Phi_K(t)}$ is the time average of $\Phi_K(t)$ over the temporal pulse profile, $I_0 = I(0, 0, 0)$ is peak pulse intensity on the z axis at the linear focal point, and A_τ is a time-averaged factor that depends on the pulse shape and the temporal response time of the nonlinearity [9]. For a nonlinear refractive index that responds and decays much faster than the pulsewidth, $A_\tau = 1/\sqrt{2}$ for a Gaussian pulse and $A_\tau = 2/3$ for a sech^2 pulse.

This normalized transmittance has a peak and a valley. The sign of Φ_{K0}, which is also the sign of n_2, is determined by the relative location of the peak and the valley, as discussed above and shown in Fig. 12.4(b). In general, the value of Φ_{K0}, thus the value of n_2, can be found by theoretical fitting of the measured normalized transmittance as a function of z for the given value of the aperture transmittance S that is used in the experiment. For a small Kerr phase shift, however, the value of Φ_{K0} can be found from the peak and the valley of the normalized transmittance. In this case, the peak and the valley are located at the same distance of about $0.86 z_R$ on the two sides from the focal point, with a separation of $\Delta z_{\text{pv}} = |z_\text{p} - z_\text{v}| \approx 1.72 z_R$. The difference of the values of the normalized transmittance at the peak and the valley is a function of the aperture transmittance S. Based on numerical fitting, it is found that this difference can be approximated, within a 2% accuracy for $|\Phi_{K0}| < \pi$, by the relation [8, 9]:

$$\Delta T_{pv} = T_p - T_v = T(z_p) - T(z_v) = 0.406(1 - S)^{0.25} |\Phi_{K0}|. \tag{12.69}$$

By measuring the difference, ΔT_{pv}, of the peak and the valley values of the normalized transmittance, the value of $|\Phi_{K0}|$ can be found from (12.69). Then the absolute value of n_2 can be found from (12.68); meanwhile, the sign of n_2 is determined by the relative locations of the peak and the valley in the transmittance curve, as shown in Fig. 12.4(b). By measuring the separation Δz_{pv} between the locations of the peak and the valley, the validity of using (12.69) can be verified.

The results obtained above are valid only for purely refractive nonlinearity of the third order. For other refractive nonlinearity, such as that of the fifth order, similar analysis can be carried out but the relations presented above are changed. In the case that the nonlinearity of the material includes both nonlinear refractive index and nonlinear absorption, the normalized transmittance of the closed-aperture Z scan has to be fitted with a theoretical model to find the nonlinear refractive index. Nevertheless, the nonlinear absorption coefficient β can be independently found from an open-aperture Z scan, as discussed below. The value of β thus obtained can then be used to fit the normalized transmittance of the closed-aperture Z scan to extract the value of Φ_{K0} as a single unknown parameter.

EXAMPLE 12.3

The nonlinear refractive index, n_2, of the BaF_2 crystal is measured through the closed-aperture Z scan by using a Gaussian laser pulse at the wavelength of $\lambda = 532$ nm, which has a full width at half-maximum pulsewidth of $\Delta t_{ps} = 27$ ps and a pulse energy of $U_{ps} = 2.0$ μJ [8]. The BaF_2 sample has a thickness of $l = 2.5$ mm; its absorption loss is negligible. The Gaussian laser beam is focused to a waist spot size of $w_0 = 18$ μm. The on-axis transmittance is measured under the condition that the aperture is nearly completely closed so that $S \approx 0$. The peak and valley values of the normalized transmittance are found to be $T_p = 1.019$ and $T_v = 0.985$, and it is found that the valley appears before the peak as the sample is moved along the z axis. Find the nonlinear refractive index of BaF_2 from the measured data.

Solution

With $T_p = 1.019$ and $T_v = 0.985$, we have $\Delta T_{pv} = T_p - T_v = 0.034$. By using (12.69) with $S \approx 0$, we find that

$$|\Phi_{K0}| = \frac{\Delta T_{pv}}{0.406} = \frac{0.034}{0.406} = 0.084.$$

Because the valley of the normalized transmittance appears before the peak, we know that $n_2 > 0$, such that $\Phi_{K0} = 0.084 > 0$.

For a Gaussian laser pulse with $\Delta t_{ps} = 27$ ps and $U_{ps} = 2.0$ μJ, its peak power is

$$P_0 = \frac{2\sqrt{\ln 2}}{\sqrt{\pi}} \frac{U_{ps}}{\Delta t_{ps}} = \frac{2\sqrt{\ln 2}}{\sqrt{\pi}} \times \frac{2 \times 10^{-6}}{27 \times 10^{-12}} \text{W} = 69.6 \text{ kW}.$$

Because it is a Gaussian beam that is focused to a spot size of $w_0 = 18$ μm, it has a spot area of $\pi w_0^2/2$ at its focal point. Thus, its on-axis peak intensity is

$$I_0 = \frac{P_0}{\pi w_0^2/2} = \frac{2 \times 69.6 \times 10^3}{\pi \times (18 \times 10^{-6})^2} \text{ W m}^{-2} = 1.37 \times 10^{14} \text{ W m}^{-2}.$$

For a Gaussian pulse, $A_\tau = 1/\sqrt{2}$. Because the absorption loss is negligible, we have $l_{\text{eff}} = l = 2.5$ mm. Then, by using (12.68), we find the nonlinear refractive index:

$$n_2 = \frac{\Phi_{K0}\lambda}{2\pi A_\tau I_0 l_{\text{eff}}} = \frac{0.084 \times 532 \times 10^{-9}}{2\pi \times (1/\sqrt{2}) \times 1.37 \times 10^{14} \times 2.5 \times 10^{-3}} \text{ m}^2 \text{ W}^{-1} = 2.94 \times 10^{-20} \text{ m}^2 \text{ W}^{-1}.$$

12.3.2 Open-Aperture Z Scan

The open-aperture Z scan is used to measure the nonlinear absorption coefficient, which is the two-photon absorption coefficient β in the case of the third-order nonlinearity. For the open-aperture Z scan, the aperture in the far field is removed, effectively making $S = 1$. Thus, all of the power of the laser beam that is transmitted through the sample is measured by the photodetector. Figure 12.5(a) shows the schematic setup of the open-aperture Z scan.

The transmittance is again normalized to the linear transmittance that is measured with negligible nonlinear absorption by placing the sample far away from the focal point. Because the total power of the transmitted beam is measured, the transmittance is not varied by the change in the beam divergence caused by nonlinear refraction from n_2, but is varied only as a result of nonlinear absorption from β. Therefore, the open-aperture Z scan provides independent measurement of the nonlinear absorption coefficient. The typical features of the normalized transmittance for the open-aperture Z scan as a function of z are plotted in Fig. 12.5(b). The transmittance curve is symmetric with respect to the focal point at $z = 0$.

Similar to the closed-aperture Z scan, the open-aperture Z scan is usually performed by using laser pulses to take advantage of the high peak intensity of a pulse. Also similar to the case of the closed-aperture Z scan, the measured signal has to be time-averaged over the pulse duration. Therefore, the normalized transmittance also depends on the temporal pulse shape and the temporal response of the nonlinear absorption process [9]. In the case of two-photon absorption that responds and decays much faster than the pulsewidth, the normalized transmittance of the open-aperture Z scan by using a CW beam or a squared pulse is

$$T_{\text{CW}}(z, S = 1) = \sum_{n=1}^{\infty} \frac{\left[-q_0(z,0)\right]^{n-1}}{n} \approx 1 - \frac{q_0(z,0)}{2} = 1 - \frac{\beta I_0 l_{\text{eff}}}{2(1 + z^2/z_{\text{R}}^2)}, \tag{12.70}$$

and that by using a Gaussian pulse is

$$T_{\text{G}}(z, S = 1) = \sum_{n=1}^{\infty} \frac{\left[-q_0(z,0)\right]^{n-1}}{n^{3/2}} \approx 1 - \frac{q_0(z,0)}{2\sqrt{2}} = 1 - \frac{\beta I_0 l_{\text{eff}}}{2\sqrt{2}(1 + z^2/z_{\text{R}}^2)}, \tag{12.71}$$

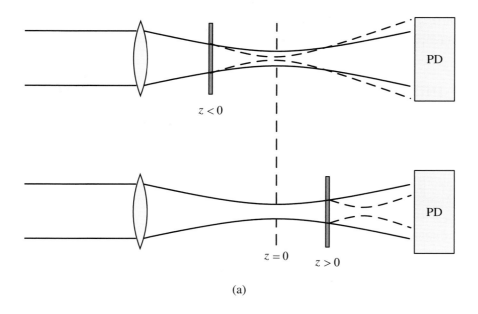

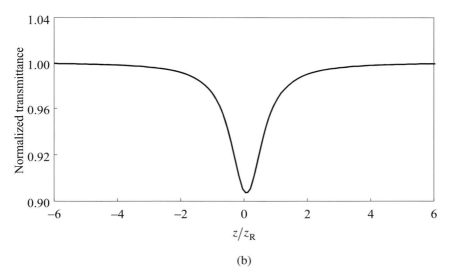

(b)

Figure 12.5 (a) Schematic setup for the open-aperture Z-scan measurement. (b) Normalized transmittance for $\beta > 0$, measured as a function of the sample position with respect to the linear focal point, which is taken to be $z = 0$.

where

$$q_0(z, 0) = \frac{\beta I_0 l_{\mathrm{eff}}}{1 + z^2/z_R^2} \tag{12.72}$$

and $I_0 = I(0, 0, 0)$ is peak pulse intensity on the z axis at the linear focal point, as defined in (12.68). By fitting the measured data of the normalized transmittance with (12.70) for a CW beam or a square pulse, or with (12.71) for a Gaussian pulse, the value of the two-photon absorption coefficient β can be determined.

12.4 SELF-INDUCED SPECTRAL BROADENING

In the case of an optical pulse, its temporally varying intensity profile leads to a temporally varying Kerr phase. Our interest in this section is the consequence of this temporally varying Kerr phase. Because a temporally varying phase leads to a frequency shift, self-phase modulation results in the spectral broadening of an optical pulse, which was first experimentally observed by Brewer in 1967 [13] and theoretically interpreted by Shimizu in the same year [14].

For simplicity, here we ignore the possible transverse spatial variations of the pulse by assuming that it has a plane wavefront so that there is no self-focusing or self-defocusing. To clearly see the effect of spectral broadening caused by self-phase modulation, we assume that the pulse initially has only one carrier frequency at ω_0 – that is, the pulse initially does not have frequency chirping. Then, the phase of the pulse as a function of space and time is, from (12.38),

$$\varphi(\mathbf{r}, t) = kl - \omega_0 t = \varphi(t) = \frac{\omega_0}{c}\left[n_0 + n_2 I(t)\right]l - \omega_0 t = \varphi_0 + \varphi_{\mathrm{K}}(t) - \omega_0 t, \quad (12.73)$$

where $\varphi_0 = n_0 \omega_0 l / c$ is the intensity-independent linear phase shift and $\varphi_{\mathrm{K}}(t)$ is the temporally varying Kerr phase:

$$\varphi_{\mathrm{K}}(t) = \frac{\omega_0}{c} n_2 l I(t). \quad (12.74)$$

The instantaneous frequency of an optical wave is generally found as

$$\omega(t) = -\frac{\partial \varphi(t)}{\partial t}. \quad (12.75)$$

For a monochromatic wave that has a constant frequency of ω_0, its temporal phase variation is $\varphi(t) = -\omega_0 t$, as expected. For an optical pulse that has an intensity-dependent, temporally varying phase of (12.73), we find that

$$\omega(t) = \omega_0 - \frac{\partial \varphi_{\mathrm{K}}(t)}{\partial t} = \omega_0\left(1 - \frac{n_2 l}{c}\frac{\mathrm{d}I(t)}{\mathrm{d}t}\right). \quad (12.76)$$

Thus, self-phase modulation causes the instantaneous carrier frequency of a pulse to vary with the temporal slope of the pulse intensity profile. Because $\mathrm{d}I/\mathrm{d}t \neq 0$ for a pulse, except at its peak, the frequencies at different times during the pulse shift away from its center frequency at ω_0. Thus, the effect of temporal self-phase modulation is to broaden the spectrum of the pulse by generating new frequencies on both the leading and the trailing edges of the pulse, as shown in Fig. 12.6. In the case that $n_2 > 0$, the frequencies generated in the leading half of the pulse are red-shifted frequencies that are lower than the center frequency, whereas those generated in the

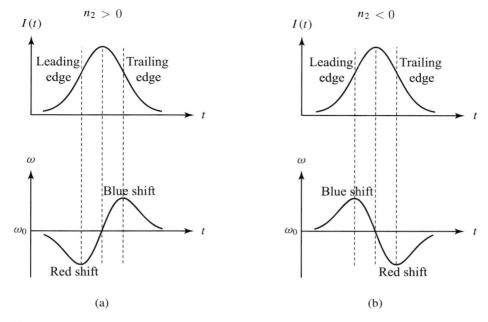

Figure 12.6 Spectral broadening caused by self-phase modulation of an optical pulse. (a) In the case that $n_2 > 0$, the frequencies generated in the leading half of the pulse are red-shifted frequencies lower than the center frequency, whereas those generated in the trailing half of the pulse are blue-shifted higher frequencies. (b) In the case that $n_2 < 0$, the frequencies generated in the leading half of the pulse are blue-shifted frequencies higher than the center frequency, whereas those generated in the trailing half of the pulse are red-shifted lower frequencies.

trailing half of the pulse are blue-shifted higher frequencies. In the case that $n_2 < 0$, the frequencies generated in the leading half of the pulse are blue-shifted frequencies that are higher than the center frequency, whereas those generated in the trailing half of the pulse are red-shifted lower frequencies. Compared to the original pulse that has a single carrier frequency of ω_0, the spectrum of the pulse is broadened because additional frequency components that vary with the pulse profile are generated.

Note that though the pulse spectrum is broadened by self-phase modulation, the pulsewidth is not changed by self-phase modulation alone because the pulse still has the same temporal intensity profile:

$$\left| \mathbf{E}(l, t) \right|^2 = \left| \mathcal{E}(t) e^{ikl - i\omega_0 t} \right|^2 = \left| \mathcal{E}(t) e^{i\varphi(\mathbf{r}, t)} \right|^2 = \left| \mathcal{E}(t) \right|^2 = \frac{I(t)}{2c\epsilon_0 n_0}. \qquad (12.77)$$

Temporally varying phase modulation changes the frequency composition of the pulse, but it does not change the intensity profile of the pulse. In other words, a broad spectrum does not necessarily lead to a shorter pulse, nor to a longer pulse. Nonetheless, a shorter pulse does require a broader spectrum. A broad spectrum is a necessary but not sufficient condition for a shorter pulse. Temporal self-phase modulation of an optical pulse thus provides only the necessary condition. With a broadened spectrum, a shorter pulse can be accomplished by *pulse compression* techniques, including *chirp reduction* and temporal *soliton formation*. This approach is one of the primary techniques for generating an optical pulse as short as a few femtoseconds, which can have a spectrum as broad as the entire visible spectrum, or even broader. A pulse that has a broadened spectrum from self-phase modulation can also spread in time due to group-velocity dispersion, thus resulting in *temporal broadening*, if the sign of the group-velocity dispersion D is the same as the nonlinear refractive index n_2 that causes the spectral broadening. The phenomena of temporal pulse compression and temporal pulse broadening that are caused by group-velocity dispersion will be further discussed in Chapter 18.

12.5 ALL-OPTICAL MODULATORS AND SWITCHES

As mentioned in Subsection 11.3.1, phase modulation on an optical wave can be translated into other forms of modulation, including polarization modulation and amplitude modulation, by proper arrangements. This concept applies to any form of phase modulation irrespective of its physical mechanism. It is also true for self-phase modulation and cross-phase modulation that were respectively discussed in Subsections 12.1.1 and 12.1.2.

The optical-field-dependent birefringence caused by the optical Kerr effect can be used for polarization modulation of an optical wave. Such polarization modulation can be either self-induced in a one-beam interaction or cross-induced in a two-beam interaction. Phase modulation or polarization modulation can be translated into amplitude modulation by a proper arrangement. For simplicity, we consider the interactions in an isotropic medium. The same principle applies to all-optical polarization modulators using anisotropic crystals.

We have already seen from the discussions in Subsection 12.1.1 that for one-beam self-phase modulation in an isotropic medium, the induced $\mathbf{P}^{(3)}$ and the optical field $\mathbf{E}$ have the same polarization state if the optical field is linearly polarized. This is also true for a circularly polarized optical field. Therefore, the optical Kerr effect does not change the polarization state of a linearly or circularly polarized optical wave that propagates alone in an isotropic medium. The situation is different for an elliptically polarized optical wave in a one-beam self-interaction, as well as for a linearly or circularly polarized optical wave in a two-beam interaction. In a one-beam self-interaction of an elliptically polarized optical wave, the polarization state of the induced $\mathbf{P}^{(3)}$ is different from that of the optical field $\mathbf{E}$, causing the polarization state of the optical field to change. The result is the phenomenon of ellipse rotation as discussed in Subsection 12.1.1 because the axes of the ellipse defined by the tip of the elliptically polarized optical field continue to rotate in space as the wave propagates through the nonlinear medium.

As seen in Subsection 12.1.2, cross-phase modulation by a linearly polarized pump wave creates a linear birefringence that is given in (12.26). Therefore, in the interaction of two linearly polarized optical waves, polarization modulation on one wave by the other through the optical Kerr effect is possible if the polarizations of the two waves are neither parallel nor perpendicular to each other. As mentioned in Subsection 12.1.2, the optical beam being modulated is called the signal or the probe, and that creating the modulation is called the pump. In the measurement of

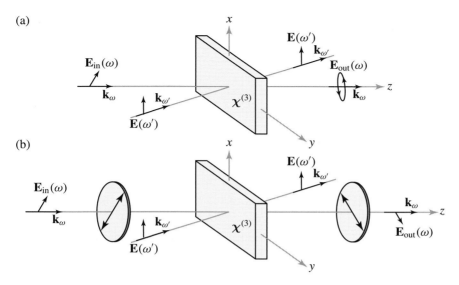

Figure 12.7 (a) All-optical polarization modulator and (b) all-optical amplitude modulator.

the optical Kerr effect through cross-modulation, the modulated wave is usually called the probe, whereas it is often called the signal in the context of a modulator. In an isotropic medium, the coordinate axes can be arbitrarily chosen. With a signal beam at a frequency of ω and a pump beam at a frequency of ω', we choose the xy plane to be that defined by the two linearly polarized field vectors $\mathbf{E}(\omega)$ and $\mathbf{E}(\omega')$, and the x axis to be in the direction of $\mathbf{E}(\omega')$, as shown in Fig. 12.7 (a). While the signal beam propagates in the z direction, the pump beam propagates in a direction on the yz plane, which may or may not be collinear with the propagation direction of the signal beam, as also shown in Fig. 12.7(a).

The optical-field-dependent birefringence that is caused by the pump wave and seen by the signal wave is described by $\Delta\epsilon(\omega, \mathbf{E})$, given in (5.36). In a practical application, the intensity of the signal beam is much lower than that of the pump beam: $I(\omega) \ll I(\omega')$. Therefore, the first term on the right-hand side of (5.36), which accounts for the self-modulation of the signal beam, can be neglected in comparison to the second term, which accounts for the cross-modulation on the signal by the pump. Then, $\Delta\epsilon(\omega, \mathbf{E})$ can be approximated by that given in (12.21). With $\mathbf{E}(\omega') \parallel \hat{x}$ and $\mathbf{E}(\omega)$ in the xy plane, as expressed in (12.22), we then have $\Delta\epsilon_{xx}(\omega) = \Delta\epsilon_{\parallel}(\omega)$ and $\Delta\epsilon_{yy}(\omega) = \Delta\epsilon_{\perp}(\omega)$, as given in (12.24), which result in $n_x(\omega) = n_{\parallel}(\omega)$ and $n_y(\omega) = n_{\perp}(\omega)$ for the signal beam, as given in (12.25). Because $n_{\parallel} \neq n_{\perp}$, the linearly polarized pump wave generates a field-dependent linear birefringence of $B = n_x - n_y = n_{\parallel} - n_{\perp}$, given in (12.26), for the signal wave. If the signal beam has a field of $\mathbf{E}(\omega) = (\hat{x}\mathcal{E}_x + \hat{y}\mathcal{E}_y)e^{-i\omega t}$ at the input surface of a nonlinear medium that has a thickness of l, its field at the output is

$$\mathbf{E}(\omega) = (\hat{x}\mathcal{E}_x e^{i\Delta\varphi} + \hat{y}\mathcal{E}_y)e^{ik^y l - i\omega t}, \tag{12.78}$$

where $k^y = n_y \omega / c$ and

$$\Delta\varphi = (k^x - k^y)l = \frac{B\omega}{c}l = \frac{3\pi}{c\epsilon_0 n_0^2 \lambda}\left(\chi_{1111}^{(3)\prime} - \chi_{1122}^{(3)\prime}\right)lI(\omega') \tag{12.79}$$

is the phase retardation of the y component with respect to the x component of the signal field. Because this phase retardation is linearly proportional to the pump intensity, the polarization state of the signal beam at the output can be modulated by varying the pump intensity if $\mathbf{E}(\omega)$ is neither parallel nor perpendicular to $\mathbf{E}(\omega')$, so that both $\mathcal{E}_x$ and $\mathcal{E}_y$ have nonvanishing values for $\mathbf{E}(\omega)$ to sense the linear birefringence that is induced by the pump field.

In the above, a linearly polarized pump beam is considered. The linearly polarized pump beam causes a linear birefringence, which can be used to modulate the polarization of a signal beam. By comparison, a circularly polarized pump beam causes a circular birefringence, as discussed Subsection 12.1.2. This circular birefringence due to cross-phase modulation can also be used to modulate the polarization of a signal beam. If a signal beam is linearly polarized, this circular birefringence results in the rotation of the direction of the linear polarization of the signal beam.

In comparison to the electro-optic polarization modulators discussed in Subsection 11.3.2, the only difference is that the all-optical polarization modulator discussed here is controlled by a pump optical beam rather than by a voltage. Other than this difference, these two types of polarization modulators have the same function and serve the same purpose.

As seen in Subsection 11.3.3, an amplitude modulator can be easily constructed by placing a polarization modulator between two polarizers. This approach is also applicable to the construction of an all-optical amplitude modulator by using an all-optical polarization modulator, as illustrated in Fig. 12.7(b). An all-optical amplitude modulator and an electro-optic amplitude modulator have the same transmission characteristics, which were discussed in Subsection 11.3.3, if they are set up in the same manner. When an ultrashort optical pulse is used as the pump beam, an all-optical amplitude modulator can function as an ultrafast *optical gate*, or an ultrafast *all-optical switch*, for switching the signal beam within a very short time.

12.6   GUIDED-WAVE ALL-OPTICAL MODULATORS AND SWITCHES

All practical guided-wave all-optical modulators and switches are of the refractive type based on the optical Kerr effect. Most of them require only one optical frequency for their operation, though multiple waveguide modes might be involved. Therefore, in this section, we consider all-optical modulators and switches that operate on only one optical frequency. The behavior of a single-frequency guided-wave device based on the optical Kerr effect is governed by the nonlinear coupled-mode equation given in (7.76):

$$\pm \frac{\mathrm{d}A_\nu}{\mathrm{d}z} = \sum_\mu \mathrm{i}\, \kappa_{\nu\mu} A_\mu \mathrm{e}^{\mathrm{i}(\beta_\mu - \beta_\nu)z} + \mathrm{i} \sum_{\mu,\xi,\zeta} \sigma_{\nu\mu\xi\zeta} A_\mu A_\xi A_\zeta^* \mathrm{e}^{\mathrm{i}(\beta_\mu + \beta_\xi - \beta_\zeta - \beta_\nu)z}, \qquad (12.80)$$

where $\kappa_{\nu\mu}$ is the linear coupling coefficient caused by any perturbations to the waveguide other than nonlinear optical perturbations, and

$$\sigma_{\nu\mu\xi\zeta} = \omega C_{\nu\mu\xi\zeta} = 3\omega\epsilon_0 \int\limits_{-\infty}^{\infty} \int\limits_{-\infty}^{\infty} \hat{\boldsymbol{\mathcal{E}}}_\nu^* \cdot \boldsymbol{\chi}^{(3)} \vdots \hat{\boldsymbol{\mathcal{E}}}_\mu \hat{\boldsymbol{\mathcal{E}}}_\xi \hat{\boldsymbol{\mathcal{E}}}_\zeta^* dxdy. \tag{12.81}$$

12.6.1 Self-Phase Modulation

In the simplest situation that a waveguide mode $\boldsymbol{\mathcal{E}}_\nu$ at a particular optical frequency ω is not coupled to any other frequencies or any other modes at the same frequency, (12.80) reduces to the simple form:

$$\frac{dA_\nu}{dz} = i\sigma_{\nu\nu\nu\nu}A_\nu|A_\nu|^2, \tag{12.82}$$

where

$$\sigma_{\nu\nu\nu\nu} = \omega C_{\nu\nu\nu\nu} = 3\omega\epsilon_0 \int\limits_{-\infty}^{\infty} \int\limits_{-\infty}^{\infty} \hat{\boldsymbol{\mathcal{E}}}_\nu^* \cdot \boldsymbol{\chi}^{(3)} \vdots \hat{\boldsymbol{\mathcal{E}}}_\nu \hat{\boldsymbol{\mathcal{E}}}_\nu \hat{\boldsymbol{\mathcal{E}}}_\nu^* dxdy. \tag{12.83}$$

Because $\boldsymbol{\chi}^{(3)}$ is real for a device that is based on the purely refractive optical Kerr effect, the nonlinear coefficient $\sigma_{\nu\nu\nu\nu}$ is also a real quantity. It is then clear from (12.82) that only the phase, but not the magnitude, of A_ν varies with z. Therefore, the mode power, $P_\nu = |A_\nu|^2$, is a constant that is independent of z. In this case, the solution of (12.82) can be easily obtained:

$$A_\nu(z) = A_\nu(0) \exp\left(i\sigma_{\nu\nu\nu\nu}P_\nu z\right) = A_\nu(0) \exp\left(i\beta_\nu^{NL} z\right), \tag{12.84}$$

where

$$\beta_\nu^{NL} = \sigma_{\nu\nu\nu\nu}P_\nu \tag{12.85}$$

is a power-dependent nonlinear modification on the propagation constant.

Clearly, the consequence of the optical Kerr effect on an individual waveguide mode is an effective propagation constant that is a function of the mode power:

$$\beta_\nu^{eff} = \beta_\nu^L + \beta_\nu^{NL} = \beta_\nu + \sigma_{\nu\nu\nu\nu}P_\nu, \tag{12.86}$$

where $\beta_\nu^L = \beta_\nu$ is the power-independent linear propagation constant of the mode. This effect leads to the self-phase modulation of the mode field over a distance of l in the waveguide:

$$\varphi_\nu^{NL} = \beta_\nu^{NL}l = \sigma_{\nu\nu\nu\nu}P_\nu l = \sigma_{\nu\nu\nu\nu}|A_\nu|^2 l, \tag{12.87}$$

which is linearly proportional to the mode power.

For a waveguide that is fabricated in an isotropic medium, such as silica glass, $\chi_{xxxx}^{(3)} = \chi_{yyyy}^{(3)} = \chi_{zzzz}^{(3)} = \chi_{1111}^{(3)}$. Then, $\sigma_{\nu\nu\nu\nu}$ defined in (12.83) takes the form:

$$\sigma_{\nu\nu\nu\nu} = 3\omega\epsilon_0 \chi_{1111}^{(3)\prime} \int\limits_{-\infty}^{\infty} \int\limits_{-\infty}^{\infty} \left|\hat{\boldsymbol{\mathcal{E}}}_\nu\right|^4 dxdy. \tag{12.88}$$

It is then convenient to define an effective area for a waveguide mode that is involved in a third-order nonlinear process as

$$
\mathcal{A}_v^{\text{eff}} = \frac{\left[\int\limits_{-\infty}^{\infty} \int\limits_{-\infty}^{\infty} \left| \hat{\mathcal{E}}_v \right|^2 dx dy \right]^2}{\int\limits_{-\infty}^{\infty} \int\limits_{-\infty}^{\infty} \left| \hat{\mathcal{E}}_v \right|^4 dx dy} = \left(\frac{\omega \mu_0}{2 \beta_v} \right)^2 \frac{1}{\int\limits_{-\infty}^{\infty} \int\limits_{-\infty}^{\infty} \left| \hat{\mathcal{E}}_v \right|^4 dx dy},
\tag{12.89}
$$

where the orthonormality relation given in (7.10) is used.[2] Then, by using (12.86), (12.88), and (12.89), we find that the power-dependent effective index of the waveguide mode can be expressed as

$$
n_v^{\text{eff}} = n_v + n_{2v} \frac{P_v}{\mathcal{A}_v^{\text{eff}}},
\tag{12.90}
$$

where $n_v^{\text{eff}} = c \beta_v^{\text{eff}} / \omega$, $n_v = c \beta_v / \omega$, and

$$
n_{2v} = \frac{3 \chi_{1111}^{(3)\prime}}{4 c \epsilon_0 n_v^2}.
\tag{12.91}
$$

Clearly, (12.90) has the form of (12.2), and (12.91) has the form of (12.3). Then the Kerr phase change that is caused by the self-phase modulation of a mode field can be expressed in the form of (12.4) as

$$
\varphi_v^{\text{K}}(\mathbf{r}, t) = \beta_v^{\text{NL}} l = \frac{\omega}{c} n_{2v} \frac{P_v}{\mathcal{A}_v^{\text{eff}}} l = \frac{2 \pi n_{2v}}{\lambda} \frac{P_v}{\mathcal{A}_v^{\text{eff}}} l.
\tag{12.92}
$$

Note that (12.91) is valid only for a mode in a waveguide that is made of a noncrystalline isotropic material. For a mode in a waveguide based on a crystalline material, such as GaAs or LiNbO$_3$, (12.92) can still be used, but (12.91) is not generally valid because the nonlinear refractive index n_{2v} in this situation is a function of the mode field polarization direction with respect to the principal axes of the crystal.

12.6.2 Two-Mode Interaction

Guided-wave all-optical modulators and switches function on the same basic principle as guided-wave electro-optic modulators and switches by transforming a differential phase shift between two waveguide modes into an amplitude modulation, except that the required phase shift is controlled by the optical power in the waveguide structure rather than by an externally applied voltage. There are two basic approaches to transforming a differential phase shift into an amplitude modulation: by interference or by phase-sensitive coupling.

---

[2] The orthonormality relation in (7.10) is strictly accurate only for TE modes. It is used here as an approximation for other types of modes. In a weakly guiding waveguide, it is a good approximation.

The operation of most devices involves only two modes of either the same waveguide or two separate waveguides. The basic formulation for the interaction of two waveguide modes based on the optical Kerr effect was already discussed in Subsection 7.2.3. For a device that functions on two waveguide modes, a and b, at the same frequency of ω, the total field is $\mathbf{E}(\mathbf{r}, t) = \mathbf{E}(\mathbf{r}) \exp(-i\omega t)$ with

$$\mathbf{E}(\mathbf{r}) = A(z)\hat{\mathcal{E}}_a(x, y)e^{i\beta_a z} + B(z)\hat{\mathcal{E}}_b(x, y)e^{i\beta_b z}, \tag{12.93}$$

which is given in (7.78). The coupled-mode equations for such a single-frequency, two-mode device are those given in (7.79) and (7.80):

$$\pm \frac{dA}{dz} = i\kappa_{aa}A + i\kappa_{ab}Be^{i(\beta_b - \beta_a)z} + i\sigma_{aaaa}\left|A\right|^2 A + \text{nonlinear cross terms}, \tag{12.94}$$

$$\pm \frac{dB}{dz} = i\kappa_{bb}B + i\kappa_{ba}Ae^{i(\beta_a - \beta_b)z} + i\sigma_{bbbb}\left|B\right|^2 B + \text{nonlinear cross terms}, \tag{12.95}$$

where $\kappa_{aa}, \kappa_{ab}, \kappa_{bb}$, and κ_{ba} are the linear coupling coefficients of the two waveguide modes, and $\sigma_{aaaa} = \omega C_{aaaa}$ and $\sigma_{bbbb} = \omega C_{bbbb}$ are the nonlinear coupling coefficients as defined in (12.83). Only the nonlinear terms that have the coefficients of σ_{aaaa} and σ_{bbbb} are explicitly expressed in the coupled equations given above. These two terms represent self-phase modulation for modes a and b, respectively. In general, $\sigma_{aaaa} \neq \sigma_{bbbb}$ because two different waveguide modes generally do not have the same self-phase modulation. The characteristics of the nonlinear cross terms were already discussed in Subsection 7.2.3. The nonlinear cross terms are generally much smaller than the direct nonlinear terms of the coefficients σ_{aaaa} and σ_{bbbb}. In most applications, the nonlinear cross terms can be ignored.

For a device that is solely based on interference, $\kappa_{ab} = \kappa_{ba} = 0$, and the nonlinear cross terms also vanish. In this case, there is generally no direct power exchange between the two modes. The nonlinear differential phase shift between the two modes controls the interference condition, thus turning an optical-power-dependent phase change into amplitude modulation or switching. For a device that is based on coupling, $\kappa_{ab} \neq 0$ and $\kappa_{ba} \neq 0$. The function of modulation or switching is then a result of direct exchange of power between the two modes. In such a device, the power-dependent differential phase shift controls the effective coupling coefficient between the two modes through its influence on the phase matching between them.

12.6.3 Nonlinear Optical Mode Mixers

A *nonlinear mode mixer* is a simple all-optical switch that is based on the power-dependent interference effect between two modes in a multimode waveguide, such as the TE_0 and TE_1 modes of a slab waveguide. Two different modes in an unperturbed waveguide are orthogonal to each other in the absence of nonlinear effects. Even when the optical Kerr effect is present, significant direct coupling between them occurs only when the power in the waveguide reaches a critical level. Below this critical power level, the optical Kerr effect leads to sufficient self-phase modulation in each individual mode, but no significant cross-phase modulation or power exchange between the mutually orthogonal modes. For a nonlinear *two-mode mixer* operating in this regime, $\kappa_{aa} = \kappa_{bb} = \kappa_{ab} = \kappa_{ba} = 0$, and the nonlinear cross-interaction between the two modes can also be ignored. Consequently, both (12.94) and (12.95) reduce to the form of (12.82) with the solution for each mode having the form of (12.84). Therefore, the total field in the two-mode mixer is

$$\mathbf{E}(\mathbf{r}) = A(0)\hat{\mathcal{E}}_a(x,y)e^{i\beta_a^{\text{eff}}z} + B(0)\hat{\mathcal{E}}_b(x,y)e^{i\beta_b^{\text{eff}}z}$$
$$= \left[A(0)\hat{\mathcal{E}}_a(x,y) + B(0)\hat{\mathcal{E}}_b(x,y)e^{i(\beta_b^{\text{eff}} - \beta_a^{\text{eff}})z} \right] e^{i\beta_a^{\text{eff}}z}. \tag{12.96}$$

For a mode mixer that has a length of l, the total differential phase shift between the two modes over the length of the device is

$$\Delta\varphi = \left(\beta_b^{\text{eff}} - \beta_a^{\text{eff}} \right) l = \Delta\varphi_{\text{L}} + \Delta\varphi_{\text{K}}, \tag{12.97}$$

where $\Delta\varphi_{\text{L}} = (\beta_b - \beta_a)l$ is the *linear differential phase shift* due to modal dispersion and $\Delta\varphi_{\text{K}} = (\beta_b^{\text{NL}} - \beta_a^{\text{NL}})l$ is the *nonlinear differential phase shift* due to the different Kerr phase shifts caused by the difference in the self-phase modulation of the two different modes. For a given device of a fixed length, the value of $\Delta\varphi_{\text{L}}$ is fixed, but that of $\Delta\varphi_{\text{K}}$ varies with the powers in the modes. Therefore, the total phase difference, $\Delta\varphi$, between the two modes can be controlled by the power that is coupled into the waveguide. Even when that power is evenly divided between the two modes, there is still a power-dependent differential phase shift between the two modes because the Kerr phase change caused by self-phase modulation for a waveguide mode is also a function of the mode-dependent effective area $\mathcal{A}_\nu^{\text{eff}}$, as seen in (12.92).

Figure 12.8 illustrates the principle of a nonlinear two-mode mixer. In this example, the power that is launched into the waveguide is equally divided between the two modes, such as TE_0 and TE_1 modes, so that $P_a = P_b = P/2$ and $A(0) = B(0)$ at the input end. Because the two modes have different field distributions, the total field is asymmetrically distributed, with its peak on one side of the waveguide. The linear differential phase shift in this example is chosen to be $\Delta\varphi_{\text{L}} = 2n\pi$, where n is an integer. At low power levels when the power-dependent nonlinear differential Kerr phase shift $\Delta\varphi_{\text{K}}$ is negligibly small, the field distribution at the output end is the same as that at the input end, as shown in Fig. 12.8(a). At a power level of P_π for which $\Delta\varphi_{\text{K}} = \pi$, the total differential phase shift is $\Delta\varphi = 2(n+1)\pi$. Then, the peak of the total field at the output end is switched to the other side of the waveguide, as shown in Fig. 12.8(b). In this manner, a nonlinear mode mixer functions as a power-dependent *all-optical switch* to switch the optical power from one spatial location to another.

A nonlinear mode mixer can take the form of a two-mode slab waveguide or that of a two-mode channel waveguide. In the latter case, both the input and the output ends of the mode mixer can be connected to Y-junction waveguides for all-optical switching of the optical power between separate waveguides, as shown in Fig. 12.9. Such a device also functions as a *nonlinear mode sorter*.

12.6.4 All-Optical Mach–Zehnder Interferometers

A nonlinear mode mixer functions as a nonlinear interferometric device only when the optical power in the waveguide is kept below the critical power level to prevent direct coupling of the

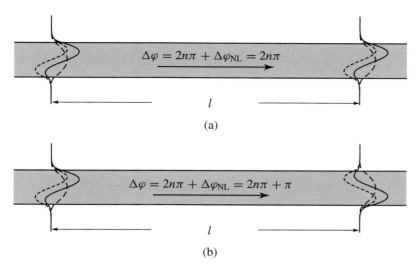

Figure 12.8 Power-dependent field distribution characteristics of a nonlinear mode mixer with a linear differential phase shift of $2n\pi$ (a) at low power levels when the nonlinear phase shift is negligible, and (b) at a power level of P_π when the nonlinear differential phase shift is π. The long dashed and short dashed curves respectively show fields of two individual modes, and the solid curve represents the total field of the two modes.

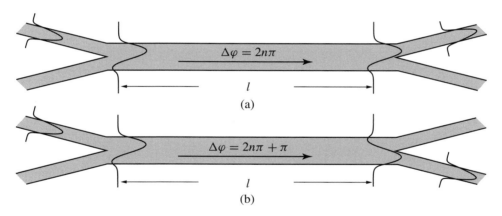

Figure 12.9 Mode mixer with Y-junction waveguides at its input and output ends for all-optical switching between separate waveguides.

modes. This limitation is caused by the fact that the two modes overlap in space while propagating codirectionally. It can be avoided in a nonlinear interferometer that consists of two separate arms, such as one in the form of a Mach–Zehnder interferometer, as shown in Fig. 12.10. There are a few significant differences between a nonlinear Mach–Zehnder interferometer and a nonlinear mode mixer:

1. Both arms of an all-optical Mach–Zehnder interferometer are generally single-mode waveguides.
2. At any power level, there is no cross-modulation between the fields in the two separate arms of the interferometer.

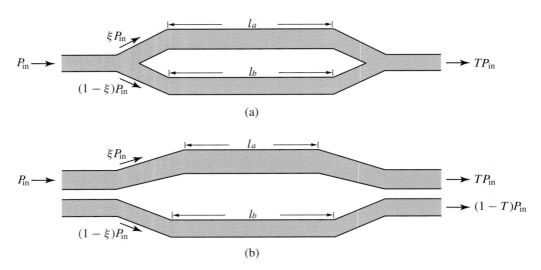

Figure 12.10 Single-input, all-optical Mach–Zehnder interferometers (a) using two Y-junction waveguides and (b) using two directional couplers for beam splitting at the input end and beam combining at the output end. The two arms of an all-optical interferometer are generally not balanced.

3. The two fields that are combined at the output end of the interferometer can experience different propagation distances because the lengths of the two arms do not have to be the same.

An all-optical Mach–Zehnder interferometer is based on the same principle as the electro-optic Mach–Zehnder interferometer discussed in Subsection 11.4.1, except that the differential phase shift $\Delta\varphi$ between its two arms is controlled by the optical power rather than by an applied electric field. An all-optical Mach–Zehnder interferometer can have a single input channel, as shown in Fig. 12.10, or three input channels, as shown in Fig. 12.11.

Figure 12.10 shows two possible structures of single-input, all-optical Mach–Zehnder interferometers. The beam-splitting and beam-combining couplers at the input and output ends, respectively, of an all-optical Mach–Zehnder interferometer can be either Y-junction waveguides, as shown in Fig. 12.10(a), or directional couplers, as shown in Fig. 12.10(b). The two arms of an all-optical interferometer are not required to be identical. Therefore, in general, the linear differential phase shift is $\Delta\varphi_\mathrm{L} = \beta_b l_b - \beta_a l_a$, and the nonlinear differential Kerr phase shift can be expressed as

$$\Delta\varphi_\mathrm{K} = \beta_b^\mathrm{NL} l_b - \beta_a^\mathrm{NL} l_a = \frac{2\pi}{\lambda}\left(n_{2b}\frac{P_b}{\mathcal{A}_b^\mathrm{eff}} l_b - n_{2a}\frac{P_a}{\mathcal{A}_a^\mathrm{eff}} l_a \right), \tag{12.98}$$

where l_a and l_b are the lengths of the two separate arms, respectively. We see that a power-dependent nonlinear differential Kerr phase shift can be obtained only when the two arms are not balanced, due to unbalanced excitation or physical asymmetry between them. With unbalanced excitation, $P_a \neq P_b$. Physical asymmetry exists when the waveguides that form the two arms have different lengths, $l_a \neq l_b$, or different effective areas, $\mathcal{A}_a^\mathrm{eff} \neq \mathcal{A}_b^\mathrm{eff}$, or different values of nonlinearity, $n_{2a} \neq n_{2b}$, or any combination of them.

To facilitate the possibility of unbalanced excitation, the Y-junction waveguides or directional couplers that are used in an all-optical Mach–Zehnder interferometer are not necessarily 3-dB couplers. For a given device, however, the beam-splitting coupler at the input end and the beam-combining coupler at the output end are usually identical couplers with a fixed power-splitting ratio of $\xi : (1 - \xi)$ between the two arms, as also shown in Fig. 12.10. For an all-optical Mach–Zehnder interferometer that uses Y-junction waveguides as input and output couplers, the power transmittance is

$$T = 1 - 2\xi(1 - \xi)(1 - \cos \Delta\varphi), \tag{12.99}$$

where $\Delta\varphi = \Delta\varphi_\mathrm{L} + \Delta\varphi_\mathrm{K}$ is the total differential phase shift. For one that uses directional couplers as input and output couplers, the power transmittance through the same channel is

$$T = 1 - 2\xi(1 - \xi)(1 + \cos \Delta\varphi). \tag{12.100}$$

As discussed above, balanced excitation with $\xi = 1/2$ is feasible only when the two arms of the interferometer are physically asymmetric. Such physical asymmetry leads to a nonvanishing linear differential phase shift, $\Delta\varphi_\mathrm{L} \neq 0$, which acts as a bias phase shift. By properly adjusting the asymmetry between the two arms, the value of this bias phase shift can be chosen for a desired operating point of the device.

Figure 12.11 shows the structure of a three-input, symmetric all-optical Mach–Zehnder interferometer that uses Y-junction waveguides for its input and output ports [15]. This device consists of three input channels that are fed into a symmetric Mach–Zehnder interferometer that has two arms of equal lengths. The data signal is sent through the central channel c, and the control signals are fed into either channel a or b, or both. The data signal wave is orthogonally polarized with respect to the control signals to avoid interference between them. A polarizer at the output end allows only the polarization of the data signal to pass. The interaction between the data signal and the control signals is only through cross-phase modulation. By using ultrashort optical pulses for the data and control signals, this interferometer functions as an *ultrafast optical gate*.

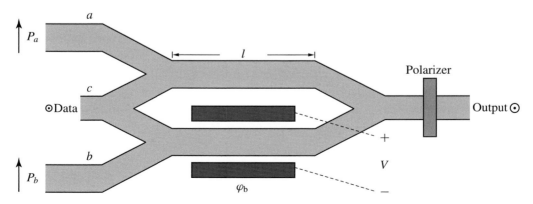

Figure 12.11 Three-input, symmetric all-optical Mach–Zehnder interferometer using Y-junction waveguides. The bias voltage V can provide a bias phase difference of φ_b between the two arms.

Because a data signal that is sent through channel c is equally split between the two arms of the interferometer, nonlinear phase shifts caused by self-phase modulation of the data signal in the two arms cancel. The net differential nonlinear phase shift is caused by the cross-phase modulation that is imposed by any control signals on the data signal. This differential nonlinear Kerr phase shift has exactly the form of (12.98) with $n_{2a} = n_{2b} = n_2$, $l_a = l_b = l$, and $\mathcal{A}_a^{\text{eff}} = \mathcal{A}_b^{\text{eff}} = \mathcal{A}_{\text{eff}}$ for a symmetric Mach–Zehnder interferometer:

$$\Delta\varphi_{\text{K}} = \frac{2\pi}{\lambda}\frac{n_2 l}{\mathcal{A}_{\text{eff}}}(P_b - P_a), \tag{12.101}$$

where n_2 is the nonlinear refractive index due to cross-phase modulation between orthogonally polarized waves. Though the two arms of a symmetric Mach–Zehnder interferometer are equal in length, it is still possible to introduce a linear phase difference between them by a bias voltage if the device is fabricated on an electro-optic material. For a symmetric Mach–Zehnder interferometer using Y-junction waveguides, the transmittance of the data signal is that given in (12.99) with $\xi = 1/2$, which reduces to the simple form:

$$T = \cos^2\frac{\Delta\varphi}{2}. \tag{12.102}$$

An all-optical interferometer has many useful applications. Like an electro-optic interferometer, it can be used as an amplitude modulator or, when accompanied by a directional coupler instead of a Y-junction waveguide at the output end, as a switch. Unlike an electro-optic interferometer, however, its function is completely controlled by the input optical power alone. Therefore, there are some unique applications of an all-optical interferometer that are not possible with an electro-optic interferometer. For instance, with unbalanced excitation in an all-optical interferometer that has symmetric arms, it is possible to shape an optical pulse by taking advantage of the fact that the power-dependent transmittance of the device now varies across the envelope of the pulse. Pulse shortening can be achieved if the maximum transmittance occurs at the peak of the pulse, while the wings of the pulse have very low transmittance. All-optical Mach–Zehnder interferometers can be made to perform certain unique functions, such as *optical logic*, *optical sampling*, and *optical on–off switching*.

Here we have ignored the losses and the dispersion of a waveguide. In reality, a waveguide has both linear losses, mainly from scattering and impurity absorption, and nonlinear losses, from both two-photon and three-photon absorption processes. These losses increase the switching power of a device while reducing its extinction ratio between on and off states. When a device is operated with short optical pulses, the dispersion of a waveguide can broaden the pulses and introduce an additional linear phase shift in the pulses. The consequences are also an increase in the switching energy and a reduction in the extinction ratio.

EXAMPLE 12.4
A three-input, symmetric all-optical Mach–Zehnder interferometer as shown in Fig. 12.11 consists of AlGaAs channel waveguides fabricated on a GaAs substrate along the [110]

direction on the (001) crystallographic plane. The data signal launched into channel c is a TM-like mode that is mainly polarized in the [001] direction. A control signal is launched into channel a as a TE-like mode that is mainly polarized in the $[1\bar{1}0]$ direction. No control signal is launched into channel b. Both data and control signals are at the wavelength of $\lambda = 1.55\ \mu m$. The length of both arms of the interferometer is $l = 2$ cm, and the effective area of the channel waveguide in each arm is $\mathcal{A}_{\text{eff}} = 6\ \mu m^2 = 6 \times 10^{-12}\ m^2$. The nonlinear refractive index characterizing cross-phase modulation between TE-like and TM-like modes in this AlGaAs waveguide at $\lambda = 1.55\ \mu m$ is $n_2 = 1.3 \times 10^{-17}\ m^2\ W^{-1}$. No linear bias phase is applied to either arm of the device. Ignoring all possible linear and nonlinear losses, find the power of the control signal that is needed for this device to function as an all-optical on–off switch. If the control signal is in the form of an optical pulse of $\Delta t_{\text{ps}} = 1$ ps pulsewidth, what is the switching energy of the control pulse?

Solution

For the device to function as an all-optical on–off switch, both the on state, with a transmittance of $T = 1$, and the off state, with a transmittance of $T = 0$, have to be accessible by varying the power of the control signal. Because there is no linear phase bias, the total differential phase shift of the device is contributed solely by the nonlinear effect; thus, $\Delta\varphi = \Delta\varphi_K$. Because no control signal is launched into channel b, $P_b = 0$. From (12.102), we then find that the minimum nonlinear differential phase shift required for $T = 1$ is $\Delta\varphi_K = 0$ and that required for $T = 0$ is $\Delta\varphi_K = -\pi$ or $\Delta\varphi_K = \pi$. Therefore, we find from (12.101) that the on state can be reached by simply making $P_a = 0$ so that $\Delta\varphi_K = 0$. By setting $\Delta\varphi_K = -\pi$ for (12.101), we find the power of the control signal required to reach the off state:

$$P_a = \frac{\lambda \mathcal{A}_{\text{eff}}}{2n_2 l} = \frac{1.55 \times 10^{-6} \times 6 \times 10^{-12}}{2 \times 1.3 \times 10^{-17} \times 2 \times 10^{-2}}\ W = 17.88\ W.$$

If the control signal is in the form of an optical pulse of $\Delta t_{\text{ps}} = 1$ ps pulsewidth, the device is in the on state with $T = 1$ in the absence of a control pulse. The device can be switched to the off state with a control pulse of a switching energy of $U_{\text{ps}} = P_a \Delta t_{\text{ps}} = 17.88$ pJ.

Clearly, a waveguide Mach–Zehnder interferometer operated with a CW beam at 17.88 W is not practical, but it is practical with an ultrashort pulse that has a peak power of 17.88 W, such as the one of 1 ps pulsewidth considered here. For this reason, the control signal for this device is generally in the form of ultrashort laser pulses, though the data signal can be of any waveform.

12.6.5 Nonlinear Optical Loop Mirrors

A *nonlinear optical loop mirror* [16, 17], is a folded Mach–Zehnder interferometer in the so-called *Sagnac configuration*, as shown in Fig. 12.12. The basic device shown in Fig. 12.12(a) consists of a single-mode waveguide loop, such as a single-mode optical fiber or a single-mode semiconductor waveguide, that is closed with a four-port directional coupler. The two paths of opposite propagation directions in the loop are equivalent to the two arms of an interferometer. The single coupler, which has a power-splitting ratio of $\xi : (1 - \xi)$, serves as both the

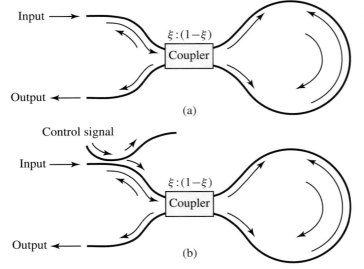

Figure 12.12 Nonlinear optical loop mirrors: (a) single-input configuration without a control signal and (b) two-input configuration with a control signal.

power-splitting input coupler and the power-combining output coupler. An input field is split into two contra-propagating fields that travel through exactly the same loop path but in opposite directions before recombining at the coupler to form the output of the device.

The optical field launched into the device can be a short pulse that has a spatial span much shorter than the loop length. Then the interaction between contradirectionally propagating pulses in the loop is negligible so that only the self-phase modulation of each individual pulse needs to be considered. It can also be a very long pulse or a CW wave that fills up the entire loop. Then the cross-phase modulation between contradirectionally propagating waves needs to be considered as well. Because of the exact symmetry between the two contradirectional paths, $\Delta\varphi_L = 0$ irrespective of the operating condition. It can be shown that for both cases discussed here, we have

$$\Delta\varphi = \Delta\varphi_K = (1 - 2\xi)\frac{2\pi n_2}{\lambda}\frac{P_{in}}{\mathcal{A}_{eff}}l, \tag{12.103}$$

where P_{in} is the input power launched into the device and l is the length of the loop. The transmittance, T, of the device is that given in (12.100), with $\Delta\varphi = \Delta\varphi_K$ given above. The device also has a reflectance of $R = 1 - T$ back to the original input port. In the linear regime at low power levels, the device functions as a mirror with $R = 4\xi(1 - \xi)$ and $T = 1 - 4\xi(1 - \xi)$. In the nonlinear regime at a high power level, the device functions as a nonlinear mirror with power-dependent reflectance and transmittance due to the dependence of $\Delta\varphi_K$ on the input power.

Similar to the Mach–Zehnder interferometer, a nonlinear optical loop mirror can also accept a control signal to switch the data signal. Figure 12.12(b) shows a two-input configuration for such a purpose. More sophisticated configurations and two-wavelength operations are also possible [17]. With a control signal, a nonlinear optical loop mirror can perform such functions as optical switching, sampling, multiplexing, and demultiplexing.

There are several advantages of using the nonlinear optical loop mirror as an all-optical interferometric device over the conventional all-optical Mach–Zehnder interferometer with two separate arms. Because the two contra-propagating fields in a nonlinear optical loop mirror travel over exactly the same path in opposite directions, they experience exactly the same linear effects, which cancel out when the two fields are combined in returning to the coupler. Therefore, the device is stable against external perturbations and does not require interferometric alignment. This unique characteristic allows a very long fiber of the order of kilometers to be used for a nonlinear optical loop mirror to function with sufficient self-phase modulation at a low optical power level, making it a truly practical all-optical device. Because the response and relaxation of optical Kerr nonlinearity in a silica fiber are nearly instantaneous, a nonlinear fiber loop mirror is also ideal for many applications that use ultrashort optical pulses. The precise match in length of the contradirectional paths in this device ensures precise coincidence of the returning pulses, which is a daunting task with a conventional interferometer of separate arms, considering the fact that the path length can be as long as a few kilometers while the pulses can be shorter than 1 ps.

EXAMPLE 12.5

A single-input nonlinear optical fiber loop mirror of the configuration shown in Fig. 12.12(a) consists of a single-mode fiber that has a loop length of $l = 100$ m and an effective cross-sectional area of $\mathcal{A}_{\text{eff}} = 3 \times 10^{-11}$ m^2 for an optical wave at the $\lambda = 1.55$ μm wavelength. The self-phase modulation nonlinear refractive index of this fiber is $n_2 = 3.2 \times 10^{-20}$ m^2 W^{-1}. At low input power levels, this loop mirror has a linear transmittance of $T = 25\%$. Find the lowest input power that is required for it to have a transmittance of 100%.

Solution

With a low-power transmittance of $T = 25\% = 1/4$, we find by solving $T = 1 - 4\xi$ $(1 - \xi) = 1/4$ that $\xi = 1/4$ for the power-splitting ratio of the coupler in the device. By plugging $\xi = 1/4$ and $T = 1$ into (12.100), we find that the nonlinear phase shift required for $T = 1$ at a high power level is a solution of the condition that $1 + \cos \Delta\varphi = 0$. Therefore, $\Delta\varphi = (2n + 1)\pi$ for any integer n. From (12.103), we see that $P_{\text{in}} \propto \Delta\varphi$. The lowest power required for $T = 100\%$ can be obtained by plugging $\Delta\varphi = \pi$ and $\xi = 1/4$ into (12.103) to find that

$$P_{\text{in}} = \frac{\lambda \mathcal{A}_{\text{eff}}}{n_2 l} = \frac{1.55 \times 10^{-6} \times 3 \times 10^{-11}}{3.2 \times 10^{-20} \times 100} \text{ W} = 14.53 \text{ W}.$$

This power is too high for this fiber device to be practical if the input is a CW signal. It is not a problem if the input signal consists of very short pulses. For instance, an average power of only 1.453 mW is required if the input signal is made up with pulses that have a pulsewidth of

1 ps at a repetition rate of 100 MHz. For this reason, nonlinear optical loop mirrors are generally operated with very short laser pulses.

12.6.6 Nonlinear Directional Couplers

The coupling efficiency of a directional coupler can be varied by varying the phase mismatch or the coupling coefficients between the two waveguides that form the directional coupler. For the electrically modulated directional coupler discussed in Subsection 11.4.2, the coupling coefficient is a function of an externally applied voltage that induces changes in the refractive index of the waveguide material through the Pockels effect. For an all-optical nonlinear directional coupler based on the optical Kerr effect [18], the coupling coefficient can be varied by varying the value or the distribution of the optical power that is launched into the device. A nonlinear directional coupler can be formed by using two parallel waveguides that are fabricated in a nonlinear crystal, such as GaAs or $LiNbO_3$, as shown in Fig. 12.13(a). It can also be formed by using a dual-core optical fiber, as shown in Fig. 12.13(b). The advantage of using a dual-core fiber is that a coupler of a very long interaction length of the order of kilometers can be easily realized to make practical use of the small optical nonlinearity of an optical fiber. In the following, we consider for simplicity only symmetric directional couplers in which the two waveguide channels are identical. Asymmetric nonlinear directional couplers have similar characteristics.

For a symmetric nonlinear directional coupler that is formed by two identical single-mode waveguides, we have $\beta_a = \beta_b$ and $\kappa_{aa} = \kappa_{bb}$. Therefore, the effective linear propagation constant of each individual waveguide mode is $\beta = \beta_a + \kappa_{aa} = \beta_b + \kappa_{bb}$. In addition, $\kappa_{ab} = \kappa_{ba}^* \equiv \kappa$, which is real and positive, and $\sigma_{aaaa} = \sigma_{bbbb} \equiv \sigma$, which is real but can be either positive or negative, depending on the sign of $\chi^{(3)}$ of the optical Kerr medium. The coupled equations given in (12.94) and (12.95) for two-mode interaction can be simplified for a symmetric nonlinear directional coupler as

$$\frac{d\widetilde{A}}{dz} = i\kappa\widetilde{B} + i\sigma\left|\widetilde{A}\right|^2\widetilde{A} + \text{nonlinear cross terms,} \tag{12.104}$$

$$\frac{d\widetilde{B}}{dz} = i\kappa\widetilde{A} + i\sigma\left|\widetilde{B}\right|^2\widetilde{B} + \text{nonlinear cross terms,} \tag{12.105}$$

where $\widetilde{A} = Ae^{-i\kappa_{aa}z}$ and $\widetilde{B} = Be^{-i\kappa_{bb}z}$. The terms that are characterized by the coupling coefficient κ represent linear coupling between the two modes. The terms that are characterized by σ represent the self-phase modulation of each individual mode. The nonlinear cross terms, which are not explicitly spelled out because of their complexity, contribute to nonlinear coupling between the two modes. In general, the nonlinear cross terms, though not completely negligible, are much smaller than the terms that represent linear coupling and self-phase modulation in each equation. Indeed, the nonlinear coupling contributed by the nonlinear cross terms is not necessary for the functioning of a nonlinear directional coupler. The basic operation principle of a nonlinear directional coupler is that the self-phase modulation of each individual mode

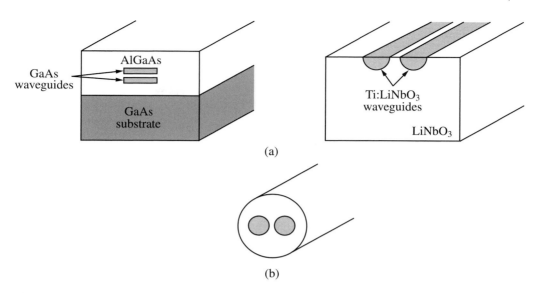

Figure 12.13 Nonlinear directional couplers formed by (a) two parallel waveguides and (b) a dual-core optical fiber.

creates a power-dependent differential phase shift that leads to a power-dependent phase mismatch between the two modes. Consequently, the coupling efficiency of the nonlinear directional coupler would become power dependent even if the only coupling were the linear coupling characterized by the linear coefficient κ. The direct nonlinear coupling contributed by the nonlinear cross terms acts as an additional perturbation, which changes the detailed quantitative characteristics of a nonlinear directional coupler. The general characteristics of a nonlinear directional coupler can be fully understood without considering the nonlinear cross terms.

We consider only the simple case that the nonlinear cross terms are neglected. We also assume that the input optical power is initially launched into only waveguide a so that $P_a(0) = P_{\text{in}}$ and $P_b(0) = 0$. Under these assumptions, the coupling efficiency of a nonlinear coupler that has an interaction length of l is found to be

$$\eta = \frac{P_b(l)}{P_{\text{in}}} = \frac{1}{2}\left[1 - \text{cn}(2\kappa l, m)\right] = \frac{1}{2}\left[1 - \text{cn}\left(2\kappa l, \frac{\sigma}{4\kappa}P_{\text{in}}\right)\right], \tag{12.106}$$

where

$$m = \frac{\sigma}{4\kappa}P_{\text{in}} = \frac{P_{\text{in}}}{P_{\text{c}}} \tag{12.107}$$

is an index that characterizes the level of the input power with respect to a critical power level of P_{c} that is defined as

$$P_{\text{c}} = \frac{4\kappa}{\sigma} = \frac{2\kappa\lambda\,\mathcal{A}_{\text{eff}}}{\pi n_2}, \tag{12.108}$$

and $\text{cn}(z, m)$ is a Jacobi elliptic function defined by the relation:

$$z = \int_{x}^{1} \frac{dt}{(1 - t^2)^{1/2}(1 - m^2 + m^2 t^2)^{1/2}} = \text{cn}^{-1}(x, m). \tag{12.109}$$

For the symmetric coupler under consideration, $P_a(l) + P_b(l) = P_{\text{in}}$, and the power transmittance through the input channel is

$$T = \frac{P_a(l)}{P_{\text{in}}} = 1 - \eta = \frac{1}{2}\left[1 + \text{cn}(2\kappa l, m)\right] = \frac{1}{2}\left[1 + \text{cn}\left(2\kappa l, \frac{\sigma}{4\kappa}P_{\text{in}}\right)\right]. \tag{12.110}$$

It can be clearly seen from (12.106) that the coupling efficiency of a nonlinear coupler is a function of the input power to the device. Figure 12.14 shows the coupling efficiency as a function of the interaction length l, normalized to the linear coupling length of $l_c^{\text{PM}} = \pi/2\kappa$, at various input power levels that are characterized by different values of the index m. In the limit of a very low input power, $P_{\text{in}} \ll P_c$ and $m \approx 0$, the coupling efficiency reduces to that of the phase-matched linear directional coupler, $\eta = (1 - \cos 2\kappa l)/2 = \sin^2 \kappa l$ because $\text{cn}(2\kappa l, 0) = \cos 2\kappa l$, and the coupling length is just l_c^{PM}.

As the input power increases, a power-dependent phase mismatch between the two waveguide channels is generated by the power-dependent differential phase shift. At a relatively low input power such that $P_{\text{in}} < P_c$ and $m < 1$, this power-dependent phase mismatch has the effect of slowing down the power transfer between the two channels. This phase mismatch is reduced as more power is transferred and is later even reversed as more than 50% of the input power is transferred. The nonlinear directional coupler thus acts like a reversed-$\Delta\beta$ coupler. Complete switching of power with $\eta = 1$ to reach the cross state still occurs, but the coupling length is longer than the linear coupling length and it increases as the input power increases. These effects can be observed from the curve for $m = 0.9$ in Fig. 12.14.

At a high input power such that $P_{\text{in}} > P_c$ and $m > 1$, the initial phase mismatch is so large that the power transfer never reaches the 50% point for the phase mismatch to be reversed. Therefore, the coupling efficiency oscillates, but $\eta < 1/2$ for any device length. The cross state cannot be reached at such a high power level, as can be seen in Fig. 12.14 from the curves for $m = 1.1$ and $m = 2$.

At the critical power level that $P_{\text{in}} = P_c$ and $m = 1$, the coupling efficiency stays at $\eta = 1/2$ indefinitely after 50% of the input power is transferred. This state is unstable as any perturbation caused by noise or fluctuations in the input power can tip this balance between the two channels.

Figure 12.15 shows the coupling efficiency as a function of the input power P_{in}, normalized to the critical power P_c, for a symmetric nonlinear directional coupler that has a fixed length of $l = l_c^{\text{PM}}$, known as the *half-beat-length coupler*, and another of $l = 2l_c^{\text{PM}}$, known as the *beat-length coupler*. For the half-beat-length coupler, which starts with a linear coupling efficiency of $\eta = 1$ at a very low input power level, the coupling efficiency remains high until the input power approaches the level of P_c, when it drops and remains low for all high power levels above P_c.

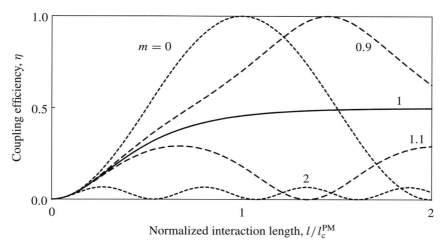

Figure 12.14 Coupling efficiency of a symmetric nonlinear directional coupler as a function of the interaction length l, normalized to the linear coupling length of $l_c^{PM} = \pi/2\kappa$, at various input power levels that are represented by different values of the index $m = P_{in}/P_c$.

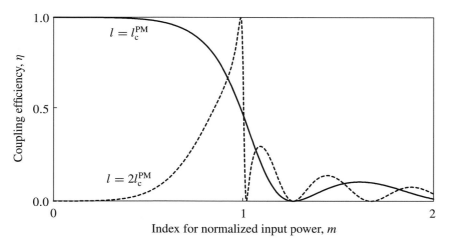

Figure 12.15 Coupling efficiency as a function of index m, which represents the input power P_{in} normalized to the critical power P_c, for two symmetric nonlinear directional couplers of fixed lengths $l = l_c^{PM}$, shown in the solid curve, and $l = 2l_c^{PM}$, shown in the dashed curve.

For the beat-length coupler, which starts with a linear coupling efficiency of $\eta = 0$ at a very low power level, only a very narrow power range exists for high coupling efficiencies with a peak value of $\eta = 1$.

In the above discussions, only symmetric directional couplers are considered, and the effect of nonlinear cross terms in (12.104) and (12.105) are ignored. There exists a general analytical solution in the form of elliptic functions for the coupled nonlinear differential equations even when the structure of the coupler is asymmetric and the nonlinear cross terms are considered. The primary effect of the nonlinear cross terms is to cause a change in the value of P_c, which depends on the strength of the nonlinear cross coupling between the two waveguide modes. For an asymmetric coupler, the initial linear phase mismatch leads to power-dependent characteristics that are nonreciprocal with respect to the detuning between the two channels.

EXAMPLE 12.6

A half-beat-length nonlinear directional coupler of the structure shown in Fig. 12.13(a) for a TE-like mode has a length of $l = 1.5$ cm. It consists of two parallel AlGaAs channel waveguides on a GaAs substrate with the same structural parameters as the AlGaAs waveguides described in Example 12.4. At the $\lambda = 1.55$ μm wavelength, the nonlinear refractive index characterizing self-phase modulation for the TE-like mode in the waveguide is $n_2 = 1.5 \times 10^{-17}$ m^2 W^{-1}. Find the critical power of the device.

Solution

Because the device is a half-beat-length coupler, we have $l_c^{PM} = l = 1.5$ cm. Therefore, the coupling coefficient is $\kappa = \pi/2l_c^{PM} = \pi/2l$. By plugging this relation into (12.108), we find the critical power:

$$P_c = \frac{\lambda \mathcal{A}_{eff}}{n_2 l} = \frac{1.55 \times 10^{-6} \times 6 \times 10^{-12}}{1.5 \times 10^{-17} \times 1.5 \times 10^{-2}} \text{ W} = 41.3 \text{ W}.$$

If the device is operated with a short pulse of $\Delta t_{ps} = 1$ ps, like that considered in Example 12.4, then the critical pulse energy is 41.3 pJ.

Problem Set

12.1.1 In this problem, we consider the field-dependent nonlinear refractive index and birefringence in an isotropic medium due to self-phase modulation caused by the optical Kerr effect, as discussed in Subsection 12.1.1.

(a) Show that if an optical wave is linearly polarized, its polarization state remains unchanged under self-phase modulation caused by the optical Kerr effect. Verify that the wave sees an intensity-dependent index of refraction that can be expressed in the form of $n = n_0 + n_2 I$, where n_2 is that given in (12.3).

(b) Show that if an optical wave is circularly polarized, its polarization state remains unchanged under self-phase modulation caused by the optical Kerr effect. Verify that the wave sees an intensity-dependent index of refraction that can be expressed in the form of $n = n_0 + n_2 I$, where n_2 is that given in (12.6).

(c) Show that if an optical wave is elliptically polarized, its polarization state does not remain unchanged under self-phase modulation caused by the optical Kerr effect. The wave does not see an intensity-dependent index of refraction that can be expressed in the form of $n = n_0 + n_2 I$. Instead, verify that the field sees a field-dependent circular birefringence B_c, as given in (12.17).

12.1.2 In this problem, we consider the field-dependent nonlinear refractive index and birefringence in an isotropic medium due to cross-phase modulation caused by the optical Kerr effect, as discussed in Subsection 12.1.2.

(a) Show that a linearly polarized pump wave causes a field-dependent linear birefringence as given in (12.26) through cross-phase modulation seen by a probe wave.

(b) Show that a circular polarized pump wave causes a field-dependent circular birefringence as given in (12.32) through cross-phase modulation seen by a probe wave. Verify that this field-dependent circular birefringence vanishes when Kleinman's symmetry condition is valid.

(c) Show that an elliptically polarized pump wave causes both field-dependent linear birefringence and field-dependent circular birefringence through cross-phase modulation by verifying (12.36) and (12.37) under the condition of general elliptical polarization. Verify that the results reduce to those found in (a) in the special case that the elliptical polarization of the pump wave reduces to linear polarization. Verify that the results reduce to those found in (b) in the special case that the elliptical polarization of the pump wave reduces to circular polarization.

12.1.3 The principal axes of a cubic crystal are $\hat{x}$, $\hat{y}$, and $\hat{z}$. It has a length of l in the z direction. All of the optical waves being considered are polarized in the xy plane and propagate in the z direction over a distance of l. The nonvanishing elements of the third-order nonlinear susceptibility of the cubic crystal have the forms: $\chi^{(3)}_{1111} > \chi^{(3)}_{1122}$, $\chi^{(3)}_{1212}$, $\chi^{(3)}_{1221} > 0$. The crystal is stressed to become slightly birefringent in its linear optical property so that $\delta n = n_{0y} - n_{0x} > 0$ and $n_0 \gg \delta n$, but its nonlinear optical susceptibilities are not changed by the stress.

(a) A beam at a frequency of ω_0 is launched into the crystal. At the input location, $z = 0$, it is linearly polarized at an angle of θ with respect to the x axis. Its output polarization state at $z = l$ can be used to deduce the stress-induced birefringence in the crystal. If we take $\theta = 45°$ and find that the beam is circularly polarized at $z = l$, what is the minimum value of δn?

(b) Because of the stress-induced birefringence, the polarization state of the output beam depends sensitively on the angle θ. It is possible to remove this problem by compensating for the stress-induced birefringence with a field-dependent birefringence through cross-phase modulation caused by another strong pump beam at a frequency of ω that is different from ω_0. If a pump beam that is linearly polarized in the x direction is launched, what is the minimum required intensity of this pump beam for the optical field at the frequency ω_0 to be always linearly polarized at $z = l$, no matter what value θ has?

(c) If the pump beam at ω is linearly polarized in the y direction instead, what is its minimum required intensity to accomplish the effect in (b)?

12.2.1 For a Gaussian beam that has a given power of P and a given beam waist radius of w_0, the effective focal length of a Kerr lens depends on the location z of the Kerr lens with respect to the beam waist, which is located at the origin of the z axis, $z = 0$. Therefore, both the beam waist radius w_{0K} and the divergence of the beam after the Kerr lens vary with the location of the Kerr lens. The effect of sending a Gaussian beam through a thin lens of a focal length f can be described by the relation:

$$\frac{1}{q'} = \frac{1}{q} + \frac{1}{f},$$
(12.111)

where $q(z) = z - iz_R$ and $q'(z) = z - iz_R'$ are the complex radii of curvature of the Gaussian beam immediately before and after the thin lens that is at the location z. The Rayleigh range, z_R, of the Gaussian beam is given in (12.44): $z_R = \pi n_0 w_0^2/\lambda$.

(a) Find $f_K(z)$ and $w_{0K}(z)$ as a function of the location z of the Kerr lens.

(b) Plot $f_K(z)/f_K(0)$ and $w_{0K}(z)/w_0$ as functions of z/z_R for the three values of $f_K(0)/z_R = 0.1, 1$, and 10.

12.2.2 A Nd:YAG laser delivers picosecond pulses that have a pulsewidth of $\Delta t_{ps} = 28$ ps at the wavelength of $\lambda = 1.064$ μm at a pulse repetition rate of 1 kHz. The average power of the laser beam in the Nd:YAG rod is $\overline{P} = 2$ W. The laser beam has a fundamental Gaussian beam profile that has a waist spot size of w_0, which is located at one surface of the Nd:YAG laser rod that has a length of l. YAG has a linear refractive index of $n_0 = 1.818$ at $\lambda = 1.064$ μm and a nonlinear refractive index of $n_2 = 6.2 \times 10^{-20}$ m^2 W^{-1} [19].

(a) Find the critical power for catastrophic self-focusing of this Gaussian laser beam in the YAG crystal.

(b) If the laser beam has a waist spot size of $w_0 = 500$ μm, what is the largest length l of the Nd:YAG rod that is limited by catastrophic self-focusing?

(c) If the length of the Nd:YAG rod is $l = 10$ cm, what is the smallest spot size that the laser beam can have as it enters the rod?

(d) The laser is frequency doubled to generate the second harmonic at $\lambda_{2\omega} = 532$ nm through intracavity second-harmonic generation at a high efficiency to generate second-harmonic pulses that have a pulsewidth of $\Delta t_{ps} = 20$ ps and an average power of $\overline{P}_{2\omega} = 1$ W. The refractive index of YAG at $\lambda_{2\omega} = 532$ nm is $n_0 = 1.838$. Answer the questions in (a)–(c) for the second-harmonic beam.

References

[1] G. A. Askaryan, "Effects of the gradient of a strong electromagnetic beam on electrons and atoms," *Soviet Physics JETP*, vol. 15, pp. 1088–1090, 1962.

[2] R. W. Boyd, S. G. Lukishova, and Y. R. Shen, eds., *Self-Focusing: Past and Present – Fundamentals and Prospects*. New York: Springer, 2009.

[3] R. Y. Chiao, E. Garmire, and C. H. Townes, "Self-trapping of optical beams," *Physical Review Letters*, vol. 13, pp. 479–482, 1964.

[4] V. I. Talanov, "Self-focusing of electromagnetic waves in nonlinear media," *Radiophysics*, vol. 8, pp. 254–257, 1964.

[5] P. L. Kelley, "Self-focusing of optical beams," *Physical Review Letters*, vol. 15, pp. 1005–1008, 1965.

[6] M. I. Weinstein, "Nonlinear Schrödinger equations and sharp interpolation estimates," *Communications in Mathematical Physics*, vol. 87, pp. 567–576, 1983.

[7] G. Fibich and A. L. Gaeta, "Critical power for self-focusing in bulk media and in hollow waveguides," *Optics Letters*, vol. 25, pp. 335–337, 2000.

[8] M. Sheik-Bahae, A. A. Said, and E. W. Van Stryland, "High-sensitivity, single-beam n_2 measurements," *Optics Letters*, vol. 14, pp. 955–957, 1989.

[9] M. Sheik-Bahae, A. A. Said, T. Wei, D. J. Hagan, and E. W. V. Stryland, "Sensitive measurement of optical nonlinearities using a single beam," *IEEE Journal of Quantum Electronics*, vol. 26, pp. 760–769, 1990.

[10] T. Xia, D. J. Hagan, M. Sheik-Bahae, and E. W. Van Stryland, "Eclipsing Z-scan measurement of $\lambda/10^4$ wave-front distortion," *Optics Letters*, vol. 19, pp. 317–319, 1994.

[11] M. Sheik-Bahae, J. Wang, R. DeSalvo, D. J. Hagan, and E. W. Van Stryland, "Measurement of nondegenerate nonlinearities using a two-color Z scan," *Optics Letters*, vol. 17, pp. 258–260, 1992.

[12] J. Wang, M. Sheik-Bahae, A. A. Said, D. J. Hagan, and E. W. Van Stryland, "Time-resolved Z-scan measurements of optical nonlinearities," *Journal of the Optical Society of America B*, vol. 11, pp. 1009–1017, 1994.

[13] R. G. Brewer, "Frequency shifts in self-focused light," *Physical Review Letters*, vol. 19, pp. 8–10, 1967.

[14] F. Shimizu, "Frequency broadening in liquids by a short light pulse," *Physical Review Letters*, vol. 19, pp. 1097–1100, 1967.

[15] A. Lattes, H. Haus, F. Leonberger, and E. Ippen, "An ultrafast all-optical gate," *IEEE Journal of Quantum Electronics*, vol. 19, pp. 1718–1723, 1983.

[16] N. J. Doran and D. Wood, "Nonlinear-optical loop mirror," *Optics Letters*, vol. 13, pp. 56–58, 1988.

[17] K. J. Blow, N. J. Doran, B. K. Nayar, and B. P. Nelson, "Two-wavelength operation of the nonlinear fiber loop mirror," *Optics Letters*, vol. 15, pp. 248–250, 1990.

[18] S. M. Jensen, "The nonlinear coherent coupler," *IEEE Transactions on Microwave Theory and Techniques*, vol. 30, pp. 1568–1571, 1982.

[19] P. Kabaciński, T. M. Kardaś, Y. Stepanenko, and C. Radzewicz, "Nonlinear refractive index measurement by SPM-induced phase regression," *Optics Express*, vol. 27, pp. 11018–11028, 2019.

13 Stimulated Raman and Brillouin Scattering

13.1 RAMAN AND BRILLOUIN SCATTERING PROCESSES

Both Raman and Brillouin scattering can be spontaneous or stimulated. *Spontaneous Raman scattering and spontaneous Brillouin scattering are linear optical processes, whereas stimulated Raman scattering and stimulated Brillouin scattering are third-order nonlinear optical processes.* Stimulated Raman scattering was discovered in 1962 by Woodbury and Ng [1] and was interpreted and studied, also in 1962, by Eckhardt *et al.* [2]. Stimulated Brillouin scattering was first experimentally demonstrated in solids by Chiao *et al.* [3] and in liquids by Garmire and Townes [4], both in 1964.

13.1.1 Differences of Spontaneous and Stimulated Processes

Both the spontaneous and the stimulated processes of Raman and Brillouin scattering involve material excitation. The excitations for Raman scattering are localized short-range excitations, such as the optical phonons in a solid-state material, the rotational and vibrational modes of a molecule, and electronic transitions. The resonance frequency of a Raman excitation falls in the range of 1–100 THz, most often within 10–100 THz. The excitations for Brillouin scattering are long-range excitations, including acoustic phonons, polaritons, and magnetons. The frequency of a Brillouin excitation is typically in the range of 1–50 GHz, which is in the hypersonic region.

Because a Raman or Brillouin scattering process involves the excitation of a material, its efficiency depends on the initial and the final states of the material that are associated with the excitation. For a spontaneous process, only the population of the initial state matters, not that of the final state. For a stimulated process, however, the efficiency depends on the *population difference* between the initial and the final states; therefore, both the populations of the initial and the final states matter. For the purpose of illustration, we consider the spontaneous and the stimulated Raman scattering processes in the following.

In spontaneous Raman scattering, the incoming light consists of only a pump wave at a frequency of ω_p, as shown in Fig. 13.1(a) for the Stokes process of spontaneous Raman scattering. The spontaneous scattering process simultaneously generates two signals at the down-shifted Stokes frequency of $\omega_S = \omega_p - \Omega_R$ and the up-shifted anti-Stokes frequency of $\omega_{AS} = \omega_p + \Omega_R$ because there is always some population in the ground state and some in the excited state of the Raman excitation, which has a resonance frequency of Ω_R. The power ratio of the two signals is proportional to their population ratio. For a material that is in thermal equilibrium at a temperature T, the population distribution of its states is governed by the Boltzmann distribution for a molecular or atomic system, or the Fermi distribution for an

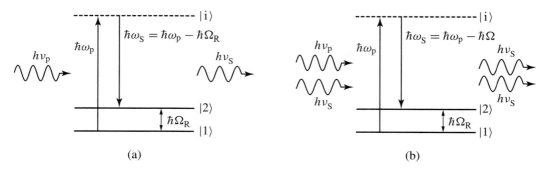

Figure 13.1 Stokes processes of (a) spontaneous Raman scattering and (b) stimulated Raman scattering. In each figure, the horizontal solid lines represent real states. The dashed line represents the virtual state.

electronic system. Therefore, the Stokes and anti-Stokes signals for a system that follows the Boltzmann distribution have the power ratio:

$$\frac{P_S}{P_{As}} = \frac{\omega_S^4}{\omega_{AS}^4} \exp\left(\frac{\hbar\Omega_R}{k_B T}\right), \tag{13.1}$$

where the frequency ratio ω_S^4/ω_{AS}^4 accounts for the frequency dependence of the scattering efficiency and k_B is the Boltzmann constant. This relation is generally used in *spontaneous Raman spectroscopy* for identifying a material through its characteristic Raman resonance frequency Ω_R, and for measuring its temperature T. Note again that spontaneous Raman scattering is a linear optical process.

By contrast, in stimulated Raman scattering, the incoming light has to consist of two frequencies, one of the pump frequency at ω_p and another of either the Stokes frequency at $\omega_S = \omega_p - \Omega$ or the anti-Stokes frequency at $\omega_{AS} = \omega_p + \Omega$. Different from spontaneous Raman scattering, however, the Stokes frequency ω_S and the anti-Stokes frequency ω_{AS} cannot be both generated or both amplified by stimulated Raman scattering because the efficiency of stimulated Raman scattering depends not only on the population distribution but also on the population difference between the ground state and the excited state of the Raman excitation. Only the Stokes signal gets amplified if the ground state is more populated, whereas only the anti-Stokes signal is amplified if the excited state is more populated. Because the ground state is always more populated than an excited state when a material is in thermal equilibrium with its surroundings, stimulated Raman scattering generally amplifies the Stokes frequency, as shown in Fig. 13.1(b). Therefore, the input consists of both the pump beam at ω_p, which can be any frequency higher than Ω_R, and the Stokes signal at $\omega_S = \omega_p - \Omega$, where Ω is close to Ω_R, within the linewidth of the Raman resonance, but it does not have to be exactly Ω_R, as is also illustrated in Fig. 13.1(b).

Note that stimulated Raman scattering is a third-order nonparametric nonlinear process that requires the pre-existence of a Stokes signal besides the pump. The Stokes photons can be supplied as part of the input, as in the case of Raman amplification, or they can be first generated through the linear process of spontaneous Raman scattering discussed above, as in the case of Raman generation. For this reason, only an optical beam at the pump frequency is supplied to a Raman generator; however, it is important to remember that the initial Stokes seed signal required for stimulated Raman scattering to take place needs to be generated through spontaneous Raman scattering. When the seed Stokes signal is supplied for Raman amplification, its frequency difference Ω from the pump frequency does not have to be exactly in resonance with the resonance frequency Ω_R of the excitation, but only has to be within the Raman linewidth for a reasonable Raman gain. When the seed Stokes signal is generated through spontaneous Raman scattering, its frequency difference from the pump frequency is determined by the resonance frequency Ω_R of the Raman excitation with a linewidth of the Raman linewidth.

One way to view the stimulated Raman scattering process is to consider the consequence of the interaction between the pump wave and the pre-existing Stokes wave. The two waves interact to generate a beat at their frequency difference, Ω. If Ω is in resonance with a Raman excitation at a resonance frequency of Ω_R, the beat signal is resonantly amplified. In quantum-mechanical terms, this process amplifies the excitation, creating many optical phonons at the frequency of Ω, which take energy away from the pump photons to reduce their photon energy to that of the Stokes signal, thus amplifying the Stokes signal. In classical terms, this process creates oscillations of the material at the frequency of Ω, which scatters the pump wave into the Stokes frequency, thus amplifying the Stokes signal. In any event, it can be seen that the stimulated scattering process is *resonantly enhanced* because the beat frequency Ω that is generated by the pump and the seed Stokes is in resonance, or near resonance, with the resonance frequency Ω_R of the Raman excitation. By comparison, the spontaneous scattering process is not resonantly enhanced because it does not generate a beat to be in resonance with Ω_R. Because of the resonant enhancement, the Stokes signal that is amplified through stimulated Raman scattering is many orders of magnitude larger than that generated through spontaneous Raman scattering. This favorable feature makes stimulated Raman scattering very efficient for many applications, including high-efficiency spectroscopy, optical signal processing, and Raman amplification and generation of coherent optical waves at Raman-shifted frequencies.

In a Raman Stokes process, the Raman medium is excited from the ground state, $|1\rangle$, to the excited state, $|2\rangle$. This Raman transition does not take place directly from $|1\rangle$ to $|2\rangle$, but via an intermediate state, $|i\rangle$, as shown in Fig. 13.1. The intermediate state can be a virtual state, as indicated by the horizontal dashed line in Fig. 13.1, or a real state. In any event, it is essential for the Raman transition, even when it is virtual. It makes the selection rules for the Raman transition from $|1\rangle$ to $|2\rangle$ via $|i\rangle$ different from those for the direct optical transition from $|1\rangle$ to $|2\rangle$ without passing through $|i\rangle$. For example, if the material is centrosymmetric, such as a centrosymmetric molecule, a direct optical transition from $|1\rangle$ to $|2\rangle$ that is caused by electric dipole interaction requires that the two states $|1\rangle$ and $|2\rangle$ have opposite parities. By contrast, the selection rules for the Raman transition from $|1\rangle$ to $|2\rangle$ via $|i\rangle$ require that $|1\rangle$ and $|2\rangle$ have the same parity because $|1\rangle$ and $|i\rangle$ are required to have opposite parities, and $|i\rangle$ and $|2\rangle$ also have opposite parities. For this reason, between the two states $|1\rangle$ and $|2\rangle$, Raman transition and direct optical transition are mutually exclusive in a centrosymmetric material. Therefore, Raman spectroscopy and linear optical spectroscopy are complementary to one another because they probe different states in a material.

In the above, we discussed the differences between spontaneous and stimulated Raman scattering processes. The general concepts apply to spontaneous and stimulated Brillouin scattering processes as well. The fundamental difference between a Raman process and a Brillouin process, no matter whether they are spontaneous or stimulated, is their different modes of material excitation, and the accompanying different excitation frequencies. This difference further leads to the different conditions for the propagation of their Stokes signals, as discussed below.

13.1.2 Stimulated Processes

The nonparametric processes of stimulated Raman scattering and stimulated Brillouin scattering both cause a shift of the optical frequency, leading to a loss for the pump beam and a gain for a Stokes beam if the material is not originally excited or a gain for an anti-Stokes beam if it is excited. On the positive side, these processes can be utilized for optical frequency conversion and optical signal amplification. On the negative side, however, they also place some serious limitations on the performance of certain optical devices and systems. As discussed above, in most applications using a Raman or Brillouin process, a material is initially in its normal state without being excited to population inversion. Therefore, we shall consider only the Stokes process in the following discussions.

For both stimulated Raman scattering and stimulated Brillouin scattering, the Stokes interaction is characterized by the imaginary part of a complex third-order nonlinear susceptibility that has the form of $\chi^{(3)}(\omega_S = \omega_S + \omega_p - \omega_p)$, with $\omega_S = \omega_p - \Omega$. The nonlinear polarization for the Stokes signal is that given in (5.55):

$$\mathbf{P}^{(3)}(\omega_S) = 6\epsilon_0\boldsymbol{\chi}^{(3)}(\omega_S = \omega_S + \omega_p - \omega_p) \vdots \mathbf{E}(\omega_S)\mathbf{E}(\omega_p)\mathbf{E}^*(\omega_p). \tag{13.2}$$

This nonlinear polarization leads to an effective susceptibility of χ_R, given in (6.53), for the stimulated Raman Stokes signal and, similarly, χ_B for the Brillouin Stokes:

$$\chi_R = \hat{e}_S^* \cdot \boldsymbol{\chi}^{(3)}(\omega_S = \omega_S + \omega_p - \omega_p) \vdots \hat{e}_S\hat{e}_p\hat{e}_p^*, \tag{13.3}$$

$$\chi_B = \hat{e}_S^* \cdot \boldsymbol{\chi}^{(3)}(\omega_S = \omega_S + \omega_p - \omega_p) \vdots \hat{e}_S\hat{e}_p\hat{e}_p^*. \tag{13.4}$$

The Stokes susceptibility has the symmetry property given in (6.49):

$$\begin{aligned}
\chi_{ijkl}^{(3)}(\omega_S = \omega_S + \omega_p - \omega_p) &= \chi_{klij}^{(3)*}(\omega_p = \omega_p + \omega_S - \omega_S) \\
&= \chi_{jilk}^{(3)}(\omega_S = \omega_S + \omega_p - \omega_p) \\
&= \chi_{lkji}^{(3)*}(\omega_p = \omega_p + \omega_S - \omega_S).
\end{aligned} \tag{13.5}$$

By following the discussions in Subsection 6.3.1 while using the relation given in (13.5), we find that the imaginary part, χ_R'' for stimulated Raman scattering or χ_B'' for stimulated Brillouin scattering, is responsible for the amplification of the Stokes signal when it has a negative value.

The Stokes signal and the pump wave can be in either the same or different polarization states. The effective nonlinear susceptibility and, consequently, the Raman or Brillouin gain factor depend on the polarization states of these two waves and on the property of the material. This dependence is similar to that discussed in Subsection 12.1.2 on the polarization states of the two interacting waves in cross-phase modulation. For example, for the interactions in an isotropic medium, if the Stokes and the pump waves are linearly polarized in the same direction, we have

$$\chi_R'' = \chi_{1111}^{(3)''} \quad \text{and} \quad \chi_B'' = \chi_{1111}^{(3)''}. \tag{13.6}$$

If the two waves are linearly polarized in mutually perpendicular directions, we have

$$\chi_R'' = \chi_{1122}^{(3)''} \quad \text{and} \quad \chi_B'' = \chi_{1122}^{(3)''}. \tag{13.7}$$

If the two waves are circularly polarized in the same sense, we have

$$\chi_R'' = \frac{1}{2} \left(\chi_{1111}^{(3)''} + \chi_{1122}^{(3)''} + \chi_{1212}^{(3)''} - \chi_{1221}^{(3)''} \right) = \chi_{1122}^{(3)''} + \chi_{1212}^{(3)''}, \tag{13.8}$$

and similarly for χ_B''. If the two waves are circularly polarized in opposite senses, we have

$$\chi_R'' = \frac{1}{2} \left(\chi_{1111}^{(3)''} + \chi_{1122}^{(3)''} - \chi_{1212}^{(3)''} + \chi_{1221}^{(3)''} \right) = \chi_{1122}^{(3)''} + \chi_{1221}^{(3)''}, \tag{13.9}$$

and similarly for χ_B''.

Being nonparametric, the Raman and Brillouin processes are automatically phase matched. The corresponding material excitation in an interaction picks up any phase mismatch between the interacting optical waves. Indeed, a Raman or Brillouin Stokes process can be viewed as a parametric interaction among a pump wave, a Stokes wave, and a material excitation wave. The material excitation wave is characterized by a frequency of Ω and a wavevector of $\mathbf{K}$. From this viewpoint, it is easy to see that a Stokes interaction is governed by these conditions:

$$\omega_S = \omega_p - \Omega, \tag{13.10}$$

$$\mathbf{k}_S = \mathbf{k}_p - \mathbf{K}. \tag{13.11}$$

Clearly, *phase matching among the pump wave, the Stokes wave, and the material excitation wave is required*, but it is automatically achieved when the pump wave generates a material excitation that allows a Raman or Brillouin process to occur. Figure 13.2 illustrates the relations among the three interacting waves in a Stokes process.

As mentioned above, the fundamental difference between the Raman and the Brillouin processes lies in the different modes of material excitation for the two processes. This difference leads to very different considerations for these two processes. One important difference results from the phase-matching condition given in (13.11). Because a Raman excitation is a localized, short-range excitation, it can cover a large range of $\mathbf{K}$ vectors for a given frequency Ω. In other words, the Fourier transform of a localized oscillation leads to a broad spectrum in the momentum space, meaning that $\mathbf{K}$ is independent of Ω. For this reason, the Raman Stokes vector, $\mathbf{k}_S$, for a given pump vector of $\mathbf{k}_p$ can be in any direction. This means that the phase-matching condition

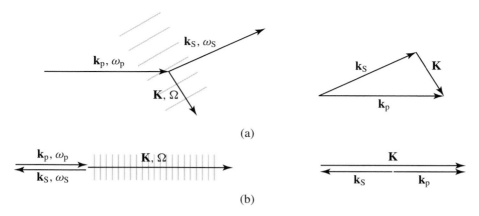

Figure 13.2 Generation of a Stokes optical wave and a material excitation wave by a pump optical wave with phase-matching condition in (a) a Raman Stokes process and (b) a Brillouin Stokes process. For the Raman process, the Stokes wave can be generated in any direction. For the Brillouin process, the Stokes wave is generated only in the backward direction.

given in (13.11) allows the Raman Stokes signal to be generated in any direction. By contrast, however, this is not true for the Brillouin Stokes signal. Because a Brillouin excitation is a long-range excitation, the value and the direction of its $\mathbf{K}$ vector are restricted and are closely related to the value of the frequency Ω. This strong dispersive relation between $\mathbf{K}$ and Ω makes satisfying the phase-matching condition given in (13.11) for a given excitation frequency Ω, given in (13.10), highly dependent on the property of the material. For a nonbirefringent material, $|\mathbf{k}_S| = k_S = \omega_S n(\omega_S)/c$, $|\mathbf{k}_p| = k_p = \omega_p n(\omega_p)/c$, and $|\mathbf{K}| = K = \Omega/v_a$, where v_a is the velocity of the acoustic wave. Because $|\mathbf{k}_S| \approx |\mathbf{k}_p|$, and $|\mathbf{K}|$ of a hypersonic wave is of the same order of magnitude as $|\mathbf{k}_S|$ and $|\mathbf{k}_p|$ due to the fact that $v_a \approx 10^{-5}c$, the only phase-matched direction for Brillouin scattering in a nonbirefringent medium is the backward direction shown in Fig. 13.2(b). Other directions, including the forward scattering direction, are possible for Brillouin scattering in a birefringent crystal under certain conditions.

13.2 RAMAN GAIN

Because an excitation that is responsible for Raman scattering is associated with a transition at the molecular or atomic level, the Raman frequency shift is determined by the resonance frequency, Ω_R, of the Raman transition. This Raman frequency is an intrinsic property of a material and is independent of the frequency of the pump optical wave. Such an excitation is also nondispersive because it is localized, as discussed above. Because a nondispersive excitation can take any wavevector $\mathbf{K}$ independently of its frequency Ω, the phase-matching condition given in (13.11) is satisfied for any combination of $\mathbf{k}_p$ and $\mathbf{k}_S$ independently of the frequency condition given in (13.10). Consequently, Raman scattering in all directions has the same frequency shift that is specific to a given material. Spontaneous Raman scattering has a nearly isotropic emission pattern in all directions, whereas stimulated Raman scattering predominantly occurs in the forward and backward directions due to the fact that a stimulated signal grows in strength as the interaction length increases. The Raman frequency shift, which is usually quoted per centimeter, for most materials is typically in the range of 300–3,000 cm^{-1} for $f_R/c = \Omega_R/2\pi c$, equivalent to 10–100 THz for f_R (1 cm^{-1} is equivalent to 30 GHz).

When a material is initially in its ground state of a Raman transition, the effective Raman susceptibility defined in (13.3) has a negative imaginary part: $\chi_R'' < 0$. This situation leads to a gain for the Raman Stokes signal at the expense of the pump wave. From (6.56), we find that the *Raman gain factor* for the Stokes signal is given by the relation:

$$\tilde{g}_R = -\frac{3\omega_S}{c^2\epsilon_0 n_{S,z} n_{p,z}}\chi_R'', \tag{13.12}$$

which has a positive value when $\chi_R'' < 0$. The unit of $\tilde{g}_R$ is meters per watt. The real and imaginary parts of the Raman susceptibility for a single Raman resonance frequency at Ω_R are shown in Fig. 13.3(a), and the corresponding Raman gain factor is shown in Fig. 13.3(b). With the Raman gain factor defined in (13.12), the coupled intensity equations for the amplification of the Stokes signal given in (6.56) and (6.57) can be expressed in the forms:

$$\frac{dI_S}{dz} = \tilde{g}_R I_p I_S, \tag{13.13}$$

$$\frac{dI_p}{dz} = -\frac{\omega_p}{\omega_S}\tilde{g}_R I_S I_p. \tag{13.14}$$

The Raman susceptibility is only very weakly dependent on the individual optical frequencies, ω_p and ω_S, but it is a strong function of the frequency difference, $\Omega = \omega_p - \omega_S$, with a resonance at the Raman resonance frequency, $\Omega = \Omega_R$, as seen in Fig. 13.3(a). In the simple

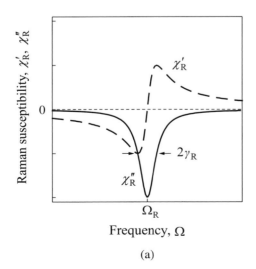
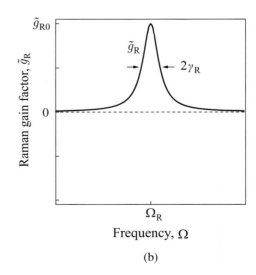

(a)

(b)

Figure 13.3 (a) Raman susceptibility χ_R and (b) Raman gain factor $\tilde{g}_R$ as functions of the frequency difference $\Omega = \omega_p - \omega_S$ between the pump and the Stokes waves.

case that there is only one Raman resonance frequency in a material, χ_R as a function of Ω has a Lorentzian lineshape as that of the linear susceptibility given in (1.167), but with the signs of both the real and the imaginary parts changed for $\chi_R'' < 0$, as shown in Fig. 13.3(a). Therefore, according to (13.12), the corresponding Raman gain factor has the form:

$$\widetilde{g}_R = \widetilde{g}_{R0} \frac{\gamma_R^2}{(\Omega - \Omega_R)^2 + \gamma_R^2} = \widetilde{g}_{R0} \frac{\gamma_R^2}{(\omega_p - \omega_S - \Omega_R)^2 + \gamma_R^2}, \tag{13.15}$$

where $\widetilde{g}_{R0}$ is the *peak Raman gain factor* and γ_R is the relaxation constant of the Raman excitation. As is shown in Fig. 13.3(b), this Raman gain factor has a full width at half-maximum linewidth given by the relation:

$$\Delta\Omega_R = 2\gamma_R, \quad \text{or} \quad \Delta f_R = \gamma_R/\pi. \tag{13.16}$$

Note that both the Raman resonance frequency f_R and the Raman spectral linewidth Δf_R are independent of the pump and the Stokes optical frequencies, but as seen from (13.12) the peak Raman gain factor varies linearly with the Stokes optical frequency: $\widetilde{g}_{R0} \propto \omega_S \propto 1/\lambda_S$.

The response time of a Raman process is measured by the time constant, $\tau_R = \gamma_R^{-1}$, which is the relaxation time of the Raman excitation, such as an optical phonon or a molecular vibration. The typical Raman response time ranges from a few hundred picoseconds for molecules through a few picoseconds for crystalline solids to tens of femtoseconds for amorphous solids such as glasses. Accordingly, the Raman linewidth Δf_R ranges from a few gigahertz to the order of 10 THz, depending on the property of a material. A Raman process can efficiently respond only to an optical signal that has a bandwidth narrower than the Raman linewidth. For an optical pulse, this means that its pulsewidth has to be greater than the Raman response time for the Raman process to efficiently take place. Therefore, depending on the specific material used, it is possible for a Raman device to function in steady state or in quasi-steady state with optical signals ranging from CW waves to picosecond or even sub-picosecond optical pulses. When an optical signal varies faster than the Raman relaxation time, the interaction is characterized by *transient stimulated Raman scattering* with a reduced Raman gain, among other features that are different from those of steady-state stimulated Raman scattering. In the following section, we shall consider only Raman devices that operate in the steady-state regime.

Forward and backward Raman interactions have the same Raman gain factor. However, the value of $\widetilde{g}_{R0}$ is a function of the polarization states of the pump and the Stokes waves because the effective Raman susceptibility χ_R that is defined in (13.3) depends on $\hat{e}_p$ and $\hat{e}_S$, as discussed in Subsection 13.1.2. In an isotropic medium, the maximum value of $\widetilde{g}_{R0}$ is found when the pump and the Stokes waves are linearly polarized in the same direction. Therefore, special attention has to be paid to the polarization states of the optical waves throughout the course of interaction in evaluating the efficiency of a Raman process.

As an example, Fig. 13.4 shows the spectral dependence of the Raman gain factor of fused silica glass measured at a pump optical wavelength of 1 μm for pump and Stokes waves that are linearly polarized in the same direction. This spectrum is very broad, and it does not have the ideal Lorentzian lineshape, because there are many closely clustered Raman resonances in such an amorphous solid material. This Raman spectral shape remains more or less the same

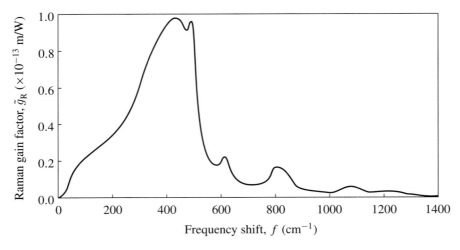

Figure 13.4 Spectrum of the Raman gain factor of fused silica measured at a pump wavelength of 1 μm for pump and Stokes waves that are linearly polarized in the same direction. Note that 1 cm^{-1} = 30 GHz. (Adapted from Stolen, R. H., "Nonlinearity in fiber transmission," *Proceedings of the IEEE*, vol. 68, pp. 1232–1236, 1980.)

for other pump wavelengths, but its peak value scales with the Stokes wavelength as $\tilde{g}_{R0} \approx (1 \times 10^{-13}/\lambda_S)\, \mathrm{m\,W^{-1}}$, where λ_S is measured in micrometers. The Raman gain factor of fused silica is relatively small compared to those of many molecular substances, such as benzene and CS$_2$. Many amorphous glass materials, such as GeO$_2$, B$_2$O$_3$, and P$_2$O$_5$, which are commonly used to dope silica fibers, also have peak Raman gain factors that are 5–10 times that of pure silica, with corresponding frequency shifts ranging from 400 to 1400 cm^{-1}. In particular, the peak Raman gain factor of GeO$_2$ at a frequency shift of 420 cm^{-1} is 9.2 times that of pure silica. Therefore, the peak value, the frequency shift corresponding to the spectral peak, and the spectral shape of the Raman gain factor of a particular optical fiber all depend on the type and concentration of the dopants in the fiber.

EXAMPLE 13.1

An optical wave at the 1.55 μm wavelength propagates in a silica optical fiber that has a peak Raman gain factor as described in the text above at a Raman frequency shift of 460 cm^{-1}. If the intensity of this optical wave is sufficiently high to generate a Raman Stokes signal, what is the wavelength of the Stokes signal? What is the Raman frequency shift in hertz? What is the Raman gain factor for this Stokes signal?

Solution

Because $\omega_S = \omega_p - \Omega$, the Stokes wavelength can be found by using the relation:

$$\frac{1}{\lambda_S} = \frac{1}{\lambda_p} - \frac{f_R}{c}.$$

The Raman frequency shift quoted per centimeter is actually f_R/c in the above relation. For the fiber considered here, we have $f_R/c = 460\,\text{cm}^{-1} = 4.6 \times 10^4\,\text{m}^{-1}$. Therefore, the wavelength of the Stokes signal is

$$\lambda_S = \left(\frac{1}{1.55 \times 10^{-6}} - 4.6 \times 10^4\right)^{-1}\text{m} = 1.669\,\mu\text{m}.$$

Because $1\,\text{cm}^{-1} = 30\,\text{GHz}$, the Raman frequency shift is $f_R = 13.8\,\text{THz}$ for $f_R/c = 460\,\text{cm}^{-1}$. From the text, $\widetilde{g}_{R0} \approx (1 \times 10^{-13}/\lambda_S)\,\text{m W}^{-1}$ for fused silica. Therefore, the Raman gain factor at this wavelength is

$$\widetilde{g}_{R0} = \frac{1 \times 10^{-13}}{1.669}\,\text{m W}^{-1} = 5.99 \times 10^{-14}\,\text{m W}^{-1}.$$

13.3 RAMAN AMPLIFICATION AND GENERATION

13.3.1 Raman Amplifiers

The Raman gain can be utilized to amplify an optical signal at a Stokes frequency of ω_S through the process of stimulated Raman scattering by choosing a proper pump wave at the frequency of $\omega_p = \omega_S + \Omega$ with a frequency shift of Ω that is within the Raman gain spectrum, ideally at the gain-peak frequency of Ω_R. Because stimulated Raman scattering has the same gain factor in the forward and backward directions, the pump wave and the Stokes signal wave can propagate either codirectionally, as shown in Fig. 13.5(a), or contradirectionally, as shown in Fig. 13.5(b), in a Raman amplifier. However, though the gain factor is the same for the two configurations, a Raman amplifier of a contradirectional configuration would have little or no efficiency for short optical pulses because of the short interaction length between the pump and the Stokes pulses that propagate in opposite directions. Here we consider only Raman amplification in a codirectional configuration. The general formulation and characteristics for Raman amplification in a contradirectional configuration are similar to those for Brillouin amplification, which will be discussed in Subsection 13.5.1.

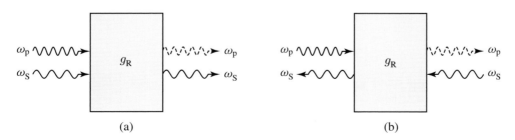

Figure 13.5 (a) Codirectional Raman amplifier and (b) contradirectional Raman amplifier. For Raman amplification, $\omega_p = \omega_S + \Omega$.

By following (13.13) and (13.14) and allowing for the existence of linear absorption loss in a medium, we have the coupled equations for Raman amplification in a codirectional configuration:

$$\frac{dI_S}{dz} + \alpha_S I_S = \tilde{g}_R I_p I_S, \tag{13.17}$$

$$\frac{dI_p}{dz} + \alpha_p I_p = -\frac{\omega_p}{\omega_S} \tilde{g}_R I_S I_p, \tag{13.18}$$

with given values of $I_S(0)$ and $I_p(0)$ at the input end, $z = 0$, of the amplifier as the initial conditions. The parameters α_S and α_p are the linear absorption coefficients of the medium at the Stokes and the pump frequencies, respectively. Because $\omega_p \approx \omega_S \gg \Omega$ in most practical situations, $\alpha_S \approx \alpha_p$. The coupled equations in (13.17) and (13.18) have exact analytical solutions when $\alpha_S = \alpha_p = \alpha$. For a forward Raman amplifier of a length l in the case that $\alpha_S = \alpha_p = \alpha$, the exact intensities of the Stokes and pump waves at the output end, $z = l$, are

$$I_S(l) = \frac{\omega_p I_S(0) + \omega_S I_p(0)}{\omega_p I_S(0) + \omega_S I_p(0) \exp\left(-g_R l_{\text{eff}}\right)} I_S(0) \exp\left(-\alpha l\right), \tag{13.19}$$

$$I_p(l) = \frac{\omega_S I_p(0) + \omega_p I_S(0)}{\omega_S I_p(0) + \omega_p I_S(0) \exp\left(g_R l_{\text{eff}}\right)} I_p(0) \exp\left(-\alpha l\right), \tag{13.20}$$

where g_R is the *Raman gain coefficient*, defined as

$$g_R = \tilde{g}_R \left[I_p(0) + \frac{\omega_p}{\omega_S} I_S(0) \right], \tag{13.21}$$

and l_{eff} is the effect length for Raman amplification, defined as

$$l_{\text{eff}} = \frac{1 - e^{-\alpha l}}{\alpha}. \tag{13.22}$$

In the amplification of a weak signal when the depletion of the pump intensity due to Raman interaction can be neglected, a simple approximate solution can be obtained by ignoring the term on the right-hand side of (13.18). This approximate solution can be found for the general situation that $\alpha_S \neq \alpha_p$; it is not necessary to assume that $\alpha_S = \alpha_p$. Then, for a Raman amplifier of a length l, we have the signal at the output end of the amplifier given by the relation:

$$I_S(l) = I_S(0) \exp\left(g_R l_{\text{eff}} - \alpha_S l\right), \tag{13.23}$$

where g_R is the Raman gain coefficient that is reduced from that defined in (13.21) under the condition that $I_S(0) \ll I_p(0)$:

$$g_R = \tilde{g}_R I_p(0) = -\frac{3\omega_S}{c^2 \epsilon_0 n_{S,z} n_{p,z}} \chi_R'' I_p(0), \tag{13.24}$$

and l_{eff} is the effective interaction length of Raman amplification given as

$$l_{\text{eff}} = \frac{1 - e^{-\alpha_p l}}{\alpha_p} \tag{13.25}$$

for the general situation that $\alpha_S \neq \alpha_p$. Note that the Raman gain coefficient increases with the pump intensity. Such dependence on an optical intensity is characteristic of an optical gain that is contributed by a nonlinear optical process. The amplification factor, or the *Raman amplifier gain*, for the Stokes signal in the case of negligible pump depletion is then given as

$$G_R = \frac{I_S(l)}{I_S(0) \exp(-\alpha_S l)} = \exp(g_R l_{\text{eff}}) = \exp\left[\tilde{g}_R I_p(0) l_{\text{eff}}\right]. \tag{13.26}$$

For a given Raman amplifier of a fixed length, the amplifier gain can be controlled by varying the pump intensity.

EXAMPLE 13.2

A silica optical fiber that has the Raman gain characteristics as described in Example 13.1 is used as a fiber Raman amplifier for codirectional amplification of an optical signal at the Stokes wavelength of $\lambda_S = 1.55$ µm. The input power of the signal is -15 dBm, and the desired output power is 0 dBm. The fiber has an absorption coefficient of $\alpha = 0.2$ dB km^{-1} at this signal wavelength and a length of $l = 25$ km. Its core has an effective cross-sectional area of $\mathcal{A}_{\text{eff}} = 5 \times 10^{-11}$ m^2. What is the pump wavelength for the largest Raman gain? What is the required pump power if the absorption coefficient at the pump wavelength is the same as that at the signal wavelength?

Solution

Because the Raman gain spectrum of a silica optical fiber is very broad, it is in general only necessary to pick a pump wavelength such that the signal wavelength falls within the Raman gain spectral range of the pump. However, to have the largest Raman gain, we need to properly choose a pump wavelength so that the Raman gain peak appears at the signal wavelength. From Example 13.1, we have $f_R/c = 460$ cm$^{-1} = 4.6 \times 10^4$ m^{-1} at the peak of the Raman spectrum. Therefore, the pump wavelength for the largest Raman gain is

$$\lambda_p = \left(\frac{1}{1.55 \times 10^{-6}} + 4.6 \times 10^4\right)^{-1} \text{m} = 1.4468 \text{ µm}.$$

The Raman gain factor at $\lambda_S = 1.55$ µm has the peak value:

$$\tilde{g}_R = \frac{1 \times 10^{-13}}{1.55} \text{ m W}^{-1} = 6.45 \times 10^{-14} \text{ m W}^{-1}.$$

With $\alpha = 0.2$ dB km$^{-1} = 0.046$ km^{-1} and $l = 25$ km, we have

$$l_{\text{eff}} = \frac{1 - e^{-0.046 \times 25}}{0.046} \text{ km} = 14.86 \text{ km}.$$

Because $P_S^{in} = I_S(0)\mathcal{A}_{eff}$ and $P_S^{out} = I_S(l)\mathcal{A}_{eff}$, we find from (13.26) that the required Raman amplifier gain in decibels is

$$G_R = P_S^{out}(dBm) - P_S^{in}(dBm) + \alpha\,(dB\,km^{-1})l(km) = 20\,dB.$$

Therefore, $G_R = 20\,dB = 100$. By identifying the pump power as $P_p = I_p(0)\mathcal{A}_{eff}$ and using (13.26), we find the required pump power:

$$P_p = \frac{\mathcal{A}_{eff}}{l_{eff}}\frac{\ln G_R}{\widetilde{g}_R} = \frac{5 \times 10^{-11} \times \ln 100}{14.86 \times 10^3 \times 6.45 \times 10^{-14}}\,W = 240\,mW.$$

Note that by using (13.26) to obtain G_R and P_p in the above, we have implicitly assumed that the depletion of the pump power due to its conversion to the signal power is negligible. This assumption is clearly valid here because the pump power obtained under such an assumption is 240 mW, while the output signal is only 0 dBm, which is 1 mW. Stimulated Brillouin scattering has to be suppressed in a Raman amplifier because it can deplete the pump power for Raman amplification. Because the Brillouin gain factor decreases with the linewidth of the pump, as discussed in Section 13.4, an optical source of a linewidth that is large enough to suppress stimulated Brillouin scattering is normally used for pumping a Raman amplifier. Multimode semiconductor lasers can serve such a purpose for pumping fiber Raman amplifiers.

13.3.2 Raman Generators

When a medium that has a Raman susceptibility is excited with an optical beam, spontaneous Raman scattering that generates incoherent Stokes and anti-Stokes emission in all directions always occurs, though the emission might be weak. In a Raman amplifier where an input signal is coherently amplified, such incoherent spontaneous emission contributes to the noise of the amplifier. In the absence of an input signal, however, the ubiquitous spontaneous Raman emission can be the seed for the generation of a Stokes or anti-Stokes wave through stimulated amplification under the right conditions. A Raman generator is normally used for the generation of the Stokes wave at the down-shifted Stokes frequency of $\omega_S = \omega_p - \Omega_R$, with the medium initially in its ground state. A Raman generator can simply be a Raman amplifier without an input signal but with a pump of a sufficient intensity for significant power conversion from the pump frequency to the Stokes frequency in a single pass through the medium. In this case, the seed Stokes signal comes from the initial spontaneous Raman emission, as discussed in Section 13.1.

In a Raman generator, the Stokes wave grows from stimulated amplification of the spontaneous Stokes emission. Because spontaneous Stokes emission occurs along the entire length of the generator, the total Stokes power at the output is the result of the cumulative amplification of all spontaneous Stokes emission over the length of the generator. A detailed analysis that takes into account such cumulative amplification can be carried out. For forward Raman generation, the net result is equivalent to treating the generator as an amplifier with the injection of an effective Stokes signal of $I_S^{eff}(0)$ at $z = 0$ while ignoring all of the spontaneous Stokes emission in the generator. For backward generation, it is equivalent to the injection of an effective Stokes signal of $I_S^{\text{eff}}(l)$ at $z = l$ while ignoring all of the spontaneous Stokes emission. The values of the effective signals depend on the Raman characteristics, particularly $\widetilde{g}_{R0}$ and Δf_R, of the medium, as well as on the pump intensity. Besides, due to the difference between the forward and the backward interactions in the geometric relation of the pump and the Stokes waves, the effective signal $I_S^{\text{eff}}(l)$ for backward generation is significantly smaller than the effective signal $I_S^{\text{eff}}(0)$ for forward generation at a given pump intensity in a given medium. This difference leads to a higher threshold for backward Raman generation than that for forward Raman generation. As a result, *only a Stokes wave in the forward direction is generated in a Raman generator*. No backward generation occurs. Note that a Raman amplifier can be backward because the input Stokes signal can be sent in the backward direction. This option is not available to a Raman generator because it does not take an input Stokes signal.

Because significant power conversion from the pump wave to the Stokes wave is desired in the application of a Raman generator, pump depletion cannot be neglected. If we assume for simplicity that $\alpha_S = \alpha_p = \alpha$ and consider the fact that $I_p(0) \gg I_S^{\text{eff}}(0)$ for a Raman generator, the complete solutions, given in (13.19) and (13.20), of the coupled equations in (13.17) and (13.18) leads to the relation:

$$\frac{I_S(l)}{I_p(l)} \approx \frac{I_S^{\text{eff}}(0)}{I_p(0)} \exp\left[\widetilde{g}_R I_p(0) l_{\text{eff}}\right] = \frac{I_S^{\text{eff}}(0)}{I_p(0)} G_R. \tag{13.27}$$

The *threshold of a Raman generator* can be defined as the condition for $I_S(l) = I_p(l)$. Then, the threshold amplification factor is obtained at the *Raman threshold*:

$$G_R^{\text{th}} = \exp\left[\widetilde{g}_R I_p^{\text{th}}(0) l_{\text{eff}}\right] = \frac{I_p^{\text{th}}(0)}{I_{S,\text{th}}^{\text{eff}}(0)}. \tag{13.28}$$

The physical meaning of this relation is that *the Raman threshold is reached when stimulated amplification of the spontaneous Stokes emission brings the Stokes intensity at the output to the same level as that of the pump intensity*. Therefore, the threshold pump intensity for forward Raman Stokes generation in a medium of a length l can be calculated by using the relation:

$$I_p^{\text{th}}(0) \approx \frac{\ln G_R^{\text{th}}}{\widetilde{g}_R l_{\text{eff}}}. \tag{13.29}$$

Because the value of $I_S^{\text{eff}}(0)$ depends on the characteristics of the medium, the value of G_R^{th} is also a function of the characteristics of the medium. The same relation of (13.29) is also valid for the Raman threshold of backward generation, with a different value for G_R^{th}. For example, $\ln G_R^{\text{th}} \approx 16$ for forward Raman Stokes generation in a single-mode silica fiber [5]. For backward Raman Stokes generation in a single-mode silica fiber, $\ln G_R^{\text{th}} \approx 20$ [5]. Backward Raman Stokes generation normally does not occur because it has a much higher threshold than forward generation.

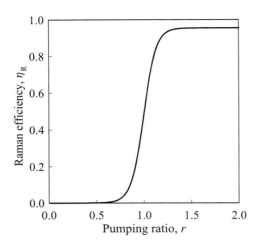

Figure 13.6 Conversion efficiency of a silica fiber Raman generator, for which $\ln G_R^{th} \approx 16$, as a function of the pumping ratio. The pump wavelength is taken to be $\lambda_p = 1\,\mu m$, and the Raman frequency shift at the gain peak is $f_R/c = 460\,cm^{-1}$ for a Stokes wavelength of $\lambda_S = 1.0482\,\mu m$. The maximum conversion efficiency for this example is $\eta_R^{max} = \omega_S/\omega_p = \lambda_p/\lambda_S = 95.4\%$ for a negligible absorption loss.

In the case that $\alpha_S = \alpha_p = \alpha$, a simple relation for calculating the conversion efficiency of a Raman generator can be obtained by assuming that $I_p(0) \gg I_S^{eff}(0) = I_{S,th}^{eff}(0)$ for any pump intensity. Then, the *Raman conversion efficiency* from the pump to the Stokes is found to be

$$\eta_R = \frac{I_S(l)}{I_p(0)} = \frac{\omega_S}{\omega_p} \frac{\exp{(-\alpha l)}}{1 + (\omega_S/\omega_p)r(G_R^{th})^{1-r}}, \tag{13.30}$$

where $r = I_p(0)/I_p^{th}(0)$ is the *pumping ratio* with respect to the threshold pump intensity. Figure 13.6 shows the Raman conversion efficiency of a Raman generator as a function of the pumping ratio. The threshold of a Raman generator is very sharp, as seen in Fig. 13.6. Below the threshold, the conversion efficiency η_R quickly approaches zero as the pumping ratio is reduced, but it quickly approaches its maximum value of $(\omega_S/\omega_p)\exp{(-\alpha l)}$ above the threshold. Therefore, (13.30) can be used to find the Raman conversion efficiency quite accurately for any value of the pumping ratio r irrespective of the assumption used in obtaining it. By using (13.30) to calculate the Raman Stokes generation in a single-mode silica fiber, it is found that a reduction in the pump intensity by 1 dB below the threshold reduces the output Stokes intensity by more than 10 dB, but an increase in the pump intensity by 1 dB above the threshold causes the conversion of photons from the pump to the Stokes to be more than 98% complete.

If the pump intensity is many times above the threshold intensity and the linear absorption loss is negligible, complete conversion of photons from the pump to the Stokes occurs within a very short distance from the input end. This first Stokes wave at the frequency of $\omega_{S1} = \omega_p - \Omega_R$ can then serve as a pump to generate the second Stokes wave at the frequency of $\omega_{S2} = \omega_{S1} - \Omega_R = \omega_p - 2\Omega_R$. This cascading process continues until the waves reach the end of the generator. Therefore, with proper choices of the generator length and the pump intensity, a high-order Stokes wave can be generated at a frequency that is down-shifted from the pump frequency by an integral multiple of the Raman frequency. However, such complete conversion of photons from the pump to the first Stokes and from a low-order Stokes to a high-order Stokes is possible only for CW waves or very long optical pulses. For short optical pulses,

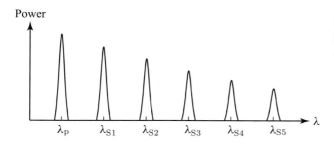

Figure 13.7 Multiple orders of Raman Stokes generated by high-power ultrashort optical pulses. This figure shows five orders of Stokes.

the conversion of photons from one order to another is normally not complete due to temporal walk-off between the interacting pulses of different wavelengths caused by group-velocity dispersion in the medium, but the generation of multiple Stokes orders is still possible with high-intensity pulses, as shown in Fig. 13.7.

Sometimes, in addition to the Stokes wave, an anti-Stokes wave at the up-shifted anti-Stokes frequency of $\omega_{AS} = \omega_p + \Omega_R$ can also be generated through Stokes–anti-Stokes coupling and/or parametric four-wave mixing with the pump if the required phase-matching conditions for such parametric processes are satisfied. In practical applications, however, a Raman generator is normally used as a nonparametric frequency converter to convert the optical power of a high-frequency pump wave to a low-frequency Stokes wave.

EXAMPLE 13.3

The fiber Raman amplifier described in Example 13.2 can be used as a fiber Raman generator for a Stokes signal at $\lambda_S = 1.55$ μm without an input signal at this wavelength by raising the pump power at $\lambda_p = 1.4468$ μm. Find the threshold pump power for this fiber Raman generator.

Solution

By identifying $P_p^{th} = I_p^{th}(0)\mathcal{A}_{eff}$ and using $\ln G_R^{th} = 16$, we have, from (13.29),

$$P_p^{th} = \frac{\mathcal{A}_{eff}}{l_{eff}} \frac{\ln G_R^{th}}{\tilde{g}_R} = \frac{16\mathcal{A}_{eff}}{\tilde{g}_R \, l_{eff}}$$

for the threshold pump power of a fiber Raman generator. By using the parameters obtained in Example 13.2, we find that

$$P_p^{th} = \frac{16 \times 5 \times 10^{-11}}{6.45 \times 10^{-14} \times 14.86 \times 10^3} \text{ W} = 835 \text{ mW}.$$

When the pump power is below P_p^{th}, very little power is converted to the Stokes signal in a Raman generator. When the pump power exceeds P_p^{th} at a certain level, it is completely converted to the Stokes with a maximum power conversion efficiency of $\eta_R^{max} = (\omega_S/\omega_p)\exp(-\alpha l) = (\lambda_p/\lambda_S)\exp(-\alpha l)$. If the pump power continues to increase, the power is converted to a successively higher order of Stokes at the output.

13.3.3 Raman Oscillators

The Raman gain can be utilized to construct a Raman oscillator by providing optical feedback to a Raman generator, such as by placing the Raman generator in a resonant optical cavity, as shown in Fig. 13.8. The optical cavity shown in Fig. 13.8(a) is optically pumped by a pump beam at the frequency of ω_p; it resonates and emits at the first Stokes frequency of $\omega_S = \omega_p - \Omega_R$. Because the Raman gain coefficient g_R increases linearly with the pump intensity, and thus with the pump power, for a fixed resonator configuration, it can be linearly increased by increasing the pump power. When the power of the pump beam is sufficiently high for the Raman gain coefficient to compensate for all losses of the optical cavity, the Raman oscillator reaches its threshold for oscillation, thus making the oscillator into a *Raman laser* that emits the Stokes signal as the laser output, as shown in Fig. 13.8(a).

It is also possible to construct a Raman laser that emits at a high-order Stokes frequency of $\omega_{Sn} = \omega_p - n\Omega_R$. For this purpose, the optical resonator is designed to resonate at all orders of the Stokes frequencies, $\omega_{S1}, \cdots, \omega_{Sn}$, but only the nth-order Stokes at ω_{Sn} is emitted through the resonator, as shown in Fig. 13.8(b). A Stokes signal can serve as the pump for generating the Stokes signal of the next order through a cascading process. A high-order Stokes oscillator has a higher threshold than a low-order Stokes oscillator because there is an additional loss associated with Raman excitation at the frequency of Ω_R for each higher order. With a sufficiently high power for the pump beam, the threshold of a Raman laser that emits at a desired order of the Stokes frequency can be reached.

Compared to an ordinary laser, a Raman laser has the flexibility of its emission frequency because the emission is not based on population inversion, nor is it in resonance with a transition between two energy levels. For a desired emission frequency at ω_{Sn} and with a given Raman gain medium that has a Raman resonance frequency of Ω_R, it is only necessary to find a sufficiently strong optical pump source of a frequency at $\omega_p = \omega_{Sn} + n\Omega_R$ for some integer n.

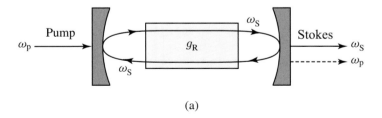

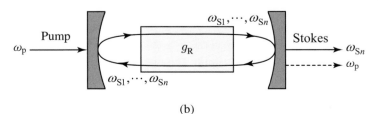

(b)

Figure 13.8 Schematic diagrams of (a) a Raman oscillator that resonates and emits at the first-order Stokes frequency and (b) a Raman oscillator that resonates at multiple orders of Stokes frequencies but only emits at the nth-order Stokes frequency.

Similar to an optical parametric oscillator, a Raman oscillator utilizes a nonlinear suscepti-bility for optical gain. Different from an optical parametric oscillator, however, a Raman oscillator is based on a third-order nonparametric process that uses the imaginary part of $\chi^{(3)}$, but an optical parametric oscillator is based on a second-order parametric process that uses the real part of $\chi^{(2)}$. For this reason, phase matching is automatic for a Raman oscillator, whereas an optical parametric oscillator requires phase matching. Frequency tuning for a Raman oscillator can be accomplished by simply tuning the pump frequency. By comparison, frequency tuning for an optical parametric oscillator is normally accomplished through tuning the phase-matching condition.

A very important characteristic of the Raman process is that it can efficiently respond only to an optical signal that has a bandwidth narrower than the Raman linewidth, as discussed in Section 13.2. For this reason, when a Raman oscillator is pumped with a short laser pulse, the Raman gain decreases as the pump pulsewidth decreases, thus increasing the Raman laser threshold. At a certain point, the Raman laser stops working because its threshold becomes too high to be reached. Therefore, a picosecond or femtosecond Raman laser is not practical. By comparison, the real part of $\chi^{(2)}$ for an optical parametric oscillator responds to optical excita-tion on the sub-femtosecond time scale. Therefore, an optical parametric oscillator readily responds to the pumping by ultrashort pulses, including femtosecond pulses. The threshold of a synchronously pumped optical parametric oscillator becomes lower as the pump pulsewidth is reduced because the peak power of the pulse is increased with the reduced pulsewidth, thus increasing the parametric gain coefficient. Raman oscillators are not useful for generating ultrashort laser pulses at the Stokes frequencies, but synchronously pumped optical parametric oscillators can efficiently generate ultrashort laser pulses at the parametric frequencies.

13.4 BRILLOUIN GAIN

For Brillouin scattering, the relevant excitation is a long-range acoustic wave, which has a linear dispersion relation between the magnitude of its wavevector and its frequency: $K = \Omega/v_a$. For Brillouin Stokes scattering, the conditions given in (13.10) and (13.11) are the same as those for the first-order down-shifted Bragg diffraction of an optical wave by an acoustic wave, except that in Brillouin scattering the acoustic wave is generated by the pump optical wave whereas in acousto-optic Bragg diffraction the acoustic wave is externally applied to the medium. The amount of frequency shift in Brillouin scattering is a function of the pump optical frequency and the scattering angle, θ, between $\mathbf{k}_S$ and $\mathbf{k}_p$. In general, the Brillouin frequency shift is a few orders of magnitude smaller than the pump and the Stokes optical frequencies. For Brillouin scattering in a nonbirefringent medium, the approximation $k_S \approx k_p$ is valid. Then, by using (13.10) and (13.11) together with the dispersion relation of the acoustic wave, we find the angle-dependent frequency shift:

$$\Omega = 2v_a k_p \sin\frac{\theta}{2} = \frac{2v_a n \omega_p}{c} \sin\frac{\theta}{2}, \tag{13.31}$$

where n is the index of refraction at the pump optical frequency ω_p, v_a is the acoustic velocity in the medium, and $\theta = \theta_S - \theta_p$ is the scattering angle of the Stokes wavevector $\mathbf{k}_S$ with respect to the pump wavevector $\mathbf{k}_p$. For a forward-scattering Stokes signal, $\mathbf{k}_S$ is parallel to $\mathbf{k}_p$ with $\theta_S = \theta_p$ so that $\theta = 0$; for a backward Stokes signal, $\mathbf{k}_S$ is anti-parallel to $\mathbf{k}_p$ with $\theta_S = \theta_p + \pi$ so that $\theta = \pi$. We see that, very differently from Raman scattering, Brillouin scattering does not have a constant frequency shift in all directions. In particular, there is no Brillouin Stokes scattering in the forward direction because Ω given in (13.31) vanishes for $\theta = 0$. Spontaneous Brillouin scattering appears in all other directions with a frequency shift that varies with the scattering angle. Stimulated Brillouin scattering occurs predominantly in the backward direction with a maximum frequency shift, known as the *Brillouin frequency*, which is determined by the phase-matching condition given in (13.11) to be

$$\Omega_B = \frac{n v_a}{c}(\omega_p + \omega_S) = \frac{2 n v_a / c}{1 + n v_a / c}\omega_p = \frac{2 n v_a / c}{1 - n v_a / c}\omega_S \approx \frac{2 n v_a}{c}\omega, \tag{13.32}$$

where we have used the fact that $\omega_p \approx \omega_S \approx \omega \gg \Omega_B$, or

$$f_B = \frac{\Omega_B}{2\pi} = \frac{2 n v_a}{\lambda}. \tag{13.33}$$

With a pump beam in the optical spectral region, the Brillouin frequency f_B falls in the hypersonic region, typically in the range of 1–50 GHz for a large variety of materials.

The *Brillouin gain factor* of a material can be expressed in a form similar to that of the Raman gain factor. For backward interaction at a frequency of $\Omega = \omega_p - \omega_S$ that is near the Brillouin frequency Ω_B, the Brillouin gain factor has the frequency dependence:

$$\tilde{g}_B = \tilde{g}_{B0}\frac{\gamma_B^2}{(\Omega - \Omega_B)^2 + \gamma_B^2} = \tilde{g}_{B0}\frac{\gamma_B^2}{(\omega_p - \omega_S - \Omega_B)^2 + \gamma_B^2}. \tag{13.34}$$

This Brillouin gain factor has a Lorentzian lineshape, which is plotted in Fig. 13.9(a). It has a full width at half-maximum linewidth of $\Delta\Omega_B = 2\gamma_B$, or $\Delta f_B = \gamma_B / \pi$, which is associated with a relaxation time of $\tau_B = \gamma_B^{-1}$ for the acoustic excitation that is responsible for the Brillouin process. Because the Brillouin response time of a common material is typically of the order of nanoseconds, the Brillouin linewidth Δf_B is typically in the range of 10 MHz to 1 GHz. Therefore, a Brillouin device does not respond efficiently to very short optical pulses, nor to any optical wave that has a spectral width in the gigahertz range or above. For a pump optical wave that has a Lorentzian spectral shape with a full width at half-maximum linewidth of Δv_p, the *peak Brillouin gain factor* of a medium scales as

$$\tilde{g}_{B0} = \frac{\Delta f_B}{\Delta f_B + \Delta v_p}\tilde{g}_{B0}^{max}, \tag{13.35}$$

as plotted in Fig. 13.9(b), where $\tilde{g}_{B0}^{max}$ is the peak Brillouin gain factor for an idealistic CW pump wave of a zero linewidth with $\Delta v_p = 0$. Clearly, when $\Delta v_p \gg \Delta f_B$, the peak Brillouin gain factor is greatly reduced. The Brillouin gain factor has other characteristics that are different from

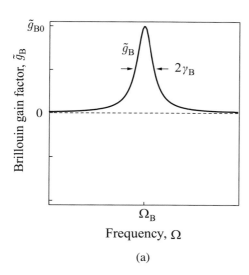
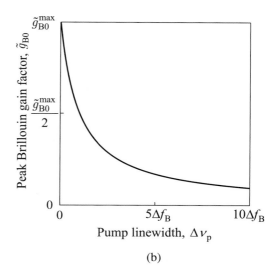

Figure 13.9 (a) Brillouin gain factor $\widetilde{g}_B$ as a function of the frequency difference $\Omega = \omega_p - \omega_S$ between the pump and the Stokes signal. (b) Peak Brillouin gain factor $\widetilde{g}_{B0}$ as a function of the pump linewidth $\Delta\nu_p$.

those of the Raman gain factor due to the fact that the Brillouin frequency is dictated by the phase-matching condition of (13.11). We have seen in (13.32) that $\Omega_B \propto \omega$. In addition, $\gamma_B \propto \omega^2$, but $\widetilde{g}_{B0}$ is independent of the optical frequency. For fused silica, $f_B \approx (17.3/\lambda)$ GHz and $\Delta f_B \approx (38.4/\lambda^2)$ MHz, where the optical wavelength λ is measured in micrometers, and $\widetilde{g}_{B0}^{max} = 4.5 \times 10^{-11}$ m W^{-1}.

EXAMPLE 13.4

For the optical wave at the 1.55 μm wavelength propagating in a silica optical fiber as described in Example 13.1, what are the Brillouin frequency shift, the Brillouin linewidth, and the peak Brillouin gain factor if the optical wave has a linewidth of 1 MHz? What is the peak Brillouin gain factor if the optical wave has a linewidth of 100 MHz?

Solution

According to the characteristics of fused silica described in the text, the Brillouin frequency shift is $f_B = (17.3/1.55)$ GHz $= 11.16$ GHz, and the Brillouin linewidth is $\Delta f_B = (38.4/1.55^2)$ MHz $= 15.98$ MHz. Though $\widetilde{g}_{B0}^{max} = 4.5 \times 10^{-11}$ m W^{-1} for the silica fiber is quite independent of the optical wavelength, the peak Brillouin gain factor varies with the linewidth $\Delta\nu_p$ of the optical wave according to (13.35). Therefore, the peak Brillouin gain factor is $\widetilde{g}_{B0} = (15.98/16.98) \times 4.5 \times 10^{-11}$ m W$^{-1} = 4.23 \times 10^{-11}$ m W^{-1} if the optical wave has a narrow linewidth of $\Delta\nu_p = 1$ MHz. If the linewidth of the optical wave is increased to $\Delta\nu_p = 100$ MHz, the peak Brillouin gain is reduced to $\widetilde{g}_{B0} = (15.98/115.98) \times 4.5 \times 10^{-11}$ m W$^{-1} = 6.2 \times 10^{-12}$ m W^{-1}. Further increase in the linewidth of the pump optical wave further reduces the peak Brillouin gain.

13.5 BRILLOUIN AMPLIFICATION AND GENERATION

13.5.1 Brillouin Amplifiers

In this subsection, we discuss the characteristics of backward amplifiers, in which the Stokes signal propagates contradirectionally to the pump wave. Specifically, we consider the Brillouin amplifiers because they are generally backward amplifiers. Nevertheless, a Raman amplifier can be either forward or backward. The analysis and the results obtained in this subsection apply equally well to backward Raman amplifiers.

The Brillouin gain in a medium can be utilized to amplify an optical signal at a frequency that is down-shifted from the pump frequency by an amount equal to the Brillouin frequency. Due to the fundamental differences between the Raman and the Brillouin processes, as discussed in Section 13.1, the Brillouin amplifiers have a few characteristics that are very different from those of the Raman amplifiers:

1. Only the contradirectional configuration shown in Fig. 13.10 is acceptable for a Brillouin amplifier because there is no forward Brillouin scattering.
2. The Brillouin linewidth is relatively narrow. Therefore, a Brillouin amplifier is useful only for the amplification of narrowband signals, whereas a Raman amplifier can be used for broadband signals or short-pulse signals because of the large Raman linewidth.
3. The peak Brillouin gain factor, $\widetilde{g}_{B0}$, of a solid or liquid medium, or a high-pressure gaseous medium, is usually much larger than the peak Raman gain factor, $\widetilde{g}_{R0}$, of the same medium. Therefore, a Brillouin amplifier usually requires a much lower pump intensity than a Raman amplifier needs to have the same amplification factor for the signal.

Because of the contradirectional configuration of a Brillouin amplifier, the signal propagates in the $-z$ direction while the pump propagates in the z direction. Therefore, Brillouin amplification is described by the coupled equations:

$$-\frac{dI_S}{dz} + \alpha_S I_S = \widetilde{g}_B I_p I_S, \tag{13.36}$$

$$\frac{dI_p}{dz} + \alpha_p I_p = -\frac{\omega_p}{\omega_S} \widetilde{g}_B I_S I_p, \tag{13.37}$$

with an input pump intensity of $I_p(0)$ at $z = 0$ and an input signal intensity of $I_S(l)$ at $z = l$ given as the boundary conditions. The exact solution for this backward amplification differs from that for the forward amplification. It can be found when $\alpha_S = \alpha_p = 0$. For a backward Brillouin

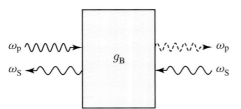

Figure 13.10 Contradirectional Brillouin amplifier. A Brillouin amplifier cannot take the codirectional configuration. For Brillouin amplification, $\omega_p = \omega_S + \Omega_B$.

amplifier of a length l in the case that $\alpha_S = \alpha_p = 0$, the intensity of the Stokes signal at its output end, $z = 0$, and that of the pump wave at its output end, $z = l$, are given by the implicit solutions:

$$I_S(0) = \frac{\omega_S I_p(0)}{\omega_S I_p(0) + \omega_p I_S(l) - \omega_p I_S(0)} I_S(l) \exp(g_B l) = \frac{I_p(0)}{I_p(l)} I_S(l) \exp(g_B l), \quad (13.38)$$

$$I_p(l) = \frac{\omega_p I_S(l)}{\omega_p I_S(l) + \omega_S I_p(0) - \omega_S I_p(l)} I_p(0) \exp(g_B l) = \frac{I_S(l)}{I_S(0)} I_p(0) \exp(g_B l), \quad (13.39)$$

where g_B is the *Brillouin gain coefficient* defined as

$$g_B = \tilde{g}_B \left[I_p(0) - \frac{\omega_p}{\omega_S} I_S(0) \right] = \tilde{g}_B \left[I_p(l) - \frac{\omega_p}{\omega_S} I_S(l) \right]. \quad (13.40)$$

In the application of an optical amplifier for the amplification of a weak signal, however, there is little pump depletion due to nonlinear Brillouin interaction. Then, the right-hand side of (13.37) can be ignored to obtain the solution for the output Stokes intensity of the signal at $z = 0$. This approximate solution can be found for the general situation that $\alpha_S \neq \alpha_p$; it is not necessary to assume that $\alpha_S = \alpha_p = 0$ as above. We have

$$I_S(0) = I_S(l) \exp(g_B l_{\text{eff}} - \alpha_S l), \quad (13.41)$$

where l_{eff} is the effective interaction length of the same form as that defined in (13.25) and g_B is the Brillouin gain coefficient that results from that defined in (13.40) under the condition that $I_S(0) \ll I_p(0)$:

$$g_B = \tilde{g}_B I_p(0) = -\frac{3\omega_S}{c^2 \epsilon_0 n_{S,z} n_{p,z}} \chi_B'' I_p(0). \quad (13.42)$$

Therefore, in the case of negligible pump depletion, the amplification factor of a Brillouin amplifier, or the *Brillouin amplifier gain*, is

$$G_B = \frac{I_S(0)}{I_S(l) \exp(-\alpha_S l)} = \exp(g_B l_{\text{eff}}) = \exp\left[\tilde{g}_B I_p(0) l_{\text{eff}} \right]. \quad (13.43)$$

EXAMPLE 13.5

If the fiber Raman amplifier described in Example 13.2 is turned into a Brillouin amplifier for the same input signal and the same desired output signal, what should the pump wavelength be? If the pump wave has a linewidth of 100 MHz, what is the required pump power?

Solution

From Example 13.4, we know that $f_B = 11.16$ GHz. Therefore, $f_B/c = 37.2 \, \text{m}^{-1}$, and the pump wavelength is

$$\lambda_p = \left(\frac{1}{1.55 \times 10^{-6}} + 37.2\right)^{-1} \text{m} = 1.5499 \text{ μm}.$$

The pump wavelength is very close to the signal wavelength because of the small Brillouin frequency shift. We find from Example 13.4 that the Brillouin gain factor for this amplifier is $\widetilde{g}_B = 6.2 \times 10^{-12} \text{ m W}^{-1}$ because the pump has a linewidth of 100 MHz. Because a Brillouin amplifier functions only in the contradirectional configuration, we identify $P_S^{in} = I_S(l)\mathcal{A}_{eff}$ and $P_S^{out} = I_S(0)\mathcal{A}_{eff}$. Then we find from (13.43) that the required Brillouin amplifier gain in decibels is

$$G_B = P_S^{out}(\text{dBm}) - P_S^{in}(\text{dBm}) + \alpha(\text{dB km}^{-1})l(\text{km}) = 20 \text{ dB},$$

which is the same as the Raman amplifier gain in Example 13.2. Therefore, $G_B = 100$. From (13.43), we find by identifying the pump power as $P_p = I_p(0)\mathcal{A}_{eff}$ that

$$P_p = \frac{\mathcal{A}_{eff}}{l_{eff}} \frac{\ln G_B}{\widetilde{g}_B} = \frac{5 \times 10^{-11} \times \ln 100}{14.86 \times 10^3 \times 6.2 \times 10^{-12}} \text{ W} = 2.5 \text{ mW}.$$

By comparing this pump power with the pump power of $P_p = 240$ mW found in Example 13.2 for the Raman amplifier, we find that for the same amplifier gain, $G_B = G_R$, the pump power required for a Brillouin amplifier is scaled from that for a Raman amplifier by a factor of $P_p^B/P_p^R = \widetilde{g}_R/\widetilde{g}_B$. Because $\widetilde{g}_B$ is larger than $\widetilde{g}_R$ by about two orders of magnitude in this example, the pump power is reduced by as much. Note that in using (13.43) to obtain G_B and P_p in the above, we have implicitly assumed that the depletion of the pump power due to its conversion to the signal power is negligible. This assumption is not valid here because the pump power obtained under such an assumption is 2.5 mW but the output signal is 1 mW. A more detailed analysis with the effect of pump depletion taken into consideration is required to obtain the accurate result.

13.5.2 Brillouin Generators

Similar to the situation in a Raman generator, the emission from spontaneous Brillouin scattering can also seed the generation of a Brillouin Stokes frequency in the presence of a pump above a *Brillouin threshold* but in the absence of an input Stokes signal. Besides the fundamental differences in terms of the frequency shift and the generation efficiency, an important difference between a Brillouin generator and a Raman generator is that the Brillouin Stokes wave is generated only in the backward direction, whereas the Raman Stokes is generated only in the forward direction.

As discussed above, for backward generation, the net result of the cumulative backward amplification of spontaneous emission over the entire length of interaction is equivalent to the injection of an effective backward-propagating Stokes signal $I_S^{eff}(l)$ at $z = l$. By considering the physical implication of the threshold amplification factor given in (13.28) for a Raman generator, the Brillouin threshold can be defined as the condition that stimulated Brillouin amplification of the spontaneous Brillouin Stokes emission brings the Stokes intensity to the level of the pump intensity. Because the effective Stokes signal at $z = l$ is related to the pump intensity at $z = l$, we then find the threshold amplification factor for a Brillouin generator:

$$G_B^{th} = \exp\left[\tilde{g}_B I_p^{th}(0) l_{eff}\right] = \frac{I_p^{th}(l)}{I_{S,th}^{eff}(l)}, \tag{13.44}$$

where $I_p^{th}(0)$ and $I_p^{th}(l)$ are the input pump intensity at $z = 0$ and the remaining pump intensity at $z = l$, respectively, at the threshold of the Brillouin generator. Therefore, the threshold pump intensity for Brillouin Stokes generation in a medium of a length l is

$$I_p^{th}(0) = \frac{\ln G_B^{th}}{\tilde{g}_B \, l_{eff}}. \tag{13.45}$$

The value of G_B^{th} is a function of the characteristics of the medium and is generally larger than that of G_R^{th} of Raman generation for the same medium, primarily because of the fact that the Brillouin Stokes is generated in the backward direction whereas the Raman Stokes is generated in the forward direction. For example, $\ln G_B^{th} \approx 21$ for Brillouin Stokes generation in a single-mode silica fiber [5].

The conversion efficiency of a Brillouin generator from the pump to the Stokes is measured in terms of an intensity reflectivity defined as

$$R_B = \frac{I_S(0)}{I_p(0)}. \tag{13.46}$$

In the case that $\alpha_S = \alpha_p = 0$, the value of R_B can be found from the transcendental relation:

$$R_B = (G_B^{th})^{r(1-R_B)-1} \tag{13.47}$$

under the approximation that $\omega_S \approx \omega_p$ for Brillouin scattering in the optical region, where $r = I_p(0)/I_p^{th}(0)$ is the *pumping ratio* of the pump intensity with respect to the threshold pump intensity. Because of the large value of G_B^{th}, the relation in (13.47) indicates a sharp threshold for Brillouin generation, as shown in Fig. 13.11. Below the Brillouin threshold, $r < 1$, we see that R_B quickly approaches zero. Above the threshold, we find that R_B varies with pump intensity approximately as

$$R_B \approx 1 - \frac{1}{r} = 1 - \frac{I_p^{th}(0)}{I_p(0)}, \quad \text{for } r > 1. \tag{13.48}$$

This relation is also shown in Fig. 13.11 as the dashed curve. It leads to the important conclusion that

$$I_p(l) = I_p^{th}(0) \quad \text{if} \quad I_p(0) \gg I_p^{th}(0). \tag{13.49}$$

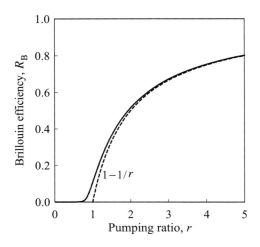

Figure 13.11 Conversion efficiency of a silica fiber Brillouin generator, for which $\ln G_{\mathrm{B}}^{\mathrm{th}} \approx 21$, as a function of the pumping ratio. The exact relation given in (13.47) is shown as the solid curve. Shown as the dashed curve is the approximate relation given in (13.48).

Therefore, when the input pump intensity exceeds the threshold pump intensity of a lossless Brillouin generator, the transmitted intensity is clamped at the level of the threshold pump intensity. The excess above the threshold is converted to the Stokes frequency and is reflected back to the input end. This characteristic allows very efficient Brillouin generation, but it also sets a very important limitation on the level of the optical power that can be transmitted through an optical system. In particular, in a fiber-optic transmission system, the generation of the Brillouin Stokes in the optical fiber severely limits the transmission power level of the system.

EXAMPLE 13.6

The fiber Brillouin amplifier described in Example 13.5 becomes a fiber Brillouin generator for a Stokes signal at $\lambda_{\mathrm{S}} = 1.55$ μm without an input signal at this wavelength if the pump power at $\lambda_{\mathrm{p}} = 1.5499$ μm is raised above a threshold level. Find the threshold pump power for this fiber Brillouin generator if the linewidth of the pump is 100 MHz. What is the threshold pump power if the linewidth of the pump is only 1 MHz?

Solution

By identifying $P_{\mathrm{p}}^{\mathrm{th}} = I_{\mathrm{p}}^{\mathrm{th}}(0)\mathcal{A}_{\mathrm{eff}}$ and using $\ln G_{\mathrm{B}}^{\mathrm{th}} \approx 21$, we have, from (13.45),

$$P_{\mathrm{p}}^{\mathrm{th}} = \frac{\mathcal{A}_{\mathrm{eff}}}{l_{\mathrm{eff}}} \frac{\ln G_{\mathrm{B}}^{\mathrm{th}}}{\tilde{g}_{\mathrm{B}}} = \frac{21\mathcal{A}_{\mathrm{eff}}}{\tilde{g}_{\mathrm{B}} l_{\mathrm{eff}}}$$

for the threshold pump power of a fiber Brillouin generator. By using the parameters obtained in Example 13.5 with $\tilde{g}_{\mathrm{B}} = 6.2 \times 10^{-12}$ m W^{-1} for a pump wave that has a linewidth of 100 MHz, we find that

$$P_{\mathrm{p}}^{\mathrm{th}} = \frac{21 \times 5 \times 10^{-11}}{6.2 \times 10^{-12} \times 14.86 \times 10^3} \, \mathrm{W} = 11.4 \, \mathrm{mW}.$$

If the pump has a narrow linewidth of only 1 MHz, we have $\tilde{g}_{\mathrm{B}} = 4.23 \times 10^{-11}$ m W^{-1} from Example 13.4. Then the threshold pump power is reduced to

$$P_\text{p}^\text{th} = \frac{21 \times 5 \times 10^{-11}}{4.23 \times 10^{-11} \times 14.86 \times 10^3} \text{ W} = 1.67 \text{ mW}.$$

The Brillouin threshold pump power can be substantially increased if the linewidth of the pump is large. Because the power that remains in the pump is clamped at the Brillouin threshold, with the rest reflected back to the input end, suppressing the Brillouin Stokes generation by sufficiently increasing the Brillouin threshold is essential for the operation of a Raman amplifier, as discussed in Example 13.2, as well as for the operation of a Raman generator.

13.5.3 Brillouin Oscillators

The Brillouin gain can also be utilized to construct a Brillouin oscillator by providing optical feedback through a resonant optical cavity in a manner similar to a Raman oscillator. As given in (13.42), the Brillouin gain coefficient g_B is linearly proportional to the pump intensity, and thus proportional to the pump power in a fixed oscillator configuration. With a pump beam at an optical frequency of ω_p, a Brillouin oscillator resonates and emits at the first Stokes frequency of $\omega_\text{S} = \omega_\text{p} - \Omega_\text{B}$ when the power of the pump beam is sufficiently high for the Brillouin gain coefficient g_B to compensate for all losses of the optical cavity, thus making the oscillator into a *Brillouin laser* that emits the Stokes signal as the laser output. These basic characteristics of a Brillouin laser are similar to those of a Raman laser discussed in Subsection 13.3.3. However, there are some fundamental differences between a Brillouin laser and a Raman laser.

A Brillouin Stokes signal is only generated and amplified in the backward direction, which is opposite to the propagation direction of its pump wave. To clearly illustrate this relationship, a Brillouin laser that has a ring cavity configuration and emits only the first-order Stokes signal is shown in Fig. 13.12(a). In this configuration, the Stokes and the pump waves are contra-propagating within the ring cavity, and the laser output at the Stokes frequency is emitted through the input port in the backward direction. By pumping a Brillouin laser at a sufficiently high pump power level, it is possible for multiple orders of Stokes signals to reach their respective oscillation thresholds through a cascading process, with a low-order Stokes signal serving as the pump for the next order. When this cascading process takes place, all odd-order Stokes signals that oscillate in the optical cavity are emitted in the backward direction opposite to the input pump at ω_p, and all even-order Stokes that oscillate in the optical cavity are emitted in the forward direction, as shown in Fig. 13.12(b).

Because the Brillouin frequency f_B typically falls in the range of 1–50 GHz, which is very small compared to the optical pump frequency that is of the order of 100–1,000 THz, the Stokes output of a Brillouin laser has a small frequency shift with respect to its pump frequency. The Brillouin gain spectrum also has a narrow linewidth Δf_B, which is typically in the range of 10 MHz to 1 GHz. Because the peak Brillouin gain factor quickly decreases with an increasing spectral width of the pump wave, as expressed in (13.35) and shown in Fig. 13.9(b), a pump wave that has a narrow linewidth of $\Delta\nu_\text{p}$ that is smaller than Δf_B is necessary for a decent Brillouin gain. When $\Delta\nu_\text{p}$ is larger than Δf_B, the threshold of a Brillouin laser quickly increases because of the decrease in the peak Brillouin gain factor. For this reason,

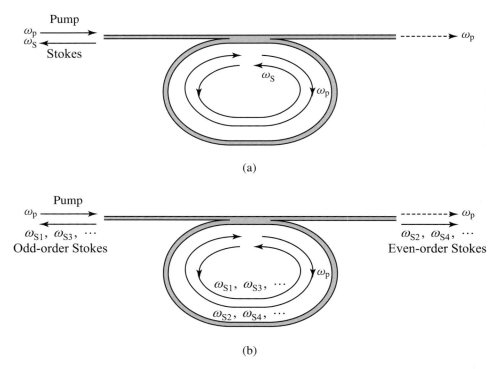

Figure 13.12 Schematic diagrams of (a) a Brillouin ring oscillator that resonates and emits at the first-order Stokes frequency and (b) a Brillouin ring oscillator that resonates and emits at multiple orders of Stokes frequencies. Odd-order Stokes signals are emitted in the backward direction, which is opposite to that of the pump wave. Even-order Stokes are emitted in the forward direction, which is the same direction as that of the pump wave.

an efficient Brillouin laser has to be pumped with a narrow-linewidth CW pump beam. Short optical pulses are not useful for pumping a Brillouin laser, although they have high peak powers.

13.6 COMPARISON OF RAMAN AND BRILLOUIN DEVICES

Because Raman and Brillouin gains exist in the same medium and both Raman Stokes and Brillouin Stokes can grow from spontaneous emission, these two processes compete with each other for the same pump power source. The one that has a lower threshold pump intensity quickly monopolizes the pump power, thus prohibiting the other from occurring. Because $\widetilde{g}_{B0}$ is usually much larger than $\widetilde{g}_{R0}$ in the same medium, Brillouin generation usually dominates, although G_B^{th} is larger than G_R^{th}. However, because the Brillouin gain has a very narrow linewidth, the threshold pump intensity for Brillouin generation increases very quickly when the pump wave has a linewidth exceeding Δf_B. Therefore, Brillouin generation dominates only when the pump has a narrow linewidth, whereas Raman generation dominates when the linewidth of the pump is larger than the Brillouin linewidth. The pump power required for a Raman amplifier is generally lower than the threshold pump power of a Raman generator, and that for a Brillouin amplifier is lower than the threshold of a Brillouin generator. However, because the Brillouin gain factor can be a few orders of magnitude larger than the Raman gain factor, stimulated Brillouin scattering can easily occur well below the power required for a Raman amplifier. The consequences of stimulated Brillouin scattering in a Raman amplifier include significant reduction of the Raman gain by depleting the pump power, generation of noise, and distortion of the signal waveform. It is therefore necessary to suppress stimulated Brillouin scattering in a Raman amplifier by, for example, using a pump of a sufficiently broad linewidth. Similar considerations apply to Raman lasers and Brillouin lasers.

Many gases, such as H_2, N_2, O_2, Ar, and Xe, have useful Raman and Brillouin gains and frequency shifts for practical applications. A gas for such applications is normally contained in a high-pressure cell, often called a *Raman cell* or a *Brillouin cell*, depending on its intended application. One significant difference between a gaseous medium and a liquid or solid medium is that $\widetilde{g}_{R0}$, Δf_R, $\widetilde{g}_{B0}$, and Δf_B of a gaseous medium depend on the density of the molecules in the medium, which can be varied by varying the gas pressure in a cell of a fixed length and a fixed volume. The value of $\widetilde{g}_{R0}$ scales linearly with the density of the gas molecules at low pressures until it saturates at a certain pressure. By comparison, the value of $\widetilde{g}_{B0}$ scales quadratically with the density of the gas molecules. Therefore, for a given gaseous medium, $\widetilde{g}_{R0}$ can be larger than $\widetilde{g}_{B0}$ at low pressures, but $\widetilde{g}_{B0}$ eventually becomes larger than $\widetilde{g}_{R0}$ at a sufficiently high pressure. In addition, $\widetilde{g}_{B0}$ and Δf_B also depend on temperature.

Because of the narrow linewidth of the typical Brillouin gain spectrum, which is of the order of 10 MHz to 1 GHz, a useful pump for a Brillouin device is practically limited to a CW laser beam that has a narrow linewidth. This limitation does not apply to a Raman device because the Raman gain spectrum is much broader, typically ranging from a few gigahertz to tens of terahertz. Correspondingly, the response time τ_R of a Raman process ranges from a few hundred picoseconds to tens of femtoseconds, depending on the material. For this reason, pumping with short laser pulses to take advantage of their high peak powers is feasible for many Raman devices. However, because the Raman frequency shift Ω_R can be a sizable fraction of the pump frequency ω_p, the dispersion between the pump and the Stokes frequencies cannot be ignored. In particular, when the pump pulses are ultrashort laser pulses of pulsewidths on the picosecond or femtosecond time scale, a pump pulse at ω_p and a Stokes pulse at $\omega_S = \omega_p - \Omega_R$ can quickly lose interaction by temporally walking away from each other due to group-velocity dispersion. Therefore, two factors have to be considered for a transient Raman process that involves short laser pulses: the response time of the Raman process, and the temporal walk-off between the pump and the Stokes pulses caused by group-velocity dispersion in the medium. In the following, we give a summary for the scenarios that take place with different pump sources.

With continuous pumping by using a CW optical wave, there is no limit on the interaction length for both forward and backward stimulated Raman scattering. Both occur. Because of the large spectral linewidth of the typical Raman gain factor, practically any CW laser has a narrower linewidth than that of the Raman gain factor. Raman amplification can take place in either the forward or the backward direction, but Raman generation takes place only in the forward direction because it has a lower threshold than backward generation. High-order

Raman Stokes can be generated if the pump intensity is sufficiently high for a given interaction length. For Brillouin amplification, the CW pump wave must have a linewidth that is narrower than the spectral width of the Brillouin gain factor. Otherwise, the Brillouin gain quickly diminishes as the pump linewidth increases. Brillouin amplification in a nonbirefringent medium only takes place in the backward direction. If the pump power is sufficiently high, high-order Brillouin Stokes can successively occur in alternating directions through a cascading process, with all odd-order Stokes propagating in the direction opposite to that of the pump wave and all even-order Stokes propagating in the same direction as that of the pump.

When pumping with a nanosecond pulse, such as that generated by a Q-switched laser, the pulsewidth is generally larger than the Raman response time – that is, the spectral width of the pulse is generally smaller than the linewidth of the Raman gain factor. Therefore, the situation is similar to stimulated Raman scattering with CW pumping, except that the temporal overlap of the pump and Stokes pulses is limited by the pulse duration in the case of backward pumping, particularly when the Raman medium is very long, such as in the case of a long optical fiber. Forward stimulated Raman scattering has little limitation with nanosecond pulse pumping, but backward stimulated Raman scattering may be substantially reduced due to a finite interaction length between the contra-propagating pump and Stokes pulses, which is determined by the pulse duration. High-order Stokes can be generated in a medium of a given length of the Raman medium if the pump intensity is sufficiently high. For stimulated Brillouin scattering, the efficiency is generally reduced because of the linewidth of a nanosecond pulse unless the Brillouin linewidth is broader than the pulse spectral width, which is possible when the pulse is relatively long, such as one of the order of tens or hundreds of nanoseconds.

When the pump beam is a picosecond pulse or a train of picosecond optical pulses, such as those generated by a mode-locked laser, the pulsewidth can be larger or smaller than the Raman relaxation time, depending on the specific Raman medium. For most Raman materials and the typical picosecond pulse, $\Delta t_{\mathrm{ps}} > \tau_{\mathrm{R}}$ so that stimulated Raman scattering still takes place. Forward stimulated Raman scattering may be reduced by a limited interaction length between the codirectionally propagating pump and Stokes pulses because of the temporal walk-off between the pump and the Stokes pulses due to group-velocity dispersion. High-order Stokes can be generated in a medium of a certain length if the pump intensity is sufficiently high. Multiple orders of Stokes can be seen if higher-order Stokes are generated. Backward stimulated Raman scattering is hardly possible because of the very short temporal overlap between the contra-propagating pump and Stokes pulses. Stimulated Brillouin scattering is generally not possible with picosecond pumping because the spectral width of a picosecond pulse is larger than the Brillouin spectral width.

As the pump pulses get into the femtosecond regime, the pulsewidth starts to be comparable to or smaller than the Raman response time for most Raman materials so that the Raman gain factor is reduced. In addition to the factors that limit the Raman efficiency for a picosecond pulse, the response time of the material is now the most important limiting factor. Backward stimulated Raman scattering is not possible for a femtosecond pulse because the interaction length between the contra-propagating pump and Stokes pulses is simply too short. When the pulse duration approaches the material response time, the forward Raman efficiency begins to be significantly reduced. No stimulated Raman scattering in either forward or backward direction can take place if the pulse duration is shorter than the material response time. The Raman response time is of the order of tens of femtoseconds to a few picoseconds for most materials. For a silica optical fiber, which has a very broad Raman spectrum, as shown in Fig. 13.4, it is around 76 fs [6]. Stimulated Brillouin scattering does not occur with femtosecond pulse pumping because the spectral width of a femtosecond pulse is many orders of magnitude larger than that of the Brillouin gain.

Problem Set

13.3.1 In this problem, Raman amplification in a codirectional configuration is considered in the case that the pump and the Stokes waves have the same absorption coefficient such that $\alpha_p = \alpha_S = \alpha$.

(a) When the medium has a nonzero absorption coefficient that $\alpha \neq 0$, the Manley–Rowe relation given in (6.59) is no longer valid. Instead, show, by using the coupled equations given in (13.17) and (13.18), that (6.59) is replaced by the relation:

$$\frac{d}{dz}\left(\frac{I_S}{\omega_S}e^{\alpha z}\right) = -\frac{d}{dz}\left(\frac{I_p}{\omega_p}e^{\alpha z}\right). \tag{13.50}$$

Therefore,

$$\frac{I_S(z)}{\omega_S}e^{\alpha z} + \frac{I_p(z)}{\omega_p}e^{\alpha z} = \frac{I_S(0)}{\omega_S} + \frac{I_p(0)}{\omega_p}, \tag{13.51}$$

which is a constant independent of z.

(b) By using the relation in (13.51), show that the coupled equations given in (13.17) and (13.18) have the exact analytical solutions given in (13.19) and (13.20).

(c) Show that, for any value of α, the pump and the Stokes signal intensities have the relation:

$$\frac{I_S(l)}{I_p(l)} = \frac{I_S(0)}{I_p(0)}\exp\left(g_R l_{\text{eff}}\right). \tag{13.52}$$

13.3.2 By using the relations in (13.27) and (13.28), and by taking the realistic assumption that $I_p(0) \gg I_S^{\text{eff}}(0) = I_{S,\text{th}}^{\text{eff}}(0)$, show that the efficiency of a Raman Stokes generator of a length l in the case that $\alpha_p = \alpha_S = \alpha$ is that given by (13.30).

13.3.3 Because of the typical long interaction length in an optical fiber, a nonlinear optical phenomenon such as stimulated Raman scattering can become important even though the nonlinear susceptibility might not be very large. The Raman spectra of oxide glasses, such as various silica, germania, and phosphorous glasses, that are used in the fabrication of optical fibers show a broad band of frequencies rather than discrete Raman lines because of the amorphous nature of glasses. Consider a nonbirefringent germania-doped silica fiber that has a Raman spectral peak at a frequency shift of $440\,\text{cm}^{-1}$.

(a) A high-power pulsed laser beam at the wavelength of $\lambda = 1$ μm is sent through such a fiber that is long enough to generate up to the fifth-order Stokes signal without completely depleting the pump laser power. What is the expected spectrum of the output at the exit end of the fiber? Identify the wavelengths of the peaks in the spectrum.

(b) Are anti-Stokes lines expected to be seen? Explain.

(c) Consider only the coupling of the pump beam and the first-order Stokes signal. Write down the equations in the slowly varying amplitude approximation to describe the wave propagation while ignoring the effects of optical-field-induced birefringence and absorption losses. Does phase matching need to be considered for this coupling? Why?

(d) If the depletion of the pump beam is negligible throughout the entire fiber, show that Raman amplification of the first-order Stokes signal is

$$\frac{P_S(l)}{P_S(0)} = \exp\left(\frac{\widetilde{g}_R P_p l}{\mathcal{A}_{\text{eff}}}\right), \tag{13.53}$$

where P_p is the input pump power, l is the length of the fiber, $\mathcal{A}_{\text{eff}}$ is the effective cross-sectional area of the fiber core, and $\widetilde{g}_R$ is the Raman gain factor.

(e) If the pump power is subject to attenuation due to linear absorption in the fiber but is not subject to appreciable depletion due to stimulated Raman scattering, how should the expression in (13.53) be modified for Raman amplification in this situation?

13.3.4 Nonlinear optical effects, such as stimulated Raman and Brillouin scattering, can be troublesome problems that limit the capability of fiber-optic communication systems. However, they can also be used in certain situations to our advantage. For example, optical amplifiers based on stimulated Raman gain have been developed for amplifying optical signals in fiber communication systems. We consider this application in this problem. The amplifier consists of an optical fiber that has a length of l, pumped by an optical beam at a frequency of ω_p from the left input end at $z = 0$, as shown in Fig. 13.13. The optical signal has a carrier frequency of ω_S that matches the first Raman Stokes frequency down-shifted from ω_p for $\omega_S = \omega_p - \Omega_R$, where Ω_R is the peak Raman resonance frequency. It has been shown that if the pump is a CW beam, bidirectional amplification of the signal is possible. Assume that the attenuation coefficient α of the fiber is the same at the pump and signal frequencies.

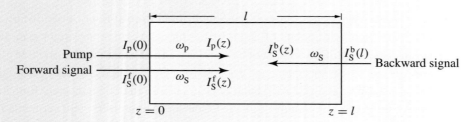

Figure 13.13 Bidirectional Raman amplification.

(a) Consider the pump and the forward-propagating signal only. Ignore the dispersion in the refractive index between ω_p and ω_S. Show that the intensities of the Stokes signal and the pump are described by the coupled differential equations given in (13.17) and (13.18), respectively.

(b) Assume that the fiber is lossless with $\alpha = 0$, but consider pump depletion. Find the total Stokes signal gain $G_R^f = I_S^f(l)/I_S^f(0)$.

(c) Assume no pump depletion due to the Raman effect, but consider fiber attenuation for both the pump and the signal. Find the total Stokes signal gain as defined in (b).

(d) Now consider the amplification of the backward-propagating signal. The total Stokes signal gain for this backward-propagating signal is defined as $G_R^b = I_S^b(0)/I_S^b(l)$, where $I_S^b(l)$ is the input intensity of the backward-propagating signal and $I_S^b(0)$ is its output intensity. It can be seen that if the fiber absorption loss is ignored, G_R^b for backward amplification is the same as G_R^f for forward amplification found in (b). The situation is less clear when the fiber absorption loss is considered. Find the Stokes signal gain G_R^b for backward amplification under the same assumptions taken in (c), and compare it to the forward Stokes signal gain obtained in (c).

13.3.5 A germania-doped silica fiber has an effective core area of $\mathcal{A}_{eff} = 2.8 \times 10^{-11}$ m^2. Its absorption loss at the optical wavelength of 1.064 μm is about 1 dB km^{-1}, and its group-velocity dispersion at this wavelength is about 40 ps km^{-1} nm^{-1}. It has a strong Raman gain peak at a frequency shift of 460 cm^{-1}. A train of optical pulses at the 1.064 μm wavelength at a repetition rate of 76 MHz is sent into the fiber. Stimulated Raman Stokes signals are observed. The Raman gain factor under these conditions is $\widetilde{g}_R \approx 1 \times 10^{-13}$ m W^{-1}. The threshold for stimulated Raman generation is at about

$$\ln G_R^{th} = \frac{\widetilde{g}_R P_{pk}^{th} l_{eff}}{\mathcal{A}_{eff}} = 16, \qquad (13.54)$$

where P_{pk} is the peak power of the pulses and l_{eff} is the effective interaction length of the stimulated Raman scattering process.

(a) What is the wavelength of the first Raman Stokes signal?

(b) Show that for pulses in the range of about 10 ps to 1 ns at the same 76-MHz repetition rate, the average power $\overline{P}_{th}$ of the pulse train for the Raman threshold is independent of pulse duration. What is the value of this threshold average power, $\overline{P}_{th}$?

(c) It is experimentally observed that $\overline{P}_{th}$ changes when the pulses become as short as 3 ps. Do you think it increases or decreases? Give a possible explanation.

(d) What do you expect to happen when the pulses get substantially shorter, say, down to about 100 fs?

(e) What is P_{th} when the input is a CW beam instead of a pulse train? What is the minimum length of the fiber for stimulated Raman generation to reach its threshold in this case?

13.5.1 In this problem, we consider Brillouin amplification that occurs only in a contradirectional configuration. The results obtained in the following apply to contra-directional Raman amplification as well if $\widetilde{g}_B$ is replaced by $\widetilde{g}_R$. We consider only the case that $\alpha_p = \alpha_S = 0$, so that exact analytical solutions can be found for the coupled equations given in (13.36) and (13.37).

(a) Show that

$$\frac{I_p(z)}{\omega_p} - \frac{I_S(z)}{\omega_S} = \frac{I_p(0)}{\omega_p} - \frac{I_S(0)}{\omega_S}, \tag{13.55}$$

which is a constant independent of z.

(b) By using the relation in (13.55), show that

$$\frac{I_S(0)}{I_p(0)} = \frac{I_S(l)}{I_p(l)} \exp\left(g_B l\right), \tag{13.56}$$

where g_B is that given in (13.40).

(c) With given values of $I_p(0)$ and $I_S(l)$ as the boundary conditions for a contradirectional amplifier, the solutions for $I_S(0)$ and $I_p(l)$ cannot be explicitly expressed in terms of $I_p(0)$ and $I_S(l)$. Show, by using (13.55) and (13.56), that $I_S(0)$ and $I_p(l)$ can be implicitly expressed as given in (13.38) and (13.39).

13.5.2 By using the relations in (13.44), (13.56), and (13.40), show that the reflectivity, R_B, of a Brillouin generator defined in (13.46) can be found from the relation in (13.47) in the case that $\alpha_p = \alpha_S = 0$.

13.6.1 A single-mode silica optical fiber has an attenuation coefficient of $0.3\,\mathrm{dB\,km}^{-1}$ and an effective cross-sectional area of $50\ \mu\mathrm{m}^2$ at the $1.3\ \mu\mathrm{m}$ optical wavelength. A CW optical beam at this wavelength is launched into the fiber. The Raman gain peak of this fiber appears at a frequency shift of $460\ \mathrm{cm}^{-1}$. The Raman and Brillouin gain factors have the characteristics described in Sections 13.2 and 13.4 for silica fibers.

(a) If the fiber has a length of 100 km and the optical beam has a linewidth of 10 MHz, what are the critical powers of the beam that reach the Raman and the Brillouin thresholds, respectively, in this fiber? What is the maximum power of the beam that can be transmitted through this fiber?

(b) How do the answers to the questions in (a) vary if the fiber length varies between 1 and 100 km, but the linewidth of the beam remains at 10 MHz? Plot them as functions of the fiber length.

(c) How do the answers to the questions in (a) vary if the fiber length is fixed at 100 km but the linewidth of the optical beam varies between 1 MHz and 100 GHz? Plot them as functions of the linewidth of the optical beam.

13.6.2 Suppression of stimulated Brillouin scattering in a Raman amplifier or generator can be accomplished by using a pump that has a sufficiently large linewidth to raise the Brillouin threshold pump power. Find the respective pump linewidths that are required to suppress the competition from stimulated Brillouin scattering for (a) the fiber Raman amplifier described in Example 13.2 and (b) the fiber Raman generator described in Example 13.3.

References

[1] E. J. Woodbury and W. K. Ng, "Ruby laser operation in the near IR," *Proceedings of the IRE*, vol. 50, p. 2367, 1962.

[2] G. Eckhardt, R. W. Hellwarth, F. J. McClung, S. E. Schwarz, D. Weiner, and E. J. Woodbury, "Stimulated Raman scattering from organic liquids," *Physical Review Letters*, vol. 9, pp. 455–457, 1962.

[3] R. Y. Chiao, C. H. Townes, and B. P. Stoicheff, "Stimulated Brillouin scattering and coherent generation of intense hypersonic waves," *Physical Review Letters*, vol. 12, pp. 592–595, 1964.

[4] E. Garmire and C. H. Townes, "Stimulated Brillouin scattering in liquids," *Applied Physics Letters*, vol. 5, pp. 84–86, 1964.

[5] R. G. Smith, "Optical power handling capacity of low loss optical fibers as determined by stimulated Raman and Brillouin scattering," *Applied Optics*, vol. 11, pp. 2489–2494, 1972.

[6] R. H. Stolen, J. P. Gordon, W. J. Tomlinson, and H. A. Haus, "Raman response function of silica-core fibers," *Journal of the Optical Society of America B*, vol. 6, pp. 1159–1166, 1989.

14 Multiphoton Absorption

14.1 ABSORPTION PROCESSES

An optical transition from a lower energy level, $|1\rangle$, to an upper energy level, $|2\rangle$, can be accomplished through a linear optical process by absorbing one photon that has an energy of $\hbar\omega_1 = E_2 - E_1$, as shown in Fig. 14.1(a), where E_1 is the energy of level $|1\rangle$ and E_2 is the energy of level $|2\rangle$. It may otherwise be accomplished through a nonlinear optical process by *simultaneously* absorbing two photons, as shown in Fig. 14.1(b), or n photons, as shown in Fig. 14.1(c), for $n > 2$. *Two-photon absorption* was originally predicted in 1931 by Göppert-Mayer in her doctoral dissertation [1]. It was first experimentally observed by Kaiser and Garrett in 1961 [2].

For a nonlinear process of multiphoton absorption, the multiple photons that are simultaneously absorbed can have either the same photon energy, as in the cases shown in Figs. 14.1(b) and (c), or different energies. The total energy of the photons involved in a multiphoton absorption process equals the energy separation, $E_2 - E_1$, between the energy levels $|1\rangle$ and $|2\rangle$. In the experiments on two-photon absorption, two photons of the same energy or different energies have been used, as discussed in the following section. Multiphoton absorption experiments for $n > 2$, where more than two photons are absorbed, are generally based on a single laser beam of a frequency ω_n that satisfies the condition $n\hbar\omega_n = E_2 - E_1$, as shown in Fig. 14.1(c).

In any event, as can be seen from Figs. 14.1(b) and (c), a multiphoton absorption involves $n - 1$ intermediate states between the initial state $|1\rangle$ and the final state $|2\rangle$. Any of the intermediate states can be either a virtual state or a real state. When an intermediate state is a real state, the nonlinear susceptibility of the multiphoton absorption is enhanced by the resonance between this state and $|1\rangle$, and that between it and $|2\rangle$, as well as that between it and any other possible real state. Nonetheless, all photons that are absorbed in a multiphoton absorption process are simultaneously absorbed in one step, but not in two or multiple separate steps, even when one or multiple intermediate states are real states.

The selection rules for different orders of multiphoton absorption are different, and they are different from those for linear optical absorption. Therefore, they provide complementary spectroscopic tools for probing a material. The difference between two-photon absorption and one-photon absorption is similar to the difference between Raman Stokes transition and one-photon absorption, as discussed in Section 13.1. In a centrosymmetric material, they are mutually exclusive for transitions between a given pair of energy levels because one-photon absorption takes place only between two states of opposite parities, but two-photon absorption takes place only between two states of the same parity.

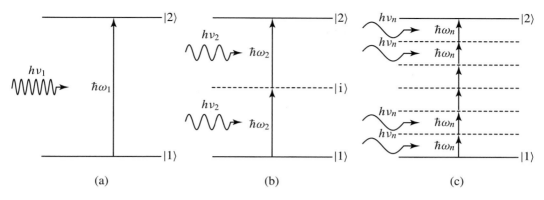

Figure 14.1 Optical transitions through (a) a linear optical process of one-photon absorption, (b) a nonlinear optical process of two-photon absorption, and (c) a nonlinear process of n-photon absorption.

A multiphoton absorption process is characterized by the imaginary part of a nonlinear optical susceptibility. For n-photon absorption, it is the $(2n-1)$th-order susceptibility: $\chi^{(2n-1)\prime\prime}$. Therefore, two-photon absorption is characterized by $\chi^{(3)\prime\prime}$, as discussed in Subsection 5.3.9, and three-photon absorption is characterized by $\chi^{(5)\prime\prime}$. Because it is associated with the imaginary part of a nonlinear susceptibility, a multiphoton absorption process is nonparametric, thus automatically phase matched.

14.2 TWO-PHOTON ABSORPTION

The simplest form of multiphoton absorption is two-photon absorption. It is a third-order nonparametric nonlinear optical process that is generally characterized by the imaginary part of a third-order nonlinear susceptibility of the form $\chi^{(3)}(\omega_1 = \omega_1 + \omega_2 - \omega_2)$. The two photons that are simultaneously absorbed in this process can have either different photon energies or the same photon energy. Therefore, there can be either one beam of a single frequency or two beams of the same or different frequencies in an experiment on two-photon absorption.

One characteristic that is distinctly different between a multiphoton absorption process and linear one-photon absorption is that multiphoton absorption is intensity dependent. Consequently, the transmittance through a medium is also intensity dependent. Another important characteristic of a multiphoton absorption process is that the photon energy that is required for enabling the transition from a lower energy level, $|1\rangle$, to an upper energy level, $|2\rangle$, is only a fraction of the transition energy $E_2 - E_1$. Being the lowest order of multiphoton absorption, two-photon absorption is intensity dependent, and the energy of each photon involved in the two-photon absorption process is a fraction of $E_2 - E_1$. In the case that the two photons have the same energy of $\hbar\omega$, it is half the transition energy: $\hbar\omega = (E_2 - E_1)/2$. Therefore, the wavelength of the photon is twice that of the photon for the same transition through linear one-photon absorption. The combination of these two characteristics makes it possible to localize two-photon absorption, and any other multiphoton absorption, by focusing a laser beam inside a two-photon absorption medium. The intensity-dependent two-photon absorption is minimal outside the focal region because the light intensity there is low. Because the single-photon energy is only half of the transition energy, the intensity-independent linear absorption is negligible if other possible optical resonances at the optical frequency are avoided. Therefore, two-photon absorption takes place primarily in the focal region. By tightly focusing the beam, this active region for two-photon absorption can be smaller than the linear diffraction limit. By using an ultrashort laser pulse, high laser intensity is also localized in time. Two-photon absorption at the peak of the pulse is further enhanced because of the high peak intensity of an ultrashort laser pulse. These characteristics have many practical applications, such as optical power limiting, two-photon microscopy, micro- and nano-fabrication of three-dimensional structures, biomedical imaging, and photodynamic therapy.

There are a few different scenarios for two-photon absorption. In the following, a few of the typical scenarios are discussed.

14.2.1 One Beam without Linear Absorption

We first consider the simplest scenario that there is only one incident beam at a frequency of ω, as shown in Fig. 14.2(a), and the two photons that are absorbed have the same photon energy of $\hbar\omega$, as shown in Fig. 14.3. We also consider the case that the intermediate state $|i\rangle$ is a virtual state, as shown in Fig. 14.3(a), so that there is no one-photon transition from $|1\rangle$ to $|i\rangle$, nor from $|i\rangle$ to $|2\rangle$. Meanwhile, other possible linear absorption at the frequency of ω is negligible.

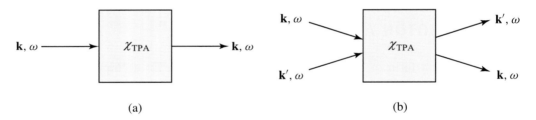

(a)   (b)

Figure 14.2 Schematics of (a) one-frequency, one-beam two-photon absorption and (b) noncollinear one-frequency, two-beam two-photon absorption.

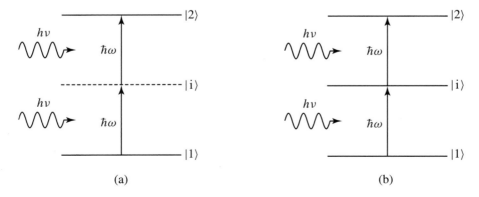

(a)   (b)

Figure 14.3 One-frequency two-photon absorption with two photons of the same photon energy. The intermediate state is (a) a virtual state and (b) a real state.

The nonlinear polarization for one-frequency, one-beam two-photon absorption is that given in (5.54):

$$\mathbf{P}^{(3)}(\omega) = 3\epsilon_0\chi^{(3)}(\omega = \omega + \omega - \omega) \vdots \mathbf{E}(\omega)\mathbf{E}(\omega)\mathbf{E}^*(\omega). \tag{14.1}$$

By using the coupled-wave equation given in (6.22) for the field amplitude $\mathcal{E}$ of the single beam at ω, we find the single coupled-wave equation:

$$\frac{d\mathcal{E}}{dz} = \frac{i3\omega}{2cn_{\omega,z}}\chi_{\text{TPA}}\mathcal{E}|\mathcal{E}|^2, \tag{14.2}$$

where $n_{\omega,z} = ck_z/\omega$ and

$$\chi_{\text{TPA}} = \hat{e}^* \cdot \chi^{(3)}(\omega = \omega + \omega - \omega) \vdots \hat{e}\hat{e}\hat{e}^* \tag{14.3}$$

is the effective nonlinear susceptibility for one-frequency two-photon absorption, and $\hat{e}$ is the unit polarization vector of the field $\mathbf{E}(\omega)$. By using the relation $I = 2c\epsilon_0 n_{\omega,z}|\mathcal{E}|^2$ from (6.31) for the light intensity, (14.2) can be transformed into an equation for single-frequency, single-beam two-photon absorption in terms of the light intensity:

$$\frac{dI}{dz} = -\beta I^2, \tag{14.4}$$

where

$$\beta = \frac{3\omega}{2c^2\epsilon_0 n_{\omega,z}^2}\chi_{\text{TPA}}'' \tag{14.5}$$

is the *two-photon absorption coefficient*. With an input intensity of $I_0 = I(0)$ at $z = 0$, (14.4) can be readily solved at $z = l$ for a *two-photon absorber* of a length l:

$$I(l) = \frac{I_0}{1 + \beta I_0 l}. \tag{14.6}$$

One useful application of two-photon absorption is *power limiting* for the protection of highly sensitive photodetectors or expensive optical instruments. As discussed in Subsection 12.2.1, an optical power limiter can be implemented by using a Kerr lens based on the parametric optical Kerr effect. The power-limiting function of a limiter based on the optical Kerr effect is activated by the intensity-dependent expansion of the transverse beam profile. By comparison, the power-limiting function of a limiter based on the nonparametric two-photon absorption is facilitated by the intensity-dependent saturation of the transmitted optical power. From (14.6), we find that for an optical power limiter of a length l that is based on two-photon absorption, the transmitted intensity is limited to a maximum intensity level that is given as

$$I(l) \xrightarrow{I_0 \to \infty} I_{\text{TPA}}^{\text{sat}} = \frac{1}{\beta l}. \tag{14.7}$$

This *saturation intensity* level can be predetermined by properly choosing the length l of a power limiter that is made of a given material of a known two-photon absorption coefficient β.

14.2.2 One Beam with Linear Absorption

Linear absorption cannot always be ignored. For example, when the intermediate state $|i\rangle$ is a real state, as shown in Fig. 14.3(b), linear optical transition from $|1\rangle$ to $|i\rangle$, and subsequently from $|i\rangle$ to $|2\rangle$, is possible together with two-photon absorption from $|1\rangle$ to $|2\rangle$, when the parity of the intermediate state $|i\rangle$ is opposite to that of state $|1\rangle$ and state $|2\rangle$. To account for the possible linear absorption, the equation given in (14.4) for single-frequency, single-beam two-photon absorption has to be modified as

$$\frac{dI}{dz} = -\alpha I - \beta I^2, \tag{14.8}$$

where

$$\alpha = \frac{\omega}{cn} \chi''_{\text{res}}(\omega) \tag{14.9}$$

is the linear absorption coefficient and $\chi''_{\text{res}}(\omega)$ is the imaginary part of the resonant susceptibility for the linear transition at the frequency of ω.

With an input intensity of $I_0 = I(0)$ at $z = 0$, (14.8) can be solved at $z = l$ for a two-photon absorber of a length l:

$$I(l) = \frac{\alpha}{\alpha + \beta I_0(1 - e^{-\alpha l})} I_0 e^{-\alpha l}. \tag{14.10}$$

By taking the limit that $\alpha \to 0$, it can be easily verified that (14.10) reduces to (14.6), as expected. In the limit that $\beta \to 0$, we find that (14.10) reduces to the linear attenuation relation that $I(l) = I_0 e^{-\alpha l}$, as is also expected. Therefore, the relation given in (14.10) expresses the general spatial intensity evolution of an optical wave that is attenuated by both one-photon absorption and two-photon absorption at the same time as it propagates through a medium.

Two-photon absorption in the presence of linear absorption can still be utilized for optical power limiting, but the saturation intensity is dependent on both the linear absorption coefficient α and the two-photon absorption coefficient β. For such an optical power limiter that has a length of l, we find from (14.10) that the transmitted intensity is limited to a maximum intensity level that is given as

$$I(l) \xrightarrow{I_0 \to \infty} I_{\text{TPA}}^{\text{sat}} = \frac{\alpha}{\beta(e^{\alpha l} - 1)}. \tag{14.11}$$

From (14.11), it is clear that for effective optical power limiting based on two-photon absorption, it is important that linear optical attenuation is minimized such that $\alpha l < 1$. Otherwise, the optical power would be primarily attenuated by the intensity-independent linear absorption process, which only diminishes the transmitted optical power without providing a power-limiting function.

EXAMPLE 14.1

Silicon has a two-photon absorption coefficient of $\beta = 2\,\mathrm{cm\,GW^{-1}} = 2 \times 10^{-11}\,\mathrm{m\,W^{-1}}$ at the wavelength of $\lambda = 1.064\,\mathrm{\mu m}$ [3]. The refractive index of silicon at this wavelength is $n = 3.5516$. Because this wavelength is very close to the bandgap wavelength of $\lambda_\mathrm{g} = 1.110\,\mathrm{\mu m}$ of silicon at room temperature, the linear absorption coefficient is very small and is dependent on the impurity concentration. A silicon wafer has a thickness of $l = 1\,\mathrm{mm}$ and a linear absorption coefficient of $\alpha = 10\,\mathrm{cm^{-1}} = 10^3\,\mathrm{m^{-1}}$. A Gaussian laser beam of a train of pulses that have a pulsewidth of $\Delta t_\mathrm{ps} = 20\,\mathrm{ps}$ and a repetition rate of $f_\mathrm{ps} = 1\,\mathrm{kHz}$ at an average power of $\overline{P}_0 = 20\,\mathrm{mW}$ is focused to a spot size of $w_0 = 50\,\mathrm{\mu m}$ on the silicon wafer. (a) By ignoring linear absorption, find the transmitted power of the beam. Compare it to the saturation power. (b) By including linear absorption, find the transmitted power of the beam. Compare it to the saturation power.

Solution

We first calculate the Rayleigh range of the beam and find it to be much larger than the thickness of the silicon wafer:

$$z_\mathrm{R} = \frac{\pi n w_0^2}{\lambda} = \frac{\pi \times 3.5516 \times (50 \times 10^{-6})}{1.064 \times 10^{-6}}\,\mathrm{m} = 2.62\,\mathrm{cm} \gg l.$$

Therefore, we can take the cross-sectional area of the focused beam to be a constant through the thickness of the silicon sample:

$$\mathcal{A} = \frac{\pi w_0^2}{2} = \frac{\pi (50 \times 10^{-6})}{2}\,\mathrm{m}^2 = 3.93 \times 10^{-9}\,\mathrm{m}^2.$$

We then find the peak intensity of the incident laser pulses at the location of the silicon sample:

$$I_0 = \frac{P_\mathrm{pk}}{\mathcal{A}} = \frac{\overline{P}_0}{\Delta t_\mathrm{ps} f_\mathrm{ps} \mathcal{A}} = \frac{20 \times 10^{-3}}{20 \times 10^{-12} \times 1 \times 10^3 \times 3.93 \times 10^{-9}}\,\mathrm{W\,m^{-2}} = 2.54 \times 10^{14}\,\mathrm{W\,m^{-2}}.$$

(a) By ignoring linear absorption, the transmitted intensity and the saturation intensity are found by using (14.6) and (14.7), respectively:

$$I(l) = \frac{I_0}{1 + \beta I_0 l} = \frac{2.54 \times 10^{14}}{1 + 2 \times 10^{-11} \times 2.54 \times 10^{14} \times 1 \times 10^{-3}}\,\mathrm{W\,m^{-2}} = 4.18 \times 10^{13}\,\mathrm{W\,m^{-2}},$$

$$I_\mathrm{TPA}^\mathrm{sat} = \frac{1}{\beta l} = \frac{1}{2 \times 10^{-11} \times 1 \times 10^{-3}}\,\mathrm{W\,m^{-2}} = 5 \times 10^{13}\,\mathrm{W\,m^{-2}}.$$

Therefore, the average transmitted power and the average saturation power are, respectively,

$$\overline{P}(l) = I(l)\Delta t_\mathrm{ps} f_\mathrm{ps} \mathcal{A} = 4.18 \times 10^{13} \times 20 \times 10^{-12} \times 1 \times 10^3 \times 3.93 \times 10^{-9}\,\mathrm{W} = 3.29\,\mathrm{mW},$$

$$\overline{P}_{\text{TPA}}^{\text{sat}} = I_{\text{TPA}}^{\text{sat}} \Delta t_{\text{ps}} f_{\text{ps}} \mathcal{A} = 5 \times 10^{13} \times 20 \times 10^{-12} \times 1 \times 10^3 \times 3.93 \times 10^{-9} \text{ W} = 3.93 \text{ mW}.$$

We find that $\overline{P}_0 > \overline{P}_{\text{TPA}}^{\text{sat}} > \overline{P}(l)$ because $I_0 > I_{\text{TPA}}^{\text{sat}} > I(l)$, as expected.

(b) When linear absorption is considered, $\alpha l = 10^3 \times 1 \times 10^{-3} = 1$. Then, the transmitted intensity and the saturation intensity are found by using (14.10) and (14.11), respectively:

$$
\begin{aligned}
I(l) &= \frac{\alpha}{\alpha + \beta I_0 (1 - e^{-\alpha l})} I_0 e^{-\alpha l} \\
&= \frac{10^3 \times 2.54 \times 10^{14} \times e^{-1}}{10^3 + 2 \times 10^{-11} \times 2.54 \times 10^{14} \times (1 - e^{-1})} \text{ W m}^{-2} \\
&= 2.22 \times 10^{13} \text{ W m}^{-2},
\end{aligned}
$$

$$I_{\text{TPA}}^{\text{sat}} = \frac{\alpha}{\beta(e^{\alpha l} - 1)} = \frac{10^3}{2 \times 10^{-11} \times (e - 1)} \text{ W m}^{-2} = 2.91 \times 10^{13} \text{ W m}^{-2}.$$

Therefore, the transmitted average power and the saturation power are, respectively,

$$\overline{P}(l) = I(l) \Delta t_{\text{ps}} f_{\text{ps}} \mathcal{A} = 2.22 \times 10^{13} \times 20 \times 10^{-12} \times 1 \times 10^3 \times 3.93 \times 10^{-9} \text{ W} = 1.75 \text{ mW}.$$

$$\overline{P}_{\text{TPA}}^{\text{sat}} = I_{\text{TPA}}^{\text{sat}} \Delta t_{\text{ps}} f_{\text{ps}} \mathcal{A} = 2.91 \times 10^{13} \times 20 \times 10^{-12} \times 1 \times 10^3 \times 3.93 \times 10^{-9} \text{ W} = 2.29 \text{ mW}.$$

We find that $\overline{P}_0 > \overline{P}_{\text{TPA}}^{\text{sat}} > \overline{P}(l)$ because $I_0 > I_{\text{TPA}}^{\text{sat}} > I(l)$, as expected. By comparing these results to those found in (a), we see that both the transmitted power and the saturation power are substantially reduced by linear absorption, though the linear absorption coefficient is small. The reason is that the two-photon absorption coefficient is also small and the incident intensity is such that $\beta I_0 = 5.08 \times 10^3 \text{ m}^{-1} = 5.08\alpha$, which is only five times the linear absorption coefficient.

14.2.3 Two Beams without Linear Absorption

Two-photon absorption can also be implemented with two beams. The two beams can have the same frequency but different propagation directions, as shown in Fig. 14.2(b). This arrangement is uncommon and often unnecessary, because when the two photons have the same frequency one beam is sufficient, as discussed in the preceding two subsections. Furthermore, in this case, each beam also undergoes two-photon absorption as it propagates through the medium before the two beams spatially overlap. For these reasons, we shall not further discuss two-photon absorption with two beams of the same frequency.

The situation is quite different for two-photon absorption with two beams of different frequencies, $\omega_1 \neq \omega_2$, such that $\hbar\omega_1 + \hbar\omega_2 = E_2 - E_1$, as shown in Figs. 14.4 and 14.5. These two beams can propagate collinearly in the same direction, as shown in Fig. 14.4(a), or noncollinearly in different directions, as shown in Fig. 14.4(b). Because $\omega_1 \neq \omega_2$ and $\hbar\omega_1 + \hbar\omega_2 = E_2 - E_1$, it is clear that $2\hbar\omega_1 \neq E_2 - E_1$ and $2\hbar\omega_2 \neq E_2 - E_1$. Therefore, unlike the situation of two beams of the same frequency discussed above, each beam of the two

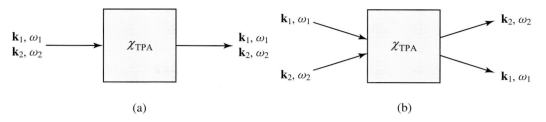

(a)   (b)

Figure 14.4 Schematics of (a) collinear two-frequency two-photon absorption and (b) noncollinear two-frequency two-photon absorption.

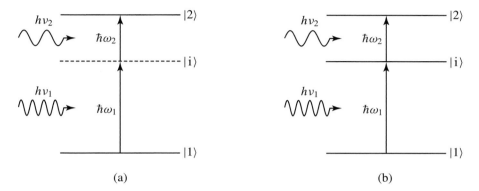

(a)   (b)

Figure 14.5 Two-frequency two-photon absorption with two photons of different photon energies. The intermediate state is (a) a virtual state and (b) a real state.

different frequencies by itself does not undergo two-photon absorption. Only when they overlap spatially and temporally, such as in the cases of short optical pulses, does two-photon absorption take place. This condition allows the control of two-photon absorption by controlling the spatial or temporal overlap of the two beams of different frequencies. This flexibility is useful in designing a practically useful optical system.

The general formulation for two-photon absorption with two frequencies at ω_1 and ω_2 was already discussed in Subsection 6.3.2. In the absence of linear absorption, the intensities of the two beams undergoing two-photon absorption are governed by the coupled equations given in (6.65) and (6.66),

$$\frac{dI_1}{dz} = -\frac{3\omega_1}{c^2\epsilon_0 n_{1,z} n_{2,z}} \chi''_{TPA} I_1 I_2 = -\beta_1 I_1 I_2, \tag{14.12}$$

$$\frac{dI_2}{dz} = -\frac{3\omega_2}{c^2\epsilon_0 n_{1,z} n_{2,z}} \chi''_{TPA} I_1 I_2 = -\beta_2 I_1 I_2, \tag{14.13}$$

where χ_{TPA} is the effective nonlinear susceptibility for two-frequency two-photon absorption given in (6.62):

$$\chi_{\text{TPA}} = \hat{e}_1^* \cdot \boldsymbol{\chi}^{(3)}(\omega_1 = \omega_1 + \omega_2 - \omega_2) \vdots \hat{e}_1 \hat{e}_2 \hat{e}_2^* = \hat{e}_2^* \cdot \boldsymbol{\chi}^{(3)}(\omega_2 = \omega_2 + \omega_1 - \omega_1) \vdots \hat{e}_2 \hat{e}_1 \hat{e}_1^*,$$

(14.14)

and

$$\beta_1 = \frac{3\omega_1}{c^2 \epsilon_0 n_{1,z} n_{2,z}} \chi_{\text{TPA}}'' \quad \text{and} \quad \beta_2 = \frac{3\omega_2}{c^2 \epsilon_0 n_{1,z} n_{2,z}} \chi_{\text{TPA}}'' \qquad (14.15)$$

are the *two-photon absorption coefficients* for the two frequencies at ω_1 and ω_2, respectively. Note that β_1 and β_2 of the two-frequency two-photon absorption are formally twice that of the one-frequency two-photon absorption coefficient β as defined in (14.5) because of the six-fold frequency permutation for $\mathbf{P}^{(3)}(\omega_1)$ and $\mathbf{P}^{(3)}(\omega_2)$, given in (5.52) and (5.53), versus the three-fold permutation for $\mathbf{P}^{(3)}(\omega)$, given in (5.54).

As discussed in Subsection 6.3.2, the coupled equations of (14.12) and (14.13) lead to the Manley–Rowe relation given in (6.68):

$$\frac{d}{dz}\left(\frac{I_1}{\omega_1}\right) = \frac{d}{dz}\left(\frac{I_2}{\omega_2}\right) = -\frac{3}{c^2 \epsilon_0 n_{1,z} n_{2,z}} \chi_{\text{TPA}}'' I_1 I_2. \qquad (14.16)$$

By integrating this Manley–Rowe relation and applying the input condition of $I_{1,0} = I_1(0)$ and $I_{2,0} = I_2(0)$, we have

$$\frac{I_1(z)}{\omega_1} - \frac{I_2(z)}{\omega_2} = \frac{I_{1,0}}{\omega_1} - \frac{I_{2,0}}{\omega_2}. \qquad (14.17)$$

By using the relation in (14.17), the exact analytical solutions can be found for the coupled equations of (14.12) and (14.13) at $z = l$ for a two-photon absorber of a length l:

$$I_1(l) = \frac{\omega_2 I_{1,0} - \omega_1 I_{2,0}}{\omega_2 I_{1,0} - \omega_1 I_{2,0} \exp(-al)} I_{1,0}, \qquad (14.18)$$

$$I_2(l) = \frac{\omega_1 I_{2,0} - \omega_2 I_{1,0}}{\omega_1 I_{2,0} - \omega_2 I_{1,0} \exp(al)} I_{2,0}, \qquad (14.19)$$

where

$$a = \frac{3}{c^2 \epsilon_0 n_{1,z} n_{2,z}} (\omega_2 I_{1,0} - \omega_1 I_{2,0}) = \left(\frac{\beta_1 \beta_2}{\omega_1 \omega_2}\right)^{1/2} (\omega_2 I_{1,0} - \omega_1 I_{2,0}). \qquad (14.20)$$

14.2.4 Two Beams with Linear Absorption

When linear absorption is present, the coupled equations given in (14.12) and (14.13) for two-photon absorption with two frequencies at ω_1 and ω_2 have to be modified to account for the linear attenuation:

$$\frac{dI_1}{dz} = -\alpha_1 I_1 - \beta_1 I_1 I_2, \qquad (14.21)$$

$$\frac{dI_2}{dz} = -\alpha_2 I_2 - \beta_2 I_1 I_2, \tag{14.22}$$

where

$$\alpha_1 = \frac{\omega_1}{cn_1} \chi''_{\mathrm{res}}(\omega_1) \quad \text{and} \quad \alpha_2 = \frac{\omega_2}{cn_2} \chi''_{\mathrm{res}}(\omega_2) \tag{14.23}$$

are the linear absorption coefficients at the frequencies ω_1 and ω_2, respectively. The coupled equations of (14.21) and (14.22) have exact analytical solutions when $\alpha_s = \alpha_p = \alpha$. With the input condition of $I_{1,0} = I_1(0)$ and $I_{2,0} = I_2(0)$ for the two waves, the analytical solutions found at $z = l$ for a two-photon absorber of a length l have the forms:

$$I_1(l) = \frac{\omega_2 I_{1,0} - \omega_1 I_{2,0}}{\omega_2 I_{1,0} - \omega_1 I_{2,0} \exp\left(-al_{\mathrm{eff}}\right)} I_{1,0} \exp\left(-\alpha l\right), \tag{14.24}$$

$$I_2(l) = \frac{\omega_1 I_{2,0} - \omega_2 I_{1,0}}{\omega_1 I_{2,0} - \omega_2 I_{1,0} \exp\left(al_{\mathrm{eff}}\right)} I_{2,0} \exp\left(-\alpha l\right), \tag{14.25}$$

where a is given in (14.20) and l_{eff} is the effect length for two-photon absorption given by the relation:

$$l_{\mathrm{eff}} = \frac{1 - e^{-\alpha l}}{\alpha}. \tag{14.26}$$

14.3 THREE-PHOTON ABSORPTION

Three-photon absorption is a fifth-order nonparametric nonlinear optical process that is characterized by the imaginary part of a nonlinear susceptibility of the form $\chi^{(5)}(\omega_1 = \omega_1 + \omega_2 - \omega_2 + \omega_3 - \omega_3)$. Three photons are simultaneously absorbed while the material makes a transition from a lower energy level, $|1\rangle$, to an upper energy level, $|2\rangle$, with $E_2 - E_1 = \hbar\omega_1 + \hbar\omega_2 + \hbar\omega_3$. In general, the three photons can be all of different frequencies, or two of the same frequency and the third being different, or all three of the same frequency. In practice, three-photon absorption is experimentally carried out by using a single laser beam, as schematically shown in Fig. 14.6(a). The three photons have the same photon energy under the condition that $E_2 - E_1 = 3\hbar\omega$, as shown in Fig. 14.6(b). In the rest of this section, only this scenario is considered.

The selection rules for three-photon absorption are different from those for two-photon absorption. Therefore, they do not take place at the same time between a given pair of energy levels, $|1\rangle$ and $|2\rangle$. However, two-photon absorption for the single laser beam at the frequency ω might happen if one of the two intermediate states for three-photon absorption is a real state. For example, if the intermediate state $|j\rangle$ is a real state, two-photon absorption with $E_j - E_1 = 2\hbar\omega$ might occur for a transition from $|1\rangle$ to $|j\rangle$ at the same time, while three-photon absorption for a transition from $|1\rangle$ to $|2\rangle$ takes place with $E_2 - E_1 = 3\hbar\omega$. Linear absorption at the frequency ω for the single laser beam might also happen, particularly when the intermediate state $|i\rangle$ is a real state.

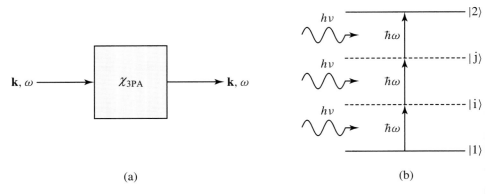

Figure 14.6 (a) Schematic of one-beam three-photon absorption. (b) One-frequency three-photon absorption with three photons of the same photon energy.

14.3.1 One Beam without Linear Absorption

The nonlinear polarization for one-frequency, one-beam three-photon absorption is

$$\mathbf{P}^{(5)}(\omega) = 10\epsilon_0\chi^{(5)}(\omega = \omega + \omega + \omega - \omega - \omega) \vdots \mathbf{E}(\omega)\mathbf{E}(\omega)\mathbf{E}(\omega)\mathbf{E}^*(\omega)\mathbf{E}^*(\omega). \quad (14.27)$$

By using the coupled-wave equation given in (6.22) for the field amplitude $\mathcal{E}$ of the single beam at ω, we find the single coupled-wave equation:

$$\frac{d\mathcal{E}}{dz} = \frac{i5\omega}{2cn_{\omega,z}}\chi_{3PA}\mathcal{E}|\mathcal{E}|^4, \quad (14.28)$$

where $n_{\omega,z} = ck_z/\omega$ and

$$\chi_{3PA} = \hat{e}^* \cdot \chi^{(5)}(\omega = \omega + \omega + \omega - \omega - \omega) \vdots \hat{e}\hat{e}\hat{e}\hat{e}^*\hat{e}^* \quad (14.29)$$

is the effective nonlinear susceptibility for one-frequency three-photon absorption, and $\hat{e}$ is the unit polarization vector of the field $\mathbf{E}(\omega)$. By using the relation $I = 2c\epsilon_0 n_{\omega,z}|\mathcal{E}|^2$ from (6.31) for the light intensity, (14.28) can be transformed into an equation for single-beam three-photon absorption in terms of the light intensity:

$$\frac{dI}{dz} = -\gamma I^3, \quad (14.30)$$

where

$$\gamma = \frac{5\omega}{4c^3\epsilon_0^2 n_{\omega,z}^3}\chi''_{3PA} \quad (14.31)$$

is the *three-photon absorption coefficient*. With an input intensity of $I_0 = I(0)$ at $z = 0$, (14.30) can be solved at $z = l$ for a *three-photon absorber* of a length l:

$$I(l) = \frac{I_0}{(1 + 2\gamma I_0^2 l)^{1/2}}. \tag{14.32}$$

Being a nonlinear absorption process, three-photon absorption also has the power-limiting effect. For an optical power limiter that has a length of l, we find from (14.32) that the transmitted intensity is limited to a maximum intensity level that is given as

$$I(l) \xrightarrow{I_0 \to \infty} I_{3PA}^{sat} = \frac{1}{(2\gamma l)^{1/2}}. \tag{14.33}$$

By comparing this saturation intensity for three-photon absorption with that for two-photon absorption found in (14.7), we find that the saturation behavior of three-photon absorption is different from that of two-photon absorption.

14.3.2 One Beam with Linear Absorption

When linear absorption is present, the equation given in (14.30) has to be modified as

$$\frac{dI}{dz} = -\alpha I - \gamma I^3, \tag{14.34}$$

where α is the linear absorption coefficient given in (14.9). With an input intensity of $I_0 = I(0)$ at $z = 0$, (14.34) can be solved at $z = l$ for a three-photon absorber of a length l in the presence of linear absorption:

$$I(l) = \left[\frac{\alpha}{\alpha + \gamma I_0^2 (1 - e^{-2\alpha l})} \right]^{1/2} I_0 e^{-\alpha l}. \tag{14.35}$$

By taking the limit that $\alpha \to 0$, it can be easily verified that (14.35) reduces to (14.32), as expected. In the limit that $\gamma \to 0$, we find that (14.35) reduces to the linear attenuation relation that $I(l) = I_0 e^{-\alpha l}$, as is also expected.

Three-photon absorption in the presence of linear absorption can also be utilized for optical power limiting. In this case, the saturation intensity is dependent on both the linear absorption coefficient α and the three-photon absorption coefficient γ. For such an optical power limiter that has a length of l, we find from (14.35) that the transmitted intensity is limited to a maximum intensity level that is given as

$$I(l) \xrightarrow{I_0 \to \infty} I_{3PA}^{sat} = \left[\frac{\alpha}{\gamma(e^{2\alpha l} - 1)} \right]^{1/2}. \tag{14.36}$$

It is clear that for effective optical power limiting based on three-photon absorption, it is important that linear optical attenuation is minimized such that $\alpha l < 1$.

In the unlikely situation that two-photon absorption takes place together with three-photon absorption in the presence of linear absorption for the same optical beam, the intensity of the beam is governed by the equation:

$$\frac{dI}{dz} = -\alpha I - \beta I^2 - \gamma I^3. \tag{14.37}$$

Problem Set

14.2.1 The characteristics of one-beam two-photon absorption are considered in this problem. The two-photon absorber has a length of l in the z direction. The input intensity of the optical beam is $I_0 = I(0)$ at $z = 0$. The two-photon absorption coefficient is β.

  (a) In the absence of linear absorption, show that the transmitted intensity of the optical beam at $z = l$ is that given in (14.6).

  (b) In the presence of linear absorption with a linear absorption coefficient of α, show that the transmitted intensity of the optical beam is that given in (14.10).

  (c) Show that the relation obtained in (b) for (14.10) reduces to that obtained in (a) for (14.6) in the absence of linear absorption. Show also that it reduces to the relation of linear attenuation in the presence of only linear absorption without two-photon absorption.

14.2.2 In this problem, we consider two-photon absorption for two collinear beams at two different frequencies of ω_1 and ω_2, with $\omega_1 \neq \omega_2$. The two-photon absorber has a length of l in the z direction. The input intensities of the two optical beams are $I_{1,0} = I_1(0)$ and $I_{2,0} = I_2(0)$, respectively, at $z = 0$. The two-photon absorption coefficients at these two frequencies are β_1 and β_2, respectively.

  (a) In the absence of linear absorption, show that the transmitted intensities of the two optical beams at $z = l$ are those given in (14.18) and (14.19).

  (b) In the presence of linear absorption with a linear absorption coefficient of $\alpha_1 = \alpha_2 = \alpha$ for both beams, find the Manley–Rowe relations. Show that in this case, the transmitted intensities of the two optical beams are those given in (14.24) and (14.25).

14.3.1 The characteristics of one-beam three-photon absorption are considered in this problem. The three-photon absorber has a length of l in the z direction. The input intensity of the optical beam is $I_0 = I(0)$ at $z = 0$. The three-photon absorption coefficient is γ.

  (a) In the absence of linear absorption, show that the transmitted intensity of the optical beam at $z = l$ is that given in (14.32).

  (b) In the presence of linear absorption with a linear absorption coefficient of α, show that the transmitted intensity of the optical beam is that given in (14.35).

  (c) Show that the relation obtained in (b) for (14.35) reduces to that obtained in (a) for (14.32) in the absence of linear absorption. Show also that it reduces to the relation of linear attenuation in the presence of only linear absorption without three-photon absorption.

References

[1] M. Göppert-Mayer, "Über Elementarakte mit zwei Quantensprüngen," *Annalen der Physik*, vol. 401, pp. 273–294, 1931.

[2] W. Kaiser and C. G. B. Garrett, "Two-photon excitation in CaF_2:Eu^{2+}," *Physical Review Letters*, vol. 7, pp. 229–231, 1961.

[3] A. D. Bristow, N. Rotenberg, and H. M. van Driel, "Two-photon absorption and Kerr coefficients of silicon for 850–2200 nm," *Applied Physics Letters*, vol. 90, p. 191104, 2007.

15 Optical Saturation

15.1 ABSORPTION SATURATION AND GAIN SATURATION

For *one-photon optical absorption,* as shown in Fig. 15.1(a), or *optical amplification,* as shown in Fig. 15.1(b), of an optical field, resonant transitions between two energy levels, $|1\rangle$ and $|2\rangle$, are involved such that the frequency ω of the optical field is in resonance with the transition frequency – that is, $\omega \approx \omega_{21}$, where $\omega_{21} = (E_2 - E_1)/\hbar$. The interaction between the medium and the optical wave results in net optical absorption if the lower energy level, $|1\rangle$, is more populated than the upper energy level, $|2\rangle$, such that $N_1 > N_2$. It results in net optical amplification if the upper energy level is more populated than the lower energy level, such that $N_2 > N_1$. The optical medium is generally a multilevel system, but the interaction that is directly associated with one-photon optical absorption or optical amplification takes place only between the two energy levels that are in resonance with the optical frequency. Therefore, we only consider the parameters that are associated with these two energy levels, as defined in Subsection 1.7.1 for a system of two levels.

The time-domain *resonant susceptibility* that is responsible for the one-photon absorption process is that given in (1.160):

$$\chi_{\mathrm{res}}(t; t') = \frac{2}{\epsilon_0 \hbar} p_{12} p_{21} \left[N_1(t - t') - N_2(t - t') \right] \sin \omega_{21} t' e^{-\gamma_{21} t'}. \tag{15.1}$$

For this time-domain expression of $\chi_{\mathrm{res}}(t; t')$, the time variable t is the general time variable, and t' is the response time that measures the time delay of the material response with respect to the optical excitation. It is the delayed material response at time t, for a time delay of t', to an excitation field of $E(t - t')$ at an earlier time of $t - t'$. The frequency-domain resonant susceptibility is obtained by taking the Fourier transform on the response time t', not on the general time variable t. The complication is that the population densities, N_1 and N_2, are functions of the excitation optical field $E(t - t')$; therefore, they also appear in (15.1) as functions of $t - t'$. As mentioned in Section 1.7.1, this dependence of the population densities on the excitation field leads to optical nonlinearity because the electric susceptibility tensor χ is generally a function of the electric field E as $\chi(t; t'; E)$. In the case of resonant susceptibility discussed in this chapter, this field dependence results in optical saturation. We can formally express the resonant susceptibility in the frequency domain as $\chi_{\mathrm{res}}(t; \omega; E)$, where E is the complex field in the frequency domain. The time dependence of $\chi_{\mathrm{res}}(t; \omega; E)$ accounts for the possibility of the temporal variations of the material property, whereas the frequency dependence accounts for the response of the material to the optical excitation.

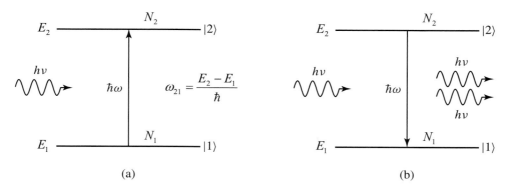

Figure 15.1 Resonant transitions through one-photon processes between two energy levels. (a) Optical absorption under the condition that $N_1 > N_2$. (b) Optical amplification under the condition that $N_2 > N_1$.

By using $\chi_{\mathrm{res}}(t; \omega; \mathbf{E})$, we can express the electric polarization at the frequency ω as

$$\mathbf{P}(\omega) = \epsilon_0 \chi_{\mathrm{res}}(t; \omega; \mathbf{E}) \cdot \mathbf{E}(\omega) = \epsilon_0 \chi_{\mathrm{res}}(t; \omega; \mathbf{E}) \cdot \hat{e}\mathcal{E}(\omega)e^{ikz}, \tag{15.2}$$

assuming that the wave propagates in the z direction with $\mathbf{k} = k\hat{z}$. Then, by using the coupled-wave equation given in (6.22), we have

$$\frac{d\mathcal{E}}{dz} = \frac{i\omega}{2cn_{\omega,z}}\chi_{\mathrm{res}}\mathcal{E}, \tag{15.3}$$

where $\chi_{\mathrm{res}}(t; \omega; \mathbf{E}) = \hat{e}^* \cdot \chi_{\mathrm{res}}(t; \omega; \mathbf{E}) \cdot \hat{e}$. By using the relation for the light intensity, $I = 2c\epsilon_0 n|\mathcal{E}|^2$, we can transform (15.3) into an equation for the attenuation of the light intensity through optical absorption:

$$\frac{dI}{dz} = -\frac{\omega}{cn_{\omega,z}}\chi_{\mathrm{res}}''I = -\alpha I, \tag{15.4}$$

where

$$\alpha(t; \omega; \mathbf{E}) = \frac{\omega}{cn_{\omega,z}}\chi_{\mathrm{res}}''(t; \omega; \mathbf{E}) \tag{15.5}$$

is the *one-photon absorption coefficient*. When $\chi_{\mathrm{res}}'' > 0$, the medium attenuates the light intensity by absorbing light with $\alpha > 0$. When $\chi_{\mathrm{res}}'' < 0$, we find that $\alpha < 0$. In this case, the medium amplifies the light intensity. Therefore, we can define a positive *gain coefficient* as $g = -\alpha$. Then, (15.4) can be expressed as

$$\frac{dI}{dz} = -\frac{\omega}{cn_{\omega,z}}\chi_{\mathrm{res}}''I = gI, \tag{15.6}$$

for the growth of the light intensity through optical amplification. Because $g = -\alpha$ by definition, we also find from (15.5) that

$$g(t; \omega; \mathbf{E}) = -\frac{\omega}{cn_{\omega, z}} \chi''_{\text{res}}(t; \omega; \mathbf{E}). \tag{15.7}$$

The absorption coefficient α and the gain coefficient g are respectively defined as the *attenuation rate* in space and the *growth rate* in space of the light intensity as light propagates: $I = I_0 \exp(-\alpha z)$ and $I = I_0 \exp(gz)$. As discussed in Subsection 4.2.3, they can also be defined through the imaginary part of the propagation constant as given in (4.39):

$$k = \beta + i\frac{\alpha}{2} = \beta - i\frac{g}{2}. \tag{15.8}$$

By using the relation that $k^2 = \omega^2 \mu_0 \epsilon = \omega^2 \mu_0 \epsilon_0 (n^2 + \chi_{\text{res}})$, we can also obtain the relation between α and χ''_{res} given in (15.5), and that between g and χ''_{res} given in (15.7) through the definitions of α and g given in (15.8).

Whether an optical medium attenuates an optical wave through optical absorption or amplifies it with an optical gain depends on the population difference between the two energy levels that are in resonance with the optical frequency. In the normal state that the medium is in thermal equilibrium with its surroundings, the population distribution is the equilibrium thermal distribution determined by temperature, as discussed in Subsection 1.5.2. Then the lower energy level $|1\rangle$ is more populated than the upper energy level $|2\rangle$ such that $\rho_{11} > \rho_{22}$ and $N_1 > N_2$. In this situation, the optical medium has a positive absorption coefficient such that $\alpha \propto N_1 - N_2 > 0$. In the case that the optical medium is pumped by an external pumping source to a state of *population inversion* in which the upper energy level $|2\rangle$ is more populated than the lower energy level $|1\rangle$, then $\rho_{22} > \rho_{11}$ and $N_2 > N_1$ for the optical medium to have a positive gain coefficient such that $g \propto N_2 - N_1 > 0$. From this discussion, it is clear that $g = -\alpha$. The medium attenuates light when $\alpha > 0$ and $g < 0$; it amplifies light when $\alpha < 0$ and $g > 0$.

Absorption saturation, in the case that the medium has a positive absorption coefficient, or *gain saturation*, in the case that the medium has a positive gain coefficient, occurs as the light intensity increases. Gain saturation is also called *gain compression*. The root cause of absorption saturation is that the population in the upper energy level increases as electrons in the lower energy level make transitions to the upper level by absorbing light, thus decreasing the population in the lower level while increasing that in the upper level. Because the total population is finite and fixed, this process leads to a decrease in the population difference $N_1 - N_2$, thus resulting in absorption saturation. A similar but opposite scenario takes place for gain saturation if the system is in a state of population inversion.

When perturbation expansion is valid, the resonant susceptibility can be approximated by the *first-order* susceptibility $\chi^{(1)}(t)$ given in (1.164):

$$\chi^{(1)}_{\text{res}}(t) = \frac{2}{\epsilon_0 \hbar} p_{12} p_{21} \left(N_1^{(0)} - N_2^{(0)} \right) \sin \omega_{21} t \, e^{-\gamma_{21} t}, \tag{15.9}$$

where t is the response time and $N_1^{(0)}$ and $N_2^{(0)}$ are the population densities when the material is in thermal equilibrium with its surroundings. This time-domain linear susceptibility leads to the frequency-domain linear susceptibility given in (1.165), and then (1.166) under the rotating-wave approximation:

$$\chi_{\text{res}}^{(1)}(\omega) = \frac{1}{\epsilon_0 \hbar} p_{12} p_{21} \left(N_1^{(0)} - N_2^{(0)} \right) \left(\frac{1}{\omega + \omega_{21} + \mathrm{i}\gamma_{21}} - \frac{1}{\omega - \omega_{21} + \mathrm{i}\gamma_{21}} \right)$$

$$\approx \frac{1}{\epsilon_0 \hbar} p_{12} p_{21} \left(N_1^{(0)} - N_2^{(0)} \right) \left(-\frac{1}{\omega - \omega_{21} + \mathrm{i}\gamma_{21}} \right) \tag{15.10}$$

$$= \frac{1}{\epsilon_0 \hbar} p_{12} p_{21} \left(N_1^{(0)} - N_2^{(0)} \right) \mathcal{L}(\omega; \omega_{21}, \gamma_{21}),$$

where

$$\mathcal{L}(\omega; \omega_{21}, \gamma_{21}) = -\frac{1}{\omega - \omega_{21} + \mathrm{i}\gamma_{21}} = -\frac{\omega - \omega_{21}}{(\omega - \omega_{21})^2 + \gamma_{21}^2} + \mathrm{i} \frac{\gamma_{21}}{(\omega - \omega_{21})^2 + \gamma_{21}^2} \tag{15.11}$$

is the complex Lorentzian lineshape function that is given in (1.167) and shown in Fig. 1.3. Because of resonant transition, the rotating-wave approximation can be taken here. Note that $\chi_{\text{res}}^{(1)}(\omega)$ as expressed in (15.10) is independent of the excitation optical field. From (15.5) and (15.7), we find the field-independent absorption coefficient and gain coefficient as

$$\alpha_0(\omega) = \frac{\omega}{c n_{\omega, z}} \chi_{\text{res}}^{(1)\prime\prime}(\omega) \propto N_1^{(0)} - N_2^{(0)} \tag{15.12}$$

and

$$g_0(\omega) = -\frac{\omega}{c n_{\omega, z}} \chi_{\text{res}}^{(1)\prime\prime}(\omega) \propto N_2^{(0)} - N_1^{(0)}, \tag{15.13}$$

respectively, where $\chi_{\text{res}}^{(1)}(\omega) = \hat{e}^* \cdot \boldsymbol{\chi}_{\text{res}}^{(1)}(\omega) \cdot \hat{e}$. The absorption coefficient α_0 is the linear absorption coefficient, known as the *unsaturated absorption coefficient*, and the gain coefficient g_0 is the linear gain coefficient, known as the *unsaturated gain coefficient*, because they are independent of the optical intensity. They are the values measured at a very low light intensity that does not cause appreciable changes in the population densities.

Because saturation necessarily occurs at resonance with two energy levels and its effect can significantly change the population distribution, the perturbation approach taken for the series expansion as done above to reach (15.10) is not valid at a sufficiently high intensity. Instead, a full analysis of the resonant transition process has to be carried out [1]. At the very least, even for a simple two-level system, the dependence of N_1 and N_2 on the excitation optical field has to be found by solving the coupled equations given in (1.153)–(1.156) for the density matrix elements to find ρ_{11} and ρ_{22}, for $N_1 = \rho_{11} N_{\text{total}}$ and $N_2 = \rho_{22} N_{\text{total}}$, while solving for ρ_{21} and ρ_{21} at the same time to find $\chi(t; t'; E)$ through $P(t)$. Such an analysis results in an intensity-dependent absorption coefficient and an intensity-dependent gain coefficient characterized by the relations:[1]

$$\alpha = \frac{\alpha_0}{1 + I/I_{\text{sat}}} \quad \text{and} \quad g = \frac{g_0}{1 + I/I_{\text{sat}}}, \tag{15.14}$$

---

[1] The absorption coefficient described by (15.14) is that for a homogeneously broadened medium. For an inhomogeneously broadened medium, the relation is $\alpha = \alpha_0 / (1 + I/I_{\text{sat}})^{1/2}$.

where α_0 is the unsaturated absorption coefficient given in (15.12), g_0 is the unsaturated gain coefficient given in (15.13), I is the intensity of the optical field, and I_{sat} is known as the *saturation intensity*. The intensity-dependent α and g in (15.14) are known as *saturated absorption coefficient* and *saturated gain coefficient*, respectively. The unsaturated absorption coefficient α_0 is independent of the optical intensity; it is the largest absorption coefficient when the system has the populations of $N_1^{(0)}$ and $N_2^{(0)}$ in thermal equilibrium. For an optical system such that the resonant transition frequency, ω_{21}, is in the optical region, $N_1^{(0)} \gg N_2^{(0)} \approx 0$ at room temperature; then $N_1^{(0)} \approx N_{total}$ and $\alpha_0 \propto N_{total}$. The unsaturated gain coefficient g_0 is also independent of the optical intensity; it is the largest gain coefficient for a given set of N_1 and N_2 in population inversion, which depends on the degree of pumping. If the pumping is so strong that the population in the lower energy level is almost completely depleted such that $N_1 \approx 0$, then $N_2 \approx N_{total}$ and $g_0 \propto N_{total}$. The saturation intensity is [1]

$$I_{sat} = \frac{\hbar\omega}{\tau_s \sigma_e}, \qquad (15.15)$$

where τ_s is an effective *saturation lifetime* of the effective population relaxation from the upper energy level and σ_e is the emission cross-section of the stimulated emission transition from the upper energy level to the lower energy level, which is proportional to the absorption cross-section, σ_a, of the absorption transition from the lower energy level to the upper energy level. Both τ_s and σ_e depend on the structures and parameters of the energy levels that are involved in the resonant transition process [1].

By using (15.5), we can find through (15.10) and (15.11) that under the rotating-wave approximation, the resonant susceptibility in the case of optical absorption can be expressed as

$$\chi_{res}(\omega) \approx -\frac{cn\alpha(\omega_{21})}{\omega_{21}} \frac{\gamma_{21}}{\omega - \omega_{21} + i\gamma_{21}} = -\frac{cn}{\omega_{21}} \frac{\alpha_0(\omega_{21})}{1 + I/I_{sat}} \frac{\gamma_{21}}{\omega - \omega_{21} + i\gamma_{21}}, \qquad (15.16)$$

where $\alpha(\omega_{21})$ and $\alpha_0(\omega_{21})$ are respectively the saturated and unsaturated absorption coefficients at the resonance frequency; and, by using (15.7) in the case of optical gain, as

$$\chi_{res}(\omega) \approx \frac{cng(\omega_{21})}{\omega_{21}} \frac{\gamma_{21}}{\omega - \omega_{21} + i\gamma_{21}} = \frac{cn}{\omega_{21}} \frac{g_0(\omega_{21})}{1 + I/I_{sat}} \frac{\gamma_{21}}{\omega - \omega_{21} + i\gamma_{21}}, \qquad (15.17)$$

where $g(\omega_{21})$ and $g_0(\omega_{21})$ are respectively the saturated and unsaturated gain coefficients at the resonance frequency. In general, these expressions have to be used for saturation processes. They are valid even when the optical intensity I is comparable to or larger than the saturation intensity I_{sat}.

Only in the case that $I < I_{sat}$, can the relations in (15.14) be expanded as

$$\alpha = \alpha_0 \left[1 - \frac{I}{I_{sat}} + \left(\frac{I}{I_{sat}}\right)^2 - \left(\frac{I}{I_{sat}}\right)^3 + \cdots \right] \quad \text{and} \quad g = g_0 \left[1 - \frac{I}{I_{sat}} + \left(\frac{I}{I_{sat}}\right)^2 - \left(\frac{I}{I_{sat}}\right)^3 + \cdots \right].$$

$$(15.18)$$

By using these expansions for the resonant susceptibilities that are given in (15.16) and (15.17), we can find, from the intensity-independent leading terms of these expansions, the linear susceptibility, $\chi_{res}^{(1)}$, as

$$\chi_{res}^{(1)}(\omega) \approx -\frac{cn\alpha_0(\omega_{21})}{\omega_{21}}\frac{\gamma_{21}}{\omega - \omega_{21} + i\gamma_{21}} = \frac{cng_0(\omega_{21})}{\omega_{21}}\frac{\gamma_{21}}{\omega - \omega_{21} + i\gamma_{21}}. \qquad (15.19)$$

In the case that the light intensity is low, such that $I(\omega) \ll I_{sat}$, perturbation expansion is valid. Then, we can generalize the relation that is given in (5.51) as

$$\chi_{res}(\omega) = \chi_{res}^{(1)}(\omega) + \frac{3\chi_{res}^{(3)}(\omega)}{2c\epsilon_0 n}I(\omega). \qquad (15.20)$$

From the intensity-dependent second terms of the expansions in (15.18), we can use (15.20) to find the third-order nonlinear susceptibility, $\chi_{res}^{(3)}$, of the resonant transition as

$$\chi_{res}^{(3)}(\omega) \approx \frac{2c^2\epsilon_0 n^2\alpha_0(\omega_{21})}{3\omega_{21}I_{sat}}\frac{\gamma_{21}}{\omega - \omega_{21} + i\gamma_{21}} = -\frac{2c^2\epsilon_0 n^2 g_0(\omega_{21})}{3\omega_{21}I_{sat}}\frac{\gamma_{21}}{\omega - \omega_{21} + i\gamma_{21}}. \qquad (15.21)$$

Note that (15.19) and (15.20) are equivalent to the results of perturbation expansion as discussed in the preceding section. They are valid only when $I \ll I_{sat}$, so that the expansions given in (15.18) are valid and the high-order terms in these expansions are negligible. We see from (15.19) and (15.21) that $\chi_{res}^{(1)\prime\prime}$ and $\chi_{res}^{(3)\prime\prime}$ have opposite signs, as expected for a saturation process.

15.2 SATURABLE ABSORBERS

A saturable absorber has an absorption coefficient that decreases with increasing light intensity, as described by the intensity-dependent absorption coefficient α that is given in (15.14). In general, this general form of α for absorption saturation has to be used rather than the perturbation forms in terms $\chi_{res}^{(1)\prime\prime}$ and $\chi_{res}^{(3)\prime\prime}$ because the light intensity can easily exceed the saturation intensity in a practical application of a saturable absorber.

The propagation of an optical wave through a saturable absorber that has an absorption coefficient as given in (15.14) is governed by the equation:

$$\frac{dI}{dz} = -\alpha I = -\frac{\alpha_0}{1 + I/I_{sat}}I. \qquad (15.22)$$

With an input light intensity of $I_{in} = I(0)$ at $z = 0$, (15.22) can be integrated to obtain the solution for $I(z)$ in terms of a transcendental relation:

$$I(z)e^{I(z)/I_{sat}} = I(0)e^{I(0)/I_{sat}}e^{-\alpha_0 z}, \qquad (15.23)$$

which can also be expressed as

$$I(z) = I(0)\exp\left[\frac{I(0) - I(z)}{I_{\text{sat}}} - \alpha_0 z\right]. \tag{15.24}$$

The transmittance of an optical wave through a saturable absorber that has a thickness of l is

$$T = \frac{I_{\text{out}}}{I_{\text{in}}} = \frac{I(l)}{I(0)}. \tag{15.25}$$

By using (15.24) for $z = l$, we find the transcendental relation for the transmittance:

$$T = T_0 \exp\left[\frac{I_{\text{in}}}{I_{\text{sat}}}(1 - T)\right], \tag{15.26}$$

where

$$T_0 = \exp\left(-\alpha_0 l\right) \tag{15.27}$$

is the *unsaturated transmittance*, also known as the *small-signal transmittance*. The transmittance that is given by (15.26) is plotted in Fig. 15.2 as a function of the input light intensity, normalized to the saturation intensity, for a few different values of $\alpha_0 l$ represented in terms of T_0.

According to (15.26), we find that $1 > T \geq T_0$ because $\alpha_0 > 1$, as expected for a saturable absorber and seen in Fig. 15.2. From (15.24) and (15.25), we find that when the input light intensity is very low such that $I_{\text{in}} \ll I_{\text{sat}}$,

$$T \xrightarrow{I_{\text{in}} \ll I_{\text{sat}}} T_0 = \exp\left(-\alpha_0 l\right). \tag{15.28}$$

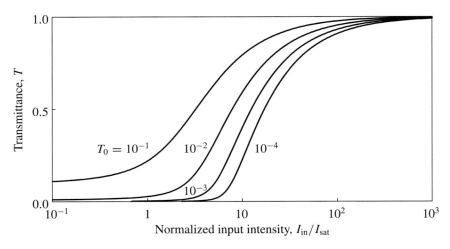

Figure 15.2 Transmittance of an optical wave through a saturable absorber that has a thickness of l and an unsaturated absorption coefficient of α_0 as a function of the input light intensity normalized to the saturation intensity. The curves are plotted for different values of $\alpha_0 l$ in terms of $T_0 = \exp\left(-\alpha_0 l\right)$.

As Fig. 15.2 shows, the optical transmittance through a saturable absorber increases nonlinearly as the input intensity is increased, and it approaches unity at high input intensities. In a particular application of a saturable absorber, the value of $\alpha_0 l$ has to be properly chosen for a desired difference between the maximum transmittance at a high intensity and the minimum transmittance at a low intensity. The *modulation depth* of a saturable absorber is defined as

$$m = \frac{T_{\max} - T_{\min}}{T_{\max}} = \frac{T_{\max} - T_0}{T_{\max}}.$$

(15.29)

It is often expressed as a percentage.

In the above discussions, we have considered saturable absorbers of the *transmissive type*. For this type, the transmittance of a saturable absorber increases when the absorption of an optical beam that is transmitted through the absorber is saturated at a high intensity. Some saturable absorbers are of the *reflective type*. For this type, the reflectance of a saturable absorber increases when the attenuation of an optical beam that is reflected by the absorber is saturated at a high intensity. The modulation depth of a reflective-type saturable absorber is

$$m = \frac{R_{\max} - R_{\min}}{R_{\max}}.$$

(15.30)

One example of the reflective type is the *semiconductor saturable absorber mirror* (SESAM) [2], which often contains a layer of semiconductor quantum-well saturable absorber on top of a Bragg reflector.

The recovery time of a saturable absorber is the decay time of the absorber from its excited state after it absorbs a photon, which is the population relaxation time τ_2 of the upper energy level, $|2\rangle$. In practical applications, a saturable absorber is often categorized as either a *fast absorber* or a *slow absorber* according to whether it has a short recovery time, thus a fast recovery from saturation, or a long recovery time, thus a slow recovery from saturation. Note that fast or slow refers to how fast a saturable recovers from being saturated; it does not refer to how fast or how slow the absorber is saturated. In other words, it refers to the recovery time but not the response time. The response time of a saturable absorber to an optical excitation is generally very fast and can be considered instantaneous for most applications. Nonetheless, fast or slow is a relative concept that requires a reference with which the recovery time can be measured. The reference is the duration of the excitation optical field. For a CW excitation, all saturable absorber can be considered fast because the interaction reaches a steady state. When the excitation comes from an optical pulse, a fast saturable absorber is one that has a recovery time shorter than the pulse duration so that the saturation of the absorber follows the temporal variations of the pulse intensity by reaching the steady state almost instantaneously. By contrast, a slow absorber is one that has a recovery time longer than the pulse duration so that the saturation of the absorber does not follow the instantaneous variations of the pulse intensity. Instead, the slow absorber is saturated by the integrated intensity of the pulse over time. Therefore, a saturable absorber that is fast for a certain application can be slow for another application. For example, a saturable absorber can be fast for a Q-switched laser that delivers nanosecond pulses, but slow for a mode-locked laser that generates picosecond pulses.

Similarly, a fast absorber for a picosecond mode-locked laser can become a slower absorber for a femtosecond mode-locked laser. For this reason, the same saturable absorber may be either a fast absorber or a slow absorber, depending on the durations of the pulses with which it is used. One important concept to keep in mind is that a fast absorber is not necessarily better or worse than a slower absorber. Indeed, slow saturable absorbers are used in most of the highly effective mode-locked lasers. Whether a certain saturable absorber is suitable for a particular application depends on many considerations, as well as on many other parameters of the system.

A saturable absorber can be used as a *spatial light filter*, which blocks low-intensity stray light or background optical noise but transmits a high-intensity signal beam. In this kind of application, a saturable absorber cleans up the spatial irregularities of an optical beam. A saturable absorber can be used as an *optical pulse discriminator*, which transmits optical pulses of intensities above a certain threshold and suppresses those below it. A saturable absorber is also commonly used as a *passive Q switch* in a *Q*-switched laser or as a *passive mode locker* in a mode-locked laser for the generation of short laser pulses. The saturable absorber in this kind of application functions as a *passive optical switch* in the time domain. It is switched open by the rising intensity of a laser pulse, and it closes through its own relaxation after the passing of the pulse. Therefore, the population relaxation time of a saturable absorber is also an important factor to consider in its application as a *Q* switch or a mode locker. Similar to its use as a spatial light filter, a saturable absorber can also be used as a *temporal light filter* outside a laser cavity to clean up the temporal irregularities and noise of an optical pulse. In this respect, it can also be used as a *thresholding device* for optical signal processing.

15.3 SATURABLE AMPLIFIERS

Any medium that provides an optical gain can be used to amplify an optical signal. Depending on the physical mechanism that is responsible for the optical gain, there are two different categories of optical amplifiers: *nonlinear optical amplifiers* and *laser amplifiers*. The optical gain of a nonlinear optical amplifier is associated with a nonlinear optical process in a nonlinear medium, whereas the gain of a laser amplifier results from a one-photon resonant transition of a medium that is pumped to population inversion. Nonlinear optical amplifiers can be parametric, such as the optical parametric amplifiers discussed in Section 10.5, or nonparametric, such as the Raman and Brillouin amplifiers, discussed in Subsections 13.3.1 and 13.5.1, respectively.

Every amplifier saturates at a high input signal power. The reason is clear. At a given pumping level, the pump power that is supplied to an amplifier is fixed. If the gain also remains fixed without saturation, the power of the output signal increases proportionally as the power of the input signal increases. This rate of increase cannot continue as the input signal power continues to increase because the output signal power would eventually exceed the pump power that is supplied to the amplifier if it could, thus violating conservation of energy. Therefore, *the general concept of gain saturation applies to every optical amplifier irrespective of its gain mechanism*. All amplifiers share similar saturation characteristics. Nonetheless, the detailed behavior of an optical amplifier depends on its physical mechanism and parameters. In this section, the general characteristics of laser amplifiers are addressed. We consider only continuously pumped laser amplifiers operating in the steady state. Not considered here are pulsed laser amplifiers that require transient dynamical analysis, including those for regenerative amplification of ultrashort laser pulses and those using transient pumping for high-power amplification of giant laser pulses. Nonlinear optical amplifiers are not considered here either.

We consider single-pass, traveling-wave laser amplifiers. Such an amplifier does not have a resonant optical cavity; therefore, the optical signal being amplified passes through it only once as a traveling wave. An amplifier can be pumped in many different ways, but the most commonly employed techniques are *electrical pumping* and *optical pumping*. The pumping arrangement can be *transverse* or *longitudinal*. For electrical pumping, a transverse pumping arrangement is more convenient and is most often used, though a longitudinal pumping arrangement is also possible. For optical pumping, both transverse and longitudinal pumping arrangements can be easily implemented. However, for an optically pumped amplifier that has a long length but a relatively small absorption coefficient at the pump frequency, such as a fiber amplifier, the longitudinal pumping arrangement is much more efficient than the transverse pumping arrangement. Longitudinal optical pumping can be arranged as unidirectional forward, unidirectional backward, or bidirectional. In the following analysis, we assume that pumping is transversely uniform but allow for the possibility that it may vary in the longitudinal z direction, which is often the case for a long amplifier. Then the unsaturated gain coefficient does not vary across the transverse profile of the signal beam, but it may vary along the longitudinal beam path such that it can be expressed as $g_0(z)$.

In general, the photons that are emitted by spontaneous emission also contribute to the output of an optical amplifier. Furthermore, they are also amplified by the optical gain. Though they are noise photons, they do change the total output power of the amplifier. For simplicity, here we ignore the contribution of these noise photons. By using (15.6) and (15.14), the amplification of the intensity, I, of an optical signal that propagates through an optical amplifier can be described as

$$\frac{\mathrm{d}I}{\mathrm{d}z} = g(z)I = \frac{g_0(z)}{1 + I/I_{\mathrm{sat}}}I, \tag{15.31}$$

where $g_0(z)$ is the unsaturated gain coefficient, which can be spatially varying in the longitudinal direction as discussed above, and I_{sat} is the saturation intensity of the gain medium as defined in (15.15).

As mentioned above, we assume transverse uniformity but consider the possibility of longitudinal nonuniformity by taking the unsaturated gain coefficient $g_0(z)$ to be only a function of z. Such a longitudinally nonuniform gain distribution is a common scenario in an amplifier under longitudinal optical pumping because of pump absorption by the gain medium. In the following discussions, we assume for simplicity that not only the gain is transversely uniform but the signal beam is also collimated throughout the length of the amplifier such that divergence of the beam is negligible. This assumption allows us to consider the power, P, of the optical signal as $P = I\mathcal{A}$, where $\mathcal{A}$ is the cross-sectional area of the signal beam. Then (15.31) can be converted to an equation for the signal power:

$$\frac{\mathrm{d}P}{\mathrm{d}z} = g(z)P = \frac{g_0(z)}{1 + P/P_{\mathrm{sat}}}P, \tag{15.32}$$

where $P_{\text{sat}} = I_{\text{sat}}A$ is the *saturation power*. By integrating (15.32), we obtain the relation:

$$\frac{P(z)}{P(0)} \exp\left[\frac{P(z) - P(0)}{P_{\text{sat}}}\right] = \exp\int_0^z g_0(z)\,dz, \tag{15.33}$$

where $P(0)$ is the input power of the signal beam at $z = 0$. It can be seen from (15.33) that when $P(z) \ll P_{\text{sat}}$, the power of the optical signal grows exponentially with distance. As $P(z)$ approaches the value of P_{sat}, its growth slows down. Eventually, the signal grows only linearly with distance when $P(z) \gg P_{\text{sat}}$.

The power gain of an optical signal that is amplified by an optical amplifier of a length l is defined as

$$G = \frac{P_{\text{out}}}{P_{\text{in}}} = \frac{P(l)}{P(0)}, \tag{15.34}$$

where $P_{\text{in}} = P(0)$ and $P_{\text{out}} = P(l)$ are the input and output powers of the signal, respectively. By taking $z = l$ for the relation in (15.33) for an amplifier of a length l, a transcendental relation for the power gain of the signal is found:

$$G = G_0 \exp\left[\frac{P_{\text{in}}}{P_{\text{sat}}}(1 - G)\right], \tag{15.35}$$

where G_0 is the *unsaturated power gain*, or the *small-signal power gain*. For a single pass through the amplifier, G_0 is given by the relation:

$$G_0 = \exp\int_0^l g_0(z)\,dz. \tag{15.36}$$

Figure 15.3 shows the amplifier gain as a function of the input signal power for a few different values of the unsaturated power gain. By comparing (15.26) for the transmittance of a saturable absorber with (15.35) for the gain of a saturable amplifier, we find that they have the same form. However, $T_0 < 1$ so that $T < 1$ for all signal intensities, whereas $G_0 > 1$ so that $G > 1$ for all signal powers. This difference leads to the difference between Fig. 15.2 and Fig. 15.3.

According to (15.35), $G_0 \geq G > 1$ because $g_0 > 0$ for an amplifier. For a weak optical signal such that $P_{\text{in}} < P_{\text{out}} \ll P_{\text{sat}}$, the power gain is simply the small-signal power gain:

$$G \xrightarrow{P_{\text{in}} < P_{\text{out}} \ll P_{\text{sat}}} G_0. \tag{15.37}$$

If the signal power approaches or even exceeds the saturation power of the amplifier, the relation in (15.35) clearly indicates that $G < G_0$ because of gain saturation. In this situation, the overall gain, G, can be found by solving (15.35) when the values of P_{in} and P_{sat}, as well as that of G_0, are given.

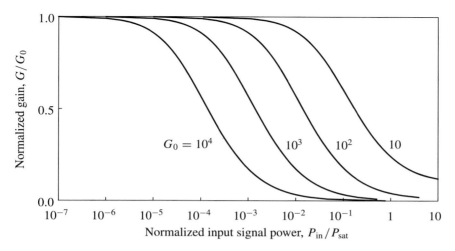

Figure 15.3 Gain, normalized to the unsaturated gain as G/G_0, of an optical amplifier as a function of the input signal power, normalized to the saturation power as P_{in}/P_{sat}, for a few different values of the unsaturated power gain.

In the above, we have assumed transverse uniformity in the profiles of the gain and the signal beam. In reality, the transverse intensity profile of a signal beam is never uniform, even though the gain profile might be quite uniform. The transverse spatial variation of the optical intensity of the signal beam results in modified saturation characteristics. In particular, because the intensity is usually higher at the beam center, the gain medium that is located in the transverse center of the beam is saturated at a lower signal power than that located in the outer parts of the beam profile. One consequence of this uneven saturation is that the transverse spatial profile of the amplified output signal may be different from that of the input signal.

As discussed in the preceding section for the saturable absorbers, the recovery time of a saturable absorber matters in the interaction of a saturable absorber with an optical pulse. Similarly, an optical gain medium also has a finite recovery time, and this recovery time matters in the amplification of optical pulses. Among the various types of optical amplifiers, the optical parametric amplifiers have the fastest recovery time, almost instantaneous for even the shortest femtosecond pulses, because its gain does not come from any resonant transition and the medium does not store energy. The significance and consequences of the finite temporal response, thus finite bandwidth, of the Raman and Brillouin amplifiers were already discussed in Section 13.6. In the case of a laser gain medium that involves a resonant transition, as shown in Fig. 15.1(b), the temporal variations in the populations of the energy levels are limited by the population relaxation times of the energy levels. Therefore, the gain is not able to respond instantly to the change in the pumping source or that in the signal power. Some laser gain media, such as the semiconductor lasers, have very short relaxation times of the order of nanoseconds or picoseconds, whereas some others, such as the solid-state lasers, have long relaxation times of the order of microseconds or milliseconds. One important consequence of the different gain relaxation times is the difference in the saturation intensity. Nonetheless, according to (15.15), the saturation intensity depends on both the emission cross-section σ_e and the saturation lifetime τ_s. Therefore, a gain medium that has a small relaxation time does not necessarily have a higher saturation intensity than one that has a large relaxation time. These parameters also have significant consequences for the dynamics of a laser.

15.4 LASER OSCILLATION

Laser action is a highly nonlinear process, though it usually is not discussed in the context of nonlinear optics. It is highly nonlinear because *one condition for stable laser oscillation is that the saturated optical gain of the laser medium is clamped at a fixed level after the laser reaches its threshold for oscillation*. Indeed, any system that has a threshold is a nonlinear system. Figure 15.4 shows the typical characteristics of the output power of a single-mode laser as a function of the pump power. It shows that before a laser reaches its threshold, it has a very low output power, which is contributed by spontaneous emission and is very close to zero on the scale shown in Fig. 15.4. Indeed, once the gain medium of a laser is pumped so that its upper laser level begins to be populated, it emits spontaneous photons regardless of whether population inversion is reached or nor, or whether the laser is oscillating or not. Clearly, the output power of a laser below the threshold is not exactly zero because spontaneous photons are already emitted from the laser before the laser reaches its threshold. This power of the spontaneous emission is proportional to the rate at which the upper laser level is populated. Because the laser output below the threshold consists of only spontaneous emission, its power increases linearly with the pump power. Therefore, before a laser reaches its oscillation threshold, it behaves as a linear optical device, very much like a light-emitting diode. Though this spontaneous emission power is incoherent and is generally small for a practical laser when it oscillates above the threshold, it is significant for a laser that is pumped below the threshold. Above the threshold, it is the major source of incoherent noise for the coherent field of the laser output.

Once a laser reaches its threshold, laser oscillation starts. The behavior and characteristics of the laser change dramatically. The output of the laser above the threshold mainly consists of photons that are coherently amplified through stimulated emission. For an ideal laser, the saturated optical gain of the laser medium is clamped at its threshold value by the effect of gain saturation as the pump power increases above the threshold, whereas the unsaturated optical gain increases linearly with the pump power. Though this is not exactly true for a realistic laser,

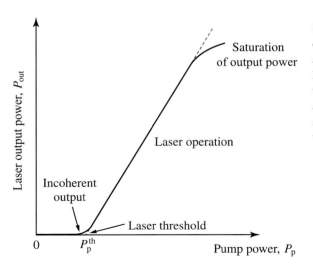

Figure 15.4 Typical characteristics of the output power of a single-mode laser as a function of the pump power. Note that the laser gain is saturated once the laser reaches its threshold, before its output is saturated at a much higher pumping level. It remains saturated for all pump powers higher than the threshold pump power.

the saturated optical gain of a practical laser is still clamped near the threshold value, while the unsaturated optical gain increases with the pump power. The consequence is that the populations of the upper and lower laser levels remain unchanged as the pump power increases above the threshold. Therefore, the power of spontaneous emission does not increase with the pump power above the threshold; it is clamped at the level that is reached at the laser threshold. On the other hand, the coherent laser output power increases linearly with the pump power above the threshold, as seen in Fig. 15.4. Upon reaching the threshold, the optical output of the laser also shows dramatic spectral narrowing that accompanies the start of laser oscillation. Because the coherent output power from stimulated emission increases with the pump power but the spontaneous emission power is fixed at the threshold level, the linewidth of an oscillating laser mode continues to narrow with the increasing laser power as the laser is pumped higher and higher above the threshold. These are the unique characteristics that distinguish a laser from other types of light sources, such as fluorescent light emitters and luminescent light sources. However, a real laser does not have exactly such ideal characteristics, mainly because of the presence of spontaneous emission and nonlinearities in the gain medium. As the pump power increases to a sufficiently high level, the unsaturated gain coefficient of the gain medium cannot continue to increase linearly with the pump power because of the depletion of the ground-level population. Therefore, we can expect the output power of a laser not to continue its linear dependence on the pump power but to increase less than linearly with the pump power at high pumping levels, as is also shown in Fig. 15.4

To clearly illustrate the condition and characteristics of gain saturation that are related to laser oscillation, we consider for simplicity a laser oscillator that has the structure of a *Fabry–Perot cavity*, as shown in Fig. 15.5. The longitudinal direction of this cavity along which the laser field propagates is taken to be the z direction. The laser cavity has a physical length of l, and it contains a gain medium of a length l_g. The *overlap factor* between the gain medium and the laser mode intensity distribution, also known as the *gain filling factor*, is defined as

$$\Gamma = \frac{\underset{\text{active}}{\iiint} |\mathbf{E}|^2 \, dxdydz}{\underset{\text{laser mode}}{\iiint} |\mathbf{E}|^2 \, dxdydz} \approx \frac{\mathcal{V}_{\text{active}}}{\mathcal{V}_{\text{mode}}}. \tag{15.38}$$

For the Fabry–Perot cavity shown in Fig. 15.5, $\Gamma \approx l_g/l$. The *round-trip optical path length*, l_{RT}, is the optical length for the laser field to make a round-trip inside the cavity. It depends on the specific structure of the cavity. For this Fabry–Perot cavity,

$$l_{\text{RT}} = 2nl_g + 2n_0(l - l_g) = 2\Gamma nl + 2(1 - \Gamma)n_0 l = 2\bar{n}l, \tag{15.39}$$

where n_0 is the refractive index in the cavity but outside the gain medium, n is the refractive index of the gain medium, and

$$\bar{n} = \Gamma n + (1 - \Gamma)n_0 \tag{15.40}$$

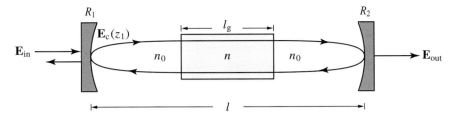

Figure 15.5 Laser with a Fabry–Perot cavity.

is the weighted average index of refraction throughout the laser cavity. The time it takes for an intracavity field to complete one round-trip in the cavity is called the *cavity round-trip time*:

$$T = \frac{\text{round-trip optical path length}}{c} = \frac{l_{\text{RT}}}{c}. \tag{15.41}$$

The general concepts discussed below apply equally well to laser oscillators that have other structures, such as one of a ring cavity, a Bragg reflector cavity, or a distributed-feedback cavity. However, the relations given above for the structural parameters have to be properly modified for a specific optical cavity.

Consider an intracavity field, $\mathbf{E}_c(z)$, at any point z on the longitudinal axis inside an optical cavity. When it completes a round-trip back to the starting position z, it is modified by a *complex amplification or attenuation factor*, a, to become $a\mathbf{E}_c(z)$. The factor a can be generally expressed as

$$a = G \exp(i\varphi_{\text{RT}}), \tag{15.42}$$

where G is the *round-trip gain factor for the field amplitude*, equivalent to the *power gain in one single pass through a linear Fabry–Perot cavity*, and φ_{RT} is the *round-trip phase shift* for the intracavity field. Both G and φ_{RT} have real values, and $G \geq 0$. If $G > 1$, the intracavity field is amplified. If $G < 1$, the intracavity field is attenuated.

We first consider the resonant characteristics of a passive optical cavity. A passive cavity cannot generate or amplify an optical field. In order to maintain a resonant intracavity field in such a cavity, it is necessary to constantly inject an input optical field, $\mathbf{E}_{\text{in}}$, to the cavity. As shown in Fig. 15.5, the forward-traveling component of the intracavity field at the location z_1 just inside the cavity next to the injection point is the sum of the transmitted input field and the fraction of the intracavity field that returns after one round-trip through the cavity:

$$\mathbf{E}_c(z_1) = t_{\text{in}}\mathbf{E}_{\text{in}} + a\mathbf{E}_c(z_1), \tag{15.43}$$

where t_{in} is the complex transmission coefficient for the input field. The transmitted output field, $\mathbf{E}_{\text{out}}$, is proportional to the intracavity field: $\mathbf{E}_{\text{out}} \propto \mathbf{E}_c(z_1)$. From (15.43), we find that

$$\mathbf{E}_{\text{out}} \propto \mathbf{E}_c(z_1) = \frac{t_{\text{in}}}{1-a}\mathbf{E}_{\text{in}}. \tag{15.44}$$

A laser is a coherent optical oscillator, and the basic function of an oscillator is to generate a coherent signal through resonant oscillation without an input signal. Therefore, no external optical field is injected into the optical cavity for laser oscillation. A practical laser device can be constructed by placing an optical gain medium inside an optical resonator. The gain medium provides amplification to the intracavity optical field, while the resonator provides optical feedback. The intracavity optical field has to grow from the field that is generated by spontaneous emission from the intracavity gain medium. When steady-state oscillation is reached, the coherent laser field at any given location inside the cavity has to become a constant of time in both phase and magnitude. In the model shown in Fig. 15.5, the situation of steady-state laser oscillation requires that $\mathbf{E}_{in} = 0$ and $\mathbf{E}_c(z_1) = $ constant $\neq 0$ so that $\mathbf{E}_{out} \neq 0$. Therefore, from (15.44), the *condition for steady-state laser oscillation* is

$$a = G \exp{(i\,\varphi_{RT})} = 1. \tag{15.45}$$

For a given laser mode that has the compound transverse mode indices of *mn* and the longitudinal mode index of *q*, this condition can be satisfied for the laser mode to oscillate only if the *gain condition*

$$G_{mnq} = 1 \tag{15.46}$$

and the *phase condition*

$$\varphi_{mnq}^{RT} = 2q\pi, \tag{15.47}$$

are simultaneously fulfilled. Note that both G_{mnq} and φ_{mnq}^{RT} are frequency dependent, and the indices *m*, *n*, and *q* are integers. The gain condition given in (15.46) indicates that there is a threshold gain, thus a corresponding threshold pumping level, for the oscillation of a laser mode. When the *mnq* mode is below the threshold, $G_{mnq} < 1$ so that it does not oscillate. When the *mnq* mode is above the threshold, it oscillates with its gain factor clamped at $G_{mnq} = 1$, as required by the gain condition. The phase condition given in (15.47) determines the oscillation frequency of a laser mode. For simplicity of the notation, in the following we consider a single-mode laser so that the mode indices *mnq* are dropped.

For the Fabry–Perot laser shown in Fig. 15.5, the round-trip field gain factor is

$$G = R_1^{1/2} R_2^{1/2} \exp{(gl_g - \bar{\alpha}l)} = R_1^{1/2} R_2^{1/2} \exp{\left[(\Gamma g - \bar{\alpha})l\right]}, \tag{15.48}$$

where R_1 and R_2 are the reflectivities of the two mirrors of the cavity, as shown in Fig. 15.5, *g* is the saturated gain coefficient of the gain medium, and $\bar{\alpha}$ is the spatially averaged *distributed loss* of the laser mode. When the laser reaches its *oscillation threshold*, the gain condition given in (15.46) indicates that $G_{th} = 1$. Therefore, from (15.48), the *threshold gain coefficient*, g_{th}, is given by the relation:

$$g_{th}l_g = \bar{\alpha}l - \ln\sqrt{R_1 R_2} \quad \text{or} \quad \Gamma g_{th} = \bar{\alpha} - \frac{1}{l}\ln\sqrt{R_1 R_2}. \tag{15.49}$$

The power that is required to pump a laser to its threshold is called the *threshold pump power*, P_p^{th}. Because the threshold gain coefficient as given in (15.49) is mode dependent, the threshold pump power is also mode dependent. The threshold pump power of a laser mode is found by calculating the power that is required for the gain medium to have an unsaturated gain coefficient equal to the threshold gain coefficient of the mode:

$$g_0 = g_{th} \tag{15.50}$$

for spatially uniform pumping in the longitudinal direction, or

$$\int_0^{l_g} g_0(z)dz = g_{th}l_g \tag{15.51}$$

for longitudinally nonuniform pumping. The threshold gain coefficient g_{th} given in (15.49) has the unit of m^{-1}; it is the spatial growth rate and is determined by the spatial loss rate $\bar{\alpha}$.

The threshold condition can also be expressed in terms of the temporal photon growth rate, g, and the temporal loss rate, γ_c, of the cavity. The net amplification factor of the intracavity field energy, or that of the photon number, in a round-trip through the cavity is G^2. By using (15.48), we can express it as

$$G^2 = \exp\left(2\Gamma gl - \gamma_c T\right) = \exp\left[(\Gamma g - \gamma_c)T\right], \tag{15.52}$$

where γ_c is the *cavity decay rate*, Γg is the *intracavity energy growth rate*, or *intracavity photon growth rate*, and T is the round-trip time of the cavity defined in (15.41). A *photon lifetime* for the decay time constant of the passive optical cavity can be defined as

$$\tau_c = \frac{1}{\gamma_c}. \tag{15.53}$$

By comparing (15.48) and (15.52), we find that

$$g = \frac{2gl}{T} = \frac{cg}{\bar{n}} \tag{15.54}$$

and

$$\gamma_c = \frac{1}{T}\left(2\bar{\alpha}l - \ln R_1 R_2\right) = \frac{c}{\bar{n}}\left(\bar{\alpha} - \frac{1}{l}\ln\sqrt{R_1 R_2}\right). \tag{15.55}$$

By using (15.49) and (15.55), we find that

$$\gamma_c = \Gamma\frac{2g_{th}l}{T} = \Gamma\frac{cg_{th}}{\bar{n}}. \tag{15.56}$$

From (15.54) and (15.56), we find that the *laser threshold condition* can be simply expressed as

$$\Gamma g_{th} = \gamma_c. \tag{15.57}$$

The energy growth rate Γg and the cavity decay rate γ_c are specific to a laser mode; they are respectively known as the *gain parameter* and the *loss parameter* of a laser mode. Thus, Γg_{th} is the *threshold gain parameter*. When a laser mode is pumped above its oscillation threshold, its gain parameter is clamped by gain saturation at its threshold value, which is equal to its loss parameter, as indicated by (15.57).

The growth rate Γg and the decay rate γ_c respectively refer to the growth and decay of the intracavity photons. The spatially averaged *intracavity photon density* of an oscillating mode is

$$S = \frac{\bar{n}I}{chv},\tag{15.58}$$

where I is the spatially averaged intracavity intensity of the laser mode and hv is the photon energy. Because the *saturated gain parameter g* is directly proportional to the saturated gain coefficient g, according to (15.54), we find by using (15.14) and (15.58) the relation:

$$g = \frac{g_0}{1 + S/S_{sat}},\tag{15.59}$$

where

$$g_0 = \frac{cg_0}{\bar{n}}\tag{15.60}$$

is the *unsaturated gain parameter* and

$$S_{sat} = \frac{\bar{n}I_{sat}}{chv} = \frac{\bar{n}}{c\tau_s\sigma_e}\tag{15.61}$$

is the *saturation photon density*.

When a laser oscillates above its threshold in the steady state, the value of Γg is clamped at its threshold value of γ_c, as indicated in (15.57). Therefore, by setting $\Gamma g = \gamma_c$, the intracavity photon density of an oscillating laser mode can be found as

$$S = \left(\frac{\Gamma g_0}{\gamma_c} - 1\right)S_{sat} = (r-1)S_{sat}, \quad \text{for} \quad r \geq 1.\tag{15.62}$$

The *dimensionless pumping ratio, r*, represents that a laser is pumped at r times its threshold:

$$r = \frac{\Gamma g_0}{\gamma_c} = \frac{g_0}{g_{th}} = \frac{g_0}{g_{th}}.\tag{15.63}$$

The output power of the laser mode can be found as

$$P_{out} = V_{mode}\,Shv\gamma_{out} = (r-1)V_{mode}\,S_{sat}\,hv\gamma_{out} = (r-1)P_{out}^{sat},\tag{15.64}$$

where γ_{out} is the output coupling rate of the laser cavity and P_{out}^{sat} is the *saturation output power* defined as $P_{out}^{sat} = V_{mode}\,S_{sat}\,hv\gamma_{out}$. Note that P_{out}^{sat} is not the power at which the laser output saturates; it is the output power when the laser is pumped at a level that is twice its threshold with $r = 2$. From (15.64), it is clearly seen that the output power of a laser mode increases linearly with the pumping level above the threshold because of the saturation of the laser gain at its threshold level.

EXAMPLE 15.1

A Nd:YAG microchip laser, shown in Fig. 15.6, is made of a thin Nd:YAG crystal that has a thickness of $l = 500$ μm. The two parallel surfaces of the crystal are coated for $R_1 = 100\%$ and $R_2 = 99.7\%$ at the 1.064 μm laser wavelength to form the laser cavity, but for $R_1 = R_2 = 0$ at the 808 nm pump wavelength to allow only a single pass of the pump beam. At the laser wavelength of $\lambda = 1.064$ μm, the refractive index of Nd:YAG is $n = 1.818$. The emission cross-section is $\sigma_e = 3.1 \times 10^{-23}$ m², and the fluorescence lifetime is $\tau_2 = 240$ μs. The distributed loss of the laser cavity is found to be $\bar{\alpha} = 0.5$ m⁻¹. (a) Find the round-trip optical path length and the round-trip time. (b) What are the cavity decay rate and the photon lifetime? (c) Find the threshold gain coefficient and the threshold gain parameter of this laser. (d) Find the saturation photon density. (e) The laser has a circular Gaussian mode that has a spot size of $w_0 = 200$ μm inside the cavity. Find the saturation output power P_{out}^{sat} of the laser. What is the output power of the laser if it is pumped at a pumping ratio of $r = 2.5$?

Solution

(a) This is a Fabry–Perot laser cavity that has a filling factor of $\Gamma = 1$. Therefore, $\bar{n} = n = 1.818$. Then the round-trip optical path length is found by using (15.39) as

$$l_{RT} = 2\bar{n}l = 2 \times 1.818 \times 500 \text{ μm} = 1.818 \text{ mm}.$$

The round-trip time is found by using (15.41):

$$T = \frac{l_{RT}}{c} = \frac{1.818 \times 10^{-3}}{3 \times 10^8} \text{ s} = 6.06 \text{ ps}.$$

(b) The cavity decay rate is found by using (15.55):

$$\gamma_c = \frac{c}{\bar{n}}\left(\bar{\alpha} - \frac{1}{l}\ln\sqrt{R_1 R_2}\right) = \frac{3 \times 10^8}{1.818} \times \left(0.5 - \frac{\ln\sqrt{1 \times 0.997}}{500 \times 10^{-6}}\right) \text{ s}^{-1} = 5.78 \times 10^8 \text{ s}^{-1}.$$

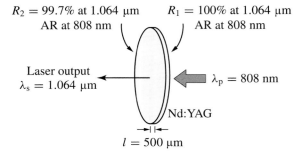

$R_2 = 99.7\%$ at 1.064 μm
AR at 808 nm

$R_1 = 100\%$ at 1.064 μm
AR at 808 nm

Laser output
$\lambda_s = 1.064$ μm

$\lambda_p = 808$ nm

Nd:YAG

$l = 500$ μm

Figure 15.6 Nd:YAG microchip laser. AR: anti-reflection.

The photon lifetime is

$$\tau_c = \frac{1}{\gamma_c} = \frac{1}{5.78 \times 10^8} \text{ s} = 1.73 \text{ ns.}$$

(c) The threshold gain coefficient is found by using (15.49):

$$g_{th} = \frac{1}{\Gamma}\left(\bar{\alpha} - \frac{1}{l}\ln\sqrt{R_1 R_2}\right) = \frac{1}{1} \times \left(0.5 - \frac{\ln\sqrt{1 \times 0.997}}{500 \times 10^{-6}}\right) \text{ m}^{-1} = 3.5 \text{ m}^{-1}.$$

The threshold gain parameter is found by using (15.57):

$$\Gamma g_{th} = \gamma_c = 5.78 \times 10^8 \text{ s}^{-1}.$$

(d) For Nd:YAG, which is a four-level system, $\tau_s \approx \tau_2 = 240$ μs. The saturation photon density is found by using (15.61):

$$S_{sat} = \frac{\bar{n}}{c\tau_s \sigma_e} = \frac{1.818}{3 \times 10^8 \times 240 \times 10^{-6} \times 3.1 \times 10^{-23}} \text{ m}^{-3} = 8.15 \times 10^{17} \text{ m}^{-3}.$$

(e) For a circular Gaussian mode that has a spot size of w_0, the mode area is $A_{mode} = \pi w_0^2/2$. Therefore, the mode volume is

$$V_{mode} = A_{mode}l = \frac{\pi w_0^2}{2}l = \frac{\pi \times (200 \times 10^{-6})^2}{2} \times 500 \times 10^{-6} \text{ m}^3 = 3.14 \times 10^{-11} \text{ m}^3.$$

The photon energy at the laser wavelength of $\lambda = 1.064$ μm is $h\nu = (1.2398/1.064)$ eV $= 1.165$ eV $= (1.165 \times 1.6 \times 10^{-19})$ J. For $R_1 = 100\%$ and $R_2 = 99.7\%$, the laser has an output only from the second mirror with an output coupling rate of

$$\gamma_{out} = \gamma_{out,2} = -\frac{c}{\bar{n}l}\ln\sqrt{R_2} = -\frac{3 \times 10^8}{1.818 \times 500 \times 10^{-6}}\ln\sqrt{0.997} \text{ s}^{-1} = 4.96 \times 10^8 \text{ s}^{-1}.$$

The saturation output power P_{out}^{sat} of the laser is found from the statement below (15.64) as

$$\begin{aligned}P_{out}^{sat} &= V_{mode}S_{sat}h\nu\gamma_{out} \\ &= 3.14 \times 10^{-11} \times 8.15 \times 10^{17} \times 1.165 \times 1.6 \times 10^{-19} \times 4.96 \times 10^8 \text{ W} \\ &= 2.37 \text{ mW.}\end{aligned}$$

For a pumping ratio of $r = 2.5$, the output power is found by using (15.64) as

$$P_{out} = (r-1)P_{out}^{sat} = (2.5 - 1) \times 2.37 \text{ mW} = 3.56 \text{ mW.}$$

15.4.1 Homogeneous Saturation

In a homogeneous system, all of the atoms that participate in a resonant optical inter-action associated with the two energy levels $|1\rangle$ and $|2\rangle$ are indistinguishable. Their responses to an optical field are characterized by the same resonance frequency ω_{21} and the same relaxation constant γ_{21}. All active atoms have the same absorption spectrum and the same emission spectrum. In such a homogeneous system, the physical mechanisms that contribute to the linewidth of the transition affect all atoms equally. Spectral broadening due to such mechanisms is called *homogeneous broadening*. When absorption saturation or gain saturation occurs in a homogeneous system, all atoms are saturated at the same degree. Consequently, the spectral shape of the system is not affected by *homogeneous saturation*.

In the preceding sections, we have considered for simplicity only homogeneous systems that have a single resonance frequency at ω_{21} and a relaxation constant of γ_{21}. As expressed in (15.12) and (15.13), the spectral characteristics of optical absorption and emission due to a resonant transition in a homogeneously broadened medium are described by the Lorentzian lineshape function of $\chi_{\mathrm{res}}^{(1)\prime\prime}(\omega)$, which has the spectral shape of the imaginary part, $\mathcal{L}''(\omega; \omega_{21}, \gamma_{21})$, of the complex Lorentzian, $\mathcal{L}(\omega; \omega_{21}, \gamma_{21})$, that is given in (15.11) and shown in Fig. 1.3. A *lineshape function*, $\hat{g}(\omega)$, or $\hat{g}(v) = 2\pi\hat{g}(\omega)$, is generally normalized as

$$\int_0^\infty \hat{g}(\omega)\,\mathrm{d}\omega = \int_0^\infty \hat{g}(v)\mathrm{d}v = 1. \tag{15.65}$$

By using this normalization condition, we find the normalized Lorentzian lineshape function for a homogeneously broadened system:

$$\hat{g}_\mathrm{h}(\omega) = \frac{1}{\pi}\frac{\gamma_{21}}{(\omega - \omega_{21})^2 + \gamma_{21}^2} = \frac{1}{2\pi}\frac{\Delta\omega_\mathrm{h}}{(\omega - \omega_{21})^2 + (\Delta\omega_\mathrm{h}/2)^2}, \tag{15.66}$$

which has a full width at half-maximum of $\Delta\omega_\mathrm{h} = 2\gamma_{21}$, or

$$\hat{g}_\mathrm{h}(v) = \frac{1}{2\pi}\frac{\Delta v_\mathrm{h}}{(v - v_{21})^2 + (\Delta v_\mathrm{h}/2)^2}, \tag{15.67}$$

where $\Delta v_\mathrm{h} = \gamma_{21}/\pi$ is the full width at half-maximum of $\hat{g}_\mathrm{h}(v)$. The Lorentzian lineshape is shown in solid curves as the normalized function $\hat{g}_\mathrm{h}(v)$ in Fig. 15.7(a) and with a normalized peak value of unity in Fig. 15.7(b).

In a homogeneously broadened laser, all modes that occupy the same spatial gain region compete for the gain from the population inversion in the same group of active atoms. The mode that has the largest gain reaches the threshold first and starts oscillating. Because the optical gain of this mode is clamped at the threshold value, the entire gain curve supported by this group of atoms saturates. Because this oscillating mode is normally the one that has a longitudinal mode frequency that is closest to the gain peak and a transverse mode pattern that has the lowest loss,

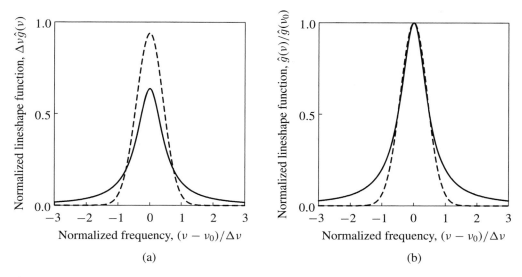

Figure 15.7 Normalized Lorentzian (solid curves) and Gaussian (dashed curves) lineshape functions of the same full width at half-maximum with (a) a normalized area as defined in (15.65) and (b) a normalized peak value. For the Lorentzian lineshape, $v_0 = v_{21}$ and $\Delta v = \Delta v_h$. For the Gaussian lineshape, $\Delta v = \Delta v_{inh}$.

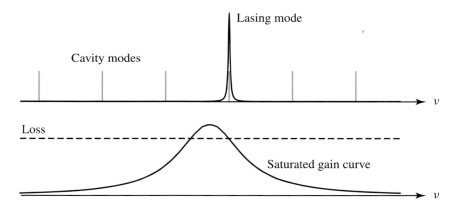

Figure 15.8 Homogeneous gain saturation. In a homogeneously broadened laser, only one longitudinal mode that has a frequency closest to the gain peak oscillates. The entire gain curve is saturated such that the gain at the single lasing frequency remains at the loss level.

the gain curve is saturated in such a manner that its value at this longitudinal mode frequency is clamped at the threshold value of the transverse mode that has the lowest threshold gain coefficient among all transverse modes. If the gain peak does not happen to coincide with this mode frequency, it still lies above the threshold when the gain curve is saturated, as shown in Fig. 15.8. Nevertheless, all other longitudinal modes belonging to this transverse mode have frequencies away from the gain peak. Therefore, even with increased pumping, they do not have a sufficient gain to reach the threshold because the entire gain curve shared by these modes is saturated, as illustrated in Fig. 15.8. Because the gain curve is saturated below their threshold levels, other transverse modes that are supported solely by this group of saturated, homogeneously broadened atoms do not have the opportunity to oscillate, either. Nevertheless, as different transverse modes have different spatial field distributions, a high-order transverse mode may draw its gain from a spatial gain region outside the region that is saturated by a low-order mode. Therefore, when the pumping level is increased, a high-order transverse mode may still reach its relatively high threshold for oscillation after a low-order transverse mode of a low threshold already oscillates. Consequently, for a homogeneously broadened CW laser in steady-state oscillation, only one among all of the longitudinal modes belonging to the same transverse mode can reach the threshold to oscillate, but it is possible for more than one transverse mode to simultaneously oscillate at a high pumping level. Note that this conclusion does not hold true for a pulsed laser. It is possible for multiple longitudinal modes all belonging to the same transverse mode to simultaneously oscillate in a pulsed laser even when the gain medium is homogeneously broadened.

15.4.2 Inhomogeneous Saturation

A resonant transition can be further broadened by inhomogeneous broadening if certain physical mechanisms exist that do not affect all active atoms equally, causing energy levels $|1\rangle$ and/or $|2\rangle$ to shift differently among different groups of atoms. The resulting inhomogeneous shifts of the resonance frequency contribute to additional broadening of the transition spectrum on top of the original homogeneous broadening. Spectral broadening due to such inhomogeneous mechanisms is called *inhomogeneous broadening*. When absorption saturation or gain saturation occurs in an inhomogeneous system, different groups of active atoms saturate differently. Consequently, the spectral shape of the system is changed by *inhomogeneous saturation*.

To account for the inhomogeneous shifts of the resonance frequency, we use v_k to represent the shifted resonance frequency of a homogeneous group k that has a normalized Lorentzian lineshape function of $\hat{g}_h(v, v_k)$ instead of $\hat{g}_h(v, v_{21})$. The distribution of the active atoms in the system is described by a probability density function of $p(v_k)$ that has the properties:

$$p(v_k) \geq 0 \quad \text{and} \quad \int_0^\infty p(v_k)\mathrm{d}v_k = 1. \tag{15.68}$$

The probability that the resonance frequency of a given atom falls in the range between v_k and $v_k + \mathrm{d}v_k$ is $p(v_k)\mathrm{d}v_k$. Then, the overall spectral lineshape of the inhomogeneously broadened transition is

$$\hat{g}_{\mathrm{inh}}(v) = \int_0^\infty p(v_k)\hat{g}_h(v, v_k)\mathrm{d}v_k. \tag{15.69}$$

The overall lineshape function that is obtained from (15.69) depends on the degree of inhomogeneous broadening in comparison to the homogeneous broadening of the atoms. Mathematically, it depends on the spread of the distribution of the probability density function $p(v_k)$ in comparison to the homogeneous linewidth Δv_h.

One possibility for inhomogeneous broadening is the existence of different isotopes, which have slightly different resonance frequencies for a given resonant transition. Another mechanism for inhomogeneous broadening is the Doppler effect in a gaseous medium at a low pressure. Another very common cause is the random distribution of active impurity atoms doped in a solid host. The inhomogeneous frequency shifts caused by these mechanisms are usually randomly distributed, resulting in a Gaussian functional distribution for $p(v_k)$. In an extremely inhomogeneously broadened system, the spread of this distribution dominates the homogeneous linewidth. Then, the transition is characterized by a normalized Gaussian lineshape function:

$$\hat{g}_{inh}(v) = \frac{2(\ln 2)^{1/2}}{\pi^{1/2}\Delta v_{inh}} \exp\left[-4\ln 2 \frac{(v-v_0)^2}{\Delta v_{inh}^2}\right],$$ (15.70)

where v_0 is the center frequency and Δv_{inh} is the full width at half-maximum of the inhomogeneously broadened spectral distribution. In terms of the angular frequency, the normalized Gaussian lineshape function is

$$\hat{g}_{inh}(\omega) = \frac{2(\ln 2)^{1/2}}{\pi^{1/2}\Delta\omega_{inh}} \exp\left[-4\ln 2 \frac{(\omega-\omega_0)^2}{\Delta\omega_{inh}^2}\right],$$ (15.71)

where $\omega_0 = 2\pi v_0$ and $\Delta\omega_{inh} = 2\pi\Delta v_{inh}$. This Gaussian lineshape function is shown as the dashed curves in Fig. 15.7 in comparison to the normalized Lorentzian lineshape function of the same full width at half-maximum.

In an inhomogeneously broadened laser, there are different groups of active atoms in the same spatial region of the gain medium. Each group saturates independently. Two modes that occupy the same spatial gain region do not compete for the same group of atoms if the separation of their frequencies is larger than the homogeneous linewidth of each group of atoms. When one longitudinal mode reaches its threshold and oscillates, only the gain coefficient around its frequency within the homogeneous linewidth is saturated, the gain coefficient at other frequencies continues to increase with increased pumping. As the pumping level increases, other longitudinal modes successively reach their respective thresholds and oscillate. As a result, at a sufficiently high pumping level, multiple longitudinal modes belonging to the same transverse mode can simultaneously oscillate. The saturation of the gain coefficient around each of the frequencies of these oscillating modes, but not across the entire gain curve, creates the effect of *spectral hole burning* in the gain curve of an inhomogeneously broadened laser medium, as illustrated in Fig. 15.9. Different transverse modes also saturate independently in an inhomogeneously broadened medium if their longitudinal mode frequencies are sufficiently separated. Therefore, an inhomogeneously broadened laser can also oscillate in multiple transverse modes. The most outstanding characteristics of inhomogeneous saturation as

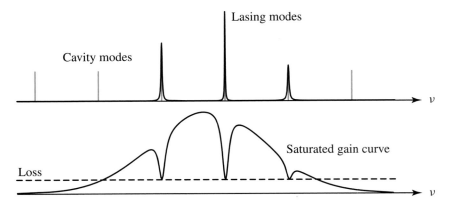

Figure 15.9 Spectral hole burning effect in inhomogeneous gain saturation. In an inhomogeneously broadened laser, multiple longitudinal modes simultaneously oscillate at a sufficiently high pumping level. The gain at each lasing frequency is saturated at the loss level.

compared to homogeneous saturation are spectral hole burning and multi-longitudinal-mode CW oscillation in steady state.

Problem Set

15.1.1 By using (15.12) and (15.13) through (15.10) and (15.11), followed by using (15.5) and (15.14), show that under the rotating-wave approximation, the resonant susceptibility can be expressed as (15.16) and (15.17) in the case of optical absorption and optical gain, respectively.

15.1.2 Show that when $I \ll I_{sat}$ so that the intensity-dependent absorption coefficient and gain coefficient can be expanded as in (15.18), $\chi^{(1)}_{res}$ and $\chi^{(3)}_{res}$ can be defined, and they have the forms of (15.19) and (15.21), respectively.

15.2.1 It is possible to use $\chi^{(3)}$ processes to shorten optical pulses. This objective can be accomplished through the processes that involve either $\chi^{(3)\prime}$ or $\chi^{(3)\prime\prime}$. Because the temporal pulsewidth of a pulse is related to its spectral width through the Fourier transform, it is generally necessary to broaden the pulse spectrum when we start with a transform-limited pulse. However, broadening the spectrum does not automatically result in a shortened pulse before chirping in the pulse is removed.

  (a) Show that a $\chi^{(3)\prime}$ process, such as self-phase modulation, broadens the pulse spectrum but does not by itself shorten the pulsewidth.

  (b) What else is needed for shortening the pulse by using a $\chi^{(3)\prime}$ process?

  (c) Show that a $\chi^{(3)\prime\prime}$ process, such as absorption saturation, can reduce the pulsewidth without the help of other processes.

  (d) Does a $\chi^{(3)\prime\prime}$ process broaden the pulse spectrum? Discuss mathematically.

15.4.1 Single-frequency CW lasers are very useful in many applications. Consider both homogeneously and inhomogeneously broadened lasers. Discuss how a CW laser in steady-state oscillation can be made to oscillate in only one frequency.

References

[1] J. M. Liu, *Photonic Devices*. Cambridge: Cambridge University Press, 2005.

[2] U. Keller, K. J. Weingarten, F. X. Kartner, D. Kopf, B. Braun, I. D. Jung, *et al.*, "Semiconductor saturable absorber mirrors (SESAM's) for femtosecond to nanosecond pulse generation in solid-state lasers," *IEEE Journal of Selected Topics in Quantum Electronics*, vol. 2, pp. 435–453, 1996.

16 Optical Bistability

16.1 CONDITIONS FOR OPTICAL BISTABILITY

Bistability is a phenomenon that has two stable states under one condition. A bistable device has two possible stable output values for one input condition. Because of this binary feature, bistable devices can be used for many digital operations, such as switches, memories, registers, and flip-flops. Bistable electronic circuits and devices have become indispensable components in a wide range of applications that require the storage of binary information. Bistable optical devices can be important for their applications as optical logic, optical memories, and analog-to-digital converters in optical signal processing systems [1]. In addition, they can also be used as optical pulse discriminators and optical power limiters.

The output parameter of a bistable device is a multivalued function of its input parameter. Any system that has such a multivalued characteristic is by definition a nonlinear system. Therefore, optical nonlinearity is required for a bistable optical device. Optical nonlinearity alone is not sufficient for bistability, however. In general, the characteristics of an optical beam that propagates through a nonlinear medium vary nonlinearly but also *monotonically* with the beam intensity. Bistability is not possible with monotonic nonlinearity alone because a monotonic characteristic does not lead to a multivalued dependence of the output on an input parameter. The required nonmonotonic characteristics for optical bistability can be made possible only with proper feedback.

The necessary conditions for optical bistability are optical nonlinearity and positive feedback. Depending on whether the optical nonlinearity that is responsible for the bistable function comes from the real or the imaginary part of a nonlinear susceptibility, a bistable optical device can be classified as either *dispersive* or *absorptive*. The dispersive type is also known as the *refractive* type. In some devices this distinction is not clear, however, because both dispersive and absorptive nonlinear mechanisms may be present. Depending on the type of feedback, a bistable optical device can also be classified as either *intrinsic* or *hybrid*. For an intrinsic bistable device, both the interaction and the feedback are optical. For a hybrid bistable device, electrical feedback is used to modify the optical interaction, thereby creating an artificial optical nonlinearity. Optical bistability was first proposed as a bistable injection laser by Lasher in 1964 [2]. Passive optical bistability was first proposed and observed in an absorptive medium by Szöke *et al.* in 1969 [3]. Optical bistability of the dispersive type was first demonstrated by McCall *et al.* in 1975 [4, 5]. Hybrid optical bistability was first demonstrated by Smith *et al.* in 1977 by using an electro-optic modulator in a Fabry–Perot optical resonator [6].

The optical characteristic that shows bistability can be any feature of an optical field, such as its phase, its polarization, its amplitude (i.e., its intensity), its frequency (i.e., its wavelength), etc. More sophisticated bistability features are possible. For example, a laser that is subjected to optical injection can have two alternative stable states of being locked or unlocked under one injection condition. A multimode laser can have two different stable spatial patterns under one operating condition. The most common bistable optical devices are based on *intensity bistability*. For this reason, we consider intensity bistability in this chapter except for the discussion on laser bistability in Section 16.5, where we will consider polarization bistability of a laser.

Figure 16.1(a) shows a generic characteristic for intensity bistability of a bistable optical device. For each input intensity within the range between $I_{\text{in}}^{\text{down}}$ and $I_{\text{in}}^{\text{up}}$, there are three values for the output intensity. Only the two values that lie on the upper and the lower branches of the curve are stable output values. The one that lies on the middle branch is unstable because the middle branch has a negative slope: $dI_{\text{out}}/dI_{\text{in}} < 0$.

When the input intensity is gradually increased from zero, the output intensity traces the lower branch of the curve until the input intensity reaches the up-transition point at $I_{\text{in}}^{\text{up}}$, where the output makes a sudden jump to the upper branch. Once the system is in a state that lies on the upper branch, it can be brought back to the lower branch only when the input intensity is lowered to the down-transition point at $I_{\text{in}}^{\text{down}}$. If the input intensity is set at a value within the bistable region between $I_{\text{in}}^{\text{down}}$ and $I_{\text{in}}^{\text{up}}$, the output can be in either stable state depending on the history of the system. With a proper external excitation, it can be switched from one of the two stable states to the other. Otherwise, it indefinitely stays in one stable state.

We see from Fig. 16.1(a) that the slope of the characteristic curve for bistability changes sign at both up- and down-transition points. This fact can be exploited to find the condition for bistability and the transition points. Though I_{out} is a multivalued function of I_{in}, I_{in} is a single-valued function of I_{out}. Therefore, it is convenient to express I_{in} as a function of I_{out}, as shown in Fig. 16.1(b). From the curve shown in Fig. 16.1(b), we find that the condition for the existence

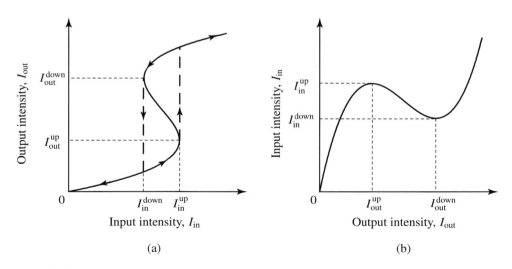

(a)   (b)

Figure 16.1 Generic characteristic for intensity bistability (a) plotted with the output intensity, I_{out}, as a function of the input intensity, I_{in}, and (b) plotted with I_{in} as a function of I_{out}.

of a bistable region is the existence of a region of negative slope, $dI_{in}/dI_{out} < 0$, between two regions of positive slope. Because both I_{in} and I_{out} are real and positive quantities, this condition can be fulfilled only when the relation

$$\frac{dI_{in}}{dI_{out}} = 0 \tag{16.1}$$

has *two nondegenerate real and positive solutions*. These two solutions correspond to the two transition points, $(I_{in}^{up}, I_{out}^{up})$ and $(I_{in}^{down}, I_{out}^{down})$, as can be seen by an examination of Figs. 16.1(a) and (b).

In principle, it is possible to construct bistable optical devices by using a variety of nonlinear effects while properly choosing the device parameters. Any optical nonlinearity can be utilized. In practice, however, the nonlinear optical media that are most commonly used for passive bistable devices are either nonabsorptive Kerr media, for the *dispersive type*, or saturable absorbers, for the *absorptive type*. The required optical feedback can also take many different forms.

A passive bistable optical device of the *intrinsic type* can be constructed by placing a nonlinear optical medium inside an optical cavity. In principle, any optical cavity that provides the needed optical feedback is applicable. A simple bistable optical device of the intrinsic type can be formed by placing a dispersive or absorptive nonlinear optical medium inside a Fabry–Perot cavity, or an *etalon*, as shown in Fig. 16.2(a), or inside a ring cavity, as shown in Fig. 16.2 (b). The mirrors of the cavity provide the needed optical feedback to the nonlinear optical

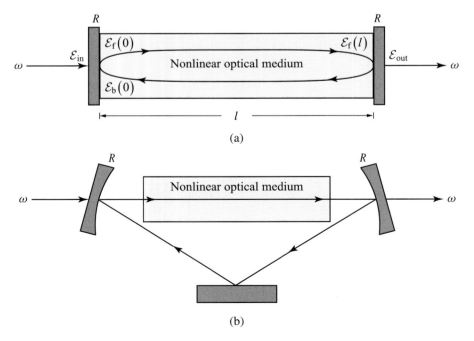

(a)

(b)

Figure 16.2 Intrinsic bistable optical devices using optical feedback in the configurations of (a) a Fabry–Perot cavity and (b) a ring cavity. In principle, the nonlinear optical medium can be any nonlinear optical material.

interaction. The only difference between the two configurations in Fig. 16.2 is that the optical wave in a Fabry–Perot cavity travels through the nonlinear medium twice in each round-trip and forms a standing wave pattern, but the wave in a ring cavity is a traveling wave that travels through the nonlinear medium only once in each round-trip. Otherwise, the basic principle and the characteristics of optical bistability are the same for the two configurations. Other optical feedback configurations for intrinsic bistable optical devices are based on the same concept. The nonlinear medium in Fig. 16.2 can be a second-order nonlinear material, a third-order one, or a high-order one. In principle, any nonlinear optical material can be used to construct a bistable optical device.

In the following three sections, we will consider passive bistable devices that have the configuration of the Fabry–Perot cavity shown in Fig. 16.2(a). For simplicity, we assume that the mirrors of the Fabry–Perot cavity are identical and lossless. The mirrors have a real reflection coefficient of r, which can be either positive or negative, but their transmission coefficient t can be complex because of the finite thickness of the mirrors. The intensity reflectance is $R = r^2$, and the intensity transmittance is $T = |t|^2 = 1 - R$, for $R + T = 1$. We also assume for simplicity that the nonlinear medium fills up the entire space inside the Fabry–Perot cavity that has a length of l. We ignore transverse spatial variations by considering only plane optical waves.

When the steady state is reached, the amplitude of the forward-traveling field, $\mathcal{E}_f$, and the amplitude of the backward-traveling field, $\mathcal{E}_b$, inside the cavity satisfy the relations at the input end, $z = 0$:

$$\mathcal{E}_f(0) = t\mathcal{E}_{in} + r\mathcal{E}_b(0), \tag{16.2}$$

$$\mathcal{E}_b(0) = r\mathcal{E}_f(0)e^{i2kl - \alpha l}, \tag{16.3}$$

where k and α are the propagation constant and the absorption coefficient, respectively, in the medium. At the output end, $z = l$, we have the relation:

$$\mathcal{E}_{out} = t\mathcal{E}_f(l) = t\mathcal{E}_f(0)e^{ikl - \alpha l/2}. \tag{16.4}$$

By using these relations, we find that

$$\mathcal{E}_{out} = \frac{t^2 e^{ikl - \alpha l/2}}{1 - r^2 e^{i2kl - \alpha l}} \mathcal{E}_{in}, \tag{16.5}$$

which gives the relation between the input and the output intensities:

$$I_{out} = \frac{(1 - R)^2 e^{-\alpha l}}{(1 - Re^{-\alpha l})^2 + 4Re^{-\alpha l} \sin^2 kl} I_{in}. \tag{16.6}$$

For intensity bistability, it is necessary that there are two values of the output intensity I_{out} for a given value of the input intensity I_{in} within a certain range of the input intensity. From (16.6), it can be seen that this is possible only when one of the three parameters, k, α, and R, varies nonlinearly with the light intensity. In the case that k varies nonlinearly with the light intensity, the optical bistability is of the dispersive type, which will be discussed in Section 16.2. In the case that α varies nonlinearly with the light intensity, the optical bistability is of the absorptive type, which will be discussed in Section 16.3. In the case that R varies nonlinearly with the light intensity through an electronic feedback, the optical bistability is of the hybrid type, which will be discussed in Section 16.4.

16.2 DISPERSIVE OPTICAL BISTABILITY

We first consider dispersive bistability in a Fabry–Perot cavity filled with a nonlinear medium that has an intensity-dependent index of refraction due to the optical Kerr effect. For simplicity, we ignore the standing wave pattern in the cavity and take the average intracavity intensity as $I_c \approx I_f + I_b \approx 2I_{out}/(1 - R)$. The intensity-dependent index of refraction is

$$n = n_0 + n_2 I_c \approx n_0 + \frac{2n_2 I_{out}}{1 - R}. \tag{16.7}$$

Then, the total phase shift over a round-trip in the cavity can be expressed as

$$2kl = \frac{2n_0 \omega l}{c} + \frac{4n_2 \omega l}{c(1 - R)} I_{out} = 2m\pi + \varphi, \tag{16.8}$$

where m is a properly chosen integer such that

$$\varphi = \varphi_0 + \varphi_2 I_{out} \tag{16.9}$$

for $|\varphi_0| < \pi$ and

$$\varphi_2 = \frac{4n_2 \omega l}{c(1 - R)} = \frac{8\pi n_2 l}{\lambda(1 - R)}. \tag{16.10}$$

Note that φ_0 is a bias phase that can be chosen at will by slightly varying the cavity length l for a given optical frequency ω or by varying the optical frequency for a fixed cavity length.

For the device under consideration, we can rearrange (16.6) as

$$\frac{I_{out}}{I_{in}} = \frac{F^2/F_0^2}{1 + 4(F^2/\pi^2) \sin^2(\varphi/2)}, \tag{16.11}$$

where

$$F = \frac{\pi \sqrt{Re^{-\alpha l}}}{1 - Re^{-\alpha l}} \tag{16.12}$$

is the *finesse* of a generic lossy Fabry–Perot cavity, and

$$F_0 = \frac{\pi \sqrt{R}}{1 - R} \tag{16.13}$$

is the finesse of a lossless Fabry–Perot cavity. The relation described by (16.11) has resonance peaks at $\varphi = 0, \pm 2\pi, \pm 4\pi, \ldots$, each of which has the same characteristics as those of the peak shown in Fig. 16.3.

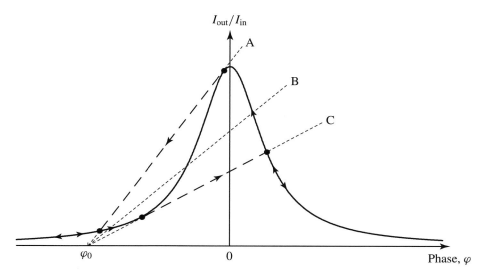

Figure 16.3 Graphic illustration of the bistable characteristics of a dispersive bistable device that has a Fabry–Perot cavity. The peaked curve is the transmission characteristic of the Fabry–Perot cavity at a resonance peak. Other resonance peaks are far away and are not shown. Line A represents the input intensity at $I_{\text{in}}^{\text{down}}$. Line B represents an input intensity in the bistable region. Line C represents the input intensity at $I_{\text{in}}^{\text{up}}$. The arrows indicate the possible directions of movement on the curve and on the lines as the value of the phase varies.

For φ being intensity dependent as given in (16.9), the system can exhibit intensity bistability under proper conditions. A graphic solution can be obtained by expressing (16.9) in the form of

$$\frac{I_{\text{out}}}{I_{\text{in}}} = \frac{\varphi - \varphi_0}{\varphi_2 I_{\text{in}}} \tag{16.14}$$

and plotting it as straight lines for various values of I_{in} to find the intersections between these lines and the curve that represents (16.11). An example of the graphic solution for $\varphi_0 < 0$ and $n_2 > 0$ is shown in Fig. 16.3. It can be seen from this illustration that up-transition corresponds to line C, which is tangent to the curve at its heel, whereas down-transition is described by line A, which is tangent to the curve near its peak.

Analytical solution is possible if $|\varphi| < 1$ so that $\sin^2(\varphi/2) \approx \varphi^2/4$. Then, by combining (16.9) and (16.11), we have

$$\frac{F^2}{F_0^2} I_{\text{in}} = \left[1 + \frac{F^2}{\pi^2}(\varphi_0 + \varphi_2 I_{\text{out}})^2\right] I_{\text{out}}. \tag{16.15}$$

By demanding that the equation $dI_{\text{in}}/dI_{\text{out}} = 0$ have two nondegenerate, real and positive solutions for I_{out}, we find that the conditions for dispersive bistability under the assumption that $|\varphi| < 1$ are

$$\varphi_0 n_2 < 0 \tag{16.16}$$

and

$$|\varphi_0| > \frac{\sqrt{3}\pi}{F}. \tag{16.17}$$

Once the conditions for bistability are satisfied, the up- and down-transition points, as well as the region of bistability, are found by using the two nondegenerate solutions for I_{out}. Figure 16.4 shows the characteristics of this dispersive nonlinear device for a few different values of the characteristic parameter $a = F^2\varphi_0^2/\pi^2$ under the condition that (16.16) is satisfied. It can be seen that bistability exists only when

$$a = \frac{F^2\varphi_0^2}{\pi^2} > 3, \tag{16.18}$$

so that (16.17) is satisfied.

There is a threshold input intensity, $I_{\mathrm{in}}^{\mathrm{th}}$, that is required for bistability to be possible in a bistable optical device of the dispersive type. This threshold input intensity is

$$I_{\mathrm{in}}^{\mathrm{th}} = \frac{\sqrt{3}}{9}\frac{F_0^2}{F^3}\frac{\lambda(1-R)}{|n_2|l}. \tag{16.19}$$

This *bistability threshold input intensity* corresponds to the switching input intensity for $a = 3$, as seen in Fig. 16.4. For $a < 3$, the device never exhibits bistability no matter how high the input intensity is, as can be seen from the curve for $a = 1$ in Fig. 16.4. For $a > 3$, bistability can be observed when the input intensity is in the range:

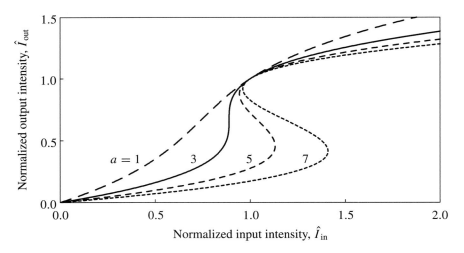

Figure 16.4 Characteristics of a dispersive nonlinear device that consists of an optical Kerr medium in a Fabry–Perot cavity, where $\hat{I}_{\mathrm{in}} = -F^2\varphi_2 I_{\mathrm{in}}/(F_0^2\varphi_0)$ and $\hat{I}_{\mathrm{out}} = -\varphi_2 I_{\mathrm{out}}/\varphi_0$. Bistability exists only when $\varphi_2/\varphi_0 < 0$ and $a = F^2\varphi_0^2/\pi^2 > 3$. The solid curve represents the threshold condition for bistability at $a = 3$. The bistability threshold input intensity is $\hat{I}_{\mathrm{in}}^{\mathrm{th}} = 8/9$ for $a = 3$.

$$I_{in}^{down} = \frac{(a+9)\sqrt{a} - (a-3)\sqrt{a-3}}{12\sqrt{3}} I_{in}^{th} < I_{in} < \frac{(a+9)\sqrt{a} + (a-3)\sqrt{a-3}}{12\sqrt{3}} I_{in}^{th} = I_{in}^{up}.$$

$$(16.20)$$

Both the lower bound and the upper bound of the bistability input intensity given in (16.20) increase when the value of a increases, as can be seen in Fig. 16.4. Meanwhile, the range of the input intensity for bistability also increases with the value of the parameter a as

$$\Delta I_{in}^{bistability} = I_{in}^{up} - I_{in}^{down} = \frac{(a-3)\sqrt{a-3}}{6\sqrt{3}} I_{th}.$$

$$(16.21)$$

As seen in Figs. 16.1 and 16.4, in order to switch the output intensity from the lower branch to the upper branch of the bistable I_{out} versus I_{in} curve, it is necessary to increase the input intensity above I_{in}^{up}. Therefore, if the input intensity is limited to a level below the upper bound given in (16.20), the device can never reach its switching point even when both F and $|\varphi_0|$ are made sufficiently large so that $a > 3$. This situation can be seen in Fig. 16.4 on the curves for $a = 5$ and $a = 7$ in the range of low input intensities below the upper switching point. When the input intensity exceeds this level, the device can enter its bistable regime for a properly chosen value of $a > 3$. However, the minimum input intensity that is required for a given device to operate properly in its bistable states increases as the value of $|\varphi_0|$ and, correspondingly, the value of the parameter a for a given finesse F increase.

Ideally for a dispersive bistable device, the medium should be completely lossless. In practice, however, there are always some absorption or scattering losses in the medium. Such losses in a dispersive device are usually very small, so that $\alpha l \ll 1$. Though the losses are small, they can reduce the finesse of the cavity, thus significantly increasing the threshold input intensity that is required for the operation of the bistable device, as can be easily seen from (16.19) when the value of the finesse F of the optical cavity is reduced by an optical loss.

EXAMPLE 16.1

A vertical cavity InGaAsP bistable device consists of an active InGaAsP layer that has a length of $l = 1$ μm between highly reflective distributed Bragg reflector mirrors of $R = 99\%$. It operates at a wavelength of $\lambda = 1.55$ μm, which is close to the bandgap wavelength, $\lambda_g = 1.49$ μm, of the InGaAsP layer. The nonlinear refractive index at this wavelength is found to be $n_2 = -9 \times 10^{-11}$ m^2 W^{-1}. (a) Find the bistability threshold input intensity of this device, assuming that the medium is lossless. If an optical beam of a circular Gaussian profile is focused to a spot size of $w_0 = 20$ μm on the device, what is the bistability threshold input power? (b) The medium is found to have an absorption coefficient of $\alpha = 1.5 \times 10^4$ m^{-1}. Accounting for this loss, what are the realistic threshold input intensity and the threshold input power of the device?

Solution

(a) For $R = 99\%$, we find that the finesse of the cavity without loss is

$$F_0 = \frac{\pi\sqrt{0.99}}{1 - 0.99} = 312.6.$$

By using (16.19) with $F = F_0$, we find the bistability threshold input intensity:

$$I_{in}^{th} = \frac{\sqrt{3} \times 1.55 \times 10^{-6} \times (1 - 0.99)}{9 \times 312.6 \times 9 \times 10^{-11} \times 1 \times 10^{-6}} \, \text{W m}^{-2} = 106 \, \text{kW m}^{-2}.$$

For a spot size of $w_0 = 20 \, \mu\text{m}$, the bistability threshold input power is

$$P_{in}^{th} = \frac{\pi w_0^2}{2} I_{in}^{th} = \frac{\pi \times (20 \times 10^{-6})^2}{2} \times 106 \times 10^3 \, \text{W} = 66.6 \, \mu\text{W}.$$

(b) With an absorption coefficient of $\alpha = 1.5 \times 10^4 \, \text{m}^{-1}$, we find that $e^{-\alpha l} = 0.9851$ for $l = 1 \, \mu\text{m}$. This amounts to a small single-pass loss of only 1.49%, but the finesse of the cavity is significantly reduced to

$$F = \frac{\pi\sqrt{0.99 \times 0.9851}}{1 - 0.99 \times 0.9851} = 125.3.$$

From (16.19), we see that the threshold input intensity of the bistable device that has such a lossy cavity is increased by a factor of $F_0^3/F^3 = 15.5$ over that obtained above for a lossless cavity. Therefore, the realistic threshold input intensity of this device is $I_{th} = 15.5 \times 106$ kW m^{-2} = 1.64 MW m^{-2}, and the realistic threshold input power is $P_{th} = 15.5 \times 66.6 \, \mu\text{W} = 1.03$ mW. We see from this example that the loss in the cavity has a very significant effect on increasing the threshold of a dispersive bistable device.

16.3 ABSORPTIVE OPTICAL BISTABILITY

For a purely absorptive bistable device, we consider a Fabry–Perot cavity that is filled with a saturable absorber. The absorption coefficient is

$$\alpha = \frac{\alpha_0}{1 + I_c/I_{sat}} = \frac{\alpha_0}{1 + 2I_{out}/I_{sat}(1 - R)}, \tag{16.22}$$

where the relation that $I_c \simeq 2I_{out}/(1 - R)$ for the average intracavity intensity mentioned in the text above (16.7) is used. The real part of the index of refraction is assumed to be independent of the light intensity. Therefore, the round-trip phase shift is a constant. We fix it at $2kl = 2m\pi$, which corresponds to a resonance peak of the Fabry–Perot cavity and can be done by tuning the cavity length at a given optical frequency. For a useful device, the total absorption has to be small in order to reduce the loss. Therefore, we consider only the limit that $\alpha l \ll 1$.

Under the conditions described above, (16.6) becomes

$$\frac{I_{out}}{I_{in}} \approx \frac{(1 - R)^2}{(1 - R + R\alpha l)^2} = \frac{1}{[1 + R\alpha l/(1 - R)]^2}. \tag{16.23}$$

The characteristics of this device are obtained by solving (16.22) and (16.23) together. A graphic solution is not necessary because the analytic solution is relatively simple.

By using (16.22), we can express the relation in (16.23) in the form:

$$I_{in} = \left[1 + \frac{R\alpha_0 l}{1 - R} \frac{1}{1 + 2I_{out}/I_{sat}(1 - R)} \right]^2 I_{out}. \tag{16.24}$$

By demanding that the equation $dI_{in}/dI_{out} = 0$ have two nondegenerate, real and positive solutions for I_{out}, we find the condition for absorptive bistability:

$$C_0 = \frac{R\alpha_0 l}{1 - R} > 8. \tag{16.25}$$

The transition points and the bistable region are found from the two nondegenerate solutions. Figure 16.5 shows the characteristics of this absorptive nonlinear device for a few different values of the characteristic parameter $C_0 = R\alpha_0 l/(1 - R)$. It can be seen that bistability exists only when $C_0 > 8$. No bistability is possible for $C_0 < 8$.

There is a threshold input intensity, I_{in}^{th}, that is required for bistability to be possible in a bistable optical device of the absorptive type. This threshold input intensity is

$$I_{in}^{th} = \frac{27}{2}(1 - R)I_{sat}. \tag{16.26}$$

This bistability threshold input intensity corresponds to the switching input intensity for $C_0 = 8$, as seen in Fig. 16.5. For $C_0 < 8$, the device never shows bistability no matter how high the input intensity is, as can be seen from the curve for $C_0 = 4$ in Fig. 16.5. For $C_0 > 8$, bistability can be observed when the input intensity is in the range:

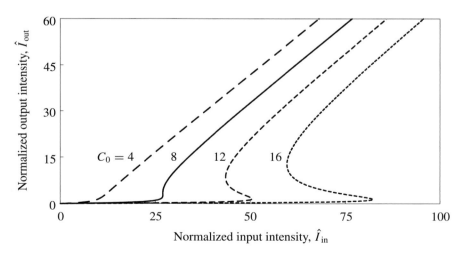

Figure 16.5 Characteristics of an absorptive nonlinear device with a saturable absorber in a Fabry–Perot cavity, where $\hat{I}_{in} = 2I_{in}/[(1 - R)I_{sat}]$ and $\hat{I}_{out} = 2I_{out}/\left[(1 - R)I_{sat}\right]$. Bistability exists only when $C_0 > 8$. The solid curve represents the threshold condition for bistability at $C_0 = 8$. The bistability threshold input intensity is $\hat{I}_{in}^{th} = 27$ for $C_0 = 8$.

$$I_{\text{in}}^{\text{down}} = \frac{C_0^2 + 20C_0 - 8 - (C_0 - 8)\sqrt{C_0^2 - 8C_0}}{216} I_{\text{in}}^{\text{th}} < I_{\text{in}}$$

$$< \frac{C_0^2 + 20C_0 - 8 + (C_0 - 8)\sqrt{C_0^2 - 8C_0}}{216} I_{\text{in}}^{\text{th}} = I_{\text{in}}^{\text{up}}.$$

(16.27)

Both the lower bound and the upper bound of the bistability input intensity given in (16.27) increase when the value of C_0 increases, as can be seen in Fig. 16.5. Meanwhile, the range of the input intensity for bistability also increases with the value of the parameter C_0:

$$\Delta I_{\text{in}}^{\text{bistability}} = I_{\text{in}}^{\text{up}} - I_{\text{in}}^{\text{down}} = \frac{(C_0 - 8)\sqrt{C_0^2 - 8C_0}}{108} I_{\text{in}}^{\text{th}}.$$

(16.28)

If the input intensity is limited to a level below the bistability threshold, the device can never reach its switching point even when $C_0 > 8$. This situation can be seen in Fig. 16.5 on the curves for $C_0 = 12$ and $C_0 = 16$.

16.4 HYBRID OPTICAL BISTABILITY

For intrinsic bistability of both the dispersive and the absorptive types, as discussed in the preceding two sections, both the nonlinearity and the feedback are optical. By comparison, the feedback mechanism of hybrid bistability generally consists of an optoelectronic component that converts the optical output to an electrical feedback signal, whereas the nonlinearity can be optical, electrical, or electro-optic. In general, a hybrid bistable optical device does not necessarily have to be constructed with an optical resonator as is generally done for an intrinsic bistable optical device. Many configurations are possible for hybrid bistability, including a ring cavity such as that shown in Fig. 16.2(b), a nonlinear optical element without a cavity, a single-waveguide mode coupler, or two coupled waveguides without a cavity.

Figure 16.6 shows two general schematics of hybrid bistability as examples. A Fabry–Perot optical resonator that consists of an optical medium is considered in Fig. 16.6. A beam splitter at the output port of the system directs a split beam to a photodetector that generally has a linear response. The photodetector generates a photocurrent, which is converted to a signal voltage, v_s. This voltage is sent into an electronic circuit. With a possible bias voltage, V_0, as a freely adjustable parameter, the voltage at the input end of the electronic circuit is

$$V = V_0 + v_s = V_0 + \mathcal{R}I_{\text{out}},$$

(16.29)

where $\mathcal{R}$ is the effective responsivity of the photodetector in the unit of V W^{-1}m^2. This relation can be expressed as

$$\frac{I_{\text{out}}}{I_{\text{in}}} = \frac{V - V_0}{\mathcal{R}I_{\text{in}}}.$$

(16.30)

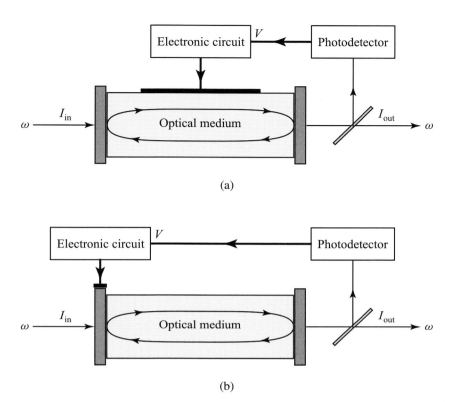

Figure 16.6 Schematics of hybrid optical bistability. The output intensity is detected by a photodetector to generate a voltage or current signal that is proportional to the output intensity. The signal provides the necessary feedback through an electronic circuit. This electronic feedback signal is used to control (a) the property of the optical medium in a Fabry–Perot resonator or (b) the configuration or structure, such as the position or the reflectivity of one mirror, of the resonator.

The electronic circuit converts its input voltage V into a control signal that modulates the optical system. The control signal can be a voltage for electro-optic control. It can be a current for current injection – for example, to a semiconductor to control the optical property of the semiconductor. The control signal can be applied on the optical medium inside the resonator to vary its optical property, as shown in Fig. 16.6(a), or it can be applied on one of the mirrors of the resonator to vary the optical configuration of the resonator, as shown in Fig. 16.6(b). The necessary nonlinearity can be implemented in the electronic circuit, in the optical medium, on the resonator cavity, or a combination of them. In any event, the outcome is that the ratio of the output optical intensity versus the input optical intensity of the system is a nonlinear function of the voltage V:

$$\frac{I_{\text{out}}}{I_{\text{in}}} = f(V). \tag{16.31}$$

For a practical hybrid system, a desirable nonlinear function $f(V)$ can be chosen by design and by proper implementation.

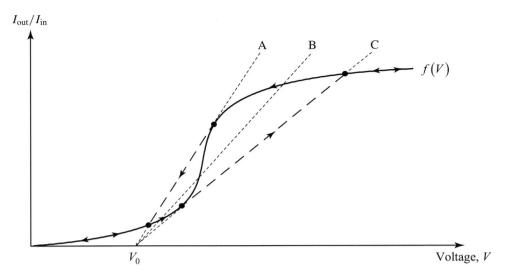

Figure 16.7 Graphic illustration of the bistable characteristics of a hybrid bistable device that is implemented with a feedback voltage signal. The solid curve, $f(V)$, is the ratio of the output intensity versus the input intensity, $I_{\text{out}}/I_{\text{in}}$, as a function of the control voltage, V, which is the output from the feedback electronic circuit shown in Fig. 16.6. This relation provides the needed nonlinearity for hybrid bistability. The straight lines represent the linear response of the photodetector to the output intensity for three different values of the input intensity. V_0 represents a possible bias voltage. Line A represents the input intensity at $I_{\text{in}}^{\text{down}}$. Line B represents an input intensity in the bistable region. Line C represents the input intensity at $I_{\text{in}}^{\text{up}}$. The arrows indicate the possible directions of movement on the curve and on the lines as the value of the voltage varies. A similar concept applies to a hybrid bistable device that has a feedback current signal.

The bistable characteristics of a hybrid bistable device that is implemented with a feedback voltage signal are illustrated in Fig. 16.7. In this figure, the relation given in (16.31) with the nonlinear function $f(V)$ is plotted as the solid curve. The linear relation given in (16.30) is plotted for three different values of the input intensity as three lines: line A for $I_{\text{in}} = I_{\text{in}}^{\text{down}}$, line B for $I_{\text{in}}^{\text{up}} > I_{\text{in}} > I_{\text{in}}^{\text{down}}$, and line C for $I_{\text{in}} = I_{\text{in}}^{\text{up}}$.

16.5 LASER BISTABILITY

The intrinsic bistability, of the dispersive and the absorptive types, and the hybrid bistability that have been discussed in the preceding three sections are all based on passive systems that do not amplify the input optical wave. Optical bistability is also possible with active systems. One example is to use a gain medium for bistability because an optical gain medium is generally a nonlinear optical medium, as discussed in Section 15.3. In particular, *a laser oscillator satisfies the necessary conditions for optical bistability*. Its gain medium serves as the required nonlinear optical medium, and its optical resonator provides the necessary optical feedback. Indeed, many bistable optical phenomena have been observed in lasers. Laser bistability can take many different forms. Besides the most common form of optical bistability that is intensity bistability, laser bistability can appear as phase bistability, wavelength bistability, polarization bistability, or spatial-mode bistability.

Because the output of a laser is defined by the oscillating mode or modes of a laser, the physical mechanism of laser bistability is generally the coupling and competition between two laser modes. In the following, we consider polarization bistability that results from the coupling and competition between two polarization modes. Polarization switching in a semiconductor laser was first observed in 1984 [7], and polarization bistability in a semiconductor laser was first observed in 1985 [8], both by Chen and Liu in InGaAsP/InP lasers. For an edge-emitting semiconductor laser, the two polarization modes are the TE mode, with its electric field transverse to its propagation direction and parallel to the junction plane of the laser, and the TM mode, with its magnetic field transverse to its propagation direction and parallel to the junction plane of the laser. For a vertical cavity surface-emitting laser (VCSEL), the polarization of the fundamental transverse mode is randomly oriented; thus, polarization switching and bistability in a VCSEL can be complicated. When a square waveguide is used to define the transverse modes of a VCSEL, two orthogonally polarized polarizations modes can be defined for clear polarization switching and bistability.

To ensure that there are only two polarization modes involved, we consider a single-transverse mode laser that supports only TE_{00} and TM_{00} modes [7, 8]. Each polarization mode consists of only one spatial mode, but it can consist of multiple longitudinal modes. The total power of a polarization mode is not affected by the internal competition and power partition among its longitudinal modes of the same polarization mode. Because TE and TM modes are orthogonal to each other, there is no coupling between the phases of the two polarization modes. The two polarization modes couple and compete with each other through saturation of the laser gain, which depends only on the photon densities of the two laser modes, but not on the phases of the mode fields. Though we consider polarization bistability between TE_{00} and TM_{00} modes of a semiconductor laser, the general concepts are also applicable to other types of laser bistability that depend only on the photon densities of the two laser modes, but not on the phases of the mode fields.

We first consider the unsaturated and saturated gain parameters of a semiconductor laser that has no competing modes and oscillates in only one laser mode. The unsaturated gain parameter of a semiconductor laser medium varies approximately linearly with the injection carrier density. It can be expressed to first order as [9]

$$g_0 = A(N - N_{tr}), \tag{16.32}$$

where g_0 is the unsaturated gain parameter for the photon growth rate in the unit of s^{-1}, which is related to the unsaturated gain coefficient g_0 as $g_0 = cg_0/\bar{n}$, defined in (15.60); A is the *linear gain constant* of the semiconductor gain medium; N is the density of the *excess carrier density* that is injected into the gain medium; and N_{tr} is the *transparency carrier density*. For $N < N_{tr}$, the semiconductor medium shows net optical absorption with $\alpha_0 > 0$ so that $g_0 < 0$ and $g_0 < 0$; for $N > N_{tr}$, it exhibits net optical gain with $g_0 > 0$ and $g_0 > 0$. The saturated gain parameter g of a laser takes the form of (15.59):

$$g = \frac{g_0}{1 + S/S_{\text{sat}}} = \frac{A(N - N_{\text{tr}})}{1 + S/S_{\text{sat}}}, \tag{16.33}$$

where S is the intracavity photon density and S_{sat} is the saturation photon density defined in (15.61). In the case that $S/S_{\text{sat}} < 1$, the saturated gain parameter can be approximated to first order as

$$g \approx g_0(1 - \xi S) = A(N - N_{\text{tr}})(1 - \xi S), \quad \text{for} \quad \xi S < 1, \tag{16.34}$$

where

$$\xi = \frac{1}{S_{\text{sat}}} \tag{16.35}$$

is the *saturation coefficient* of the laser mode.

In the presence of the coupling and competition between TE and TM modes, the gain parameters of the two polarization modes are different, and gain saturation comes from both modes. The unsaturated gain parameters of the TE and TM modes are, respectively,

$$g_{\text{E0}} = A_{\text{E}}(N - N_{\text{E}}^{\text{tr}}) \tag{16.36}$$

and

$$g_{\text{M0}} = A_{\text{M}}(N - N_{\text{M}}^{\text{tr}}). \tag{16.37}$$

The saturated gain parameters of the TE and TM modes are, respectively,

$$g_{\text{E}} = \frac{g_{\text{E0}}}{1 + S_{\text{E}}/S_{\text{EE}}^{\text{sat}} + S_{\text{M}}/S_{\text{EM}}^{\text{sat}}} = \frac{A_{\text{E}}(N - N_{\text{E}}^{\text{tr}})}{1 + S_{\text{E}}/S_{\text{EE}}^{\text{sat}} + S_{\text{M}}/S_{\text{EM}}^{\text{sat}}} \tag{16.38}$$

and

$$g_{\text{M}} = \frac{g_{\text{M0}}}{1 + S_{\text{M}}/S_{\text{MM}}^{\text{sat}} + S_{\text{E}}/S_{\text{ME}}^{\text{sat}}} = \frac{A_{\text{M}}(N - N_{\text{M}}^{\text{tr}})}{1 + S_{\text{M}}/S_{\text{MM}}^{\text{sat}} + S_{\text{E}}/S_{\text{ME}}^{\text{sat}}}, \tag{16.39}$$

where S_{E} and S_{M} are the intracavity photon densities of the TE and TM modes, respectively; $S_{\text{EE}}^{\text{sat}}$ and $S_{\text{MM}}^{\text{sat}}$ are the *self-saturation photon densities* of the TE and TM modes, respectively; and $S_{\text{EM}}^{\text{sat}}$ and $S_{\text{ME}}^{\text{sat}}$ are the *cross-saturation photon densities* between the two polarization modes. In the case that $S_{\text{E}}/S_{\text{EE}}^{\text{sat}} < 1$, $S_{\text{M}}/S_{\text{EM}}^{\text{sat}} < 1$, $S_{\text{M}}/S_{\text{MM}}^{\text{sat}} < 1$, and $S_{\text{E}}/S_{\text{ME}}^{\text{sat}} < 1$, the saturated gain parameters of the TE and TM modes can be approximated to first order as

$$g_{\text{E}} \approx g_{\text{E0}}(1 - \xi_{\text{EE}}S_{\text{E}} - \xi_{\text{EM}}S_{\text{M}}) = A_{\text{E}}(N - N_{\text{E}}^{\text{tr}})(1 - \xi_{\text{EE}}S_{\text{E}} - \xi_{\text{EM}}S_{\text{M}}) \tag{16.40}$$

and

$$g_{\text{M}} \approx g_{\text{M0}}(1 - \xi_{\text{MM}}S_{\text{M}} - \xi_{\text{ME}}S_{\text{E}}) = A_{\text{M}}(N - N_{\text{M}}^{\text{tr}})(1 - \xi_{\text{MM}}S_{\text{M}} - \xi_{\text{ME}}S_{\text{E}}), \tag{16.41}$$

where

$$\xi_{\text{EE}} = \frac{1}{S_{\text{EE}}^{\text{sat}}} \quad \text{and} \quad \xi_{\text{MM}} = \frac{1}{S_{\text{MM}}^{\text{sat}}} \tag{16.42}$$

are the *self-saturation coefficients* of the TE and TM modes, respectively; and

$$\xi_{EM} = \frac{1}{S_{EM}^{sat}} \quad \text{and} \quad \xi_{ME} = \frac{1}{S_{ME}^{sat}} \tag{16.43}$$

are the *cross-saturation coefficients* between the TE and the TM modes. Each of these saturation parameters is proportional to a corresponding element of the third-order nonlinear resonant susceptibility $\chi_{res}^{(3)}$ of the gain medium [10], as can be seen from (15.21).

The coupling and competition between two orthogonal modes of a semiconductor laser can be modeled with coupled rate equations. In our example the two modes are TE and TM modes, and the coupled equations for the laser take the forms:

$$\frac{dS_E}{dt} = -\gamma_E^c S_E + \Gamma_E g_E S_E, \tag{16.44}$$

$$\frac{dS_M}{dt} = -\gamma_M^c S_M + \Gamma_M g_M S_M, \tag{16.45}$$

$$\frac{dN}{dt} = \frac{J}{ed} - \frac{N}{\tau_s} - g_E S_E - g_M S_M, \tag{16.46}$$

where γ_E^c and γ_M^c are the cavity decay rates for the TE and TM modes, respectively, Γ_E and Γ_M are the gain filling factors, as defined in (15.38), J is the injection current density in the active region of the laser, d is the thickness of the active region, e is the electronic charge, and τ_s is the spontaneous carrier relaxation time. Figure 16.8 shows as an example the cross-sectional structure of an edge-emitting InGaAsP/InP that exhibits polarization bistability under proper conditions [10]. As is seen in the figure, the active region is much smaller than the transverse cross-sectional area of the fundamental TE and TM modes so that both polarizations oscillate in

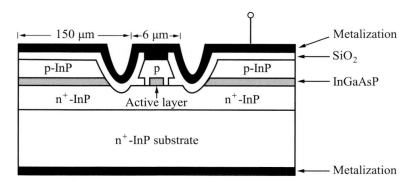

Figure 16.8 Cross-sectional structure of an edge-emitting InGaAsP/InP laser that exhibits polarization bistability under proper operating conditions. For clarity, different layers and regions are not proportionally scaled. (Adapted from B. M. Yu and J. M. Liu, "Polarization-dependent gain, gain nonlinearities, and emission characteristics of internally strained InGaAsP/InP semiconductor lasers," *Journal of Applied Physics*, vol. 69, pp. 7444–7459, 1991.)

single transverse modes, though they may each oscillate in multiple longitudinal modes. For this reason, $\Gamma_E \ll 1$ and $\Gamma_M \ll 1$. In (16.44) and (16.45), the noise terms that account for the fluctuations in the photon densities and in the carrier density that are respectively caused by spontaneous emission and electronic noise are neglected to simplify the equations because they do not have direct consequences on the primary bistability characteristics of the laser.

The condition that $S_E/S_{EE}^{sat} < 1$, $S_M/S_{EM}^{sat} < 1$, $S_M/S_{MM}^{sat} < 1$, and $S_E/S_{ME}^{sat} < 1$ for approximating the saturated gain parameters as given in (16.40) and (16.41) is usually valid for a polarization-bistable laser operating in the typical operating condition. Then, by using (16.40) and (16.41), the three coupled equations given in (16.44)–(16.46) can be approximated as [10, 11]

$$\frac{dS_E}{dt} = -\frac{S_E}{\tau_E^c} + \Gamma_E A_E (N - N_E^{tr})(1 - \xi_{EE}S_E - \xi_{EM}S_M)S_E, \tag{16.47}$$

$$\frac{dS_M}{dt} = -\frac{S_M}{\tau_M^c} + \Gamma_M A_M (N - N_M^{tr})(1 - \xi_{MM}S_M - \xi_{ME}S_E)S_M, \tag{16.48}$$

$$\frac{dN}{dt} = \frac{J}{ed} - \frac{N}{\tau_s} - A_E(N - N_E^{tr})(1 - \xi_{EE}S_E - \xi_{EM}S_M)S_E$$

$$-A_M(N - N_M^{tr})(1 - \xi_{MM}S_M - \xi_{ME}S_E)S_M. \tag{16.49}$$

where $\tau_E^c = 1/\gamma_E^c$ and $\tau_M^c = 1/\gamma_M^c$ are the photon lifetimes, as defined in (15.53). These three coupled nonlinear equations can lead to very sophisticated dynamics because they have three degrees of freedom. They are nonlinear coupled equations that have to be numerically solved. For our purpose here, we only examine the conditions for bistable operation between the TE and the TM modes.

Two general conditions have to be fulfilled for bistable operation:

1. the existence of two stable states for a given set of operating parameters; and
2. the two stable states are mutually exclusive and are switchable.

The first condition is clearly necessary because bistability requires two stable states. However, it is not sufficient because the two stable states might coexist but not be switchable from one to the other. By analyzing the coupled equations given in (16.47) and (16.48) while neglecting the nonlinear gain terms that are characterized by the self- and cross-saturation coefficients, *the condition for the existence of two stable states*, one for the stable oscillation of the TE mode and the other for the stable oscillation of the TM mode, for a given injection current density is found as [11, 12]

$$[1 + \Gamma_E \tau_E^c A_E (N_E^{tr} - N_M^{tr})][1 + \Gamma_M \tau_M^c A_M (N_M^{tr} - N_E^{tr})] \geq 1. \tag{16.50}$$

By analyzing (16.47) and (16.48) without neglecting the nonlinear gain terms, *the condition for switching* between the two stable TE and TM states is found as [10–12]

$$\xi_{EM}\xi_{ME} > \xi_{EE}\xi_{MM}. \tag{16.51}$$

This condition indicates that cross-saturation has to be stronger than self-saturation for the two stable TE and TM states to be mutually exclusive but switchable. It is also found that *the*

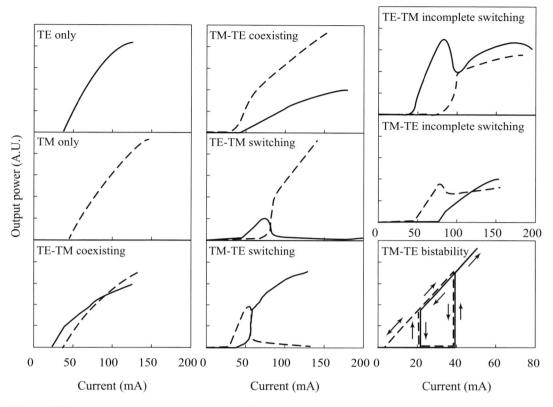

Figure 16.9 Experimentally measured TE (solid curves) and TM (dashed curves) emission characteristics of an InGaAsP/InP laser: TE-only emission, TM-only emission, TE–TM coexisting, TM–TE coexisting, TE–TM switching, TM–TE switching, TE–TM incomplete switching, TM–TE incomplete switching, and TM–TE bistability. (This figure is plotted by using data from B. M. Yu and J. M. Liu, "Polarization-dependent gain, gain nonlinearities, and emission characteristics of internally strained InGaAsP/InP semiconductor lasers," *Journal of Applied Physics*, vol. 69, pp. 7444–7459, 1991.)

condition for the coexistence of TE and TM emission without switching between them is that self-saturation is stronger than cross-saturation [10]:

$$\xi_{EE}\xi_{MM} > \xi_{EM}\xi_{ME}. \tag{16.52}$$

Various emission characteristics of the polarization modes of an InGaAsP/InP laser under different operating conditions are experimentally observed [10] and shown in Fig. 16.9. The experimentally measured TM–TE bistability characteristics are also shown in Fig. 16.9. Each of these experimentally measured emission characteristics, including the bistability, can be obtained by numerical simulation of the coupled equations described in (16.47)–(16.49) with properly chosen laser parameters [10].

Problem Set

16.2.1 A dispersive bistable optical device consists of a Fabry–Perot cavity filled with a purely dispersive optical Kerr medium. Consider its operation under the condition that $|\varphi| < 1$.

(a) Show that the conditions for bistability in such a device are those given in (16.16) and (16.17).

(b) Show that the threshold input intensity, I_{in}^{th}, required for this device to be able to reach its bistability is that given in (16.19).

16.2.2 From the characteristics of a dispersive bistable optical device as shown in Fig. 16.4, we find that when operating in its bistable regime with $a > 3$ as given in (16.18), the device has an up-transition point at an input intensity of I_{in}^{up} and a down-transition point at I_{in}^{down}. The bistability range is $\Delta I_{in}^{bistability} = I_{in}^{up} - I_{in}^{down}$. Consider its operation under the condition that $|\varphi| < 1$ and $a > 3$.

(a) Show that the down-transition point and the up-transition point occur at those given in (16.20) so that the bistability range is that given in (16.21).

(b) Show that at the up-transition point, the output intensity of the device makes the following jump from a low level to a high level:

$$\left(I_{out}^{up}\right)_{low} = \frac{\sqrt{3}}{8}\left(2\sqrt{a} - \sqrt{a-3}\right)\frac{F^2}{F_0^2}I_{in}^{th} \Rightarrow \left(I_{out}^{up}\right)_{high} = \frac{\sqrt{3}}{4}\left(\sqrt{a} + \sqrt{a-3}\right)\frac{F^2}{F_0^2}I_{in}^{th}.$$

(16.53)

(c) Show that at the down-transition point, the output intensity of the device makes the following jump from a high level to a low level:

$$\left(I_{out}^{down}\right)_{high} = \frac{\sqrt{3}}{8}\left(2\sqrt{a} + \sqrt{a-3}\right)\frac{F^2}{F_0^2}I_{in}^{th} \Rightarrow \left(I_{out}^{down}\right)_{low} = \frac{\sqrt{3}}{4}\left(\sqrt{a} - \sqrt{a-3}\right)\frac{F^2}{F_0^2}I_{in}^{th}.$$

(16.54)

16.2.3 It is desired that the dispersive bistable device described in Example 16.1 be operated with a bistability input power range of $\Delta P_{in}^{bistability} = 1$ mW between the up-transition and the down-transition points.

(a) What value of the biased phase φ_0 should be chosen?

(b) What are the required input powers at the two transition points?

(c) What are the output powers at the two transition points?

16.2.4 Find the threshold intensity for the dispersive bistable device described in Example 16.1 if the reflectivities of both mirrors are increased to $R = 99.5\%$. What is the threshold power if the Gaussian input beam is still focused to a spot size of $w_0 = 20$ μm? Compare the results with those found in Example 16.1.

16.3.1 An absorptive bistable optical device consists of a Fabry–Perot cavity filled with a saturable absorber. Consider its operation under the condition that $2kl = 2m\pi$ and $al \ll 1$.

(a) Show that the condition for bistability in such a device is $C_0 > 8$, as given in (16.25).

(b) Show that the threshold input intensity, I_{in}^{th}, required for this device to be able to reach its bistability is that given in (16.26).

16.3.2 From the characteristics of an absorptive bistable optical device, as shown in Fig. 16.5, we find that when operating in its bistable regime with $C_0 > 8$ as given in (16.25), the device has an up-transition point at an input intensity of I_{in}^{up} and a down-transition point at I_{in}^{down}. The bistability range is $\Delta I_{in}^{bistability} = I_{in}^{up} - I_{in}^{down}$. Consider its operation under the condition that $C_0 > 8$.

(a) Show that the down-transition point and the up-transition point occur at those given in (16.27) so that the bistability range is that given in (16.28).

(b) Show that at the up-transition point, the output intensity of the device makes the following jump from a low level to a high level:

$$\left(I_{out}^{up}\right)_{low} = \frac{C_0 - 2 - \sqrt{C_0^2 - 8C_0}}{54} I_{in}^{th} \Rightarrow \left(I_{out}^{up}\right)_{high} = \frac{C_0^2 - 4C_0 - 8 + C_0\sqrt{C_0^2 - 8C_0}}{216} I_{in}^{th}.$$

(16.55)

(c) Show that at the down-transition point, the output intensity of the device makes the following jump from a high level to a low level:

$$\left(I_{out}^{down}\right)_{high} = \frac{C_0 - 2 + \sqrt{C_0^2 - 8C_0}}{54} I_{in}^{th} \Rightarrow \left(I_{out}^{down}\right)_{low} = \frac{C_0^2 - 4C_0 - 8 - C_0\sqrt{C_0^2 - 8C_0}}{216} I_{in}^{th}.$$

(16.56)

16.3.3 An absorptive bistable device has the structure of a vertical cavity that consists of a doped GaAs semiconductor saturable absorber layer between two symmetric GaAs/AlGaAs distributed Bragg reflector mirrors that have the same reflectivity, R. The unsaturated absorption coefficient is $\alpha_0 = 8 \times 10^4 \, m^{-1}$ at the wavelength of $\lambda = 850$ nm.

(a) If the length of the absorber layer is $l = 0.5 \, \mu m$, what is the required reflectivity R for bistability?

(b) If the reflectivity of the distributed Bragg reflector mirrors is limited to $R = 99\%$, what should the length of the absorber layer be in order for the device to function bistably?

References

[1] J. M. Liu and Y. C. Chen, "Digital optical signal processing with polarization-bistable semiconductor lasers," *IEEE Journal of Quantum Electronics*, vol. 21, pp. 298–306, 1985.

[2] G. J. Lasher, "Analysis of a proposed bistable injection laser," *Solid-State Electronics*, vol. 7, pp. 707–716, 1964.

[3] A. Szöke, V. Daneu, J. Goldhar, and N. A. Kurnit, "Bistable optical element and its applications," *Applied Physics Letters*, vol. 15, pp. 376–379, 1969.

[4] S. L. McCall, H. M. Gibbs, G. G. Churchill, and T. N. C. Venkatesan, "Optical transistor and bistability," *Bulletin of American Physical Society*, vol. 20, p. 636, 1975.

[5] H. M. Gibbs, S. L. McCall, and T. N. C. Venkatesan, "Differential gain and bistability using a sodium-filled Fabry–Perot interferometer," *Physical Review Letters*, vol. 36, pp. 1135–1138, 1976.

[6] P. W. Smith and E. H. Turner, "A bistable Fabry-Perot resonator," *Applied Physics Letters*, vol. 30, pp. 280–281, 1977.

[7] Y. C. Chen and J. M. Liu, "Direct polarization switching in semiconductor lasers," *Applied Physics Letters*, vol. 45, pp. 604–606, 1984.

[8] Y. C. Chen and J. M. Liu, "Polarization bistability in semiconductor lasers," *Applied Physics Letters*, vol. 46, pp. 16–18, 1985.

[9] J. M. Liu, *Photonic Devices*. Cambridge: Cambridge University Press, 2005.

[10] B. M. Yu and J. M. Liu, "Polarization-dependent gain, gain nonlinearities, and emission characteristics of internally strained InGaAsP/InP semiconductor lasers," *Journal of Applied Physics*, vol. 69, pp. 7444–7459, 1991.

[11] Y. C. Chen and J. M. Liu, "Polarization bistability in semiconductor laser: rate equation analysis," *Applied Physics Letters*, vol. 50, pp. 1406–1408, 1987.

[12] W. E. Lamb, "Theory of an optical maser," *Physical Review*, vol. 134, pp. A1429–A1450, 1964.

17 Generation of Laser Pulses

17.1 EQUATIONS AND PARAMETERS OF PULSED LASERS

Many techniques have been developed for the generation of laser pulses over a wide range of pulsewidths from the order of milliseconds to femtoseconds. All of them share a common feature: They all utilize some form of *optical nonlinearity* that is coupled to the *dynamics of a laser*. Therefore, the generation of a laser pulse is inherently a nonlinear optical process, though the subject is not generally discussed in the context of nonlinear optics but is considered as a subject of laser physics or laser engineering. In this chapter, we discuss the basic concepts of the primary techniques for the generation of laser pulses.

The optical nonlinearity needed for a pulsed laser can take many different forms. The most basic is the intrinsic gain saturation of the laser gain medium. Additional nonlinear optical elements are usually used. The most common nonlinear optical elements are saturable absorbers and nonlinear refractive components such as Kerr lenses. Hybrid components that are actively controlled are also used, such as electro-optic modulators and acousto-optic modulators. In principle, any device that can prompt a desired nonlinear response from the laser can be utilized to generate laser pulses through proper design and implementation.

Laser pulses are generated either individually as independent pulses, through the *transient dynamics* of a laser, or regeneratively as a train of pulses, through the steady-state *regenerative dynamics* of a laser. The generation of individual pulses through the techniques of gain switching and Q switching is based on transient dynamics. By contrast, mode locking, which generates a train of pulses, is based on regenerative dynamics.

The requirements and the optimum conditions for the operation of a gain-switched or Q-switched laser that is transiently pulsed are very different from those for the operation of a mode-locked laser that is regeneratively pulsed. The most basic difference is that a *transiently pulsed laser* is a lumped device that generates a pulse longer than the cavity round-trip time through transient dynamics, whereas a *regeneratively pulsed laser* is a distributed device that generates a train of repetitive pulses of a pulsewidth smaller than the cavity round-trip time through steady-state dynamics. This difference is illustrated in Fig. 17.1. In a transiently pulsed laser, the photon energy is distributed throughout the laser cavity, and a pulse is generated through fast temporal evolution of this distributed energy. As a result, the laser can be modeled as a lumped device. In a regeneratively pulsed laser, however, the photon energy is concentrated in a short spatial span that circulates in the cavity. Therefore, a mode-locked laser cannot be modeled as a lumped device. In general, a pulse that is generated by the transient technique of gain switching or Q switching cannot be shorter than the cavity round-trip time. Therefore, it has a spatial span longer than the cavity length. By contrast, a pulse that is generated by the

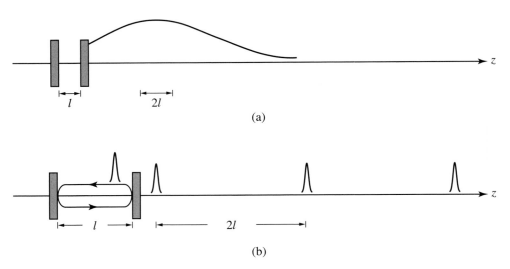

Figure 17.1 Comparison between (a) a transiently pulsed laser such as a gain-switched or Q-switched laser and (b) a regeneratively pulsed laser such as a mode-locked laser.

regenerative approach of mode locking always has a spatial span much shorter than the cavity length and a pulsewidth significantly smaller than the cavity round-trip time.

The pulsing dynamics of a laser are determined by many factors. The properties of the laser gain medium and the structural parameters of the laser cavity determine the basic forms and parameters of the equations of motion of the laser. Additional control of the laser, such as active control through an electro-optic or acousto-optic component or passive control through a saturable absorber, further modifies the laser equations.

17.1.1 Laser Gain Medium

The major characteristics of a laser gain medium are defined by the resonant susceptibility that was described in Sections 1.7 and 1.8 for the resonant interaction between the laser field and two energy levels, $|1\rangle$ and $|2\rangle$, such that the laser frequency is in resonance with the transition frequency: $\omega \approx \omega_{21}$. Figure 17.2 shows the relevant parameters of the two resonant energy levels. These parameters were already defined in Section 1.7. The *phase relaxation rate* γ_{21} characterizes the relaxation of the resonant polarization $\mathbf{P}_{res}$, as seen in (1.158); it defines a *phase relaxation time* of γ_{21}^{-1}. The *population relaxation rate* γ_2 of the upper laser level, $|2\rangle$, characterizes the relaxation of the population in the upper energy level, as defined in (1.154); it defines a *population relaxation time* for the upper energy level as $\tau_2 = \gamma_2^{-1}$. Similarly, the lower laser level, $|1\rangle$, has a population relaxation rate of γ_1 and a corresponding population relaxation time of $\tau_1 = \gamma_1^{-1}$. The population relaxation times τ_2 and τ_1 are commonly known as the *fluorescence lifetimes* of levels $|2\rangle$ and $|1\rangle$, respectively.

In general, many energy levels are involved in a laser action. However, laser gain media can be generally categorized into three basic systems: the *quasi-two-level system*, the *three-level system*, and the *four-level system*. The most significant difference among different systems is the

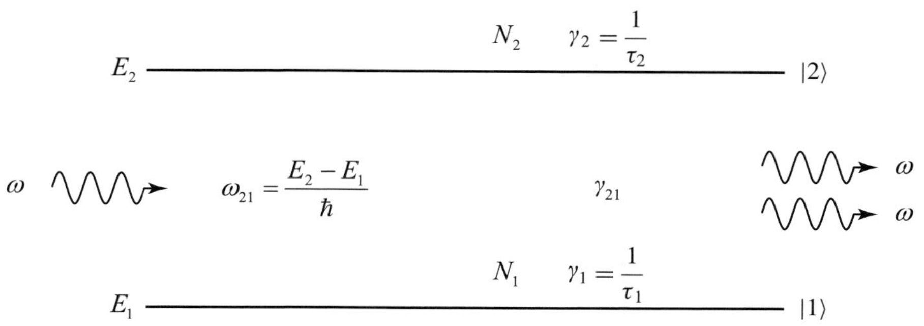

$$E_2 \quad\rule{0pt}{0pt}\hspace{2cm} N_2 \quad \gamma_2 = \frac{1}{\tau_2} \hspace{3cm} |2\rangle$$

$$\omega \quad \rightsquigarrow \qquad \omega_{21} = \frac{E_2 - E_1}{\hbar} \qquad\qquad \gamma_{21} \qquad\qquad \rightsquigarrow \omega$$
$$\rightsquigarrow \omega$$

$$E_1 \quad\rule{0pt}{0pt}\hspace{2cm} N_1 \quad \gamma_1 = \frac{1}{\tau_1} \hspace{3cm} |1\rangle$$

Figure 17.2 Parameters of the two resonant energy levels for laser action.

efficiency of pumping, which is characterized by a *bottleneck factor* β of a value in the range of $2 \geq \beta \geq 1$, depending on the system [1]. An ideal four-level system is the most efficient, with a bottleneck factor of $\beta = 1$, and an ideal three-level system is the least efficient, with a bottleneck factor of $\beta = 2$, whereas a quasi-two-level system has a bottleneck factor between the two values. The saturation of the gain medium and the dynamics of a laser depend on the system of its gain medium mainly through the bottleneck factor in the rate equation for the population inversion N [1]:

$$\frac{dN(t)}{dt} = R(t) - \frac{N(t)}{\tau_2} - \beta v\sigma_e N(t) S(t)$$
$$= R(t) - \frac{N(t)}{\tau_2} - \frac{N(t)}{\tau_2}\frac{S(t)}{S_{\text{sat}}}, \tag{17.1}$$

where R is the effective pumping rate, $v = c/\bar{n}$ is the effective phase velocity of the intracavity laser field, σ_e is the stimulated emission cross-section of the optical transition from the upper laser level to the lower laser level, S is the intracavity photon density, and

$$S_{\text{sat}} = \frac{1}{\beta v \tau_2 \sigma_e} = \frac{\bar{n}}{\beta c \tau_2 \sigma_e} = \frac{\bar{n}}{c \tau_s \sigma_e} \tag{17.2}$$

is the saturation photon density as defined in (15.61), where we can identify the *saturation lifetime of the population inversion* as $\tau_s = \beta \tau_2$. The laser gain coefficient and gain parameter are, respectively,

$$g = \sigma_e N \quad \text{and} \quad g = \frac{c}{\bar{n}} g = v\sigma_e N = \frac{N}{\beta \tau_2 S_{\text{sat}}}. \tag{17.3}$$

17.1.2 Laser Cavity

Besides the time constants γ_{21}^{-1}, τ_2, and τ_1 of the laser gain medium, the cavity round-trip time T, as defined in (15.41), and the photon lifetime τ_c, as defined in (15.53), are two important time parameters of a laser. Both of these time parameters are determined by the parameters of the laser cavity.

The cavity round-trip time is determined by the round-trip optical path length l_{RT} of the cavity. Therefore, it has direct relations with the pulsewidth of a pulse generated by a laser and, in the case of a regeneratively pulsed laser, with the repetition rate of the pulses in a pulse train. As can be seen from Fig. 17.1(a), the pulsewidth of a transiently generated pulse cannot be smaller than the cavity round-trip time; thus, it is limited on the lower side by the relation: $\Delta t_{ps} \geq T$. By contrast, as can be seen from Fig. 17.1(b), a regeneratively generated laser pulse in a train of repetitive pulses cannot have a pulsewidth larger than the cavity round-trip time, and the pulse repetition rate f_{ps} is determined by the cavity round-trip time. Therefore, $\Delta t_{ps} \ll T$ and $f_{ps} = T/n$, where n is an integer; usually, only one pulse circulates inside the laser cavity so that $n = 1$, but more than one pulse that are equally spaced in time can circulate in the cavity so that $n > 1$. The cavity round-trip time determines the frequency spacing of the longitudinal modes through the relation:

$$\Delta \nu_L = \frac{1}{T}. \tag{17.4}$$

The *quality factor Q* of the optical resonator is determined by the resonance mode frequency ω_c and the photon lifetime τ_c, and equivalently the cavity decay rate γ_c, as

$$Q = \omega_c \tau_c = \frac{\omega_c}{\gamma_c}. \tag{17.5}$$

The cavity decay rate also determines the laser threshold through the relation given in (15.57). Therefore, a laser that has a high-Q cavity has a low threshold because it has a large photon lifetime and thus a small cavity decay rate, and a low-Q cavity leads to a high laser threshold.

The intracavity laser field can be expanded as a linear superposition of the normal mode fields of the laser cavity:

$$\mathbf{E}(\mathbf{r}, t) = \sum_{mnq} \mathbf{E}_{mnq}(\mathbf{r}, t) = \sum_{mnq} A_{mnq}(t) \hat{\mathbf{E}}_{mnq}(\mathbf{r}) e^{-i\omega_{mnq}^c t}, \tag{17.6}$$

where $\hat{\mathbf{E}}_{mnq}(\mathbf{r})$ is the normalized mode field distribution of the transverse mode mn and the longitudinal mode q, $A_{mnq}(t)$ is its amplitude, and ω_{mnq}^c is its resonance mode frequency. For each mode, the laser mode field amplitude is governed by the equation:

$$\frac{dA(t)}{dt} = -\frac{\gamma_c}{2} A(t) + i(\omega - \omega_c) A(t) + \frac{\Gamma g(t)}{2} A(t) - i\delta\omega(t) A(t)$$
$$+ \frac{i\omega_c}{2\bar{\epsilon}} \left[P_{NL}(t) + P_{ext}(t) \right] e^{i\omega t} + F_{sp}(t), \tag{17.7}$$

where ω is the oscillating frequency of the laser mode, which can be different from the resonant mode frequency, ω_c, of the laser cavity; Γ is the gain filling factor defined in (15.38); $P_{NL}(t)$ is the nonlinear polarization that is contributed by a possible nonlinear optical component in the laser; $P_{ext}(t)$ is the polarization that is contributed by a possible externally controlled hybrid mechanism, such as an electro-optic or acousto-optic device; and $F_{sp}(t)$ accounts for the stochastic fluctuations of the field caused by spontaneous emission; and $\delta\omega(t)$ and $g(t)$ generally vary with time, defined as

$$\delta\omega(t) = -\Gamma\frac{\epsilon_0\omega_c}{2\bar{\epsilon}}\chi'_{\text{res}}(t) \quad \text{and} \quad g(t) = -\frac{\epsilon_0\omega_c}{\bar{\epsilon}}\chi''_{\text{res}}(t) = v\sigma_e N(t). \tag{17.8}$$

Note that in (17.7) and (17.8) we have omitted the mode indices mnq for simplicity of notation, but it is important to remember that one equation in the form of (17.7) has to be written for each oscillating mode. Furthermore, multiple oscillating modes can be coupled through $g(t)$, $\delta\omega(t)$, $P_{\text{NL}}(t)$, or $P_{\text{ext}}(t)$. As an example, for the two-mode polarization bistability discussed in Section 16.5, the two polarization modes are coupled through $g(t)$ due to gain saturation.

In the case that the phases of the mode fields are relevant to the operation of a pulsed laser, such as in the case of mode locking, field equations of the form given in (17.7) have to be used. In the case that the phases are irrelevant, the field equations can still be used, but equations in terms of the photon densities that eliminate the phases can also be used:

$$\begin{aligned}
\frac{dS(t)}{dt} &= -\gamma_c S(t) + \Gamma g(t)\left[S(t) + S_{\text{sp}}\right] + \left(\frac{dS(t)}{dt}\right)_{\text{NL}} + \left(\frac{dS(t)}{dt}\right)_{\text{ext}} + F_S(t) \\
&= -\gamma_c S(t) + \Gamma\frac{N(t)}{\beta\tau_2}\frac{\left[S(t) + S_{\text{sp}}\right]}{S_{\text{sat}}} + \left(\frac{dS(t)}{dt}\right)_{\text{NL}} + \left(\frac{dS(t)}{dt}\right)_{\text{ext}} + F_S(t),
\end{aligned} \tag{17.9}$$

where S_{sp} is the *spontaneous photon density* that is injected by spontaneous emission into the laser mode, $(dS/dt)_{\text{NL}}$ accounts for the effect of nonlinear polarization P_{NL}, $(dS/dt)_{\text{ext}}$ accounts for the contribution from P_{ext}, and $F_S(t)$ is a stochastic function of *zero mean* that characterizes the random fluctuations of the spontaneous emission photons in the oscillating mode. Note that both S_{sp} and $F_S(t)$ are associated with spontaneous emission, but they account for different effects on the photon density of the laser mode: $F_S(t)$ is a stochastic term that accounts for the fluctuations of the photon density, whereas S_{sp} is the seed for the growth of the coherent photons of the laser mode. Both terms are needed. Again the mode indices mnq is omitted in (17.9) for simplicity of notation, but it is important to remember that one equation in the form of (17.9) has to be written for each oscillating mode.

17.1.3 Nonlinear Mechanisms

Nonlinearity is necessary for the generation of a laser pulse. Except for gain switching, which only utilizes the intrinsic nonlinearity of the gain medium, an additional nonlinear mechanism is generally implemented for a pulsed laser. In principle, there is no limit of the possible nonlinear mechanisms. Common mechanisms can be categorized as passive or active, and as intracavity or extracavity; further categorization can be as optical, electrical, or mechanical, etc. A mechanism is passive if it is not actively controlled by an external signal; it is active if it is controlled by an external signal. The most common passive components are saturable absorbers and optical Kerr elements. The most common active components include electro-optic modulators and acousto-optic modulators. A modification can be made intracavity as part of the laser or extracavity as a feature outside the cavity. Most of the passive and active modifications can be made either intracavity or extracavity. Hybrid modifications that combine passive and active elements, as well as intracavity and extracavity, are often employed. When an additional modification is made to a laser, the coupled laser equations have to be accordingly modified to incorporate the effect of the modification.

For the generation of a laser pulse, the recovery speed of the controlling elements matters because of the short duration of the laser pulse. Active components are generally chosen to be much faster than the dynamical evolution of the pulsed laser. Therefore, active control is generally considered to be instantaneous without considering the delay of response.

A passive component of the refractive type, such as a Kerr lens, is usually a fast component because the optical Kerr effect of the typical dielectric material responds and recovers on the time scale of femtoseconds, which is shorter than most laser pulses. For this reason, a Kerr lens responds to the instantaneous intensity of the laser field that appears at the Kerr lens. This means that the temporally varying Kerr phase as given in (12.74) is valid for a laser pulse that has a temporal intensity profile of $I(t)$:

$$\varphi_K(t) = \frac{\omega_0}{c} n_2 l I(t). \tag{17.10}$$

For a laser pulse that has a transverse spatial profile of $I(r,t)$, where $r = (x^2 + y^2)^{1/2}$, a temporally varying focal length of the Kerr lens can be found by using (12.42):

$$\frac{1}{f_K(t)} = -a\frac{c}{\omega}\frac{d^2\varphi_K(r,t)}{dr^2}\bigg|_{r=0} = -an_2 l\frac{d^2 I(r,t)}{dr^2}\bigg|_{r=0}. \tag{17.11}$$

For an optical pulse that propagates as a circular Gaussian beam that has a beam waist of w at the location of the Kerr lens, the effective focal length of the Kerr lens follows the temporal variation of the pulse intensity, and the pulse power, as

$$f_K(t) = \frac{w^2}{4an_2 l I_0(t)} = \frac{\pi w^4}{8an_2 l P(t)}. \tag{17.12}$$

By contrast, a passive component of the absorptive type, which is commonly a saturable absorber, can be either faster or slower than the dynamical time scale of the laser. As was already mentioned in Section 15.2 a *fast saturable absorber* has a recovery time that is shorter than the duration of the laser pulse, whereas a *slow saturable absorber* has a recovery time that is longer than the duration of the laser pulse.

A saturable absorber is usually a two-level system with the initial population all in the ground state. By taking N_1^{ab} and N_2^{ab} to be the spatially averaged population densities of the absorber in the ground and excited states, respectively, we can describe the population changes of the absorber by using the rate equations:

$$\frac{dN_1^{ab}(t)}{dt} = \frac{N_2^{ab}(t)}{\tau_2^{ab}} - v\sigma_{ab}\left[N_1^{ab}(t) - N_2^{ab}(t)\right]S(t), \tag{17.13}$$

$$\frac{dN_2^{ab}(t)}{dt} = -\frac{N_2^{ab}(t)}{\tau_2^{ab}} + v\sigma_{ab}\left[N_1^{ab}(t) - N_2^{ab}(t)\right]S(t),$$ (17.14)

where σ_{ab} is the absorption cross-section of the absorber and τ_2^{ab} is the population relaxation time constant of the excited state of the absorber. Note that N_1^{ab} and N_2^{ab} are spatially averaged over the absorber volume. The population difference in the saturable absorber can be defined as

$$N_{ab}(t) = N_1^{ab}(t) - N_2^{ab}(t),$$ (17.15)

and the initial population difference is

$$N_i^{ab} = N_{ab}(0) \approx N_1^{ab}(0) = N_1^{ab}(t) + N_2^{ab}(t).$$ (17.16)

By combining the two equations in (17.13) and (17.14), the saturable absorber can then be described with a single rate equation in terms of its population difference:

$$\begin{aligned}\frac{dN_{ab}(t)}{dt} &= -\frac{N_{ab}(t) - N_i^{ab}}{\tau_2^{ab}} - 2v\sigma_{ab}N_{ab}(t)S(t) \\ &= -\frac{N_{ab}(t) - N_i^{ab}}{\tau_2^{ab}} - \frac{N_{ab}(t)}{\tau_2^{ab}}\frac{S(t)}{S_{sat}^{ab}} \\ &= -\frac{N_{ab}(t) - N_i^{ab}}{\tau_2^{ab}} - \frac{N_{ab}(t)}{\tau_2^{ab}}\frac{I(t)}{I_{sat}^{ab}},\end{aligned}$$ (17.17)

where

$$S_{sat}^{ab} = \frac{1}{2v\sigma_{ab}\tau_2^{ab}} \quad \text{and} \quad I_{sat}^{ab} = \frac{hv}{2\sigma_{ab}\tau_2^{ab}}$$ (17.18)

are, respectively, the saturation photon density and the saturation intensity of the saturable absorber.

Fast-Recovering Saturable Absorber

If the recovery time of the saturable absorber is short compared to the duration of a laser pulse, the absorber saturates only in response to the instantaneous photon density of the laser pulse. In this case of fast-recovering saturable absorber, the population difference of the saturable absorber reaches a steady state to follow the temporal variation of the pulse intensity at any moment of the pulse evolution. Then, from (17.17),

$$\frac{dN_{ab}(t)}{dt} = -\frac{N_{ab}(t) - N_i^{ab}}{\tau_2^{ab}} - \frac{N_{ab}(t)}{\tau_2^{ab}}\frac{S(t)}{S_{sat}^{ab}} = -\frac{N_{ab}(t) - N_i^{ab}}{\tau_2^{ab}} - \frac{N_{ab}(t)}{\tau_2^{ab}}\frac{I(t)}{I_{sat}^{ab}} = 0.$$ (17.19)

Thus, the instantaneous population difference of a fast-recovering saturable absorber is

$$N_{ab}(t) = \frac{N_i^{ab}}{1 + S(t)/S_{sat}^{ab}} = \frac{N_i^{ab}}{1 + I(t)/I_{sat}^{ab}}.$$ (17.20)

We see from (17.20) that a fast-recovering saturable absorber responds to the instantaneous photon density, or the instantaneous intensity, of a laser pulse.

Slow-Recovering Saturable Absorber

In case the recovery time of the saturable absorber is long compared to the duration of a laser pulse, the population difference of the saturable absorber does not reach a steady state to instantaneously follow the temporal variation of the pulse. Therefore, $dN_{ab}(t)/dt \neq 0$. However, because τ_2^{ab} is large for a slow-recovering saturable absorber, the relaxation term in (17.17) can be ignored during the entire process of pulse evolution:

$$\frac{dN_{ab}(t)}{dt} \approx -\frac{N_{ab}(t)}{\tau_2^{ab}} \frac{S(t)}{S_{sat}^{ab}} = -\frac{N_{ab}(t)}{\tau_2^{ab}} \frac{I(t)}{I_{sat}^{ab}}. \tag{17.21}$$

This equation can be integrated to obtain the relation:

$$N_{ab}(t) = N_i^{ab} \exp\left(-\frac{1}{\tau_2^{ab}} \int_{-\infty}^{t} \frac{S(t')}{S_{sat}^{ab}} dt'\right) = N_i^{ab} \exp\left(-\frac{1}{\tau_2^{ab}} \int_{-\infty}^{t} \frac{I(t')}{I_{sat}^{ab}} dt'\right). \tag{17.22}$$

Note that the lower bound of the integral in (17.22) is taken as $t = -\infty$ only to symbolically mean that the integral is evaluated from the starting point of the pulse, which may or may not be defined as $t = 0$, depending on the choice of the reference for the time variable. We see from (17.22) that a slow-recovering saturable absorber responds to the cumulative number of photons, or the cumulative energy, of a laser pulse.

17.2 GAIN SWITCHING

Gain switching is the simplest technique for the generation of short laser pulses. This technique was first proposed by Roess [2] in 1966 in terms of controlled oscillator transients for giant pulse shortening. It is a transient pulsing technique that relies on the control of the transient dynamics of a laser. To generate a short pulse through laser transient, it requires a mechanism to start the pulse by rapidly bringing the intracavity photon density to a high level and another mechanism to terminate the pulse as soon as the pulse reaches its peak. For gain switching, the mechanism that enables the fast rise of the laser pulse is fast pulsed pumping to rapidly bring the laser gain parameter, $\Gamma g(t)$, to a maximum level of Γg_{max} that is high above the laser threshold, which is defined by the loss parameter, γ_c, of the laser cavity. The mechanism that terminates the pulse is the intrinsic saturation of the gain by the laser pulse as the intracavity photon density grows to its peak value. The peak *pumping ratio* for a gain-switched laser is

$$r = \frac{\Gamma g_{max}}{\gamma_c} > 1. \tag{17.23}$$

These concepts are illustrated in Fig. 17.3. Different from a CW laser, a gain-switched laser can oscillate in only one mode or in multiple modes regardless of whether the gain medium is

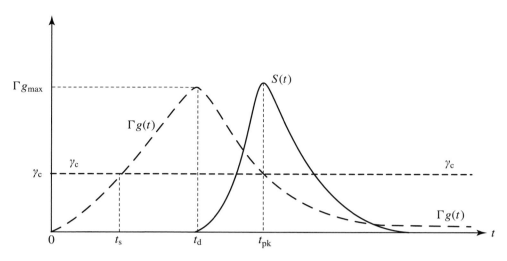

Figure 17.3 Temporal evolutions of the gain parameter and the intracavity photon density of a gain-switched laser. The laser pulse is initiated by a gain parameter that is brought high above the threshold by fast pulsed pumping. It is terminated by the saturation of the laser gain.

homogeneously or inhomogeneously broadened. For simplicity, we ignore the effects of the possible competition or coupling among multiple modes for a gain-switched laser.

For the operation of a gain-switched laser, the gain medium is pumped so fast that the population inversion builds up more rapidly than the growth of the intracavity photon density. The gain is thus raised considerably above the threshold before the laser field starts to build up from the initial noise level in the cavity. The transient dynamic that follows under the condition of an excessively high gain results in the generation of a short, powerful laser pulse.

Because the intracavity laser photon density grows exponentially in time with the net gain, for a gain-switched laser pulse to have a short risetime it is only necessary to create a large excess gain above the threshold before stimulated emission starts to reduce it by depleting the population inversion. This condition requires hard and fast pumping. Therefore, a gain-switched laser is generally pumped with short, powerful pump pulses. In addition, a lifetime, τ_2, of the upper laser level that is long compared to the duration of the pump pulse also helps in building up the excess population inversion. To make sure of a short falltime for the gain-switched pulse, first the gain has to be terminated when the photon density builds up to its peak value, and then the intracavity photons have to be quickly depleted. The first of these two conditions requires short pumping duration, while the second requires a short photon lifetime, τ_c, and thus a large photon decay rate, γ_c. Gain saturation is necessary for the generation of a short gain-switched pulse because the laser gain can be terminated even more rapidly to substantially reduce the pulse falltime. For a gain-switched laser, the degree of gain saturation depends only on the peak value of the intracavity photon density and how fast this peak value is reached, both of which depend on the pumping ratio r, the photon decay rate γ_c, and the saturation photon density S_{sat}.

The dynamical behavior of a gain-switched laser is governed by the coupled equations obtained from (17.1) and (17.9):

$$\frac{dN(t)}{dt} = R(t) - \frac{N(t)}{\tau_2} - \frac{N(t)}{\tau_2}\frac{S(t)}{S_{sat}}, \tag{17.24}$$

$$\frac{dS(t)}{dt} = -\gamma_c S(t) + \Gamma g(t)\left[S(t) + S_{sp}\right]$$
$$= -\gamma_c S(t) + \Gamma \frac{N(t)}{\beta\tau_2}\frac{\left[S(t) + S_{sp}\right]}{S_{sat}}, \tag{17.25}$$

where the two terms $(dS/dt)_{NL}$ and $(dS/dt)_{ext}$ in (17.9) are set to zero for a gain-switched laser, and the stochastic term $F_S(t)$ is ignored. The growth of the gain-switched pulse is initially seeded by a spontaneous photon per laser mode; thus, the seed photon density for each mode can be approximately taken as $S_{sp} = 1/\mathcal{V}_{mode}$, where $\mathcal{V}_{mode}$ is the mode volume. Note that S_{sp} is the seed for the growth of the coherent photons of the laser pulse. Therefore, S_{sp} sets the initial condition for (17.25) as $S(-\infty) = S_{sp}$ if we take the pulse to start at $t = -\infty$, or $S(0) = S_{sp}$ if we take the pulse to start at $t = 0$. The temporally varying gain parameter $\Gamma g(t)$ for a gain-switched laser is determined by the temporally varying pumping rate $R(t)$.

To generate an ultrashort laser pulse by gain switching, it is important to have an extremely small cavity lifetime τ_c. Because the shortest intracavity photon decay time is limited by τ_c, the tail of a gain-switched pulse cannot be shorter than τ_c. Sometimes τ_c can be smaller than the cavity round-trip time T because of a high intracavity loss or a high output-coupling loss. However, unless the small τ_c is caused by a high distributed loss, which is undesirable, the shortest pulse that can be generated by gain switching a laser that has $\tau_c < T$ is limited by T rather than by τ_c. The cause of this limitation is that it takes at least one round-trip to deplete all of the intracavity photons by output coupling through the mirrors. As illustrated in Fig. 17.1, this limitation also applies to Q switching, which is also a transient technique, but it does not apply to mode locking, which is a regenerative technique. From the above discussion, we see that the conditions for the generation of very short laser pulses by gain switching are

1. a large excess population inversion at the onset of laser oscillation;
2. a short photon lifetime τ_c; and
3. sufficient gain saturation after pulse build-up.

Successful gain switching of a laser can be accomplished by choosing

1. a very short and strong pump pulse;
2. a short laser cavity; and
3. a laser medium that has a low saturation intensity and a long fluorescence lifetime τ_2 for the upper laser level in comparison to the pump pulse duration.

For ideal gain switching, the pump pulse duration is so much shorter than τ_2 such that the pumping rate in (17.24) is effectively a delta function:

$$R(t) \rightarrow N_{max}\delta(t). \tag{17.26}$$

Then, the temporal evolution of the gain parameter can be found by integrating (17.24) with $R(t) = N_{max}\delta(t)$ so that (17.25) can be expressed as

$$\frac{dS(t)}{dt} = \Gamma g_{max} S(t) \exp\left\{ -\frac{1}{\tau_2} \int_0^t \left[1 + \frac{S(t')}{S_{sat}} \right] dt' \right\} - \gamma_c S(t)$$

$$= \Gamma g_{max} S(t) \exp\left[-\frac{1}{\tau_2} \int_0^t \frac{S(t')}{S_{sat}} dt' - \frac{t}{\tau_2} \right] - \gamma_c S(t),$$

(17.27)

where $g_{max} = N_{max}/\beta\tau_2 S_{sat}$. This equation has an approximate analytical solution of the form [3]:

$$S(t) = \frac{r \exp\left[-(t - t_{pk})/\tau_c \right]}{\exp\left[-r(t - t_{pk})/\tau_c \right] + r - 1} S_{pk}.$$

(17.28)

The peak photon density is

$$S_{pk} = \frac{\tau_2}{\tau_c} (r - \ln r - 1) S_{sat},$$

(17.29)

which occurs at the time

$$t_{pk} \approx \frac{\tau_c}{r - 1} \ln \frac{S_{pk}}{S_{sp}}$$

(17.30)

after the pump pulse.

The output power of the gain-switched laser pulse can be found as

$$P_{out}(t) = V_{mode} S(t) h\nu\gamma_{out},$$

(17.31)

with a peak power of

$$P_{pk} = \frac{\tau_2}{\tau_c} (r - \ln r - 1) P_{out}^{sat},$$

(17.32)

where γ_{out} is the output-coupling rate of the laser cavity and P_{out}^{sat} is the saturation output power that is defined as $P_{out}^{sat} = V_{mode} S_{sat} h\nu\gamma_{out}$ in (15.64). By comparing P_{pk} given in (17.32) for a gain-switched pulse with P_{out} given in (15.64) for a CW output, we find that the peak power of a gain-switched pulse is enhanced by a factor of approximately τ_2/τ_c in the ideal situation that the laser is pumped with a very short pump pulse and gain saturation is effective. For $1.2 < r < 5$, the full width at half-maximum pulsewidth, Δt_{ps}, of an ideally gain-switched pulse can be quite accurately approximated as

$$\Delta t_{ps} = \frac{2.5}{(r - \ln r - 1)^{1/2}} \tau_c,$$

(17.33)

which is obtained by approximate analytical fitting of ideally gain-switched pulses. The energy of a gain-switched pulse can be approximated as

$$U_{ps} \approx P_{pk}\Delta t_{ps}.$$ (17.34)

17.3 Q SWITCHING

Q switching is the most widely used technique for the generation of high-power giant laser pulses of short durations. The concept of *Q* switching was first proposed in 1961 by Hellwarth [4] and was subsequently demonstrated in 1962 by McClung and Hellwarth [5] in a ruby laser. Similar to gain switching, *Q* switching relies on controlling the transient dynamics of a laser to generate very short laser pulses. It also requires a mechanism to start the laser pulse by rapidly bringing the intracavity photon density to a high level and another mechanism to terminate the pulse as soon as the pulse reaches its peak. In contrast to gain switching, however, it does not require extremely fast pumping. Instead, the mechanism that enables the fast rise of the laser pulse is the rapid switching of the cavity *Q* factor from a low value in an initial *pumping phase* to a high value at the onset of the *lasing phase*. Thus, the cavity loss rate rapidly drops from a high level of γ_{cp} in the pumping phase to a low level of γ_{cl} below the initial gain parameter of Γg_i in the lasing phase, as shown in Fig. 17.4. Indeed, it is possible to pump the gain medium continuously while switching the cavity *Q* factor repetitively to generate a periodic train of

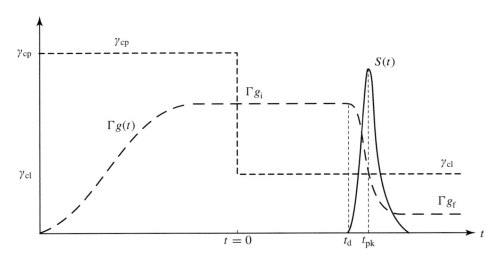

Figure 17.4 Temporal evolutions of the cavity loss rate, the gain parameter, and the intracavity photon density of an ideally *Q*-switched laser. The gain parameter is brought high in the pumping phase by maintaining a high cavity loss rate of γ_{cp} for a low cavity *Q* to prohibit premature laser oscillation, thus preventing gain depletion. The laser pulse is initiated in the lasing phase by switching down the cavity loss rate to γ_{cl} for a high cavity *Q* to lower the laser threshold below the gain parameter for an initial pumping ratio of $r > 1$. It is terminated by the saturation of the gain. This illustration is based on an initial pumping ratio of $r = 2.2$.

Q-switched pulses. Similar to the termination of a gain-switched pulse, a Q-switched pulse is terminated by the intrinsic saturation of the gain by the laser pulse as the intracavity photon density grows to its peak value. These concepts are illustrated in Fig. 17.4. The characteristics of a Q-switched laser pulse are determined by the initial *pumping ratio* at the beginning of the lasing phase:

$$r = \frac{\Gamma g_i}{\gamma_{cl}} > 1. \tag{17.35}$$

As a pulsed laser, a Q-switched laser can oscillate in only one mode or in multiple modes, regardless of whether the gain medium is homogeneously or inhomogeneously broadened. For simplicity, we ignore the effects of possible competition or coupling among multiple modes for a Q-switched laser in this section.

The principle of Q switching is based on delaying the onset of laser oscillation relative to the start of pumping to accumulate a large population inversion. This task is accomplished by reducing the laser cavity Q factor in the pumping phase to prohibit the depletion of the population inversion caused by premature laser oscillation. Upon reaching a large population inversion, the Q factor is rapidly increased to lower the laser threshold, resulting in a large excess gain above the threshold value and a burst of high-intensity short pulse that is driven by the transient dynamics of the laser. Because $Q = \omega_c/\gamma_c$ according to (17.5), modulating the cavity Q factor is equivalent to modulating the cavity loss rate γ_c. Therefore, different from a gain-switched laser, for which the cavity loss rate is kept constant, a Q-switched laser has a temporally varying cavity loss rate $\gamma_c(t)$.

In the pumping phase of a Q-switched laser, population inversion builds up without being depleted by stimulated emission. Clearly, if the pump pulse duration is much shorter than the fluorescence lifetime, τ_2, of the gain medium, gain switching can be very effective; therefore, there is no need for Q switching in this case. This is the condition discussed in the preceding section for gain switching. When the pump pulse is long, the gain grows slowly. However, given sufficient time, the gain can still accumulate to a substantial value if τ_2 is sufficiently long. In this situation, Q switching can be effectively implemented to generate a short and strong laser pulse. Therefore, a large τ_2 is even more desirable for Q switching than for gain switching. For the most efficient utilization of the pump energy, the pump duration should not be too much longer than τ_2, although it does not have to be short. Because of spontaneous relaxation, population inversion, if not depleted by laser oscillation, cannot continue to build up much longer than a period of τ_2. For a repetitively Q-switched laser under continuous pumping, this fact means that the overall efficiency of the laser drops when the pulse repetition rate is below τ_2^{-1}. The major difference between Q switching and gain switching is only in the pumping phase when $\gamma_c(t)$ is kept high in the case of Q switching. In the lasing phase, gain-switched and Q-switched lasers are driven by the same transient laser dynamics that are initiated by the initial excess population inversion. This can be seen by comparing Fig. 17.4 to Fig. 17.3.

It is clear that the conditions discussed in the preceding section for the generation of very short laser pulses by gain switching apply equally well to Q switching. Q switching differs from gain switching only in the technical aspect of how the large initial excess population inversion is achieved. Gain switching relies on a very fast and strong pump pulse to achieve a high peak gain before stimulated emission starts. This condition is not required for Q switching, as the high cavity loss of a Q-switched laser in the pumping phase prohibits the laser from oscillating. Thus, the requirement on the pump source for Q switching is less demanding than that for gain switching. The technical demand of a Q-switched laser is shifted from the pump to the Q switch. In the lasing phase, the requirements for a Q-switched laser are the same as those for a gain-switched laser. To generate a short Q-switched pulse, a small cavity lifetime τ_{cl} for the lasing phase is necessary because the tail of a Q-switched pulse cannot be shorter than τ_{cl}. The shortest pulse that can be generated by a Q-switched laser is still limited by the round-trip time T of its cavity. In order to generate a very short pulse by Q switching, it is therefore desirable to have:

1. an effective Q switch that switches the cavity Q very rapidly from a very low value to a high value at the moment the gain reaches a desired high level;
2. a short laser cavity; and
3. a laser medium that has a low saturation intensity and a long fluorescence lifetime.

Figure 17.5 shows the schematics of Q-switched lasers. The Q-switching element can be placed anywhere in the cavity. In principle, it can be any device that can effectively vary the cavity loss rate. A few common examples are shown in the figure. Depending on the technique that is used to modulate the cavity Q factor, the type of Q switching can be generally categorized as active or passive. For *active Q switching*, the Q switch is controlled by an external signal. Various techniques have been developed for active Q switching, including mechanical modulation, electro-optic modulation, acousto-optic modulation, and magneto-optic modulation. Most of the actively Q-switched lasers use Pockels cells or traveling-wave acousto-optic modulators. For *passive Q switching*, the Q switch is typically a nonlinear optical element that changes the cavity loss by responding directly to the intracavity laser intensity. No externally controlled signal is necessary. With a proper arrangement, any all-optical switch, such as a saturable absorber or a Kerr lens, can function as a passive Q switch. A saturable absorber is most commonly used. The saturable absorber can be either a fast-recovering absorber or a slow-recovering absorber.

17.3.1 Active Q Switching

For active Q switching, the cavity Q factor is switched by an externally controlled signal. Therefore, the cavity loss rate $\gamma_c(t)$ is a function of time that is varied by an external signal from a high value of γ_{cp} in the pumping phase to a low value of γ_{cl} in the lasing phase. The dynamical behavior of an actively Q-switched laser is governed by the coupled equations for gain switching given in (17.24) and (17.25), but with the constant cavity loss rate in (17.25) replaced by a temporally varying loss rate:

$$\frac{dN(t)}{dt} = R(t) - \frac{N(t)}{\tau_2} - \frac{N(t)}{\tau_2}\frac{S(t)}{S_{sat}}, \tag{17.36}$$

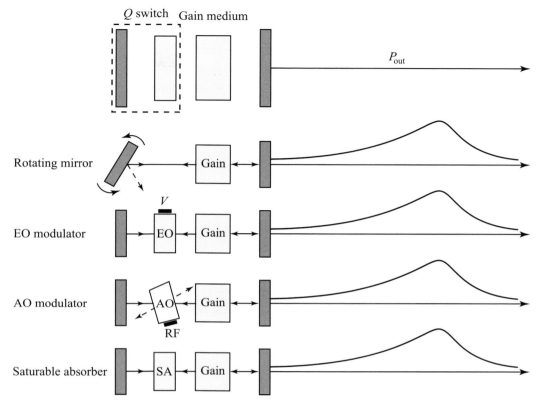

Figure 17.5 Schematics of Q-switched lasers. For active Q switching, the Q switch can be a rotating mirror, an electro-optic modulator, or a traveling-wave acousto-optical modulator. For passive Q switching, a saturable absorber is most commonly used. The saturable absorber can be either a fast-recovering absorber or a slow-recovering absorber.

$$\begin{aligned}
\frac{\mathrm{d}S(t)}{\mathrm{d}t} &= -\gamma_\mathrm{c}(t)S(t) + \Gamma g(t)\Big[S(t) + S_\mathrm{sp}\Big] \\
&= -\gamma_\mathrm{c}(t)S(t) + \Gamma \frac{N(t)}{\beta\tau_2}\frac{\big[S(t) + S_\mathrm{sp}\big]}{S_\mathrm{sat}}.
\end{aligned} \tag{17.37}$$

For ideal Q switching, $\gamma_\mathrm{c}(t)$ is switched abruptly from a constant high level of γ_cp to a constant low level of γ_cl, as shown in Fig. 17.4. As discussed in the preceding section, ideal gain switching is accomplished by pumping with a pulse that is much shorter than τ_2. Ideal Q switching is accomplished by abrupt switching at the onset of the lasing phase within a time duration that is much shorter than τ_2. In practice, because it takes a time delay, t_d shown in Fig. 17.4, after the time of switching into the lasing phase for the Q-switched pulse to build up to a significant level, the condition of ideal Q switching can be approximated by *fast Q switching*, for which the transition from the pumping phase to the lasing phase is completed within a time duration less than the time delay of the pulse build-up. In this ideal, or nearly ideal, situation of fast Q switching, the behavior of the Q-switched laser in the lasing phase is governed by the coupled equations given in (17.36) and (17.37), with $\gamma_c(t) = \gamma_{cl}$, which have the same forms as (17.24) and (17.25) for the gain-switched laser. The characteristics of the Q-switched pulse are completely determined by the initial pumping ratio r at the onset of the lasing phase. Consequently, a laser pulse that is generated by ideal Q switching has the same characteristics as a pulse that is generated by ideal gain switching for the same pumping ratio r. By taking γ_c to be γ_{cl}, the quantitative characteristics expressed in (17.27)–(17.34) for an ideally gain-switched laser pulse apply equally well to an ideally Q-switched laser pulse.

17.3.2 Passive Q Switching

For passive Q switching, the net cavity loss rate is not switched by an external signal but is changed by the intracavity photon density. Because most passively Q-switched lasers use saturable absorbers, in this subsection we only consider passive Q switching with saturable absorbers. The behavior of the laser is governed by three coupled equations, from (17.1) for the population inversion of the gain medium, from (17.17) for the saturable absorber, and from (17.9) for the intracavity photon density:

$$\frac{\mathrm{d}N(t)}{\mathrm{d}t} = R(t) - \frac{N(t)}{\tau_2} - \frac{N(t)}{\tau_2}\frac{S(t)}{S_{sat}}, \tag{17.38}$$

$$\frac{\mathrm{d}N_{ab}(t)}{\mathrm{d}t} = -\frac{N_{ab}(t) - N_i^{ab}}{\tau_2^{ab}} - \frac{N_{ab}(t)}{\tau_2^{ab}}\frac{S(t)}{S_{sat}^{ab}}, \tag{17.39}$$

$$\frac{\mathrm{d}S(t)}{\mathrm{d}t} = -\gamma_c S(t) + \Gamma g(t)\left[S(t) + S_{sp}\right] - \Gamma_{ab}a(t)S(t)$$

$$= -\gamma_c S(t) + \Gamma\frac{N(t)}{\beta\tau_2}\frac{\left[S(t) + S_{sp}\right]}{S_{sat}} - \Gamma_{ab}\frac{N_{ab}(t)}{2\tau_2^{ab}}\frac{S(t)}{S_{sat}^{ab}}, \tag{17.40}$$

where the last term in (17.40) is the $(\mathrm{d}S/\mathrm{d}t)_{NL}$ term in (17.9) that accounts for the effect of the nonlinear polarization P_{NL} of the saturable absorber; Γ_{ab} is the overlap factor of the laser field and the saturable absorber, which is defined in a way similar to (15.38) for Γ; and

$$a = v\sigma_{ab}N_{ab} = \frac{N_{ab}}{2\tau_2^{ab}S_{sat}^{ab}} \tag{17.41}$$

is the *absorption rate* of the saturable absorber. Thus, $\Gamma_{ab}a$ is the intracavity photon loss rate due to the absorption of photons by the saturable absorber. Note that γ_c in (17.40) is a constant. The total loss rate is $\gamma_c + \Gamma_{ab}a(t)$, which varies with time as the intracavity photon density $S(t)$ evolves, thus providing the Q-switching function.

With a saturable absorber in a laser cavity, the laser action is inhibited by the absorption of the absorber when the gain medium is first excited, thus encouraging the build-up of population inversion. When the gain medium is pumped to an initial population inversion such that the gain exceeds the total cavity loss including the absorber absorption, the intracavity photon density can start growing gradually. The nonlinear behavior of the absorber allows the growing photon density to saturate it, making it transparent to start the Q-switching process. In comparison to ideal, or nearly ideal, active Q switching, passive Q switching through the saturation of an absorber does not abruptly switch the cavity loss rate because it takes time for the intracavity photon density to grow after the gain first exceeds the loss.

Initially, before the absorber or the gain medium is saturated, (17.40) can be approximated as

$$\frac{dS(t)}{dt} \approx (\Gamma g_i - \Gamma_{ab} a_i - \gamma_c)S(t), \tag{17.42}$$

where Γg_i is the unsaturated initial gain, $\Gamma_{ab} a_i$ is the unsaturated initial absorption rate, and $\Gamma g_i - \Gamma_{ab} a_i - \gamma_c$ is the net initial growth rate of the intracavity photon density. Because the cavity loss is not actively switched, it is necessary that this net initial growth rate be positive for the intracavity photon density to start growing before the cavity Q is switched by the saturation of the absorber. Otherwise, the passive Q-switching action of the absorber saturation never takes place. To ensure that the absorber saturation has a large effect on the change of the cavity Q, however, it is advantageous to have $\Gamma_{ab} a_i \gg \Gamma g_i - \Gamma_{ab} a_i - \gamma_c$. Thus, *the initial condition for passive Q switching* is

$$1 \gg \frac{\Gamma g_i - \Gamma_{ab} a_i - \gamma_c}{\Gamma_{ab} a_i} > 0. \tag{17.43}$$

This condition implies that $\Gamma g_i - \gamma_c$ is only slightly larger than $\Gamma_{ab} a_i$. Consequently, the initial condition for passive Q switching requires that

$$N_i - N_{th} > \frac{\Gamma_{ab}\sigma_{ab}}{\Gamma\sigma_e} N_i^{ab} \gg N_i - N_{th} - \frac{\Gamma_{ab}\sigma_{ab}}{\Gamma\sigma_e} N_i^{ab}. \tag{17.44}$$

Fast-Recovering Saturable Absorber

A saturable absorber is considered to be fast recovering if its recovery time is short compared to the Q-switched pulse duration. A fast absorber saturates only in response to the instantaneous photon density. In this case, the population difference of the saturable absorber is in a steady state following the temporal pulse variation at any moment of the pulse evolution. Then, we use (17.20) for the instantaneous population difference of the saturable absorber to find the absorption rate of a fast-recovering absorber as

$$a(t) = \frac{a_i}{1 + S(t)/S_{sat}^{ab}}. \tag{17.45}$$

By using (17.45) for the saturable absorber to analyze the coupled equations given in (17.38) and (17.40) for the population inversion and the photon density, we find *the condition for passive Q switching with a fast saturable absorber* in terms of the condition for the effective cross-sectional ratio, ρ, of the absorber to the gain medium:

$$\rho = \frac{2\sigma_{ab}}{\beta\sigma_e} = \frac{S_{sat}}{S_{sat}^{ab}} > \frac{\Gamma g_i}{\Gamma_{ab} a_i} \frac{1}{(\Gamma g_i - \Gamma_{ab} a_i - \gamma_c)\tau_2^{ab}} > \frac{1}{(\Gamma g_i - \Gamma_{ab} a_i - \gamma_c)\tau_2^{ab}}. \tag{17.46}$$

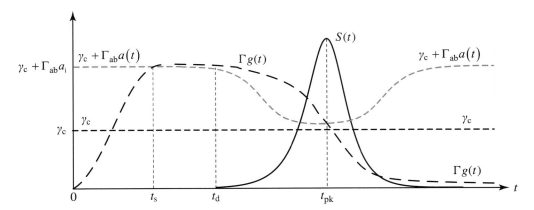

Figure 17.6 Temporal evolutions of the cavity loss rate, the gain parameter, and the intracavity photon density of a passively Q-switched laser with a fast-recovering saturable absorber. The total cavity loss rate recovers as soon as the intracavity photon density starts to drop because the recovery time of the absorber is short compared to the pulse duration. By contrast, the gain parameter does not recover within the pulse duration but stays saturated after the photon density is depleted because the fluorescence lifetime of the gain medium is long compared to the pulse duration.

Because $(\Gamma g_i - \Gamma_{ab}a_i - \gamma_c)\tau_2^{ab} \ll 1$ for the typical fast absorber, the effective transition cross-section of the absorber has to be much larger than that of the gain medium. This condition means that $S_{sat} \gg S_{sat}^{ab}$ to make sure that the absorber is sufficiently saturated for the Q-switched pulse to take off before the gain is saturated.

Figure 17.6 shows the temporal evolutions of the cavity loss rate, the gain parameter, and the intracavity photon density of a passively Q-switched laser with a fast-recovering saturable absorber. The total cavity loss rate recovers as soon as the intracavity photon density starts to drop, because the recovery time of the absorber is short compared to the pulse duration. Compared to a laser pulse generated by fast active Q switching, which has an asymmetric temporal pulse shape with a risetime generally smaller than its falltime, as shown in Fig. 17.4, a laser pulse generated by passive Q switching with a fast saturable absorber has a more symmetric shape with a long rising edge, as can be seen in Fig. 17.6. This temporal pulse shape has to be found by numerically solving the coupled equations for the laser.

Slow-Recovering Saturable Absorber

A saturable absorber is considered to be slow recovering if its recovery time is long compared to the Q-switched pulse duration. Then, we use (17.22) for the temporal variation of the absorber population difference to find the absorption rate of a slow-recovering absorber as

$$a(t) = a_i \exp\left(-\frac{1}{\tau_2^{ab}} \int_0^t \frac{S(t')}{S_{sat}^{ab}} \, dt'\right). \tag{17.47}$$

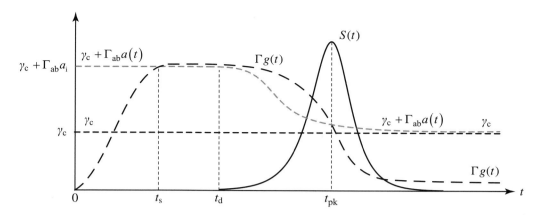

Figure 17.7 Temporal evolutions of the cavity loss rate, the gain parameter, and the intracavity photon density of a passively Q-switched laser with a slow-recovering saturable absorber. Neither the total cavity loss rate nor the gain parameter recovers within the pulse duration. Both of them stay saturated after the photon density is depleted because both the recovery time of the absorber and the fluorescence lifetime of the gain medium are long compared to the pulse duration.

By using (17.47) for the saturable absorber to analyze the coupled equations given in (17.38) and (17.40) for the population inversion and the photon density, we find *the condition for passive Q switching with a slow saturable absorber* in terms of the condition for the effective cross-sectional ratio, ρ, of the absorber to the gain medium:

$$\rho = \frac{2\sigma_{ab}}{\beta\sigma_e} = \frac{S_{sat}}{S_{sat}^{ab}} > \frac{\Gamma g_i}{\Gamma_{ab} a_i} > 1. \tag{17.48}$$

Similar to the condition given in (17.46) for a fast-recovering saturable absorber, this condition also implies that a slow-recovering saturable absorber has to be sufficiently saturated before the gain is saturated to ensure a period of net gain for the Q-switched pulse to quickly take off. By comparison to the condition in (17.46), however, this condition simply requires that the effective absorption cross-section of the slow absorber be larger than the effective stimulated emission cross-section of the gain medium. This difference is caused by the fact that both the slow absorber and the gain medium are saturated in the same manner by the integrated photon density, while the fast absorber is saturated only by the instantaneous photon density.

Figure 17.7 shows the temporal evolutions of the cavity loss rate, the gain parameter, and the intracavity photon density of a passively Q-switched laser with a slow-recovering saturable absorber. Neither the total cavity loss rate nor the gain parameter recovers within the pulse duration. Both of them stay saturated after the photon density is depleted because both the recovery time of the absorber and the fluorescence lifetime of the gain medium are long compared to the pulse duration. Similar to a laser pulse generated by passive Q switching with a fast saturable absorber, a laser pulse generated by passive Q switching with a slow saturable absorber also has a symmetric shape with a long rising edge, as can be seen in Fig. 17.7. This temporal pulse shape also has to be found by numerically solving the coupled equations for the laser.

EXAMPLE 17.1

In this example, we consider the ideal Q-switching or gain-switching operation of the Nd:YAG microchip laser that is described in Example 15.1 and shown in Fig. 15.6. The laser wavelength is $\lambda = 1.064\ \mu\text{m}$. For the gain-switching operation, all of the laser parameters, including those of the cavity and the gain medium, remain the same as those described in Example 15.1, except that the pump is an optical pulse at the pump wavelength of 808 nm. For the Q-switching operation, a Q switch introduces an additional high loss to the laser in the pumping phase, but the laser parameters in the lasing phase are the same as those for the gain-switching operation. A possible Q-switching mechanism is passive Q switching by co-doping the Nd:YAG with Cr^{4+} ions as the saturable absorber. For direct comparison with CW operation, we take the pumping ratio to be $r = 2.5$, as taken in Example 15.1 for a CW saturation output power of $P_{\text{out}}^{\text{sat}} = 2.37\ \text{mW}$. (a) What are the longitudinal mode spacing and the Q factor of this cavity? (b) What are the required conditions for the laser to be nearly ideally Q-switched? (c) Find the peak power, the pulsewidth, and the pulse energy of the ideally Q-switched laser pulse. Compare the peak power of the Q-switched pulse to that of the CW power of $P_{\text{out}} = 3.56\ \text{mW}$. (d) What are the required conditions for the laser to be nearly ideally gain switched? (e) What are the characteristics of the ideally gain-switched pulse?

Solution

(a) From Example 15.1, the cavity round-trip time is $T = 6.06\ \text{ps}$. Therefore, the longitudinal mode spacing is found by using (17.4) as

$$\Delta v_{\text{L}} = \frac{1}{T} = \frac{1}{6.06 \times 10^{-12}}\ \text{Hz} = 165\ \text{GHz}.$$

The cavity decay rate is found in Example 15.1 to be $\gamma_{\text{c}} = 5.78 \times 10^8\ \text{s}^{-1}$. Thus, the cavity Q is found by using (17.5):

$$Q = \frac{\omega_{\text{c}}}{\gamma_{\text{c}}} = \frac{2\pi c}{\lambda \gamma_{\text{c}}} = \frac{2\pi \times 3 \times 10^8}{1.064 \times 10^{-6} \times 5.78 \times 10^8} = 3.065 \times 10^6.$$

(b) Because the laser parameters in the lasing phase are the same as those of the CW laser described in Example 15.1, we have $\gamma_{\text{cl}} = \gamma_{\text{c}} = 5.78 \times 10^8\ \text{s}^{-1}$ and $\tau_{\text{cl}} = \tau_{\text{c}} = 1.73\ \text{ns}$. Two of the three conditions for ideal Q switching, namely a short laser cavity and a gain medium with a long fluorescence lifetime and a low saturation intensity, are already met by this laser. Therefore, the only requirement that has to be considered is an effective Q switch that switches the cavity from a high loss of γ_{cp} to a low loss of γ_{cl}. First, γ_{cp} has to be larger than Γg_{i}, which for a pumping ratio of $r = 2.5$ is $\Gamma g_{\text{i}} = r\gamma_{\text{cl}} = 2.5\gamma_{\text{cl}}$. Thus, an effective Q switch has to keep a high cavity decay rate at $\gamma_{\text{cp}} > 2.5\gamma_{\text{cl}} = 1.445 \times 10^9\ \text{s}^{-1}$ in the pumping phase. Next, the Q switch has to switch fast enough. Quantitatively, the Q switch has to switch the cavity loss from the high value of γ_{cp} to the low value of γ_{cl} within a time interval of Δt_{QS} that is shorter than the pulse delay time t_d, as shown in Fig. 17.4, for the process to qualify as fast Q switching.

The pulse delay time can be estimated by considering the fact that the pulse grows exponentially from a seed of spontaneous emission to the saturation photon density with a rate of $\Gamma g_i - \gamma_{cl}$. The saturation photon density is S_{sat}, which has a value of $S_{sat} = 8.15 \times 10^{17}\,\text{m}^{-3}$, found in Example 15.1 for this laser. The seed of spontaneous emission is one photon per mode, which translates into a spontaneous photon density of $1/\mathcal{V}_{mode}$, with $\mathcal{V}_{mode} = 3.14 \times 10^{-11}\,\text{m}^3$, also found in Example 15.1 for this laser. If we take t_d to be the time it takes the photon density of the oscillating laser mode to grow exponentially with a rate of $\Gamma g_i - \gamma_{cl}$ from $1/\mathcal{V}_{mode}$ to S_{sat}, then t_d can be found as

$$
\begin{aligned}
t_d &= \frac{1}{\Gamma g_i - \gamma_{cl}} \ln \frac{S_{sat}}{1/\mathcal{V}_{mode}} \\
&= \frac{\tau_{cl}}{r-1} \ln\left(S_{sat}\mathcal{V}_{mode}\right) \\
&= \frac{1.73}{2.5-1} \ln\left(8.15 \times 10^{17} \times 3.14 \times 10^{-11}\right)\,\text{ns} \\
&= 19.7\,\text{ns}.
\end{aligned}
$$

Therefore, the requirements for ideal Q switching of this laser at the given pumping ratio of $r = 2.5$ are $\gamma_{cp} > 1.445 \times 10^9\,\text{s}^{-1}$ and $\Delta t_{QS} \ll 19.7\,\text{ns}$.

(c) From Example 15.1, we find that $P_{out}^{sat} = 2.37\,\text{mW}$ and $\tau_2 = 240\,\mu\text{s}$ for this laser, and $\tau_{cl} = \tau_c = 1.73\,\text{ns}$. Therefore, from (17.32), the peak power of the ideally Q-switched pulse is

$$
\begin{aligned}
P_{pk} &= \frac{\tau_2}{\tau_{cl}}(r - \ln r - 1)P_{out}^{sat} \\
&= \frac{240 \times 10^{-6}}{1.73 \times 10^{-9}}(2.5 - \ln 2.5 - 1) \times 2.37 \times 10^{-3}\,\text{W} \\
&= 192\,\text{W}.
\end{aligned}
$$

Compared to the 3.56 mW output power of the laser in CW operation, the peak power of this Q-switched pulse is 5.42×10^4 times higher primarily because of the fact that τ_2 is five orders of magnitude larger than τ_{cl}. This demonstrates that a gain medium that has a large τ_2 makes a good Q-switched laser.

The pulsewidth is found from (17.33) to be

$$
\Delta t_{ps} = \frac{2.5}{(r - \ln r - 1)^{1/2}}\tau_{cl} = \frac{2.5}{(2.5 - \ln 2.5 - 1)^{1/2}} \times 1.73\,\text{ns} = 5.66\,\text{ns}.
$$

Compared to the cavity round-trip time of $T = 6.07\,\text{ps}$ found in Example 15.1, which sets the ultimate lower limit for the pulsewidth of a Q-switched pulse, this pulsewidth is quite large. It can be reduced by pumping the laser higher to increase the pumping ratio r and by using an output-coupling mirror of a lower reflectivity to reduce τ_{cl}.

The pulse energy is simply

$$U_{ps} \approx P_{pk} \Delta t_{ps} = 192 \times 5.66 \times 10^{-9}\, J = 1.09\, \mu J.$$

This pulse energy is not very high because of the small amount of energy that can be stored in the small gain volume of the microchip laser. To increase the Q-switched pulse energy, one must increase the volume of the gain medium as well as the mode volume of the oscillating laser field.

(d) Two of the three conditions for ideal gain switching are the same as those for ideal Q switching, which are already met by this laser. The only condition remaining to be considered is a very short and strong pump for gain switching. Whether a pump pulse is short or not is relative to the fluorescence lifetime, τ_2, of the gain medium. Because τ_2 is the relaxation time constant of the excited population in the upper laser level, the pump energy can be efficiently stored in the population inversion of the gain medium if the pump pulse duration is much smaller than τ_2. If τ_2 is much smaller than the pump pulse duration, then the pump energy cannot be efficiently stored in the population inversion of the gain medium because population relaxation during the pumping process is significant. From this discussion, we understand that for ideal gain switching, it is necessary that the pump pulse be much shorter than τ_2 of the gain medium. It does not have to be extremely short, however. A pump pulse that has a duration of $\tau_2/10$ is short enough, while one that has a duration of $\tau_2/100$ is close to an ideal delta pump pulse. For ideal gain switching of this Nd:YAG laser with $\tau_2 = 240\,\mu s$, we need a short pump pulse of a few microseconds or less in duration that has a sufficiently high energy to pump the laser to the desired pumping ratio of $r = 2.5$ in such a short duration.

(e) The characteristics of an ideally gain-switched pulse are the same as those of an ideally Q-switched pulse found in (c). The only difference is in the pump pulse. Ideal Q switching can be accomplished with a relatively long pump pulse so long as the Q switch satisfies the conditions discussed in (b).

17.4 MODE LOCKING

Mode locking is the most important technique for the generation of repetitive, ultrashort laser pulses. The principle of mode locking is very different from those of gain switching and Q switching in that it is not based on the transient dynamics of a laser. Instead, a mode-locked laser operates in a dynamical steady state.

A pulsed laser can oscillate in multiple longitudinal modes regardless of whether the gain medium is homogeneously or inhomogeneously broadened. Mode locking refers to the situation that all of the oscillating longitudinal modes of a laser have fixed relative phases. The mechanism that locks these phases is generally a nonlinear optical process. When this phase locking is accomplished, constructive interference of all of the oscillating modes results in a short pulse that circulates inside the cavity, which is regeneratively amplified by the gain medium after periodically delivering an output pulse in each round-trip through an output-coupling mirror. The mode-locking operation is accomplished by a nonlinear optical element known as the *mode locker*.

Viewed in the frequency domain, mode locking is a process that generates a train of short laser pulses by locking multiple longitudinal laser modes in phase. The function of the mode locker in the frequency domain is thus to lock the phases of the oscillating modes together through nonlinear interactions among the mode fields. In the time domain, the mode-locking process can be understood as a regenerative pulse-generating process by which a short pulse circulating inside the laser cavity is formed when the laser reaches steady state. The action of the mode locker in the time domain resembles that of a pulse-shaping optical shutter that opens periodically in synchronism with the arrival at the mode locker of the laser pulse that circulates in the cavity. Consequently, the output of a mode-locked laser is a train of regularly spaced pulses of identical pulse envelope.

A practically useful mode-locked laser generally oscillates in a very large number of longitudinal modes, in the range from a few hundred to a few million modes. The total laser field of such a laser is the sum of all oscillating modes:

$$E(t) = \sum_q \mathcal{E}_q e^{i\varphi_q} e^{-i\omega_q t}, \tag{17.49}$$

where $\mathcal{E}_q$ is taken to be a positive, real quantity representing the magnitude of the field amplitude of mode q, φ_q is the phase of the mode field, and the summation is taken over all of the oscillating longitudinal modes.

The temporal characteristics of the combined laser field described in (17.49) depend on the phase relationships among the oscillating modes, as well as on the distribution of the field magnitudes $\mathcal{E}_q$ and the frequency spacing between neighboring modes. If the phases vary randomly with time, (17.49) describes the field of a CW multimode laser, which is of no interest here. For mode locking, we consider the situation that the oscillating laser modes are equally spaced, with a longitudinal mode spacing of $\Delta\omega_L$, and all of their phases are time independent. The magnitudes and phases of the mode fields are functions of the mode frequencies. Their spectral distribution can be described by a complex spectral envelope function of $\mathcal{E}(\omega)$ as

$$\mathcal{E}_q e^{i\varphi_q} = \frac{\Delta\omega_L}{2\pi} \mathcal{E}(\omega_q - \omega_0). \tag{17.50}$$

For simplicity, we choose ω_0 to be a longitudinal mode frequency that is at or closest to the center of the spectrum such that $\omega_q = \omega_0 + q\Delta\omega_L$. The total field in (17.49) can then be transformed between the frequency domain and the time domain as

$$E(t) = \frac{\Delta\omega_L}{2\pi} \sum_{q=-\infty}^{\infty} \mathcal{E}(\omega_q - \omega_0)e^{-i\omega_q t} = e^{-i\omega_0 t} \sum_{m=-\infty}^{\infty} \mathcal{E}(t - mT), \tag{17.51}$$

where $T = 2\pi/\Delta\omega_L$ and

$$\mathcal{E}(t) = \mathcal{F}^{-1}\{\mathcal{E}(\omega)\} = \frac{1}{2\pi} \int_{-\infty}^{\infty} \mathcal{E}(\omega)e^{-i\omega t}d\omega. \tag{17.52}$$

In (17.52), $\mathcal{F}^{-1}\{\mathcal{E}(\omega)\}$ performs the inverse Fourier transform from the frequency domain to the time domain. The result in (17.52) is obtained under the assumption that the mode phases φ_q do not vary with time. This is the condition of mode locking because the oscillating modes are phase locked with fixed phase relationships when none of the mode phases vary with time.

From (17.52), we find that when mode locking is accomplished, the total laser field $E(t)$ is a periodic function of time with a period of T and a temporal profile of $\mathcal{E}(t)$. Thus, a mode-locked laser produces a train of optical pulses at a temporal pulse spacing of $T = 2\pi/\Delta\omega_\mathrm{L}$, which has a spectrum of multiple discrete mode frequencies at a spectral spacing of $\Delta\omega_\mathrm{L}$. The temporal profile $\mathcal{E}(t)$ and the spectral envelope $\mathcal{E}(\omega)$ of the mode-locked pulses are related through the Fourier-transform relationship given in (17.52). Figure 17.8 shows the spectral and temporal characteristics of the field and intensity profiles of mode-locked laser pulses.

The spectral width, $\Delta\omega_\mathrm{ps}$, of a laser pulse is defined as the full width at half-maximum of the spectral intensity distribution, $I(\omega)$, as shown in Fig. 17.8(b). Correspondingly, the temporal pulsewidth, Δt_ps, is defined as the full width at half-maximum of the temporal intensity profile, $I(t)$, of an individual pulse, as illustrated in Fig. 17.8(d). Because of the Fourier-transform relationship, given in (17.52), between the temporal field profile, $\mathcal{E}(t)$, and the spectral field profile, $\mathcal{E}(\omega)$, the temporal and spectral widths of a mode-locked laser pulse are subject to the relation

$$\Delta\nu_\mathrm{ps}\Delta t_\mathrm{ps} \geq K, \tag{17.53}$$

where $\Delta\nu_\mathrm{ps} = \Delta\omega_\mathrm{ps}/2\pi$ and K is a constant of the order of unity that depends on the pulse shape. For any pulse with a given pulse shape, the best one can hope for is $\Delta\nu_\mathrm{ps}\Delta t_\mathrm{ps} = K$. When this is accomplished, the pulse is said to be *Fourier-transform limited*, or simply *transform limited*. A transform-limited pulse is one that has the smallest pulsewidth of $\Delta t_\mathrm{ps} = K/\Delta\nu_\mathrm{ps}$ for a given pulse spectral width of $\Delta\nu_\mathrm{ps}$.

When all of the modes of a laser are locked to a common phase, we can set $\varphi_q = \varphi_0 = 0$ for all oscillating modes because a constant common phase has no physical significance. This is the ideal situation of *complete mode locking*. From (17.50), we find that the spectral envelope is a real function when $\varphi_q = 0$ for all q. This implies that $\mathcal{E}(\omega) = \mathcal{E}^*(\omega)$ and

$$\mathcal{E}(t) = \frac{1}{2\pi}\int_{-\infty}^{\infty}\mathcal{E}(\omega)\mathrm{e}^{-\mathrm{i}\omega t}\mathrm{d}\omega = \frac{1}{2\pi}\int_{-\infty}^{\infty}\mathcal{E}^*(\omega)\mathrm{e}^{-\mathrm{i}\omega t}\mathrm{d}\omega = \left[\frac{1}{2\pi}\int_{-\infty}^{\infty}\mathcal{E}(\omega)\mathrm{e}^{\mathrm{i}\omega t}\mathrm{d}\omega\right]^* = \mathcal{E}^*(-t).$$

$$\tag{17.54}$$

Therefore,

$$I(t) = I(-t) \tag{17.55}$$

for a completely mode-locked laser pulse. From the above discussions, it can be concluded that *a completely mode-locked laser pulse has a symmetric temporal intensity profile and is transform limited. It does not necessarily have a symmetric spectral intensity profile, but an asymmetric temporal pulse shape or a deviation from the transform limit signifies incomplete mode locking. The reverse is not true, however,*

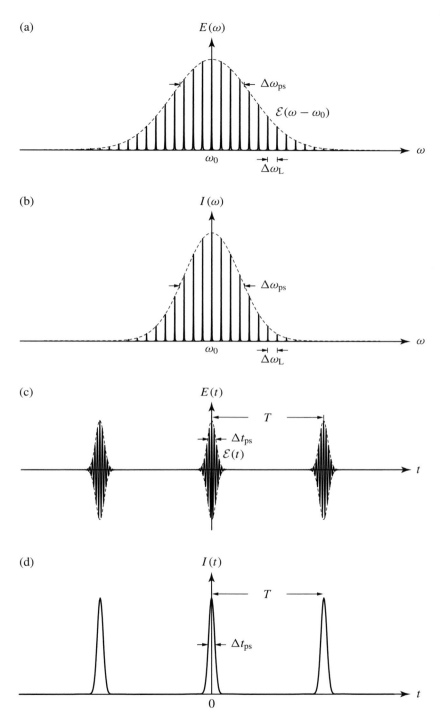

Figure 17.8 (a) Spectral field distribution, (b) spectral intensity distribution, (c) temporal field variation, and (d) temporal intensity variation of mode-locked laser pulses. For simplicity, it is assumed that the center of the spectral distribution coincides with the frequency of a longitudinal mode. Note that $\Delta\omega_{ps}$ and Δt_{ps} are defined as the full widths at half-maximum of $I(\omega)$ and $I(t)$, respectively.

because *a transform-limited pulse is not necessarily completely mode locked.* When the longitudinal laser modes are not completely locked in phase, pulses can still be formed, but with less than ideal characteristics.

A completely mode-locked pulse, being transform limited, satisfies the condition that $\Delta v_{ps} \Delta t_{ps} = K$. Because the number of oscillating modes can be estimated with the relation that

$$N \approx \frac{\Delta v_{ps}}{\Delta v_L}, \tag{17.56}$$

the temporal width of a mode-locked pulse is inversely proportional to the number of oscillating modes:

$$\Delta t_{ps} \approx \frac{K}{N \Delta v_L} = K \frac{T}{N}. \tag{17.57}$$

At a fixed longitudinal mode spacing of Δv_L, hence a fixed pulse repetition rate of $f_{ps} = 1/T = \Delta v_L$, the pulsewidth can be reduced by increasing the number of oscillating modes. It is common to expect a pulsewidth that is two to six orders of magnitude smaller than the pulse spacing in a train of mode-locked pulses. The relation in (17.57) indicates that this pulsewidth requires locking of hundreds to millions of oscillating modes.

An inhomogeneously broadened laser naturally oscillates in multiple longitudinal modes. In such a laser, the mode locker only has to lock these modes in phase to produce a train of mode-locked pulses. However, many mode-locked lasers that produce ultrashort pulses are homogeneously broadened. In the free-running steady state of a homogeneously broadened laser, only one longitudinal mode oscillates because of homogeneous saturation across the gain spectrum. Even though it is possible to force multimode oscillation in a homogeneously broadened laser when it is pulsed, the homogeneously broadened gain medium has a natural tendency to contract the spectral bandwidth of the oscillating laser field. Therefore, besides locking the phases of the oscillating laser modes together, the mode locker has the function of expanding the spectral width of the laser pulse to counteract the spectral narrowing effect of the gain medium. For this reason, a mode locker is fundamentally a nonlinear optical element.

For the pulses that are generated by a given mode-locked laser, the pulse spectral bandwidth, Δv_{ps}, is ultimately limited by the spontaneous linewidth, Δv, of the gain medium, such that $\Delta v_{ps} \leq \Delta v$, because Δv sets the limit for the gain bandwidth of the laser. By combining this limitation with the limitation set by (17.57), we find that the pulsewidth, Δt_{ps}, of a mode-locked pulse that can be generated from a given laser, regardless of whether it is homogeneously or inhomogeneously broadened, is subject to the absolute limitations:

$$T = \frac{1}{\Delta v_L} \gg \Delta t_{ps} \geq \frac{K}{\Delta v_{ps}} \geq \frac{K}{\Delta v}. \tag{17.58}$$

For most mode-locked lasers, only a fraction of the laser gain bandwidth is utilized so that Δv_{ps} is only a fraction of Δv. This fraction of bandwidth utilization depends on a number of operating parameters, including the modulation strength and the modulation frequency of the mode locker, as well as the type of mode locker that is used. Increasing this fraction is the key to reducing the temporal pulsewidth of a mode-locked pulse.

A continuously mode-locked laser is continuously pumped to deliver a steady train of short pulses at a constant average output power of $\overline{P}$, while each pulse has a high peak power of P_{pk}. Effectively, the energy of the laser output in each pulse repetition period T is concentrated within the duration of the pulsewidth Δt_{ps}. Therefore, the peak power of each pulse is enhanced over the average laser power by a factor of $T/\Delta t_{ps}$ in accordance with the relation:

$$P_{pk} = K' \frac{T}{\Delta t_{ps}} \overline{P} = K' \frac{\overline{P}}{f_{ps} \Delta t_{ps}} = \frac{K'}{K} N\overline{P}, \qquad (17.59)$$

where K' is a constant of the order of unity that depends on the pulse shape. From this relationship, we see that the enhancement of the pulse peak power over the average power is proportional to the number of locked modes.

Two pulse shapes are of most interest for mode-locked lasers. As we shall see below, *actively mode-locked pulses tend to have Gaussian shapes*, whereas *passively mode-locked pulses usually have* sech^2 *shapes*. For a Gaussian pulse, both $\mathcal{E}(\omega)$ and $\mathcal{E}(t)$ are Gaussian functions because the Fourier transform of a Gaussian function is another Gaussian function, and both its temporal intensity profile and spectral intensity profile are also Gaussian. For a sech^2 pulse, both $\mathcal{E}(\omega)$ and $\mathcal{E}(t)$ are hyperbolic secant functions because the Fourier transform of a hyperbolic secant function is another hyperbolic secant function, and both its temporal intensity profile and its spectral intensity profile are sech^2 functions. For a Gaussian pulse, $K = 2\ln 2/\pi = 0.4413$ and $K' = 2\sqrt{\ln 2}/\sqrt{\pi} = 0.9394$. For a sech^2 pulse, $K = 4\ln^2(1 + \sqrt{2})/\pi^2 = 0.3148$ and $K' = \ln(1 + \sqrt{2}) = 0.8814$.

EXAMPLE 17.2

By properly incorporating a suitable mode locker in the laser cavity, a CW Nd:YAG laser can often be mode locked with little additional loss, thus maintaining the average power while delivering a regular train of ultrashort laser pulses. A mode-locked Nd:YAG laser consists of a Nd:YAG gain medium that has a spontaneous linewidth of $\Delta\nu = 150$ GHz in a Fabry–Perot cavity that has a round-trip optical path length of $l_{RT} = 2$ m. The laser is continuously pumped to have an average output power of $\overline{P} = 2$ W. The mode locker used in this laser generates pulses of Gaussian temporal and spectral shapes. (a) What is the repetition rate of the mode-locked pulses? Does it vary with the pulsewidth or the laser output power? (b) If transform-limited pulses of $\Delta t_{ps} = 100$ ps pulsewidth are generated, how much of the bandwidth of the gain medium is utilized? How many longitudinal modes have to oscillate and be locked in phase to generate such pulses? (c) What is the peak power of the pulses? (d) What is the pulsewidth of the shortest pulses that can possibly be generated from this laser? Under what conditions can such pulses be generated?

Solution

(a) The pulse repetition rate is determined by the cavity round-trip time, T, which in turn is determined by the round-trip optical path length, l_{RT}. Therefore,

$$f_{ps} = \frac{1}{T} = \frac{c}{l_{RT}} = \frac{3 \times 10^8}{2} \text{ Hz} = 150 \text{ MHz}.$$

It does not vary with either the pulsewidth or the laser output power. We also find from this result that $T = 6.7$ ns.

(b) Because $K = 0.4413$, we have

$$\Delta\nu_{ps} = \frac{K}{\Delta t_{ps}} = \frac{0.4413}{100 \times 10^{-12}} \text{ Hz} = 4.413 \text{ GHz}$$

for transform-limited Gaussian pulses of $\Delta t_{ps} = 100$ ps pulsewidth. Because $\Delta\nu = 150$ GHz, we have $\Delta\nu_{ps}/\Delta\nu = 4.413/150 = 2.942\%$. Therefore, only 2.942% of the bandwidth of the gain medium is used. The longitudinal mode spacing is simply the same as the pulse repetition rate: $\Delta\nu_L = f_{ps} = 150$ MHz. The number of oscillating modes that are locked to generate these pulses is found from (17.56):

$$N \approx \frac{\Delta\nu_{ps}}{\Delta\nu_L} = \frac{4.413 \times 10^9}{150 \times 10^6} = 30.$$

Only 30 oscillating modes are required because the pulsewidth of 100 ps is relatively large for mode-locked pulses in a cavity that has a round-trip time of $T = 6.7$ ns.

(c) The peak power of these Gaussian pulses can be found by using (17.59) with $K' = 0.9394$:

$$P_{pk} = K'\frac{T}{\Delta t_{ps}}\overline{P} = 0.9394 \times \frac{6.67 \times 10^{-9}}{100 \times 10^{-12}} \times 2 \text{ W} = 125 \text{ W}.$$

This peak power is only about 63 times the average power because of the moderate value of the $T/\Delta t_{ps}$ ratio and the correspondingly small number of oscillating modes.

(d) The pulsewidth is ultimately limited by the condition given in (17.58). For Gaussian pulses, $K = 0.4413$. Thus, the shortest pulses that can be generated from this laser have the pulsewidth:

$$\Delta t_{min} = \frac{K}{\Delta\nu} = \frac{0.4413}{150 \times 10^9} \text{ s} = 2.942 \text{ ps}.$$

Such pulses are generated under the conditions: (1) the entire bandwidth of the laser gain medium is utilized so that $\Delta\nu_{ps} = \Delta\nu$; and (2) the pulses are transform limited so that $\Delta t_{ps}\Delta\nu_{ps} = K$. To utilize the entire bandwidth of the laser gain medium is not a simple matter. Aside from sufficiently pumping the laser to realize its entire gain bandwidth, it requires that each of the optical elements in the laser cavity, including the mirrors and the mode locker, have a bandwidth larger than $\Delta\nu$. It also requires that the mode-locking mechanism be sufficiently strong to force all modes across the entire bandwidth to oscillate and lock in phase. The generation of transform-limited pulses is not a trivial matter, either. It requires the elimination or compensation of all possible sources of dispersion in the laser while using an effective mode-locking scheme to lock all oscillating modes in phase.

As discussed in Section 17.1 and illustrated in Fig. 17.1, the fundamental difference between a transiently pulsed laser, such as a gain-switched laser or a Q-switched laser, and a regeneratively pulsed laser, such as a mode-locked laser, is that a transiently pulsed laser is a lumped device that produces an individual pulse longer than the cavity round-trip time such that $\Delta t_{ps} \geq T$, whereas a regeneratively pulsed laser is a distributed device that produces a train of pulses each much shorter than the cavity round-trip time such that $\Delta t_{ps} \ll T$. Because a regeneratively pulsed laser does not function by control of the laser transient, it does not depend on the rapid depletion of intracavity photons to generate a short pulse, as do transiently pulsed lasers. Consequently, it does not require a very short photon lifetime and a correspondingly short laser cavity. Another difference between a transiently pulsed laser and a regeneratively pulsed laser is the characteristic requirements of the gain medium. A transiently pulsed laser requires a long fluorescence lifetime and prefers to have it as long as possible. By contrast, the fluorescence lifetime τ_2 varies among different types of regeneratively pulsed lasers, but a particularly large τ_2 is not required. A *synchronously pumped laser* can successfully operate with a gain medium that has a very small τ_2 or even one that has no energy-storage mechanism, such as in the case of a synchronously pumped optical parametric oscillator [6, 7]. In certain mode-locked systems, τ_2 is preferred to be sufficiently large but not so large as to cause competition between transient oscillation and the build-up of mode-locked pulses. Therefore, the τ_2 requirement of a mode-locked laser is a sophisticated issue that depends on the specific type and the mechanism of the mode-locking operation.

Mode locking can take the form of either periodic loss modulation or periodic gain modulation. In principle, any nonlinear mirror that modulates the loss or gain of the laser in synchronism with the pulse repetition rate can function as a mode locker. The concepts of some important mode-locking techniques are illustrated in Fig. 17.9.

The location of the mode locker is not arbitrary. As shown in Fig. 17.9, the mode locker is typically placed near one end of the cavity if the laser has the configuration of a linear cavity. In the time domain, the mode locker acts as a shutter that opens only briefly once every cavity round-trip time in synchronism with the arrival of the mode-locked laser pulse that circulates inside the cavity; any noise spike or pulse that arrives at a wrong time is blocked by the mode locker. This arrangement produces a train of pulses at a repetition rate of $f_{ps} = 1/T$. It is possible to place the mode locker at a location that is l_{RT}/n from one end of a linear cavity while modulating it at a rate of n/T so that n equally spaced pulses circulate in the cavity to produce a train of pulses at a repetition rate of $f_{ps} = n/T$, where n is an integer. For example, by placing

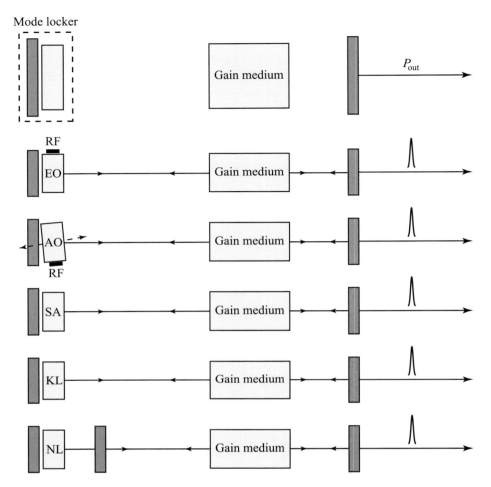

Figure 17.9 Schematics of mode-locked lasers. For active mode locking, the mode locker can be an electro-optic modulator or a standing-wave acousto-optical modulator. For passive mode locking, the mode locker can be a saturable absorber, a Kerr lens, or a coupled cavity containing a nonlinear element. The saturable absorber can be either a fast-recovering absorber or a slow-recovering absorber. Other schemes are possible for both active and passive mode locking.

the mode locker at the center of a linear cavity while modulating it at a rate of $2/T$, it is possible to produce a train of mode-locked pulses at a repetition rate of $f_{ps} = 2/T$.

Mode locking has been applied to a wide variety of laser materials to generate laser pulses with pulsewidths ranging from the order of 5 fs to the order of 1 ns. Similar to Q switching, mode locking can also be either active or passive, depending on the type of mode locker used. In an actively mode-locked laser, the mode locker is controlled by an externally applied signal. In a passively mode-locked laser, the mode locker directly responds to the optical field in the laser cavity through a nonlinear optical mechanism. For a given laser material, passive mode locking typically generates shorter pulses than active mode locking because the laser pulse modulates itself through the passive mode locker much faster than is possible with active modulation, but active mode locking often produces pulses with less fluctuation and a smaller jitter because an active mode locker is controlled by a stable external signal. Some systems combine active mode locking with passive mode locking in a form of *hybrid mode locking* to realize the advantages of both. It is also possible to combine mode locking with a transient pulsing technique. In this situation, the laser does not reach a complete steady state. An important example of this possibility is the operation of a *Q-switched mode-locked laser* by combining Q switching and mode locking. Because the transiently Q-switched pulse has a duration longer than the cavity round-trip time, the result is a finite train of equally spaced mode-locked pulses with unequal amplitudes under a Q-switched pulse envelope.

EXAMPLE 17.3

Nd:YAG lasers can undertake all modes of laser operation, including CW, gain-switching, Q-switching, and mode-locking operations. With the exception of synchronous pumping, almost all other mode-locking techniques can be successfully employed to mode lock Nd:YAG lasers, either in a pure form of CW mode locking or in a hybrid form that combines Q switching with mode locking. These being the facts, however, the microchip Nd:YAG laser with its cavity parameters described in Example 17.1 cannot be mode locked by any means. Give quantitative reasons for this problem.

Solution

First, consider the fact that the longitudinal mode spacing of this microchip laser is $\Delta v_L = 165$ GHz, as found in Example 17.1, while the entire linewidth of the Nd:YAG plate used for this laser is only $\Delta v = 150$ GHz. Although a homogeneously broadened laser can oscillate in multiple longitudinal modes when the laser is mode locked, as discussed in the text above, this microchip laser can only oscillate in a single longitudinal mode regardless of how it is being operated because $\Delta v_L > \Delta v$, not because it is homogeneously broadened. Clearly, there is no possibility of mode locking if a laser can only oscillate in one longitudinal mode.

We can see the problem from another angle in the time domain. A mode-locked pulse must have a spatial span that is much shorter than the cavity length to allow it to circulate inside the cavity as a regenerative pulse. For a laser that has a linear Fabry–Perot cavity such as the microchip laser under consideration, the mode-locked pulse has to fit loosely into the length of the cavity to allow it to circulate inside without wrapping itself up, thus having a pulsewidth that is much smaller than one-half of the cavity round-trip time: $\Delta t_{ps} \ll T/2$. From Example 15.1, we find that $T = 6.06$ ps for this laser. Therefore, any mode-locked pulse that can possibly be generated from this laser has a pulsewidth subject to the limitation that $\Delta t_{ps} \ll 3.03$ ps. However, the pulsewidth of a mode-locked pulse is also subject to the limitation given in (17.58). With $\Delta v = 150$ GHz, we have $\Delta t_{ps} \geq 2.942$ ps according to the calculation in Example 17.2 if the pulse has a Gaussian shape. These two conflicting limitations cannot be simultaneously satisfied, thus excluding any possibility of mode locking this laser.

17.4.1 Active Mode Locking

Active mode locking was first reported in 1964 by DiDomenico [8] and by Hargrove *et al.* [9]. Active mode locking can be done by modulating the frequency, known as *FM mode locking*, or the amplitude, known as *AM mode locking*, of the intracavity laser field [10–12]. In both cases, the modulation frequency is the repetition frequency of the mode-locked laser pulses. Both FM mode locking and AM mode locking generate *Gaussian pulses*, which have Gaussian temporal shapes and Gaussian spectral envelopes. FM mode locking is usually carried out by electro-optic phase modulation with an RF signal, as shown in Fig. 17.9. Being modulated at the pulse repetition frequency, which is the same as the frequency spacing of the longitudinal laser modes, the FM mode locker creates sidebands for each oscillating laser mode to injection lock the neighboring modes, thus accomplishing mode locking. For AM mode locking with loss modulation, the most commonly used technique is acousto-optic modulation with an externally applied RF signal, as also shown in Fig. 17.9. An acousto-optic modulator that is used for mode locking is different from one used for Q switching. An acousto-optic mode locker is a standing-wave Bragg diffractor that is turned on continuously and has to be modulated with an acoustic wave at half the pulse repetition frequency for a loss modulation at the pulse repetition frequency, but an acousto-optic Q switch is a traveling-wave Bragg diffractor that is modulated at a convenient frequency and is turned on and off to switch the cavity Q between different values.

In this subsection, only AM mode locking is considered. The temporal variations of the gain and loss with respect to the laser pulses for AM mode locking are shown in Fig. 17.10. For active mode locking, the population relaxation time of the gain medium is longer than the cavity round-trip time such that $\tau_2 > T$, and the laser is continuously pumped. Therefore, as shown in Fig. 17.10, the laser gain is saturated at a constant level by the averaged power of the laser. The laser pulse is temporally shortened by the nonlinear effect of loss modulation but is temporally lengthened by the finite gain bandwidth and the dispersion of the gain medium. The balance between these two opposing processes determines the steady-state pulse duration.

For a mode-locked laser in steady state oscillation, a very short pulse circulates in the laser cavity by regenerating itself, in both phase and amplitude, once every round-trip through the cavity. Because both phase and amplitude have to be considered, and because the pulse travels inside the cavity, the behavior of the laser cannot be described by the averaged photon density, as is done for transiently pulsed lasers. Instead, it has to be described by a self-consistent traveling-wave equation in terms of the intracavity laser field. For active mode locking with loss modulation, the governing equation is

$$\left[\frac{\Gamma g}{2} \left(1 + \frac{1}{\gamma_{21}^2} \frac{d^2}{dt^2} \right) - \frac{\gamma_c}{2} - \frac{\gamma_m}{2} (1 - \cos \Omega_m t) + \frac{\delta T}{T} \frac{d}{dt} \right] \mathcal{E}(t) = 0, \qquad (17.60)$$

where Γg is the gain parameter that accounts for the photon growth rate defined in Subsection 17.1.2; γ_{21} is the phase relaxation rate of the gain medium defined in Subsection 17.1.1; γ_c is the cavity photon decay rate defined in Section 17.1.2; γ_m is the photon decay rate due to loss modulation, which is related to the modulation index m and the cavity round-trip time T as $\gamma_m = m/T$; $\Omega_m = 2\pi f_m$ is the loss modulation frequency. In (17.60), the second-order time derivative term accounts for the dispersion caused by the finite bandwidth

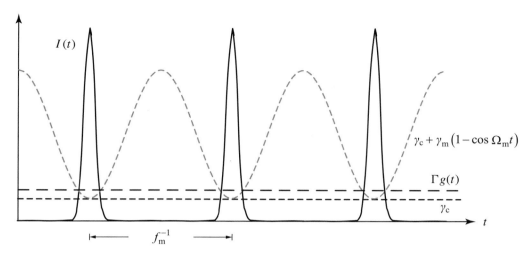

Figure 17.10 Temporal variations of the gain and the loss with respect to the laser pulses for AM mode locking.

of the gain medium, which is determined by the phase relaxation rate as $\Delta\omega_h = 2\gamma_{21}$ from (15.66). This dispersion term temporally lengthens the pulse, whereas the modulation term shortens the pulse. The *cavity detuning time*, δT, is the time delay of the pulse repetition time period, $T_{\text{ps}} = f_{\text{ps}}^{-1}$, with respect to the cavity round-trip time T:

$$\delta T = T_{\text{ps}} - T = \frac{1}{f_{\text{ps}}} - T. \tag{17.61}$$

In the case of active mode locking, the pulse repetition rate is determined by the loss modulation frequency such that $f_{\text{ps}} = f_{\text{m}}$, whereas the cavity round-trip time T is determined by the optical path of the cavity. Though they have to be closely matched, they might not be exactly the same; thus, there might be a nonzero detuning time, $\delta T \neq 0$.

Note that in (17.60), Γg, γ_c, and γ_m are multiplied by a factor of $1/2$ because they are defined as the *photon* growth and decay rates, whereas the equation is expressed for the field envelope $\mathcal{E}(t)$, not for the photon density. By multiplying the equation by a factor of 2, (17.60) can be expressed as

$$\left[\Gamma g \left(1 + \frac{1}{\gamma_{21}^2} \frac{d^2}{dt^2} \right) - \gamma_c - \gamma_m (1 - \cos \Omega_m t) + \frac{2\delta T}{T} \frac{d}{dt} \right] \mathcal{E}(t) = 0. \tag{17.62}$$

Equation (17.62) is a Mathieu equation that has periodic solutions. Because $\Delta t_{\text{ps}} \ll T \approx f_{\text{m}}^{-1}$ for a mode-locked pulse, it is generally true that $|\Omega_m t| \ll 1$ within the pulse duration. By taking the approximation that $1 - \cos \Omega_m t \approx \Omega_m^2 t^2 / 2$, (17.62) takes the form:

$$\left[\Gamma g \left(1 + \frac{1}{\gamma_{21}^2} \frac{d^2}{dt^2} \right) - \gamma_c - \frac{\gamma_m}{2} \Omega_m^2 t^2 + \frac{2\delta T}{T} \frac{d}{dt} \right] \mathcal{E}(t) = 0. \tag{17.63}$$

The solutions are Hermite–Gaussian functions. The fundamental solution is the Gaussian pulse:

$$\mathcal{E}(t) = \mathcal{E}_0 \exp\left(-\frac{t^2}{\tau^2}\right) = \mathcal{E}_0 \exp\left(-2\ln 2 \frac{t^2}{\Delta t_{\text{ps}}^2}\right), \tag{17.64}$$

where

$$\tau = \frac{1}{(\gamma_{21}\Omega_{\text{m}})^{1/2}} \left(\frac{8\Gamma g}{\gamma_m}\right)^{1/4}, \tag{17.65}$$

and

$$\Delta t_{\text{ps}} = \sqrt{2\ln 2}\,\tau \tag{17.66}$$

is the full width at half-maximum of the *intensity* profile of the pulse. We also find from (17.63) that the threshold gain parameter at which the laser is saturated under AM mode locking is higher than that for CW operation:

$$\Gamma g_{\text{th}} = \frac{1}{1 - (\Omega_{\text{m}}/\gamma_{21})(\gamma_{\text{m}}/2\Gamma g_{\text{th}})^{1/2}} \gamma_{\text{c}}. \tag{17.67}$$

The increase in this threshold gain parameter is caused by the additional loss from the AM modulation.

17.4.2 Passive Mode Locking

Passive mode locking in the form of simultaneous Q switching and mode locking with a saturable absorber was first reported in 1965 by Mocker and Collins [13] and in 1966 by DeMaria *et al.* [14]. CW passive mode locking was first reported by Ippen *et al.* in 1972 [15]. Passive mode locking can be accomplished by using a saturable absorber that is localized and placed near one end of a linear laser cavity, as shown in Fig. 17.9. Sometimes, a saturable absorber that is used for passive Q switching can also be used for passive mode locking, but in general the requirements for passive mode locking are very different from those for passive Q switching. Passive mode locking with a saturable absorber can be arranged in a ring configuration, shown in Fig. 17.11, for *colliding-pulse mode locking*. In this mode-locking scheme, there are two intracavity laser pulses that circulate in opposite directions and collide at the saturable absorber to enhance the pulse-shortening function of the saturable absorber. Passive mode locking can also be accomplished without the use of a saturable absorber by employing nonlinear refractive index changes through the real part of $\chi^{(3)}$, or even $\chi^{(2)}$, of a nonlinear optical element, as shown in Fig. 17.9.

Fast-Recovering Saturable Absorbers
A saturable absorber is considered to be fast recovering if its recovery time is short compared to the mode-locked pulse duration. A fast absorber saturates only in response to the instantaneous

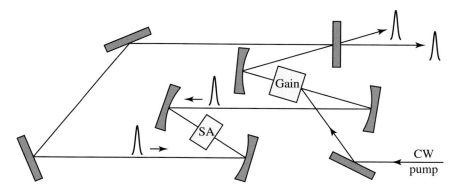

Figure 17.11 Ring cavity configuration of colliding-pulse mode locking with a saturable absorber. The spacing between the gain medium and the saturable absorber is not arbitrary but has to be one-quarter of the ring path length for the optimum operation of the system. The two contra-propagating pulses in the cavity collide at the saturable absorber for maximum absorber saturation, but the separation of their arrival times at the gain medium is maximized to minimize gain saturation.

photon density. Because a mode-locked laser pulse has a spatial span that is much shorter than the cavity length, it is not feasible to use the averaged photon density as is done for gain-switched and Q-switched laser pulses. Instead, the intensity or the power of the mode-locked laser pulse has to be used. Therefore, instead of (17.45), the absorption rate of a fast-recovering absorber for a mode-locked laser takes the form:

$$a(t) = \frac{a_i}{1 + I(t)/I_{sat}^{ab}}. \tag{17.68}$$

The temporal variations of the gain and the loss with respect to the laser pulse for passive mode locking with a fast saturable absorber are shown in Fig. 17.12.

For passive mode locking with a fast saturable absorber, the population relaxation time of the gain medium is longer than the cavity round-trip time, and the population relaxation time of the saturable absorber is shorter than the pulsewidth; therefore, $\tau_2 > T \gg \Delta t_{ps} > \tau_2^{ab}$. The laser is continuously pumped. As shown in Fig. 17.12, the laser gain is saturated at a constant level by the averaged power of the laser. The laser pulse is shortened by the nonlinear effect of the saturable absorber but is lengthened by the finite gain bandwidth and the dispersion of the gain medium. The balance between these two opposing processes determines the steady-state pulse duration.

The self-consistent traveling-wave equation for passive mode locking with a fast saturable absorber is [16–18]

$$\left[\Gamma g \left(1 + \frac{1}{\gamma_{21}^2} \frac{d^2}{dt^2} \right) - \gamma_c - \Gamma_{ab} a(t) + \frac{2\delta T}{T} \frac{d}{dt} \right] \mathcal{E}(t) = 0, \tag{17.69}$$

where Γ_{ab} is the overlap factor of the laser field and the saturable absorber, defined in (17.40), and $\Gamma_{ab} a(t)$ is the intracavity photon loss rate due to the absorption of the saturable absorber. By taking the approximation that

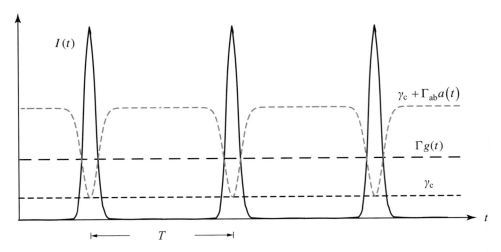

Figure 17.12 Temporal variations of the gain and the loss with respect to the laser pulse for passive mode locking with a fast saturable absorber.

$$\Gamma_{ab}a(t) = \frac{\Gamma_{ab}a_i}{1 + I(t)/I_{sat}^{ab}} \approx \Gamma_{ab}a_i - \gamma|\mathcal{E}(t)|^2 \tag{17.70}$$

with

$$\gamma|\mathcal{E}(t)|^2 = \Gamma_{ab}a_i \frac{I(t)}{I_{sat}^{ab}}, \tag{17.71}$$

(17.69) takes the form:

$$\left[\Gamma g\left(1 + \frac{1}{\gamma_{21}^2}\frac{d^2}{dt^2}\right) - \gamma_c - \Gamma_{ab}a_i + \gamma|\mathcal{E}(t)|^2 + \frac{2\delta T}{T}\frac{d}{dt}\right]\mathcal{E}(t) = 0. \tag{17.72}$$

The solution is the hyperbolic secant function:

$$\mathcal{E}(t) = \mathcal{E}_0 \operatorname{sech}\left(\frac{t}{\tau}\right) = \mathcal{E}_0 \operatorname{sech}\left[2\ln(1 + \sqrt{2})\frac{t}{\Delta t_{ps}}\right], \tag{17.73}$$

where

$$\tau = \frac{1}{\gamma_{21}}\left(\frac{2\Gamma g}{\gamma|\mathcal{E}_0|^2}\right)^{1/2} = \frac{1}{\gamma_{21}}\left(\frac{2\Gamma g}{\Gamma_{ab}a_i}\right)^{1/2}\left(\frac{I_{sat}^{ab}}{I_0}\right)^{1/2}, \tag{17.74}$$

I_0 is the peak intensity of the pulse, and

$$\Delta t_{ps} = 2\ln\left(1 + \sqrt{2}\right)\tau \tag{17.75}$$

is the full width at half-maximum of the *intensity* profile of the pulse. The intensity profile of the pulse is a sech^2 function. We also find from (17.72) that the threshold gain parameter at which the laser is saturated under mode locking by a fast absorber is higher than that for CW operation:

$$\Gamma g_{th} = \frac{\gamma_c + \Gamma_{ab} a_i}{1 + \left(\Gamma_{ab} a_i / 2\Gamma g_{th}\right)\left(I_0/I_{sat}^{ab}\right)}. \tag{17.76}$$

The increase in this threshold gain parameter is caused by the additional loss from the absorber.

For mode-locked lasers that generate ultrashort pulses of pulsewidths of the order of picoseconds or femtoseconds, the recovery times of most *real saturable absorbers* are much longer than the pulsewidths. Therefore, most real saturable absorbers that are used for passive mode locking are slow saturable absorbers. Fast saturable absorbers for passive mode locking are possible with *artificial saturable absorbers* that have very fast nonlinear optical responses. Compared to the resonant absorption processes of saturable absorbers, nonresonant reactive nonlinear processes are much faster. Indeed, parametric nonlinearities that are characterized by the real part of $\chi^{(3)}$ or $\chi^{(2)}$ are the fastest nonlinear optical processes, which usually have response times of the order of a few femtoseconds. Many concepts of artificial saturable absorbers based on self-phase modulation or cross-phase modulation have been realized for fast passive mode locking [17, 18]. Besides having very fast responses, these artificial saturable absorbers can be used over a broad spectral range and they do not dissipate the laser power. Two very important concepts belonging to this category are *additive-pulse mode locking* and *Kerr-lens mode locking*, which are illustrated in Figs. 17.13(a) and (b), respectively. For additive-pulse mode locking, the external cavity contains an optical fiber that creates intensity-dependent self-phase modulation for the pulse. The length of the external cavity is fine-tuned such that two pulses that meet at the coupling mirror are in phase at the pulse peak but are out of

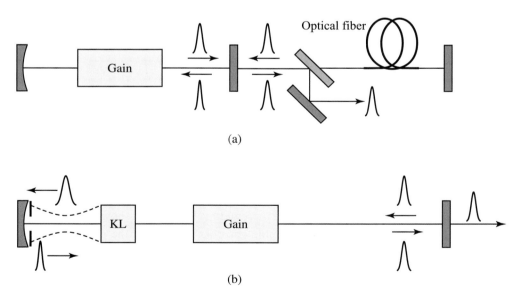

Figure 17.13 (a) Additive-pulse mode locking and (b) Kerr-lens mode locking. For additive-pulse mode locking, the external cavity contains an optical fiber that creates intensity-dependent self-phase modulation for the pulse. The pulse that is reflected by the coupling mirror is shorter than the incoming pulse. For Kerr-lens mode locking, the intensity-dependent focal length of the Kerr lens leads to a tighter focus for the pulse peak than for the pulse tails. The aperture that is placed in front of the end mirror thus transmits the pulse peak but cuts off the pulse tails, resulting in a shorter pulse that is reflected from the mirror.

phase at the pulse tails. The pulse that is reflected by the coupling mirror is shorter than the incoming pulse due to constructive interference at the pulse peak but destructive interference at the pulse tails. For Kerr-lens mode locking, the intensity-dependent focal length of the Kerr lens leads to a tighter focus for the pulse peak than for the pulse tails. The aperture that is placed in front of the end mirror thus transmits the pulse peak but cuts off the pulse tails, resulting in a shorter pulse that is reflected from the mirror.

A general master equation can be written for passive mode locking in steady state with a real or artificial fast absorber [17, 18]:

$$\left[\Gamma g - \Gamma_{ab} a_i - \gamma_c + i\psi_i + \frac{\Gamma g}{\gamma_{21}^2} \frac{d^2}{dt^2} + i \frac{D}{\bar{n}\omega} \frac{d^2}{dt^2} + \frac{2\delta T}{T} \frac{d}{dt} + (\gamma - i\delta) \left| \mathcal{E}(t) \right|^2 \right] \mathcal{E}(t) = 0. \quad (17.77)$$

This equation has the general steady-state solution [19]:

$$\mathcal{E}(t) = \mathcal{E}_0 \, \text{sech}^{(1+i\beta)} \left(\frac{t}{\tau} \right), \quad (17.78)$$

where β is a real chirp parameter. Therefore, the intensity profile of the pulse is a sech^2 function.

Slow-Recovering Saturable Absorbers

A slow-recovering saturable absorber has a recovery time that is long compared to the mode-locked pulse duration. A slow absorber is not saturated by the instantaneous photon density but by the total energy of a pulse. Therefore, instead of (17.47), the absorption rate of a slow-recovering absorber for a mode-locked laser takes the form, from (17.22),

$$a(t) = a_i \exp \left(-\frac{1}{\tau_2^{ab}} \int_{-\infty}^{t} \frac{I(t')}{I_{sat}^{ab}} dt' \right). \quad (17.79)$$

Note that the lower bound of the integral in (17.79) is taken as $t = -\infty$ only to symbolically mean that the integral is evaluated from the starting point of the pulse, which is not defined as $t = 0$ because for a mode-locked pulse $t = 0$ is usually taken to be the time at the pulse peak. The temporal variations of the gain and the loss with respect to the laser pulses for passive mode locking with a slow saturable absorber are shown in Fig. 17.14.

For passive mode locking with a slow saturable absorber, the population relaxation time of the saturable absorber is longer than the pulsewidth. However, as shown in Fig. 17.14, both the population relaxation time of the saturable absorber and that of the gain medium have to be shorter than the cavity round-trip time for the absorber and the gain medium to recover before the next pulse arrives; therefore, $T > \tau_2 \geq \tau_2^{ab} > \Delta t_{ps}$. Because $\tau_2 > \Delta t_{ps}$, the gain medium is also saturated in a manner similar to the saturation of the absorber:

$$g(t) = g_i \exp \left(-\frac{1}{\tau_2} \int_{-\infty}^{t} \frac{I(t')}{I_{sat}} dt' \right). \quad (17.80)$$

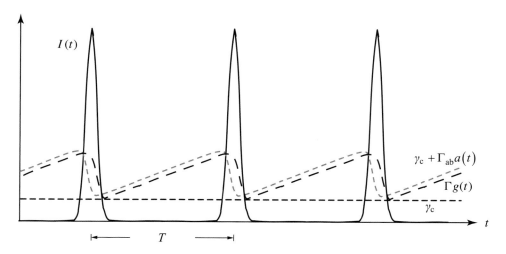

Figure 17.14 Temporal variations of the gain and the loss with respect to the laser pulse for passive mode locking with a slow saturable absorber.

The laser is continuously pumped. The laser pulse is shortened by the nonlinear effects of absorption saturation and gain saturation, as seen in Fig. 17.14, but it is lengthened by the finite gain bandwidth and the dispersion of the gain medium. The balance between these two opposing processes determines the steady-state pulse duration.

The self-consistent traveling-wave equation for passive mode locking with a slow saturable absorber is [17, 18, 20]

$$\left[\Gamma g(t) - \Gamma_{ab} a(t) - \gamma_c + \frac{\gamma_c}{\gamma_{21}^2}\frac{d^2}{dt^2} + \frac{2\delta T}{T}\frac{d}{dt}\right]\mathcal{E}(t) = 0, \tag{17.81}$$

where $g(t)$ and $a(t)$ are those given in (17.80) and (17.79), respectively, with the pulse intensity related to the pulse field envelope as $I(t) = 2c\epsilon_0 n|\mathcal{E}(t)|^2$. This equation for passive mode locking with a slow saturable absorber has a solution of the same hyperbolic secant function as that given in (17.73) for passive mode locking with a fast saturable absorber, but the pulsewidth is different:

$$\mathcal{E}(t) = \mathcal{E}_0 \operatorname{sech}\left(\frac{t}{\tau}\right) = \mathcal{E}_0 \operatorname{sech}\left[2\ln\left(1 + \sqrt{2}\right)\frac{t}{\Delta t_{ps}}\right], \tag{17.82}$$

where

$$\tau = \left[\frac{\gamma_{21}^2}{4\gamma_c}\left(4\sigma_{ab}^2\Gamma_{ab}a_i - \beta^2\sigma_e^2\Gamma g_i\right)\left(\frac{I_0}{h\nu}\right)^2\right]^{-1/4} = \left\{\frac{\gamma_{21}^2}{4\gamma_c}\left[\frac{\Gamma_{ab}a_i}{(\tau_2^{ab})^2}\left(\frac{I_0}{I_{sat}^{ab}}\right)^2 - \frac{\Gamma g_i}{(\tau_2)^2}\left(\frac{I_0}{I_{sat}}\right)^2\right]\right\}^{-1/4}, \tag{17.83}$$

I_0 is the peak intensity of the pulse, and

$$\Delta t_{\text{ps}} = 2 \ln (1 + \sqrt{2})\tau \tag{17.84}$$

is the full width at half-maximum of the *intensity* profile of the pulse. The intensity profile of the pulse is a sech^2 function. When the laser reaches the steady state, we find that the initial gain parameter is saturated at the level:

$$\Gamma g_{\text{i}} = \Gamma_{\text{ab}} a_{\text{i}} + \gamma_{\text{c}} \left(1 + \frac{3}{\gamma_{21}^2 \tau^2} \right). \tag{17.85}$$

The increase in this saturated gain parameter is caused by the additional loss from the absorber.

17.4.3 Synchronous Pumping

A very important technique, known as *synchronous pumping*, illustrated in Fig. 17.15, for generating ultrashort laser pulses can be considered as active mode locking by gain modulation. Synchronous pumping as a mode-locking technique was first reported in 1968 by Glenn et al. [21] and by Bradley and Durrant [22]. For synchronous pumping, the gain medium in the laser cavity is localized and is placed near one end of the cavity. The gain medium is periodically pumped, either optically or electrically, with a train of very short pulses at the same repetition rate as that of the periodic arrival at the gain medium of the pulse that circulates inside the laser cavity. The pulse that circulates inside the cavity delivers an output pulse every time it arrives at the output-coupling mirror. It meets with a pump pulse at the gain medium in every round-trip through the cavity so that it is regenerated. In this manner, the mode-locking condition is satisfied. Through this arrangement, the synchronously pumped laser delivers a train of mode-locked pulses at the same repetition rate as that of the pump pulses.

Figure 17.16 shows the temporal variations of the pump pulse, the gain, and the regenerated laser pulse for synchronous pumping with a gain medium that has a population relaxation time

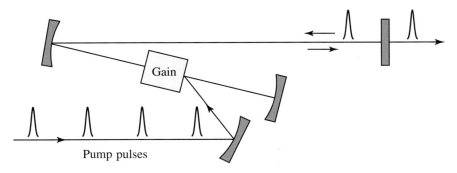

Figure 17.15 Schematic of synchronous pumping. The pump pulses are not shown to scale. In reality, they have to be equally spaced at or very close to the cavity round-trip time for the pulse that circulates inside the cavity to be synchronized to the pump pulses and to meet them at the gain medium in every round-trip through the cavity.

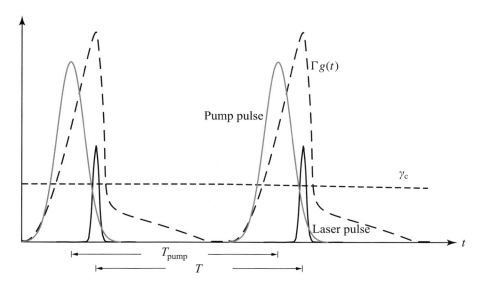

Figure 17.16 Temporal variations of the pump pulse, the gain, and the regenerated laser pulse for synchronous pumping with a gain medium that has a population relaxation time longer than the pump pulse duration. The time period between two successive pump pulses is T_{pump}, which is determined by the repetition rate of the pump pulses. The time period between two successive laser pulses is the round-trip time, T, of the laser cavity. For synchronous pumping, these two time periods have to be synchronized such that $T_{pump} \approx T$ within a very small detuning range that is generally only a small fraction of the laser pulsewidth. The regenerated laser pulse is generally shorter than the pump pulse.

longer than the pump pulse duration. The evolution of the gain within each pumping cycle of synchronous pumping is similar to that in the case of gain switching. A short and strong pump pulse quickly brings the gain above the threshold for a net gain to amplify the arriving laser pulse. The gain is depleted by the regenerated laser pulse through gain saturation, thus terminating the regenerated pulse. However, the analogy with gain switching stops here. For gain switching, each laser pulse starts from spontaneous emission noise and evolves through transient dynamics, even when the laser is repetitively gain switched. For synchronous pumping, by contrast, a pulse circulates in the laser cavity and is regeneratively amplified; it does not redevelop from noise when the laser reaches steady-state oscillation. As shown in Fig. 17.16, the time period between two successive pump pulses is T_{pump}, which is determined by the repetition rate of the pump pulses. The time period between two successive laser pulses is the round-trip time, T, of the laser cavity. For synchronous pumping, these two time periods have to be synchronized such that $T_{pump} \approx T$ within a very small detuning range that is generally only a small fraction of the laser pulsewidth. Therefore, the pulses generated through synchronous pumping are coherent, mode-locked pulses, whereas those generated by repetitive gain switching are uncorrelated, individual pulses.

Because the gain of a synchronously pumped laser is supplied by a pump pulse in synchronism with the regenerated laser pulse, the fluorescence lifetime τ_2 of the gain medium is not subject to a stringent condition. Indeed, it can be very long, as the case preferred by a gain-switched laser, or very short, or even zero, as the case for a synchronously pumped optical parametric oscillator [6, 7]. Therefore, the quantitative characteristics of the gain parameter $g(t)$

of a synchronously pumped laser depends on the specific properties of the gain medium and the pump pulse. Though not necessary for the operation of the laser, a saturable absorber can also be incorporated into a synchronously pumped laser for hybrid mode locking to further shorten the laser pulse. The characteristics of the absorption rate $a(t)$ depends on the specific properties of the saturable absorber. The behavior of a synchronously pumped laser is governed by the master equation given in (17.81), but with the parameters $g(t)$ and $a(t)$ specific to a particular system. For this reason, there is no general solution for all synchronously pumped lasers, though the shape of the pulse field envelope for a synchronously pumped dye laser has the hyperbolic secant function of a mode-locked laser with a fast saturable absorber [23]. In any event, the mode-locked pulses generated by a synchronously pumped laser generally have a pulsewidth smaller than that of the pump pulses, as is illustrated in Fig. 17.16.

Problem Set

17.2.1 Ideal gain switching with a very short pump pulse is considered. Take the time $t = 0$ to be the start of the pump pulse. The pump pulse duration is much shorter than the fluorescence lifetime τ_2 of the gain medium and the photon lifetime τ_c of the optical cavity such that the pump pulse can be approximated as a delta function, $R(t) \rightarrow N_{max}\delta(t)$, as given in (17.26).

    (a) Show, by integrating the equation for the density of the population inversion given in (17.24), that the equation for the intracavity photon density given in (17.25) can be expressed in the form of (17.27).

    (b) The photon density described by (17.27) has an approximate analytical solution as given in (17.28), which is characterized by the pumping ratio, r, the peak photon density, S_{pk}, and the time, t_{pk}, when the peak of the pulse occurs. Show that for a given value of the pumping ratio r, the values of S_{pk} and t_{pk} are given by (17.29) and (17.30), respectively.

17.3.1 In this problem, we consider laser pulses generated by gain switching and Q switching.

    (a) What is the absolutely shortest laser pulse one can generate by gain switching a given laser? What has to be done to approach that limit?

    (b) Answer the questions in (a) for Q switching instead of gain switching.

    (c) With given cavity mirrors and a given gain medium for a gain-switched or Q-switched laser, name two different approaches that can easily be taken to reduce the output laser pulsewidth.

    (d) In what situation can the same laser pulse generated by Q switching a laser be generated by gain switching the same laser so that the functioning of the Q switch becomes redundant?

17.3.2 The pulsewidth of the Q-switched pulse generated by the Q-switched Nd:YAG microchip laser considered in Example 17.1 can be reduced by increasing the pumping level while keeping the laser parameters unchanged. By so doing, the pumping ratio r is increased while the photon lifetime τ_{cl} in the lasing phase remains unchanged. When the pulsewidth changes, the peak power and the pulse energy both change accordingly. Meanwhile, the requirements on the Q switch to act as an ideal Q switch are changed. A Q-switched pulse of $\Delta t_{ps} = 4\,\text{ns}$ is desired.

(a) What is the required pumping ratio for the Q-switched pulse to have $\Delta t_{ps} = 4\,\text{ns}$?

(b) What are the peak power and the energy of this Q-switched pulse?

(c) What are the requirements on the Q switch for ideal Q switching of this laser?

17.3.3 The Nd:YAG laser described in Example 17.2 can also be gain switched or Q switched. It has a cavity round-trip time of $T = 6.7\,\text{ns}$, which is determined by the round-trip optical path length of $l_{RT} = 2\,\text{m}$. The Nd:YAG gain medium has a fluorescence lifetime of $\tau_2 = 240\,\mu\text{s}$. Without additional information on the parameters of the laser or its cavity, find the upper and lower limits of the pulsewidth of a gain-switched or Q-switched pulse that can be generated from this laser. Though the photon lifetime τ_c is not known, consider its relation with T to estimate the limitation it imposes on the pulsewidth. Also consider the required pumping ratio for effective gain switching or Q switching. What is the most likely range for the pulsewidth?

17.4.1 The basic features of mode locking are considered.

(a) What can be said about the temporal pulse characteristics of a completely mode-locked pulse no matter how it is generated?

(b) What can be said about a pulse generated by a mode-locked laser if the temporal pulse shape is asymmetric?

(c) What are the expected shapes of the temporal and spectral envelopes of laser pulses generated by active and passive mode locking, respectively?

(d) For most modes of laser operation, it is desirable that the fluorescence lifetime of the gain medium be as long as possible, but there are exceptions. Give two examples of laser operation in which a gain medium with a very long fluorescence lifetime is not desirable. What is the desirable fluorescence lifetime in each of those two situations?

17.4.2 Show that a laser pulse that has a Gaussian temporal intensity profile also has a Gaussian spectral intensity profile. Show that for such Gaussian pulses the transform-limit constant K defined in (17.53) has the value of $K = 2\ln 2/\pi = 0.4413$. Show also that for such Gaussian pulses the constant K' defined in (17.59) has the value of $K' = 2\sqrt{\ln 2}/\sqrt{\pi} = 0.9394$. Note that

$$\int_{-\infty}^{\infty} e^{-x^2}\,dx = \sqrt{\pi}. \tag{17.86}$$

17.4.3 Show that a laser pulse that has a sech^2 temporal intensity profile also has a sech^2 spectral intensity profile. Show that for such sech^2 pulses the transform-limit constant K defined in (17.53) has the value of $K = 4\ln^2(1 + \sqrt{2})/\pi^2 = 0.3148$. Show also that for such sech^2 pulses the constant K' defined in (17.59) has the value of $K' = \ln(1 + \sqrt{2}) = 0.8814$. Note that

$$\int_{-\infty}^{\infty} \text{sech}^2 x dx = 2.$$ (17.87)

17.4.4 Several types of lasers are very versatile in terms of their mode of operation. For example, by simply turning the active mode locker on or off, a Nd:YAG laser can be switched between CW operation and continuous mode-locking operation without much change in its average output power. As another example, a semiconductor laser can be biased at a constant DC level while being switched between CW or repetitive gain-switching operations, the latter of which delivers a regular train of gain-switched pulses. The relation between the peak power and the average power of the pulses of a constant pulsewidth Δt_{ps} in a pulse train of a pulse repetition rate of f_{ps} is given in (17.59). The laser beam at the fundamental frequency of ω is sent through a second-harmonic crystal to generate its second harmonic at the second-harmonic frequency of 2ω with an efficiency that is proportional to the instantaneous power of the laser such that $P_{2\omega}(t) = aP_{\omega}^2(t)$, with a being a constant. The average powers $\overline{P}_{\omega}$ and $\overline{P}_{2\omega}$ of the fundamental and the second harmonic, respectively, are monitored for both CW and pulsed operations of the laser. Clearly, $(\overline{P}_{2\omega})_{\text{pulsed}}$ is higher than $(\overline{P}_{2\omega})_{\text{CW}}$ if $(\overline{P}_{\omega})_{\text{pulsed}}$ is comparable to $(\overline{P}_{\omega})_{\text{CW}}$. Show that the pulsewidth can be found from the relation:

$$\Delta t_{ps} = A \frac{\left(\overline{P}_{2\omega}/\overline{P}_{\omega}^2\right)_{\text{CW}}}{f_{ps}\left(\overline{P}_{2\omega}/\overline{P}_{\omega}^2\right)_{\text{pulsed}}},$$ (17.88)

where A is a constant that depends on the pulse shape. Show also that $A = (2\ln 2/\pi)^{1/2} = 0.6643$ for Gaussian pulses and that $A = (2/3)\ln(1+\sqrt{2}) = 0.5876$ for sech^2 pulses. This problem describes a convenient way of measuring the pulsewidth of repetitive pulses if the pulse shape and the pulse repetition rate are both known. Note that

$$\int_{-\infty}^{\infty} \text{sech}^4 x dx = \frac{4}{3}.$$ (17.89)

(See Y. C. Chen and J. M. Liu, "Measurement of picosecond semiconductor laser pulse duration with internally generated second harmonic emission," *Applied Physics Letters*, vol. 47, pp. 662–664, 1985.)

17.4.5 By using the equation given in (17.63) for AM mode locking, show that the Gaussian pulse given in (17.64) is a solution with the pulsewidth parameter, τ, given in (17.65) so that the full width at half-maximum pulsewidth, Δt_{ps}, is that given in (17.66). Show also that the gain parameter of the laser is saturated at a threshold level given by (17.67). For simplicity, consider the case without detuning such that $\delta T = 0$.

17.4.6 By using the equation given in (17.72) for passive mode locking with a fast absorber, show that the hyperbolic secant pulse given in (17.73) is a solution with the pulsewidth parameter, τ, given in (17.74) so that the full width at half-maximum pulsewidth, Δt_{ps}, is that given in (17.75). Show also that the gain parameter of the laser is saturated at a threshold level given by (17.76). For simplicity, consider the case without detuning, such that $\delta T = 0$.

17.4.7 By using the equation given in (17.81) for passive mode locking with a slow absorber, show that the hyperbolic secant pulse given in (17.82) is a solution with the pulsewidth parameter, τ, given in (17.83) so that the full width at half-maximum pulsewidth, Δt_{ps}, is that given in (17.84). Show also that the gain parameter of the laser is saturated at a threshold level given by (17.84). For simplicity, consider the case without detuning, such that $\delta T = 0$.

References

[1] J. M. Liu, *Principles of Photonics*. Cambridge: Cambridge University Press, 2016.

[2] D. Roess, "Giant pulse shortening by resonator transients," *Journal of Applied Physics*, vol. 37, pp. 2004–2006, 1966.

[3] L. W. Casperson, "Analytic modeling of gain-switched lasers: I. Laser oscillators," *Journal of Applied Physics*, vol. 47, pp. 4555–4562, 1976.

[4] R. W. Hellwarth, "Theory of the pulsation of fluorescent light from ruby," *Physical Review Letters*, vol. 6, p. 4, 1961.

[5] F. J. McClung and R. W. Hellwarth, "Giant optical pulsations from ruby," *Journal of Applied Physics*, vol. 33, pp. 828–829, 1962.

[6] E. C. Cheung and J. M. Liu, "Theory of a synchronously pumped optical parametric oscillator in steady-state operation," *Journal of the Optical Society of America B*, vol. 7, pp. 1385–1401, 1990.

[7] E. C. Cheung and J. M. Liu, "Efficient generation of ultrashort, wavelength-tunable infrared pulses," *Journal of the Optical Society of America B*, vol. 8, pp. 1491–1506, 1991.

[8] M. J. DiDomenico, "Small-signal analysis of internal (coupling-type) modulation of lasers," *Journal of Applied Physics*, vol. 35, pp. 2870–2876, 1964.

[9] L. E. Hargrove, R. L. Fork, and M. A. Pollack, "Locking of He–Ne laser modes induced by synchronous intracavity modulation," *Applied Physics Letters*, vol. 5, pp. 4–5, 1964.

[10] D. Kuizenga and A. Siegman, "FM and AM mode locking of the homogeneous laser – Part I: theory," *IEEE Journal of Quantum Electronics*, vol. 6, pp. 694–708, 1970.

[11] D. Kuizenga and A. Siegman, "FM and AM mode locking of the homogeneous laser – Part II: experimental results in a Nd:YAG laser with internal FM modulation," *IEEE Journal of Quantum Electronics*, vol. 6, pp. 709–715, 1970.

[12] A. Siegman and D. Kuizenga, "Modulator frequency detuning effects in the FM mode-locked laser," *IEEE Journal of Quantum Electronics*, vol. 6, pp. 803–808, 1970.

[13] H. W. Mocker and R. J. Collins, "Mode competition and self-locking effects in a Q-switched ruby laser," *Applied Physics Letters*, vol. 7, pp. 270–273, 1965.

[14] A. J. DeMaria, D. A. Stetser, and H. Heynau, "Self mode-locking of lasers with saturable absorbers," *Applied Physics Letters*, vol. 8, pp. 174–176, 1966.

[15] E. P. Ippen, C. V. Shank, and A. Dienes, "Passive mode locking of the cw dye laser," *Applied Physics Letters*, vol. 21, pp. 348–350, 1972.

[16] H. A. Haus, "Theory of mode locking with a fast saturable absorber," *Journal of Applied Physics*, vol. 46, pp. 3049–3058, 1975.

[17] E. P. Ippen, "Principles of passive mode locking," *Applied Physics B*, vol. 58, pp. 159–170, 1994.

[18] H. A. Haus, "Mode-locking of lasers," *IEEE Journal of Selected Topics in Quantum Electronics*, vol. 6, pp. 1173–1185, 2000.

[19] O. E. Martinez, R. L. Fork, and J. P. Gordon, "Theory of passively mode-locked lasers including self-phase modulation and group-velocity dispersion," *Optics Letters*, vol. 9, pp. 156–158, 1984.

[20] H. Haus, "Theory of mode locking with a slow saturable absorber," *IEEE Journal of Quantum Electronics*, vol. 11, pp. 736–746, 1975.

[21] W. H. Glenn, M. J. Brienza, and A. J. DeMaria, "Mode locking of an organic dye laser," *Applied Physics Letters*, vol. 12, pp. 54–56, 1968.

[22] D. J. Bradley and A. J. F. Durrant, "Generation of ultrashort dye laser pulses by mode locking," *Physics Letters A*, vol. 27, pp. 73–74, 1968.

[23] J. M. Catherall, G. H. C. New, and P. M. Radmore, "Approach to the theory of mode locking by synchronous pumping," *Optics Letters*, vol. 7, pp. 319–321, 1982.

18 Propagation of Optical Pulses

18.1 SINGLE-MODE OPTICAL PULSE PROPAGATION

In this chapter, we discuss the characteristics of pulse propagation in an isotropic and spatially homogeneous optical Kerr medium. The analysis done and the results obtained in this chapter apply equally well to pulse propagation in a single-mode waveguide that is made of an isotropic optical Kerr material, such as a silica optical fiber. The only difference is that for the propagation in a spatially homogeneous medium considered here, the amplitude $\mathcal{E}(z,t)$ of the *pulse field envelope* is used and the propagation constant is k, whereas for the propagation in a single-mode waveguide the mode amplitude $A(z,t)$ has to be used and the propagation constant is β. The general equations for optical pulse propagation have been derived in Chapter 8.

We limit our discussion in this chapter to the cases for which the conditions for the applicability of the pulse propagation equation given in (8.58) are valid. We also ignore the shock term $(\mathrm{i}/\omega_0)\partial/\partial t$ on the right-hand side of (8.58). In Chapter 8, the possibility of a background absorption or a resonant susceptibility was not considered. For generality, however, here we include a background absorption loss, characterized by an absorption coefficient α, and a resonant susceptibility, characterized by a resonant polarization $\mathbf{P}_{\mathrm{res}}(\mathbf{r},t) = \mathcal{P}_{\mathrm{res}}(\mathbf{r},t)\exp\left(\mathrm{i}k_0 z - \mathrm{i}\omega_0 t\right)$, as expressed in (8.6). Thus, we can extend (8.58), while dropping the shock term $(\mathrm{i}/\omega_0)\partial/\partial t$, as

$$
\frac{\partial \mathcal{E}(z,t)}{\partial z} + \frac{1}{v_{\mathrm{g}}}\frac{\partial \mathcal{E}(z,t)}{\partial t} + \sum_{\infty}^{n=2}\frac{\mathrm{i}^{n-1}k_n}{n!}\frac{\partial^n \mathcal{E}(z,t)}{\partial t^n}
$$
$$
= -\frac{\alpha}{2}\mathcal{E}(z,t) + \mathrm{i}\frac{\omega_0}{2c\epsilon_0 n_0}\mathcal{P}_{\mathrm{res}}(z,t) + \mathrm{i}s\left|\mathcal{E}(z,t)\right|^2\mathcal{E}(z,t).
$$

(18.1)

18.1.1 Rate-Equation Approximation

The resonant susceptibility connects the electric field to a resonant polarization through the time-domain convolution relation given in (1.159). In general, the resonant susceptibility is a function of the electric field so that the time-domain convolution relation of (1.159) cannot be simply transformed to a direct-product relation in the frequency domain by straightforward Fourier transform. This transformation is valid, however, under the conditions for the *rate-equation approximation*:

1. The phase relaxation rate of the gain medium is much larger than the population relaxation rate, such that $\gamma_{21} \gg \gamma_2 = \tau_2^{-1}$.
2. The gain bandwidth, and thus the phase relaxation rate, of the gain medium is much larger than the spectral width of the optical pulse such that $\Delta v_h = \gamma_{21}/\pi \gg \Delta v_{ps} \approx \Delta t_{ps}^{-1}$, where Δv_h is the homogeneous gain bandwidth of the gain medium, as defined in (15.67), Δv_{ps} is the spectral width of the pulse, and Δt_{ps} is the temporal width of the pulse.

For very short optical pulses, we consider the common situation that $\tau_2 \gg \Delta t_{ps}$ such that the population density of the gain medium does not vary during the pulse duration.

When the above conditions are satisfied, the rate-equation approximation can be applied to express $\mathcal{P}_{res}(z, t)$ as a function of $N(z)$ and $\mathcal{E}(z, t)$ as

$$\mathcal{P}_{res}(z, t) = \epsilon_0 \chi_{res}(z, \omega_0)\mathcal{E}(z, t), \tag{18.2}$$

where

$$\chi_{res}(z, \omega_0) = \frac{1}{\epsilon_0 \hbar} \frac{|\boldsymbol{p}_{21} \cdot \hat{e}|^2}{\omega_0 - \omega_{21} + i\gamma_{21}} N(z) = -\frac{1}{\epsilon_0 \hbar} |\boldsymbol{p}_{21} \cdot \hat{e}|^2 N(z)\mathcal{L}(\omega_0, \omega_{21}). \tag{18.3}$$

In (18.3), $\hat{e}$ is the polarization vector of the optical field: $\mathcal{E}(z, t) = \hat{e}\mathcal{E}(z, t)$, $N = N_2 - N_1$ is the density of the population inversion of the gain medium, and $\mathcal{L}$ is the complex Lorentzian lineshape function defined in (1.167).

Under the rate-equation approximation, the gain coefficient that is contributed by the resonant susceptibility is

$$g(z, \omega_0) = -\frac{\omega_0}{cn_0}\chi_{res}''(z, \omega_0), \tag{18.4}$$

which has the same form as (15.7), and the change in the propagation constant associated with the resonant susceptibility is

$$\Delta k(z, \omega_0) = \frac{\omega_0}{2cn_0}\chi_{res}'(z, \omega_0). \tag{18.5}$$

Both $g(z, \omega_0)$ and $\Delta k(z, \omega_0)$ are directly proportional to $N(z)$, as can be seen from (18.3). Then the pulse propagation equation of (18.1) can be expressed as

$$\frac{\partial \mathcal{E}(z, t)}{\partial z} + \frac{1}{v_g}\frac{\partial \mathcal{E}(z, t)}{\partial t} + \sum_{n=2}^{\infty} \frac{i^{n-1}k_n}{n!}\frac{\partial^n \mathcal{E}(z, t)}{\partial t^n}$$
$$= -\frac{\alpha}{2}\mathcal{E}(z, t) + \frac{g(z, \omega_0)}{2}\mathcal{E}(z, t) + i\Delta k(z, \omega_0)\mathcal{E}(z, t) + is|\mathcal{E}(z, t)|^2\mathcal{E}(z, t). \tag{18.6}$$

Note that in this equation, α is the background absorption coefficient that is not spatially varying in the spatially homogeneous medium, but $g(z, \omega_0)$ is contributed by a resonant gain process of the medium, which can vary with space due to spatially varying pumping. Therefore, $\alpha \neq -g(z, \omega_0)$ because they are from different sources.

18.2 LINEAR PULSE PROPAGATION

In this subsection, we consider the propagation of an optical pulse in a linear medium where the nonlinear susceptibility does not exist, such that $P_{NL}(z,t) = 0$. Then, (18.6) reduces to a linear pulse propagation equation of the general form:

$$\frac{\partial \mathcal{E}(z,t)}{\partial z} + \frac{1}{v_g}\frac{\partial \mathcal{E}(z,t)}{\partial t} + i\frac{D}{2c\omega_0}\frac{\partial^2 \mathcal{E}(z,t)}{\partial t^2} + \sum_{n=3}^{\infty}\frac{i^{n-1}k_n}{n!}\frac{\partial^n \mathcal{E}(z,t)}{\partial t^n}$$
$$= -\frac{\alpha}{2}\mathcal{E}(z,t) + \frac{g(z,\omega_0)}{2}\mathcal{E}(z,t) + i\Delta k(z,\omega_0)\mathcal{E}(z,t). \tag{18.7}$$

This linear equation can be formally solved through Fourier transform.

By taking the Fourier transform on the time variable t, (18.7) has the form in the frequency domain:

$$\frac{\partial \mathcal{E}(z,\omega)}{\partial z} - i\frac{\omega}{v_g}\mathcal{E}(z,\omega) - i\frac{D\omega^2}{2c\omega_0}\mathcal{E}(z,\omega) - i\sum_{n=3}^{\infty}\frac{\omega^n k_n}{n!}\mathcal{E}(z,\omega)$$
$$= -\frac{\alpha}{2}\mathcal{E}(z,\omega) + \frac{g(z,\omega_0)}{2}\mathcal{E}(z,\omega) + i\Delta k(z,\omega_0)\mathcal{E}(z,\omega). \tag{18.8}$$

By integrating this equation over z, we obtain

$$\mathcal{E}(z,\omega) = L(z,\omega)\mathcal{E}(0,\omega) = G(z)T(z,\omega)\mathcal{E}(0,\omega), \tag{18.9}$$

where $L(z,\omega) = G(z)T(z,\omega)$ is the *linear propagation function* with

$$G(z) = \exp\left[-\frac{\alpha}{2}z + \int_0^z \frac{g(z,\omega_0)}{2}dz + i\int_0^z \Delta k(z,\omega_0)dz\right] \tag{18.10}$$

accounting for the *growth*, or attenuation, of the amplitude and the phase of the pulse field, and

$$T(z,\omega) = \exp\left(i\frac{\omega}{v_g}z + i\frac{D\omega^2}{2c\omega_0}z + i\sum_{n=3}^{\infty}\frac{\omega^n k_n}{n!}z\right) \tag{18.11}$$

accounting for the *transposition* of the pulse field as it propagates. Note that α, g, and Δk are independent of ω because we assume that they do not vary with time on the temporal scale of the pulse and we also ignore their dispersive nature.

By taking the inverse Fourier transform on ω for (18.9), we find the formal time-domain solution for the propagation of the pulse as a temporal convolution integral:

$$\mathcal{E}(z,t) = L(z,t) * \mathcal{E}(0,t) = G(z)T(z,t) * \mathcal{E}(0,t). \tag{18.12}$$

Therefore, with a given initial pulse field envelope of $\mathcal{E}(0,t)$, the pulse field envelope $\mathcal{E}(z,t)$ at the location z can be found once $T(z,t)$ is found by taking the inverse Fourier transform for $T(z,\omega)$. Note that $T(z,\omega)$ depends only on the linear optical properties of the medium; it is independent of the characteristics of the optical pulse. Therefore, both $G(z)$ and $T(z,t)$ are completely determined by the linear optical properties of the optical medium.

In general, $T(z,t)$ cannot be expressed in an analytical form if high-order dispersion that is characterized by the terms of $k_n = (\mathrm{d}^n k / \mathrm{d}\omega^n)_{\omega_0}$ is considered for $T(z,\omega)$, as given in (18.11). However, an analytical form can be found if only the two leading terms are considered while neglecting the high-order dispersion by taking the approximation:

$$T(z,\omega) \approx \exp\left(i\frac{z\omega}{v_g} + i\frac{Dz\omega^2}{2c\omega_0}\right) = T_v(z,\omega)T_{\mathrm{GVD}}(z,\omega),$$

(18.13)

where

$$T_v(z,\omega) = \exp\left(i\frac{z\omega}{v_g}\right),$$

(18.14)

and

$$T_{\mathrm{GVD}}(z,\omega) = \exp\left(i\frac{Dz\omega^2}{2c\omega_0}\right).$$

(18.15)

By taking the inverse Fourier transform, we find that

$$T_v(z,t) = \delta\left(t - \frac{z}{v_g}\right),$$

(18.16)

$$T_{\mathrm{GVD}}(z,t) = \left(\frac{ic\omega_0}{2\pi Dz}\right)^{1/2} \exp\left(-\frac{ic\omega_0}{2Dz}t^2\right),$$

(18.17)

and

$$T(z,t) \approx T_v(z,t) * T_{\mathrm{GVD}}(z,t) = \left(\frac{ic\omega_0}{2\pi Dz}\right)^{1/2} \exp\left[-\frac{ic\omega_0}{2Dz}\left(t - \frac{z}{v_g}\right)^2\right].$$

(18.18)

Thus, from (18.12), the pulse envelope travels in the time domain as

$$\mathcal{E}(z,t) = G(z)T(z,t) * \mathcal{E}(0,t)$$

$$\approx \left(\frac{ic\omega_0}{2\pi Dz}\right)^{1/2} \exp\left[-\frac{\alpha z}{2} + \int_0^z \frac{g(z,\omega_0)}{2}\mathrm{d}z + i\int_0^z \Delta k\,(z,\omega_0)\mathrm{d}z\right]$$

$$\times \int_{-\infty}^{\infty} \exp\left[-\frac{ic\omega_0}{2Dz}\left(t' - \frac{z}{v_g}\right)^2\right]\mathcal{E}(0,t-t')\mathrm{d}t'.$$

(18.19)

18.2.1 Linear Propagation of a Gaussian Pulse

As an example of linear pulse propagation, we consider in this subsection the propagation of a Gaussian optical pulse. We first define the general parameters of a Gaussian pulse by considering a general pulse envelope of the form:

$$\mathcal{E}(t) = \mathcal{E}_{\text{pk}} \exp\left(-\frac{t^2}{\tau_G^2}\right), \tag{18.20}$$

where $\mathcal{E}_{\text{pk}}$ is the peak field amplitude of the pulse and τ_G is the Gaussian pulsewidth parameter. The frequency-domain field profile of this pulse is

$$\mathcal{E}(\omega) = \int_{-\infty}^{\infty} \mathcal{E}(t)e^{i\omega t}dt = \sqrt{\pi}\tau_G \mathcal{E}_{\text{pk}} \exp\left(-\frac{\omega^2 \tau_G^2}{4}\right). \tag{18.21}$$

The electric field has the forms in the time and frequency domains, respectively:

$$E(t) = \mathcal{E}(t) \exp\left(-i\omega_0 t\right) = \mathcal{E}_{\text{pk}} \exp\left(-\frac{t^2}{\tau_G^2}\right) \exp\left(-i\omega_0 t\right), \tag{18.22}$$

and

$$E(\omega) = \mathcal{E}(\omega - \omega_0) = \sqrt{\pi}\tau_G \mathcal{E}_{\text{pk}} \exp\left[-\frac{(\omega - \omega_0)^2 \tau_G^2}{4}\right]. \tag{18.23}$$

The temporal intensity profile and the spectral intensity profile are, respectively,

$$I(t) = 2c\epsilon_0 n \left|E(t)\right|^2 = I_{\text{pk}} \exp\left(-\frac{2t^2}{\tau_G^2}\right), \tag{18.24}$$

and

$$I(\omega) = 2c\epsilon_0 n \left|E(\omega)\right|^2 = \pi\tau_G^2 I_{\text{pk}} \exp\left[-\frac{(\omega - \omega_0)^2 \tau_G^2}{2}\right], \tag{18.25}$$

where the peak intensity of the pulse is $I_{\text{pk}} = 2c\epsilon_0 n \left|\mathcal{E}_{\text{pk}}\right|^2$. The pulsewidth, Δt_{ps}, which is defined as the full width at half-maximum of the temporal intensity profile, $I(t)$, and the spectral width, $\Delta\omega_{\text{ps}}$, which is defined as the full width at half-maximum of the spectral intensity profile, $I(\omega)$, are respectively,

$$\Delta t_{\text{ps}} = \sqrt{2\ln 2}\,\tau_G \quad \text{and} \quad \Delta\omega_{\text{ps}} = \frac{2\sqrt{2\ln 2}}{\tau_G}. \tag{18.26}$$

We consider a Gaussian optical pulse of initial temporal and spectral field envelopes at $z = 0$ of the forms:

$$\mathcal{E}(0, t) = \mathcal{E}_0 \exp\left(-\frac{t^2}{\tau_{G0}^2}\right) \quad \text{and} \quad \mathcal{E}(0, \omega) = \sqrt{\pi}\tau_{G0} \mathcal{E}_0 \exp\left(-\frac{\omega^2 \tau_{G0}^2}{4}\right), \tag{18.27}$$

where $\mathcal{E}_0 = \mathcal{E}_{\text{pk}}(0)$ is the initial peak field amplitude of the pulse and $\tau_{G0} = \tau_G(0)$ is the initial pulsewidth parameter, both measured at $z = 0$. The initial fields in the time and frequency domains are, respectively,

$$E(0, t) = \mathcal{E}(0, t)e^{-i\omega_0 t} = \mathcal{E}_0 \exp\left(-\frac{t^2}{\tau_{G0}^2}\right)e^{-i\omega_0 t} \tag{18.28}$$

and

$$E(0, \omega) = \mathcal{E}(0, \omega - \omega_0) = \sqrt{\pi}\tau_{G0}\mathcal{E}_0 \exp\left[-\frac{(\omega - \omega_0)^2 \tau_{G0}^2}{4}\right]. \tag{18.29}$$

To focus our attention on the effect of dispersion on pulse propagation, we ignore the attenuation due to the background loss represented by α. We also ignore the amplification represented by g and the accompanying phase shift of Δk, both due to the resonant susceptibility. Thus, we take $G(z) = 1$ so that, from (18.9),

$$\mathcal{E}(z, \omega) = T(z, \omega)\mathcal{E}(0, \omega) = \sqrt{\pi}\tau_{G0}\mathcal{E}_0 \exp\left[i\frac{z\omega}{v_g} - \left(\tau_{G0}^2 - i\frac{2Dz}{c\omega_0}\right)\frac{\omega^2}{4}\right], \tag{18.30}$$

and

$$\mathcal{E}(z, t) = T(z, t) * \mathcal{E}(0, t) = \frac{\tau_{G0}}{\left(\tau_{G0}^2 - i\dfrac{2Dz}{c\omega_0}\right)^{1/2}} \mathcal{E}_0 \exp\left[-\frac{(t - z/v_g)^2}{\tau_{G0}^2 - i\dfrac{2Dz}{c\omega_0}}\right]. \tag{18.31}$$

We first examine the frequency-domain characteristics. By using (18.23) and (18.30), we find the field spectrum at the location z as

$$E(z, \omega) = \mathcal{E}(z, \omega - \omega_0)$$

$$= \sqrt{\pi}\tau_{G0}\mathcal{E}_0 \exp\left[-\frac{(\omega - \omega_0)^2 \tau_{G0}^2}{4}\right] \exp\left[i\frac{z(\omega - \omega_0)}{v_g} + i\frac{2Dz}{c\omega_0}\frac{(\omega - \omega_0)^2}{4}\right] \tag{18.32}$$

$$= \mathcal{E}(0, \omega - \omega_0) \exp\left[i\varphi(z, \omega)\right]$$

$$= E(0, \omega)\exp\left[i\varphi(z, \omega)\right],$$

where $\varphi(z, \omega)$ is a frequency-dependent phase shift that is linearly proportional to the propagation distance:

$$\varphi(z, \omega) = \left[\frac{(\omega - \omega_0)}{v_g} + \frac{2D}{c\omega_0}\frac{(\omega - \omega_0)^2}{4}\right] z. \tag{18.33}$$

From (18.32), we find that $\left|E(z, \omega)\right| = \left|E(0, \omega)\right|$. Therefore, the spectral intensity profile and the spectral width of the pulse stays unchanged as the pulse propagates:

$$I(z, \omega) = I(0, \omega) = \pi\tau_{G0}^2 I_0 \exp\left[-\frac{(\omega - \omega_0)^2 \tau_{G0}^2}{2}\right] \tag{18.34}$$

and

$$\Delta\omega_{ps}(z) = \Delta\omega_{ps}(0) = \frac{2\sqrt{2\ln 2}}{\tau_{G0}}. \tag{18.35}$$

In the time domain, we find, by using (18.22) and (18.31), that the electric field is

$$E(z,t) = \mathcal{E}(z,t)\exp(ik_0 z - i\omega_0 t)$$

$$= \frac{\tau_{G0}}{\left(\tau_{G0}^2 - i\dfrac{2Dz}{c\omega_0}\right)^{1/2}} \mathcal{E}_0 \exp\left[-\frac{(t - z/v_g)^2}{\tau_{G0}^2 - i\dfrac{2Dz}{c\omega_0}}\right] \exp(ik_0 z - i\omega_0 t)$$

$$= \frac{\tau_{G0}^{1/2}}{\tau_{G}^{1/2}} \mathcal{E}_0 \exp\left[-\frac{(t - z/v_g)^2}{\tau_G^2}\right] \exp[i\varphi(z,t)]$$

$$= \mathcal{E}_{pk}\exp\left[-\frac{(t - z/v_g)^2}{\tau_G^2}\right] \exp[i\varphi(z,t)], \tag{18.36}$$

where τ_G is the pulsewidth parameter, $\mathcal{E}_{pk}$ is the peak field amplitude, and $\varphi(z,t)$ is the spatially and temporally varying phase shift, all measured at the location z:

$$\tau_G = \left[\tau_{G0}^2 + \left(\frac{2Dz}{c\omega_0\tau_{G0}}\right)^2\right]^{1/2} = \left[1 + \left(\frac{2Dz}{c\omega_0\tau_{G0}^2}\right)^2\right]^{1/2}\tau_{G0}, \tag{18.37}$$

$$\mathcal{E}_{pk} = \frac{\tau_{G0}^{1/2}}{\tau_G^{1/2}}\mathcal{E}_0, \tag{18.38}$$

$$\varphi(z,t) = \theta(z) + k_0 z - \omega_0 t - \frac{\dfrac{2Dz}{c\omega_0}}{\tau_{G0}^4 + \left(\dfrac{2Dz}{c\omega_0}\right)^2}t^2 = \theta(z) + k_0 z - \omega_0 t - \frac{(\tau_G^2/\tau_{G0}^2 - 1)^{1/2}}{\tau_G^2}t^2,$$

$$\tag{18.39}$$

with $\theta(z) = (1/2)\tan^{-1}(2Dz/c\omega_0\tau_{G0}^2)$. From (18.36) and (18.37), we find that the pulse propagates with a group velocity of v_g, while its pulsewidth parameter τ_G increases, and its peak field amplitude $\mathcal{E}_{pk}$ decreases, with the propagation distance due to group-velocity dispersion. From (18.39), we find that the instantaneous frequency of the pulse linearly varies with time:

$$\omega(t) = -\frac{d\varphi}{dt} = \omega_0 + \gamma t, \tag{18.40}$$

where γ is the *chirp parameter*:

$$\gamma = \frac{d\omega}{dt} = \frac{\dfrac{4Dz}{c\omega_0}}{\tau_{G0}^4 + \left(\dfrac{2Dz}{c\omega_0}\right)^2} = \frac{2(\tau_G^2/\tau_{G0}^2 - 1)^{1/2}}{\tau_G^2}. \tag{18.41}$$

The pulse becomes *linearly chirped* due to group-velocity dispersion as it propagates because its instantaneous frequency as found in (18.40) is a linear function of time with a constant chirp parameter. For positive group-velocity dispersion, $D > 0$, the pulse has a *positive chirp* of $\gamma > 0$ with an increasing instantaneous frequency from its leading edge to its trailing edge. For negative group-velocity dispersion, $D < 0$, the pulse has a *negative chirp* of $\gamma < 0$ with a decreasing instantaneous frequency from its leading edge to its trailing edge. The characteristics of the pulse field in the time domain are summarized in Fig. 18.1. From (18.36), we find that the temporal intensity profile of the pulse spreads and the pulsewidth increases as the pulse propagates:

$$I(z,t) = \frac{\tau_{G0}}{\tau_G} I_0 \exp\left[-\frac{2(t - z/v_g)^2}{\tau_G^2}\right] \tag{18.42}$$

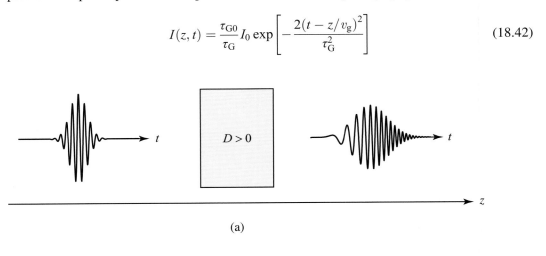

(a)

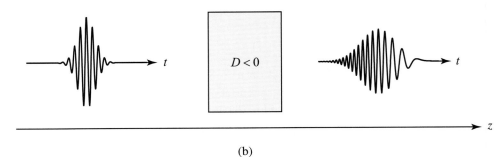

(b)

Figure 18.1 Propagation, spreading, and chirping of a Gaussian optical pulse through a medium (a) with a positive group-velocity dispersion, thus a positive chirp, and (b) with a negative group-velocity dispersion, thus a negative chirp. This figure shows the temporal variations of the electric field of an initially unchirped pulse before and after it propagates through a dispersive medium.

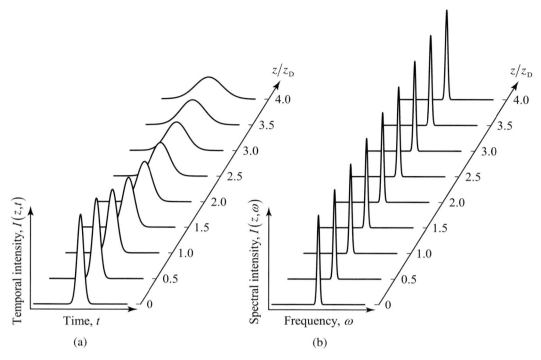

Figure 18.2 Evolutions of (a) the temporal intensity profile and (b) the spectral intensity profile of a Gaussian optical pulse as it propagates through a dispersive linear medium. In this figure, $z_D = c\omega_0\tau_{G0}^2/2D$.

and

$$\Delta t_{ps}(z) = \sqrt{2\ln 2}\,\tau_G > \Delta t_{ps}(0) = \sqrt{2\ln 2}\,\tau_{G0}. \tag{18.43}$$

From the above discussions, we conclude that *linear propagation of a pulse through a dispersive medium does not lead to any change in the pulse spectrum in the frequency domain, but it results in a temporally broadened pulse with a frequency chirp and a reduced peak amplitude.* Therefore, only the temporal intensity profile of the pulse changes, but the spectral intensity profile stays unchanged. Consequently, if a pulse is initially transform-limited, for which $\Delta t_{ps}\Delta \nu_{ps} = K$, it becomes nontransform-limited with $\Delta t_{ps}\Delta \nu_{ps} > K$ as Δt_{ps} increases while $\Delta \nu_{ps}$ remains unchanged. The evolutions of $I(z,t)$, as given in (18.42), and $I(z,\omega)$, as given in (18.34), as a pulse propagates through a dispersive linear medium are shown in Fig. 18.2.

EXAMPLE 18.1

The transmission distance and bit rate of an optical digital data transmission system are limited by the effect of dispersion of the transmission medium, such as an optical fiber. Transform-limited Gaussian pulses that have a pulsewidth of Δt_{ps0} and a spectral width of $\Delta \nu_{ps0}$, such that $\Delta t_{ps0}\Delta \nu_{ps0} = K = 2\ln 2/\pi$ are used as the input data bits at $z = 0$. Assume that the peak power of the pulses is sufficiently low to avoid nonlinear effects, such as the optical Kerr effect, such that the transmission is linear transmission. (a) Find the limitation on the transmission distance for a bit rate of f_{bit}. (b) Consider the transmission of data pulses at the wavelength of $\lambda = 1.55$ μm through a silica fiber at a bit rate of $f_{bit} = 10$ Gbit s^{-1}. Each of the pulses is a transform-limited Gaussian pulse with an initial pulsewidth of $\Delta t_{ps0} = 5$ ps. The silica fiber has a group-velocity dispersion of $D_\lambda = 17$ ps nm^{-1} km^{-1} and an absorption coefficient of $\alpha = 0.15$ dB km^{-1} at $\lambda = 1.55$ μm. Find the limitation on the transmission distance due to dispersion.

Solution

For a bit rate of f_{bit}, the time interval of each bit, represented by a pulse, is $T_{bit} = f_{bit}^{-1}$. The dimensionless coefficient D of group-velocity dispersion is related to D_λ as $D = -c\lambda D_\lambda$, from (4.84). The spectral width of the pulse measured in wavelength is related to that measured in frequency as $\Delta\nu_{ps} = c\Delta\lambda_{ps}/\lambda^2$. The full width at half-maximum pulsewidth of a Gaussian pulse is related to its pulsewidth parameter as $\Delta t_{ps} = \sqrt{2\ln 2}\tau_G$, from (17.66).

(a) The temporal pulsewidth of an initially transform-limited Gaussian pulse broadens as given in (18.37) as it propagates through a dispersive medium when nonlinear optical effects are negligible. For clear identification of each bit that is represented by a pulse, the limitation on the transmission distance is imposed by the requirement that $T_{bit} > \Delta t_{ps}$. By using the relation that $\Delta t_{ps} = \sqrt{2\ln 2}\tau_G$ for (18.37), this limitation leads to the relation:

$$\Delta t_{ps} = \left[1 + \left(\frac{4\ln 2 Dz}{c\omega_0 \Delta t_{ps0}^2}\right)^2\right]^{1/2} \Delta t_{ps0} < T_{bit} = f_{bit}^{-1}.$$

By rearranging this relation, we obtain the limitation on the transmission distance as

$$z < \frac{c\omega_0 \Delta t_{ps0}}{4\ln 2 |D| f_{bit}} (1 - f_{bit}^2 \Delta t_{ps0}^2)^{1/2} \qquad \text{in terms of } \Delta t_{ps0},$$

$$z < \frac{c\omega_0}{2\pi |D| f_{bit}\Delta\nu_{ps0}} \left[1 - \frac{4(\ln 2)^2 f_{bit}^2}{\pi^2 \Delta\nu_{ps0}^2}\right]^{1/2} \qquad \text{in terms of } \Delta\nu_{ps0}, \qquad (18.44)$$

$$z < \frac{1}{|D_\lambda| f_{bit}\Delta\lambda_{ps0}} \left[1 - \frac{4(\ln 2)^2 \lambda^4 f_{bit}^2}{\pi^2 c^2 \Delta\lambda_{ps0}^2}\right]^{1/2} \qquad \text{in terms of } \Delta\lambda_{ps0}.$$

(b) With $\lambda = 1.55$ μm and $D_\lambda = 17$ ps nm^{-1} km^{-1}, we have $\omega_0 = 2\pi c/\lambda = 1.216 \times 10^{15}$ rad s^{-1} and $D = -c\lambda D_\lambda = -7.9 \times 10^{-3}$. Therefore, for $f_{bit} = 10$ Gbit s^{-1} and $\Delta t_{ps0} = 5$ ps, we find from the first expression of (18.44) that the limitation on the transmission distance is

$$z < \frac{c\omega_0 \Delta t_{ps0}}{4\ln 2 |D| f_{bit}} (1 - f_{bit}^2 \Delta t_{ps0}^2)^{1/2} = 8.32 \text{ km}.$$

The transmission distance is limited by the temporal pulse broadening due to dispersion, but not by the attenuation due to absorption. Therefore, the absorption coefficient is irrelevant in this example.

18.2.2 Compression of Chirped Pulses

A chirped pulse can be compressed by removing the frequency chirp of the pulse through a dispersive element that has the opposite dispersion. Compression of optical pulses was first proposed in 1968 by Giordmaine *et al.* [1]. Compression of a linearly chirped pulse requires only a linearly dispersive element that introduces the opposite dispersion to eliminate the chirp. The most commonly used dispersive element for this purpose is a pair of gratings, but many other possibilities exist, such as a pair of prisms.

One important application of the compression of chirped pulses is the technique of *chirped-pulse amplification* of pulses, a concept that was originally developed for radar. This technique was demonstrated for the effective amplification of high-power optical pulses by Strickland and Mourou [2], who shared the Nobel Prize in physics in 2018 for this work. Figure 18.3 shows the schematics of this technique.

A high-power optical pulse can easily saturate the gain medium or cause damage to an optical amplifier, and to other optical components in its path, through nonlinear processes such as self-focusing and multiphoton absorption. Before the introduction of chirped-pulse amplification for

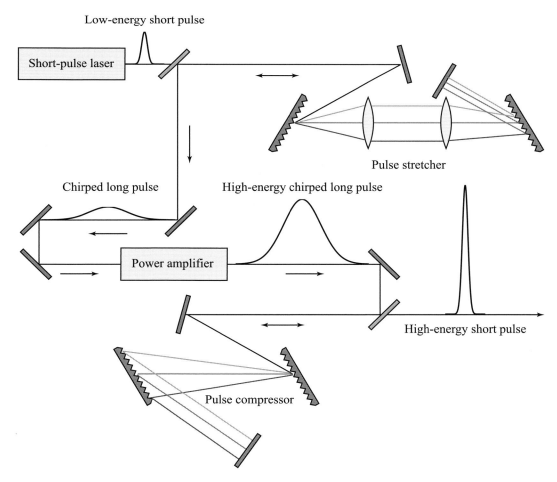

Figure 18.3 Schematics of chirped-pulse amplification for high-power optical pulses.

optical pulses in 1985, the peak power of laser pulses was limited by this problem. In chirped-pulse amplification, a laser pulse is chirped and temporally stretched to a much longer duration by using a *pulse stretcher*, which is a highly dispersive element. This process reduces the peak power to a sufficiently low level so that the stretched pulse can be safely amplified without causing the detrimental nonlinear effects. The result is an amplified high-energy pulse that has a stretched duration and a relatively low peak power. After the stretched pulse is amplified to a high energy level, the chirp in this pulse is then removed by using a *pulse compressor*, which is an optical element that has the opposite dispersion of the pulse stretcher. By removing the chirp in the pulse, the pulse compressor compresses the temporal pulsewidth of the pulse while increasing the peak power. By using highly dispersive pulse stretchers and compressors that stretch and compress pulses by 4–6 orders of magnitude, the technique of chirped-pulse amplification has led to the development of tabletop terawatt lasers that deliver femtosecond pulses of terawatt peak powers and the development of pulses of petawatt peak powers at large facilities.

18.3 TEMPORAL SELF-PHASE MODULATION

In this section, we consider the effect of self-phase modulation on the propagation of an optical pulse in an optical Kerr medium. Self-phase modulation causes spectral broadening of optical pulses. Spectral broadening caused by self-phase modulation was first experimentally observed by Brewer [3] and theoretically interpreted by Shimizu [4], both in 1967.

When an optical pulse propagates in a spatially homogeneous optical Kerr medium, the intensity-dependent Kerr phase change generally varies with both space and time, as expressed in (5.42). The transverse spatial intensity profile of the beam causes spatially varying self-phase modulation, which leads to self-focusing or self-defocusing, as discussed in Section 12.2. By contrast, the temporal intensity profile of the pulse causes temporally varying self-phase modulation, which results in self-induced spectral broadening, as discussed in Section 12.4. The possibility of self-focusing or self-defocusing creates complications for pulse propagation. These complications can be avoided by propagating an optical pulse in a single-mode waveguide, such as a single-mode optical fiber. Because there is only one transverse mode, the spatial phase change is uniform across the transverse mode profile. Meanwhile, the integrity of the spatial mode is maintained by the waveguiding effect, which prevents the mode from self-focusing or self-defocusing. In this case, the one-dimensional traveling-wave equation given in (18.1) is valid.

To focus on the effect of self-phase modulation on the propagation of an optical pulse, we assume a lossless background medium and no resonant susceptibility such that $\alpha = 0$ and $\mathcal{P}_{res} = 0$ for (18.1). We also ignore dispersion of the third and higher orders. Then, (18.1) reduces to the form of (8.52):

$$\frac{\partial \mathcal{E}(z,t)}{\partial z} + \frac{1}{v_g}\frac{\partial \mathcal{E}(z,t)}{\partial t} + i\frac{k_2}{2}\frac{\partial^2 \mathcal{E}(z,t)}{\partial t^2} = is\left|\mathcal{E}(z,t)\right|^2 \mathcal{E}(z,t), \tag{18.45}$$

where $k_2 = D/c\omega_0$ and $s = 2\omega_0\epsilon_0 n_0 n_2$, as defined in (8.50). By performing the change of variables $\zeta = z$ and $\tau = t - z/v_g$ as expressed in (8.55), we transform (18.45) to the nonlinear Schrödinger equation given in (8.56):

$$i\frac{\partial\mathcal{E}(\zeta,\tau)}{\partial\zeta} = \frac{k_2}{2}\frac{\partial^2\mathcal{E}(\zeta,\tau)}{\partial\tau^2} - s\big|\mathcal{E}(\zeta,\tau)\big|^2\mathcal{E}(\zeta,\tau). \tag{18.46}$$

In this section, we first consider the propagation of a pulse in an optical Kerr medium without dispersion such that $k_2 = 0$ but $s \neq 0$. Then, (18.46) reduces to the form:

$$i\frac{\partial\mathcal{E}(\zeta,\tau)}{\partial\zeta} = -s\big|\mathcal{E}(\zeta,\tau)\big|^2\mathcal{E}(\zeta,\tau). \tag{18.47}$$

The solution of this equation is

$$\mathcal{E}(\zeta,\tau) = \mathcal{E}(0,\tau)\exp\big[is\big|\mathcal{E}(0,\tau)\big|^2\zeta\big] = \mathcal{E}(0,\tau)\exp\Big[i\frac{\omega_0}{c}n_2I(0,\tau)\zeta\Big]. \tag{18.48}$$

Therefore,

$$\mathcal{E}(z,t) = \mathcal{E}\Big(0, t - \frac{z}{v_g}\Big)\exp\big[i\varphi_K(z,t)\big], \tag{18.49}$$

where

$$\varphi_K(z,t) = \frac{\omega_0}{c}n_2I(0, t - z/v_g)z = \frac{\omega_0}{c}n_2I(z,t)z \tag{18.50}$$

is the spatially and temporally varying Kerr phase, and

$$I(z,t) = 2c\epsilon_0 n\big|\mathcal{E}(z,t)\big|^2 = 2c\epsilon_0 n\big|\mathcal{E}(0, t - z/v_g)\big|^2 = I(0, t - z/v_g). \tag{18.51}$$

For pulse propagation over a distance of l, we find that (18.50) is identical to (12.74) with $z = l$. Thus, the discussions that follow (12.74) are valid. The consequence is spectral broadening caused by self-phase modulation, as illustrated in Fig. 12.6. By using (18.49), we have

$$E(z,t) = \mathcal{E}(z,t)\exp(ik_0z - i\omega_0t) = \mathcal{E}(0, t - z/v_g)\exp\big[i\varphi(z,t)\big], \tag{18.52}$$

so that

$$\varphi(z,t) = k_0z - \omega_0t + \varphi_K(z,t). \tag{18.53}$$

Then, by using (18.50) and (18.51), we find that the instantaneous frequency of the pulse varies nonlinearly with time as

$$\omega(z,t) = -\frac{d\varphi}{dt} = \omega_0 - \frac{d\varphi_K}{dt} = \omega_0 - \frac{\omega_0 n_2 z}{c}\frac{dI(z,t)}{dt}. \tag{18.54}$$

Therefore, the pulse becomes nonlinearly chirped with a broadened spectrum as it propagates through the optical Kerr medium even when there is no dispersion.

For a Gaussian pulse of the form given in (18.20) but with an initial pulsewidth parameter of τ_{G0}, it can be shown that its spectrum falls within a range of $\omega_{max} = \omega_0 + \omega_m \geq \omega(z,t) \geq \omega_0 - \omega_m = \omega_{min}$. As the pulse propagates through the nondispersive optical Kerr medium, we find from (18.51) that its temporal intensity profile stays unchanged, but we find from

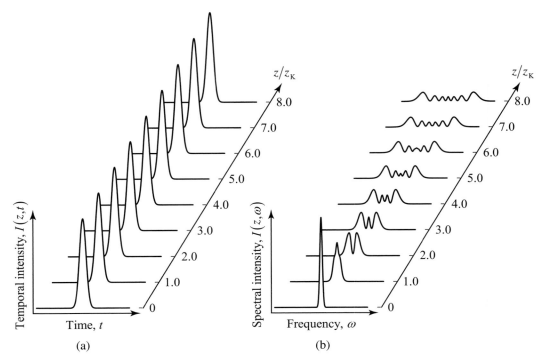

Figure 18.4 Evolutions of (a) the temporal intensity profile and (b) the spectral intensity profile of a Gaussian optical pulse as it propagates through a nondispersive optical Kerr medium. For this figure, z_K is defined in (18.56).

(18.55) that its spectral width increases linearly with the propagation distance. *In the absence of dispersion, self-phase modulation of an optical pulse generates new frequency components to broaden the spectral intensity profile of the pulse but not its temporal intensity profile.* The evolutions of $I(z, \omega)$ and $I(z, t)$ as a pulse propagates through a nondispersive optical Kerr medium are illustrated in Fig. 18.4. Because of the complex shape of the spectral intensity profile, a full width at half-maximum cannot be defined. Instead, the spectral width can be defined as the width between the maximum and the minimum frequencies:

$$\Delta\omega_{ps}(z) = \omega_{max}(z) - \omega_{min}(z) = \frac{4\omega_0 |n_2| I_0}{e^{1/2} c \tau_{G0}} z = \frac{2\pi}{\tau_{G0}} \frac{z}{z_K}, \tag{18.55}$$

where

$$z_K = \frac{\pi e^{1/2} c}{2\omega_0 |n_2| I_0}. \tag{18.56}$$

The spectral width of the pulse linearly increases with the propagation distance.

It is seen in Fig. 18.4 that the spectral intensity $I(z, \omega)$ shows a semi-periodic interference structure, with the number of peaks increasing with the propagation distance [5–7]. By examining Fig. 12.6, this feature can be understood from the fact that self-phase modulation generates each frequency twice within its spectrum at two different points of time in the pulse duration. Because the two components of the same frequency are generated at different times, they have different phases. The phase difference varies with the frequency across the spectrum, resulting in an interference pattern with peaks and valleys, as seen in Fig. 18.4. Both the spectral width and the phase of each frequency, as well as the phase difference between the two components of each frequency, in the self-phase-modulated spectrum increase with the pulse propagation distance. Thus, both the spectral width of $I(z, \omega)$ and the number of peaks and valleys in the spectrum increase with the propagation distance. By contrast, the temporal intensity profile $I(z, t)$ stays unchanged because all frequency components travel at the same velocity in the absence of dispersion.

18.4 SPECTRAL STRETCHING AND PULSE COMPRESSION

In Section 18.2, we considered pulse propagation in a dispersive linear medium, for which $D \neq 0$ but $n_2 = 0$. In Section 18.3, we considered pulse propagation in a nondispersive optical Kerr medium, for which $D = 0$ but $n_2 \neq 0$. In this and the following sections, we consider pulse propagation in a dispersive optical Kerr medium, for which $D \neq 0$ and $n_2 \neq 0$. There are two general cases: In this section, we consider the case that the group-velocity dispersion D and the nonlinear refractive index n_2 have the same sign, such that $Dn_2 > 0$. In the following section, we will consider the case that the group-velocity dispersion D and the nonlinear refractive index n_2 have different signs, such that $Dn_2 < 0$.

There are two different situations for the case that $Dn_2 > 0$: either (1) $D > 0$ and $n_2 > 0$, or (2) $D < 0$ and $n_2 < 0$. The scenarios of these two situations can be visualized by considering the self-phase modulation spectra shown in Fig. 12.6 for $n_2 > 0$ and $n_2 < 0$, respectively.

In the situation that $D > 0$ and $n_2 > 0$, we find from Fig. 12.6(a) that the frequencies generated by self-phase modulation in the leading half of the pulse are red-shifted to lower frequencies with respect to the center carrier frequency ω_0, whereas those in the trailing half of the pulse are blue-shifted to higher frequencies with respect to the center carrier frequency ω_0. In a medium that has a positive group-velocity dispersion with $D > 0$, a wave packet at a lower frequency travels faster at a higher group velocity than that at a higher frequency. Consequently, with $D > 0$ and $n_2 > 0$, the leading half of the pulse speeds up, while the trailing half slows down, with respect to the center of the pulse, as their frequencies are varied by self-phase modulation. This process results in temporal broadening of the pulse while the spectrum of the pulse is stretched by self-phase modulation. The temporal broadening and the spectral stretching continue to enhance each other as the pulse propagates further through the medium.

In the situation that $D < 0$ and $n_2 < 0$, we find from Fig. 12.6(b) that the frequencies generated by self-phase modulation in the leading half of the pulse are blue-shifted to higher frequencies with respect to the center carrier frequency ω_0, whereas those in the trailing half of the pulse are red-shifted to lower frequencies. In a medium that has a negative group-velocity dispersion with $D < 0$, a wave packet at a higher frequency travels faster at a higher group velocity than that at a lower frequency. Consequently, as in the situation with $D > 0$ and $n_2 > 0$, we also find with $D < 0$ and $n_2 < 0$ that the pulse is temporally broadened as its leading half speeds up, while its trailing half slows down, with respect to the pulse center, as their frequencies are varied by self-phase modulation.

From the above qualitative analysis, we conclude that when a pulse propagates in a dispersive optical Kerr medium that has the same sign for D and n_2, such that $Dn_2 > 0$, the pulse is temporally broadened by group-velocity dispersion as it is spectrally stretched by self-phase modulation. The two effects continue as the pulse propagates further through the medium. Over a sufficiently long propagation distance, the pulse develops into a square temporal shape that is approximately linearly chirped within the flattop duration of the square pulse [6, 8], as shown in Fig. 18.5 for the comparison between (a) the situation that $D = 0$ and $n_2 > 0$ and (b) the situation that $D > 0$ and $n_2 > 0$. In both situations for $n_2 > 0$, the central portion of the pulse has a positive chirp. In the situations for $n_2 < 0$, with either $D = 0$ or $D < 0$, the general characteristics are similar to those shown in Fig. 18.5, but the temporal evolution of the frequency is reversed, with the central portion of the pulse having a negative chirp. The propagation distance for the pulse to develop into the square pulse shape in the presence of group-velocity dispersion depends on the input temporal width and the peak power of the pulse; it is shorter for a shorter pulse and for a higher-power pulse.

Both self-phase-modulated pulses shown in Fig. 18.5 have stretched spectra. They are not transform-limited and thus can be temporally compressed. Because the central portion of each pulse is approximately linearly chirped, the pulse can be compressed with a dispersive linear element that has the opposite dispersion. The concept is illustrated in Fig. 18.6 with a pair of properly arranged gratings serving as the pulse compressor. The period and the separation of the gratings are properly chosen and adjusted to optimally cancel out most of the chirp in the pulse. It can be seen from Fig. 18.5 that the linearly chirped central portion of each of the two pulses contains most of the pulse energy, but the outer parts of the pulse are not linearly chirped.

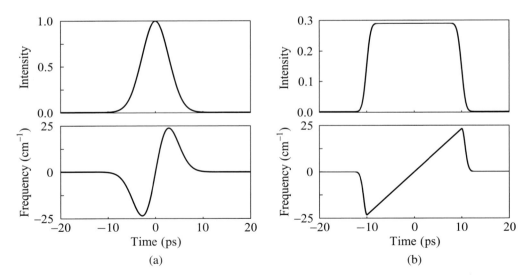

Figure 18.5 Temporal shape and frequency modulation for (a) an optical pulse with only self-phase modulation after propagation through a nondispersive optical Kerr medium with $D = 0$ and $n_2 > 0$ and (b) a square pulse developed through group-velocity dispersion and self-phase modulation after propagation through a dispersive optical Kerr medium with $D > 0$ and $n_2 > 0$. The central portion of the pulse is linearly and positively chirped. (Adapted from D. Grischkowsky and A. C. Balant, "Optical pulse compression based on enhanced frequency chirping," *Applied Physics Letters*, vol. 41, pp. 1–3, 1982.)

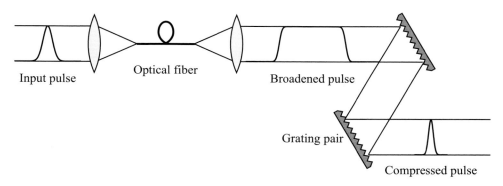

Figure 18.6 Schematics of a pulse compressor that uses an optical fiber as a dispersive self-phase modulator and a pair of gratings as the pulse compressor. The pulse is both temporally and spectrally broadened by the dispersion and the nonlinearity of the optical fiber. The grating pair compresses it by reducing its temporal width without changing its spectral width.

Therefore, the compressed pulse has some structures on its wings, though most of its energy is contained in its central peak. From Fig. 18.5, we find that the square pulse shown in Fig. 18.5(b) has more energy contained in the linearly chirped portion than the pulse shown in Fig. 18.5 (a). Therefore, the pulse resulting from linear compression of the square pulse that is generated through a dispersive optical Kerr medium has a higher central peak with less extensive wings than the one resulting from linear compression of the pulse that is generated through a nondispersive optical Kerr medium [6]. In the paper that first proposed optical pulse compression in 1968 by Giordmaine *et al.* [1], mode-locked laser pulses were electro-optically modulated with an optical frequency shifter to accomplish spectral stretching with a linear chirp. Compression of ultrashort laser pulses, however, commonly uses a passive optical Kerr medium, such as an optical fiber, for spectral stretching. The group-velocity dispersion of a silica fiber is positive, $D > 0$, for an optical wavelength shorter than 1.3 μm, and the nonlinear index is also positive, $n_2 > 0$. Therefore, the concepts of spectral stretching and temporal compression discussed in this section can be realized by using a silica fiber as the optical Kerr medium for pulses at $\lambda < 1.3$ μm [6, 9–11]. Pulse compression with large compression ratios has been successfully applied to various laser pulses.

18.5 TEMPORAL SOLITONS

In this section, we consider optical pulse propagation in a dispersive optical Kerr medium for which the group-velocity dispersion D and the nonlinear refractive index n_2 have opposite signs, such that $Dn_2 < 0$. There are two situations for this case: (1) $D < 0$ and $n_2 > 0$, or (2) $D > 0$ and $n_2 < 0$. Because $n_2 > 0$ for most optical materials, we only consider the situation that $D < 0$ and $n_2 > 0$. The general concepts and conclusions apply to the other situation that $D > 0$ and $n_2 < 0$, except that the signs of both parameters are simultaneously changed.

When an optical pulse propagates in an optical Kerr medium with $n_2 > 0$, self-phase modulation stretches the pulse spectrum by generating red-shifted frequencies in the leading half of the pulse and blue-shifted frequencies in the trailing half, as discussed in the preceding section and seen in Fig. 12.6(a) and in Fig. 18.5(a). If the group-velocity dispersion of the medium is negative, $D < 0$, then a wave packet in the red-shifted leading half travels at a lower velocity than one in the blue-shifted trailing half. Therefore, the pulse is not temporally spread. In the steady state, the effects of positive self-phase modulation and negative group-velocity dispersion reach a balance, resulting in an optical pulse of a definite shape with a finite pulsewidth, known as an *optical soliton*. Solitons appear universally in various dispersive nonlinear systems. Solitary waves have been studied since Russel made the first observation in the propagation of solitary water waves in 1834. Solitons as stable entities and their interactions in a collisionless plasma were first observed and analyzed by Zabusky and Kruskal in 1965 [12]. Optical solitons were first theoretically shown by Hasegawa and Tappert in 1973 [13] and experimentally demonstrated in optical fibers by Mollenauer *et al.* in 1980 [14]. Here, we consider only *bright temporal solitons*. *Dark temporal solitons* have been observed in optical fibers [15]. A *spatial soliton* can be formed when spatial spreading due to diffraction is balanced by self-focusing, as mentioned in Subsection 12.2.2.

The behavior of an optical pulse traveling in a dispersive optical Kerr medium with $D < 0$ and $n_2 > 0$ is governed by the nonlinear Schrödinger equation given in (18.46):

$$i\frac{\partial \mathcal{E}(\zeta,\tau)}{\partial \zeta} = \frac{k_2}{2}\frac{\partial^2 \mathcal{E}(\zeta,\tau)}{\partial \tau^2} - s\left|\mathcal{E}(\zeta,\tau)\right|^2 \mathcal{E}(\zeta,\tau), \tag{18.57}$$

where

$$k_2 = \frac{D}{c\omega_0} = \left(\frac{d^2 k}{d\omega^2}\right)_{\omega_0} < 0 \quad \text{and} \quad s = 2\omega_0\epsilon_0 n_0 n_2 > 0. \tag{18.58}$$

Depending on the parameters of the input pulse $\mathcal{E}(0,\tau)$ at $\zeta = 0$, the steady-state solution $\mathcal{E}(\zeta,\tau)$ of this equation takes one of the forms for different orders of solitons [16, 17]. The group-velocity dispersion of a silica fiber is negative, $D < 0$, for an optical wavelength longer than 1.3 μm, and the nonlinear index is positive, $n_2 > 0$. Therefore, for pulses at $\lambda > 1.3$ μm, optical solitons can be formed in silica fibers under the correct conditions [13, 14, 18–20].

18.5.1 Fundamental Soliton

The lowest-order solution, known as the *fundamental soliton*, of (18.57) is

$$\mathcal{E}_s(\zeta,\tau) = \mathcal{E}_0 \operatorname{sech}\left(\frac{\tau}{\tau_s}\right)e^{i\varphi(\zeta)}, \tag{18.59}$$

where τ_s is the pulsewidth parameter of the soliton. Thus, in terms of the original variables z and t,

$$\mathcal{E}_s(z,t) = \mathcal{E}_0 \operatorname{sech}\left(\frac{t - z/v_g}{\tau_s}\right)e^{i\varphi(z)}. \tag{18.60}$$

For (18.59) and (18.60), the amplitude $\mathcal{E}_0$ is a real quantity that satisfies the condition:

$$\mathcal{E}_0\tau_s = \left(\frac{|k_2|}{s}\right)^{1/2} = \left(\frac{|D|}{2\omega_0^2 c\epsilon_0 n_0 n_2}\right)^{1/2}, \qquad (18.61)$$

and $\varphi(z)$ is a phase given by the relation:

$$\varphi(z) = \frac{|k_2|}{2\tau_s^2}z = \frac{1}{2}s\mathcal{E}_0^2 z = \frac{|D|}{2c\omega_0\tau_s^2}z = \omega_0\epsilon_0 n_0 n_2 \mathcal{E}_0^2 z. \qquad (18.62)$$

From (18.61), we find that the fundamental soliton is not characterized by a fixed value for its amplitude $\mathcal{E}_0$ or a fixed value for its pulsewidth parameter τ_s. It is also not characterized by a fixed value for its intensity or power. Instead, it is characterized by a fixed value for its *area* defined as

$$\mathcal{A}_0 = \int_{-\infty}^{\infty} |\mathcal{E}_s(\zeta,\tau)| d\tau = \pi\mathcal{E}_0\tau_s = \pi\left(\frac{|k_2|}{s}\right)^{1/2} = \pi\left(\frac{|D|}{2\omega_0^2 c\epsilon_0 n_0 n_2}\right)^{1/2}. \qquad (18.63)$$

The amplitude $\mathcal{E}_0$ and the pulsewidth parameter τ_s of a fundamental soliton are not fixed; they can be varied under the condition that their product has a fixed value that makes the area of the pulse satisfy the condition given in (18.63). Under this condition, the amplitude and the pulsewidth parameter of the fundamental soliton are determined by the *fluence* of the pulse:

$$F_0 = \int_{-\infty}^{\infty} I(\zeta,\tau) d\tau = 2c\epsilon_0 n_0 \int_{-\infty}^{\infty} |\mathcal{E}_s(\zeta,\tau)|^2 d\tau = 4c\epsilon_0 n_0 \mathcal{E}_0^2\tau_s = 2I_0\tau_s, \qquad (18.64)$$

where $I_0 = 2c\epsilon_0 n_0 \mathcal{E}_0^2 = |D|/(\omega_0^2 n_2\tau_s^2)$ is the peak intensity of the fundamental soliton pulse.

Because the area $\mathcal{A}_0$ is a constant that is only determined by the parameters of the medium and the carrier frequency of the pulse, by using (18.63) and (18.64), we find that

$$\mathcal{E}_0 = \frac{\pi}{4c\epsilon_0 n_0}\frac{F_0}{\mathcal{A}_0} = \left(\frac{\omega_0^2 n_2}{8|D|c\epsilon_0 n_0}\right)^{1/2} F_0 \qquad (18.65)$$

and

$$\tau_s = \frac{4c\epsilon_0 n_0}{\pi^2}\frac{\mathcal{A}_0^2}{F_0} = \frac{2|D|}{\omega_0^2 n_2 F_0}. \qquad (18.66)$$

Therefore, once the parameters of the medium, including D and n_2, and the carrier frequency, ω_0, of the pulse are given, the area of the fundamental soliton has a fixed value that is determined by (18.63), but the fluence of the pulse is a free parameter that can be varied. The amplitude $\mathcal{E}_0$ and the pulsewidth parameter τ_s of the fundamental soliton are determined by the fluence F_0 of the pulse. To keep the pulse area fixed at the constant value given in (18.63), the pulse amplitude increases proportionally with the fluence, as seen in (18.65); meanwhile, the pulsewidth decreases inversely with the fluence, as seen in (18.66).

As seen from (18.60), the fundamental soliton has a field envelope of the hyperbolic secant shape, and thus an intensity profile of the sech^2 shape. As a fundamental soliton propagates through the medium, its temporal shape stays unchanged; only its phase increases with the propagation distance, as seen in (18.62). A soliton is robust because it is a mode of the nonlinear Schrödinger equation. When two fundamental solitons collide, either because they propagate in opposite directions or because they have different carrier frequencies, they interfere with each other as optical waves do, but each of them regains its original profile when they separate after the collision. Only their phases are changed by the collision. Therefore, a soliton has intrinsic integrity, like a particle.

When an optical pulse that has the parameters matching those of a fundamental soliton is launched into an optical Kerr medium, such as an optical fiber, the pulse maintains both its amplitude and pulsewidth as it propagates; only its phase changes according to (18.62). If the parameters of the launched pulse do not match those of a soliton, neither a fundamental soliton nor a high-order soliton as discussed below, the pulse evolves into a soliton of the closest order and a temporally spreading background. This spreading background is a *dispersive wave* because its spectral components at different frequencies travel at different group velocities. Unlike a soliton, for which the effect of group-velocity dispersion is balanced by the effect of optical nonlinearity, a dispersive wave is spread by the effect of group-velocity dispersion that is uncompensated by optical nonlinearity.

EXAMPLE 18.2

A fundamental soliton pulse that has a full width at half-maximum pulsewidth of $\Delta t_{ps0} = 5$ ps and a carrier wavelength of $\lambda = 1.55$ μm is transmitted through a silica optical fiber that has the parameters given in Example 18.1. At this wavelength, the silica fiber has the following parameters: the refractive index is $n_0 = 1.444$, the nonlinear refractive index is $n_2 = 2.4 \times 10^{-20}$ m^2 W^{-1}, the group-velocity dispersion is $D = -c\lambda D_\lambda = -7.9 \times 10^{-3}$, and the effective core radius is $r = 4.5$ μm. Find the peak intensity and the peak power of this soliton pulse. Then find its fluence and its energy.

Solution
The effective area of the fiber core is

$$\mathcal{A} = \pi r^2 = \pi \times 4.5^2 \ \mu\text{m}^2 = 6.36 \times 10^{-11} \ \text{m}^2.$$

A fundamental soliton pulse has the sech field profile and the sech^2 intensity profile. With a full width at half-maximum pulsewidth of $\Delta t_{ps0} = 5$ ps, the pulsewidth parameter of this pulse is, from (17.75),

$$\tau_s = \frac{\Delta t_{ps0}}{2 \ln (1 + \sqrt{2})} = \frac{5}{2 \ln (1 + \sqrt{2})} \ \text{ps} = 2.84 \ \text{ps}.$$

For $\lambda = 1.55$ μm, $\omega_0 = 2\pi c/\lambda = 1.216 \times 10^{15}$ rad s^{-1}, as found in Example 18.1. By using (18.61), we find the peak intensity of this fundamental soliton pulse:

$$I_0 = 2c\epsilon_0 n_0 \mathcal{E}_0^2$$

$$= \frac{|D|}{\omega_0^2 n_2 \tau_s^2}$$

$$= \frac{7.9 \times 10^{-3}}{(1.216 \times 10^{15})^2 \times 2.4 \times 10^{-20} \times (2.84 \times 10^{-12})^2} \ \text{W m}^{-2}$$

$$= 27.6 \ \text{GW m}^{-2}.$$

Therefore, the peak power of the pulse is

$$P_0 = I_0 \mathcal{A} = 27.6 \times 10^9 \times 6.36 \times 10^{-11} \ \text{W} = 1.76 \ \text{W}.$$

By using (18.64), we find the fluence of this fundamental soliton pulse:

$$F_0 = 2I_0 \tau_s = 2 \times 27.6 \times 10^9 \times 2.84 \times 10^{-12} \ \text{J m}^{-2} = 157 \ \text{mJ m}^{-2}.$$

Therefore, the pulse energy is

$$U_{\text{ps0}} = F_0 \mathcal{A} = 157 \times 10^{-3} \times 6.36 \times 10^{-11} \ \text{J} = 10 \ \text{pJ}.$$

18.5.2 High-Order Solitons

Multiple fundamental solitons can be superimposed to form a *high-order soliton* [16, 17] of the input field:

$$\mathcal{E}(0, \tau) = N \mathcal{E}_s(0, \tau) = N \mathcal{E}_0 \operatorname{sech}\left(\frac{\tau}{\tau_s}\right) = \mathcal{E} \operatorname{sech}\left(\frac{\tau}{\tau_s}\right), \tag{18.67}$$

where N is an integer, with $N = 1$ for the fundamental soliton, and $\mathcal{E} = N \mathcal{E}_0$. This soliton has an area of

$$\mathcal{A} = N \mathcal{A}_0 \tag{18.68}$$

and a fluence of

$$F = N^2 F_0, \tag{18.69}$$

where $\mathcal{A}_0 = \pi \mathcal{E}_0 \tau_s = \pi(|k_2|/s)^{1/2}$ as seen in (18.63), and $F_0 = 2I_0 \tau_s = 4c\epsilon_0 n_0 \mathcal{A}_0 \mathcal{E}_0 / \pi$, as seen in (18.64) and (18.65). The area $\mathcal{A}_0$ is a constant that is completely determined by the parameters of the medium; the fluence F_0 is not a constant, but it remains unchanged if $\mathcal{E}_0$ remains unchanged.

As seen in the preceding subsection, the temporal shape of a fundamental soliton, with $N = 1$, stays unchanged as the pulse propagates; thus, the spectrum is also not changed. This is not true, however, for a high-order soliton with $N \geq 2$. A high-order soliton propagates in a complex manner that is characterized by both spectral and temporal compression and splitting, followed by subsequent recovery to the original pulse shape and spectrum after a characteristic propagation distance of ζ_0, or z_0 in terms of the original spatial variable, known as the *soliton period*. Thus, both the temporal shape and the spectrum of a high-order soliton vary periodically as the high-order soliton propagates. The soliton period ζ_0 of this periodic evolution is determined by the condition that

$$\varphi(\zeta_0) = \frac{\pi}{4}, \tag{18.70}$$

such that

$$\zeta_0 = \frac{\pi \tau_s^2}{2|k_2|} = \frac{\pi c \omega_0 \tau_s^2}{2|D|} \quad \text{and} \quad z_0 = \zeta_0. \tag{18.71}$$

A soliton is characterized by the pulsewidth parameter τ_s in time and by the soliton period ζ_0 in space; and its amplitude is defined by its area $\mathcal{A} = N \mathcal{A}_0$ for an Nth-order soliton, with

$$\mathcal{E}_0 = \frac{\mathcal{A}_0}{\pi \tau_s} \tag{18.72}$$

as defined in (18.63) for the fundamental soliton. Therefore, the temporal and spatial evolution of a fundamental soliton, with $N = 1$, is completely described as

$$\mathcal{E}_{N=1}(\zeta, \tau) = \mathcal{E}_s(\zeta, \tau) = \left(\frac{\mathcal{A}_0}{\pi \tau_s}\right) \operatorname{sech}\left(\frac{\tau}{\tau_s}\right) \exp\left(i \frac{\pi}{4} \frac{\zeta}{\zeta_0}\right). \tag{18.73}$$

The temporal and spatial evolution of a second-order soliton, with $N = 2$, can be described by the analytical relation [17]:

$$\mathcal{E}_{N=2}(\zeta, \tau) = 4\left(\frac{\mathcal{A}_0}{\pi \tau_s}\right) \frac{\cosh\left(\frac{3\tau}{\tau_s}\right) + 3\cosh\left(\frac{\tau}{\tau_s}\right) \exp\left(i \frac{2\pi\zeta}{\zeta_0}\right)}{\cosh\left(\frac{4\tau}{\tau_s}\right) + 4\cosh\left(\frac{2\tau}{\tau_s}\right) + 3\cos\left(\frac{2\pi\zeta}{\zeta_0}\right)} \exp\left(i \frac{\pi}{4} \frac{\zeta}{\zeta_0}\right). \tag{18.74}$$

The temporal and spatial evolution of a third-order soliton, with $N = 3$, can be described by the analytical relation [21]:

$$\mathcal{E}_{N=3}(\zeta, \tau) = 3\left(\frac{\mathcal{A}_0}{\pi \tau_s}\right) \frac{G_3(\zeta, \tau)}{H_3(\zeta, \tau)} \exp\left(i \frac{\pi}{4} \frac{\zeta}{\zeta_0}\right), \tag{18.75}$$

where

$$G_3(\zeta, \tau) = 2\cosh\left(\frac{8\tau}{\tau_s}\right) + 32\cosh\left(\frac{2\tau}{\tau_s}\right) + \left[16\cosh\left(\frac{6\tau}{\tau_s}\right) + 36\cosh\left(\frac{4\tau}{\tau_s}\right)\right] \exp\left(i \frac{2\pi\zeta}{\zeta_0}\right)$$

$$+ \left[20\cosh\left(\frac{4\tau}{\tau_s}\right) + 80\cosh\left(\frac{2\tau}{\tau_s}\right)\right] \exp\left(i \frac{6\pi\zeta}{\zeta_0}\right) \tag{18.76}$$

$$+ 5\exp\left(-i \frac{4\pi\zeta}{\zeta_0}\right) + 45\exp\left(i \frac{4\pi\zeta}{\zeta_0}\right) + 20\exp\left(i \frac{8\pi\zeta}{\zeta_0}\right)$$

and

$$H_3(\zeta, \tau) = \cosh\left(\frac{9\tau}{\tau_s}\right) + 9\cosh\left(\frac{7\tau}{\tau_s}\right) + 64\cosh\left(\frac{3\tau}{\tau_s}\right) + 36\cosh\left(\frac{\tau}{\tau_s}\right)$$

$$+ 36\cosh\left(\frac{5\tau}{\tau_s}\right)\cos\left(\frac{2\pi\zeta}{\zeta_0}\right) + 20\cosh\left(\frac{3\tau}{\tau_s}\right)\cos\left(\frac{6\pi\zeta}{\zeta_0}\right) + 90\cosh\left(\frac{\tau}{\tau_s}\right)\cos\left(\frac{4\pi\zeta}{\zeta_0}\right).$$

$$(18.77)$$

It is clear that for $\zeta = 0$, $\mathcal{E}_{N=1}(\zeta, \tau)$ as given in (18.73) reduces to (18.67) with $N = 1$. It can be shown that for $\zeta = 0$, $\mathcal{E}_{N=2}(\zeta, \tau)$ as given in (18.74) reduces to (18.67) with $N = 2$, and $\mathcal{E}_{N=3}(\zeta, \tau)$ as given in (18.75) reduces to (18.67) with $N = 3$.

Figure 18.7 shows the evolutions of the temporal and spectral intensity profiles of a fundamental soliton. Figures 18.8 and 18.9 respectively show the periodic evolutions of the temporal and spectral intensity profiles of second-order and third-order solitons [14, 20]. In these figures, we plot only the profiles of the temporal and spectral intensity distributions by normalizing them as $\hat{I}_N(z, t) = I_N(z, t)/N^2$ and $\hat{I}_N(z, \omega) = I_N(z, \omega)/N^2$. For a fundamental soliton, $N = 1$, which is plotted in Fig. 18.7, both $\hat{I}_1(z, t)$ and $\hat{I}_1(z, \omega)$ do not vary with z because

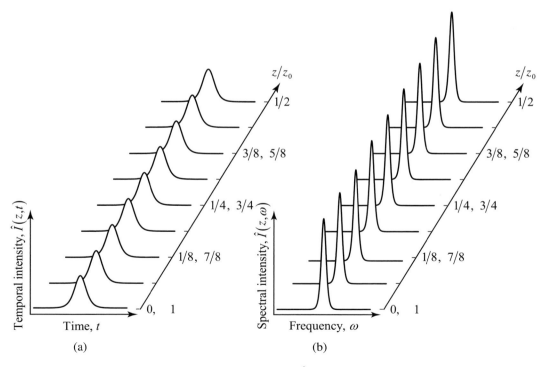

Figure 18.7 Evolutions of (a) the temporal intensity profile, $\hat{I}_1(z, t)$, and (b) the spectral intensity profile, $\hat{I}_1(z, \omega)$, of a fundamental soliton, $N = 1$. In this figure, z_0 is the soliton period defined in (18.71). Only half a period is plotted because the evolutions in the second half-period are the reverse of those in the first half-period.

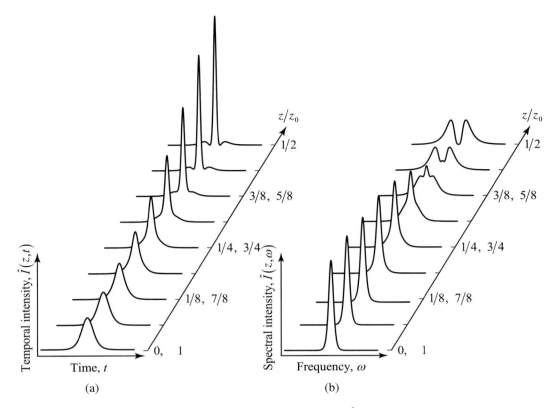

Figure 18.8 Periodic evolutions of (a) the temporal intensity profile, $\hat{I}_2(z, t)$, and (b) the spectral intensity profile, $\hat{I}_2(z, \omega)$, of a second-order soliton, $N = 2$. In this figure, z_0 is the soliton period defined in (18.71). Only half a period is plotted because the evolutions in the second half-period are the reverse of those in the first half-period.

$\hat{I}_1(z, t) = I_1(z, t) = I_0 \operatorname{sech}^2(t/\tau_s)$ and $\hat{I}_1(z, \omega) = I_1(z, \omega) = (\pi \tau_s)^2 I_0 \operatorname{sech}^2 [\pi(\omega - \omega_0)\tau_s/2]$, where $I_0 = 2c\epsilon_0 n_0 \mathcal{E}_0^2$ is the peak intensity of the fundamental soliton, as defined in (18.64), and ω_0 is the center carrier frequency of the soliton pulse. For a second-order soliton, $N = 2$, which is plotted in Fig. 18.8, and a third-order soliton, $N = 3$, which is plotted in Fig. 18.9, both $\hat{I}_N(z, t)$ and $\hat{I}_N(z, \omega)$ periodically vary with z, with a soliton period of z_0. At $z = 0$, $\hat{I}_N(0, t) = I_1(0, t) = I_0 \operatorname{sech}^2(t/\tau_s)$ and $\hat{I}_N(0, \omega) = I_1(0, \omega) = (\pi \tau_s)^2 I_0 \operatorname{sech}^2 [\pi(\omega - \omega_0)\tau_s/2]$ for both second-order and third-order solitons, as seen in Figs. 18.8 and 18.9.

Only half a period is plotted in each figure because the evolutions in the second half-period are the reverse of those in the first half-period. It is seen from Fig. 18.7 that both the temporal and the spectral intensity profiles of a fundamental soliton remain unchanged as the fundamental soliton propagates. By comparison, it is seen from Figs. 18.8 and 18.9 that both the temporal and the spectral intensity profiles of a high-order soliton, with $N > 1$, vary periodically with a soliton period of z_0, defined in (18.71), as the soliton propagates. Though a high-order soliton undergoes complex evolution in each period, its evolution in the first stage of each period is always characterized by spectral broadening caused by self-phase

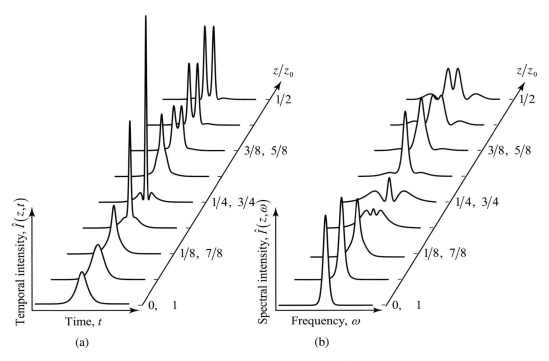

Figure 18.9 Periodic evolutions of (a) the temporal intensity profile, $\hat{I}_3(z, t)$, and (b) the spectral intensity profile, $\hat{I}_3(z, \omega)$, of a third-order soliton, $N = 3$. In this figure, z_0 is the soliton period defined in (18.71). Only half a period is plotted because the evolutions in the second half-period are the reverse of those in the first half-period.

modulation and temporal pulsewidth compression due to negative group-velocity dispersion. The periodic pulse narrowing effect of the temporal evolution of a high-order soliton becomes more significant for a higher order. It can be utilized for optical pulse compression of a large compression ratio [19].

If the input does not have exactly the correct hyperbolic secant shape or its area is not exactly an integral multiple of that of a fundamental soliton, then the field consists of a soliton part and a nonsoliton part. The initial evolution as the pulse propagates through the optical Kerr medium is complicated. Over a long propagation distance, the soliton part evolves as a soliton of the nearest order, whereas the nonsoliton part evolves into a dispersive wave and gradually spreads out.

Consider, for example, an input field

$$\mathcal{E}_{\text{in}}(0, \tau) = (N + a)\mathcal{E}_{\text{s}}(0, \tau) = (N + a)\mathcal{E}_0 \operatorname{sech}\left(\frac{\tau}{\tau_{\text{s}}}\right) = \mathcal{E}_{\text{in}} \operatorname{sech}\left(\frac{\tau}{\tau_{\text{s}}}\right), \qquad (18.78)$$

where $-1/2 < a < 1/2$ and $\mathcal{E}_{\text{in}} = (N + a)\mathcal{E}_0$. This input has an initial area of

$$\mathcal{A}_{\text{in}} = \pi \mathcal{E}_{\text{in}} \tau_{\text{s}} = (N + a)\mathcal{A}_0, \qquad (18.79)$$

and an initial fluence of

$$F_{in} = 4c\epsilon_0 n_0 \mathcal{E}_{in}^2 \tau_s = (N + a)^2 F_0, \tag{18.80}$$

where $\mathcal{A}_0 = \pi\mathcal{E}_0\tau_s = \pi(|k_2|/s)^{1/2}$ as seen in (18.63), and $F_0 = 2I_0\tau_s = 4c\epsilon_0 n_0 \mathcal{A}_0\mathcal{E}_0/\pi$ as seen in (18.64) and (18.65). The area $\mathcal{A}_0$ is a constant that is completely determined by the parameters of the medium; the fluence F_0 is not a constant but it remains unchanged if $\mathcal{E}_0$ remains unchanged. In the discussions here, $\mathcal{E}_0$ is taken to be a fixed reference value. Therefore, both $\mathcal{A}_0$ and F_0 are constants.

Over a long propagation distance, this field evolves into an Nth-order soliton because $|a| < 1/2$. Therefore, the area approaches the value given in (18.68):

$$\mathcal{A}_\infty = \pi\mathcal{E}_\infty\tau_\infty = N\mathcal{A}_0. \tag{18.81}$$

However, the fluence does not approach the value given in (18.69). Because the dispersive wave that contains the nonsoliton part of the field spreads out, both the field amplitude and the temporal width of the pulse are changed so that the fluence asymptotically approaches the value [17]:

$$F_\infty = 4c\epsilon_0 n_0 \mathcal{E}_\infty^2 \tau_\infty = N(N + 2a)F_0 = \left[1 - \frac{a^2}{(N + a)^2}\right] F_{in}. \tag{18.82}$$

By using (18.81) and (18.82), we find the initial field amplitude and pulsewidth parameter for the periodic evolution of the Nth-order soliton when it reaches the long-term steady state:

$$\mathcal{E}_\infty = (N + 2a)\mathcal{E}_0 = \frac{N + 2a}{N + a}\mathcal{E}_{in}, \tag{18.83}$$

$$\tau_\infty = \frac{N}{N + 2a}\tau_s. \tag{18.84}$$

EXAMPLE 18.3

Consider the transmission of soliton pulses at a carrier wavelength of $\lambda = 1.55$ μm through a silica fiber that has the parameters given in Example 18.1. From Example 18.1, we find that a fundamental soliton pulse that has a full width at half-maximum pulsewidth of $\Delta t_{ps0} = 5$ ps has a pulse energy of $U_{ps0} = 10$ pJ. High-order solitons can be excited by raising the energy of an input pulse while keeping its pulsewidth at $\Delta t_{ps}^{in} = \Delta t_{ps0} = 5$ ps. (a) With an input pulse that has twice the energy of the fundamental soliton, $U_{ps}^{in} = 2U_{ps0} = 20$ pJ, what is the order of the soliton when it reaches long-term steady state? What are its energy and its pulsewidth parameter in the long-term periodic evolution? How much energy is dissipated from the pulse as a dispersive wave? (b) With an input pulse that has twice the energy of the fundamental soliton, $U_{ps}^{in} = 3U_{ps0} = 30$ pJ, what is the order of the soliton when it reaches its long-term steady state? What are its energy and its pulsewidth parameter in the long-term periodic evolution? How much energy is dissipated from the pulse as a dispersive wave? (c) What is the required input pulse energy for the excitation of a second-order soliton that keeps the same initial pulsewidth as $\Delta t_{ps}^{in} = \Delta t_{ps0} = 5$ ps?

Solution

From (17.75), $\Delta t_{ps} = 2 \ln (1 + \sqrt{2})\tau_s$. Therefore, as found in Example 18.2, $\tau_s = 2.84$ ps for $\Delta t_{ps0} = 5$ ps. Because the pulsewidth is kept unchanged for the input pulse while the energy of the pulse is changed, we find the relation between the energy and the field amplitude of the input pulse:

$$U_{ps}^{in} = F_{in}\mathcal{A} = 4c\epsilon_0 n_0 \mathcal{E}_{in}^2 \tau_s \mathcal{A} \propto \mathcal{E}_{in}^2.$$

Therefore, by keeping $\tau_s = 2.84$ ps constant for the input pulse, we have the relation for the field amplitude of the input pulse:

$$\mathcal{E}_{in} = \sqrt{\frac{U_{ps}^{in}}{U_{ps0}}} \mathcal{E}_0.$$

We also have the relation for the long-term asymptotic pulse energy:

$$U_{ps\infty}^N = F_\infty \mathcal{A} = N(N + 2a)F_0 \mathcal{A} = N(N + 2a)U_{ps0}.$$

(a) With $U_{ps}^{in} = 2U_{ps0} = 20$ pJ, we find that

$$\mathcal{E}_{in} = \sqrt{2}\mathcal{E}_0 = (N + a)\mathcal{E}_0 \quad \Rightarrow \quad 0.5 < N + a = \sqrt{2} < 1.5.$$

In this case, because $0.5 < N + a < 1.5$, we find that

$$N = 1 \quad \text{and} \quad a = \sqrt{2} - 1.$$

Therefore, the pulse is a fundamental soliton when it reaches the long-term steady state. The long-term value of the pulse energy is

$$U_{ps\infty}^{N=1} = N(N + 2a)U_{ps0} = 1 \times \left[1 + 2\left(\sqrt{2} - 1\right)\right]U_{ps0} = \left(2\sqrt{2} - 1\right)U_{ps0} = 18.3\,\text{pJ},$$

and the long-term value of the pulsewidth parameter is

$$\tau_\infty^{N=1} = \frac{N}{N + 2a}\tau_s = \frac{1}{1 + 2(\sqrt{2} - 1)}\tau_s = \frac{1}{2\sqrt{2} - 1}\tau_s = 1.55\,\text{ps}.$$

The energy that is dissipated from the pulse as a dispersive wave is

$$U_{ps}^{in} - U_{ps\infty}^{N=1} = 1.7\,\text{pJ}.$$

(b) With $U_{ps}^{in} = 3U_{ps0} = 30$ pJ, we find that

$$\mathcal{E}_{in} = \sqrt{3}\mathcal{E}_0 = (N + a)\mathcal{E}_0 \quad \Rightarrow \quad 1.5 < N + a = \sqrt{3} < 2.5.$$

In this case, because $1.5 < N + a < 2.5$, we find that

$$N = 2 \quad \text{and} \quad a = \sqrt{3} - 2.$$

Therefore, the pulse is a second-order soliton when it reaches the long-term steady state. The long-term value of the pulse energy is

$$U_{\mathrm{ps}\infty}^{N=2} = N(N + 2a)U_{\mathrm{ps0}} = 2 \times \left[2 + 2(\sqrt{3} - 2)\right]U_{\mathrm{ps0}} = 2 \times \left(2\sqrt{3} - 2\right)U_{\mathrm{ps0}} = 29.3 \text{ pJ},$$

and the long-term value of the pulsewidth parameter is

$$\tau_{\infty}^{N=2} = \frac{N}{N + 2a}\tau_{\mathrm{s}} = \frac{2}{2 + 2(\sqrt{3} - 2)}\tau_{\mathrm{s}} = \frac{1}{\sqrt{3} - 1}\tau_{\mathrm{s}} = 3.88 \text{ ps}.$$

The energy that is dissipated from the pulse as a dispersive wave is

$$U_{\mathrm{ps}}^{\mathrm{in}} - U_{\mathrm{ps}\infty}^{N=1} = 0.7 \text{ pJ}.$$

(c) To keep the same pulsewidth of $\Delta t_{\mathrm{ps}} = 5$ ps, thus $\tau_{\infty}^{N=2} = \tau_{\mathrm{s}} = 2.84$ ps for the long term, it is necessary that $a = 0$, as can be seen from (18.84). For a second-order soliton, we then need $N = 2$ and $a = 0$ for the input pulse. Therefore, the required input pulse energy is

$$U_{\mathrm{ps}}^{N=2} = N^2 U_{\mathrm{ps0}} = 2^2 U_{\mathrm{ps0}} = 40 \text{ pJ}.$$

18.6 MODULATION INSTABILITY

One issue related to nonlinear wave propagation is *modulation instability* [22], which is characterized by an exponential growth of a small perturbation to the wave. Modulation instability usually occurs in the same parameter range where solitons can be formed. Therefore, it can be the precursor of soliton formation. Modulation instability can take place in space, resulting in self-focusing or spatial solitons, or in time, resulting in temporal oscillations or temporal solitons. Here, we consider only temporal modulation instability.

The phenomenon of modulation instability can be qualitatively seen. When a CW optical wave is modulated by a small perturbation, either by its noise or by an external signal, its intensity fluctuates with time. If the wave propagates in an optical Kerr medium that has a positive nonlinear refractive index, $n_2 > 0$, the temporal location that has a slightly increased intensity has a slightly larger refractive index, which leads to a lower group velocity. Optical wave packets are thus attracted to this region of low group velocity in a manner much like *temporal self-focusing*. If the medium has a negative group-velocity dispersion, $D < 0$, the localized growth of light intensity is first enhanced but then limited by a diffraction mechanism in a manner that is similar to the formation of temporal solitons. The final temporal pattern of modulation instability is reached when the nonlinear effect is balanced by the diffraction effect.

It is clear from this discussion that the characteristics of modulation instability depend on the linear and nonlinear optical parameters of the medium and on the intensity of the optical wave.

Modulation instability for an optical wave that propagates in an optical Kerr medium [23] can be analyzed by using the nonlinear Schrödinger equation given in (18.57) because it is a phenomenon of nonlinear wave propagation even when the input is a CW wave. Here we consider the case that $n_2 > 0$ such that $s = 2\omega_0\epsilon_0 n_0 n_2 > 0$ according to (18.58). With a CW input, (18.57) has a solution of the form:

$$\mathcal{E}(\zeta, \tau) = \mathcal{E}_0 \exp\left(is\mathcal{E}_0^2\zeta\right),\tag{18.85}$$

where the amplitude $\mathcal{E}_0$ is taken to be a real quantity. To analyze the stability of this solution, we consider a small perturbation of $\rho(\zeta, \tau)$ to its amplitude:

$$\mathcal{E}(\zeta, \tau) = \left[\mathcal{E}_0 + \rho(\zeta, \tau)\right] \exp\left(is\mathcal{E}_0^2\zeta\right).\tag{18.86}$$

By plugging (18.86) into (18.57), we find the equation for $\rho(\zeta, \tau)$ by keeping only the linear terms:

$$i\frac{\partial\rho}{\partial\zeta} = \frac{k_2}{2}\frac{\partial^2\rho}{\partial\tau^2} - s\mathcal{E}_0^2(\rho + \rho^*).\tag{18.87}$$

The general solution of this equation has the form:

$$\rho(\zeta, \tau) = \mathcal{E}_- \exp\left[-i(K\zeta - \Omega\tau)\right] + \mathcal{E}_+ \exp\left[i(K\zeta - \Omega\tau)\right].\tag{18.88}$$

By plugging (18.88) into (18.87), we obtain two homogeneous equations:

$$\left(K + \frac{k_2}{2}\Omega^2 + s\mathcal{E}_0^2\right)\mathcal{E}_- + s\mathcal{E}_0^2\mathcal{E}_+^* = 0,\tag{18.89}$$

$$s\mathcal{E}_0^2\mathcal{E}_- + \left(-K + \frac{k_2}{2}\Omega^2 + s\mathcal{E}_0^2\right)\mathcal{E}_+^* = 0.\tag{18.90}$$

The condition for nontrivial solutions of $\mathcal{E}_-$ and $\mathcal{E}_+$ to exist is that the propagation constant K satisfy the relation:

$$K^2 = \frac{k_2}{2}\Omega^2\left(\frac{k_2}{2}\Omega^2 + 2s\mathcal{E}_0^2\right).\tag{18.91}$$

Because $\Omega^2 > 0$, $\mathcal{E}_0^2 > 0$, and $s > 0$, the two values of K are both real when the medium has positive group-velocity dispersion with $D > 0$ such that $k_2 > 0$. In this case, the CW solution given in (18.85) is stable against a small perturbation because $\rho(\zeta, \tau)$ varies periodically but does not grow as the wave propagates. For the CW solution to become unstable as the wave propagates, it is necessary that K have an imaginary part that has a negative value so that $\rho(\zeta, \tau)$ grows exponentially with the propagation distance. When the medium has negative group-velocity dispersion with $D < 0$ such that $k_2 < 0$, the two values of K from (18.91) become purely imaginary, and one of them has a negative value because the two values of K are complex conjugates, if the frequency is below a *critical frequency*:

$$\Omega < \Omega_c = 2\left(\frac{s}{|k_2|}\right)^{1/2}\mathcal{E}_0 = 2\omega_0\left(\frac{n_2}{|D|}\right)^{1/2}\left(\frac{P_0}{\mathcal{A}}\right)^{1/2}, \tag{18.92}$$

where P_0 is the power of the CW optical beam, $\mathcal{A}$ is the cross-sectional area of the beam, and ω_0 is the carrier frequency of the optical wave, as defined in the preceding sections. Alternatively, this relation indicates that a perturbation at a frequency of Ω grows exponentially and becomes unstable if the power of the optical beam is above a *critical power*:

$$P_0 > P_c = c\epsilon_0 n_0 \frac{|k_2|}{2s}\Omega^2\mathcal{A} = \frac{|D|}{4n_2}\frac{\Omega^2}{\omega_0^2}\mathcal{A}. \tag{18.93}$$

When the condition given in (18.92), or equivalently that in (18.93), is satisfied, the perturbations at the shifted frequencies of $\omega_0 - \Omega$ and $\omega_0 + \Omega$ both grow exponentially with a *power gain coefficient* of

$$g(\pm\Omega) = 2\left|\text{Im } K\right| = 2|K| = |k_2|\Omega(\Omega_c^2 - \Omega^2)^{1/2} = \frac{|D|}{c\omega_0}\Omega(\Omega_c^2 - \Omega^2)^{1/2}. \tag{18.94}$$

In terms of the power of the beam, the gain coefficient can be expressed as

$$g(P_0) = \frac{2s}{c\epsilon_0 n_0 \mathcal{A}}P_c^{1/2}(P_0 - P_c)^{1/2} = \frac{4\omega_0 n_2}{c\mathcal{A}}P_c^{1/2}(P_0 - P_c)^{1/2}. \tag{18.95}$$

It can be seen from (18.94) that for a given value of Ω_c, thus a given pump power P_0, there is a maximum value for $g(\pm\Omega)$. By contrast, from (18.95), it can also be seen that for a given value of P_c, thus a given frequency shift Ω or $-\Omega$, $g(P_0)$ continues to increase with P_0 without a maximum value.

For a given pump power P_0, thus a given value of Ω_c, the maximum gain coefficient occurs at the shifted frequencies of $\omega_0 - \Omega_{\text{max}}$ and $\omega_0 + \Omega_{\text{max}}$, with

$$\Omega_{\text{max}} = \frac{\Omega_c}{\sqrt{2}} \tag{18.96}$$

and a maximum gain of

$$g_{\text{max}} = \frac{|k_2|}{2}\Omega_c^2 = 2s\mathcal{E}_0^2 = \frac{2\omega_0 n_2}{c\mathcal{A}}P_0. \tag{18.97}$$

In terms of g_{max} and Ω_c, the power gain spectrum, $g(\pm\Omega)$, can be expressed in the normalized form:

$$g(\pm\Omega) = 2g_{\text{max}}\frac{\Omega}{\Omega_c}\left(1 - \frac{\Omega^2}{\Omega_c^2}\right)^{1/2}. \tag{18.98}$$

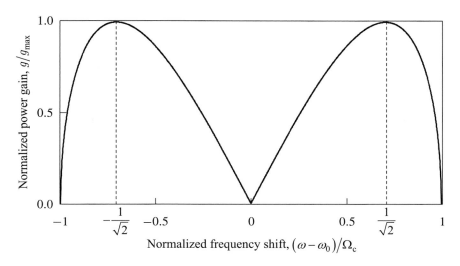

Figure 18.10 Gain spectrum of modulation instability as a function of the frequency shift from the carrier frequency ω_0.

The normalized power gain spectrum as a function of the frequency shift, normalized to the critical frequency shift, is shown in Fig. 18.10. It is symmetric with respect to the center frequency so that the two sidebands at the frequencies of $\omega_0 - \Omega$ and $\omega_0 + \Omega$ have the same gain coefficient. This spectrum is identical to Fig. 10.10 for the normalized four-wave mixing power gain because (18.98) is identical to (10.109).

In the above, modulation instability is analyzed in the time domain by considering the stability of a perturbation using the nonlinear Schrödinger equation. We find that the instability occurs only under the condition of a negative group-velocity dispersion for a positive nonlinear refractive index. Similar analysis for a negative nonlinear refractive index yields the condition of a positive group-velocity dispersion. Therefore, modulation instability occurs when the group-velocity dispersion and the nonlinear refractive index have opposite signs, such that $Dn_2 < 0$. As discussed in the preceding section, this is the same condition for the formation of temporal solitons. We also find that modulation instability can occur only for a perturbation frequency shift that is smaller than the critical frequency shift, and the two sidebands of up-shifted and down-shifted frequencies have the same gain coefficient.

In the frequency domain, the characteristics discussed above indicate that modulation instability is a partially degenerated four-wave mixing process that involves the three frequencies of the center frequency at ω_0 and the two sideband frequencies at $\omega_0 - \Omega$ and $\omega_0 + \Omega$. Phase matching for this parametric process is made possible by the intensity-dependent index of refraction. The growth of the two sidebands occurs when two photons at ω_0 are simultaneously annihilated to generate two photons at $\omega_0 - \Omega$ and $\omega_0 + \Omega$ through the third-order parametric process $\omega_0 + \omega_0 \rightarrow (\omega_0 - \Omega) + (\omega_0 + \Omega)$, which is facilitated by $\chi^{(3)}\left(\omega_0 - \Omega = \omega_0 + \omega_0 - (\omega_0 + \Omega)\right)$ and $\chi^{(3)}\left(\omega_0 + \Omega = \omega_0 + \omega_0 - (\omega_0 - \Omega)\right)$. The conditions that $Dn_2 < 0$ and $\Omega < \Omega_c$ for it to take place are the conditions for phase matching this parametric process of four-wave mixing to be possible. From this picture, it can be understood that the two sidebands have the same gain coefficient because $\Omega \ll \omega_0$, so that $\chi^{(3)}(\omega_0 - \Omega) \approx \chi^{(3)}(\omega_0 + \Omega)$. It can also be seen that the frequency shift is limited to a range below the critical frequency because phase matching is not possible for a large frequency shift due to dispersion. Because phase matching depends on the intensity-dependent contribution to the refractive index, the range of positive gain increases with the power of the beam. This four-wave mixing analysis has been presented in Section 10.7. It can be seen that (18.94) and (18.95) obtained in this section are respectively identical to (10.103) and (10.105) obtained in Section 10.7. Therefore, the normalized gain spectrum of modulation instability as shown in Fig. 18.10 is the same as the normalized four-wave mixing gain spectrum shown in Fig. 10.10 because (10.109) and (18.98) are identical.

Problem Set

18.2.1 Show that the linear pulse propagation equation given in (18.7) has the solution as given in (18.9) in the frequency domain and that as given in (18.19) in the time domain if the dispersion terms of the third and higher orders are neglected.

18.2.2 Consider the propagation of an optical pulse in a linear but dispersive medium. Ignore the dispersion terms of the third and higher orders. Show, by using an initially trans-form-limited Gaussian pulse, that as the pulse propagates through the medium its spectral width remains unchanged but its temporal pulsewidth broadens while its peak amplitude decreases. Show also that the pulse becomes linearly chirped as it propagates. Find the time–bandwidth product, $\Delta t_{ps}\Delta\nu_{ps}$, of the pulse as a function of the propagation distance.

18.2.3 A digital data transmission system uses a silica optical fiber that has a group-velocity dispersion of $D_\lambda = 17$ ps nm^{-1} km^{-1} at the transmission wavelength of $\lambda = 1.55$ μm, as described in Example 18.1. Assume that the effects of both optical attenuation and optical nonlinearity can be neglected.

    (a) If transform-limited Gaussian pulses that have an initial pulsewidth of $\Delta t_{ps0} = 5$ ps are used for the data bits, what is the highest bit rate for a transmission distance of at least 10 km?

    (b) At a bit rate of $f_{bit} = 10$ Gbit s^{-1} and a transmission distance of at least 10 km, what is the limitation on the initial pulsewidth?

18.3.1 Consider the propagation of an optical pulse in a nondispersive optical Kerr medium. Show that as the pulse propagates through the medium both its temporal pulsewidth and its peak amplitude remain unchanged, but its spectral width broadens and the pulse becomes nonlinearly chirped. Show that an initially transform-limited Gaussian pulse has a broadened spectrum as given by (18.55) as it propagates. Find its time–bandwidth product, $\Delta t_{ps}\Delta\nu_{ps}$, as a function of the propagation distance.

18.5.1 Show that under the condition given in (18.58), the fundamental soliton given in (18.59) is a solution of the nonlinear Schrödinger equation given in (18.57), with the phase shift varying linearly with the propagation distance, as given in (18.62), and the pulse area as a constant of propagation, as given in (18.63). Show also that the field amplitude of the soliton varies linearly with the input fluence of the pulse, whereas the pulsewidth parameter varies inversely with the input fluence.

18.5.2 Because the area of a fundamental soliton pulse is a constant that is completely determined by the carrier frequency of the pulse and the parameters of the optical medium, the peak intensity, peak power, fluence, and energy of a fundamental soliton at a given carrier frequency scale with the pulsewidth. Consider the fundamental soliton at $\lambda = 1.55$ μm that propagates in the silica optical fiber, which has the parameters described in Example 18.2.

(a) What are the peak intensity, peak power, fluence, and energy of a fundamental soliton in this fiber if the pulsewidth is reduced by half to $\Delta t_{ps0} = 2.5$ ps? Compare them to the values found for $\Delta t_{ps0} = 5$ ps in Example 18.2.

(b) What are the peak intensity, peak power, fluence, and energy of a fundamental soliton in this fiber if the pulsewidth is doubled to $\Delta t_{ps0} = 10$ ps? Compare them to the values found for $\Delta t_{ps0} = 5$ ps in Example 18.2.

18.5.3 Show that for $\zeta = 0$, $\mathcal{E}_{N=2}(\zeta, \tau)$ as given in (18.74) reduces to (18.67) with $N = 2$.

18.5.4 Show that for $\zeta = 0$, $\mathcal{E}_{N=3}(\zeta, \tau)$ as given in (18.75) reduces to (18.67) with $N = 3$.

18.5.5 The fundamental soliton that is considered in Examples 18.2 and 18.3 at a carrier wavelength of $\lambda = 1.55$ μm has a full width at half-maximum pulsewidth of $\Delta t_{ps0} = 5$ ps and a pulse energy of $U_{ps0} = 10$ pJ. High-order solitons can be excited by raising the energy of the input pulse while keeping its pulsewidth unchanged at $\Delta t_{ps0} = 5$ ps. The order of the soliton that is excited depends on this input pulse energy. When the excited soliton reaches the long-term steady state, both its energy and its pulsewidth parameter reach their long-term asymptotic values. Based on the parameters of the fundamental soliton given above, find the range of the input pulse energy that will excite a second-order soliton. Then, find the corresponding values of the long-term asymptotic values of the energy and the pulsewidth parameter of the second-order soliton.

18.6.1 By using the coupled equations given in (18.89) and (18.90) for the amplitudes of the perturbation field, show that modulation instability can occur for a given pump power only at frequency shifts below a critical frequency as given in (18.92), or for a given frequency shift only above a critical pump power as given in (18.93). Then show that the power gain coefficient takes the form of (18.94) in terms of the frequency shift, and the form of (18.95) in terms of the pump power. Show also that for a given pump power, the maximum gain coefficient occurs at $\pm \Omega_{max}$, as given in (18.96), for a maximum value of g_{max} that is given in (18.97), but the gain coefficient does not have a maximum value for a given frequency shift as the pump power increases.

References

[1] J. Giordmaine, M. Duguay, and J. Hansen, "Compression of optical pulses," *IEEE Journal of Quantum Electronics*, vol. 4, pp. 252–255, 1968.

[2] D. Strickland and G. Mourou, "Compression of amplified chirped optical pulses," *Optics Communications*, vol. 56, pp. 219–221, 1985.

[3] R. G. Brewer, "Frequency shifts in self-focused light," *Physical Review Letters*, vol. 19, pp. 8–10, 1967.

[4] F. Shimizu, "Frequency broadening in liquids by a short light pulse," *Physical Review Letters*, vol. 19, pp. 1097–1100, 1967.

[5] R. H. Stolen and C. Lin, "Self-phase-modulation in silica optical fibers," *Physical Review A*, vol. 17, pp. 1448–1453, 1978.

[6] D. Grischkowsky and A. C. Balant, "Optical pulse compression based on enhanced frequency chirping," *Applied Physics Letters*, vol. 41, pp. 1–3, 1982.

[7] P. L. François, "Nonlinear propagation of ultrashort pulses in optical fibers: total field formulation in the frequency domain," *Journal of the Optical Society of America B*, vol. 8, pp. 276–293, 1991.

[8] H. Nakatsuka, D. Grischkowsky, and A. C. Balant, "Nonlinear picosecond-pulse propagation through optical fibers with positive group velocity dispersion," *Physical Review Letters*, vol. 47, pp. 910–913, 1981.

[9] C. V. Shank, R. L. Fork, R. Yen, R. H. Stolen, and W. J. Tomlinson, "Compression of femtosecond optical pulses," *Applied Physics Letters*, vol. 40, pp. 761–763, 1982.

[10] B. Nikolaus and D. Grischkowsky, "12× pulse compression using optical fibers," *Applied Physics Letters*, vol. 42, pp. 1–2, 1983.

[11] B. Nikolaus and D. Grischkowsky, "90-fs tunable optical pulses obtained by two-stage pulse compression," *Applied Physics Letters*, vol. 43, pp. 228–230, 1983.

[12] N. J. Zabusky and M. D. Kruskal, "Interaction of 'solitons' in a collisionless plasma and the recurrence of initial states," *Physical Review Letters*, vol. 15, pp. 240–243, 1965.

[13] A. Hasegawa and F. Tappert, "Transmission of stationary nonlinear optical pulses in dispersive dielectric fibers: I. Anomalous dispersion," *Applied Physics Letters*, vol. 23, pp. 142–144, 1973.

[14] L. F. Mollenauer, R. H. Stolen, and J. P. Gordon, "Experimental observation of picosecond pulse narrowing and solitons in optical fibers," *Physical Review Letters*, vol. 45, pp. 1095–1098, 1980.

[15] A. M. Weiner, J. P. Heritage, R. J. Hawkins, R. N. Thurston, E. M. Kirschner, D. E. Leaird, *et al.*, "Experimental observation of the fundamental dark soliton in optical fibers," *Physical Review Letters*, vol. 61, pp. 2445–2448, 1988.

[16] V. E. Zakharov and A. B. Shabat, "Exact theory of two-dimensional self-focusing and one-dimensional self-modulation of wave in nonlinear media," *Soviet Physics JETP*, vol. 34, p. 8, 1972.

[17] J. Satsuma and N. Yajima, "B. Initial value problems of one-dimensional self-modulation of nonlinear waves in dispersive media," *Progress of Theoretical Physics Supplement*, vol. 55, pp. 284–306, 1974.

[18] A. Hasegawa and Y. Kodama, "Signal transmission by optical solitons in monomode fiber," *Proceedings of the IEEE*, vol. 69, pp. 1145–1150, 1981.

[19] L. F. Mollenauer, R. H. Stolen, J. P. Gordon, and W. J. Tomlinson, "Extreme picosecond pulse narrowing by means of soliton effect in single-mode optical fibers," *Optics Letters*, vol. 8, pp. 289–291, 1983.

[20] R. H. Stolen, L. F. Mollenauer, and W. J. Tomlinson, "Observation of pulse restoration at the soliton period in optical fibers," *Optics Letters*, vol. 8, p. 3, 1983.

[21] J. Gutierrez Vega, S. Lopez-Aguayo, and J. Ochoa-Ricoux, "Exploring the behavior of solitons on a desktop personal computer," *Revista Mexicana de Física*, vol. 52, pp. 28–36, 2006.

[22] V. E. Zakharov and L. A. Ostrovsky, "Modulation instability: the beginning," *Physica D: Nonlinear Phenomena*, vol. 238, pp. 540–548, 2009.

[23] A. Hasegawa and W. Brinkman, "Tunable coherent IR and FIR sources utilizing modulational instability," *IEEE Journal of Quantum Electronics*, vol. 16, pp. 694–697, 1980.

Supercontinuum Generation

19.1 SUPERCONTINUUM GENERATION IN OPTICAL FIBERS

Supercontinuum generation is a nonlinear optical process that produces a broad continuous spectrum, often spanning over an octave, when an intense laser beam of an initially narrow bandwidth propagates through a nonlinear medium. It was first observed in 1970 by Alfano and Shapiro through picosecond excitation of bulk glass and crystals [1, 2]. Given a sufficiently high laser power, supercontinuum generation can be observed in any material, including air and water. In general, many nonlinear processes are involved, including self-phase modulation, cross-phase modulation, four-wave mixing, modulation instability, self-focusing, stimulated Raman scattering, soliton dynamics, and dispersive-wave generation. The specific nonlinear optical processes that are involved depend on the optical properties of the material and on the wavelength and the temporal characteristics of the laser beam. For example, self-focusing is usually involved in supercontinuum generation in a bulk medium, but not in an optical waveguide such as an optical fiber. As another example, when an optical fiber is pumped with femtosecond pulses at a wavelength in the spectral region of negative group-velocity dispersion, supercontinuum generation strongly depends on soliton dynamics. By contrast, if the group-velocity dispersion at the excitation wavelength is positive, the spectral broadening is mainly caused by self-phase modulation but not by soliton dynamics.

Before the emergence of optical fibers, supercontinuum was generated in bulk solids, liquids, or gases. It required high-power laser beams, generally ultrashort laser pulses that were highly focused to reach the necessary high intensities. Supercontinuum generation in a bulk material is a complicated process that involves the coupling of spatial and temporal effects. The effect of diffraction limits the interaction length and, consequently, the efficiency and the quality of the generated supercontinuum. In contrast to supercontinuum generation in a bulk material, super-continuum generation in an optical fiber involves only temporal effects because the transverse mode characteristics of the optical field are determined by the waveguiding structure of the optical fiber.

In 1976, Lin and Stolen first accomplished supercontinuum generation in an optical fiber by injecting a 10-nanosecond Q-switched dye laser pulse into a 20-meter silica fiber [3]. The usage of optical fibers greatly facilitated the development of supercontinuum generation, and the subsequent development of optical fiber technology further accelerated the research and development of supercontinuum generation across the broad spectral range from the deep-ultraviolet to the mid-infrared. Because of the waveguiding effect, both the pump beam and the generated supercontinuum can propagate over a long distance in an optical fiber without diverging due to diffraction or self-focusing due to Kerr lensing. Thus, an optical fiber provides the favorable

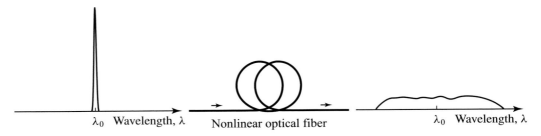

Figure 19.1 Schematic illustration of supercontinuum generation in a nonlinear optical fiber.

combination of both high intensity and long interaction length for efficient supercontinuum generation. Figure 19.1 illustrates supercontinuum generation in a nonlinear optical fiber.

Silica fibers are the most well developed and widely used optical fibers, but they have a few limitations for supercontinuum generation. They have a relatively narrow optical transmission window that is limited to wavelengths longer than 200 nm but shorter than 2.4 μm, a small nonlinear coefficient, and a spectral region of negative group-velocity dispersion that is limited to wavelengths longer than 1.3 μm. Zero group-velocity dispersion appears at 1.284 μm for pure silica, but the dispersion of the waveguide structure of a conventional optical fiber shifts it to a longer wavelength. Many specialty fibers are developed for supercontinuum generation in the diverse spectral regions from the vacuum-ultraviolet to the mid-infrared.

The introduction of *photonic crystal fibers* [4] has led to accelerated development of supercontinuum in the near-ultraviolet, visible, and near-infrared spectral regions [5]. Most of them are made of silica glass, but their properties are very different from those of conventional silica fibers. There are two basic types of photonic crystal fibers: the *solid-core photonic crystal fiber* and the *hollow-core photonic crystal fiber*, as shown in Figs. 19.2(a) and (b), respectively. The solid-core photonic crystal fibers are used for supercontinuum generation in the spectral range from the near-ultraviolet to the near-infrared, whereas the hollow-core photonic crystal fibers can be used to extend the spectral range to the vacuum-ultraviolet wavelengths close to 100 nm. Photonic crystal fibers have a few favorable properties, including the flexibility to design their dispersion properties, an enhanced nonlinear coefficient due to strong mode confinement, and single-mode propagation over a broad spectral range. The dispersion property of a photonic crystal fiber can be varied by suitable design; of particular importance for supercontinuum generation is that the *zero-dispersion wavelength* of a photonic crystal fiber that is made of silica can be shifted to a wavelength much shorter than the intrinsic 1.284 μm zero-dispersion wavelength of silica. This flexibility makes it possible for a photonic crystal fiber to have a negative group-velocity dispersion coefficient, thus the possibility of soliton formation, across the near-infrared and visible regions. A photonic crystal fiber has a reduced effective mode area, which effectively enhances the optical Kerr coefficient compared to a conventional silica fiber. Another important property of a properly designed photonic crystal fiber is that it can be single mode in all wavelengths [6]. Though a photonic crystal fiber used for supercontinuum generation is not designed for such *endless single-mode property*, its fundamental mode is robust to the excitation of high-order modes. Thus, supercontinuum generated in a photonic

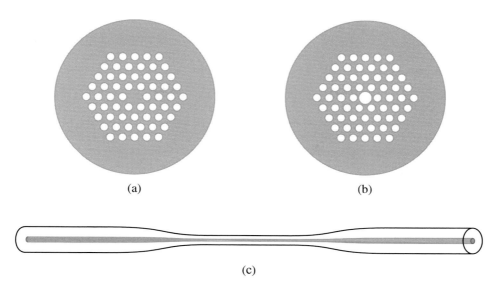

Figure 19.2 Structures of (a) a solid-core photonic crystal fiber, (b) a hollow-core photonic crystal fiber, and (c) a tapered fiber.

crystal fiber is consistently observed in the fundamental mode [5]. In spite of these favorable properties, however, solid-core photonic crystal fibers made of silica are still limited to the spectral range from 200 nm to 2.4 µm because of the high absorption of silica in the ultraviolet and infrared spectral regions.

Essentially identical dispersion and nonlinearity characteristics as those of the photonic crystal fibers can be obtained simply by appropriate tapering of a standard optical fiber, as shown in Fig. 19.2(c). The fundamental cause for the favorable dispersion and nonlinearity properties of photonic crystal fibers and *tapered fibers* is the reduction of the effective core diameter. The tight mode confinement resulting from the reduction of the core diameter shifts the zero-dispersion point to a short wavelength and enhances the optical nonlinearity.

Many types of infrared fibers are available for extending the supercontinuum spectra into the mid-infrared beyond the 2.4 µm limit of silica fibers. Highly germania-doped or pure germania-core fibers have low-loss transmission extending to the 3.0 µm wavelength, and they have higher nonlinearity and larger Raman response than pure silica fibers. They are also compatible with silica fibers, making them easy to fuse with silica fibers. Some fluoride fibers are useful for mid-infrared supercontinuum generation. The most commonly used ones are ZBLAN fibers, which have the composition of ZrF_4–BaF_2–LaF_3–AlF_3–NaF, and indium fluoride fibers, which is made of InF_3. These fluoride fibers have low-loss transmission windows extending to about the 5.0 µm wavelength. Beyond this wavelength, various types of chalcogenide fibers are available, which are various compositions of sulfides, selenides, or tellurides of arsenic or germanium. The transmission window of a chalcogenide fiber depends on its composition. Some of these fibers have transmission windows over the 20 µm wavelength, making mid-infrared supercontinuum covering a spectrum up to 20 µm possible. Photonic crystal fibers made of the infrared glasses are also available. They have the favorable properties of photonic crystal fibers, as discussed in the preceding paragraph, and the extended infrared transmission windows of infrared glasses, as discussed above.

19.2 GENERALIZED NONLINEAR SCHRÖDINGER EQUATION

The nonlinear Schrödinger equation developed in Section 8.3.2 and given in (8.56) for the propagation in a homogeneous medium, or (8.57) for the propagation in a single-mode waveguide, accounts for the second-order dispersion and the nonlinear response of an optical Kerr medium. As seen in Sections 18.3–18.6, it has successfully described many phenomena of nonlinear optical wave propagation, including spectral broadening caused by self-phase modulation, soliton dynamics, and modulation instability. However, it is not sufficient to describe all of the linear and nonlinear optical processes that are involved in supercontinuum generation, not even for supercontinuum generation in an optical fiber where spatially dependent nonlinear effects such as self-focusing can be ignored.

A few important considerations set supercontinuum generation apart from the nonlinear wave propagation phenomena that were described in Sections 18.3–18.6. One is the extremely broad spectrum of a supercontinuum, which requires the inclusion of high-order dispersion beyond the second order. The other is the intrapulse stimulated Raman scattering. This process causes self-frequency shifting as an optical pulse propagates; thus, it is difficult to handle in the frequency domain. Instead, the effect of intrapulse stimulated Raman scattering can be treated by considering the delayed response of the nonlinear medium in the time domain [7, 8]. Furthermore, the *self-steeping effect* of the propagating pulse caused by the dispersion of the Kerr and Raman nonlinearities has to be considered in a complete model [8–10]. By adding a term for the possible background absorption to (8.63), the generalized nonlinear Schrödinger equation that accounts for all of the above effects is [5, 9–12]:

$$
\frac{\partial A(z,t)}{\partial z} + \frac{1}{v_g}\frac{\partial A(z,t)}{\partial t} + \sum_{n=2}^{\infty} \frac{i^{n-1}\beta_n}{n!}\frac{\partial^n A(z,t)}{\partial t^n}
$$
$$
= -\frac{\alpha}{2}A(z,t) + i\sigma\left(1 + \frac{i}{\omega_0}\frac{\partial}{\partial t}\right)\left[A(z,t)\int_{-\infty}^{\infty} R(t')|A(z,t-t')|^2 \mathrm{d}t'\right].
\tag{19.1}
$$

Under the *slowly evolving wave approximation*, this generalized nonlinear Schrödinger equation is valid for an optical pulse as short as a single optical cycle and for the extremely broad spectrum of a supercontinuum [5, 9]. Note that here we consider pulse propagation in a single-mode waveguide because supercontinuum generation is usually carried out in a single-mode optical fiber.

In (19.1), the time derivative terms on the left-hand side account for linear dispersion including high-order dispersion. The time derivative on the right-hand side, which has the coefficient of i/ω_0, accounts for the self-steeping effect due to the dispersion of the nonlinear effects, including self-phase modulation and cross-phase modulation due to the optical Kerr effect and stimulated Raman scattering. The function $R(t)$ in the integral on the right-hand side describes the response of the nonlinear medium, including the instantaneous response of the Kerr nonlinearity and the delayed response of the Raman process:

$$R(t) = (1 - f_R)\delta(t) + f_R h_R(t), \tag{19.2}$$

where $1 - f_R$ is the fraction of the contribution from the instantaneous Kerr nonlinearity, f_R is the fraction of the contribution from the delayed Raman nonlinearity, and $h_R(t)$ is the Raman response function [7]. Both f_R and $h_R(t)$ are the characteristic properties of a particular nonlinear material. For silica, $f_R = 0.18$ [7].

By performing the change of variables as expressed in (8.55), the generalized nonlinear Schrödinger equation given in (19.1) is converted to the form:

$$
\begin{aligned}
&\frac{\partial A(\zeta, \tau)}{\partial \zeta} + \sum_{n=2}^{\infty} \frac{i^{n-1} \beta_n}{n!} \frac{\partial^n A(\zeta, \tau)}{\partial \tau^n} \\
&= -\frac{\alpha}{2} A(\zeta, \tau) + i\sigma \left(1 + \frac{i}{\omega_0} \frac{\partial}{\partial \tau} \right) \left[A(\zeta, \tau) \int_{-\infty}^{\infty} R(\tau') \left| A\left(\zeta, \tau - \tau'\right) \right|^2 d\tau' \right].
\end{aligned}
\tag{19.3}
$$

The generalized nonlinear Schrödinger equation as given in (19.1), or equivalently in (19.3), cannot be analytically solved in general. Instead, it is numerically solved with an input of $A(0, t)$ for (19.1), or $A(0, \tau)$ for (19.3), as the initial condition. The numerical solution is commonly carried out by using the split-step Fourier scheme. This scheme takes Fourier transform back and forth between the time domain and the frequency domain. In each numerical step, the linear and nonlinear terms of the equation are separately integrated by treating each linear term in the frequency domain and each nonlinear term in the time domain.

19.3 SUPERCONTINUUM GENERATION WITH FEMTOSECOND PULSES

The processes involved in supercontinuum generation and the characteristics of the generated supercontinuum both depend on the combination of the parameters of the pump pulse, including its peak power, wavelength, and pulsewidth, and on the properties of the optical fiber, including its dispersion and nonlinearity. The power of the pulse has to be sufficiently high to sustain the necessary nonlinear processes for supercontinuum generation. Above a certain level, increasing the pump power simply increases the total power of the supercontinuum, with only a slight increase in its spectral width. The pump wavelength has to be properly chosen so that the generated supercontinuum falls within the low-loss transmission window of the fiber. The pulsewidth is an important factor because it determines the nonlinear processes that are involved. Femtosecond pump pulses are considered in this section; longer pulses will be considered in the following section. By pumping with a femtosecond pulse of a peak power of the order of kilowatts, a broadband supercontinuum can be generated with a short high-nonlinearity fiber of a length of centimeters or a few tens of centimeters.

The relation of the pump wavelength with respect to the zero-dispersion wavelength of the optical fiber is also an important factor to consider. Propagation of an ultrashort pulse in an optical fiber near the zero-dispersion wavelength allows for an interaction over a large length at a high peak power, leading to dramatic nonlinear effects. Because the nonlinear refractive index, n_2, of an optical fiber is positive, soliton formation is possible only in the spectral region of negative group-velocity dispersion. Consequently, the ability for a photonic crystal fiber or a tapered fiber to have its zero-dispersion wavelength shifted to a wavelength that is much shorter than the intrinsic zero-dispersion wavelength of the fiber material is an important feature for supercontinuum generation. For a silica fiber, this flexibility allows for efficient super-continuum generation throughout the visible and the near-infrared regions, as discussed in Section 19.1. For an infrared fiber, it allows for shifting the zero-dispersion wavelength to a convenient near-infrared or mid-infrared wavelength. For this reason, in both this section and the following section, we assume supercontinuum generation with this flexibility to choose a desired zero-dispersion wavelength by design.

19.3.1 Dependence on the Pump Wavelength

The nonlinear optical processes that are involved in the generation of a supercontinuum, and the spectral and temporal features of the supercontinuum, strongly depend on the pump wavelength, λ_0, especially its relation to the zero-dispersion wavelength, λ_{ZDW}, of an optical fiber. For an optical fiber, positive group-velocity dispersion is found in the spectral region where the optical wavelength is shorter than the zero-dispersion wave-length, $\lambda < \lambda_{ZDW}$, whereas negative group-velocity dispersion is found where the wave-length is longer than the zero-dispersion wavelength, $\lambda > \lambda_{ZDW}$. As mentioned above, $n_2 > 0$ for an optical fiber. Therefore, soliton dynamics are possible in the spectral region of negative group-velocity dispersion, but not in the spectral region of positive group-velocity dispersion. To facilitate the following discussions, numerically simulated spectral and temporal features of the supercontinua that are generated by pumping a 15-cm photonic crystal fiber of $\lambda_{ZDW} = 780$ nm with a 50-fs pulse of 10 kW peak power at various pump wavelengths are shown in Fig. 19.3 [5].

When the pump wavelength lies in the spectral region of positive group-velocity dispersion such that $\lambda_0 < \lambda_{ZDW}$, the dominant spectral broadening mechanism is simply self-phase modulation because soliton formation is not possible in this spectral region. This spectral broadening is also accompanied by the temporal broadening of the pulse because of positive group-velocity dispersion, as discussed in Section 18.4. If λ_0 is far from λ_{ZDW} such that the entire supercontinuum spectrum resides in the region of positive group-velocity dispersion, as is the case for $\lambda_0 = 600$ nm in Fig. 19.3, the spectral and the temporal profiles broaden by the combined effects of self-phase modulation and linear dispersion. The decrease in the peak power caused by the temporal broadening of the pulse quickly reduces the effect of self-phase modulation, thus limiting the spectral broadening. With λ_0 still in the region of positive group-velocity dispersion but approaching λ_{ZDW}, as are the cases for $\lambda_0 = 650$ nm and $\lambda_0 = 700$ nm in Fig. 19.3, self-phase modulation is still responsible for the spectral broadening in the region of positive group-velocity dispersion, and linear dispersion still causes the temporal profile to broaden, but some optical energy is transferred to the long-wavelength region of negative group-velocity dispersion. As the energy in this region increases, soliton dynamics start to take

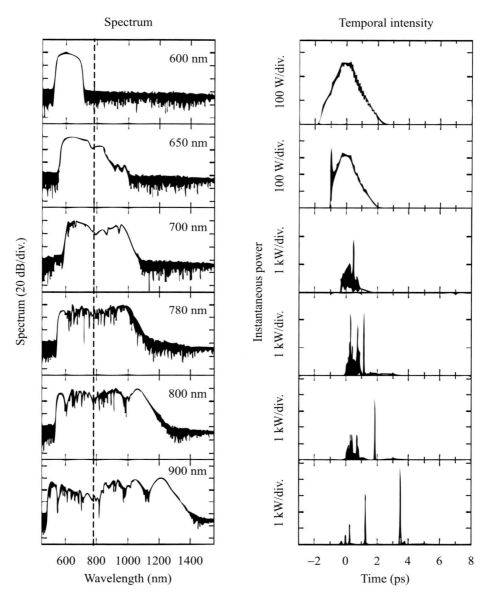

Figure 19.3 Numerically simulated spectral and temporal features of the supercontinua that are generated by pumping a photonic crystal fiber at various pump wavelengths. The length of the fiber is 15 cm. The pump wavelength for each case is indicated in the panel of the spectrum. The peak power of the input pump pulse is 10 kW, and the full width at half-maximum pulsewidth is 50 fs. The dashed line indicates the zero-dispersion wavelength of the fiber at 780 nm. (This figure is plotted by using data from J. M. Dudley, G. Genty, and S. Coen, "Supercontinuum generation in photonic crystal fiber," *Reviews of Modern Physics*, vol. 78, pp. 1135–1184, 2006.)

action, resulting in further spectral broadening on the long-wavelength side of the spectrum but narrowing of the temporal profile with narrow soliton peaks, as seen for the case of $\lambda_0 = 700$ nm.

When the pump wavelength is in the region of negative group-velocity dispersion such that $\lambda_0 > \lambda_{ZDW}$, soliton dynamics play dominant roles. Clear spectral and temporal soliton features are already seen for pumping at λ_{ZDW} with $\lambda_0 = \lambda_{ZDW} = 780$ nm. In this spectral region, the soliton dynamics evolve from the excitation of a high-order soliton, followed by fission of this high-order soliton, and continuous frequency down-shifting of each ejected fundamental soliton to longer wavelengths by stimulated Raman scattering in the case that the pump pulsewidth is larger than the Raman response time. Because of the large spectral width of the femtosecond soliton, frequency down-shifting takes place within the pulse through intrapulse stimulated Raman scattering. High-order solitons are susceptible to perturbations by high-order linear dispersion and by other nonlinear effects. The most important step of the evolution of soliton dynamics for supercontinuum generation is the *soliton fission* process, whereby a higher-order soliton is perturbed and broken up to eject a series of lower-amplitude fundamental solitons. Each ejected fundamental soliton is continuously frequency down-shifted by intrapulse stimulated Raman scattering, thus breaking away by propagating at a lower group velocity due to the negative group-velocity dispersion. In this manner, the combined effect of soliton dynamics and stimulated Raman scattering results in a series of narrow-width soliton peaks in the temporal profile, as seen in Fig. 19.3 for $\lambda_0 = 780$ nm, 800 nm, and 900 nm.

Besides causing soliton fission, the third-order linear dispersion plays an important role in *dispersive-wave generation*. Simultaneously with the ejection of each fundamental soliton in the soliton fission process, the third-order linear dispersion causes some optical energy to be transferred to the spectral region of positive group-velocity dispersion through phase-matched dispersive-wave generation [13]. Thus, the spectrum is broadened on the long-wavelength side in the region of negative group-velocity dispersion by the combined effect of soliton fission and stimulated Raman scattering, and on the short-wavelength side in the region of positive group-velocity dispersion by dispersive-wave generation. As seen in Fig. 19.3, a longer pump wavelength leads to a broader spectrum and more distinct soliton peaks in both the spectral and the temporal profiles. In particular, a series of clearly distinct soliton peaks are seen in the temporal profile for $\lambda_0 = 900$ nm. In this series, the fundamental soliton that is first ejected has the highest peak. It also has the largest time delay because it undergoes the largest Raman frequency down-shift, thus propagating at the lowest group velocity.

19.3.2 Dependence on the Pump Pulsewidth

When the pump wavelength is in the spectral region of positive group-velocity dispersion with $\lambda_0 < \lambda_{ZDW}$, different pump pulsewidths do not cause significant differences in the processes that are involved because the dominant process is self-phase modulation. The only difference is that a broader spectrum is generated with a shorter pump pulse if the energy of the pulse is the same.

When the pump wavelength is in the region of negative group-velocity dispersion with $\lambda_0 > \lambda_{ZDW}$, significant differences can be seen for different pump pulsewidths. There are three fundamental reasons:

1. The dominant process in this spectral region is soliton fission.
2. Soliton dynamics and the related process of dispersive-wave generation are more significant for shorter pulses.
3. The significance of stimulated Raman scattering diminishes as the pulsewidth is reduced below the Raman response time, which is around 76 fs for a silica fiber [7].

Figure 19.4 shows numerically simulated spectral and temporal features of the supercontinua that are generated by pumping a photonic crystal fiber of $\lambda_{ZDW} = 780$ nm with three different pump pulsewidths of 20 fs, 100 fs, and 500 fs at $\lambda_0 = 835$ nm [5]. For the two cases of pumping with 20-fs and 100-fs pulses, the signatures of soliton fission and dispersive-wave generation are clearly seen. Between these two cases, it is also clear that spectral broadening on the long-wavelength side and temporal slowing down of solitons, both caused by frequency down-shifting due to stimulated Raman scattering, are much more significant for pumping with the 100-fs pulse, which is longer than the Raman response time, and are substantially reduced for pumping with the 20-fs pulse, which is shorter than the Raman response time. For pumping with the 500-fs pulse, the initial spectral broadening develops from noise through the process of modulation instability, which can be viewed as four-wave mixing in the frequency domain. Subsequently, soliton dynamics and dispersive-wave generation take place after modulation instability breaks the pulse up. Stimulated Raman scattering further broadens the spectrum on the long-wavelength side.

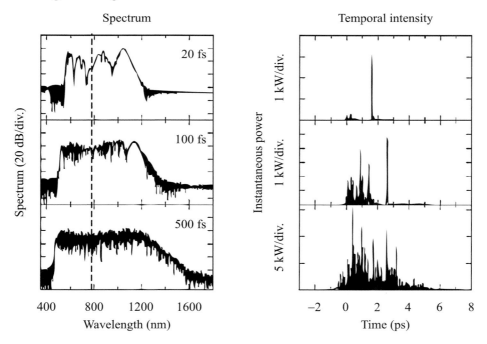

Figure 19.4 Numerically simulated spectral and temporal features of the supercontinua that are generated by pumping a photonic crystal fiber with three different pump pulsewidths. The length of the fiber is 15 cm. The pump wavelength is 835 nm in the spectral region of negative group-velocity dispersion. The peak power of the input pump pulse is 10 kW. The full width at half-maximum pump pulsewidth for each case is indicated in the panel of the spectrum. The dashed line indicates the zero-dispersion wavelength of the fiber at 780 nm. (This figure is plotted by using data from J. M. Dudley, G. Genty, and S. Coen, "Supercontinuum generation in photonic crystal fiber," *Reviews of Modern Physics*, vol. 78, pp. 1135–1184, 2006.)

19.3.3 Primary Nonlinear Processes

The primary nonlinear optical processes that are involved in supercontinuum generation with femtosecond pump pulses are summarized in Fig. 19.5.

For pumping at a wavelength shorter than the zero-dispersion wavelength, $\lambda_0 < \lambda_{ZDW}$, as shown in Fig. 19.5(a), spectral broadening is mainly caused by self-phase modulation in the spectral region, where the fiber has positive group-velocity dispersion. If the pump wavelength is not very far from the zero-dispersion wavelength and the pump power is sufficiently high, some optical energy is transferred to the region of negative group-velocity dispersion to initiate soliton dynamics and stimulated Raman scattering on the long-wavelength side of the supercontinuum.

For pumping at a wavelength longer than the zero-dispersion wavelength, $\lambda_0 > \lambda_{ZDW}$, as shown in Fig. 19.5(b), the dominant processes that are responsible for spectral broadening in the spectral region where the fiber has negative group-velocity dispersion are soliton fission and stimulated Raman scattering. Optical energy is transferred to the spectral region of positive group-velocity dispersion through the process of dispersive-wave generation to broaden the supercontinuum on the short-wavelength side. The third-order linear dispersion plays an important role in the processes of soliton fission and dispersive-wave generation

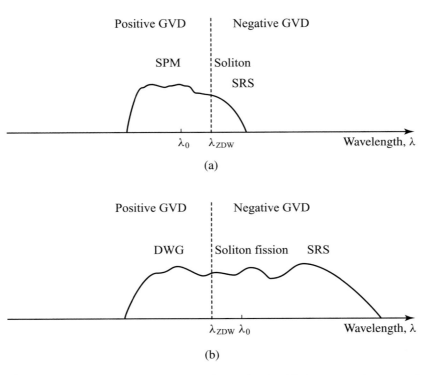

Figure 19.5 Primary nonlinear processes that are involved in femtosecond supercontinuum generation for (a) pumping at a wavelength of λ_0 in the spectral region of positive group-velocity dispersion with $\lambda_0 < \lambda_{ZDW}$ and (b) pumping at a wavelength of λ_0 in the spectral region of negative group-velocity dispersion with $\lambda_0 > \lambda_{ZDW}$. ZDW: zero-dispersion wavelength. GVD: group-velocity dispersion. SPM: self-phase modulation. SRS: stimulated Raman scattering. DWG: dispersive-wave generation.

The effect of stimulated Raman scattering diminishes as the pulsewidth of the pump pulse becomes smaller than the Raman response time. By contrast, the effects of modulation instability and four-wave mixing start to appear as the pump pulsewidth is increased to the order of a few hundred femtoseconds.

19.4 SUPERCONTINUUM GENERATION WITH LONG PULSES

Broadband supercontinuum can be generated with long laser pulses, ranging from picoseconds to nanoseconds, and even with CW laser beams. Indeed, the first supercontinuum was generated with picosecond excitation [1, 2], and the first supercontinuum generation in an optical fiber was pumped with a 10-ns dye laser pulse [3]. In comparison to the processes that take place in supercontinuum generation with femtosecond-pulse pumping, the processes involved with long-pulse pumping are quite different. As shown in Fig. 19.5(a), for femtosecond-pulse pumping in the region of positive group-velocity dispersion with $\lambda_0 < \lambda_{ZDW}$, the dominant nonlinear process is self-phase modulation. For long-pulse pumping, however, self-phase modulation is not able to generate a sufficiently broad spectrum because the spectral width is proportional to the slope of the pulse intensity, as expressed in (12.76) and shown in Fig. 12.6, and is thus inversely proportional to the pulsewidth for a given peak intensity. Instead, as discussed below, the dominant process for long-pulse pumping with $\lambda_0 < \lambda_{ZDW}$ is stimulated Raman scattering. As shown in Fig. 19.5(b), for femtosecond-pulse pumping in the region of negative group-velocity dispersion with $\lambda_0 > \lambda_{ZDW}$, the dominant nonlinear process is soliton fission. Because the soliton period is proportional to the square of the pulse duration, as expressed in (18.71), the soliton period for a long pump pulse can easily exceed the fiber length unless the long pulse is first broken up into very short pulses by other processes, such as modulation instability. Therefore, the effect of soliton dynamics is significantly reduced as the pump pulsewidth approaches the picosecond range and is completely diminished for a nanosecond pulse. As discussed below, the dominant process for long-pulse pumping with $\lambda_0 > \lambda_{ZDW}$ is four-wave mixing, or modulation instability if viewed in the time domain, which is seeded by noise. By pumping with a long laser pulse of a peak power of the order of hundreds of watts or kilowatts, a long fiber of the order of meters or tens of meters is necessary for supercontinuum generation.

19.4.1 Dependence on the Pump Wavelength

Though the processes and the characteristics are very different from supercontinuum generation with femtosecond-pulse pumping, the processes involved in supercontinuum generation with long-pulse pumping, and the spectral and temporal features of the supercontinuum, also strongly depend on the pump wavelength, λ_0, and its relation to the zero-dispersion wavelength, λ_{ZDW}, of an optical fiber. Figure 19.6 shows numerically simulated spectral and temporal features of the supercontinua that are generated by pumping a 2-m photonic crystal fiber of $\lambda_{ZDW} = 780$ nm with a 20-ps pulse of 500 W peak power at various pump wavelengths [5].

When the pump wavelength is in the spectral region of positive group-velocity dispersion such that $\lambda_0 < \lambda_{ZDW}$, the dominant spectral broadening mechanism is stimulated Raman

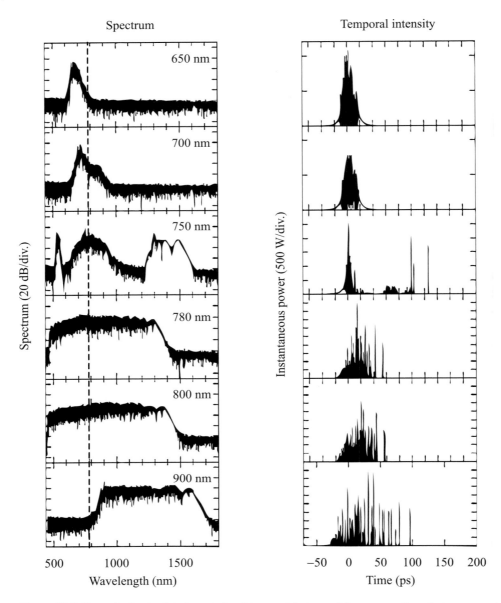

Figure 19.6 Numerically simulated spectral and temporal features of the supercontinua that are generated by pumping a photonic crystal fiber at various pump wavelengths. The length of the fiber is 2 m. The pump wavelength for each case is indicated in the panel of the spectrum. The peak power of the input pump pulse is 500 W, and the full width at half-maximum pulsewidth is 20 ps. The dashed line indicates the zero-dispersion wavelength of the fiber at 780 nm. (This figure is plotted by using data from J. M. Dudley, G. Genty, and S. Coen, "Supercontinuum generation in photonic crystal fiber," *Reviews of Modern Physics*, vol. 78, pp. 1135–1184, 2006.)

scattering because self-phase modulation is not effective for long pulses, as discussed above. Modulation instability facilitated by the second-order dispersion, which is equivalent to four-wave mixing phase matched through the second-order dispersion, is also not possible in the region of positive group-velocity dispersion, as discussed in Section 18.6 and in Subsection 10.7.2. Different from the intrapulse Raman process for femtosecond pulses, the spectral width of a long pulse is not sufficiently broad for intrapulse stimulated Raman scattering in a silica fiber, which has a Raman frequency shift of 13.2 THz. Instead, the spectrum is broadened by cascaded Raman processes, but the spectral width is limited, as is seen in the spectra for $\lambda_0 = 650$ nm and $\lambda_0 = 700$ nm in Fig. 19.6. In the time domain, the temporal modulation caused by these Raman processes breaks the pulse up into short pulses. Four-wave mixing with a pump wavelength in the region of positive group-velocity dispersion with $\lambda_0 < \lambda_{ZDW}$ is possible only by phase matching through high-order dispersion, as mentioned in Subsection 10.7.2, and the two sideband frequencies are discrete frequencies that are far separate from the pump frequency [5]. For $\lambda_0 = 650$ nm and $\lambda_0 = 700$ nm, the pump wavelength is too far from λ_{ZDW} for four-wave mixing to take place. For pumping at a wavelength that is close to λ_{ZDW}, four-wave mixing can have a sufficient gain to generate two sidebands, as seen in the spectrum for $\lambda_0 = 750$ nm in Fig. 19.6. The processes of stimulated Raman scattering and cross-phase modulation further broadens these sidebands, and soliton-like processes can occur in the region of negative group-velocity dispersion of the spectrum, as seen in the spectral and temporal profiles of the supercontinuum for this pump wavelength.

When the pump wavelength is in the region of negative group-velocity dispersion such that $\lambda_0 > \lambda_{ZDW}$, the dominant process is four-wave mixing, which is modulation instability in the time domain. As discussed in Section 18.6, modulation instability occurs for a negative group-velocity dispersion with the same gain bandwidth for both sidebands without a gap. Because the pump pulse is too long to initiate soliton dynamics, the spectrum is first broadened by modulation instability, which is initiated by noise. Once the pulse is broken up into short pulses, soliton dynamics can take action to broaden the spectrum, as seen in the spectral and temporal profiles for $\lambda_0 = 780$ nm, 800 nm, and 900 nm in Fig. 19.6. Subsequent stimulated Raman scattering and cross-phase modulation further broaden the spectrum. For $\lambda_0 = 780$ nm and 800 nm, dispersive-wave generation broadens the spectrum on the short-wavelength side in the region of positive group-velocity dispersion. Pumping too far from λ_{ZDW} results in a reduced spectral width because of the difficulty of phase matching for dispersive-wave generation, as seen in the spectrum for $\lambda_0 = 900$ nm.

19.4.2 Dependence on the Pump Pulsewidth

When pumping in the spectral region of positive group-velocity dispersion with $\lambda_0 < \lambda_{ZDW}$, different pump pulsewidths do not cause significant differences in the processes that are involved because the dominant process is stimulated Raman scattering, followed by four-wave mixing and cross-phase modulation. These processes are not critically dependent on the pulse duration for long pulses. Consequently, there are also no significant differences in the spectral and temporal characteristics for different pulsewidths.

When pumping in the spectral region of negative group-velocity dispersion with $\lambda_0 > \lambda_{ZDW}$, the initial process is four-wave mixing, or modulation instability, which is not critically dependent on the pulsewidth for a long pulse. As mentioned above, this four-wave mixing process also takes place in the case of pumping with $\lambda_0 < \lambda_{ZDW}$. The major difference for the

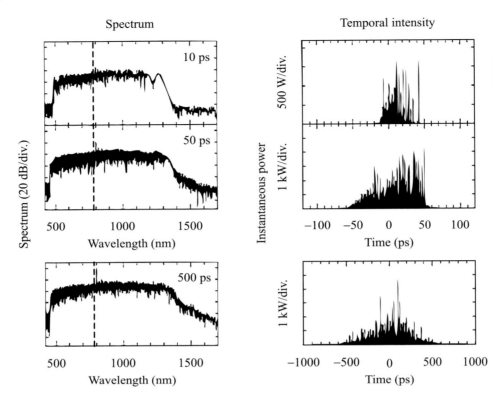

Figure 19.7 Numerically simulated spectral and temporal features of the supercontinua that are generated by pumping a photonic crystal fiber with three different pump pulsewidths. The length of the fiber is 2 m. The pump wavelength is 800 nm in the spectral region of negative group-velocity dispersion. The peak power of the input pump pulse is 500 W. The full width at half-maximum pump pulsewidth for each case is indicated in the panel of the spectrum. The dashed line indicates the zero-dispersion wavelength of the fiber at 780 nm. (This figure is plotted by using data from J. M. Dudley, G. Genty, and S. Coen, "Supercontinuum generation in photonic crystal fiber," *Reviews of Modern Physics*, vol. 78, pp. 1135–1184, 2006.)

case of pumping with $\lambda_0 > \lambda_{ZDW}$ is seen in the subsequent soliton dynamics after the long pulse is broken up into short pulses. Figure 19.7 shows numerically simulated spectral and temporal features of the supercontinua that are generated by pumping a photonic crystal fiber of $\lambda_{ZDW} = 780$ nm with three different pump pulsewidths of 10 ps, 50 ps, and 100 ps at $\lambda_0 = 800$ nm [5]. It can be seen that the spectral and temporal features of solitons are more distinct for pumping with a shorter pulse.

19.4.3 Primary Nonlinear Processes

The primary nonlinear optical processes that are involved in supercontinuum generation with long pump pulses are summarized in Fig. 19.8.

For pumping at a wavelength shorter than the zero-dispersion wavelength, $\lambda_0 < \lambda_{ZDW}$, as shown in Fig. 19.8(a), spectral broadening is mainly caused by cascaded stimulated Raman

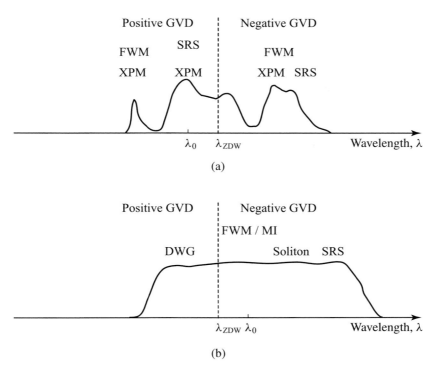

Figure 19.8 Primary nonlinear processes that are involved in long-pulse supercontinuum generation for (a) pumping at a wavelength of λ_0 in the spectral region of positive group-velocity dispersion with $\lambda_0 < \lambda_{ZDW}$ and (b) pumping at a wavelength of λ_0 in the spectral region of negative group-velocity dispersion with $\lambda_0 > \lambda_{ZDW}$. ZDW: zero-dispersion wavelength. GVD: group-velocity dispersion. SRS: stimulated Raman scattering. FWM: four-wave mixing. XPM: cross-phase modulation. MI: modulation instability. DWG: dispersive-wave generation.

scattering in the spectral region where the fiber has positive group-velocity dispersion. The spectrum can be further broadened by cross-phase modulation. If the pump wavelength is not very far from the zero-dispersion wavelength, two discrete sidebands can be generated by four-wave mixing that is phase matched through high-order dispersion. These sidebands are subsequently broadened by cross-phase modulation and stimulated Raman scattering.

For pumping at a wavelength longer than the zero-dispersion wavelength, $\lambda_0 > \lambda_{ZDW}$, as shown in Fig. 19.5(b), the dominant process that is responsible for the spectral broadening in the region of negative group-velocity dispersion is four-wave mixing (i.e., modulation instability) that is phase matched by the negative group-velocity dispersion and initiated by noise. This process also breaks up the long pulse into many short pulses, which can evolve into solitons. The spectrum is subsequently broadened by soliton dynamics and stimulated Raman scattering on the long-wavelength side, and by dispersive-wave generation on the short-wavelength side if the pump wavelength is not very far from λ_{ZDW}.

Except for the soliton dynamics, which are more significant for a shorter pump pulse, the primary nonlinear processes involved do not depend significantly on the pump pulsewidth for pumping with a long pulse.

References

[1] R. R. Alfano and S. L. Shapiro, "Emission in the region 4000 to 7000 Å via four-photon coupling in glass," *Physical Review Letters*, vol. 24, pp. 584–587, 1970.

[2] R. R. Alfano and S. L. Shapiro, "Observation of self-phase modulation and small-scale filaments in crystals and glasses," *Physical Review Letters*, vol. 24, pp. 592–594, 1970.

[3] C. Lin and R. H. Stolen, "New nanosecond continuum for excited-state spectroscopy," *Applied Physics Letters*, vol. 28, pp. 216–218, 1976.

[4] P. Russell, "Photonic crystal fibers," *Science*, vol. 299, pp. 358–362, 2003.

[5] J. M. Dudley, G. Genty, and S. Coen, "Supercontinuum generation in photonic crystal fiber," *Reviews of Modern Physics*, vol. 78, pp. 1135–1184, 2006.

[6] T. A. Birks, J. C. Knight, and P. S. J. Russell, "Endlessly single-mode photonic crystal fiber," *Optics Letters*, vol. 22, pp. 961–963, 1997.

[7] R. H. Stolen, J. P. Gordon, W. J. Tomlinson, and H. A. Haus, "Raman response function of silica-core fibers," *Journal of the Optical Society of America B*, vol. 6, pp. 1159–1166, 1989.

[8] K. J. Blow and D. Wood, "Theoretical description of transient stimulated Raman scattering in optical fibers," *IEEE Journal of Quantum Electronics*, vol. 25, pp. 2665–2673, 1989.

[9] T. Brabec and F. Krausz, "Nonlinear optical pulse propagation in the single-cycle regime," *Physical Review Letters*, vol. 78, pp. 3282–3285, 1997.

[10] P. V. Mamyshev and S. V. Chernikov, "Ultrashort-pulse propagation in optical fibers," *Optics Letters*, vol. 15, pp. 1076–1078, 1990.

[11] A. L. Gaeta, "Nonlinear propagation and continuum generation in microstructured optical fibers," *Optics Letters*, vol. 27, pp. 924–926, 2002.

[12] D. Hollenbeck and C. D. Cantrell, "Multiple-vibrational-mode model for fiber-optic Raman gain spectrum and response function," *Journal of the Optical Society of America B*, vol. 19, pp. 2886–2892, 2002.

[13] I. Cristiani, R. Tediosi, L. Tartara, and V. Degiorgio, "Dispersive wave generation by solitons in microstructured optical fibers," *Optics Express*, vol. 12, pp. 124–135, 2004.

APPENDIX A

Symbols and Notations

A.1 FIELDS

Field vectors and their scalar magnitudes are represented by using a consistent system of symbols and fonts. All vectors except for unit vectors are represented in bold-face fonts, whereas all scalar quantities are represented in nonbold fonts. This system is illustrated in the following by using the electric field as an example.

A.1.1 Real Fields

All real fields are defined in the real space and time domain only. All real field vectors are represented in the italic bold capital Roman font, such as

$$\boldsymbol{E}(\mathbf{r}, t) \tag{A.1}$$

for the real electric field vector. Other real field vectors are $\boldsymbol{H}(\mathbf{r}, t)$, $\boldsymbol{D}(\mathbf{r}, t)$, $\boldsymbol{B}(\mathbf{r}, t)$, $\boldsymbol{P}(\mathbf{r}, t)$, $\boldsymbol{M}(\mathbf{r}, t)$, $\boldsymbol{J}(\mathbf{r}, t)$, $\boldsymbol{A}(\mathbf{r}, t)$, and $\boldsymbol{S}(\mathbf{r}, t)$. Two real scalar fields are $\rho(\mathbf{r}, t)$ and $\varphi(\mathbf{r}, t)$. Except for the current density, all real field vectors are always represented in the vector form without separate symbols defined for their scalar magnitudes. The scalar magnitude of $\boldsymbol{J}$ is represented as J.

A.1.2 Complex Fields

All complex field vectors are represented in the nonitalic bold capital Roman font. All complex field vectors in the real space and time domain are defined in relation to their respective real field vectors, such as $\mathbf{E}(\mathbf{r}, t)$ defined in (1.37) for the complex electric field vector:

$$\boldsymbol{E}(\mathbf{r}, t) = \mathbf{E}(\mathbf{r}, t) + \mathbf{E}^*(\mathbf{r}, t) = \mathbf{E}(\mathbf{r}, t) + \text{complex conjugate.} \tag{A.2}$$

Other complex field vectors defined in a similar manner are $\mathbf{H}(\mathbf{r}, t)$, $\mathbf{D}(\mathbf{r}, t)$, $\mathbf{B}(\mathbf{r}, t)$, $\mathbf{P}(\mathbf{r}, t)$, $\mathbf{M}(\mathbf{r}, t)$, $\mathbf{J}(\mathbf{r}, t)$, and $\mathbf{A}(\mathbf{r}, t)$. Two similarly defined complex scalar fields are $\rho(\mathbf{r}, t)$ and $\varphi(\mathbf{r}, t)$. The complex Poynting vector is defined differently, as given in (4.8) and (4.9):

$$\mathbf{S} = \mathbf{E} \times \mathbf{H}^* \quad \text{so that} \quad \overline{\boldsymbol{S}} = \mathbf{S} + \mathbf{S}^*. \tag{A.3}$$

The scalar magnitude of a complex field vector is represented in the nonbold italic capital Roman font, such as E for the magnitude of $\mathbf{E}$:

$$\mathbf{E} = E\hat{e}, \tag{A.4}$$

where $\hat{e}$ is the unit vector of $\mathbf{E}$. Other scalar field magnitudes represented in a similar manner are H, D, B, P, and M. No scalar complex current density at an optical frequency is used; the scalar J represents the magnitude of a real current density vector $\mathbf{J}$ at DC or a low frequency. No scalar magnitude of the complex vector potential $\mathbf{A}$ is used. No scalar magnitude of the complex Poynting vector $\mathbf{S}$ is used.

A.1.3 Complex Field Amplitudes

The slowly varying amplitude vector of a complex field vector is represented in the bold capital script font, such as $\boldsymbol{\mathcal{E}}$ for the slowly varying amplitude of $\mathbf{E}$. It is defined as the slow variation of the field envelope on its carrier frequency through the relation

$$\mathbf{E}(\mathbf{r}, t) = \boldsymbol{\mathcal{E}}(\mathbf{r}, t) \exp\left(i\mathbf{k} \cdot \mathbf{r} - i\omega t\right), \tag{A.5}$$

as expressed in (1.45) for the electric field. Other slowly varying field amplitude vectors defined in a similar manner are $\boldsymbol{\mathcal{H}}, \boldsymbol{\mathcal{D}}, \boldsymbol{\mathcal{B}}, \boldsymbol{\mathcal{P}}$, and $\boldsymbol{\mathcal{M}}$, but not all of them are used in the text. No slowly varying field amplitude is defined for the vector potential or the Poynting vector.

The scalar magnitude of a slowly varying field amplitude vector is represented in the nonbold capital script font, such as $\mathcal{E}$ for the magnitude of $\boldsymbol{\mathcal{E}}$:

$$\boldsymbol{\mathcal{E}} = \mathcal{E}\hat{e}. \tag{A.6}$$

Other scalar magnitudes of slowly varying field amplitudes represented in a similar manner are $\mathcal{H}, \mathcal{D}, \mathcal{B}, \mathcal{P}$, and $\mathcal{M}$, but not all of them are used in the text.

A.1.4 Mode Fields

Complex mode field vectors are represented as $\mathbf{E}_\nu(\mathbf{r}, t)$ and $\mathbf{H}_\nu(\mathbf{r}, t)$ with their scalar magnitudes represented as $E_\nu(\mathbf{r}, t)$ and $H_\nu(\mathbf{r}, t)$, respectively, where the subscript index ν represents a compound mode index such as m or mn for waveguide modes, or mnq for Gaussian modes. The vectorial field profiles of a waveguide mode characterize the transverse spatial distributions of the mode fields. A waveguide mode field profile is a function of the transverse spatial coordinates only. The vectorial waveguide mode field profiles are represented as $\boldsymbol{\mathcal{E}}_\nu(x, y)$ and $\boldsymbol{\mathcal{H}}_\nu(x, y)$, or $\boldsymbol{\mathcal{E}}_\nu(\phi, r)$ and $\boldsymbol{\mathcal{H}}_\nu(\phi, r)$, with their scalar magnitudes represented as $\mathcal{E}_\nu(x, y)$ and $\mathcal{H}_\nu(x, y)$, or $\mathcal{E}_\nu(\phi, r)$ and $\mathcal{H}_\nu(\phi, r)$. Normalized vectorial mode field patterns, defined in (7.9), are represented as $\hat{\boldsymbol{\mathcal{E}}}_\nu(x, y)$ and $\hat{\boldsymbol{\mathcal{H}}}_\nu(x, y)$, or $\hat{\boldsymbol{\mathcal{E}}}_\nu(\phi, r)$ and $\hat{\boldsymbol{\mathcal{H}}}_\nu(\phi, r)$. Gaussian modes are represented by using similar symbols, but they are functions of both transverse and longitudinal spatial coordinates, x, y, and z.

A.2 VECTORS AND TENSORS

All vectors are represented in bold-face, with the exceptions of unit vectors, and their magnitudes are represented with corresponding symbols in nonbold fonts. A vector is also represented in the form of a 3 × 1 column matrix. Besides the field vectors and their magnitudes described in the preceding section, we have

$$\mathbf{k}, \;\; k; \;\; \mathbf{K}, \;\; K; \;\; \mathbf{r}, \;\; r; \;\; \Delta\mathbf{k}, \;\; \Delta k.$$

Scalar physical operators are represented with a hat. Vector and tensor physical operators are represented in bold-face with a hat. Density-matrix operators are represented without a hat.

$$\hat{H} = [H_{ab}], \;\; \hat{\boldsymbol{p}} = [p_{ab}], \;\; \hat{\boldsymbol{m}} = [m_{ab}], \;\; \hat{\boldsymbol{Q}} = [Q_{ab}], \;\; \hat{\boldsymbol{P}} = [P_{ab}], \;\; \rho = [\rho_{ab}], \;\; \rho^{(n)} = [\rho_{ab}^{(n)}].$$

All tensors and transformation matrices are represented in bold-face or in terms of their elements with subscript indices. Second-order tensors and transformation matrices are also represented in the form of 3 × 3 square matrices. The tensors used include

$$[r_{ijk}], \qquad\qquad [s_{ijkl}], \qquad \boldsymbol{\chi} = [\chi_{ij}], \qquad \boldsymbol{\chi}^{(1)} = [\chi_{ij}^{(1)}], \qquad \boldsymbol{\chi}^{(2)} = [\chi_{ijk}^{(2)}],$$

$$\boldsymbol{\chi}^{(3)} = [\chi_{ijkl}^{(3)}], \qquad \boldsymbol{\epsilon} = [\epsilon_{ij}], \qquad \Delta\boldsymbol{\epsilon} = [\Delta\epsilon_{ij}], \qquad \boldsymbol{\eta} = [\eta_{ij}], \qquad \Delta\boldsymbol{\eta} = [\Delta\eta_{ij}].$$

The transformation matrices used in the text include

$$\mathbf{T}, \;\; \widetilde{\mathbf{T}}.$$

A.3   FOURIER-TRANSFORM PAIRS

The same symbol is used for a quantity in the real space and its counterpart in the momentum space, or one in the time domain and its counterpart in the frequency domain. The difference is indicated by expressing a quantity as a function of $\mathbf{r}$ or $\mathbf{k}$, or as a function of t or ω. Note that the unit of a quantity is multiplied by a length unit of a meter each time one of the three spatial dimensions is transformed from the real space to the momentum space, and is multiplied by a time unit of a second when the quantity is transformed from the time domain to the frequency domain. For example, the electric field $E(\mathbf{r}, t)$ in the real space and time domain has the unit of volts per meter (V m^{-1}), but $E(\mathbf{k}, t)$ has the unit of volt-square-meters (V m^2), $E(\mathbf{r}, \omega)$ has the unit of volt-seconds per meter (V s m^{-1}), and $E(\mathbf{k}, \omega)$ has the unit of volt-second-square-meters (V s m^2).

A.4   SPECIAL NOTATIONS

A few special notations are used to label symbols for special meanings.

A.4.1   Unit Vectors and Normalized Quantities

Unit vectors are denoted with a hat on top of a symbol. The unit vectors that appear in the text are

$$\hat{e}, \ \hat{k}, \ \hat{n}, \ \hat{r}, \ \hat{u}, \ \hat{x}, \ \hat{y}, \ \hat{z}, \ \hat{X}, \ \hat{Y}, \ \hat{Z}.$$

Normalized quantities are also denoted with a hat on top of a symbol. The normalized mode field profiles appear in both vector and scalar forms:

$$\hat{\boldsymbol{\mathcal{E}}}_v, \ \hat{\mathcal{E}}_v, \ \hat{\boldsymbol{\mathcal{H}}}_v, \ \hat{\mathcal{H}}_v.$$

Other normalized quantities that appear in the text include

$$\hat{\eta}_{\mathrm{SH}}, \ \hat{I}_{\mathrm{in}}, \ \hat{I}_{\mathrm{out}}, \ \hat{g}(v), \ \hat{T}_{\mathrm{FP}}.$$

A.4.2 Modified Quantities

A quantity that is modified from the original quantity in some manner is denoted with a tilde on top of a symbol. Modified quantities that appear in the text include

$$\widetilde{A}, \ \widetilde{B}, \ \widetilde{g}_{\mathrm{B}}, \ \widetilde{g}_{\mathrm{B0}}, \ \widetilde{g}_{\mathrm{R}}, \ \widetilde{g}_{\mathrm{R0}}, \ \Delta\widetilde{\boldsymbol{\epsilon}}, \ \Delta\widetilde{\epsilon}, \ \widetilde{\kappa}.$$

A.4.3 Average Values

The spatial average, temporal average, weighted average, or mean value of a quantity is denoted with a bar on top of a symbol, such as

$$\overline{k}, \ \overline{n}, \ \overline{P}, \ \overline{P}_{\mathrm{th}}, \ \overline{P}_{2\omega}, \ \overline{I}, \ \overline{I}_{\mathrm{TPA}}^{\mathrm{sat}}, \ \overline{\boldsymbol{S}}, \ \overline{W}_{\mathrm{p}}, \ \overline{\alpha}.$$

A.5   SUBSCRIPTS AND SUPERSCRIPTS

Various fonts and notations are used for subscripts and superscripts. They include numerals, the mathematical font, the Greek font, coordinate symbols, and the Roman font. Bare numerals, mathematic font letters, and Greek letters that represent indices or variables are normally used only for subscripts. One exception is the labeling as superscripts of the coordinate indices i, j, k, and l, in (3.11)–(3.13). Roman letters and some special notations that have literal meanings can appear either as subscripts or as superscripts.

A.5.1 Numerals

Bare numerals are used only for subscripts. The following four numbers have special meanings in a proper context:

0 base value $(\alpha_0, \ m_0)$, constant value $(P_0, \ S_0)$, free-space value $(\epsilon_0, \ \mu_0)$, center value $(v_0, \ \omega_0)$, unsaturated value $(g_0, \ g_0)$, equilibrium value $(n_0, \ p_0)$, beam waist (w_0), or static field $(\boldsymbol{E}_0, \ \boldsymbol{H}_0)$;

1 parameters for waveguide core $(n_1, N_1, D_1, k_1, h_1)$ or
  parameters for the lower laser level $|1\rangle$ (E_1, N_1, R_1);
2 parameters for waveguide substrate $(n_2, N_2, D_1, k_2, \gamma_2)$ or
  parameters for the upper laser level $|2\rangle$ (E_2, N_2, R_2);
3 parameters for waveguide cover $(n_3, N_3, D_3, k_3, \gamma_3)$ or
  parameters for the energy level $|3\rangle$.

Note that the same symbol can have different meanings in different contexts. For example, n_2 in nonlinear optics also represents the coefficient of intensity-dependent index change defined in (5.38).

The numbers 1, 2, and 3 are also used as subscripts to represent the orthogonal coordinates of a general three-dimensional spatial coordinate system. The numbers 1 through 6 are also used as subscripts representing double indices to label tensor elements under the index contraction rule defined in (3.32):

xx	yy	zz	yz, zy	zx, xz	xy, yx
1	2	3	4	5	6

A numeral in the superscript is always placed in parentheses so that it is never confused with an exponent. It represents a perturbation order or the order of an interaction process. For example, $\chi^{(1)}$ is a linear susceptibility, $\chi^{(2)}$ is a second-order nonlinear susceptibility, $\chi^{(3)}$ is a third-order nonlinear susceptibility, and so forth.

A.5.2 Mathematic and Greek Subscripts

Mathematic and Greek fonts are used only for subscripts. They represent variable indices with the following well-defined meanings:

a, b, c, d general indices or general mode indices;
i, j, k, l integers or coordinate indices;
m, n, p, q integers or frequency component indices;
m, n transverse mode indices, each labeling a spatial dimension;
q longitudinal mode index or diffraction order;
α, β contracted indices representing double coordinate indices;
μ, ν, ξ, ζ compound transverse mode indices, each representing a mode.

Some Greek subscripts do not represent indices or variables, but express literal meanings. These include:

$$\beta, \chi, \lambda, \lambda/2, \lambda/4, \pi, \pi/2, \omega, 2\omega, 3\omega.$$

A.5.3 Coordinate Labels

General orthogonal spatial coordinates are labeled as 1, 2, and 3. Specific coordinates include the rectilinear coordinates (x, y, z), the cylindrical coordinates (r, ϕ, z), and the spherical coordinates (r, θ, ϕ). One set of special rectilinear coordinates (X, Y, Z) is used for the new principal axes $\hat{X}$, $\hat{Y}$, and $\hat{Z}$ of a crystal transformed under the Pockels effect. Two orthogonal unit vectors, $\hat{e}_+$ and $\hat{e}_-$, are used for left- and right-circular polarizations, respectively. Two special symbols are also used to represent directions: $\perp$ for perpendicular and $\parallel$ for parallel.

Coordinate labels generally appear as subscripts with commonly accepted meanings, with two exceptions. One exception appears in (3.11)–(3.13), as mentioned above. The other exception takes place when labeling a propagation constant k and the corresponding wavevector $\mathbf{k}$ of an optical field that represents a particular polarization normal mode. Because k_x conventionally represents the x component of the $\mathbf{k}$ vector, meaning $k_x = \mathbf{k} \cdot \hat{x}$, the propagation constant of an x-polarized optical field that can propagate in any direction perpendicular to $\hat{x}$ is represented as k^x in order to avoid confusion. To be consistent, the corresponding wavevector is labeled as $\mathbf{k}^x$. Thus, $k^x = n_x \omega/c \neq k_x$, and $\mathbf{k}^x = k^x \hat{k}$ where $\hat{k} \perp \hat{x}$. Such superscript coordinate labeling for k and $\mathbf{k}$ applies only to the following:

$$k^x, \quad k^y, \quad k^z, \quad k^X, \quad k^Y, \quad k^Z, \quad k^+, \quad k^-,$$
$$\mathbf{k}^x, \quad \mathbf{k}^y, \quad \mathbf{k}^z, \quad \mathbf{k}^X, \quad \mathbf{k}^Y, \quad \mathbf{k}^Z, \quad \mathbf{k}^+, \quad \mathbf{k}^-.$$

A.5.4 Roman Labels

All superscript and subscript labels in the Roman font have literal meaning. A given Roman label can appear either as a subscript or as a superscript, depending on the convenience of the situation, with exactly the same meaning. Among all subscript and superscript labels, only Roman labels have such flexibility. With only a few exceptions for avoiding confusion, the conventional rules for abbreviations are largely followed: (1) abbreviations for common words are in lower case, with the exceptions of E for TE, M for TM, L for longitudinal or linear, and T for transverse; (2) abbreviations for proper nouns are in upper case; and (3) acronyms are all in upper case. The Roman labels used in this book are listed below.

Label	Meaning
3PA	three-photon absorption
a	absorption, acoustic, aperture
ab	absorber
active	active region
AS	anti-Stokes
b	background, backward, bias
B	Bohr, Boltzmann, Bragg, Brillouin
bistability	bistability
bit	bit
c	carrier, cavity, center, characteristic, circular, conduction band, coupling, critical, cutoff

cl	cavity in lasing phase $(\gamma_{cl},\ \tau_{cl})$
coh	coherence
cp	cavity in pumping phase (γ_{cp})
CW	continuous wave
d	delay, diffraction, donor
D	Dirac, Doppler
DFG	difference-frequency generation
down	downward transition
DWG	dispersive-wave generation
e	electrical, electrode, electron, emission, extraordinary
E	TE mode
ED	electric dipole
EE	TE–TE
eff	effective
EM	TE–TM
EQ	electric quadrupole
ext	external
f	fast, forward
F	Fermi
FP	Fabry–Perot
FSR	free spectral range
FWM	four-wave mixing
g	bandgap, gain, group
G	Gaussian
GVD	group-velocity dispersion
h	hole, homogeneous
HH	high-order harmonic
high	high level
i	incidence, initial, intrinsic, ionization, intermediate
I	type I
II	type II
in	input
ind	induced
inh	inhomogeneous
int	interaction
K	Kerr
l	orbital, lasing
L	linear, longitudinal
low	low level
m	magnetization, modulation
M	TM mode
max	maximum

MD	magnetic dipole
ME	TM–TE
MI	modulation instability
min	minimum
MM	TM–TM
mode	laser mode
n	n-type
NL	nonlinear
o	ordinary
OA	optical axis
OPA	optical parametric amplification
OPG	optical parametric generation
OPO	optical parametric oscillation, optical parametric oscillator
out	output
p	p-type, phase, ponderomotive, polarization, pump
pk	peak
PM	phase matched
ps	pulse
pv	peak to valley
Q	quasi-phase matched
QS	Q switching
r	recombination, reflection, response
R	Rayleigh, Raman
rand	random
res	resonant
RT	round-trip
s	saturation, scattered, soliton, signal, slow, spin
S	Schrödinger, Stokes
sat	saturation
SF	self-focusing, sum frequency
SFG	sum-frequency generation
SH	second harmonic
SHG	second-harmonic generation
sp	spontaneous
SPM	self-phase modulation
SBS	stimulated Brillouin scattering
SRS	stimulated Raman scattering
t	transmission
T	transverse
TE	transverse electric
th	thermal, threshold
THG	third-harmonic generation

TM	transverse magnetic
total	total
TPA	two-photon absorption
tr	transparency
up	upward transition
v	valence band
XPM	cross-phase modulation
ZDW	zero-dispersion wavelength

APPENDIX B
SI Metric System

Table B.1 **SI base units**

Quantity	Name	Symbol
Length	meter	m
Mass	kilogram	kg
Time	second	s
Electric current	ampere	A
Temperature	kelvin	K
Amount of substance	mole	mol
Luminous intensity	candela	cd

Table B.2 **SI derived units**

Quantity	Name	Symbol	Equivalent
Plane angle	radian	rad	$m\,m^{-1} = 1$
Solid angle	steradian	sr	$m^2\,m^{-2} = 1$
Frequency	hertz	Hz	s^{-1}
Force	newton	N	$kg\,m\,s^{-2}$
Pressure	pascal	Pa	$N\,m^{-2}$
Energy	joule	J	$kg\,m^2\,s^{-2}$
Power	watt	W	$J\,s^{-1}$
Electric charge	coulomb	C	$A\,s$

Table B.2 *(cont.)*

Quantity	Name	Symbol	Equivalent
Electric potential	volt	V	$J\,C^{-1}$, $W\,A^{-1}$
Magnetic flux	weber	Wb	V s
Magnetic flux intensity	tesla	T	$Wb\,m^{-2}$
Resistance	ohm	Ω	$V\,A^{-1}$
Conductance	siemens	S	$A\,V^{-1}$, Ω^{-1}
Capacitance	farad	F	$C\,V^{-1}$
Inductance	henry	H	$Wb\,A^{-1}$
Luminous flux	lumen	lm	cd sr
Illuminance	lux	lx	$lm\,m^{-2}$

Source: R. A. Nelson, "Guide for metric practice," *Physics Today*, BG15–BG16, August, 2002.

Table B.3 **Metric prefixes**

Name	Symbol	Factor
Exa	E	10^{18}
Peta	P	10^{15}
Tera	T	10^{12}
Giga	G	10^{9}
Mega	M	10^{6}
Kilo	k	10^{3}
Hecto	h	10^{2}
Deca	da	10
Unit		1

Table B.3 (*cont.*)

Name	Symbol	Factor
Deci	d	10^{-1}
Centi	c	10^{-2}
Milli	m	10^{-3}
Micro	μ	10^{-6}
Nano	n	10^{-9}
Pico	p	10^{-12}
Femto	f	10^{-15}
Atto	a	10^{-18}

APPENDIX C

Fundamental Physical Constants

Table C.1 **Physical constants**

Quantity	Symbol	Value	Unit
Speed of light in free space	c	$2.997\,924\,58 \times 10^8$	$\mathrm{m\,s^{-1}}$
Magnetic permeability of free space	μ_0	$4\pi \times 10^{-7}$	$\mathrm{H\,m^{-1}}$
		$1.256\,637\,061\,4 \times 10^{-6}$	$\mathrm{H\,m^{-1}}$
Electric permittivity of free space	ϵ_0	$8.854\,187\,817 \times 10^{-12}$	$\mathrm{F\,m^{-1}}$
Impedance of free space $(\mu_0/\epsilon_0)^{1/2}$	Z_0	$376.730\,313\,461$	Ω
Planck constant	h	$6.626\,068\,765\,2 \times 10^{-34}$	$\mathrm{J\,s}$
		$4.135\,667\,271\,6 \times 10^{-15}$	$\mathrm{eV\,s}$
Planck constant $h/2\pi$	$\hbar$	$1.054\,571\,596\,8 \times 10^{-34}$	$\mathrm{J\,s}$
		$6.582\,118\,892\,6 \times 10^{-16}$	$\mathrm{eV\,s}$
Elementary charge	e	$1.602\,176\,462\,6 \times 10^{-19}$	C
Electron rest mass	m_0	$9.109\,381\,887\,2 \times 10^{-31}$	kg
Proton rest mass	m_p	$1.672\,621\,581\,3 \times 10^{-27}$	kg
Atomic mass unit	m_u	$1.660\,538\,731\,3 \times 10^{-27}$	kg
Boltzmann constant	k_B	$1.380\,650\,324 \times 10^{-23}$	$\mathrm{J\,K^{-1}}$
		$8.617\,342\,15 \times 10^{-5}$	$\mathrm{eV\,K^{-1}}$
Thermal energy at $T = 300$ K	$k_\mathrm{B}T$	$2.585\,202\,645 \times 10^{-2}$	eV
Photon constant $hc = \lambda h\nu$	hc	$1.239\,841\,86 \times 10^{-6}$	$\mathrm{eV\,m}$

Source: P. J. Mohr and B. N. Taylor, "The fundamental physical constants," *Physics Today*, BG6–BG13, August, 2002.

Fourier-Transform Relations

According to the discussions in Chapter 1, we define the Fourier transform between the time domain and the frequency domain in terms of the angular frequency as follows:

$$E(\omega) = \mathcal{F}\{E(t)\} = \int_{-\infty}^{\infty} E(t)e^{i\omega t}\,dt \tag{D.1}$$

and

$$E(t) = \mathcal{F}^{-1}\{E(\omega)\} = \frac{1}{2\pi} \int_{-\infty}^{\infty} E(\omega)e^{-i\omega t}\,d\omega. \tag{D.2}$$

In terms of the real frequency $v = \omega/2\pi$, we have

$$E(v) = \int_{-\infty}^{\infty} E(t)e^{i2\pi v t}\,dt \tag{D.3}$$

and

$$E(t) = \int_{-\infty}^{\infty} E(v)e^{-i2\pi v t}\,dv. \tag{D.4}$$

The Fourier-transform relations for common functions encountered in the description of various waveforms are listed in Table D.1. In this table, the Heaviside step function $H(x)$ is defined as

$$H(x) = \begin{cases} 1, & \text{if } x \geq 0, \\ 0 & \text{if } x < 0; \end{cases} \tag{D.5}$$

the rectangular function $\Pi(x)$ is defined as

$$\Pi(x) = \begin{cases} 1, & \text{if } |x| \leq 1/2, \\ 0 & \text{if } |x| > 1/2; \end{cases} \tag{D.6}$$

and the triangular function $\Lambda(x)$ is defined as

$$\Lambda(x) = \begin{cases} 1 - |x|, & \text{if } |x| \leq 1, \\ 0 & \text{if } |x| > 1. \end{cases} \tag{D.7}$$

Table D.1 **Fourier-transform relations**

Function form	$E(t)$	$E(\omega)$	Function form
Gaussian	e^{-t^2/τ^2}	$\sqrt{\pi}\tau e^{-\omega^2\tau^2/4}$	Gaussian
Hyperbolic secant	$\operatorname{sech}\dfrac{t}{\tau}$	$\pi\tau\operatorname{sech}\dfrac{\pi\omega\tau}{2}$	Hyperbolic secant
Infinite impulse sequence	$\displaystyle\sum_m \delta\left(\dfrac{t}{\tau}-m\right)$	$\displaystyle\tau\sum_n \delta\left(\dfrac{\omega\tau}{2\pi}-n\right)$	Infinite impulse sequence
Complex exponential	$e^{-i\omega_0 t}$	$2\pi\delta(\omega-\omega_0)$	Delta
Double-sided exponential	$e^{-|t/\tau|}$	$\dfrac{2\tau}{1+\omega^2\tau^2}$	Lorentzian
Single-sided exponential	$e^{-t/\tau}H(t)$	$\dfrac{\tau}{1-i\omega\tau}$	Complex Lorentzian
Rectangular	$\Pi\left(\dfrac{t}{\tau}\right)$	$\tau\dfrac{\sin(\omega\tau/2)}{\omega\tau/2}$	Sinc
Triangular	$\Lambda\left(\dfrac{t}{\tau}\right)$	$\tau\dfrac{\sin^2(\omega\tau/2)}{(\omega\tau/2)^2}$	Sinc^2
Convolution	$f(t) * g(t)$	$f(\omega)g(\omega)$	Product
Product	$f(t)g(t)$	$\dfrac{1}{2\pi}f(\omega) * g(\omega)$	Convolution
Complex conjugate	$f^*(t)$	$[f(-\omega)]^*$	

The convolution integral is defined as

$$f(x) * g(x) = \int_{-\infty}^{\infty} f(x-x')g(x')\mathrm{d}x' = \int_{-\infty}^{\infty} f(x')g(x-x')\mathrm{d}x'. \tag{D.8}$$

By using the Fourier-transform relation between $f(t) * g(t)$ and $f(\omega)\,g(\omega)$ and that between $f^*(t)$ and $f^*(-\omega)$ shown in Table D.1, some useful relations can be obtained:

$$\text{Correlation theorem}: \quad \int_{-\infty}^{\infty} f^*(t)g(t+\tau)\mathrm{d}t = \frac{1}{2\pi}\int_{-\infty}^{\infty} f^*(\omega)g(\omega)e^{-i\omega\tau}\mathrm{d}\omega, \tag{D.9}$$

$$\text{Autocorrelation theorem}: \quad \int_{-\infty}^{\infty} f^*(t)f(t+\tau)\mathrm{d}t = \frac{1}{2\pi}\int_{-\infty}^{\infty} \left|f(\omega)\right|^2 e^{-i\omega\tau}\mathrm{d}\omega, \tag{D.10}$$

$$\text{Power theorem}: \quad \int_{-\infty}^{\infty} f^*(t)g(t)\mathrm{d}t = \frac{1}{2\pi} \int_{-\infty}^{\infty} f^*(\omega)g(\omega)\mathrm{d}\omega, \tag{D.11}$$

$$\text{Parseval's theorem}: \quad \int_{-\infty}^{\infty} \left| f(t) \right|^2 \mathrm{d}t = \frac{1}{2\pi} \int_{-\infty}^{\infty} \left| f(\omega) \right|^2 \mathrm{d}\omega. \tag{D.12}$$

By using (D.3) and (D.4), Parseval's theorem can be written as

$$\int_{-\infty}^{\infty} \left| E(t) \right|^2 \mathrm{d}t = \int_{-\infty}^{\infty} \left| E(v) \right|^2 \mathrm{d}v = \frac{1}{2\pi} \int_{-\infty}^{\infty} \left| E(\omega) \right|^2 \mathrm{d}\omega. \tag{D.13}$$

INDEX